Zoological Monographs

Volume 6

Series Editors

The book series *Zoological Monographs* is devoted to the study of the animal kingdom and covers all aspects of animals, both living and extinct, including taxonomy, systematics and evolution, biodiversity and biogeography, ecology, morphology and anatomy, physiology and behavior. The volumes in the series either discuss the biology of a single animal group, approach a topic from an interspecies level, or present recent methodologies. The series is intended for graduate students, university teachers, scientists and professionals.

More information about this series at http://www.springer.com/series/15188

Fernando Pérez-Miles
Editor

New World Tarantulas

Taxonomy, Biogeography and Evolutionary Biology of Theraphosidae

Springer

Editor
Fernando Pérez-Miles
Facultad de Ciencias, Sección Entomología
Universidad de la República
Montevideo, Uruguay

ISSN 2523-3904 ISSN 2523-3912 (electronic)
Zoological Monographs
ISBN 978-3-030-48643-3 ISBN 978-3-030-48644-0 (eBook)
https://doi.org/10.1007/978-3-030-48644-0

This Springer imprint is published by the registered company Springer Nature Switzerland AG.
The registered company address is: Gewerbestrasse 11, 6330 Cham, Switzerland

To tarantula taxonomists who bravely encouraged themselves to approach the study of a particularly difficult group.
To my colleagues, tarantula breeders and students for their multiple teachings and for encouraging me to face the challenge of this book.
To the loves of my life, especially to my children Lucia, Mateo, and Flavia.

Preface

Theraphosid tarantulas impressed naturalists since early times due to their large size, spectacular appearance, and extraordinary behaviors. However, the study of their taxonomy and especially their biology slowly increased until the second half of the twentieth century. The book of Baerg 1958 was one of the first contributions to the biology of North American theraphosids. Since the 1960s, starting with the papers of Argentinean arachnologists Gerschman de Pikelin and Schiapelli, the knowledge of the taxonomy of Neotropical tarantulas becomes more rigorous and some genera and subfamilies were seriously reviewed. The first cladistic analysis of Mygalomorphae done by Raven 1985 was a landmark in the taxonomy of the group and stimulated several colleagues to study the phylogeny of several mygalomorph families including Theraphosidae.

During the last few decades, many researchers interested in Theraphosidae from all over the world made relevant contributions in the study of tarantulas, mainly in taxonomical and phylogenetic aspects but also in biological aspects. The development of tarantula hobby also contributed to informal but valuable observations on breeding and other aspects of biology.

I prefer to use the name tarantula for the theraphosid spiders although it was originally used for lycosids because tarantula was widely extended in scientific and popular language for Theraphosidae, and I think the language is alive and dynamic.

I have been working on the taxonomy and biology of tarantulas for more than 40 years. This experience gave me the opportunity and privilege to know most specialists in the world, and I invited most of them to participate in this book. I trust their expertise and knowledge shared in this book could constitute a good motivation for students and researchers to continue developing studies in this fascinating group.

Montevideo, Uruguay — Fernando Pérez-Miles

Acknowledgments

I am grateful to many colleagues for their time and effort in the revision of early versions of the manuscripts of the chapters improving the quality of this book: Anita Aisenberg, Nelson Ferretti, Oscar Francke, Caroline Fukushima, Richard Gallon, Jose Paulo Guadanucci, Brent Hendrixon, Martín Hüsser, Facundo Labarque, Witold Lapinsky, Jorge Mendoza, Juan J. Morrone, Carlos Perafán, Martín Ramírez, Paul Selden, Danniela Sherwood, Cara Shillington, Joseph Spagna, Claudio Veloso, Rick West, and Jonas Wolff.

I also thank all the colleagues who generously contributed images to the book; they are recognized in the figure captions.

Contents

Chapter 1
Introduction to the Theraphosidae

Fernando Pérez-Miles

Abstract The mygalomorph spiders of the family Theraphosidae, commonly named as tarantulas, are one of the most famous and diverse groups of arachnids, which include the largest spider species in the world. Theraphosidae contains almost 1000 species from all continents, except Antarctica and includes burrowing, terrestrial and arboreal taxa with diverse ecological adaptations and natural histories. Since the description of the family in 1869 their systematics was largely based on the study of morphological characters, and many authors agreed in the chaotic situation of their taxonomy. First phylogenetic studies were also based on morphological characters and molecular studies only started in the twenty-first century. Most authors recognize 12–13 subfamilies in the world; 5 of them occur in the New World. The most diverse subfamily, Theraphosinae includes about a half of the known species of tarantulas. In this chapter we introduce the phylogenetic position of the Theraphosidae within the Araneae, the general characteristics of the tarantulas including taxonomical, evolutionary and biological aspects of the group. We analyze Theraphosidae diagnostic characters and the affinities with other families and discuss the intrafamilial relationships and subfamilial characteristics focusing in New World groups.

1.1 Introduction

Until now almost 50,000 species of spiders are described grouped into 128 families (World Spider Catalog 2020). All of them are carnivorous, some specialized as web builders, others live in burrows or are wandering species using circumstantial shelters. The order Araneae, which include the spiders, is divided into three suborders: the Mesothelae, the Araneomorphae, and the Mygalomorphae. The Mesothelae constitute a small group of spiders from Asia which exhibit segmented abdomen. The Araneomorphae constitute the largest group of spiders including about 90% of the species and characterized by the presence of vertical chelicerae opossing each other. Mygalomorphae is the sister group of the Araneomorphae and is

F. Pérez-Miles (✉)
Entomología, Facultad de Ciencias, UDELAR, Montevideo, Uruguay
e-mail: myga@fcien.edu.uy

F. Pérez-Miles (ed.), *New World Tarantulas*, Zoological Monographs 6,
https://doi.org/10.1007/978-3-030-48644-0_1

characterized by the presence of the parallel alignment of the chelicerae; this group includes the tarantula family Theraphosidae.

Theraphosidae are mostly large, hairy spiders, that live more or less hidden in burrows, under stones or in silken retreats on the ground or on trees (Fig. 1.1). They occur mainly in tropical and subtropical regions and occupy habitats from sea level up to more than 4000 m (Ferretti et al. 2018) and are also found in caves at depths down to 800 m (Mendoza and Francke 2018). Theraphosidae are commonly called tarantulas although this common name was originally used also to name the wolf-spiders (Lycosidae). In several parts of Latin-America they are also called araña pollito (bird spider, chicken spider), araña mono (monkey spider); araña pica caballo (horse biting spider), and in Portuguese aranha caranguejeira or caragueijo (crabspider). Despite their large size and fearsome aspect, giving them a bad reputation, the great majority of tarantula species are not poisonous spiders. The bite can be painful due to the mechanic action of large chelicerae but the envenomation effect from the bite is usually similar to a wasp sting. In the last decades, the popularity of tarantulas as exotic pets increased considerably, mainly in developed countries.

Fig. 1.1 Habitus of Theraphosidae, (**a**) *Theraphosa blondi*, (**b**) *Psalmopoeus irminia* (Photos: R. West)

Consequently, the pet trade has exacerbated threats by overexploitation of natural populations (Costa and Pérez-Miles 2007; Aisenberg and Pérez-Miles 2016; Luddecke et al. 2018). In fact, several species from both the Old and New World were included in the IUCN red list as being in different status of risk; some of them protected by conservation actions as CITES (IUCN 2017) and national laws.

1.1.1 What Is a Theraphosidae?

Theraphosid tarantulas are usually large and hairy spiders. In fact, the type of setae that they have on their legs are fundamental for their classification as Theraphosidae. The diagnostic characters of the family are the presence of claw tufts together with dense tarsal scopulae (Fig. 1.2). Exceptionally, the genus *Agnostopelma* Pérez-Miles and Weinmann (2010) lacks the scopulae on posterior legs and an undescribed troglobic genus of Theraphosinae lacks scopulae on all legs (Perafán pers. comm.). Spiders of the family Barychelidae, closely related to the Theraphosidae, also have claw tufts and scopulae (usually less dense than Theraphosidae) and, in some genera, restricted to forelegs. According to Guadanucci (2014), Barychelidae has few clavate trichobothria in a group located on mid dorsal tarsi whereas in Theraphosidae the trichobothria are large and form lines or rows (Fig. 1.3). A typical barychelid has claw tufts, weak scopula, short apical segment of posterior lateral spinnerets (PLS), maxilla without projection and few tricobothria in a group; while a typical theraphosid presents claw tufts, dense scopula, apical segment of PLS digitiform, maxilla with anterior projection and tricobohria large and arranged in lines. Between both families there is a range of variation in morphology that makes difficult to clearly separate some barychelids from theraphosids. In this book we follow the most conservative criterion, using Theraphosidae *sensu stricto*, including only spiders that fit

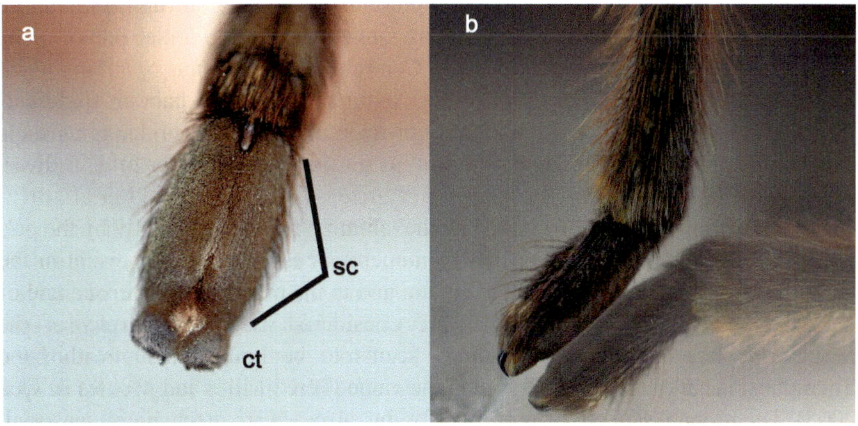

Fig. 1.2 Tarsal scopula (sc) and claw tufts (ct) of the tarantula *Grammostola anthracina*; (**a**) ventral view, (**b**) lateral view of tarsus on a glass surface

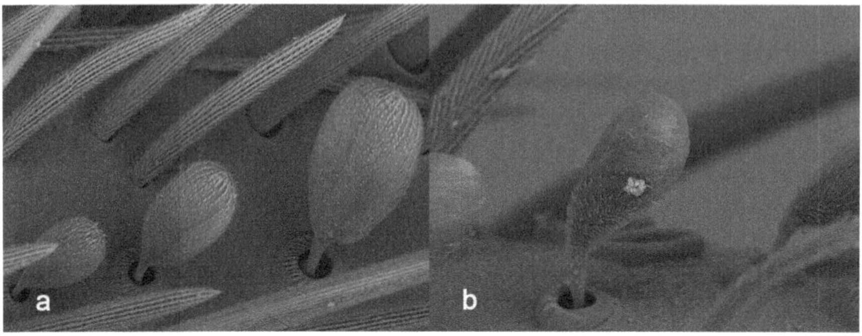

Fig. 1.3 Tarsal tricobotriae, (**a**) *Trichopelma* sp. (Theraphosidae), (**b**) *Neodiplothele martinsi* (Barychelidae) (Photos: JP Guadanucci)

with typical characters mentioned above. Most anatomical and morphological characteristics are similar to those found in other spiders. Characteristic morphological features consistent with other mygalomorphs are the parallel chelicerae (orthognath), two pairs of book lungs, absence of anterior spinnerets, palpal coxae similar to leg coxae, and presence of cuspules on labium and palpal coxae.

1.2 Morphology

Theraphosids are spiders with the carapace oval to subrectangular, of variable shape (Figs. 1.4 and 1.5). They usually have eight eyes in an elevated ocular tubercle; in some cave tarantulas (e.g., some species of *Hemirrhagus*) eyes can be reduced or absent, as well as the tubercle. As in other spiders, dorsally in the carapace, a transverse depression is present: the thoracic fovea. The tarantula fovea is usually procurved or straight. The spider fovea represents an internal invagination as an apodeme for the attachment of the sucking stomach muscle (dorsal dilator). The fovea presents a spherical process in the Central American genus *Sphaerobothria* and also could present a horn-like process as in some African baboon spiders as *Ceratogyrus* spp., in which the horn may provide an increased surface for muscle attachment (ring-shaped apodeme) as well as an increased area for midgut diverticula (West 1986).

Ventrally the mouth is surrounded by the labium behind and laterally by the palpal coxae, also named maxillae. Few to numerous cuspules can be present in the anterior edge or other position of the labium and in the prolateral proximal angle of maxillae (Fig. 1.6). These cuspules are considered as an apomorphy of the Mygalomorphae and their morphology seems to be homogeneous within the Theraphosidae and have limited systematic value (Pérez-Miles and Montes de Oca 2005). The probable functions proposed for the cuspules are mechanical, sensorial, and glandular. Cutler and Vuilliomenet (2001) proposed sensorial or glandular

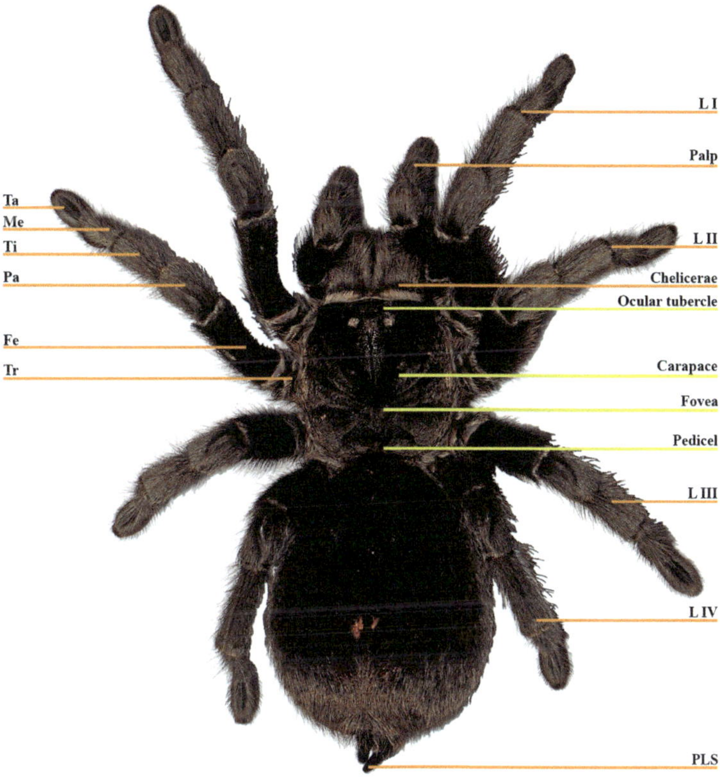

Fig. 1.4 Morphology of a Theraphosidae, dorsal view. *Fe* femur, *L* leg, *Me* metatarsus, *Ta* tarsus, *Ti* tibia, *Tr* trochanter, *Pa* patella, *PLS* posterior lateral spinnerets

function in *Aphonopelma seemani*, due to a pore observed in the apical region of the cuspule. Pérez-Miles and Montes de Oca (2005) proposed that cuspules could help in prey retention by opposing the backward force of the chelicerae. Posterior to the labium and separated by the labiosternal junction, there is the sternum with more or less conspicuous sigillae which represent apodemes. Sigillae are usually present in three pairs and could be marginal or submarginal.

Appendages In theraphosids the basal cheliceral segment usually has two rows of denticles on promargin and retromargin, respectively. The number and size of these denticles can vary in different species. Basal segments of chelicerae, palps and legs can present tegumentary modifications on prolateral and/or retrolateral faces; they are called stridulatory organs. Stridulatory organs are usually placed between the chelicerae on prolateral face, between chelicerae (retrolateral face) and palps (prolateral coxa); between basal segments of palps and legs I (retrolateral and prolateral faces, respectively) or between basal segments of legs. There are different types of stridulatory organs, that can be formed by bacilliform setae (most Selenocosmiinae), plumose setae (some Theraphosinae and most Harpactirinae), paddle- or spike-

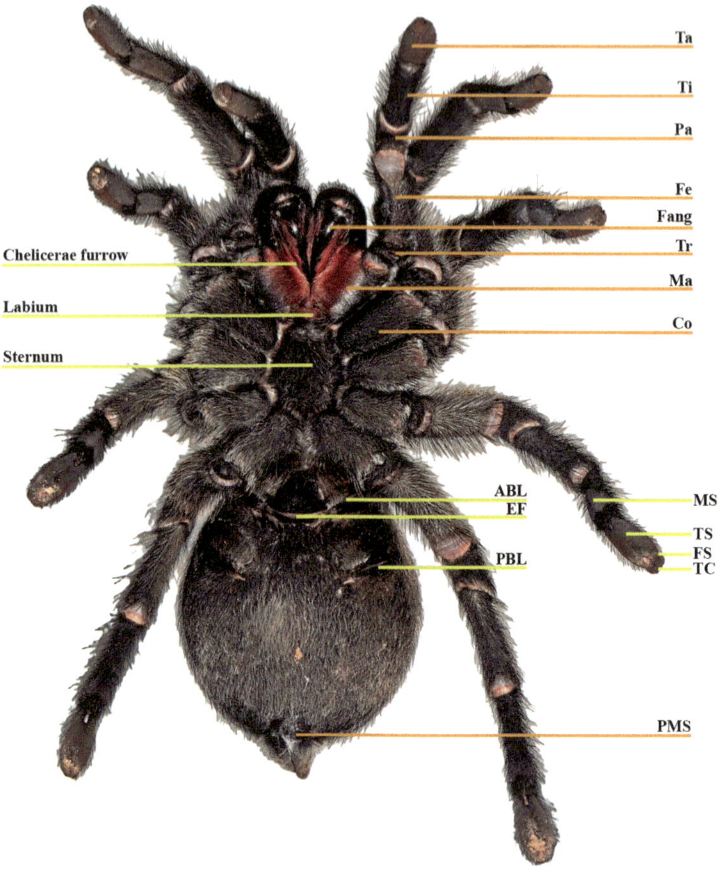

Fig. 1.5 Morphology of a Theraphosidae, ventral view. *ABL* anterior booklung, *Co* coxa, *EF* epigastric furrow, *FS* claw tufts, *Ma* maxilla, *Ms* metatarsal scopula, *PBL* postrior booklungs, *PMS* posterior median spinnerets, *Ts* tarsal scopula, *Tc* tarsal claw, other abbreviations as in Fig. 1.4

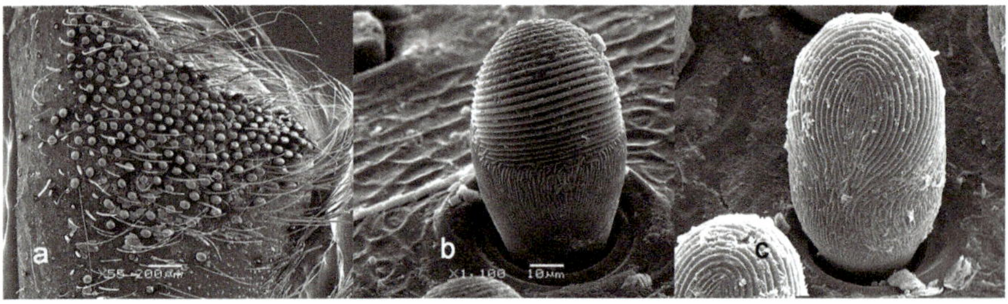

Fig. 1.6 Maxillary and labial cuspules of *Acanthoscurria cordubensis*. (**a**) Right maxillae (palpal coxa) showing cuspules, ventral view, (**b**) close-up of a labial cuspule (anterior view), (**c**) close-up of a labial cuspule (posterior view)

shaped setae (most Eumenophorinae), curved paddle setae (all Ornithoctoninae), and short modified setae (all Thrigmopoeinae) (Raven 1985; Gallon 2002; Bertani et al. 2008). Stridulatory setae present on New World genera of Theraphosinae were thoroughly studied in a recent article by Lima and Guadanucci (2019). These authors describe four different types of stridulatory setae present on coxae and/or trochantera and discuss function and evolution of such organs (See Chap. 11). In Old World tarantulas stridulatory organs can be more complex, involving diverse morphological types of setae, sometimes forming a lyra, and they are especially important to separate some subfamilies.

Palps are the second pair of appendages in spiders, they have one segment less than legs (lacking metatarsus) and have mainly sensitive functions. The palps are involved in prey capture and holding. In males, the apex of the palp is modified into sexual copulatory organs, the papal bulbs. Theraphosid palpal bulbs are very sclerotized (not expandable) as in other mygalomorphs and in some subfamilies as Theraphosinae have conspicuous keels (Fig. 1.7). Because of the heavy sclerotization, the intake and expulsion of sperm is not activated by hydraulic forces but rather by glandular mechanisms (see Chap. 14). The morphology of male palpal bulb and nested morphology (keels, apophysis, etc.) are important sources of taxonomic characters to diagnose different categories such as subfamilies, genera, and species. The theraphosid palpal bulb is relatively simplified compared to other spiders (Kraus 1978, 1984) and two parts can be recognized: the proximal subtegulum and the distal part including the tegulum and embolous (Fig. 1.7). The subtegulum is extended in the subfamily Theraphosinae, which was considered as a synapomorphy of the group (Raven 1985). Although theraphosid bulbs are not as complex as those of most araneomorphs, the study of Bertani (2000) proposed many

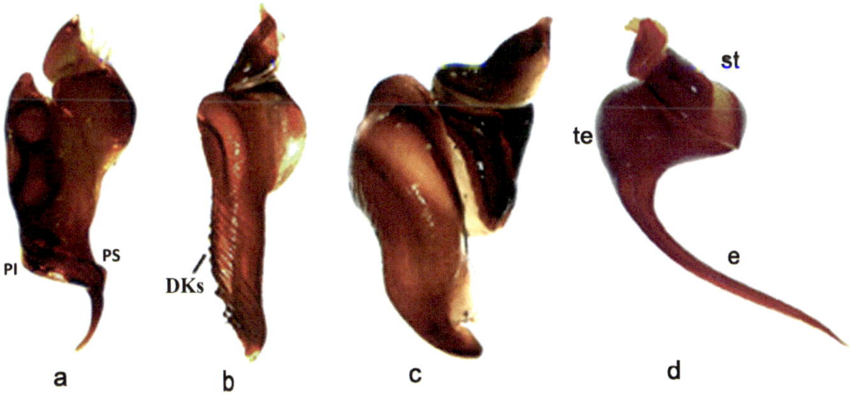

Fig. 1.7 Male palpal bulbs as examples of some New World Subfamilies (prolateral views). (**a**) *Tmesiphantes crassifemur* (Theraphosinae); (**b**) *Kankuamo marquezi* (Theraphosinae); (**c**) *Schismatothele weimanni* (Schismatothelinae); (**d**) *Avicularia merianae* (Aviculariinae). *DKs* dorsal supernumerary keels, *PI* prolateral superior keel, *PS* prolateral inferior keel, *st* subtegulum, *te* tegulum, *e* embolous. Photographs are not in scale. (**c**: after Valencia-Cuellar et al. (2019) and **d**: after Fukushima and Bertani (2017); with permission of the authors)

homologies among palpal bulb keels and main structures for Theraphosinae. The prolateral superior keel (PS) and prolateral inferior keel (PI) are widespread in most Theraphosinae genera whereas apical, subapical keels and other structures are present in some groups (Bertani 2000). Several other genera were described, and some of them with palpal bulb structures that obviously could not be considered in the homology hypothesis proposed by Bertani (2000). An extreme of these features are the supernumerary keels present in *Kankuamo* Perafán, Galvis and Pérez-Miles (Perafán et al. 2016). A new analysis of palpal feature homology seems to be necessary to include genera described in last decades.

Legs Scopulae and claw tufts setae, present in all Theraphosidae spiders, are adhesive and are involved in prey capture and locomotion, mainly during climbing (see Chap. 12). In Aviculariinae the scopulae are laterally extended giving the appearance of wider tarsi than other leg segments; this feature is probably related with the arboreal habits of most species in the subfamily. In some species the scopula is entire, formed by more or less homogeneous setae or medially divided by a stripe of longer conical setae; such division is usually wider in hindlegs (see Chap. 12).

In most New World theraphosid genera, males have a specialized apophysis on distal ventral or proventral anterior tibiae (Fig. 1.8). This tibial apophysis clasps the female chelicerae during copulation and are variable in morphology; they can have one (e.g., *Acanthoscurria, Schizopelma*), two (most genera), or three branches (some *Bonnetina*) or can occasionally be absent (e.g., *Metriopelma*, *Theraphosa blondi*, *Nhandu carapoensis*, several Ischnocolinae and Old-world genera, among others). In those with paired branches they can be convergent, straight, or divergent and can have their bases more or less fused. In Aviculariinae tarantulas tibial apophyses can be formed by grouped megaspines and can be present only in first pair of legs or in first and second pair (e.g., *Iridopelma*).

Tarantula legs are covered by varied types of setae, tricobothria, and spines. In Aviculariinae spines are scarce or absent whereas they are usually present in the other subfamilies, and the pattern and disposition of spines can be useful for taxonomy. The tricobothria morphology and arrangement patterns were studied in

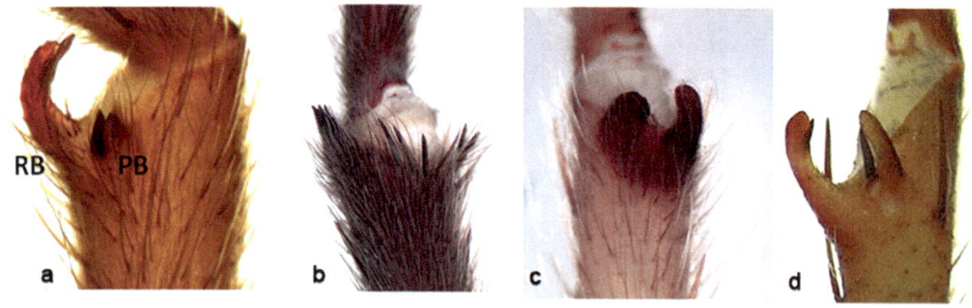

Fig. 1.8 Male tibial apophysis, (**a**) *Tmesiphantes uru*, two unequal branches, (**b**) *Acanthoscurria cordubensis* (ex *A. suina*), only one branch, (**c**) *Hapalopus formosus*, two branches fused at the bases, (**d**) *Schismatothele olsoni*, two separated branches (**d**: after Valencia-Cuellar et al. (2019), with permission of the author). *RB* retrolateral branch, *PB* prolateral branch

detail by Guadanucci et al. (2017) in Theraphosidae and Barychelidae spiders (see Chap. 11). The abundance of spines usually shows anterior posterior gradations, being more abundant in forelegs than in hindlegs.

Opisthosoma In theraphosids, the opisthosoma has a thinner cuticle in comparison with other body parts and similar to other spiders, this condition allows it to expand during prey ingestion and when females produce eggs. The opisthosoma is usually oval and can vary considerably in size and color pattern; most species have uniform dark coloration but several genera have conspicuous patterns and colorations as: *Avicularia, Antillena, Brachypelma, Caribena, Chromatopelma, Cyriocosmus, Davus, Hapalopus, Iridopelma, Megaphobema, Neoholothele, Pamphobeteus, Pachistopelma, Psalmopoeus, Typhochlaena, Xenesthis, Ybyrapora.*

A unique characteristic of the New World subfamilies Aviculariinae and Theraphosinae is the presence of urticating setae on the dorsal abdomen used for defense; also, the Avicularinae *Ephebopus* has urticating setae on palps. Abdominal urticating setae can be transferred to the target by contact in some *Avicularia* or released to the air by the friction of hindlegs against the abdomen in other theraphosids. Urticating setae can be thrown as a reaction to the perturbing animal (active defense) or incorporated to egg-sac and/or molting mat (passive defense). There are several setae types described by Cooke et al. (1972), Marshall and Uetz (1990a, b), Pérez-Miles (1998), and Perafán et al. (2016). Detailed information on urticating setae morphology, biology, and evolution of tarantulas is given in Chap. 9.

Ventrally the opisthosoma shows two pairs of book-lungs; a transverse furrow can be seen between the anterior pair of lungs: the epigastric furrow. In the center of the epigastric furrow is located the genital opening, that in males serves to expel the sperm drop during the sperm induction maneuver (see Chap. 14). The epigastric furrow in males is also surrounded by a special type of spigots (epiandrous apparatus) that contribute to build a part of the sperm web. In adult females the epigastric furrow has a conspicuous darker and slightly raised crescent, that connects with the genitalia. Internally, female genitalia consist of seminal receptacles, the spermathecae (Fig. 1.9), that are connected with the external uterus.

The male palpal bulb delivers sperm into the spermathecae during copulation, which is defined as sperm transference. In tarantulas the copulatory duct is at the same time the fertilization duct, as in the Haplogynae. The shape, size, and characteristics of the spermathecae can vary, and are considered useful tools as taxonomic characters, since the contributions by Schiapelli and de Pikelin (1962). The receptacle wall is usually covered by glands that can be related with sperm nutrition (De Carlo 1973). The spermathecae is derived from the ectoderm, and thus the cuticular lining of spermathecae is shed with the exuvium.

Theraphosid spermathecae comprise one or two receptacles that can show different degrees of sclerotization. Exceptionally, *Sickius longibulbi* and *Encyocratella olivacea* are the only tarantula species in which females lack spermathecae (Bertani and Silva 2002; Gallon 2003, 2005); in the absence of spermathecae, sperm storage is undertaken by the oviducts and uterus and consequently a short interval between copulation and oviposition is expected.

Fig. 1.9 Female spermathecae as examples of some Theraphosidae. (**a**) *Schizopelma bicarinatum* (single, only one receptacle), (**b**) *Pamphobeteus* sp. (two receptacles, basally fused), (**c**) *Grammostola* sp. (two separated receptacles). (**a**: after Gabriel (2016), with permission of the author)

It is usually difficult or almost impossible to differentiate males from females in juvenile individuals, and it is also somehow difficult to distinguish adult females from all juveniles. Only adult males are clearly recognizable because of the presence of palpal organs, and in several species also by tibial spurs. Large juvenile males have on the anterior edge of the epigastric furrow a group of specialized spigots called epiandrous apparatus, which females lack. These spigots are involved in the production of aditional silk for the sperm web (Costa and Pérez-Miles 2002). These structures can be used to distinguish between sexes in juveniles, but sometimes are not easy to see mainly in dark species; a good amount of light and high magnification is usually helpful. In some species it is possible to observe a different setae arrangement on the anterior edge of the epigastric furrow; males have a narrow semicircular darker area whereas females have a wider dark area and the epigastric furrow is more arched. Also the anterior pair of booklungs are more separated in females than in males.

Another more accurate way to differentiate males from females is through the examination of the inner face of the exuviae, if available. Spermathecae cuticle remains attached to the females' molt and they can be observed inside the exuviae next to the epigastric furrow.

Tarantulas have two pairs of spinnerets on the posterior extreme of the opisthosoma; the posterior median spinnerets (PLS) have three segments, with the apical segment digitiform and longer than median; and the posterior median spinnerets (PMS), are short undivided.

1.3 Tarantulas: Numbers and Taxonomy

Theraphosidae is the largest family of the largest sized spiders among the infraorder Mygalmorphae. The family includes 985 species in 147 genera; they usually occur in tropical and subtropical regions, most of them in the Neotropics (World Spider Catalog 2020). Largest species of the family (*Theraphosa* spp.) reach about 30 cm of legspan and weight more than 100 g, but there are also small species in which adults measure only a few centimeters.

The arrangement of theraphosid tarantulas in subfamilies has been a matter of controversy (Figs. 1.10 and 1.11) but the latest classifications (Luddecke et al. 2018; Foley et al. 2019) include 12–13 subfamilies. Aviculariinae was considered endemic to the New World until West et al. (2008) analyzed the phylogeny of the group, and representatives of *Heteroscodra* and *Stromatopelma* from Africa that were previously classified in the subfamily Stromatopelminae, nested inside the clade Aviculariinae. Fukushima and Bertani (2017) exhaustively studied the Aviculariinae

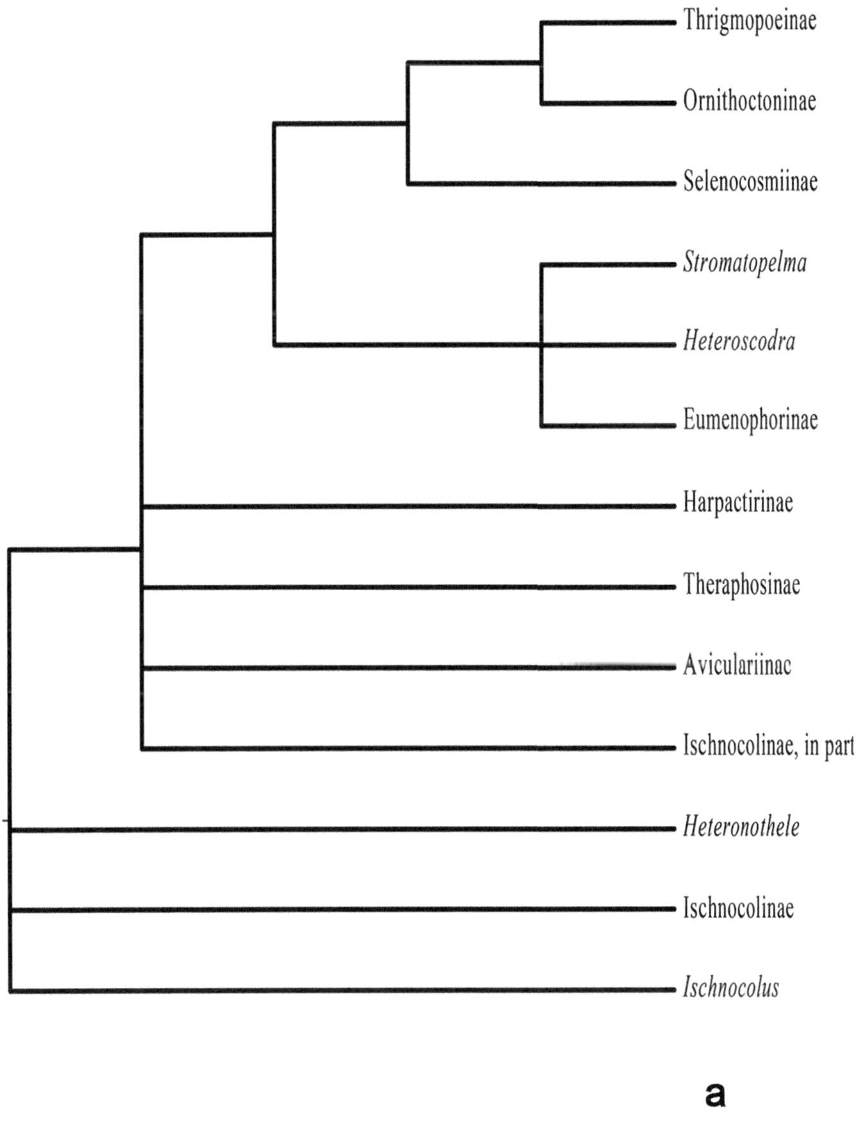

Fig. 1.10 Phylogenetic proposals for the Theraphosidae. (**a**) Raven (1985), (**b**) Guadanucci (2014)

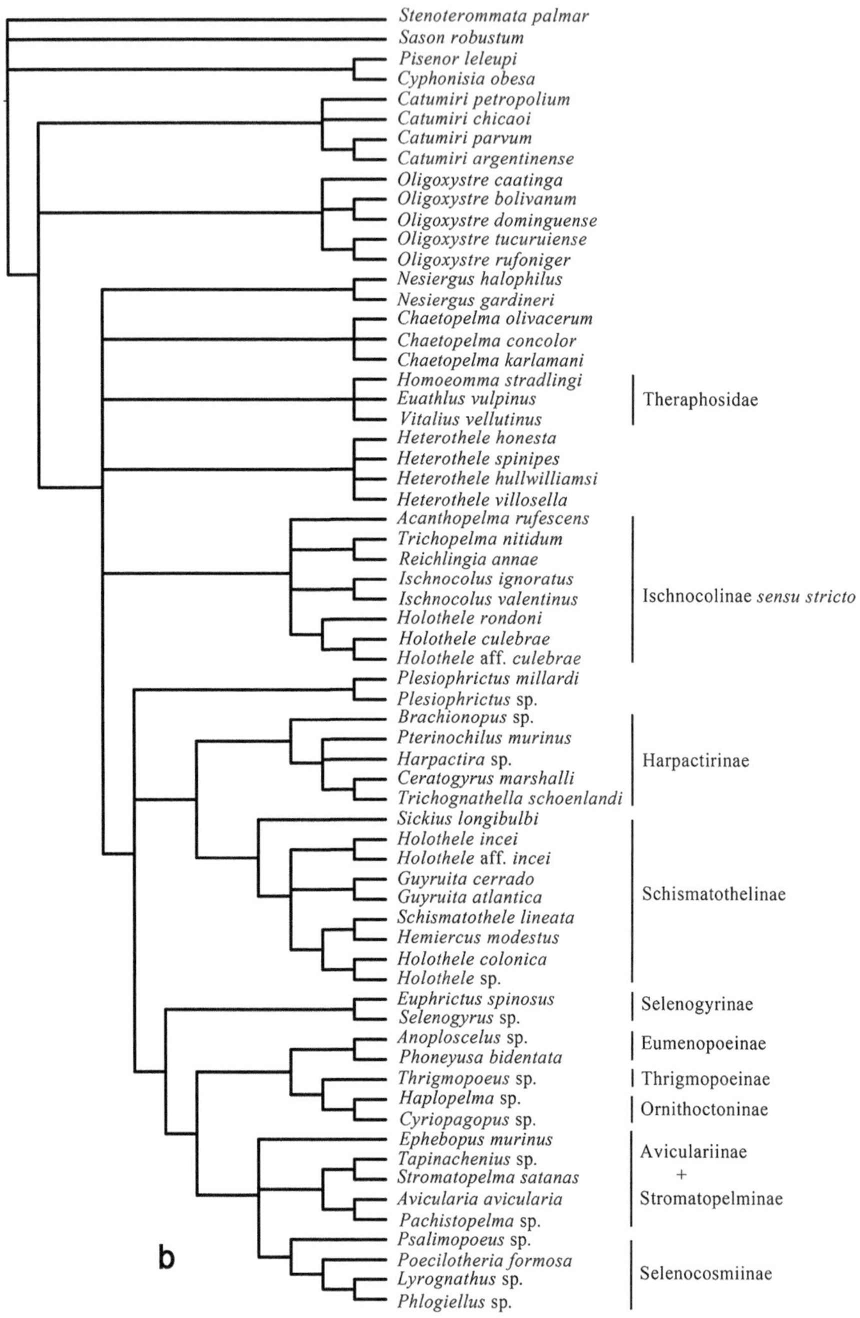

Fig. 1.10 (continued)

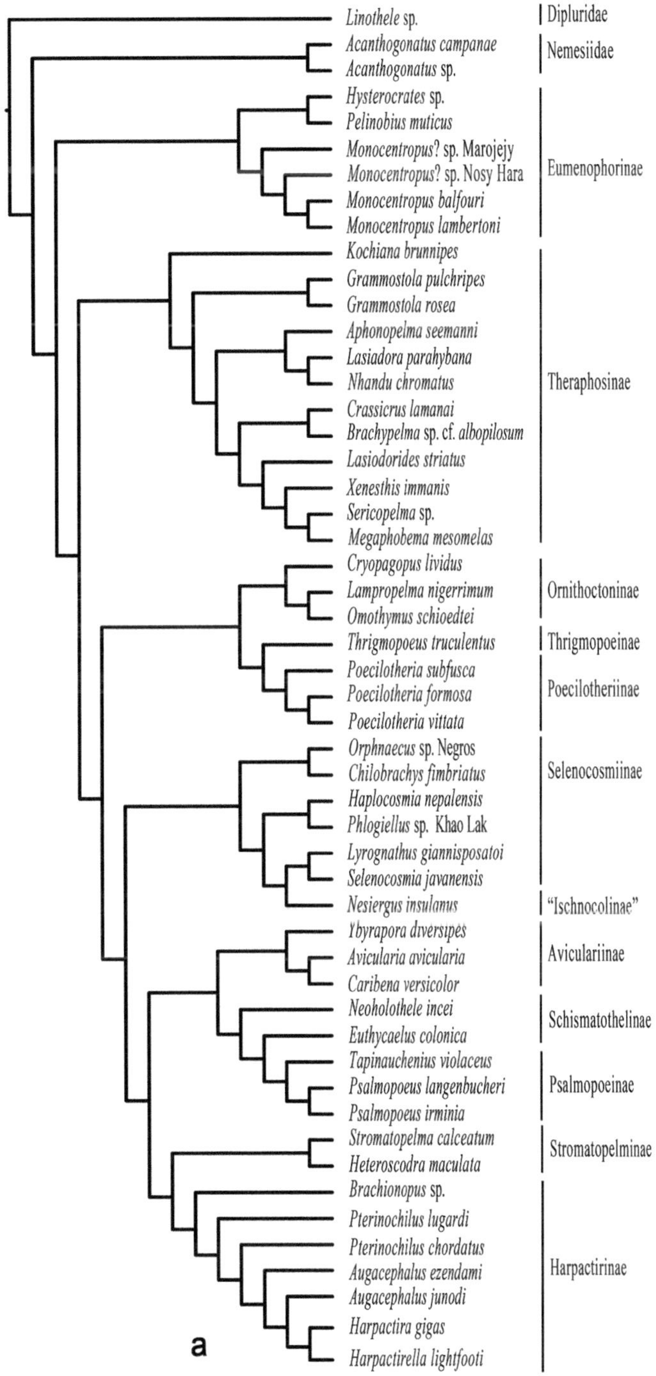

Fig. 1.11 Phylogenetic proposals for the Theraphosidae. (**a**) Luddecke et al. (2018), (**b**) Turner et al. (2018), (**c**) Foley et al. (2019)

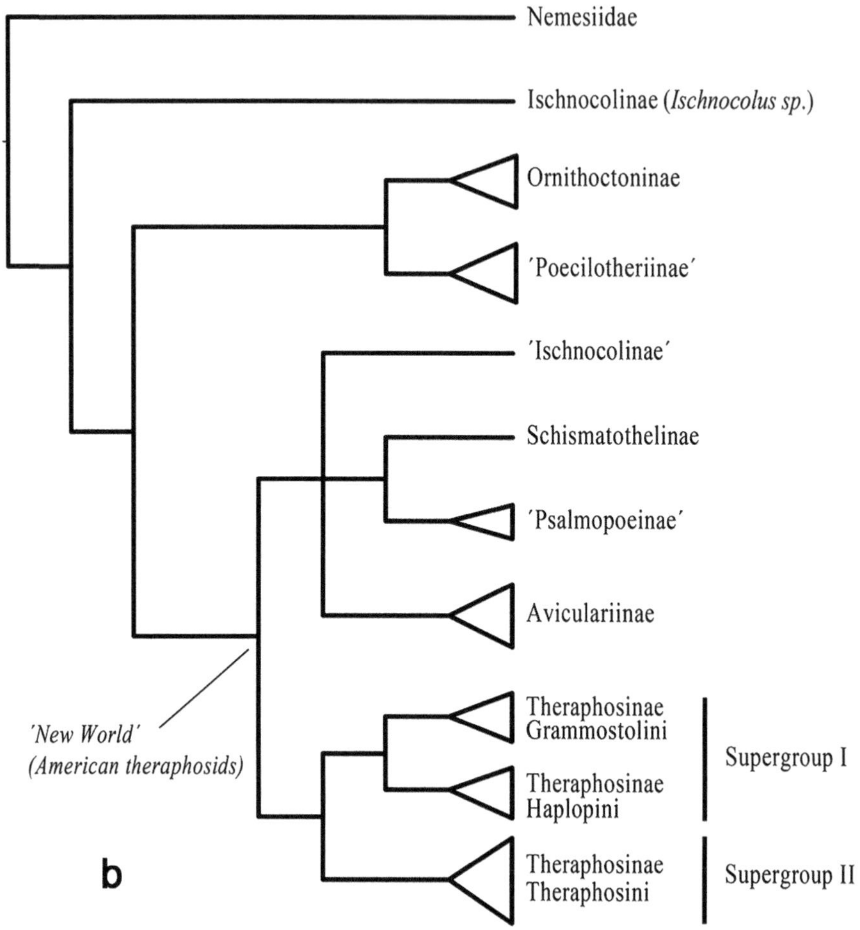

Fig. 1.11 (continued)

and in their morphologically based cladogram the subfamily also included *Heteroscodra* and *Stromatopelma*, as well as *Psalmopoeus*, *Tapinauchenius*, and *Ephebopus*. In the study of Guadanucci (2014) *Stromatopelma satan* was recovered in the same clade with the aviculariines with *Avicularia* and *Ephebopus*. In that study, Guadanucci (2014) revised intrafamilial relationships, and as a consequence, established Schismatothelinae including some genera previously considered as Ischnocolinae.

Recently, Turner et al. (2018) analyzed intrafamilial and generic relationships focussing on Theraphosinae using mitochondrial DNA and accepted, at least as informal groups, "Poecilotherinae" and "Psalmopoeinae." Luddecke et al. (2018) reanalyzed the intrafamilial phylogeny using nuclear and mitochondrial sequences and gave important insights to the taxonomy of theraphosid subfamilies. The

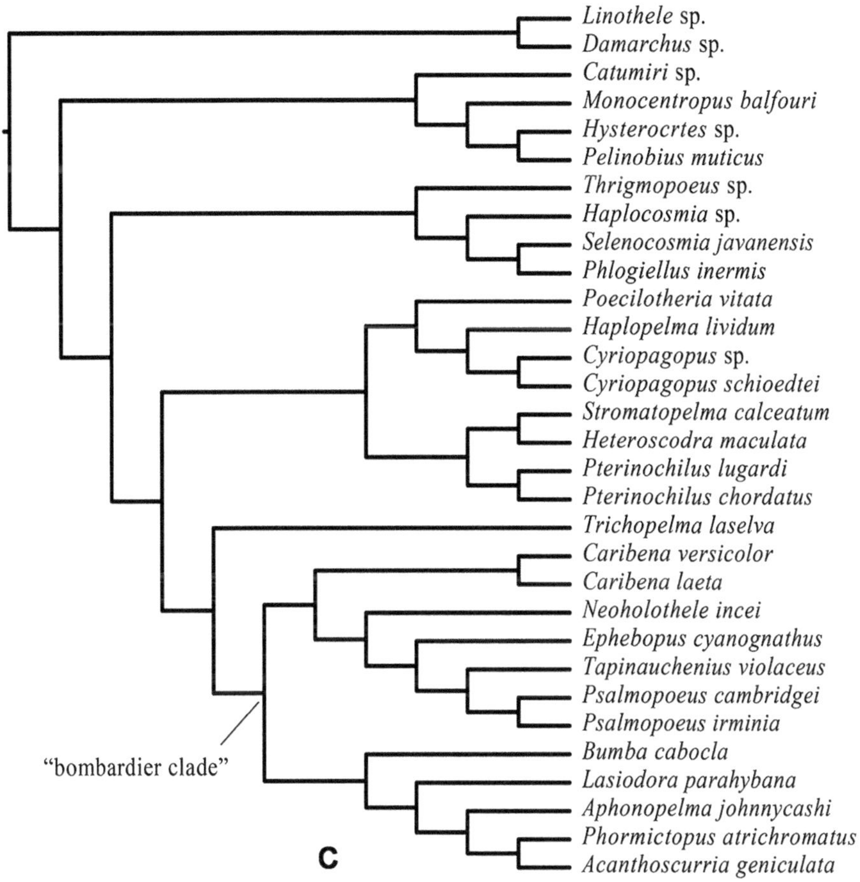

Fig. 1.11 (continued)

analysis of Luddecke et al. (2018) supports the monophyly of Poecilotherinae, Psalmopoeinae, and Stromatopelminae. As *Stromatopelma* and *Heteroscodra* from Africa were transferred back from Aviculariinae to Stromatopelminae, then Avicularinae remains endemic to the New World. In the analysis of Luddecke et al. (2018) Schismatothelinae was rendered paraphyletic and related to Psalmopoeinae whereas the "Ischnocolinae" *Nesiergus* was nested in Selenocosmiinae, but the authors said "...a more comprehensive sampling is necessary to confirm these results." The study of Guadanucci (2014) had an exhaustive sampling of Ischnocolinae sensu lato; for this reason I prefer to maintain Schismatothelinae as valid until new studies are performed (see Chap. 3 for further information).

More recently Foley et al. (2019) performed the first phylogenomic multigene study with representatives of several theraphosid subfamilies. Besides Old-World tarantulas, they concluded that Ischnocolinae, as suggested by Guadanucci (2014) represent multiple independent theraphosid taxa. Considering Ischnocolinae sensu lato, Foley et al. (2019) found at least three lineages: *Catumiri* as one of the earliest branching of Theraphosidae; *Trichopelma laselva* considered as Ischnocolinae sensu stricto and *Neoholothele incei,* formely in Ischnocolinae and now in Schismatothelinae. These authors proposed the "bombardier clade" (in reference to their ability of throwing urticating setae) for the group which comprises: ((Aviculariinae) (*Neoholothele incei* (Psalmopoeinae)) (Theraphosinae)). Foley et al. (2019) also agree with the monophyly and validity of Psalmopoeinae supported in Luddecke et al. (2018) and in Turner et al. (2018), including the genus *Ephebopus* in this subfamily, as propossed by Hüsser (2018). A relationship between Aviculariinae (sensu lato) and Theraphosinae, was earlier suggested by Pérez-Miles et al. (1996) based on behavioral evidence, and never followed until now; the "bombardier clade" is somehow reminiscent of that proposal. Furthermore, Foley et al. (2019) found that with the exception of *Catumiri* all Neotropical representatives included in their study resulted in a monophyletic clade.

Although the molecular studies cited above greatly increased the evidences about the phylogeny of Theraphosidae and important limitation is the fragmentary taxonomic sampling of each study. The development of integrative studies including an exhaustive taxonomic sampling throughout the diversity of tarantulas, is fundamentally important.

Summarizing, five subfamilies are currently considered as present in the New World: Aviculariinae, Ischnocolinae, Psalmopoeinae, Schismatothelinae, and Theraphosinae.

1.3.1 Key to New World Theraphosidae Subfamilies

1.- Abdominal urticating setae present (See Chap. 9)..2
- Abdominal urticating setae absent...3

2.- One or two types of urticating setae, types: I, III, IV, VI, VII (except some cave dwelling species), male palpal bulb with keels and subtegulum extended ....Theraphosinae (Chap. 5)
- Type II urticating setae, tarsal scopula laterally extended, male palpal bulb without conspicuous keels and subtegulum not extended; most species arboreal, legs with few or no spines...Aviculariinae (Chap. 4)

3.- Tarsal scopula extended in combination with the absence of abdominal urticating hairs. Lyriform stridulatory organs on prolateral face of palpal coxa (*Psalmopoeus, Pseudoclamoris*), reduced or absent (*Tapinauchenius*) or type V urticating setae present on prolateral palpal femur *Ephebopus*); males always with double tibial apophysis, legs with few spines.........Psalmopoeinae (Chap. 4)
- Tarsal scopula not extended, urticating setae absent, legs usually spinose.......... ....Ischnocolinae and Schismatothelinae (Chap. 3)

1.4 Biology

Most tarantula species live in burrows or crevices on the ground but others are arboreal and live in silken retreats on trees. They are usually sit-and-wait predators with nocturnal habits but during reproductive periods it is possible to observe wandering males during daylight searching for females (Costa and Pérez-Miles 2002). Adult females remain inside their retreats most of the time, leaving the burrow only for prey capturing and to discard remains of cocoons or food (or etc.). They usually do not move very far away from the burrow entrance (Alvarez et al. 2016). Detailed information about tarantula ecology is given in Chaps. 7, 8 and 10.

1.4.1 Life Cycles, Lifespan

Theraphosid tarantulas are among the longest-lived spiders; females can reach an age of 30 years in captivity (Costa and Pérez-Miles 2002). Although the mygalomorph *Gaius villosus* (Idiopidae) was recently reported as a record reaching 43 years old in the wild (Mason et al. 2018), the lifespan of Theraphosidae is high as compared to most araneomorphs which usually live 1 or 2 years. Particularly long lifespans are restricted to female tarantulas because males usually die after their first reproductive season (Costa and Pérez-Miles 2002; Klaas 2007). This sexual difference in lifespan could be explained because males have higher metabolic rates at resting and during locomotion, as reported for *Aphonopelma anax*, in which males weigh half as much as females (Shillington and Peterson 2002; Shillington 2005).

Ibler et al. (2013) compiled lifespan data for 85 theraphosid species held in captivity; they also studied patterns of correlation between lifespan and other variables taken from literature and found several interesting relations. These authors suggest that species inhabiting more predictable environments such as humid tropics or low altitudes have longer lifespans. Also large body size, low abundance, fossorial lifestyle, and aggressive behavior are associated with longer lifespans. However, some reasons underlying these relations are unclear. The mean lifespan reported by Ibler et al. (2013) is relatively shorter than that of other studies; the authors explain such differences because they raised their animals at relatively high temperatures and provided ad libitum feeding. Also this difference could be influenced because they averaged male and female lifespans together, whereas several data from the literature correspond to longer-lived females.

We observed that males of some species as *Eupalaestrus weijenberghi* and *Acanthoscurria cordubensis* (ex *A. suina*) that live about 2 months as adults in the wild, could reach almost a year in captivity (pers. obs.). This could be explained because they feed less and have a lower metabolic cost than in nature, due to limited movements and other activities.

1.4.2 Cocoons and Development

Three types of Theraphosidae cocoons or egg-sacs have been described (Gallon and Gabriel 2006): (1) mobile cocoon, (2) fixed hammock, and (3) fixed flat. A mobile cocoon is a fully detached sac, usually a subspherical or ovoid loose sac, of which the female takes care during incubation, and is present in most Theraphosidae. A fixed hammock is a sac forming a hammock-like structure, permanently attached to the female's retreat by distinct silk bands. This kind of egg-sac is found in some Eumenophorinae, Harpactirinae, Stromatopelminae, and Thrigmopoeinae. A fixed flat is a sac permanently attached to the female's retreat, which is usually discoidal, flat, and secured against a surface. A fixed flat egg-sac is only present in some Ischnocolinae and Stromatopelminae. Gallon (2003) considered the mobile egg-sac to be plesiomorphic while fixed egg-sacs evolved independently in several groups. In several theraphosids studied egg-sacs can contain between 12 and 2000 eggs (Punzo and Henderson 1999); usually clutch size is correlated with adult body size in spiders (Simpson 1995).

In the temperate region, spiderlings emerge between 30 and 75 days after oviposition (Célérier 1992; Costa and Pérez-Miles 2002) but this period should vary with temperature and probably other environmental conditions. Studies of post-embryonic development show that tarantulas have 3–4 larval stages inside the cocoon, the first of them intrachorional (Galiano 1969, 1973a, b). It is interesting to remark that the urticating setae occur in the first stage free of the cocoon (stages 4 or 5). The occurrence of different types of urticating setae during the development was studied in some species (Pérez-Miles 2002). Type III urticating setae are related with active defense, usually occuring later than other types.

Theraphosid species usually take 3–6 years to reach adulthood in humid tropical and temperate conditions, while they can take longer (up to 10 years in *Aphonopelma hentzi*) in arid zones (Baerg 1958; Stradling 1978; Trabalon and Blais 2012). As in all arthropods, spiders have to molt in order to grow in body size. In contrast with other spiders, tarantula females continue molting after maturation whereas males cease molting after reaching adulthood. During female post-mature molting spermathecae dimensions increase allometrically more than body size, which was interpreted to increase fecundity (Pérez-Miles 1989). Males are prevented from molting after adulthood due to anatomical limitations such as the basal constriction at palpal bulb. Some males that exceptionally molted after adulthood by hormonal disorders showed male palpal organ aberrations (pers. obs.).

In males, the juvenile period is shorter than in females, undergoing from 8 to 12 molts, while females usually have more than 9 juvenile molts (Trabalon and Blais 2012). Shortest male juvenile periods were found in the African species *Pterinochilus* sp. and *Stromatopelma grisepis*, lasting about 500–600 days to adulthood (Perret 1974; Célérier 1981). Several environmental factors such as temperature, food availability and photoperiod can produce variation in size increasing and instar duration (Turnbull 1962, 1965; Peck and Whitcomb 1970; Kotzman 1990).

1.5 Venom

Due to their body size and fearsome appearance, tarantulas usually have a bad reputation as being poisonous. However, this is far from true because most species' venom is not dangerous to humans. The bite could be painful due to the size of chelicerae that can penetrate deeply, but toxicity effects can be compared with a wasp sting. Most tarantulas are not aggressive and they usually try to escape rather than trying to bite. Old-world tarantulas are not so quiet, and more reactive than their New-World relatives; and their bites can be more severe. In effect species of *Pterinochilus*, *Poecilotheria*, and *Hysterocrates* can produce painful muscle cramps that can last approximately for a month (Ezendam 2007; Fuchs et al. 2014; Hofler 1996 and pers. observ.). Toxicologists have paid significant attention to tarantula venom because of the high quantities that can be extracted from large-size individuals. The venom of about 60 species of tarantula was studied (Bode et al. 2001; Escoubas and Rash 2004); some peptides found in tarantula venom seem to have potential application in medicine and industry. For example, venom of *Theraphosa blondi* is a source of curaremimetic toxins and ω-toxins of possible interest as tools in bioscientific research (Fontana et al. 2002). Spider venoms are complex mixtures of substances such as inorganic salts, small molecules such as amino acids, neurotransmitters and larger polyamines, peptides, and proteins, which work synergistically, improving venom efficiency (Santana et al. 2017). Tarantula venoms contain adenosine, histamine, and serotonin (Krug and Elston 2011). The purpose and function of spider venom is prey capture and digestion; secondarily venom could be used against predators as a deterrence tactic. A very interesting discovery demonstrates changes in venom composition during development in *Selenocosmia crassipes*. This modification in venom composition could be associated with change in the prey the spiders encounter at different life stages or to defend themselves from different predators along their lifespans (Santana et al. 2017).

1.6 Origin and Paleontology

Several authors suggested that Mygalomorphae radiated in major lineages during the Mesozoic (Dunlop et al. 2008). Eskov and Zhonshtein (1990) called the Cretaceous as the "age of the mygalomorphs" suggesting that in the Cenozoic the araneomorphs dominated the araneofauna until more recent times. Recent phylogenomic studies estimated the origin of the Theraphosidae between 34 and 71 Mya (Bond et al. 2014) and between 2 and 125 Mya (Garrison et al. 2016). Very recently Opatova et al. (2019) suggested the split between Barychelids and Theraphosidae around 106 Mya.

In fact, the geographic distribution of tarantulas with gondwanic preponderance suggests the origin of the family prior to the division of the continental plate, near to higher age estimations. However, the theraphosid fossil record is scarce and came entirely from Neogene-Recent (Dunlop et al. 2018); in their list three fossil species are recognized as Theraphosidae. The first described was found in Dominican amber and assigned to the subfamily Ischnocolinae: *Ischnocolinopsis acutus* Wunderlich 1988 (Dunlop et al. 2018). Two other species came from Chiapas amber in Mexico, an undetermined genus and species described by Dunlop et al. (2008) and *Hemirrhagus* sp. described by García-Villafuerte (2008). Probably the environmental conditions prior to the Neogene were not favorable for fossilization of tarantulas, which could explain the absence of theraphosid fossils in earlier times. Recently, Hembree (2017) developed a neoichnological study of tarantula burrows presenting criteria to recognize these spider burrows in the fossil record. This study has potential application for continental ischnofossil assemblages in order to improve paleoecological interpretations of ancient soil ecosystems and to improve the knowledge of tarantula biochron.

References

Aisenberg A, Pérez-Miles F (2016) Uruguayan tarantulas in danger: pet traders strike again. Montevideo, 1 March 2016. http://tentacletheatre.com/post/141213127835/attn-tarantula-keepers. Accessed 1 Mar 2019

Alvarez L, Perafán C, Pérez-Miles F (2016) At what time, for what distance, and for how long does the tarantula *Eupalaestrus weijenberghi* (Araneae, Theraphosidae) leave its burrow during the mating season. Arachnology 17:152–154

Baerg WJ (1958) The tarantula. University of Kansas Press, Lawrence, KS, p 88

Bertani R (2000) Male palpal bulbs and homologous features in Theraphosinae (Araneae, Theraphosidae). J Arachnol 28:29–42

Bertani R, Silva PI (2002) The first mygalomorph spider without spermathecae: *Sickius longibulbi*, with a revalidation of *Sickius* (Araneae, Theraphosidae, Ischnocolinae). J Arachnol 30:519–526

Bertani R, Fukushima CS, Silva PI (2008) Two new species of *Pamphobeteus* Pocock 1901 (Araneae: Mygalomorphae: Theraphosidae) from Brazil, with a new type of stridulatory organ. Zootaxa 1826:45–58

Bode F, Sachs F, Franz MR (2001) A peptide venom can prevent the heartbeat from losing its rhythm. Nature 409:35–36

Bond JA, Garrison NL, Hamilton CA, Godwin RL, Hedin M, Agnarsson I (2014) Phylogenomics resolves a spider backbone phylogeny and rejects a prevailing paradigm for orb web evolution. Curr Biol 24:1765–1771

Célérier ML (1981) Cycles biologiques et croissance de quelques espèces. In: Blandin P, Célérier M (eds) Les araignées des savanes de Lamto (Cote-D'Ivoire). Publications du Laboratoire de Zoologie, Ecole Normale Superieure 21, Paris, pp 209–324

Célérier ML (1992) Fifteen years experience of breeding a theraphosid spider from Ivory Coast: *Scodra griseipes* Pocock, 1897, in captivity. In: Cooper JE, Pearce-Kelly P, Williams DL (eds) Arachnida: Proceedings of a Symposium on Spiders and their Allies, London, pp 99–110

Cooke JAL, Roth VD, Miller FH (1972) The urticating hairs of theraphosid spiders. Am Mus Novit 2498:1–43

Costa FG, Pérez-Miles F (2002) Reproductive biology of uruguayan theraphosids (Araneae, Mygalomorphae). J Arachnol 30:571–587

Costa F, Pérez-Miles F (2007) Cruel and irresponsible traffic of spiders in Uruguay. Arachne 12:40

Cutler B, Vuilliomenet F (2001) Surface ultrastructure of *Aphonopelma seemanni* cuspules (Araenae: Theraphosidae). Am Arachnol 62:4

De Carlo JM (1973) Anatomia microscopica de las espermatecas de los géneros *Grammostola* y *Acanthoscurria* (Araneae, Theraphosidae). Physis 32:343–350

Dunlop JA, Harms D, Penney D (2008) A fossil tarantula (Araneae: Theraphosidae) from Miocene Chiapas amber, Mexico. Rev Ibér de Aracnol 15:9–17

Dunlop JA, Penney D, Jekel D (2018) A summary list of fossil spiders and their relatives. In: World Spider Catalog, version 19.0. Natural History Museum, Bern. http://wsc.nmbe.ch. Accessed 25 Jan 2018

Escoubas P, Rash L (2004) Tarantulas: eight-legged pharmacists and combinatorial chemists. Toxicon 43:555–574

Eskov KY, Zonstein S (1990) First Mesozoic mygalomorph spiders from the Lower Cretaceous of Siberia and Mongolia, with notes on the system and evolution of the infraorder Mygalomorphae (Chelicerata: Araneae). Neues Jahrbuch für Mineralogie, Geologie und Paläontologie, Abhandlungen 178:325–368

Ezendam T (2007) The hunt for Blue Foot. J Br Tarantula Soc 23:30–33

Ferretti N, Cavallo P, Chaparro JC, Ríos-Tamayo D, Seimon TA, West R (2018) The Neotropical genus *Hapalotremus* Simon, 1903 (Araneae: Theraphosidae), with the description of seven new species and the highest altitude record for the family. J Nat Hist 52:1927–1984

Foley S, Lüddecke T, Cheng DQ, Krehenwinkel H, Künzel S, Longhorn SJ, Wendt I, Von Wirth V, Tänzler R, Vences M, Piel WH (2019) Tarantula phylogenomics: a robust phylogeny of multiple tarantula lineages inferred from transcriptome data sheds light on the prickly issue of urticating setae evolution. Mol Phylogenet Evol. https://doi.org/10.1016/j.ympev.2019.106573

Fontana MD, Lucas HSM, Vital Brazil O (2002) Neuromuscular blocking action of the *Theraphosa blondii* spider venom. J Venom Anim Toxins 8:1–3

Fuchs J, von Dechend M, Mordasini R, Ceschi A, Nentwig W (2014) A verified spider bite and a review of the literature confirm Indian ornamental tree spiders (*Poecilotheria* species) as underestimated theraphosids of medical importance. Toxicon 77:73–77

Fukushima CS, Bertani R (2017) Taxonomic revision and cladistic analysis of *Avicularia* Lamarck, 1818 (Araneae, Theraphosidae, Aviculariinae) with description of three new aviculariine genera. Zookeys 659:1–185

Gabriel R (2016) Revised taxonomic placement of the species in the Central American genera Davus O. Pickard-Cambridge, 1892, Metriopelma Becker, 1878, and Schizopelma F. O. Pickard-Cambridge, 1897, with comments on species in related genera (Araneae: Theraphosidae). Arachnology 17:61–92

Galiano ME (1969) El desarrollo postembrionario larval de *Grammostola pulchripes* (Simon, 1891) (Araneae, Theraphosidae). Physis 29:73–90

Galiano ME (1973a) El desarrollo postembrionario larval de *Avicularia avicularia* (Linnaeus, 1758) (Araneae, Theraphosidae). Physis 32:315–327

Galiano ME (1973b) El desarrollo postembrionario larval en Theraphosidae (Araneae). Physis 32:37–46

Gallon RC (2002) Revision of the African genera *Pterinochilus* and *Eucratoscelus* (Araneae, Theraphosidae, Harpactirinae) with description of two new genera. Bull Br Arachnol Soc 12:201–232

Gallon RC (2003) A new African arboreal genus and species of theraphosid spider (Araneae, Theraphosidae, Stromatopelminae) which lacks spermathecae. Bull Br Arachnol Soc 12:405–411

Gallon RC (2005) *Ashantia* Strand, 1908 is a junior synonym of Euathlus Ausserer, 1875 (Araneae, Theraphosidae, Theraphosinae). Bull Br Arachnol Soc 13:199–201

Gallon RC, Gabriel R (2006) Theraphosidae egg-sac types. Newsl Br Arachnol Soc 106:5–10

García-Villafuerte MA (2008) Primer registro fósil del género *Hemirrhagus* (Araneae, Theraphosidae) en ámbar del Terciario, Chiapas, México. Rev Ibér Aracnol 16:43–47

Garrison NL, Rodríguez J, Agnarsson I, Coddington JA, Griswold CE, Hamilton CA, Hedin M, Kocot KM, Ledford JM, Bond J (2016) Spider phylogenomics: untangling the Spider Tree of Life. PeerJ. https://doi.org/10.7717/peerj.1719

Guadanucci JP (2014) Theraphosidae phylogeny: relationships of the 'Ischnocolinae' genera (Araneae, Mygalomorphae). Zool Scr 43:508–518. https://doi.org/10.1111/zsc.1206

Guadanucci JPL, Perafán C, Valencia-Cuellar D (2017) The genus *Holothele* Karsch, 1879: the identity of the type species (Mygalomorphae, Theraphosidae). Zoosystema 39:263–271

Hembree DI (2017) Neoichnology of tarantulas (Araneae: Theraphosidae): criteria for recognizing spider burrows in the fossil record. Palaeontol Electron 20.3.45A:1–30

Höfler T (1996) Auswirkungen eines Bisses von *Poecilotheria fasciata* (Latreille 1804). Arachnol Mag 4:8–10

Hüsser M (2018) A first phylogenetic analysis reveals a new arboreal tarantula genus from South America with description of a new species and two new species of *Tapinauchenius* Ausserer, 1871 (Araneae, Mygalomorphae, Theraphosidae). Zookeys 784:59–93

Ibler B, Michalik P, Fischer K (2013) Factors affecting lifespan in bird-eating spiders (Arachnida: Mygalomorphae, Theraphosidae) – A multi-species approach. Zool Anz 253:126–136

IUCN (2017) The IUCN red list of threatened species, version 2017-3. http://www.iucnredlist.org. Accessed 05 Dec 2017

Klaas P (2007) Vogelspinnen. Herkunft. Pflege. Arten. Eugen Ulmer, Stuttgart

Kotzman M (1990) Annual activity patterns of the Australian tarantula *Selenocosmia stirlingi* (Araneae, Theraphosidae) in an arid area. J Arachnol 18:123–130

Kraus O (1978) *Liphistius* and the evolution of spider genitalia. In: Merrett P (ed) Arachnology, Symposia of the Zoological Society of London, vol 42, pp 235–254

Kraus O (1984) Male spider genitalia: evolutionary changes in structure and function. Verhandlungen des Naturwissenschaftlichen Vereins in Hamburg N.F. 27:373–382

Krug LE, Elston DM (2011) What's eating you? Tarantulas (Theraphosidae). Cutis 87:10–12

Lima AG, Guadanucci JPL (2019) Comparative morphology of stridulating setae of Theraphosinae (Araneae, Theraphosidae). Zool Anz 283:58–68

Luddecke T, Krehenwinkel H, Canning G, Glaw F, Longhorn S, Tänzler R, Wendt I, Vences M (2018) Discovering the silk road: nuclear and mitochondrial sequence data resolve the phylogenetic relationships among theraphosid spider subfamilies. Mol Phylogenet Evol 119:63–70. https://doi.org/10.1016/j.ympev.2017.10.015

Marshall SD, Uetz GW (1990a) The pedipalpal brush of *Ephebopus* sp. (Araneae, Theraphosidae): Evidence of a new site for urticating hairs. Bull Br Arachnol Soc 8:122–124

Marshall SD, Uetz GW (1990b) Incorporation of urticating hairs into silk: a novel defense mechanism in two Neotropical tarantulas (Araneae, Theraphosidae). J Arachnol 18:143–149

Mason LD, Wardell-Johnson G, Main BY (2018) The longest-lived spider: mygalomorph dig deep, and persevere. Pac Conserv Biol 24:203–206

Mendoza JI, Francke OF (2018) Five new cave-dwelling species of *Hemirrhagus* Simon 1903 (Araneae, Theraphosidae, Theraphosinae), with notes on the generic distribution and novel morphological features. Zootaxa 4407:451–482

Opatova V, Hamilton CA, Hedin M, Montes de Oca L, Kral J, Bond JE (2019) Phylogenetic systematics and evolution of the spider infraorder Mygalomorphae using genomic scale data. Syst Biol. https://doi.org/10.1093/sysbio/syz064

Peck W, Whitcomb W (1970) Studies on the biology of a spider, *Chieracanthium inclusum* (Hentz). University of Arkansas, Agricultural Experiment Station Bulletin no. 753

Perafán C, Galvis W, Gutierrez M, Pérez-Miles F (2016) *Kankuamo*, a new theraphosid genus from Colombia (Araneae, Mygalomorphae), with a new type of urticating setae and divergent male genitalia. Zookeys 601:89–109

Pérez-Miles F (1989) Variación relativa de caracteres somáticos y genitales en *Grammostola* mollicoma (Araneae, Theraphosidae). J Arachnol 17:263–274

Pérez-Miles F (1998) Notes on the systematics of the little known theraphosid spider *Hemirrhagus cervinus,* with a description of a new type of urticating hair. J Arachnol 26:120–123

Pérez-Miles F (2002) The occurrence of abdominal urticating hairs during development in Theraphosinae (Araneae, Theraphosidae): phylogenetic implications. J Arachnol 30:316–320

Pérez-Miles F, Montes de Oca L (2005) Surface ultrastructure of labial and maxillary cuspules in eight species of Theraphosidae (Araneae). J Arachnol 33:43–49
Pérez-Miles F, Weinmann D (2010) *Agnostopelma*: a new genus of tarantula without a scopula on leg IV (Araneae: Theraphosidae: Theraphosinae). J Arachnol 38:104–112
Pérez-Miles F, Lucas SM, da Silva P, Bertani R (1996) Systematic revision and cladistic analysis of Theraphosinae (Araneae: Theraphosidae). Mygalomorph 1:33–62
Perret BA (1974) The venom of the East African spider *Pterinochilus* sp. Toxicon 12:303–310
Punzo F, Henderson L (1999) Aspects of the natural history and behavioural ecology of the tarantula spider *Aphonopelma hentzi* (Girard, 1854) (Orthognatha, Theraphosidae). Bull Br Arachnol Soc 11:121–128
Raven R (1985) The spider infraorder Mygalomorphae (Araneae): cladistics and systematics. Bull Am Mus Nat Hist 182:1–180
Santana RC, Perez D, Dobson J, Panagides N, Raven R, Nouwens A, Jones A, King GF, Fry BG (2017) Venom profiling of a population of the theraphosid spider *Phlogius crassipes* reveals continuous ontogenetic changes from juveniles through adulthood. Toxins 9:116. https://doi.org/10.3390/toxins9040116
Schiapelli RD, Pikelin BSG (1962) Importancia de las espermatecas en la sistemática de las arañas del suborden Mygalomorphae (Araneae). Physis 23:69–75
Shillington C (2005) Inter-sexual differences in resting metabolic rates in the Texas tarantula, *Aphonopelma anax*. Comp Biochem Physiol 142A:439–445
Shillington C, Peterson CC (2002) Energy metabolism of male and female tarantulas (*Aphonopelma anax*) during locomotion. J Exp Biol 205:2909–2914
Simpson MR (1995) Convariation of spider egg and clutch size: the influence of foraging and parental care. Ecology 76:795–800
Stradling DJ (1978) The growth and maturation of the “tarantula” *Avicularia avicularia* L. Zool J Linn Soc 62:291–303
Trabalon M, Blais C (2012) Juvenile development ecdysteroids and hemolymph level of metabolites in the spider *Brachypelma albopilosum* (Theraphosidae). J Exp Zool 317:236–247
Turnbull AL (1962) Quantitative studies of the food of *Linyphia triangularis* Clerck (Araneae; Lyniphiidae). Canad J Entomol 91:1233–1249
Turnbull AL (1965) Effects of prey abundance on the development of the spider *Agelenopsis potteri* (Blackwall) (Araneae: Agelenidae). Can J Entomol 97:141–147
Turner SP, Longhorn SJ, Hamilton CA, Gabriel R, Pérez-Miles F, Vogler AP (2018) Re-evaluating conservation priorities of New World tarantulas (Araneae: Theraphosidae) in a molecular framework indicates nonmonophyly of the genera, *Aphonopelma* and *Brachypelma*. Syst Biodivers 16:89–107
Valencia-Cuéllar D, Perafán C, Guerrero RJ, Guadanucci JPL (2019) Schismatothelinae spiders (Araneae, Mygalomorphae, Theraphosidae) from Colombia: four new species and an approach to their diversity. Zootaxa 4545:548–562
West RC (1986) Ceratogyrus. J Br Tarantula Soc 1:79–80
West RC, Marshall SD, Fukushima CS, Bertani R (2008) Review and cladistic analysis of the Neotropical tarantula genus *Ephebopus* Simon 1892 (Araneae: Theraphosidae) with notes on the Aviculariinae. Zootaxa 1849:35–58
World Spider Catalog (2020) World Spider Catalog, version 20.5. Natural History Museum, Bern. http://wsc.nmbe.ch. Accessed 28 Sep 2020

Chapter 2
A Molecular Approach to the Phylogeny of Theraphosidae and Their Kin

Stuart J. Longhorn and Chris A. Hamilton

Abstract Molecular data are increasingly helping inform tarantula evolutionary history. This includes redefining taxonomic groups at many levels, from clarifying species limits and matching sexes to elucidating the boundaries of genera and higher taxonomic ranks. We initially overview early molecular studies with tarantulas, before more closely looking at later developments with either a focus on questions around the population–species interface, or on aspects of the broader phylogeny. In both, we consider the gene fragments used for insights, but also introduce the role that newer high-throughput sequencing can play to expand the scope of such datasets. We then move into other approaches offered by the age of genomics, with some focus on mitogenomics versus nuclear genomics. Here we discuss some useful aspects of genomes such as gene arrangements that may be treated as "rare-events" to resolve intractable systematic questions. We then overview transcriptomic methods versus target capture approaches, each of which provide increasingly powerful methods for new insights. Finally, we speculate on where additional taxon sampling is needed to resolve the tarantula phylogeny, before concluding how existing studies now form a solid baseline for future projects, in particular on biogeography or the evolution of body size, venom, or other interesting comparative questions.

2.1 Introduction

Spiders are one of the oldest terrestrial arthropod lineages, dating back 400 million years to the Devonian (Garrison et al. 2016). With over 48,000 species (World Spider Catalog 2019), they constitute one of the most diverse radiations of predators on the planet (see Chap. 8), inhabiting almost every terrestrial ecosystem (Wheeler et al. 2017). One of the most recognizable and charismatic spider lineages is the

S. J. Longhorn (✉)
Arachnology Research Association, Birmingham, UK

C. A. Hamilton
Department of Entomology, Plant Pathology, & Nematology, University of Idaho, Moscow, ID, USA

F. Pérez-Miles (ed.), *New World Tarantulas*, Zoological Monographs 6,
https://doi.org/10.1007/978-3-030-48644-0_2

family Theraphosidae (tarantulas). These are a derived subgroup of Mygalomorphae—the broader lineage that also includes Purse-web spiders, various types of Trapdoor spiders, plus many other groups. As with many spiders, our understanding about the evolutionary relationships of tarantulas and their kin has remained poorly understood until the recent, rapid progress in available data. This is due to advances in studies looking at diverse aspects such as anatomy and physiology, but also the increased availability of molecular data, which have together dramatically improved our knowledge of the Spider Tree of Life.

In general, spider researchers were slow to adopt the use of molecular data to answer evolutionary questions, including those relating to systematics. The latter aspect can be defined as a field encompassing taxonomy and classification, but also phylogenetics—a topic that looks to clarify the evolutionary history and relationships of taxa. Historically, systematics largely relied on the evaluation of morphology to establish the foundations of our modern understanding. In the 1990s, the first molecular-focused studies began to shed light on spider systematics (e.g., after Huber et al. 1993; Gillespie et al. 1994), followed by the first large-scale studies around the start of the new millennium. More recently, molecules have become an increasingly dominant source of data in spider systematics (see Agnarsson et al. 2013), leading to fresh insights about the evolutionary history of many spider lineages. Progress in the study of tarantulas and their kin with molecular data has been particularly slow, but recent years have finally begun to see an explosion of new studies. The decreasing costs of gathering sequence data has democratized its availability, leading to increased use of molecules to aid species descriptions or assist in the definition of higher groupings (generic level etc), while others have focused on using molecules to resolve even wider systematic issues, such as the redefinition of major lineages like subfamilies.

2.2 Overview of Developments in Molecular Studies of Spiders, Especially Tarantulas

During the dawn of the age of molecular systematics, the potential for certain molecules to give insights on systematic questions was often recognized soon after the molecules themselves were first characterized (e.g., Zuckerkandl and Pauling 1965). In the earliest years, some studies achieved insights by evaluating the shared presence or absence of gross features in focal biochemicals such as enzyme proteins. This was often achieved by methods akin to those of morphological taxonomists to evaluate physical similarities and differences. Soon though, the focus of molecular studies shifted toward comparing fine scale variation within sequences, particularly nucleic acids of DNA or RNA, but also directly from proteins as amino acids (e.g., after Zuckerkandl and Pauling 1965). Accordingly, each structural position within any given molecular sequence from one individual is compared against the estimated equivalent position in another individual. For DNA and RNA, the positions typically refer to individual nucleotides, with counts given as sites (or base pairs—b.p.), even if the nucleotides are protein coding. For proteins themselves, the positions refer to individual amino acids (a.a), again called sites. Regardless of the type of molecular sequence used, when large numbers of sites are compared simultaneously between multiple individuals, some of the variation can be informative to mathematical formulae at the core of a computational phylogenetic analysis. The output can be presented as a branching diagram—or rather a phylogeny—as a hypothesis about how the individuals (often exemplifying different taxa) are plausibly related to each other. Results can be informative to help revise taxonomy, such as to clarify which individuals may together form a natural group, and perhaps even correspond to a formal taxonomic rank (for example, a species, genus, tribe, subfamily, etc).

It was another 20 years or so after the basic framework of molecular systematics was established before studies began broadly evaluating diverse animals. An important early paper compared a short repetitive molecule called 5S rRNA across several animals alongside a focus on Fungi (Huysmans et al. 1983). This molecule forms part of a core cellular structure present in most cells called the ribosome. Soon after, several arthropods (i.e., the group for spiders plus more distant relatives like insects, crustaceans etc.) were studied. One of the few animals used at this nascent stage was a North American tarantula recorded as *Eurypelma californicum*,[1] where the 5S rRNA sequence was determined alongside those from a handful of other arthropods. This was done by reading the messenger RNA (rather than DNA as most modern methods) in a labor-intensive approach yielding a mere 120 b.p. of sequence data (Hendriks et al. 1986). The resultant phylogenetic analyses (considered crude by modern standards) reflected some expected taxonomic groups anticipated by older morphological studies, but other questionable aspects led these authors to conclude that their data were too short for enough character variation, particularly across such a deep evolutionary timescale (Hendriks et al. 1986). Follow-up studies then began a quest for greater sources of variation from longer molecular sequences, again predominantly those associated with the ribosome. For genes such as for the nuclear encoded small ribosomal subunit (i.e., 18S rRNA in animals, see Fig. 2.1a left), multiple sequencing reads were soon combined as overlapping pieces, leading to a major study that compared about 1000 b.p. of 18S rRNA across several animals (Field et al. 1988). The same year, another study generated the first 18S sequence from a tarantula—again using "*E. californicum*" (Hendriks et al. 1988). This study used an alternative method of first digesting DNA into many fragments, before selecting only those matching pieces of this focal molecule. Once many fragments were sequenced, their overlaps were determined to give a total read of 1814 b.p., nearly the entire length of the complete molecule. When compared against other

[1] The name "*Eurypelma californicum*" was used in over 100 studies mainly of biochemistry, with the likely identity being *Aphonopelma hentzi* (Nentwig 2012).

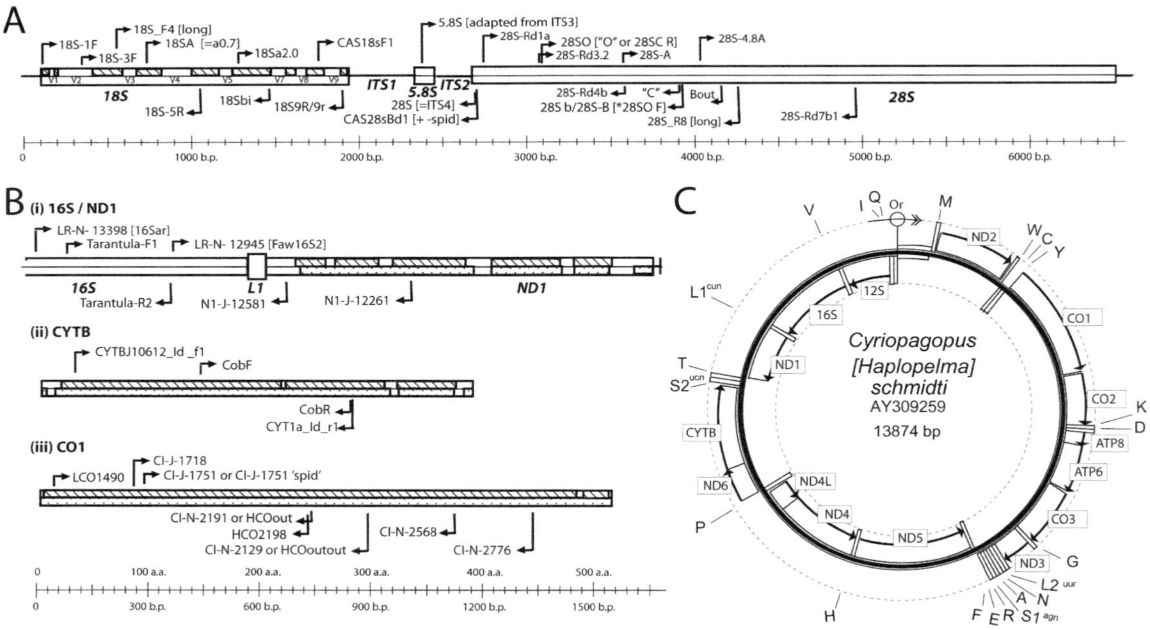

Fig. 2.1 (**a**) The rRNA cistron in the nuclear genome of tarantulas and their kin based on combined fragments of various *Aphonopelma* spp. and select other spiders. Sections are named according to standard nomenclature, so that 18S, 5.8S and 28S (boxed regions) are transcribed regions of rRNA, while ITS1 and ITS2 are highly variable internal transcribed spacers. For 18S, the more conserved domains are indicated (hashed boxes above midline) with variable domains in between (labeled V1–V9). Locations of select PCR primers used for various spiders are shown. Notably, some confusion exists in the literature on the identity of 28S primers "O" and "C", herein given as in original sources. Scale in 100 b.p. increments. (**b**) Select genes from the mitochondrial genome, where for PCGs the relative conservation of amino acids is shown using GBlock analysis as per Masta et al. (2009) under either higher stringency (hashed boxes above midline) or lower stringency (dotted blocks below). Again the position of select primer sequences are indicated. Scale in 30 b.p./10 a.a increments. (**c**) Published structure of the complete mitochondrial genome for a tarantula, showing the relative size and arrangement of genes and other notable features, where single letter codes represent tRNAs and Or stands for the Origin of replication

such near-complete 18S sequences from diverse organisms, these gave a combined character set with almost double the number of analysed sites than any such previous studies. The resultant molecular phylogeny was found to recover many taxonomic groups anticipated by other studies, and helped solidify the utility of molecular data for systematics (Hendriks et al. 1988).

Meanwhile, novel methods were devised for direct amplification of selected gene regions via the "Polymerase Chain Reaction"—abbreviated to PCR (Mullis and Faloona 1987). This method was quickly seen as easily automatable, using a pair of short custom-made sequences called primers that bind to complimentary sites on the desired nucleic acid molecule, with each of a pair binding to a rather conserved region either side of the selected fragment. Then, a polymerase enzyme adds nucleotides between the primer pair to form new chains that match the as yet unknown intervening DNA. Essentially, multiple copies are then amplified by repeated thermal cycles (i.e., heating and cooling) of these core components in a biochemical mixture. This results in massive numbers of replicates of the selected fragment that enables them to be later sequenced in a separate reaction, which until recently was largely based on the related method of Sanger sequencing (Sanger and Coulson 1975). Over the next years, systematists began using PCR to target mitochondrial DNA (mtDNA), which in animals forms a separate circular genome to that of the cell nucleus. The mitochondrial genome (hereafter mitogenome) occurs in most animal cells with a high copy number that facilitates PCR amplification, and typically has a very conserved gene content, including two further rRNA-coding genes named 12S and 16S (See Fig. 2.1b, c). A significant early study used PCR to successfully amplify three distinct regions of the mitogenome from diverse animals to demonstrate the utility of this approach for systematics (Kocher et al. 1989). This included 386 b.p. of 12S rRNA from a tarantula, becoming the first published mitochondrial sequence for this spider family, but from an unknown species. The utility of these methods was then supported by others sequencing a shorter fragment (about 250 b.p.) of this same molecule from two "true spiders" (i.e., Araneomorphae), which were compared against diverse animals (Croom et al. 1991; Ballard et al. 1992). Soon, in one of the earliest molecular systematic studies of spiders, other 12S sequences were compared across multiple species in the same genus (Gillespie et al. 1994). Around this time, other researchers used PCR-based methods to investigate the utility of another region of the mitochondrial genome for spider systematics. In a study focused on the araneomorph family Ctenidae, a fragment from 16S rRNA (of about 450 b.p.) was compared across diverse spiders (Huber et al. 1993). As with many earlier studies, a tarantula was also included in that study—therein specified as an unknown *Psalmopoeus* sp.—but its sequence data was not integrated into their phylogenetic analysis. In the following years, many distinct regions of spider mitogenomes became accessible to PCR following development of a large array of primers (after Folmer et al. 1994; Simon et al. 1994). Subsequent studies on spiders and other taxa exploited these primers (and others) to amplify various mitochondrial regions including from protein coding genes (PCGs)—the most notable being the Cytochrome Oxidase subunit 1 (= CO1). This gene often became the go-to mitochondrial sequence (see sections below) in later years, due in part to its ubiquitousness across animals as it encodes an essential protein. That said, the conserved presence of the same 37 genes across mitogenomes of diverse animals also facilitated studies with other mitochondrial fragments. However, due to differences in the arrangement of certain mitochondrial genes being found early on in several taxa, some researchers began to investigate whether the order of these genes could be informative to systematics, rather than just variation between sequences themselves (after Boore et al. 1995). Further exploration was enabled by advances in PCR techniques with new protocols (such as "long-PCR" after Roehrdanz (1995), see below) and the development of high-fidelity enzymes.

A little earlier, building on the near complete nuclear 18S rRNA sequence for a tarantula (as discussed above) plus the success of initial systematic studies with this region; other researchers began using PCR-based methods to generate 18S from diverse arthropods (although some direct RNA sequencing was also still employed initially). With the addition of a second 18S sequence from *"Rhechostica chalcodes,"* this became the first genetic fragment (totaling about 700 b.p.) compared between two tarantulas (Turbeville et al. 1991). Due in part to the larger size of this fragment than most other gene sequences compared at the time, this same region became repeatedly targeted in many later studies. This and adjacent regions of nuclear rRNA (termed the rRNA cistron, Fig. 2.1a) have often been preferred due to their highly repetitive nature where tandem repeats of the entire 18S-28S region facilitated PCR amplification. Large sections of 18S could also be easily amplified from multiple overlapping fragments as PCR-based protocols improved, then combined per taxa. Sequence data from different genes began to be published together in combined "multigene" analyses (i.e., after Wheeler et al. 1993), in particular with other nuclear rRNA fragments. For example, parts of the adjacent 28S rRNA were used to investigate the broadest arthropod relationships (Friedrich and Tautz 1995), or later those of diverse arachnids (Wheeler and Hayashi 1998). Systematic insights on spiders were also gained from 28S rRNA alone, for example revealing support for the monophyly of the araneomorph subgroup Entelegyne, using a North American tarantula *Aphonopelma sp.* for the broadest comparison (Hausdorf 1999). To move away from dependence on rRNA and generate additional molecular characters, other studies around this time shifted investigation onto various protein-coding genes (PCGs) of the nuclear genome for their potential to inform spider systematics. One of the earliest of such studies examined the potential use of Polyubiquitin (Wheeler et al. 1993). Here, multiple gene-copies likely facilitated PCR amplification, although this was only successful for a few arthropods and the adopted fragment had a short (<250 b.p.) length. Soon other nuclear encoded PCGs such as EF1α (as 364 b.p.) were generated for systematic questions again first using a tarantula to represent mygalomorphs (Regier and Schultz (1997) with *Aphonopelma chalcodes*), while another early PCG was Histone H3 (see Colgan et al. 1998). A wealth of molecular studies followed in subsequent years using an even broader selection of gene fragments to investigate diverse questions about spider systematics (see below, also Agnarsson et al. 2013).

Below, we give an overview of more recent developments in molecular systematic studies of spiders and their allies, with focus on mygalomorphs, and where possible specifically with tarantulas (family Theraphosidae). We initially focus on two approaches that increasingly diverged around the turn of the millennium, divided into: (i) Studies primarily concerned with closely related allies such as between species; and (ii) Studies often concerned about broader relationships, such as between genera, subfamilies or wider.

The first approach typically relates to species delineation, where a fragment of CO1 has become the go-to genetic region to help rapidly distinguish species, delimit species boundaries, or uncover lineages of novel genetic diversity. This may be referred to as DNA barcoding (e.g., Hebert et al. 2003a, b). Using CO1, investigators can take advantage of its relatively conserved nature (i.e., protein-coding sequence), while capitalizing on faster evolving 'silent' substitutions in the third codon position. Due to the degeneracy in the genetic code, these third codon positions contain information important at the species-level split, relative to uncertainties caused by saturation (Simon et al. 1994; Folmer et al. 1994). By sequencing this same region (or same few regions) from ever-larger numbers of taxa (i.e., a more taxa approach), fragments of CO1 have become increasingly useful in separating or even identifying species of tarantulas (Graham et al. 2015; Hamilton et al. 2011, 2014, 2016a; Hendrixson et al. 2013, 2015; Hüsser 2018; Montes de Oca et al. 2016; Petersen et al. 2007; Mendoza and Francke 2017; Ortiz and Francke 2017; Ortiz et al. 2018), and also more broadly across spiders (Astrin et al. 2006; Barrett and Hebert 2005; Blagoev et al. 2013, 2016; Castalanelli et al. 2014; Hedin and Carlson 2011; Kuntner and Agnarsson 2011; Lopardo and Uhl 2014; Miller et al. 2013; Rix et al. 2017b; Robinson et al. 2009; Satler et al. 2011).

The second approach often relates to broader groupings of taxa, where systematic studies have tended to focus on insights from multiple gene regions simultaneously (i.e., a more genes approach) which are compared across taxa of concern. Usefully, the various included fragments can be representative of scattered genomic locations, each potentially subject to different selective pressures that can enhance overall information content once combined. However, region selection has rarely been driven by utility, but rather accessibility. This has often caused tough a priori choices for a systematist about which of many potential gene fragments to adopt in a study. The inability of any preselected fragments to resolve systematic questions at multiple hierarchical levels due lack of historical signal, incongruence between fragments and confounding factors such as paralogy, have led to uncertainties about some aspects of phylogenies with this approach. That said, the combined resolution from multiple fragments has often led to significant progress in our understanding of spider systematics, with valuable insights at multiple levels, including within tarantulas (Hüsser 2018; Lüddecke et al. 2018; Ortiz et al. 2018; Turner et al. 2018; etc).

Finally, in both above approaches, we note that the field of molecular systematics has recently shifted toward the more commonplace adoption of increasingly cost effective high-throughput sequencing (HTS) via adoption of next-generation sequencing (NGS) technologies (e.g., see Brewer et al. 2014 for review). By necessity there is some overlap in the studies detailed below, which are not always mutually exclusive.

2.2.1 Clarification of Species Limits

The earliest published molecular study to explicitly focus on systematic questions for any mygalomorph was an assessment of population variation in the Californian trapdoor spider *Aptostichus simus* (Bond et al. 2001). As with many earlier molecular studies, this relied on rRNA sequences (here from mitochondrial 16S) rather than PCGs. Comparisons of rRNA sequences between different mygalomorphs from multiple genera soon followed, initially for three *A. simus* individuals against single representatives of a few other genera (Smith and Bond 2003). In follow-up studies, focus often shifted to comparisons of the CO1 gene across multiple individuals. The earliest example with mygalomorphs involved another genus of Californian trapdoor spiders, namely *Apomastus,* where CO1 sequences helped differentiate an undescribed 'cryptic species' that until then could not be recognized morphologically with either females or juveniles (Bond 2004). The switch to CO1 reflected a movement for DNA sequences to take a core role in taxonomy, and even provide a scaffold for species identification and delimitation (after Tautz et al. 2002; Hebert et al. 2003a, b), thereafter often called "DNA barcoding." Both then and since, an investigator has been faced with choices about which of several possible gene-fragments to exploit for a new molecular systematic study. The rapid growth of potential "molecular markers" around the new millennium became of increased concern to molecular systematists (i.e., after Caterino et al. 2000), but the higher information potential of larger and/or varied gene fragments led others to simply suggest "it seems advisable to use more than one sequence region for assigning taxonomic status" (Tautz et al. 2003, p.72). However, several factors meant that it was either difficult or unfeasible for many systematists to obtain molecular sequences back then (and still for many today) due to the specialist infrastructure required, as well as technical skills and expense. Given this, proposals for systematists to routinely sequence multiple fragments were especially unrealistic, especially to those in less developed countries. Against this, calls for the unifying role that may be played by global adoption of a single short DNA region (i.e., CO1) as the molecular "bio-identification system for animals" became increasingly dominant. This "DNA Barcoding" approach had core claims that a fragment of CO1 could provide a "molecular-based species identification" (after Hebert et al. 2003a, b, e.g., Barrett and Hebert 2005). Many criticisms soon followed, such as fears over the abandonment of morphology (see Hamilton et al. 2014 for overview) or that mistaken identifications could arise, especially when the genetic databases lacked the most appropriate comparison data or contained errors.

Conceptual advances in "DNA barcoding" over the next years focused on questions about species delimitation (e.g., Pons et al. 2006) including several studies of araneomorphs that generally showed CO1 could be a useful aid in species identification and delimitation, and particularly in cases where the genetic variation among populations was considerably less than between closely related species—the "barcode gap" (e.g., Astrin et al. 2006; Barrett and Hebert 2005; Blagoev et al. 2013, 2016; Castalanelli et al. 2014; Kuntner and Agnarsson 2011; Lopardo and Uhl 2014; Miller et al. 2013; Robinson et al. 2009). For mygalomorphs, the study of genetic variation and species delimitation has been carried out across several families, again often relying on CO1, or at least in part combined with other gene regions. These [following the scheme of Opatova et al. (2019)] include studies of: (1) Anamidae, *Hesperonatalius* (Castalanelli et al. 2017) or various genera (Harvey et al. 2018); (2) Antrodiaetidae *Aliatypus* (Hedin and Carlson 2011; Satler et al. 2011, 2013; Starrett et al. 2018) and *Antrodiaetus* (Hendrixson and Bond 2005, 2007, 2009; Starrett and Hedin 2007; Hedin et al. 2013); (3) Atracidae, *Atrax* and *Hadronyche* (Beavis et al. 2011); (4) Ctenizidae, *Cyrtocarenum* (Kornilios et al. 2016); (5) Euctenidizae, *Apomastus* (Bond 2004; Bond et al. 2006), *Aptostichus* (Bond and Stockman 2008; Bond 2012), *Promyrmekiaphila* (Stockman and Bond 2007) and *Myrmekiaphila* (Bailey et al. 2010); (6) Halonoproctidae, *Ummidia* (Opatova et al. 2013, 2016); (7) Idiopidae, multiple genera (Rix et al. 2017a), *Blakistonia* (Harrison et al. 2018), *Bungulla* (Rix et al. 2018b), *Cataxia* (Rix et al. 2017b), *Gaius* (Rix et al. 2018a), *Titanidiops* (Opatova and Arnedo 2014a); (8) Macrothelidae, *Macrothele* (Arnedo and Ferrández 2007); (9) Microhexuridae, *Microhexura* (Hedin et al. 2015); (10) Migidae, *Moggridgea* (Cooper et al. 2011) and that plus *Bertmainius* (Harrison et al. 2017); (11) Nemesiidae, *Calisoga* (Leavitt et al. 2015), *Iberesia* and *Nemesia* (Mora et al. 2017); (12) Theraphosidae, *Aphonopelma* (Hamilton et al. 2011, 2014, 2016a; Wilson et al. 2012; Hendrixson et al. 2013, 2015; Graham et al. 2015), *Brachypelma* (Petersen et al. 2007; Longhorn et al. 2007; Mendoza and Francke 2017), *Bonnetina* (Ortiz and Francke 2016, 2017) plus several allied genera (Ortiz et al. 2018), and *Grammostola* (Montes de Oca et al. 2016).

While many of these aforementioned studies on mygalomorphs have successfully applied molecular data to help make informed decisions about species definitions, several have found that extensive genetic sub-structuring can make it hard to distinguish between fragmented populations and independent evolutionary lineages (e.g., Opatova and Arnedo 2014a; Leavitt et al. 2015). This can be related to the tendency of many mygalomorphs to naturally have populations sub-divided into clustered aggregations. As a result, decisions about revised species limits that incorporate molecular data should be made in an informed manner that considers the group's wider taxonomy, ecology, biogeography, and population genetics in an integrative decision making process (after Bond and Stockman 2008; e.g., Hamilton et al. 2016a). One important aspect of the above wealth of molecular studies is a strong bias towards either Australian or North American taxa. The power of molecular data to aid species delineation and provide context for revision of Australian mygalomorphs was clearly shown in an influential analysis of multiple families, where divergence in CO1 between putative species was around 9.5% (Castalanelli et al. 2014). Here, the same data also showed great utility to associate otherwise unidentifiable juveniles and females with their corresponding adult males—the latter being vital for morphological species definition yet poorly represented (i.e., by only 5% of their specimens). Another notable study used CO1 in combination with other genes to reveal a remarkable case of probable long-distance oceanic dispersal to Australia in Migidae (Harrison et al. 2017). Other important studies have combined morphological and molecular insights for revision of several Idiopidae genera, together revealing a remarkable degree of endemism in the Southwest of the continent (e.g., Rix et al. 2018a, b). However, there are no comparable studies yet published for Australian tarantulas, but these are likely on the horizon. Meanwhile in North America, studies on various mygalomorphs, especially those in California, have revealed the high potential for endemism due to genetic fragmentation linked to limited dispersal in multiple families (Bond et al. 2001, 2006; Bond 2004; Stockman and Bond 2007; Bond and Stockman 2008; Hedin and Carlson 2011; Satler et al. 2011, 2013; Hedin et al. 2013; Hamilton et al. 2014; Hendrixson et al. 2015; Leavitt et al. 2015; Starrett et al. 2018). Several important studies have been published on North American tarantulas, which have increasingly used integrative methodological approaches. Particularly notable are studies on the genus *Aphonopelma,* culminating with an integrative revision of all species found within the United States (Hamilton et al. 2016a, b and references therein), including a comprehensive molecular sampling resulting in the most heavily DNA barcoded spider genus at the time (n = 1032 specimens). Another set of important studies focused on the *Bonnetina* from Mexico, using multiple lines of evidence to arrive at their preferred "integrative delineation" (Ortiz and Francke 2016, 2017). For other North American lineages, a growing framework of molecular data now exists for *Brachypelma* (Mendoza and Francke 2017; Turner et al. 2018). In addition to aiding formal species delimitation within this genus of conservation importance (see Chap. 16), here sequence data can form a framework for molecular identification to inform stakeholders involved in trade, e.g., to facilitate decision making by law enforcement officials on the identity of traded livestock (Petersen et al. 2007; Longhorn et al. 2007). Outside of North America, a selection of CO1 barcodes has already been published for many natural variants of several *Grammostola* spp. from Uruguay (Montes de Oca et al. 2016). Again, in addition to aiding formal delineation of species, such sequence data now provides a valuable molecular framework against which other specimens of dubious origins can be compared, with potential to at least clarify the country of origin and even the precise geographic region.

Recently, remarkable developments in various next-generation sequencing (NGS) technologies have led researchers to investigate their potential to yield faster, more efficient and cheaper molecular species delineation. One important limitation of the older Sanger sequencing technology has been the relatively short reads of only a few hundred base pairs. There can also be significant logistical effort and expense when each read (from every gene fragment and taxon) is sequenced in isolation—which is not particularly cost-effective when scaled up (Shokralla et al. 2015). An important development in NGS (or high-throughput sequencing) has been the ability to pool samples from multiple sources—either extracted DNA or amplified fragments—before bioinformatically differentiating them. For example, Shokralla et al. (2015) showed that pooled DNA extracts from arthropods could be differentiated after sequencing and used to discriminate species. Using relatively short reads from Illumina sequencing, it was possible for them to match around 75% of 1010 individual insects (and allied arthropods) in a mixed array of specimens to morphological identifications. For large-scale DNA barcoding projects, these authors calculated that their method equated to a 27% reduction in time, and a 79% reduction in costs compared to Sanger sequencing—where they highlighted that a high failure rate is not uncommon. Elsewhere, the characterization of CO1 barcodes from mixtures of DNA has been termed "metabarcoding" (Yu et al. 2012). The short read lengths of early NGS technologies (mostly of only a few hundred b.p.) limited the number of possible applications, but increasingly long-read technologies like SMRT or Nanopore sequencing may play an increasingly important role in species delimitation (Krehenwinkel et al. 2019). These technologies are capable of reads of several thousand base pairs, greater than typical for Sanger technologies (i.e., the entirety of fragments often used for CO1 barcodes at around 650 b.p.), but increasingly at a fraction of the cost per base compared to such older methods. Multiple PCR products can also be pooled prior to read generation—even originating from different source specimens—as shown by the simultaneously sequencing of CO1 barcodes from two sets of *c.* 500 insects with Nanopore technology (Srivathsan et al. 2018). For spiders, Krehenwinkel et al. (2019) used a dual indexing approach on PCR amplified fragments of nuclear rRNA (ITS1–5.8S–ITS2) from a broad selection of araneomorphs, before being subject to Nanopore sequencing. Here, later analyses of these "long rDNA amplicons" allowed species delimitation between multiple species of *Tetragnatha* from Hawaii, and also recovered excellent support at several hierarchical levels in phylogenetic analyses. The essence of the approach is that PCR primers used in amplification carry a 15 base pair index. When two primers are used in combination, a unique pair of indexes can be incorporated into each amplification product, that together later can bioinformatically assign the output sequences back to its original source. It is therefore possible to sequence any PCR amplified material from different taxa in parallel, proved each has a different pair of indexes. Given the successful results of aforementioned studies, we can anticipate that the exploitation of similar technologies for mygalomorphs such as tarantulas will likely be an extremely powerful aid to their species delineation in the near future.

2.2.2 Broader Phylogenetic Scheme

Since the turn of the millennium, the bulk of molecular studies focused on broader taxonomic relationships of spiders (i.e., genus level and beyond) have been limited to a handful of genetic fragments that could be amplified by PCR and sequenced by established Sanger technology (see also Agnarsson et al. 2013). These fragments have increasingly been published in ever-larger combined multigene sets. This approach has been fruitful toward advancing our understanding of broader spider relationships alongside important evolutionary questions, including for some tarantulas. Much earlier, this approach was first successfully used for spiders of the araneomorph family Salticidae (Hedin and Maddison 2001; Maddison and Hedin 2003), followed by studies within other allied families, or to clarify relationships between them (e.g., Spagna and Gillespie 2008; see Agnarsson et al. 2013 for overview). The most widely employed genetic fragments for such studies included mitochondrial PCG for CO1 (see above) and various rRNA fragments—particularly 16S, 18S, and 28S. Together, these and other often-used fragments began being referred to as the "usual suspects" due to their widespread usage in spider systematics (e.g., see Agnarsson et al. 2013). Over time, this set expanded to 13 or so fragments becoming regularly used in multigene studies, although the subset used varies widely (see Wheeler et al. 2017). A few regions of the nuclear genome have been considered "usual suspects," especially portions of rRNA cistron encoding 18S (~800–1800 b.p.) and 28S (~800–2700 b.p.), both of which are often sequenced from multiple overlapping fragments (see Fig. 2.1a). Other parts of the rRNA cistron such as the two internal transcribed spacers have been used for spider systematics, such as ITS1 (~450 b.p. in known tarantulas), ITS2 (~300 b.p. as previous), but these can be of limited utility for broader comparisons as they may vary hugely in size across taxa. However, in tarantulas, both spacers can be easily amplified together with PCR as the complete ITS–5.8S–ITS2 is not especially large (at ~800 b.p.). Elsewhere, a select few nuclear encoded PCGs have been included in the "usual suspects," such as actin (~900–1300 b.p.), EF1γ (~600–1200 b.p.), H3 (~200–325 b.p.), and wingless (~350 b.p.). Most other "usual suspects" are from the mitochondrial genome (see Fig. 2.1b, c), in particular rRNA-coding fragments of 12S (~400 b.p.) and 16S (~250–1200 b.p.), plus fragments of select PCGs such as ND1 (~350 b.p.), CO1 (~200–1200 b.p.), and CYTB (~650 b.p.), or less commonly CO2 (~400 b.p.).

Many of the aforementioned "usual suspects" have been repeatedly used in molecular systematic studies of spiders due to their practical feasibility (i.e., their ability to be amplified and sequenced with relative ease), regardless of the fragments' phylogenetic informativeness. When phylogenetic information content of the sequences themselves is poor, the recovered fragments can fail to contain suitable variation (i.e., information) for the focal systematic questions, and so do a poor job at recovering support for key nodes on a resultant phylogeny (Agnarsson et al. 2013; Dimitrov et al. 2017; Wheeler et al. 2017). Despite such concerns, several molecular studies on mygalomorphs—including some for tarantulas—have successfully used a subset of these "usual suspects" to greatly inform our knowledge of systematics. Building on the success of earlier studies with various rRNA (e.g., Smith and Bond 2003), fragments of the two largest nuclear rRNA (namely 18S and 28S) were sequenced from a broad sample of mygalomorphs to help revise classification (Hedin and Bond 2006). Here, Barychelidae was found to be the most plausible sister group to tarantulas for first time with molecular data, although the limits of monophyly for both families remained unclear due to limited taxon sampling. Molecular data also indicated that Paratropididae are more distant relatives, despite prior views to the contrary from morphology (Hedin and Bond 2006; Bond et al. 2012). An important practical note is that significant laboratory effort was needed for sequencing the relatively long rRNA fragments used here, where lengths of 1700 b.p. (for 18S) and 1900 b.p. (for 28S) exceeded feasible Sanger reads (of around 750 b.p. per primer). This required a nested series of internal reactions at extra effort and cost (see below for "primer walking"). Else, these authors reported that while their 18S sequences had little variation in length between different mygalomorphs, it was extreme in their 28S, which necessitated a "multifaceted alignment approach that incorporated both different alignment methods and an array of alignment parameters" (Hedin and Bond 2006, p.458, also see Bond and Hedin 2006).

Due to the paucity of nuclear PCGs being used around this time, fragments from a select few nuclear-encoded PCGs were specifically evaluated across a broad sample of mygalomorphs, such as EF1γ (Ayoub et al. 2007). Still, one or both of the aforementioned nuclear rRNA genes continued to be used alongside such PCGs in dedicated multigene studies. For example in family Antrodiaetidae, a combination of both 18S and 28S alongside CO1 and later EF1γ were used to establish a well-supported phylogeny of *Antrodiaetus* (Hendrixson and Bond 2007, 2009). Similarly, the combination of 18S and 28S, plus EF1γ, helped to resolve uncertainties about the family Hexathelidae—supporting its polyphyly alongside revised placement of the *Macrothele* (Opatova and Arnedo 2014b). Elsewhere, revision of other mygalomorph genera such as *Ummidia* again relied on 28S and EF1γ (Opatova et al. 2013), as well as H3 sequences. In an important move toward consensus, insights from morphology together with 18S, 28S, and EF1γ helped resolve earlier uncertainties about many aspects of the broadest relationships within mygalomorphs (Bond 2012). Here, the combination of multigene molecular data with morphology helped reinforce earlier concerns about the non-monophyly of certain families (in particular Nemesiidae and Cyrtaucheniidae), leading authors to revise several genera under a new family Euctenizidae. As for earlier findings with rRNA, their combined analysis again provided strong support for Barychelidae as the sister group to the tarantulas (i.e., Theraphosidae), although sampling of both families was again small. More recently, adoption of NGS has begun to further inform such broader aspects of mygalomorph relationships alongside other aspects of spider systematics (see below, e.g., after Bond et al. 2014). Yet regardless, several multigene studies have continued to use one or more of the "usual suspects" to make significant progress in resolving the phylogenetic relationships of various mygalomorphs. Following Opatova et al. (2019), these include important studies on the broader scheme within several revised mygalomorph families, namely Anamidae (Harvey et al. 2018), Halonoproctidae (Opatova et al. 2013, 2016), Idiopidae (Opatova and Arnedo 2014a; Rix et al. 2017a), Macrothelidae (Opatova and Arnedo 2014b), Migidae (Harrison et al. 2017), and Nemesiidae (Mora et al. 2017).

For tarantulas, several publications have recently emerged using molecular data from some "usual suspects." Particular focus has been given to their broadest relationships, such as between subfamilies and tribes. The first, by Turner et al. (2018), used only a single amplified fragment for 16S/ND1 (of ~1000 b.p.), where fragments of two "usual suspects" were co-amplified. These authors focused on the most speciose subfamily Theraphosinae (see Chap. 5), most notably finding support for three tribes, but also importantly indicating the non-monophyly of two focal genera, namely *Brachypelma* and *Aphonopelma*. Around the same time, Lüddecke et al. (2018) used a more comprehensive sampling of five fragments to create a combined multigene set of around 3500 b.p. encompassing six genes. Here, about half their molecular data was of mitochondrial origin: namely from CO1 (285 b.p.) and 12S/16S rRNA (1463 b.p.). The other half was nuclear: namely from 18S (1024 b.p.), 28S (522 b.p.), and H3 (204 b.p.). When analysed alone, each fragment only provided weak support at many nodes, which in some cases (e.g., CO1) could have been due to the lack of informative variation due to a relatively short length. 18S was found to have an especially idiosyncratic pattern of molecular evolution, both across sites and between several taxa—i.e., varying from highly conserved across many taxa at many sites, while a few taxa or sites were highly divergent. Yet, when the multiple fragments were analyzed together, the multigene data provided a useful combined set for new insights about relationships between tarantula subfamilies (Lüddecke et al. 2018). One important finding was a clear biogeographic pattern, where major clades corresponded to groups of taxa distributed on particular continents (see Chap. 6). Several well-supported nodes also provided confidence about the validity of some previously contentious subfamilies such as Indian Poecilotheriinae, American Psalmopoeinae, or the African Stromatopelminae. For the latter, there was strong support for the African Harpactirinae as their closest relatives, rather than others such as American Aviculariinae (i.e., Fukushima and Bertani 2017; after West et al. 2008; etc). As also found by Turner et al. (2018), Psalmopoeinae were resolved as closely related to Schismatothelinae—a recently proposed American subfamily derived from partial dismemberment of "Ischnocolinae" (Guadanucci 2014)—rather than, i.e., with Aviculariinae (see Chap. 4). These findings were echoed in another multigene analysis in the same year (Hüsser 2018) based on fragments from CO1 (~700 b.p.), 16S (~270 b.p.), and 28S (of ~800 b.p.). Here, combined molecular data plus morphological insights again supported Psalmopoeinae as a unique subfamily, but also importantly included the first molecular data from *Ephebopus*, finding strong support for its inclusion in this subfamily (Hüsser 2018). Soon after, another informative multigene study by Ortiz et al. (2018) was published, again using several of the "usual suspects." This focused on a broad selection of genera in the dominant new-world subfamily Theraphosinae allied to *Bonnetina*. Here, a set of six genetic fragments were combined to give ~4800 b.p., including CO1 (~950 b.p.), ITS1 (~450–700 b.p.) and EF1γ (~650 b.p.) from the 'usual suspects'. Importantly, these authors developed three additional nuclear fragments, namely MID1IP1 (~550 b.p.), MRPL44 (~830 b.p.), and I3568 (903 b.p.). Like EF1γ, these three latter genes were anticipated to be single-copy, but potential for problematic allelic variation led them to use a modified amplification approach (nested PCR) to maximize specificity (see Ortiz et al. 2018). The resultant phylogeny from this multigene dataset was mostly well-resolved with strong support for most groups, and the first estimated ages for some key historical events in the evolution of Theraphosinae (see Chap. 5). For example, they proposed the origin of this subfamily in the late Cretaceous ~92 million years ago (=Ma), with later divergence into three tribes in a relatively short time close to the Cretaceous–Tertiary boundary (68–63 Ma). Their focal genus *Bonnetina* was estimated to have originated in the early Miocene (~23 Ma), and was recovered as a well-supported group (except for one species transferred to an allied genus). Alongside this was useful exploration of variation in some diagnostic morphological attributes relative to their phylogeny, but also of molecular variation for their chosen fragments. On their dated scheme, if each fragment was analyzed individually, CO1 and ITS1 were more informative than any of the four other PCG fragments across the whole topology (i.e., from very recent splits to earlier), but together the latter four were more informative for earlier divergences (Ortiz et al. 2018).

Overall for tarantulas, the burgeoning wealth of multigene molecular studies reflects a general drive of molecular systematists towards ever-larger multigene datasets (both in total sites and scope of taxon sampling) for addressing questions about phylogeny. This of course means that both the required time, effort and expense grows significantly when using labor-intensive Sanger sequencing approaches. To overcome this, studies focused on Australian mygalomorphs (and particularly the family Idiopidae) have made important advances by simplifying sequencing methodology for such multigene approaches. For example, Rix et al. (2017a) adopted a parallel-tagged amplicon sequencing (TAS) approach (Bybee et al. 2011) to form an impressive multigene set of up to 18 genes (from 16 fragments) for a trial set of 16 arachnids (mostly spiders, with 12 mygalomorphs). This involved pooling the various PCR fragments for each specimen during library preparation, then also pooling those of different taxa before applying NGS on the Illumina MiSeq platform. Subsequently, the phylogenetic informativeness of each gene (which included several of the "usual suspects") was assessed. From the resultant rankings, a preferred subset of 12 fragments were next amplified across a more diverse sample of focal taxa, namely of >100 Australian Idiopidae. These were again sequenced together by the TAS approach. Later Rix et al. (2017b) again successfully used this approach with only two fragments in a study focused on *Cataxia* (Idiopidae). Around this same time, Krehenwinkel et al. (2018) presented what was heralded as a "flexible, simple and cost-efficient means to perform multilocus phylogenetic analysis" for a trial sets of either 42 arthropods or 79 spiders (from six families). Here, several "usual suspects" were co-amplified with multiplex PCR to efficiently gather set of eight genes. Then, an additional round of amplification (indexing PCR) introduced adaptors and dual indexes that allowed bioinformatic identification of all fragments after being simultaneously sequenced. Together, these latter studies have successfully demonstrated how it is feasible to co-amplify and sequence multigene datasets from spiders both quickly and at reasonable cost. Adoption of this or other such developing techniques for systematic studies of tarantulas and their allies will likely revolutionize data gathering of the "usual suspects" and such fragments.

2.3 Expansion into Genomic-Scale Sequencing Initiatives

In sections below, we provide an overview of other developments in molecular studies, again where possible focusing on the relevance to systematics, and in particular for mygalomorph spiders and especially tarantulas. Regardless of taxa, advances in various NGS technologies now make it increasingly cost-effective to expand molecular projects into high-throughput sequencing to characterize large numbers of genes for a greater proportion of biodiversity. In the following sections, we divide these approaches into: (i) Mitochondrial genomics (aka Mitogenomics) and (ii) Nuclear genomics.

The first approach exploits the moderate amounts of sequences from multigene sections of mitochondrial genome, which totals around 14,000 sites (i.e., 14,000 b.p.) in arthropods (Timmermans et al. 2010). This scale of molecular data can be considered as a compromise between more traditional approaches such as PCR of preselected fragments that at most generally result in only a few thousand sites for phylogenetic matrices, against other 'omics approaches that can provide many thousands. The lack of efficient methods for routine amplification of either partial or complete mitochondrial genomes from divergent taxa has been a significant hurdle to the widespread use of mitogenomic approaches (e.g., Briscoe et al. 2013), but more recent adoption of NGS technologies are now providing new avenues of investigation.

The second approach exploits the comparatively huge amounts of sequence data available in nuclear genomes. These may vary massively in size and complexity even between closely related species of spider (Gregory and Shorthouse 2003; Král et al. 2013), and where sequenced assemblies may also vary in quality and completion (Garb et al. 2018). These factors, alongside the current small number of spiders with either genome projects underway or published (see Table 2.1) means that the

Table 2.1 Various aspects of published spider nuclear genomes.

	Estimated size (Gb)	Assembled size	Transcribed PCGs (supported by expression data)	Given genes counts
Acanthoscurria geniculata	ca 6.5	5.8 Gb	20,755[a] (2193) [31,627 in their Fig. 2c]	–
Araneus ventricosus	–	3.66 GB	23,414	29,380
Loxoscles reclusa	3.26	[Not yet released]	[Not yet released]	20,616
Latrodectus mactans	1.14	[Not yet released]	[Not yet released]	17,364
Nephila clavipes	2.49–3.45	2.44 Gb	14,025	14,025
Parasteatoda tepidariorum	1.45	1.44 Mb	27,990	27,990
Stegodyphus mimosarum	2.55–2.74	2.7 Gb	27,235 (2171) [31,693 in their Fig. 2c]	27,252

[a]Reported in the supplementary information of Sanggaard et al. (2014)

utility of nuclear genomes for systematics is currently limited, but are beginning to provide a valuable scaffold for spider systematics. In particular large sets of nuclear-encoded PCGs are being increasingly used in phylogenomic analyses—i.e., phylogenetic methods where multigene assortments of sequences are harvested and compared from a wider array of available sequence data.

2.3.1 Mitogenomics

Mitochondrial DNA has long been considered a powerful tool for evolutionary studies of animals (Moritz et al. 1987). This has led to much investigation of the phylogenetic utility of different genes encoded by mitochondrial genomes (e.g., after Simon et al. 1994, 2006). Several spider "mitogenomes" have been published since the millennium (see below), often from a single newly sequenced species alone. These each have a similar size (at around 14,500 b.p.) and near uniform gene content as those of most other animals, typically with the same set of 37 genes; 2 for rRNA (named 12S and 16S), 22 for tRNAs, and 13 PCGs (Fig. 2.1c). There is also strong functional conservation between the structures of most mitochondrial genes across animals. As a result, the incorporation of any new mitogenomic sequences into existing phylogenetic matrices can be straightforward. This relative ease of combining data may have appealed to various authors as accessible "genome-scale data" for phylogenetics. The first published spider mitogenome from the jumping spider *Habronattus oregonensis* (Masta and Boore 2004) was quickly followed by that of a Chinese tarantula "*Ornithoctonus huwena*" plus an armored trapdoor spider *Heptathela hangzhouensis* (Qiu et al. 2005). These three together represented the broadest array of spiders, with two being Opisthothelae (Araneomorphae and Mygalomorphae respectively), and the latter from Mesothelae. The first combined phylogenetic analysis of these was published the next year, alongside a partial mitogenome from *Argiope sp.* (Araneidae) and non-spiders (Hassanin 2006). Other phylogenetic analyses followed (e.g., Masta and Boore 2008; Masta et al. 2009) that together highlighted unusual molecular evolution in several lineages (see below). Yet, it was several more years before other publications added more spider mitogenomes (e.g., Liu et al. 2015; Tian et al. 2016; Wang et al. 2016a, 2016b; etc). As of early 2019, the diversity of annotated spider mitogenomes has expanded to 37 publically available, most published by just two Chinese research groups (72%), while others are unpublished in the public domain (10%). This set is dominated by 32 (86%) from Araneomorphae, of which 29 (78%) are from Entelegyne (after Zhu et al. 2019). Only two (5%) are from Mesothelae and three (8%) from Mygalomorphae. In addition to the single complete tarantula mitogenome, the other two mygalomorphs are Nemesiidae *Calisoga longitarsis* (Masta and Boore 2008; Masta et al. 2009) and the Euagridae *Phyxioschema suthepium* (Podsiadlowski et al. unpublished). These three all have slightly smaller mitogenomes than most spiders—where the araneomorph *Argyroneta aquatica* (Dictynidae) has the largest known mitogenome of any spider to date, at over 16,000 b.p. (Liu et al. 2015).

Most spider mitogenomic studies have used long-PCR to amplify mitogenomes in one or more multigene sections, which were then sequenced by well-established Sanger technologies (e.g., Zhu et al. 2019). In this approach, degenerate primers are often first used to amplify small PCR fragments from one or more conserved genes such as CO1. The resultant species-specific fragments then provide a precise template for design of custom primers that are more useful in long-PCR to amplify much longer multigene sections (after Roehrdanz 1995; Hwang et al. 2001). Those sections can then be sheared into smaller fragments (e.g., each of a few hundred b.p.) that may be cloned into vectors, and selectively sequenced (after Masta and Boore 2004, 2008; Masta et al. 2009; also see Liu et al. 2015). More recently, a few studies have attempted to produce lineage-specific primers for long-PCR that work for various closely related species (e.g., for Salticidae in Fang et al. 2016). However, most other studies have produced spider mitogenomes by comparatively intensive methods, instead relying on multiple smaller amplifications of between 4 and 14 overlapping sections (e.g., Liu et al. 2015; Tian et al. 2016). The ends these can be directly sequenced by Sanger technologies, but when larger than a few hundred b.p. a nested set of additional custom primers must also be repeatedly applied—each internal to the previous set—in a primer walking approach (see Qiu et al. 2005). Although both cost and labor intensive, this method has been successfully used on several occasions, often combined with a cloning step to facilitate sequencing (e.g., Fang et al. 2016; Wang et al. 2016a; Zhu et al. 2019).

In the earliest phylogenetic analyses of spider mitogenomes and allied arthropods, multiple genes of some taxa were found to be affected by persuasive molecular differences. At the extreme, some mitogenomes were proposed as having reversals in mutational constraints or significant codon-usage biases (after Hassanin et al. 2005). Such findings can be important for systematics, because differing patterns of nucleotide or amino acid usage across taxa can mislead phylogenetic reconstructions (after Foster and Hickey 1999). This factor (i.e., compositional heterogeneity), alongside rapid accumulation of state changes that can mask divergences (i.e., saturation) plus variation in substitution rate between taxa (i.e., heterotachy), have together caused much debate about the utility of mitogenomic data for systematics. For spiders, some dubious phylogenetic schemes have been recovered with mitogenomic data, such as failure to detect the monophyly of Araneomorphae (e.g., Zhu et al. 2019). Plausibly, this may be due to the use of poor analytical methods that failed to adequately deal with known problematic factors, as briefly outlined above, especially compositional heterogeneity. In such cases, more sophisticated phylogenetic analyses and adoption of data recoding schemes can both be useful strategies to ameliorate misleading signals (e.g., Hassanin 2006; Masta et al. 2009). If given such careful treatment, mitogenome data can provide valuable insights about spider systematics (i.e., Liu et al. 2015). Regardless of approach, several phylogenetic studies of spider mitogenomes have increasingly found strong support for many plausible taxon groupings within Araneomorphae, plus monophyly of several families (e.g., Fang et al. 2016; Tian et al. 2016; Zhu et al. 2019) that accord well with other recent studies (e.g., Wheeler et al. 2017). These results indicate the strong potential for mitogenomic data to inform spider systematics if given dense taxon sampling and careful consideration of factors that may otherwise mislead phylogeny reconstruction.

In addition to insights from the sequences themselves, mitogenomic studies have sometimes considered phylogenetic information from rare genomic changes (RGCs). These include novelties from duplications, alterations in genetic code, modified molecular folding, and most importantly changes in gene orientation and order (after Boore et al. 1995). For the latter, the translocation of tRNA-L2 (one of several tRNAs) to a novel location between CO1 and CO2 was proposed as a RGC linking Crustacea and Hexapoda as a sister group (after Boore et al. 1995). At that time, most other known arthropod mitogenomes had L2 adjacent to ND1, as later found in Mesothelae (Fig. 2.2 top), and many other arachnids (after Qiu et al. 2005; Masta et al. 2010). Yet in later years, all Opisthothelae were found to have L2 in another novel location between ND3 and tRNA-N (Fig. 2.2, after Masta and Boore 2004), since interpreted as another RGC from an independent translocation (Zhu et al. 2019). Unfortunately, further gene order changes in this region cause uncertainty about such events. One proposal is that the ancestral Opisthothelae experienced a multigene duplication in this region, then several losses alongside the translocation of L2 (Qiu et al. 2005). Other RGCs also appear to have affected mygalomorphs, such as translocation of tRNA-T and reversal of tRNA-C and -Y (see Wang et al. 2016a). Alterations are often associated to the non-coding Origin of replication (Or), which itself was proposed to have experienced a reversal during the divergence of Opisthothelae from Mesothelae (Hassanin et al. 2005). Beyond these, other RGCs have apparently occurred within araneomorphs (see Fig. 2.2 lower), for example I is reversed in three non-Entelegyne spiders, or translocated between ND6 and CYTB in Entelegyne. Other RGCs for various Entelegyne include possible inversions of L2 and N in *Argyroneta aquatica* or further rearrangements for *Agelena silvatica* (after Liu et al. 2015). In others, the region adjacent to the Or has experienced significant alterations, for example in *Trichonehila clavata* or in *Ebrachtella tricusidata* [but also see other differences in various *Tetragnatha* spp., e.g., Wang et al. (2016a) versus Tian et al. (2016)]. Given that changes associated with Or region often show complicated differences, this raises multiple questions about the potential timing and order of translocation events in those lineages. For mygalomorphs and particularly tarantulas, the small number of annotated mitogenomes means that any other such RGC novelties are yet unknown and await investigation.

In the last decade, significant progress in NGS technologies has allowed more rapid production of mitogenomic data, whereby multiple mitogenomes can now be assembled for dozens of species from a single sequencing run (Crampton-Platt et al. 2016) or discovered as "bycatch" in genomic NGS sequencing. An important exploratory study with 30 diverse beetles used long-PCR to generate a large section of about 10,000 b.p. plus three smaller sections for each (Timmermans et al. 2010). Here, multiple amplification products from each species were pooled together, then with those of different species, before being sequenced together with the 454-sequencing platform. From this mix, the complete or near complete mitogenomes were recovered for the vast majority of taxa (28 out of 30), with an average

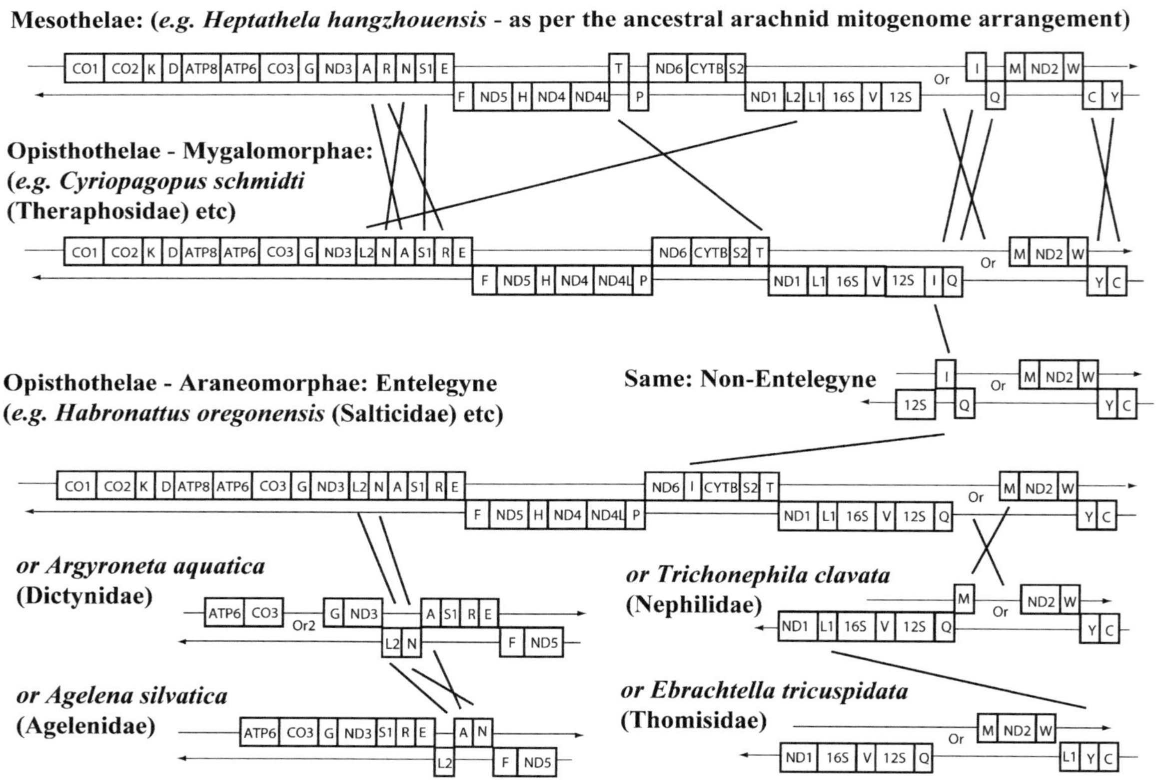

Fig. 2.2 Relative arrangements in the genes across mitochondrial genomes (mitogenomes from diverse spiders, for details of annotations see in Fig. 2.1). Lines between mitogenomes of different taxa indicate alterations in the relative gene-order and/or coding direction

assembly length of over 7000 b.p. (of about 14,500 b.p.). Subsequently, Briscoe et al. (2013) investigated how such an approach could be used for spider mitogenomes on a similar scale, using a broad taxon selection of 33 spiders from both freshly collected and ethanol stored samples. While these latter authors were able to amplify short fragments of CO1 and 16S genes in most taxa, approximately two-thirds failed to give reliable long-PCR amplifications of longer sections despite "an extensive campaign of PCR strategies and optimization" (Briscoe et al. 2013, p.5). Of the 11 taxa where long sections could be reliably amplified, this included two tarantulas *Eupalaestrus campestratus* and *Psalmopoeus cambridgei*. The amplified long-PCR sections from these and other taxa were each sheared, then those from each taxon were given a unique recognition sequence as MID-tagged libraries, before multiple products were combined for sequencing (by 454 pyrosequencing). Afterwards, the MID-tags allowed the various output reads to be bioinformatically assigned back to their specific originating taxon. However, their output (under SRX135055 in Genbank) showed a wide variation in read coverage (i.e., overlapping sequences per taxon), but problematically even the highly covered mitogenomes did allow full assembly (Briscoe et al. 2013). For example, the *E. campestratus* data contained 22 fragments with a small mean size of only 514 b.p. (209–1937 b.p.). Our own comparisons of this data found that several of these fragments were redundant in the CO1 gene and several had overlaps, so their lack of merger in the published assembly is rather surprising. Other fragments originated from a section that encompassed 16S–12S–ND1, which if allowing for small gaps (where there is no apparent overlap) would cover a span of about 3500 b.p. Similarly, their *P. cambridgei* assemblies were also highly fragmented, containing 28 fragments with a mean size of 766 b.p. (251–4477 b.p.). Here, there was one especially large assembled section, but our own comparisons suggested this may be malformed (i.e., chimeric) from two discrete multigene sections of about 2200 b.p., being comprised of one for 16S–12S and adjacent regions toward ND2, wrongly combined with another for partial CYTB, ND6, ND4L and ND4. Consequently, with recovered mitogenomes being highly fragmented and assemblies problematic, the results of Briscoe et al. (2013) highlight the challenges that may be faced in producing mitogenomes on this scale. Later studies of other spiders have shown that such an approach may be more successful if conducted on a smaller scale. Using similar methods, complete mitogenomes were generated from two species of orb-weavers *Neoscona spp.* (Wang et al. 2016b). Both were first amplified in multiple large sections by long-PCR as in many previous studies, but next, these sections were fragmented and given additional motifs to yield a paired-end tag (PET) library, before being subjected to Illumina sequencing. Unfortunately, Wang et al. (2016b) do not give details about the distribution of the typically short output reads from this method across their final combined assemblies to help understand sequencing coverage, quality, nor cost-effectiveness.

In general, researchers working on diverse organisms have increasingly recognized that even when successful, the relatively tedious process of long-range PCR amplification (especially if followed by primer walking) is poorly suited to high-throughput processing, limiting the viability of mitogenomic data for large-scale applications (i.e., see Crampton-Platt et al. 2016). To circumvent the need for long-PCR, a more straightforward approach relies on low-coverage shotgun sequencing of total DNA, whereby only the high-copy elements (such as mitochondrial sequences) and repeat regions are present in sufficient quantities to permit frequent detection and sequencing. Given this, those from the same region can often be assembled together in a process referred to as "genome skimming" (Straub et al. 2012). The related discipline of "Mitochondrial metagenomics" (MMG) (or other names, see Tang et al. 2014) specifically targets the mitochondrial fraction, and may be defined as "a methodology for shotgun sequencing of total DNA from specimen mixtures and subsequent bioinformatic extraction of mitochondrial sequences" (Crampton-Platt et al. 2016). Of course, despite being able to rapidly generate mitogenomic data on large-scale, this approach also leads to high volumes of other reads from across the nuclear genome, often of little utility and so are discarded. This has led others to search for cost-effective methods to enrich DNA extracts for mitogenomes prior to sequencing, such as simplistic alteration of purification protocols (e.g., Tilak et al. 2015; Macher et al. 2018) or more technological ones like using probes on a microarray to preferentially select mitochondrial fragments (Liu et al. 2016). Together, such developments have increasingly demonstrated the possibility of generating mitogenomic data from mixtures of specimens without indexing. However, the later re-assembly process can be problematic when different samples are very closely related (Timmermans et al. 2016). That said, although such advanced approaches to mitogenomics have not yet been published with spiders, some of these methods can hold great promise for their systematic studies, including for tarantulas.

2.3.2 Nuclear Genomics

A new era of molecular analyses has dawned with the shift into the field of comparative nuclear genomics. Already, the first tarantula genome has been published from a Brazilian tarantula *Acanthoscurria geniculata,* concurrently with that of a Velvet spider *Stegodyphus mimosarum* (Sanggaard et al. 2014). These first two spider nuclear genomes have been slowly followed by a few select others, namely a Golden silk spider *Trichonephila clavipes* (Babb et al. 2017), an American house spider *Parasteatoda tepidariorum* (Schwager et al. 2017) and a round-shouldered orb-weaver *Araneus ventricosus* (Kono et al. 2019). For most, the characterization of genes associated with silk production have figured prominently, especially in the latter study where a hybrid sequencing approach combining output from long-read Nanopore sequencing with a dense array of short reads from Illumina technologies. Here, the combination of these approaches was said to be fundamental for rigorous genome assembly during curation. This was especially the case for the silk-encoding Spidroin genes, which are typically very long (around 10 kbp) and almost entirely composed of repetitive sequences between more conserved terminal domains (Kono et al. 2019). At the time of writing, the above spider genomes are complemented by only a small handful of genomes from other arachnids and beyond (after Richards 2015), but several more spider genomes are in progress. These include both the Western black widow *Latrodectus hesperus* and the Brown recluse *Loxosceles reclusa* as early outputs within a broader goal to sequence 34 spider genomes as a subset of 5000 focal arthropods (i5K Consortium 2013). Another unrelated genome project exists for *Dolomedes plantarius,* which won its place in public vote to decide which of 25 organisms from the UK should be sequenced. This latter project forms part of the broader Darwin Tree of Life Project with ambitious aims to sequence genomes of 66,000 (Eukaryotic) species in the UK, which should incorporate a large fraction of the country's 650 or so species of spider. These include a single near relative of tarantulas, namely *Atypus affinis* in the mygalomorph family Atypidae. Beyond this, an even more formidable proposal has since been announced to merge several ongoing nuclear genome projects under the aegis of the Earth Biogenome Project (EBP). This was declared as "a moonshot for biology that aims to sequence, catalog, and characterize the genomes of all of Earth's eukaryotic biodiversity over a period of 10 years" (Lewin et al. 2018). A way towards such a lofty goal might be to put early focus on prioritizing species that represent different taxonomic families (e.g., Richards 2015). If so, then the tarantula family Theraphosidae is already represented by *A. geniculata*, although other mygalomorphs from closely allied families will likely provide useful comparisons. Due to the low socioeconomic importance of most mygalomorphs to humans (e.g., little impact on health and disease, pest ecology etc.), additional genomes from tarantula and their allies are unlikely to be high priority in the next wave of genomics initiatives. Yet, other researchers may deem them valuable, as some hold strong potential to display exciting molecular attributes of high value for biotechnology such as venom components (i.e., as per Garb et al. 2018), as well as genome evolution due to the potential for enormous sizes (i.e., *Aphonopelma* possesses an estimated 16 Gb—see Hamilton et al. 2016b). Some useful investigations of novel genes and networks involved in molecular adaptation of venoms and also silks have already been carried out in a few tarantulas using transcriptomic methods (see below). Likewise, partial genomic data for three tarantulas, namely *Cyriopagopus (Haplopelma) schmidti, Poecilotheria regalis,* and *Lasiodora parahybana* have also been generated, derived from projects interested in the microbial diversity associated with venom toxicity (e.g., Bioprojects PRJEB20997 and PRJEB4693). Given these cases, the adoption of large-scale genomic sequencing to investigate venom evolution has begun for tarantulas and their allies. Beyond this, other aspects of physiological and behavioral adaptation in tarantulas such as cold tolerance in Nearctic or high elevation taxa may be reflected in exciting genomic novelties that await discovery, or the molecular basis to adaptations for prey detection, mate recognition, etc.

The first published tarantula genome for *A. geniculata* was based on an assembly of 5.8 Gb (around 6 billion nucleotides) of the total estimated 6–6.5 Gb genome size, of which 2193 transcripts could be securely identified (Sanggaard et al. 2014, see transcriptomics). Due to a fragmented assembly resulting from both a relatively large genome size and unexpected heterogeneity, the total gene number remains unclear, but was reported as <73,821 PCGs—although only 20,755 PCGs were listed in a comparative analysis. Richards (2015) suggested there are around 12,000 PCGs (or rather clusters of orthologous genes) in the ancestral arthropod genome, but actual gene counts to date are consistently higher due to existence of variously expanded gene families and other unique sequences. Nuclear genomes for the first two araneomorphs have relatively high gene-numbers, whereby *P. trepidarium* has about 27,990 PCGs Schwager et al. (2017), similar to *S. mimosarum* with 27,235 PCGs (Sanggaard et al. 2014). For other arachnids, the gene number can be even larger, such as the Chinese scorpion *Mesobuthus martensii* with about 32,016 PCGs (Cao et al. 2013) and the Bark scorpion *Centruroides sculpuratus* with about 30,456 (Schwager et al. 2017), both far more than most other animals. At the other extreme, some arachnid nuclear genomes can be much smaller, such as the Red spider mite *Tetranychus urticae* with only about 18,414 PCGs, where extreme compaction is linked to considerable gene loss (Grbic et al. 2011).

A practical complication for systematic studies exploiting nuclear genomes is the potential existence of polyploidy, either from restricted changes such as unequal crossing with local duplication, or larger scale duplication events (likely the main reason for the large genome sizes of some mygalomorphs). The evaluated scorpions in particular appear to have experienced high gene family turnover as well as frequent duplications and associated neofunctionalization (Cao et al. 2013). Given these issues, an important consideration for any molecular systematic study interested in relationships between taxa should be that the compared sequences have diverged through speciation events (i.e., = orthologous) rather than from duplication events (= paralogous). The mistaken inclusion of paralogs into datasets can lead to confounding noise that ameliorates or masks historical signal about evolution of the focal organisms (see Kriventseva et al. 2019). To limit this, several studies have attempted to determine subsets of "core genes" within nuclear genomes that are easily identifiable, highly conserved, and single copy (or nearly so) across diverse taxa. An early but now defunct approach was CEGMA (Parra et al. 2009), which used a narrow sample of three animals, one plant and two fungi to summarize gene similarity between these first completed genomes of multicellular organisms. From these, a set of 458 "core genes" were identified as highly conserved with identifiable orthologs, but later reduced to only 248 after filtering out those with multiple copies in several species. Additionally, the detection rate of such "core genes" was proposed as a useful metric by which the completion of new genome assemblies could be assessed. For example, a survey of gene content in the first scorpion nuclear genome (for *Mesobuthus martensii*) found 99% of that initial 458 core-gene set, so indicating the relatively high completeness of its assembly (Cao et al. 2013). Several later approaches to identify orthologs have been based on reciprocal best match in pairwise comparisons (see Altenhoff et al. 2013), such as was used to identify 452 single-copy orthologs for comparison of the first two spider genomes against six other animals (Sanggaard et al. 2014). More recently, the BUSCO (Benchmarked Universal Single-copy orthologs) tool has become widely adopted for such comparisons, including with spiders (e.g., Babb et al. 2017; Schwager et al. 2017). This relies on OrthoDB, which is currently one of the largest database of orthologs, alongside maps of functional categories (Kriventseva et al. 2019). This also incorporates a comprehensive spread of taxa with published genomes, including refinement for certain subsets of taxa (such as arthropods). The BUSCO tool contains hidden Markov model (HMM) profiles from amino acid alignments of each gene, consisting of probability states for each residue, which can help identify likely orthologs from additional genomes (Waterhouse et al. 2018). These can then be summarized to discuss relative completion of genome sequencing, or help assemble gene alignments for phylogenomic analyses. In its earliest iteration, the arthropod subset BUSCO–Ar incorporated 38 taxa (mostly from insects) to yield 2676 genes present as single-copy orthologs in at least 90% of given species (Simão et al. 2015). For spiders, 2058 of these genes in the tick *Ixodes scapularis* were then used to statistically evaluate quality of the *T. clavipes* genome, where around 94% had a single-copy equivalent, so indicating a high quality of its assembly and annotation (Babb et al. 2017). Similarly, of 1610 found in the beetle *Tribolium castaneum*, 91% had an equivalent the spider *P. trepidorum*, plus further fragmented ones (Schwager et al. 2017). However, additional comparisons against various other arthropods were then used to demonstrate that a considerable percentage (5.5–7.2%) of genes in BUSCO–Ar set had duplicated copies in *P. trepidorum* (representing spiders) or *C. sculpuratus* (for scorpions), plus for several cases in both. From this, it appears that both these latter genomes have a significantly higher proportion of genes with duplicated copies than expected. This led the authors to the radical conclusion that both spiders and scorpions may have experienced a whole-genome duplication event in their common ancestor (Schwager et al. 2017). In future, nuclear genomes from other spiders, including additional tarantulas and their allies can be expected to also yield similarly exciting discoveries about their molecular evolution, and provided that meaningful historical signal can be clarified, these should play an important role to inform systematic questions.

2.4 Targeted Sequencing Approaches

In sections below, we provide an overview of another set of developments in molecular studies, again with a focus on mygalomorph systematics, especially for tarantulas. We provide an overview of different approaches that can together be considered targeted sequencing, or else as reduced representation of genomes (primarily nuclear, but can also include valuable sequence data from mitochondrial genomes). Below, these are divided into: (i) transcriptomics and (ii) target capture—namely Anchored Hybrid Enrichment and Ultra-Conserved Elements. These various approaches have only recently become used for spider systematics, alongside development of high-throughput genomic sequencing and other methodological advances. However, each has rapidly changed the field by increasing the volume and rate of data collection by several orders of magnitude (see Table 2.2 for some overview) when compared to most other earlier approaches. As a result, various "targeted sequencing" methodologies now provide molecular systematists with a powerful set

Table 2.2 Overview of relevant studies with genomic or transcriptomic data

Reference	Scope of taxa	Orthologs in largest set	Orthologs in reduced sets	Source
Sanggaard et al. (2014)	Arachnida plus outgroup	452		GENOM
Sharma et al. (2014)	Arachnida plus outgroup	3644 [1,235,915 a.a.]	1237/516/etc.	TRANSC
Fernández et al. (2014)	Arachnida plus outgroup	2647 [791,793 a.a.]	789/100/94	TRANSC
Bond et al. (2014)	Araneae plus outgroup	327 [110,808 a.a.]	128	TRANSC
Frías-López et al. (2015)	Araneae	35 [4534 a.a.]	N/a	TRANSC
Garrison et al. (2016)	Araneae	3398 [of 4934][a] [696,652 a.a.]	1699/850 etc.	TRANSC
Babb et al. (2017)	Araneae	N/a, only spidroin family	N/a	GENOM
Carlson and Hedin (2017)	Araneae	3345 [Not given]	N/a	TRANSC
Rix et al. (2017a)	Araneae plus outgroup	151 [of 1478] [74,323 a.a.]	11	TRANSC
Schwager et al. (2017)	Arthropoda	–	–	GENOM
Cheng and Piel (2018)	Araneae	2581 [822,142 a.a.]	1387/1291/646/etc.	TRANSC
Fernández et al. (2018)	Araneae	2365 [623,892 a.a.]	750/255	TRANSC
Kallal et al. (2018)	Araneidae	1780 [591,554 a.a.]	1404/641/548/299/64	TRANSC
Coddington et al. (2019)	Araneae	–	–	TRANSC
Ballesteros and Sharma (2019)	Arachnida plus outgroup	3534 [1,484,206 a.a.]	1499/882/721/98	Both
Foley et al. (2019)	Theraphosidae plus outgroup	2460 [1,096,124 a.a.]	1230/615/etc.	TRANSC

Note: "etc." refers to other variant datasets
[a]Spider-specific core ortholog group (OG) set with 4934 OGs

of tools for resolving relationships across the Spider Tree of Life at multiple hierarchical levels (e.g., Bond et al. 2014; Fernández et al. 2014, 2018; Frías-López et al. 2015; Garrison et al. 2016; Hamilton et al. 2016b; Rix et al. 2017a; Maddison et al. 2017; Starrett et al. 2017; Chamberland et al. 2018; Cheng and Piel 2018; Godwin et al. 2018; Hedin et al. 2018a, b; Kallal et al. 2018; Wood et al. 2018; Coddington et al. 2019; Kuntner et al. 2019; Foley et al. 2019; Opatova et al. 2019). Together, these approaches are providing an increasingly solid phylogenetic framework to address important evolutionary questions for spiders. Many formerly stable phylogenetic hypotheses have not survived this influx of new data, such as previous views about the monophyly of orb-weaving spiders, the relationships between primitive araneomorphs, monophyly of various mygalomorph families, or more generally, patterns of spider diversification and ages of major spider groups.

2.4.1 Transcriptome Sequencing

Rather than directly using genomic sequences (i.e., DNA), transcriptomic methods instead exploit a sampling of expressed messenger-transcripts (mRNA). These largely correspond to protein-coding genes (PCGs) as indirect representations of functional regions of genomes. For spiders, the earliest transcriptomic data (i.e., assortments of diverse transcripts) were often generated for gene discovery. An early study of importance for tarantulas was the small-scale evaluation of transcripts from hemocytes of a Brazilian tarantula *Acanthoscurria gomesiana* to characterize PCGs involved in immunity (Lorenzini et al. 2006). Another transcriptomic study generated transcripts from a North American *Aphonopelma* sp. to evaluate PCGs of skeletal muscle (Zhu et al. 2009), which later helped clarify the acquisition of muscles in animal evolution (Alamo et al. 2018). More recent initiatives with other spiders have had academic goals by comparing PCG transcripts of certain functional roles such as translation (e.g., French et al. 2014) or from certain tissues such as chemosensory structures (e.g., Frías-López et al. 2015). Others had more applied goals to characterize PCGs related to insecticide action and detoxification (e.g., Meng et al. 2015). Transcriptomic projects of spiders have also often focused on venom-associated proteins (e.g., Carlson and Hedin 2017; see Corzo and Escoubas 2003). For this, one approach has been to focus only on select transcripts from a given tissue. For example, RNA from the venom glands of the red-knee tarantula *Brachypelma smithi* was used to construct a library of all tissue-specific transcripts, before purified proteins with insecticidal activity were used to selectively sequence only their corresponding transcripts (Corzo et al. 2008). In other studies, mixed pools of unknown transcripts have more simply been sequenced from various tissues en masse, prior to detailing those of focal interest bioinformatically. Examples for tarantulas include the sequencing of a few hundred PCGs with diverse cellular roles from *Chilobrachys guangxiensis* (reported as *C. jingzhao*) alongside focal toxins (Chen et al. 2008); 41 PCGs from *Cyriopagopus schmidti* (Jiang et al. 2010); or the handful from *Pelinobius muticus* (Diego-García et al. 2010). Such studies have led to a strong taxonomic bias in spider-toxin databases, with tarantulas making up about 34.6% of all species (Herzig et al. 2019). Here, preference for tarantulas [and other mygalomorphs like the barychelid *Trittami loki* by Undheim et al. (2013)] appears driven by their large size—which can ease extraction of workable venom yields.

Recently, several gene discovery projects have exploited NGS technologies to gather comparatively large samples of spider transcriptomes via shotgun approaches, again often focused on genetic components of venoms. These have either used extracted venom, glands themselves, or indeed whole or partial animals. Projects with tarantulas include those on *Selenotholus foelschei* (Bioproject PRJEB16008), *Coremiocnemis tropix* (PRJEB15661), *Poecilotheria fasciata* (PRJNA357041), *Grammostola rosea* (PRJDB864), *Cyriopagopus spp.* (PRJNA270392, PRJNA241286, also PRJNA240354). As of early 2019, the Sequence Read Archive of the NCBI database contains 491 spider transcriptome samples from 278 species (251 > 1000 MBases), which are largely derived from venom-focused research (See Herzig et al. 2019). However, each remains rather isolated, and only specific PCGs of focal interest are carefully detailed, leaving large volumes of unevaluated sequences. Elsewhere, other transcriptomes have been generated to support nuclear genome projects, as transcripts from either the same or closely allied species can be informative to annotate assemblies by helping identifying boundaries of PCGs. For example, previously published transcripts from venom glands (Haney et al. 2014) or silk glands (Clarke et al. 2014, 2015) are helping inform the on-going genome assembly of *L. hesperus*, which currently has the largest transcriptomic data for any spider (with about 200,000 transcripts as of early 2019). Similarly, existing transcriptomes from various araneoid spiders, plus others created for the task, proved highly useful to inform the genome assembly and annotation of *A. ventricosus* (Kono et al. 2019). Another large transcriptome set exists for the tarantula *A. geniculata* with 109,122 transcripts from two large-scale projects (SRR1024075, SRR1024076). As above, these have also been generated to complement the corresponding genome assembly (Sanggaard et al. 2014, see above), but viewed apart are the largest annotated transcriptomic set in the public domain (as of early 2019) for any mygalomorph.

Regardless of the diverse reasons that transcriptomes may have been generated, their gene content can often be easily compared. Consequently, several studies have recently begun to exploit transcriptomic data for comparative analyses of spiders. A notable study by Carlson and Hedin (2017) contrasted gene diversity between transcriptomes of six araneomorphs from three families (two from each). When independently sequenced, these ranged in size from 35,817 to 146,297 possible PCGs per species. Within these, a large percentage (87–98%) of the 248 "core genes" of CEGMA (see above) were cross-detected, as well as many of PCGs of focal interest (Carlson and Hedin 2017). It logically follows that comparable transcriptomes can also be exploited to create gene matrices for systematic studies, and increasingly some spider transcriptomes have been specifically generated for this purpose. One of the first of these was for a small set of 257–262 transcripts from the woodlouse spider *Dysdera* sp., which was created alongside many insect transcriptomes (Hughes et al. 2006). This early study was fundamental to demonstrate the power of the approach, specifically that even small transcript sets could be combined in systematic studies to infer a defensible phylogeny. In subsequent years, transcriptomic data has been increasingly used in phylogenetic analyses of various organisms. These are often termed phylogenomic analyses—generally to emphasize a large number and diverse nature of PCGs in such datasets. Yet importantly, the size and gene content has varied widely in such studies (See Table 2.2). Even within a particular dataset, the represented PCGs for each taxon (i.e., occupancy) can differ widely due in part to variation in the diversity of transcripts expressed (and detected) at time of generation, but also linked to differences in sequencing intensity (i.e., coverage, redundancy), or even sequencing technologies used (that vary in read length, etc). In 2014, three systematic studies of spider transcriptomic data were published in quick succession (Bond et al. 2014, Fernández et al. 2014, Sharma et al. 2014). The one with broadest taxon scope focused on arachnid orders, but included transcriptomes from five spiders (Sharma et al. 2014), with four newly sequenced, while the fifth for *A. gomesiana* was from a previous gene-discovery project (see above). From these, a set of 3644 PCGs was assembled with moderate completion at only 36.2% occupancy—with taxa differing widely in the numbers of genes represented. The recovered phylogeny was also examined from different subsets of data, such as after selecting fewer genes with higher cross-detection across taxa (i.e., reducing set to 516 PCGs, but raising occupancy to 40%), or from fewer taxa of highest completion (1237 PCGs, occupancy 62.5%). Similarly, the effect of removal of certain genes was also carefully evaluated; such as only “slower” genes with more conserved sites. Another linked study then evaluated a further eight araneomorph transcriptomes to focus on orb-weavers (Orbiculariae), again using existing data from *A. gomesiana* to represent mygalomorphs (Fernández et al. 2014). Here, a set of 2647 PCGs with 62.5% occupancy was initially analyzed, before reduction to subsets with enhanced completion (e.g., up to 92.3% with only 94 PCGs), or more conserved genes, etc. Once carefully evaluated, the resultant phylogenies led the authors to propose that two types of orb webs (cribellate and ecribellate) had either independent origins or a more ancestral origin than previously anticipated with subsequent loses. At the same time, an independent study assembled transcriptomic data from a much denser array of 40 spiders (Bond et al. 2014). Here, the combined gene set was smaller (only 327 PCGs, occupancy 76.9%) than the initial matrices of aforementioned studies, but again completion was increased by adoption of a revised subset with higher detection (up to 91.3% occupancy from 128 PCGs). Phylogenomic analyses of this data also rejected the long-held hypothesis that orb-weavers were monophyletic—a fundamental shift in understanding about the araneomorph phylogeny. For mygalomorphs, these authors greatly expanded taxon sampling beyond previous studies, generating new transcriptomes for the tarantula *Aphonopelma johnnycashi* (therein as *A. iviei*) and *Trichopelma laselva* (therein as Barychelidae, since transferred to Theraphosidae) plus diverse representatives of several other mygalomorph families. Their resultant phylogenomic analyses gave several interesting findings, such as reflecting earlier proposals that the mygalomorph family Nemesiidae was not monophyletic, with some such as *Damarchus* sp. actually closer to tarantulas than others. Else, the family Paratropididae—which had sometimes been proposed as closely related to tarantulas—were shown as only distantly affiliated to them. Beyond this, these three aforementioned studies together presented a complementary sampling of spiders with 53 non-overlapping species, with only a single araneomorph (labeled as *Frontinella communis*) seemingly being redundantly sequenced. Around this time, transcriptomic data from another mygalomorph *Macrothele calpeiana* was also independently published (Frías-López et al. 2015), from which 35 PCGs (from the 327 set mentioned above) were compared to give a mygalomorph phylogeny that mostly mirrored previous findings. With the exception of this latter project, most of the published spider transcriptomes plus others from a further 21 species (with some genomic data of *S. mimosarum* and *A. geniculata*) were then merged to give the largest-scale phylogenomic analysis of the time (Garrison et al. 2016). For mygalomorphs, the most important aspect of this revised set was the inclusion *A. geniculata*, where a comprehensive array of genes had become available from its transcriptome (see above). Data assembly here involved a prior step where a set of 4934 candidate orthologous PCGs were evaluated across 12 select spider transcriptomes (plus one other), excluding those either too short, present in too few taxa, or with some remaining evidence of gene duplication (i.e., paralogy), before a second filtering across spiders to exclude those with lowest representation. The resultant gene-set (3398 PCGs, occupancy 61.5%) was similar to other contemporary datasets but with many more taxa. For tarantulas and their kin, the results echoed previous findings to indicate that *T. laselva* was more closely affiliated with tarantulas (*A. johnnycashi* and *A. geniculata*, respectively) than other lineages, while the broader scheme was largely consistent with other transcriptomic datasets. Other later studies have subsequently compared gene representation using the same 4934 putative "spider-specific core ortholog groups" (e.g., Carlson and Hedin 2017), or have used alternative sets of candidate orthologs and different filtering stringencies, but similarly aim to optimize sets toward genes with strong detection in the focal taxa, but little or no evidence for duplication (i.e., paralogy).

Recently, Foley et al. (2019) have generated a wealth of transcriptomic data from various tarantulas to produce the first robust phylogenomic hypothesis focused on relationships within the family. These authors exploited a slightly smaller and arguably more stringent set of 4446 putatively orthologous genes than used for diverse spiders by Garrison et al. (2016). The "core ortholog set" of Foley et al. (2019) had previously been shown to be highly conserved across arachnids (mostly spiders), and which, after further filtering had successfully informed phylogenomic analyses of Entelegyne spiders (Cheng and Piel 2018). From this, a revised set of 2460 PCGs (with 83.2% occupancy) was generated from several new tarantula transcriptomes. This study also integrated existing transcriptomic data for *A. johnnycashi* and *T. laselva* (Bond et al. 2014) as well as *A. geniculata* (Sanggaard et al. 2014) to present a broad sampling of tarantula lineages that included representatives of almost all subfamilies (Foley et al. 2019). As in other such studies, various gene-subsets were carefully analyzed, in particular removal of genes with comparably higher missing data, which raised occupancy (up to 94.6% in their smallest 615 gene set) and enhanced completion. Regardless of data subset, the resultant phylogenies were notably congruent, including strong support for the unique status of some previously questionable subfamilies, such as Poecilotheriinae and Psalmopoeinae. For the first time, the relationships between these and almost all other subfamilies were confidently inferred with molecular data. One important aspect was the strong support for a novel group of several New World lineages with transcriptomic data, therein termed the Bombardier clade, which included subfamilies Aviculariinae, Schismatothelinae, Psalmopoeinae, Theraphosinae, and potentially some "Ischnocolinae." This novel clade, and others, reflected a clear biogeographic pattern that was apparent across the preferred topology, with major groups corresponding to taxa distributed on particular continents, as seen previously in other smaller molecular datasets (e.g., Lüddecke et al. 2018; see Chap. 6). However, the Bombardier clade of Foley et al. (2019) indicated for the first time that most American tarantulas could have arisen from a single radiation, rather than at least two as previously suggested (e.g., Lüddecke et al. 2018). In particular, this topological difference has important consequences for our understanding of the evolution of defensive urticating setae during diversification of New World tarantulas (see Chap. 9), which requires additional taxon sampling to fully resolve. Another key difference from the topology of Lüddecke et al. (2018) was the placement of the Indian Thrigmopoeinae. Here, analyses of transcriptomic data in Foley et al. (2019) instead supported Thrigmopoeinae as more closely related to Central and Southeast Asian Selenocosmiinae than to Indian Poecilotheriinae and Asian Ornithoctoninae. Regardless of specifics, it is clear from combined insights that the tarantula fauna of India are closely tied with Asian lineages, rather than say African ones. For the latter, as with some previous studies, transcriptomic analyses found strong support for African Harpactirinae and Stromatopelminae as sister-groups, plus a deep division of these lineages from the Eumenophorinae. This indicates at least two radiations of African tarantulas, again reflecting earlier findings with other molecular data (see Lüddecke et al. 2018). Remaining more contentious aspects of the tarantula phylogeny are discussed below in a later section.

Finally, other advancements in methods of orthology assessment (e.g., UPhO in Ballesteros and Hormiga 2016) have led to alternative sets of orthologs being preferred in other transcriptomic studies on spiders and their allies. For example, a study of Australian trapdoor spiders filtered 1478 putative orthologs down to only 146 strict orthologs (plus a handful of other preselected PCGs) for phylogenetic analyses (Rix et al. 2017a). Elsewhere, Fernández et al. (2018) and reanalysis by Coddington et al. (2019) reduced an initial 2675 putative orthologs to resolve systematic questions across the broadest selection of spiders to date (their largest set being 2365 PCGs with 40.1% occupancy). Here, several additional spider transcriptomes (including two mygalomorphs; one new, the other from *Macrothele*) allowed investigation of lineage diversification across spiders, leading to valuable insight on the evolution of silk and web usage, but with little consequence to our understanding of tarantula or their close allies. Similarly, other recent transcriptomic studies with either a narrow taxon sampling [such as across the araneomorph family Araneidae in Kallal et al. (2018)] or much broader sampling [such as diverse arthropods in Ballesteros and Sharma (2019)] have varied widely in their gene content of preferred ortholog-sets. Overall, these differences in component data make it impossible to judge the relative strengths of the various recent studies in direct comparison. However, regardless of scope and scale of data used, most transcriptomic studies report their phylogenetic analyses are able to successfully resolve many of the previously uncertain or intractable nodes, which together demonstrate the power of the transcriptomic approach, at least for broader systematic questions.

2.4.2 Ultra-conserved Elements/Anchored Hybrid Enrichment

Target capture sequencing approaches such as Anchored Hybrid Enrichment (AHE) and Ultraconserved Elements (UCE) are another set of increasingly popular methods for collecting large amounts of phylogenetic data for non-model taxa. These alternative, yet highly similar solutions sample hundreds of orthologous loci from across the genome of non-model organisms, leading to significant changes to the numbers and types of loci available for phylogenetic inference, and provide powerful approaches for resolving relationships across the Tree of Life (Hamilton et al. 2016a, b; Branstetter et al. 2017a, b; Maddison et al. 2017; Van Dam et al. 2017; Chamberland et al. 2018; Godwin et al. 2018; Hedin et al. 2018a, b; Kuntner et al. 2019; Wood et al. 2018; Opatova et al. 2019). In spiders, AHE (after Lemmon et al. 2012) and UCE (after Faircloth et al. 2012) were both designed to recover hundreds (AHE = 585) or thousands (UCE = 1120) of unique orthologs (i.e., single-copy, phylogenetically informative markers) that are distributed across the genome. Datasets of these sizes are theoretically sufficient to resolve difficult nodes (i.e., short branch lengths relative to population size), as in the case of rapid radiations and recent divergences, while also accounting for other factors such as deep coalescence, incomplete lineage sorting, and/or hybridization (Leaché and Rannala 2011; Liu and Yu 2011; Maddison and Knowles 2006; Edwards et al. 2007; Huang et al. 2010).

Originally, the term UCE referred to a conserved DNA sequence within vertebrates, with ≥80% identity and over ≥100 bp across taxa. In their development, the ultraconserved elements—as their name suggests—held extreme conservation across taxonomic groups. UCE loci (genomic regions for phylogenetic usage) were designed as those with moderate levels of phylogenetic information, but higher numbers of recovered loci. That said, both this and the AHE approach utilize short probes to "capture" conserved regions spread throughout the genome. This typically involves using <60–120 bp nucleotide sequences (capture probes) that are complementary to target regions (probe regions) in the new taxa being studied. By comparing existing genome and transcriptome data from taxa close to those of interest, probes can be designed in more conserved anchor (or bait) regions of the unknown target genome that neighbor variable flanking regions (see Fig. 2.3 for AHE). A diverse set of probes will include those from heterogeneous genomic regions: exons, introns, intergenic, and other conserved regions. By hybridizing the probes to a DNA library, the targets of interest are isolated prior to NGS and the non-target regions of the genome are then discarded (Lemmon et al. 2012; Faircloth et al. 2012; Smith et al. 2014; Faircloth 2017). By using high-throughput NGS technologies, millions of reads can be generated to assemble large chunks of the genome (contigs) where long stretches of their DNA have been sequenced multiple times (i.e., high redundancy) to ensure accuracy through fidelity at individual sites. The resultant sequence data are often informative at both deep and shallow phylogenetic levels in non-model systems (Lemmon et al. 2012; Faircloth et al. 2012; Faircloth 2017). Additionally, when the phylogenetic informativeness of the flanking regions is investigated, the datasets can exhibit similar levels of sequence variation and

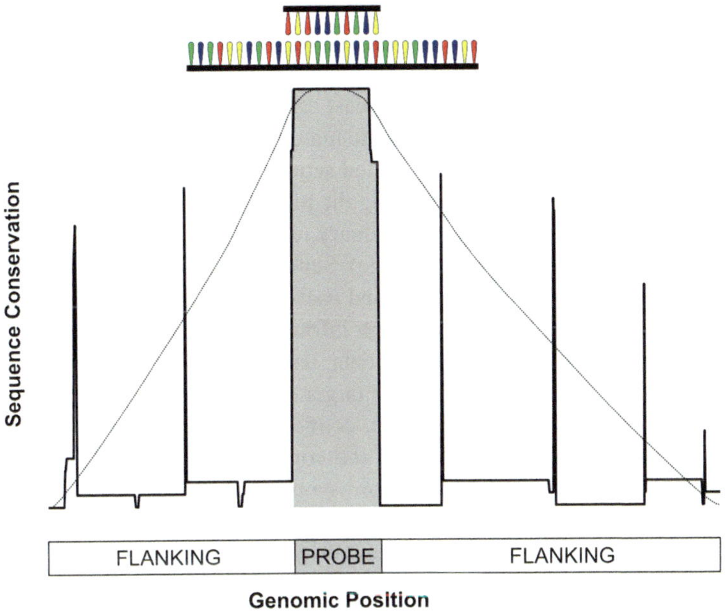

Fig. 2.3 Target capture approaches like Anchored Hybrid Enrichment and Ultraconserved Elements use probes to target specific regions (probe regions) across the nuclear genome. These regions are then sequenced out in both directions to obtain the more variable flanking region data. The conserved nature decreases as you move away from these regions, often due to insertion-deletion variation or intergenic sequences that differ between the taxa of interest. These methods then gather and utilize peaks of conservation in the flanking regions for their phylogenetic information

structure as SNP datasets, and hence produce similar levels of resolution at the species and population interface.

Both AHE and UCE methods facilitate meta-analyses, thus are accelerating resolution of various phylogenetic questions, because the same loci can be targeted across broad taxonomic groups. Unfortunately, as both AHE and UCE approaches moved into invertebrate research, a similar problem arose, namely that major differences in genome size and structure among taxa caused a rethink of the loci determination strategy. Vertebrate genomes are relatively more similar to each other than the vast diversity found across invertebrate genomes. For example, spiders even exhibit large differences in chromosome number and genome size within families (see Sanggaard et al. 2014; Král et al. 2013), an issue that has slowed the development of genomic resources for the lineage. Consequently, the genomic diversity of invertebrates has caused both approaches to incorporate much more variation into the probe regions when targeting non-vertebrates (see Hamilton et al. 2016b; Starrett et al. 2017). This has unfortunately also restricted the effectiveness of both approaches for invertebrates, with recent studies typically limited to those within a taxonomic Order or Class. It is important to note that a major benefit of both target capture approaches is that they allow us the opportunity to utilize specimens from natural history collections, provided DNA can be retrieved (Blaimer et al. 2016; Mikheyev et al. 2017; Ruane and Austin 2017; St Laurent et al. 2018). In contrast, transcriptomic methods rely almost exclusively on either fresh genetic samples (often direct from live specimens) or at least carefully stored tissue of high genetic quality. However, the ability to use genetic material from older or otherwise genetically degraded samples with these targeted sequencing approaches opens up many new avenues of investigation by allowing the placement of extinct or rare taxa into a phylogeny, or even examining evolutionary response over time to environmental pressures or changes (e.g., climate change). Several other factors also make targeted sequencing approaches both attractive and useful for systematists, particularly: (1) efficiency in non-model species; (2) high informative content of the loci; (3) high loci recovery and low levels of missing data; and (4) rapid data collection and cost effectiveness. Perhaps most importantly, target capture approaches provide mechanisms whereby different researchers can confidently and effectively use the same loci for independent projects, allowing the combination of data across studies—enabling us to constantly update and improve our understanding of the Tree of Life. One new addition to these approaches is the ability to find many of these same loci within published transcriptome sequences. While those do not possess the flanking regions, for most phylogenetic questions the probe regions will be sufficient to place taxa. Finally, while these are targeted approaches, there can also be valuable "leakage" when it comes to what reads can be recovered from high-throughput sequencing (e.g., such "leaked" reads can be used to assemble full or partial mitochondrial genomes).

For spiders, target capture approaches have now been used to answer longstanding questions regarding the relationships and diversification of several major groups (Hamilton et al. 2016a, b; Maddison et al. 2017; Chamberland et al. 2018; Godwin et al. 2018; Hedin et al. 2018a, b, 2019; Kuntner et al. 2019; Wood et al. 2018; Opatova et al. 2019). Together, these are providing a solid revised phylogenetic framework needed to address the important questions regarding spider evolution. In particular, AHE is playing an increasingly dominant role to improve our understanding of evolutionary relationships of mygalomorphs and those of tarantulas specifically. Molecular datasets with hundreds of loci from AHE have already been incredibly successful at resolving relationships both at the tips [e.g., to aid in species delineation/delimitation within the genus *Aphonopelma* in Hamilton et al. (2016a, b)] and the widest backbone of Mygalomorphae [e.g., to re-evaluate the monophyly of families in Opatova et al. (2019)]. The focused sampling of Hamilton et al. (2016b) on *Aphonopelma* showed how morphologically-defined lineages—including the identification of 14 novel species—were also genealogically exclusive with AHE data, but also confirmed that deep mitochondrial divergence and introgression otherwise obscured the understanding of evolutionary relationships within that focal genus. The much broader taxon sampling of Opatova et al. (2019) highlighted many cases of non-monophyly across the Mygalomorphae, as well as robustly estimated the historical timing of their diversification under the revised classification. Importantly, these findings aided in the elevation of five subfamilies to the family level, as well as identification of three new families. Finally, for future spider research with AHE approaches, we consider it important to note that the initial spider AHE probe set (version 1) was slightly biased in its recovery of mygalomorph taxa. Therefore, Hamilton et al. (in prep.) has refined the probe set (version 2) to capture the same loci but yield greater enrichment within araneomorph spiders. In addition, AHE data, including library preparation, enrichment, and sequencing has been traditionally generated at the Center for Anchored Phylogenomics at Florida State University (www.anchoredphylogeny.com), but in the near future, a novel open source bioinformatic pipeline will be available that will enable researchers worldwide to easily utilize AHE in their research questions much as is already available to UCE users.

2.5 Toward Resolution of the Tarantula Phylogeny

Thanks to the topology proposed by Foley et al. (2019), based on transcriptomic data, a strong molecular scaffold hypothesis now exists for tarantula evolutionary history. Against this framework, we can speculate about plausible phylogenetic affinities of select other genera for which little or no molecular data has been published (Fig. 2.4). For most American lineages, recent molecular studies appear to support or be consistent with a single large radiation [i.e., the Bombardier clade of Foley et al. (2019)]. We can anticipate that two enigmatic American genera—namely *Kankuamo* and *Hemirrhagus*—will likely be especially influential to further evaluate the robustness of this clade. Both genera have combinations of morphological features to indicate a close alliance to one or more known subfamilies, leaving uncertainties about the specific placement of either. For *Kankuamo,* placement in Theraphosinae has been supported by some morphology in phylogenetic analyses. However, unique urticating hair structures in particular can instead indicate an alternative affinity with Aviculariinae (Perafán et al. 2016). For *Hemirrhagus,* placement in the Theraphosinae has been proposed based on morphology (Pérez-Miles 1998), although has been questioned (i.e., Turner et al. 2018). Some molecular data for this latter genus has already been published by Ortiz et al. (2018) and used in a molecular phylogenetic analysis, but their taxon sampling was focused on Theraphosinae and otherwise only compared an *Ischnocolus* sp. While those authors agreed *Hemirrhagus* may be wrongly placed in Theraphosinae (i.e., as per Turner et al. 2018), their limited sampling did not allow a definitive conclusion about its broader placement. A plausible option to evaluate is whether *Hemirrhagus* is the sister-group of Theraphosinae, or if better placed elsewhere. The division of Theraphosinae into the tribes proposed by Turner et al. (2018) also needs further investigation with additional molecular data, in particular the placement of their defining genera, such as *Hapalopus* against *Theraphosa.* Insights from molecular studies are also still needed to secure the generic limits of some other subfamilies within the Bombardier clade. In particular the limits of the Schismatothelinae subfamily are in critical need of further evaluation, as well as relationships between genera against the type genus *Schismatothele,* as no molecular study to date has

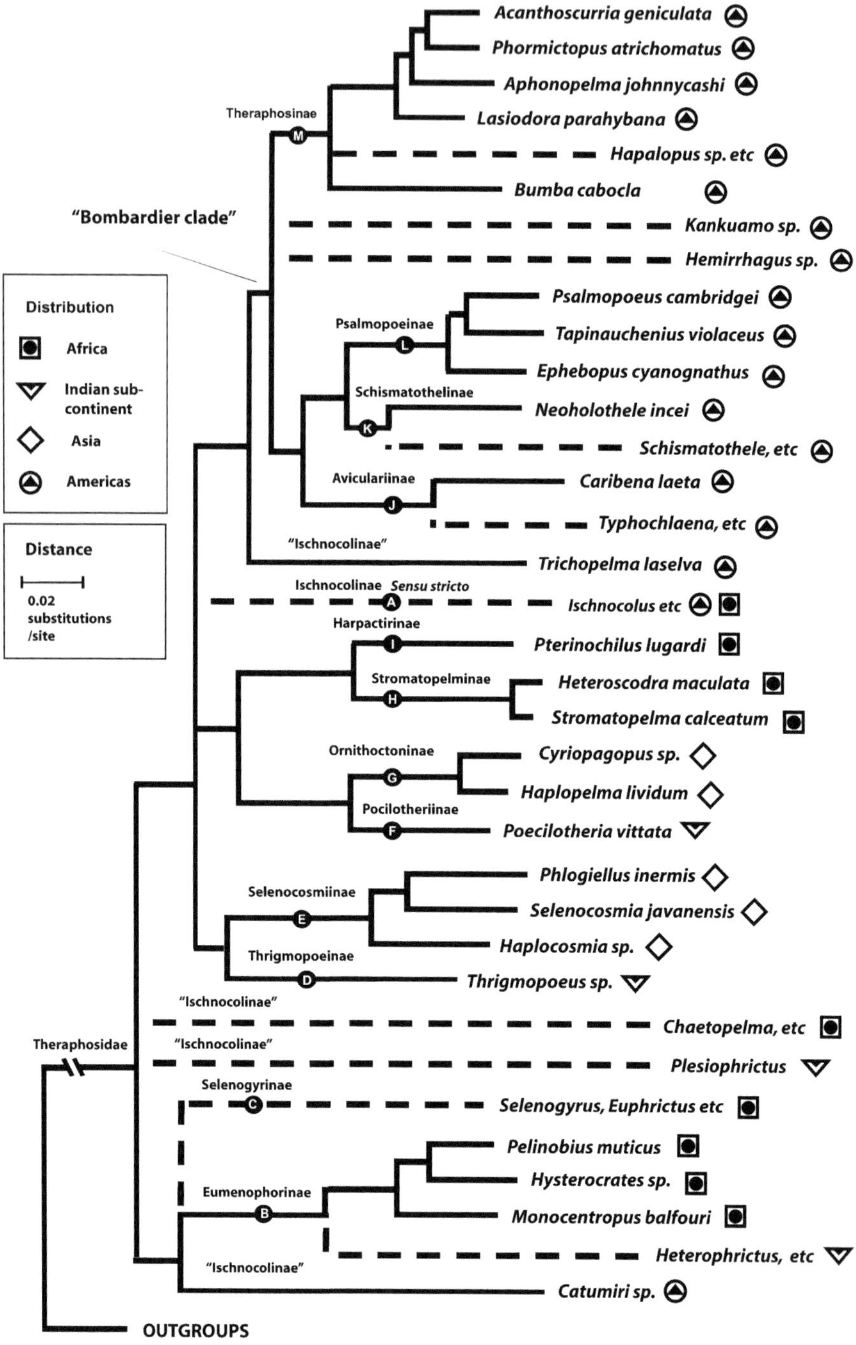

Fig. 2.4 A speculative view on the potential placement of select additional taxa within the context of the robust molecular phylogeny of the tarantula family Theraphosidae established by Foley et al. (2019). The given markers to the right of taxon names indicate geographic origins of each specific taxon (and typically also their closest allies). Nodes corresponding to named subfamilies are also marked where appropriate, labeled A–M to distinguish them

supported sub-familial monophyly with respect to Psalmopoeinae (i.e., Hüsser 2018; Lüddecke et al. 2018; Turner et al. 2018). Elsewhere for Aviculariinae, the placement of *Typhochlaena* relative to other more secure subfamily members is also worthy of investigation with molecular data. Here, several traits may link this genus with taxa outside of Aviculariinae, for example even with some taxa currently placed in the family Barychelidae (see Andre and Hüsser 2018).

More broadly, the limit of the Ischnocolinae is an important aspect of the tarantula phylogeny that has yet to be adequately addressed with molecular data, and where other character sources have proved equivocal at best (see Chap. 3). To date, the most complete morphological study indicates that a clade of several "Ischnocolinae" genera may exist to unite the Afro-European *Ischnocolus* with several American genera (Guadanucci 2014). However, this clade remains unevaluated with published molecular data, where only the small molecular dataset from Turner et al. (2018) found a distant relationships between *Ischnocolus* and diverse American tarantula species—none of which corresponded to those deemed Ischnocolinae *sensu stricto* (Guadanucci 2014). Else, the more extensive molecular AHE data from Opatova et al. (2019) found *Ischnocolus* as more closely related to Harpactirinae and Ornithoctoninae than the few other sampled tarantulas. Put together, if the Bombardier clade proposed by Foley et al. (2019) remains robust in future studies, we may speculate that several of the as yet under-evaluated American "Ischnocolinae" genera may actually be closely affiliated to this clade, perhaps most plausibly in basal positions, as seen in the placement of *Trichopelma*. If so, their addition would further support the current preferred hypothesis of one major radiation for most American tarantulas. However, we speculate that *Ischnocolus* may not be as closely allied to the other suggested American "Ischnocolinae *sensu stricto*" as proposed with morphological data, but instead have a more basal position outside of such American lineages. One option given the Afro-European distribution of *Ischnocolus* is that it can be more closely affiliated with some African lineages such as Harpactirinae/Stromatopelminae (see Fig. 2.4), as potentially indicated by the molecular data in Opatova et al. (2019). Alternatively, *Ischnocolus* may have a more basal position within the phylogeny, such as allied to African Eumenophorinae as previously proposed [see Guadanucci and Wendt (2014) for some discussion]. Another critical "Ischnocolinae" in need of phylogenetic investigation with molecular data is the Afro-European *Chaetopelma*, which, if biogeographic patterns are informative for other such unsampled groups, may be more closely related to *Ischnocolus* than recently anticipated. A similar situation can exist for the primarily Indian *Plesiophrictus,* although this may be more closely allied to Indo-Asian lineages. Other genera that require in-depth investigation with molecular data are those considered as Asian Eumenophorinae in some morphological studies (i.e., Mirza et al. 2014). In particular, several Indian genera (such as *Heterophrictus,* etc.*),* and other allied genera considered unique (e.g., *Neoheterophrictus,* etc.) may provide and important phylogeographic link with African lineages. Elsewhere, the only well-defined subfamily not included in the transcriptomic analyses of Foley et al. (2019) was the African Selenogyrinae. In the past, these have been proposed to have ancestral affinity to African Eumenophorinae, although the recovery of the American

Catumiri as close allies of the latter with transcriptomic data causes complexity around the issue. Here, the relative placement of these respective groups and other problematic "Ischnocolinae" may be greatly influenced by comparative sampling of other non-theraphosid mygalomorphs in various molecular studies, as choice of outgroup can be critical to polarize relationships of focal taxa. Unfortunately, different molecular studies of the broader tarantula relationships have used diverse mygalomorphs for comparative purposes. It will therefore be critical to conduct future genetic comparisons of Theraphosidae against Barychelidae, with a comprehensive sample of lineages from both families to evaluate monophyletic groupings, and clarify the closest sister-group and taxonomic limits of the Theraphosidae.

2.6 Closing Aspects and Future Perspective

Tarantulas and their kin have often led the way in molecular studies with spiders. As outlined above, some tarantulas were among the first arthropods (or even animals) to have partial or later near-complete fragments of their nuclear rRNA sequenced (Hendriks et al. 1986, 1988). Tarantulas were also some of the first animals compared in any molecular systematic studies (e.g., Turbeville et al. 1991), and similarly among the first in such studies exclusively on spiders (e.g., Huber et al. 1993). Later, a tarantula was the joint second spider with a completed mitochondrial genome (Qiu et al. 2005), then among the first few spiders with publically available transcriptomic sequences (Lorenzini et al. 2006). Tarantulas have also often led the way in other areas of molecular research on spiders. Here, a tarantula was the joint first spider to have a nuclear genome sequenced (Sanggaard et al. 2014), and the first group of spiders where the AHE target capture approach was successfully used (Hamilton et al. 2016a, b). Together, these cases highlight the remarkable appeal that tarantulas and their allies can have for molecular research on spiders and their kin. Studies such as these have together played vital roles in helping us understand the evolutionary history and biodiversity of tarantulas, as well as for other spiders and more distant relatives.

We suggest that a well-resolved phylogeny of the Theraphosidae, encompassing a dense sampling of all major lineages, is an easily achievable goal within the next few years. This will particularly be the case if future studies combine insights from diverse data-sources—including extensive molecular data—as integrative analyses, building on the available pool of data, expertise and interest that presently exists for the task. Given a well-resolved phylogenetic scaffold, many important evolutionary questions about these fascinating spiders can then be addressed. In particular, clarifying much of their biogeography seems an easily approachable goal, where the current outline already suggest that major groups are closely tied to geographical patterns (see Chap. 6). This may be directly related to the limited dispersal capability that tarantulas often display. An equally clear geographic pattern could also exist in many recently diverging groups, but this still needs clarifying for many. It is therefore important to use phylogeographic methods to clarify the historical processes leading to such patterns, for example to help explain the exceptional diversification of many American lineages—and particularly within the most speciose subfamily Theraphosinae (see Chap. 5). This latter group appears to have diverged long after the split of South America and Africa according to initial divergence dates from molecular data (i.e., Ortiz et al. 2018; Opatova et al. 2019), mostly since the Mesozoic–Cenozoic boundary (around 66 mya). After this time, new avenues for physical dispersal into North America began to appear, both through Central America, as well as the Caribbean. Yet scenarios of faunal-expansion and exchange between American landmasses have been little explored with any invertebrates, a task for which tarantulas appear to be ideal candidates if given a robust phylogeny. A related topic that could provide other fruitful research tied to ecology, is the investigation of adaptation of certain tarantula genera and species to extremes of climate (see Chap. 10). In particular tolerance of aridity and/or cold seem to be important questions, as both can act together for long periods in certain regions, such as the higher altitudes within South America, or only certain periods of the year such as high deserts of the USA. The relative roles of behavioral versus physiological adaptations in various tarantulas to such environmental factors will likely lead to interesting results against a robust phylogeny. Equally, useful predictions can then be made about how different lineages may respond to past and on-going environmental changes, such as increasing isolation of those with preferences for cooler regions (like higher-elevations) in response to climatic warming, or impact of extended drought on those adapted to some wetter period (e.g., breeding cues, etc.—see Chap. 14). Once such aspects are meaningfully evaluated, informed decisions about conservation actions or management plans can be made (see Chap. 16). Elsewhere, and more academically, the evolution of extreme differences in body-size can provide an exciting avenue of future research, as tarantula species can occupy the opposite ends of size extremes, often in the same geographic region. Again, a robust phylogeny is needed to provide secure insights about the polarity of such physical differences, before being related to factors that may have driven them. For a more applied topic, the variation in colors and patterns on tarantula bodies should also be re-evaluated against a well-resolved phylogeny. This can for example help to understand efficient crypsis, or conversely the development of aposematism, both of which may have findings applicable to human interests. More obviously, applied venom research may be greatly informed by a robust phylogeny, as studies of their component variation and relative efficacies can be greatly informed by knowledge of the evolutionary history. Here, a robust phylogenetic framework can help identify which unsampled taxa may be the most valuable additional targets for generating influential results.

Finally, we consider it vital to point out that even in the face of a tremendous amount of sequence data now possible from many of the approaches discussed, none of them should be considered a panacea for molecular systematics. From systematic studies of other taxa, it already appears that certain nodes in the Tree of Life may be beyond unequivocal resolution regardless of the scale of sequence data evaluated. This may be for example because the relevant phylogenetic signature is lacking, or masked by complex molecular evolution. Even with the availability of massive-scale sequencing for many tarantulas in the near future, we may anticipate that some rapid-divergence events combined with complex patterns of molecular evolution may confound systematic resolution of certain tarantula groups. Despite this, we expect that ongoing and future molecular systematic studies of tarantulas and their kin—especially if combined with other informative data—will continue to provide us with many valuable insights about the evolution and diversification of these amazing creatures.

Acknowledgments We thank Brent Hendrixson for his insightful comments that greatly improved this chapter, and Fernando Pérez-Miles for the opportunity to provide this contribution. Chris Hamilton thanks Jason Bond for his guidance and mentorship during Hamilton's PhD and partnership during development of Anchored Hybrid Enrichment in spiders. Stuart Longhorn thanks Alfried Vogler for mentorship during Longhorn's formative years, as well as Susan Masta and Davide Pisani for later informative guidance.

References

Agnarsson I, Coddington J, Kuntner M (2013) Systematics – progress in the study of spider diversity and evolution. In: Penney D (ed) Spider research in the 21st century – trends and perspectives. Siri Scientific Press, Manchester, pp 58–111

Alamo L, Pinto A, Sulbarán G, Mavárez J, Padrón R (2018) Lessons from a tarantula: new insights into myosin interacting-heads motif evolution and its implications on disease. Biophys Rev 10(5):1465–1477

Altenhoff AM, Gil M, Gonnet GH, Dessimoz C (2013) Inferring hierarchical orthologous groups from orthologous gene pairs. PLoS One 8(1):e53786

Andre C, Hüsser M (2018) About trapdoors and bridges – new insights in the little-known ecology and lifestyle of the genus *Typhochlaena* C.L. Koch 1850. J Br Tarantula Soc 32(3):3–29

Arnedo MA, Ferrández MA (2007) Mitochondrial markers reveal deep population subdivisión in the European protected spider *Macrothele calpeiana* (Walckenaer, 1805) (Araneae, Hexathelidae). Cons Genetics 8:1147–1162

Astrin JJ, Huber BA, Misof B, Klütsch CFC (2006) Molecular taxonomy in pholcid spiders (Pholcidae, Araneae): evaluation of species identification methods using CO1 and 16S rRNA. Zool Scr 35:441–457

Ayoub NA, Garb JE, Hedin M, Hayashi CY (2007) Utility of the nuclear protein-coding gene, elongation factor-1 gamma (EF-1y) for spider systematics, emphasizing family level relationships of tarantulas and their kin (Araneae: Mygalomorphae). Mol Phylogenet Evol 42(2):394–409

Babb PL, Lahens NF, Correa-Garhwal SM, Nicholson DN, Kim EJ, Hogenesch JB, Kuntner M, Higgins L, Hayashi CY, Agnarsson I, Voight BF (2017) The *Nephila clavipes* genome highlights the diversity of spider silk genes and their complex expression. Nat Genet 49(6):895–903

Bailey AL, Brewer MS, Hendrixson BE, Bond JE (2010) Phylogeny and classification of the trapdoor genus *Myrmekiaphila*: an integrative approach to evaluating taxonomic hypotheses. PLoS One 5:e12744

Ballard JWO, Olsen GJ, Faith DP, Odgers WA, Rowell DM, Atkinson PW (1992) Evidence from 12S ribosomal RNA sequences that Onychophorans are modified arthropods. Science 258:1345–1348

Ballesteros JA, Hormiga G (2016) A new orthology assessment method for phylogenomic data: unrooted phylogenetic orthology. Mol Biol Evol 33:2117–2134

Ballesteros JA, Sharma PP (2019) A critical appraisal of the placement of Xiphosura (Chelicerata) with account of known sources of phylogenetic error. Syst Biol 68(6):896–917

Barrett RDH, Hebert PDN (2005) Identifying spiders through DNA barcodes. Can J Zool 83:481–491

Beavis AS, Sunnucks P, Roswell DM (2011) Microhabitat preferences drive phylogeographic disparities in two Australian funnel web spiders. Biol J Linn Soc 104:805–819

Blagoev GA, Nikolova NI, Sobel CN, Hebert PDN, Adamowicz SJ (2013) Spiders (Araneae) of Churchill, Manitoba: DNA barcodes and morphology reveal high species diversity and new Canadian records. BMC Ecol 13:e44

Blagoev GA, DeWaard JR, Ratnasingham S, DeWaard SL, Lu L, Robertson J, Telfer AC, Hebert PDN (2016) Untangling taxonomy: a DNA barcode reference library for Canadian spiders. Mol Ecol Resour 16:325–341

Blaimer BB, Lloyd MW, Guillory WX, Brady SG (2016) Sequence capture and phylogenetic utility of genomic Ultraconserved Elements obtained from pinned insect specimens. PLoS One 11:e0161531–e0161520

Bond JE (2004) Systematics of the Californian euctenizine spider genus *Apomastus* (Araneae: Mygalomorphae: Cyrtauchenidae): the relationship between molecular and morphological taxonomy. Invertebr Syst 18:361–376

Bond JE (2012) Phylogenetic treatment and taxonomic revision of the trapdoor spider genus *Apostichus* Simon (Araneae, Mygalomorphae, Euctenizidae). ZooKeys 252:1–209

Bond JE, Hendrixson BE, Hamilton CA, Hedin M (2012) A reconsideration of the classification of the spider infraorder Mygalomorphae (Arachnida: Araneae) based on three nuclear genes and morphology. PLoS One 7:e38753

Bond JE, Hedin MC (2006) A total evidence assessment of the phylogeny of North American euctenizine trapdoor spiders (Araneae, Mygalomorphae, Cyrtaucheniidae) using Bayesian inference. Mol Phylogenet Evol 41:70–85

Bond JE, Stockman AK (2008) An integrative method for delimiting cohesion species: finding the population-species interface in a group of Californian trapdoor spiders with extreme genetic divergence and geographic structuring. Syst Biol 57:628–646

Bond JE, Hedin MC, Ramirez MG, Opell BD (2001) Deep molecular divergence in the absence of morphological and ecological change in the Californian coastal dune endemic trapdoor spider *Aptostichus simus*. Mol Ecol 10:899–910

Bond JE, Beamer DA, Lamb T, Hedin M (2006) Combining genetic and geospatial analyses to infer population extinction in mygalomorph spiders endemic to the Los Angeles region. Anim Conserv 9:145–157

Bond JE, Garrison NL, Hamilton CA, Godwin RL, Hedin M, Agnarsson I (2014) Phylogenomics resolves a spider backbone phylogeny and rejects a prevailing paradigm for orb web evolution. Curr Biol 24:1765–1771

Boore JL, Collins TM, Stanton D, Daehler LL, Brown WM (1995) Deducing the pattern of arthropod phylogeny from mitochondrial DNA re-arrangements. Nature 376:163–167

Branstetter MG, Danforth BN, Pitts JP, Faircloth BC, Ward PS, Buffington ML, Gates MW, Kula RR, Brady SG (2017a) Phylogenomic insights into the evolution of stinging wasps and the origins of ants and bees. Curr Biol 27:1019–1025

Branstetter MG, Longino JT, Ward PS, Faircloth BC (2017b) Enriching the ant tree of life: enhanced UCE bait set for genome-scale phylogenetics of ants and other Hymenoptera. Methods Ecol Evol 8:768–776

Brewer MS, Cotoras DD, Croucher PJP, Gillespie RG (2014) New sequencing technologies, the development of genomics tools, and their applications in evolutionary arachnology. J Arachnol 42:1–15

Briscoe AG, Goodacre S, Masta SE, Taylor MI, Arnedo MA, Penny D, Kenny J, Creer S (2013) Can long-range PCR be used to amplify genetically divergent mitochondrial genomes for comparative phylogenetics? A case study within spiders (Arthropoda: Araneae). PLoS One 8:e62404

Bybee SM, Bracken-Grissom H, Haynes BD, Hermansen RA, Byers RL, Clement MJ, Udall JA, Wilcox ER, Crandall KA (2011) Targeted amplicon sequencing (TAS): a scalable next-gen approach to multi-locus, multi-taxa phylogenetics. Genome Biol Evol 3:1312–1323

Cao Z, Yu Y, Wu Y, Hao P, Di Z, He Y, Chen Z, Yang W, Shen Z, He X et al (2013) The genome of *Mesobuthus martensii* reveals a unique adaptation model of arthropods. Nat Commun 4:2602

Carlson DE, Hedin M (2017) Comparative transcriptomics of Entelegyne spiders (Araneae, Entelegyne), with emphasis on molecular evolution of orphan genes. PLoS One 12:e0174102

Castalanelli MA, Teale R, Rix MG, Kennington WJ, Harvey MS (2014) Barcoding of mygalomorph spiders (Araneae: Mygalomorphae) in the Pilbara bioregion of Western Australia reveals a highly diverse biota. Invertebr Syst 28:375–385

Castalanelli MA, Huey JA, Hillyer MJ, Harvey MS (2017) Molecular and morphological evidence for a new genus of small trapdoor spiders from arid Western Australia (Araneae: Mygalomorphae: Nemesiidae: Anaminae). Invertebr Syst 31:492–505

Caterino MS, Cho S, Sperling FAH (2000) The current state of insect molecular systematics: a thriving tower of babel. Ann Rev Entomol 45:1–54

Chamberland L, McHugh A, Kechejian S, Binford GJ, Bond JE, Coddington J, Dolman G, Hamilton CA, Harvey MS, Kuntner M, Agnarsson I (2018) From Gondwana to GAARlandia: evolutionary history and biogeography of ogre-faced spiders (*Deinopis*). J Biogeogr 155:252–216

Chen J, Zhao L, Jiang L, Meng E, Zhang Y, Xiong X, Liang S (2008) Transcriptome analysis revealed novel possible venom components and cellular processes of the tarantula *Chilobrachys jingzhao* venom gland. Toxicon 52:794–806

Cheng D-Q, Piel WH (2018) The origins of the Psechridae: web-building lycosoid spiders. Mol Phylogenet Evol 125:213–219

Clarke TH, Garb JE, Hayashi CY, Haney RA, Lancaster AK, Corbett S, Ayoub NA (2014) Multi-tissue transcriptomics of the black widow spider reveals expansions, co-options, and functional processes of the silk-gland gene toolkit. BMC Genomics 15:e365

Clarke TH, Garb JE, Hayashi CY, Arensburger P, Ayoub NA (2015) Spider transcriptomes identify ancient large-scale gene duplication event potentially important in silk gland evolution. Genome Biol Evol 7:1856–1870

Coddington JA, Agnarsson I, Hamilton CA, Bond JE (2019) Spiders did not repeatedly gain, but repeatedly lost, foraging webs. PeerJ 7:e6703

Colgan DJ, McLauchlan A, Wilson GDF, Livingston SP, Edgecombe GD, Macarans J, Cassis G, Gray MR (1998) Histone H3 and U2 snRNA DNA sequences and arthropod molecular evolution. Aust J Zool 46:419–437

Cooper SJB, Harvey MS, Saint KM, Main BY (2011) Deep phylogeographic structuring of populations of the trapdoor spider *Moggridgea tingle* (Migidae) from southwestern Australia: evidence for long-term refugia within refugia. Mol Ecol 20:3219–3236

Corzo G, Escoubas P (2003) Pharmacologically active spider peptide toxins. Cell Mol Life Sci 60:2409–2426

Corzo G, Diego-García E, Clement H, Peigneur S, Odell G, Tytgat J, Possani LD, Alagón A (2008) An insecticidal peptide from the theraposid ***[sic]*** *Brachypelma smithi* spider venom reveals common molecular features among spider species from different genera. Peptides 29:1901–1908

Crampton-Platt A, Yu DW, Zhou X, Vogler AP (2016) Mitochondrial metagenomics: letting the genes out of the bottle. GigaScience 5:e15

Croom HB, Gillespie RG, Palumbi SR (1991) Mitochondrial DNA sequences coding for a portion of the RNA of the small ribosomal subunits of *Tetragnatha mandibulata* and *Tetragnatha hawaiensis* (Araneae, Tetragnathidae). J Arachnol 19:210–214

Diego-García E, Peigneur S, Waelkens E, Debaveye S, Tytgat J (2010) Venom components from *Citharischius crawshayi* spider (Family Theraphosidae): exploring transcriptome, venomics, and function. Cell Mol Life Sci 67:2799–2813

Dimitrov D, Benavides LR, Arnedo MA, Giribet G, Griswold CE, Scharff N, Hormiga G (2017) Rounding up the usual suspects: a standard target-gene approach for resolving the interfamilial phylogenetic relationships of ecribellate orb-weaving spiders with a new family-rank classification (Araneae, Araneoidea). Cladistics 33:221–250

Edwards S, Liu L, Pearl D (2007) High-resolution species trees without concatenation. PNAS 104:5936–5941

Faircloth BC (2017) Identifying conserved genomic elements and designing universal probe sets to enrich them. Mol Ecol Resour 8:1103–1112

Faircloth BC, McCormack JE, Crawford NG, Harvey MG, Brumfield RT, Glenn TC (2012) Ultraconserved elements anchor thousands of genetic markers spanning multiple evolutionary timescales. Syst Biol 61:717–726

Fang W-Y, Wang Z-L, Li C, Yang X-Q, Yu X-P (2016) The complete mitogenome of a jumping spider *Carrhotus xanthogramma* (Araneae: Salticidae) and comparative analysis in four salticid mitogenomes. Genetica 144:699–709

Fernández R, Hormiga G, Giribet G (2014) Phylogenomic analysis of spiders reveals nonmonophyly of orb weavers. Curr Biol 24:1772–1777

Fernández R, Kallal RJ, Dimitrov D, Ballesteros JA, Arnedo MA, Giribet G, Hormiga G (2018) Phylogenomics, diversification dynamics, and comparative transcriptomics across the spider tree of life. Curr Biol 28:1489–1497

Field KG, Olsen GJ, Lane DJ, Giovannoni SJ, Ghiselin MT, Raff EC, Pace NR, Raff RA (1988) Molecular phylogeny of the animal kingdom. Science 239:748–753

Foley S, Lüddecke T, Cheng D-Q, Krehenwinkel H, Künzel S, Longhorn SJ, Wendt I, von Wirth V, Tänzler R, Vences M, Piel WH (2019) Tarantula phylogenomics: a robust phylogeny of deep theraphosid clades inferred from transcriptome data sheds light on the prickly issue of urticating setae evolution. Mol Phylogenet Evol. 140:106573

Folmer O, Black M, Hoeh W, Lutz R, Vrijenhoek R (1994) DNA primers for amplification of mitochondrial cytochrome c oxidase subunit I from diverse metazoan invertebrates. Mol Mar Biol Biotech 3:294–299

Foster PG, Hickey DA (1999) Compositional bias may affect both DNA-based and protein-based phylogenetic reconstructions. J Mol Evol 48:284–290

French AS, Li AW, Meisner S, Torkkeli PH (2014) Upstream open reading frames and Kozak regions of assembled transcriptome sequences from the spider *Cupiennius salei*. Selection or chance? Gene 539:03–208

Frías-López C, Almeida FC, Guirao-Rico S, Vizueta J, Sánchez-Gracia A, Arnedo MA, Rozas J (2015) Comparative analysis of tissue-specific transcriptomes in the funnel-web spidr *Macrothele calpeiana* (Araneae, Hexathelidae). PeerJ 3:e1064

Friedrich M, Tautz D (1995) Ribosomal DNA phylogeny of the major extant arthropod classes and the evolution of myriapods. Nature 376:165–167

Fukushima CS, Bertani R (2017) Taxonomic revision and cladistic analysis of *Avicularia* Lamarck, 1818 (Araneae, Theraphosidae, Aviculariinae) with description of three new Aviculariine genera. ZooKeys 659:1–185

Garb JE, Shara PP, Ayoub NA (2018) Recent progress and prospects for advancing arachnid genomics. Curr Opin Insect Sci 25:51–57

Garrison NL, Rodriquez J, Agnarsson I, Coddington JA, Griswold CE, Hamilton CA, Hedin M, Kocot KM, Ledford JM, Bond JE (2016) Spider phylogenomics: untangling the spider tree of life. PeerJ 4:e1719

Gillespie RG, Croom HB, Palumbi SR (1994) Multiple origins of a spider radiation in Hawaii. PNAS 91:2290–2294

Godwin RL, Opatova V, Garrison NL, Hamilton CA, Bond JE (2018) Phylogeny of a cosmopolitan family of morphologically conserved trapdoor spiders (Mygalomorphae, Ctenizidae) using Anchored Hybrid Enrichment, with a description of the family, Halonproctidae Pocock 1901. Mol Phylogenet Evol 126:303–313

Graham MR, Hendrixson BE, Hamilton CA, Bond JE (2015) Miocene extensional tectonics explain ancient patterns of diversification among turret-building tarantulas (*Aphonopelma mojave*) group in the Mojave and Sonoran deserts. J Biogeogr 42:1052–1065

Grbic M, Van Leeuwen T, Clark RM, Rombauts S, Rouze P, Grbic V, Osborne EJ, Dermauw W, Ngoc PC, Ortego F et al (2011) The genome of *Tetranychus urticae* reveals herbivorous pest adaptations. Nature 479:487–492

Gregory TR, Shorthouse DP (2003) Genome sizes of spiders. J Hered 94:285–290

Guadanucci JPL (2014) Theraphosidae phylogeny: relationships of the '*Ischnocolinae*' genera (Araneae, Mygalomorphae). Zool Scr 43:508–518

Guadanucci JPL, Wendt I (2014) Revision of the spider genus *Ischnocolus* Ausserer, 1871 (Mygalomorphae: Theraphosidae: Ischnocolinae). J Nat Hist 48(7–8):387–402

Hamilton CA, Formanowicz DR, Bond JE (2011) Species delimitation and phylogeography of *Aphonopelma hentzi* (Araneae, Mygalomorphae, Theraphosidae): cryptic diversity in North American tarantulas. PLoS One 6:e26207

Hamilton CA, Hendrixson BE, Brewer MS, Bond JE (2014) An evaluation of sampling effects on multiple DNA barcoding methods leads to an integrative approach for delimiting species: a case study of the North American tarantula genus *Aphonopelma* (Araneae, Mygalomorphae, Theraphosidae). Mol Phylogenet Evol 71:79–93

Hamilton CA, Hendrixson BE, Bond JE (2016a) Taxonomic revision of the tarantula genus *Aphonopelma* Pocock, 1901 (Araneae, Mygalomorphae, Theraphosidae) within the United States. ZooKeys 560:1–340

Hamilton CA, Lemmon AR, Lemmon EM, Bond JE (2016b) Expanding anchored hybrid enrichment to resolve both deep and shallow relationships within the spider tree of life. BMC Evol Biol 16:e212

Haney RA, Ayoub NA, Clarke TH, Hayashi CY, Garb JE (2014) Dramatic expansion of the black widow toxin arsenal uncovered by multi-tissue transcriptomics and venom proteomics. BMC Genomics 15:e366

Harrison SE, Harvey MS, Cooper SJB, Austin AD, Rix MG (2017) Across the Indian Ocean: a remarkable example of trans-oceanic dispersal in an austral mygalomorph spider. PLoS One 12(8):e0180139

Harrison SE, Rix MG, Harvey MS, Austin AD (2018) Systematics of the Australian spiny trapdoor spiders of the genus *Blakistonia* Hogg (Araneae: Idiopidae). Zootaxa 4518:001–076

Harvey MS, Hillyer MJ, Main BY, Moulds TA, Raven RJ, Rix MG, Vink CJ, Huey JA (2018) Phylogenetic relationships of the Australian open-holed trapdoor spiders (Araneae: Mygalomorphae: Nemesiidae: Anaminae): multi-locus molecular analyses resolve the generic classification of a highly diverse fauna. Zool J Linn Soc 184:407–452

Hassanin A (2006) Phylogeny of Arthropoda inferred from mitochondrial sequences: strategies for limiting the misleading effects of multiple changes in pattern and rates of substitution. Mol Phylogenet Evol 38:100–116

Hassanin A, Léger N, Deutsch J (2005) Evidence for multiple reversals of asymmetric mutational constraints during the evolution of the mitochondrial genome of Metazoa, and consequences for phylogenetic inferences. Syst Biol 54:277–298

Hausdorf B (1999) Molecular phylogeny of araneomorph spiders. J Evol Biol 12:980–985

Hebert PDN, Cywinska A, Ball SL, deWaard JR (2003a) Biological identifications through DNA barcodes. P R Soc Lond Ser B Biol Sci 270:313–321

Hebert PDN, Ratnasingham S, de Waard JR (2003b) Barcoding animal life: cytochrome c oxidase subunit 1 divergences among closely related species. P R Soc Lond Ser B Biol Sci 270:S96–S99

Hedin M, Bond JE (2006) Molecular phylogenetics of the spider infraorder Mygalomorphae using nuclear rRNA genes (18S and 28S): conflict and agreement with the current system of classification. Mol Phylogenet Evol 41:454–471

Hedin M, Carlson D (2011) A new trapdoor spider species from the southern coast ranges of California (Mygalomorphae, Antrodiaetidae, *Aliatypus coylei,* sp. nov,) including consideration of mitochondrial phylogeographic structuring. Zootaxa 2963:55–68

Hedin MC, Maddison WP (2001) A combined molecular approach to phylogeny of the jumping spider subfamily Dendryphantinae (Araneae: Salticidae). Mol Phylogenet Evol 18(3):386–403

Hedin M, Starrett J, Hayashi C (2013) Crossing the uncrossable: novel trans-valley biogeographic patterns revealed in the genetic history of low-dispersal mygalomorph spiders (Antrodiaetidae, *Antrodiaetus*) from California. Mol Ecol 22:508–526

Hedin M, Carlson D, Coyle F (2015) Sky island diversification meets the multispecies coalescent – divergence in the spruce-fir moss spider (*Microhexura montivaga*, Araneae, Mygalomorphae) on the highest peaks of southern Appalachia. Mol Ecol 24:3467–3484

Hedin M, Derkarabetian S, Ramirez MJ, Vink C, Bond JE (2018a) Phylogenomic reclassification of the world's most venomous spiders (Mygalomorphae, Atracinae), with implications for venom evolution. Nat Sci Rep 8:e1636

Hedin M, Derkarabetian S, Blair J, Paquin P (2018b) Sequence capture phylogenomics of eyeless *Cicurina* spiders from Texas caves, with emphasis on US federally-endangered species from Bexar County (Araneae, Hahniidae). ZooKeys 769:49

Hedin M, Derkarabetian S, Alfaro A, Ramírez MJ, Bond JE (2019) Phylogenomic analysis and revised classification of atypoid mygalomorph spiders (Araneae, Mygalomorphae), with notes on arachnid ultraconserved element loci. PeerJ 7:e6864

Hendriks L, Huysmans E, Vandenberghe A, De Wachter R (1986) Primary structures of the 5S ribosomal RNAs of 11 arthropods and applicability of 5S RNA to the study of metazoan evolution. J Mol Evol 24:103–109

Hendriks L, Van Broeckhoven C, Vandenberghe A, Van de Peer Y, De Wachter R (1988) Primary and Secondary Structure of the 18S ribosomal RNA of the bird spider *Eurypelma californica* and evolutionary relationships among eukaryotic phyla. Eur J Biochem 177:15–20

Hendrixson BE, Bond JE (2005) Testing species boundaries in the *Antrodiaetus unicolor complex* (Araneae: Mygalomorphae: Antrodiaetidae): "paraphyly" and cryptic diversity. Mol Phylogenet Evol 36:405–416

Hendrixson BE, Bond JE (2007) Molecular phylogeny and biogeography of an ancient Holarctic lineage of mygalomorph spiders (Araneae: Antrodiaetidae: *Antrodiaetus*). Mol Phylogenet Evol 42:738–755

Hendrixson BE, Bond JE (2009) Evaluating the efficacy of continuous quantitative characters for reconstructing the phylogeny of a morphologically homogeneous spider taxon (Araneae, Mygalomorphae, Antrodiaetidae, *Antrodiaetus*). Mol Phylogenet Evol 53:300–313

Hendrixson BE, DeRussy BM, Hamilton CA, Bond JE (2013) An exploration of species boundaries in turret-building tarantulas of the Mojave Desert (Araneae, Mygalomorphae, Theraphosidae, *Aphonopelma*). Mol Phylogenet Evol 66:327–340

Hendrixson E, Gulce AV, Bond JE (2015) Integrative species delimitation and conservation of tarantulas (Araneae, Mygalomorphae, Theraphosidae) from a North American biodiversity hotspot. Insect Conser Diver 8:120–131

Herzig V, King GF, Undheim EAB (2019) Can we resolve the taxonomic bias in spider venom research? Toxicon 1:e100005

Huang H, He Q, Kubatko LS, Knowles LL (2010) Sources of error inherent in species-tree estimation: impact of mutational and coalescent effects on accuracy and implications for choosing among different methods. Syst Biol 59:573–583

Huber KC, Haider THS, Muller MW, Huber BA, Schwayen RJ, Barth FG (1993) DNA sequence data indicates the polyphyly of the family Ctenidae (Araneae). J Arachnol 21:194–201

Hughes J, Longhorn SJ, Papadopoulou A, Theodorides K, de Riva A, Meija-Chang M, Foster PG, Vogler AP (2006) Dense taxonomic EST sampling and its applications for molecular systematics of the Coleoptera (Beetles). Mol Biol Evol 23:268–278

Hüsser M (2018) A first phylogenetic analysis reveals a new arboreal tarantula genus from South America with description of a new species and two new species of *Tapinauchenius* Ausserer, 1871 (Araneae, Mygalomorphae, Theraphosidae). ZooKeys 784:59–93

Huysmans E, Dams E, Vandenberghe A, De Wachter R (1983) The nucleotide sequences of the 5S rRNA of four mushrooms and their use in studying the phylogenetic position of basidiomycetes among the eukaryotes. Nucleic Acids Res 11:2871–2879

Hwang UW, Park CJ, Yong TS, Kim W (2001) One-step PCR amplification of complete arthropod mitochondrial genomes. Mol Phylogenet Evol 26:110–120
i5K Consortium (2013) The i5K initiative: advancing arthropod genomics for knowledge, human health, agriculture, and the environment. J Hered 104:595–600
Jiang L, Zhang D, Zhang Y, Peng L, Chen J, Liang S (2010) Venomics of the spider *Ornithoctonus huwena* based on transcriptomic versus proteomic analysis. Comp Biochem Physiol D 5:81–88
Kallal RJ, Fernández R, Giribet G, Hormiga G (2018) A phylotranscriptomic backbone of the orb-weaving spider family Araneidae (Arachnida, Araneae) supported by multiple methodological approaches. Mol Phylogenet Evol 126:129–140
Kocher TD, Thomas WK, Meyer A, Edwards SV, Paabo S, Villablanca FX (1989) Dynamics of mitochondrial DNA evolution in animals: amplification and sequencing with conserved primers. PNAS 86:6196–6200
Kono N, Nakamura H, Ohtoshi R, Pedrazzoli-Moran DA, Shinohara A, Yoshida Y, Fujiwara M, Mori M, Tomita M, Arakawa K (2019) Orb-weaving spider *Araneus ventricosus* genome elucidates the spidroin gene catalogue. Sci Rep 9:e8380
Kornilios P, Thanou E, Kapli P, Parmakelis A, Chatzaki M (2016) Peeking through the trapdoor: historical biogeography of the Aegean endemic spider *Cyrtocarenum* Ausserer, 1871 with an estimation of mtDNA substitution rates for Mygalomorphae. Mol Phylogenet Evol 98:300–313
Král J, Kořínková T, Krkavcová L, Musilová J, Forman M, Ávila Herrera IM, Haddad CR, Vítková M, Henriques S, Palacios Vargas JG, Hedin M (2013) Evolution of karyotype, sex chromosomes, and meiosis in mygalomorph spiders (Araneae: Mygalomorphae). Biol J Linn Soc 109:377–408
Krehenwinkel H, Kennedy SR, Rueda A, Lam A, Gillespie RG (2018) Scaling up DNA barcoding – primer sets for simple and cost efficient arthropod systematics by multiplex PCR and Illumina amplicon sequencing. Methods Ecol Evol 9:2181–2193
Krehenwinkel H, Pomerantz A, Henderson JB, Kennedy SR, Lim JY, Swamy V, Shoobridge JD, Patel NH, Gillespie RG, Prost S (2019) Nanopore sequencing of long ribosomal DNA amplicons enables portable and simple biodiversity assessments with high phylogenetic resolution across broad taxonomic scale. GigaScience 8:giz006
Kriventseva EV, Kuznetsov D, Tegenfeldt F, Manni M, Dias R, Simão FA, Zdobnov EM (2019) OrthoDB v10: sampling the diversity of animal, plant, fungal, protist, bacterial and viral genomes for evolutionary and functional annotations of orthologs. Nucleic Acids Res 47:D807–D811
Kuntner M, Agnarsson I (2011) Biogeography and diversification of hermit spiders on Indian Ocean islands (Nephilidae: *Nephilengys*). Mol Phylogenet Evol 59:477–488
Kuntner M, Hamilton CA, Cheng R-C, Gregorič M, Lupše N, Lemmon EM, Lemmon AR, Agnarsson I, Coddington JA, Bond JE (2019) Golden orbweavers ignore biological rules: phylogenomic and comparative analyses unravel a complex evolution of sexual size dimorphism. Syst Biol 68:555–572
Leaché AD, Rannala B (2011) The accuracy of species tree estimation under simulation: a comparison of methods. Syst Biol 60:126–137
Leavitt DH, Starrett J, Westphal MF, Hedin M (2015) Multilocus sequence data reveals dozens of putative cryptic species in a radiation of endemic Californian mygalomoprph spiders (Araneae, Mygalomorphae, Nemesiidae). Mol Phylogenet Evol 91:56–67
Lemmon AR, Emme SA, Lemmon EM (2012) Anchored Hybrid Enrichment for massively high-throughput phylogenomics. Syst Biol 61:727–744
Lewin HA, Robinson GE, Kress WJ et al (2018) Earth BioGenome Project: sequencing life for the future of life. PNAS 17:4325–4333
Liu L, Yu L (2011) Estimating species trees from unrooted gene trees. Syst Biol 60:661–667
Liu M, Zhang Z, Peng Z (2015) The mitochondrial genome of the water spider *Argyroneta aquatica* (Araneae: Cybaeidae). Zool Scr 44:179–190

Liu S, Wang X, Xie L, Tan M, Li Z, Su X, Zhang H, Misof B, Kjer KM, Tang M, Niehuis O, Jiang H, Zhou X (2016) Mitochondrial capture enriches mito-DNA 100 fold, enabling PCR-free mitogenomics biodiversity analysis. Mol Ecol Resour 16:470–479

Longhorn SJ, Nicholas M, Chuter J, Vogler AP (2007) The utility of molecular markers from non-lethal DNA samples of the CITES II protected "tarantula" *Brachypelma vagans* (Araneae, Theraphosidae). J Arachnol 35:278–292

Lopardo L, Uhl G (2014) Testing mitochondrial marker efficacy for DNA barcoding in spiders: a test case using the dwarf spider genus *Oedothorax* (Araneae: Linyphiidae: Erigoninae). Invertebr Syst 28:501–521

Lorenzini DM, da Silva PI, Soares MB, Arruda P, Setubal J, Daffre S (2006) Discovery of immune-related genes expressed in hemocytes of the tarantula spider *Acanthoscurria gomesiana*. Dev Comp Immun 30:545–556

Lüddecke T, Krehenwinkel H, Canning G, Glaw F, Longhorn SJ, Tänzler R, Wendt I, Vences M (2018) Discovering the silk road: nuclear and mitochondrial sequence data resolve the phylogenetic relationships among theraphosid spider subfamilies. Mol Phylogenet Evol 119:63–70

Macher JK, Zizka V, Weigand AM, Leese F (2018) A simple centrifugation protocol for metagenomic studies increases mitochondrial DNA yield by two orders of magnitude. Methods Ecol Evol 9:1070–1074

Maddison WP, Hedin MC (2003) Jumping spider phylogeny (Araneae: Salticidae). Invertebr Syst 17:529–549

Maddison W, Knowles L (2006) Inferring phylogeny despite incomplete lineage sorting. Syst Biol 55:21–30

Maddison WP, Evans SC, Hamilton CA, Bond JE, Lemmon AR, Lemmon EM (2017) A genome-wide phylogeny of jumping spiders (Araneae, Salticidae), using anchored hybrid enrichment. ZooKeys 89:89–101

Masta SE, Boore JL (2004) The complete mitochondrial genome sequence of the spider *Habronattus oregonensis* reveals rearranged and extremely truncated tRNAs. Mol Biol Evol 21:893–902

Masta SE, Boore JL (2008) Parallel evolution of truncated transfer RNA genes in arachnid mitochondrial genomes. Mol Biol Evol 25:949–959

Masta SE, Longhorn SJ, Boore JL (2009) Arachnid relationships based on mitochondrial genomes: asymmetric nucleotide and amino acid bias affects phylogenetic analyses. Mol Phylogenet Evol 50:117–128

Masta SE, McCall A, Longhorn SJ (2010) Rare genomic changes and mitochondrial sequences provide independent support for congruent relationships among the sea spiders (Arthropoda, Pycnogonida). Mol Phylogenet Evol 57:59–70

Mendoza J, Francke O (2017) Systematic revision of *Brachypelma* red-kneed tarantulas (Araneae: Theraphosidae), and the use of DNA barcodes to assist in the identification and conservation of CITES-listed species. Invertebr Syst 31:157–179

Meng X, Zhang Y, Bao H, Liu Z (2015) Sequence analysis of insecticide action and detoxification-related genes in the insect pest natural enemy *Pardosa pseudoannulata*. PLoS One 10:e0125242

Mikheyev AS, Zwick A, Magrath MJL, Grau ML, Qiu L, Su YN, Yeates D (2017) Museum genomics confirms that the Lord Howe Island stick insect survived extinction. CURBIO 27:3157–3161.e4

Miller JA, Beentjes KK, van Helsdingen P, Ijland S (2013) Which specimens from a museum collection will yield DNA barcodes? A time series study of spiders in alcohol. ZooKeys 365:245–261

Mirza Z, Sanap RV, Bhosale H (2014) Preliminary review of Indian Eumenophorinae (Araneae: Theraphosidae) with description of a new genus and five new species from the Western Ghats. PLoS One 9:e87928

Montes de Oca L, D'Elía G, Pérez-Miles F (2016) An integrative approach for species delimitation in the spider genus *Grammostola* (Theraphosidae, Mygalomorphae). Zool Scr 45:322–333

Mora E, Paspati A, Decae AE, Arnedo MA (2017) Rafting spiders or drifting islands? Origins and diversification of the endemic trap-door spiders from the Balearic Islands, Western Mediterranean. J Biogeogr 44:924–936

Moritz C, Dowling TE, Brown WM (1987) Evolution of animal mitochondrial DNA: relevance for population biology and systematics. Ann Rev Ecol Syst 18:269–292

Mullis KB, Faloona FA (1987) Specific synthesis of DNA in vitro via a polymerase catalyzed chain reaction. Method Enzymol 155:335–350

Nentwig W (2012) The species referred to as *Eurypelma californicum* (Theraphosidae) in more than 100 publications is likely to be *Aphonopelma hentzi*. J Arachnol 40:128–130

Opatova V, Arnedo MA (2014a) Spiders on a hot volcanic roof: colonisation pathways and phylogeography of the Canary Islands endemic trap-door spider *Titanidiops canariensis* (Araneae, Idiopidae). PLoS One 9:e115078

Opatova V, Arnedo MA (2014b) From Gondwana to Europe: inferring the origins of Mediteranean Macrothele spiders (Araneae: Hexathelidae) and the limits of the faily Hexathelidae. Invertebr Syst 28:361–374

Opatova V, Bond JE, Arnedo MA (2013) Ancient origins of the Mediterranean trap-door spiders of the family Ctenizidae (Araneae, Mygalomorphae). Mol Phylogenet Evol 69:1135–1145

Opatova V, Bond JE, Arnedo MA (2016) Uncovering the role of the Western Mediterranean tectonics in shaping the diversity and distribution of the trap-door spider genus *Ummidia* (Araneae, Ctenizidae). J Biogeogr 43:1955–1966

Opatova V, Hamilton CH, Hedin M, Montes de Oca L, Král J, Bond JE (2019) Phylogenetic systematics and evolution of the spider infraorder Mygalomorphae using genomic scale data. Systematic Biology 69:671–707

Ortiz D, Francke OF (2016) Two DNA barcodes and morphology for multi-method species delimitation in *Bonnetina* tarantulas (Araneae: Theraphosidae). Mol Phylogenet Evol 101:176–193

Ortiz D, Francke OF (2017) Reconciling morphological an molecular systematics in tarantulas (Araneae: Theraphosidae): revision of the Mexican endemic genus *Bonnetina*. Zool J Linn Soc 180(4):819–886

Ortiz D, Francke OF, Bond JE (2018) A tangle of forms and phylogeny: extensive morphological homoplasy and molecular clock heterogeneity in *Bonnetina* and related tarantulas. Mol Phylogenet Evol 127:55–73

Parra G, Bradnam K, Ning Z, Keane T, Korf I (2009) Assessing the gene space in draft genomes. Nucleic Acids Res 37:289–297

Perafán C, Galvis W, Gutiérrez M, Pérez-Miles F (2016) *Kankuamo*, a new theraphosid genus from Colombia (Araneae, Mygalomorphae), with a new type of urticating setae and divergent male genitalia. ZooKeys 601:89–109

Pérez-Miles F (1998) Notes on the systematics of the little known theraphosid spider *Hemirrhagus cervinus*, with a description of a new type of urticating hair. J Arachnol 26:120–123

Petersen SD, Mason T, Akber S, West R, White B, Wilson P (2007) Species identification of tarantulas using exuviae for international wildlife law enforcement. Conserv Genet 8:497–502

Podsiadlowski L, Arabi J and Fahrein K (Unpublished) Mitochondrial genomes of three spider species. Genbank submission 31 Jan 2013

Pons J, Barraclough TG, Gomez-Zurita J, Cardoso A, Duran DP, Hazell S, Kamoun S, Sumlin WD, Vogler AP (2006) Sequence-based species delimitation for the DNA taxonomy of undescribed insects. Syst Biol 55:595–609

Qiu Y, Song D, Zhou K, Sun H (2005) The mitochondrial sequences of *Heptethela hangzhouensis* and *Ornithoctonus huwena* reveal unique gene arrangements and atypical tRNAs. J Mol Evol 60:57–71

Regier JC, Schultz JW (1997) Molecular phylogeny of the major arthropod Groups indicates Polyphyly of Crustaceans and a new hypothesis for the origins of Hexapods. Mol Biol Evol 14:902–913

Richards S (2015) It's more than stamp collecting: how genome sequencing can unify biological research. Trends Genet 31:411–421

Rix MG, Cooper SJB, Meusemann K, Klopfstein S, Harrison SE, Harvey MS (2017a) Post-Eocene climate change across continental Australia and the diversification of Australasian spiny trapdoor spiders (Idiopidae: Arbanitinae). Mol Phylogenet Evol 109:302–320

Rix MG, Bain K, Main BY, Raven RJ, Austin AD, Cooper SJB (2017b) Systematics of the spiny trapdoor spiders of the genus *Cataxia* (Mygalomorphae: Idiopidae) from south-western Australia: documenting a threatened fauna in a sky-island landscape. J Arachnol 45:395–423

Rix MG, Raven RJ, Harvey MS (2018a) Systematics of the giant trapdoor spiders of the genus *Gaius* Rainbow (Mygalomorphae: Idiopidae: Aganippini): documenting an iconic lineage of the Western Australian inland arid zone. J Arachnol 46:438–472

Rix MG, Raven RJ, Austin AD, Cooper SJB, Harvey MS (2018b) Systematics of the spiny trapdoor spider genus *Bungulla* (Mygalomorphae: Idiopidae): revealing a remarkable radiation of mygalomorph spiders from the Western Australian arid zone. J Arachnol 46:249–344

Robinson AE, Blagoev G, Hebert P, Adamowicz SJ (2009) Prospects for using DNA barcoding to identify spiders in species-rich genera. ZooKeys 16:27–46

Roehrdanz RL (1995) Amplification of complete insect mitochondrial genome in two easy pieces. Insect Mol Biol 4:169–172

Ruane S, Austin CC (2017) Phylogenomics using formalin-fixed and 100+ year-old intractable natural history specimens. Mol Ecol Resour 17:1003–1008

Sanger F, Coulson AR (1975) A rapid method for determining sequences in DNA by primed synthesis with DNA polymerase. J Mol Biol 94:441–448

Sanggaard KW, Bechsgaard JS, Fang X, Duan J, Dyrlund TF, Gupta V, Jiang X, Cheng L, Fan D, Feng Y, Han L, Huang Z, Wu Z, Liao L, Settepani V, Thøgersen IB, Vanthournout B, Wang T, Zhu Y, Funch P, Enghild JJ, Schauser L, Andersen SU, Villesen P, Schierup MH, Bilde T, Wang J (2014) Spider genomes provide insight into composition and evolution of venom and silk. Nat Commun 5:1–11

Satler JD, Starrett J, Hayashi CY, Hedin M (2011) Inferring species trees from gene trees in a radiation of California trapdoor spiders (Araneae, Antrodiatidae, *Aliatypus*). PLoS One 6:e25355

Satler JD, Carstens BC, Hedin M (2013) Multilocus species delimitation in a complex of morphologically conserved trapdoor spiders (Mygalomorphae, Antrodiatidae, *Aliatypus*). Syst Biol 62:805–823

Schwager EE, Sharma PP, Clarke T, Leite DJ, Wierschin T, Pechmann M, Akiyama-Oda Y, Esposito L, Bechsgaard J, Bilde T et al (2017) The house spider genome reveals an ancient whole-genome duplication during arachnid evolution. BMC Biol 15:e62

Sharma PP, Kaluziak ST, Perez-Porro AR, Gonzalez VL, Hormiga G, Wheeler WC, Giribet G (2014) Phylogenomic interrogation of Arachnida reveals systemic conflicts in phylogenetic signal. Mol Biol Evol 31:2963–2984

Shokralla S, Porter TM, Gibson JF, Dobosz R, Janzen DH, Hallwachs W, Golding GB, Hajibabaei M (2015) Massively parallel multiplex DNA sequencing for specimen identification using an Illumina MiSeq platform. Sci Rep 5:9687

Simão FA, Waterhouse RM, Ioannidis P, Kriventseva EV, Zdobnov EM (2015) BUSCO: assessing genome assembly and annotation completeness with single-copy orthologs. Bioinformatics 31:3210–3212

Simon C, Frati F, Beckenbach A, Crespi B, Liu H, Flook P (1994) Evolution, weighting and phylogenetic utility of mitochondrial gene sequences and a compilation of conserved polymerase chain reaction primers. Ann Entomol Soc Amer 87:651–701

Simon C, Buckley TR, Frati F, Stewart JB, Beckenbach AT (2006) Incorporating molecular evolution into phylogenetic analyses, and a new compilation of conserved polymerase chain reaction primers for animal mitochondrial DNA. Ann Rev Ecol Evol Syst 37:545–579

Smith SD, Bond JE (2003) An analysis of the secondary structure of the mitochondrial large subunit rRNA gene (16S) in spiders and its implications for phylogenetic reconstructions. J Arachnol 31:44–54

Smith BT, Harvey MG, Faircloth BC, Glenn TC, Brumfield RT (2014) Target capture and massively parallel sequencing of ultraconserved elements for comparative studies at shallow evolutionary time scales. Syst Biol 63:83–95

Spagna JC, Gillespie RG (2008) More data, fewer shifts: molecular insights into the evolution of the spinning apparatus in non-orb-weaving spiders. Mol Phylogenet Evol 46:347–368

Srivathsan A, Baloğlu B, Wang W, Tan WX, Bertrand D, AHQ N, EJH B, JJY K, Nagarajan N, Meier R (2018) A MinION(tm)-based pipeline for fast and cost-effective DNA barcoding. Mol Ecol Resour 18:1035–1049

St Laurent RA, Hamilton CA, Kawahara AY (2018) Museum specimens provide phylogenomic data to resolve relationships of sack-bearer moths (Lepidoptera, Mimallonoidea, Mimallonidae). Syst Entomol 43:729–761

Starrett J, Hedin M (2007) Multilocus genealogies reveal multiple cryptic species and biogeographical complexity in the California turret spider *Antrodiaetus riversi* (Mygalomorphae, Antrodiaetidae). Mol Ecol 16:583–604

Starrett J, Derkarabetian S, Hedin M, Bryson RW, McCormack JE, Faircloth BC (2017) High phylogenetic utility of an ultraconserved element probe set designed for Arachnida. Mol Ecol Resour 17:812–823

Starrett J, Hayashi CY, Derkarabetian S, Hedin M (2018) Cryptic elevational zonation in trapdoor spiders (Araneae, Antrodiaetidae, *Aliatypus janus* complex) from the California southern Sierra Nevada. Mol Phylogenet Evol 118:403–413

Stockman AK, Bond JE (2007) Delimiting cohesion species: extreme population structuring and the role of ecological interchangeability. Mol Ecol 16:3374–3392

Straub SCK, Parks M, Writemier K, Fishbein M, Cronn RC, Liston A (2012) Navigating the tip of the genoic iceberg: next-genetation sequencing for plant systematics. Am J Botany 99:349–364

Tang M, Tan M, Meng G, Yang S, Liu S, Song W, Li Y, Wu Q, Zhang A, Zhou X (2014) Multiplex sequencing of pooled mitochondrial genomes-a crucial step toward biodiversity analysis using mito-metagenomics. Nucleic Acids Res 42:e166

Tautz D, Arctander P, Minelli A, Thomas RH, Voger AP (2002) DNA points the way ahead in taxonomy. Nature 418:479

Tautz D, Arctander P, Minelli A, Thomas RH, Voger AP (2003) A plea for DNA taxonomy. Trends Ecol Evol 18:70–74

Tian X-X, Pan W-J, Chen L-L, Xu Y-Y, Pan H-C (2016) The complete mitochondrial genome of stretch spider *Tetragnatha maxillosa* (Araneae: Tetragnathidae). Mitochondrial DNA A DNA Mapp Seq Anal 27:3469–3470

Tilak M-K, Justy F, Debiais-Thibaud M, Botero-Castro F, Delsuc F, Douzery EJP (2015) A cost-effective straightforward protocol for shotgun Illumina libraries designed to assemble complete mitogenomes from non-model species. Conserv Genet Resour 7:37–40

Timmermans MJTN, Dodsworth S, Culverwell CL, Bocak L, Ahrens D, Littlewood DT, Pons J, Vogler AP (2010) Why barcode? High-throughput multiplex sequencing of mitochondrial genomes for molecular systematics. Nucleic Acids Res 38:e197

Timmermans MJTN, Viberg C, Martin G, Hopkins K, Vogler AP (2016) Rapid assembly of taxonomically validated mitochondrial genomes from historical insects collections. Biol J Linn Soc 117:83–95

Turbeville JM, Pfeifer DM, Field KG, Raff RA (1991) The phylogenetic status of arthropods as inferred from 18S rRNA sequences. Mol Biol Evol 8:669–686

Turner SP, Longhorn SJ, Hamilton CA, Gabriel R, Pérez-Miles F, Vogler AP (2018) Re-evaluating conservation priorities of New World tarantulas (Araneae: Theraphosidae) in a molecular framework indicates non-monophyly of the genera, *Aphonopelma* and *Brachypelma*. Syst Biodiv 16:89–107

Undheim EAB, Sunagar K, Herzig V, Kely L, Low DHW, Jackson TNW, Jones A, Kurniawan N, King GF, Ali SA, Antunes A, Ruder T, Fry BG (2013) A proteomics and transcriptomics investigation of the venom from the barychelid spider *Trittame loki* (Brush-foot trapdoor). Toxins 5:2488–2503

Van Dam MH, Lam AW, Sagata K, Gewa B, Laufa R, Balke M, Faircloth BC, Riedel A (2017) Ultraconserved elements (UCEs) resolve the phylogeny of Australasian smurf-weevils. PLoS One 12:e0188044
Wang Z-L, Li C, Fang W-Y, Yu X-P (2016a) The complete mitochondrial genome of two *Tetragnatha* spiders (Araneae: Tetragnathidae): severe truncation of tRNAs and novel gene rearrangements in Araneae. Int J Biol Sci 12:109–119
Wang Z-L, Li C, Fang WY, Yu XP (2016b) Characterization of the complete mitogenomes of two *Neoscona* spiders (Araneae: Araneidae) and its phylogenetic implications. Gene 590:298–306
Waterhouse RM, Seppey M, Simão FA, Manni M, Ioannidis P, Klioutchnikov G, Kriventseva EV, Zdobnov EM (2018) BUSCO applications from quality assessments to gene prediction and phylogenomics. Mol Biol Evol 35:543–548
West RC, Marshall SD, Fukushima CS, Bertani R (2008) Review and cladistic analysis of the Neotropical tarantula genus *Ephebopus* Simon 1892 (Araneae: Theraphosidae) with notes on the Aviculariinae. Zootaxa 1849:35–58
Wheeler WC, Hayashi CY (1998) The phylogeny of extant Chelicerate orders. Cladistics 14:173–192
Wheeler WC, Cartwright P, Hayashi CY (1993) Arthropod phylogeny: a combined approach. Cladistics 9:1–39
Wheeler WC, Coddington JA, Crowley LM, Dimitrov D, Goloboff PA, Griswold CE, Hormiga G, Prendini L, Ramírez MJ, Sierwald P, Almeida-Silva L, Alvarez-Padilla F, Arnedo MA, Benavides Silva LR, Benjamin SP, Bond JE, Grismado CJ, Hasan E, Hedin M, Izquierdo MA, Labarque FM, Ledford J, Lopardo L, Maddison WP, Miller JA, Piacentini LN, Platnick NI, Polotow D, Silva-Dávila D, Scharff N, Szűts T, Ubick D, Vink CJ, Wood HM, Zhang J (2017) The spider tree of life: phylogeny of Araneae based on target-gene analyses from an extensive taxon sampling. Cladistics 33:574–616
Wilson JS, Gunnell CF, Wahl DB, Pitts JP (2012) Testing the species limits of the tarantulas (Araneae: Theraphosidae) endemic to California's southern coast ranges, USA. Insect Conserv Div 6:365–371
Wood HM, González VL, Lloyd M, Coddington J, Scharff N (2018) Next-Generation museum genomics: phylogenetic relationships among palpimanoid spiders using sequence capture techniques (Araneae: Palpimanoidea). Mol Phylogenet Evol 127:907–918
World Spider Catalog (2019) World Spider Catalog. Version 20.0. Natural History Museum, Bern. http://wsc.nmbe.ch. Accessed Jun 2019
Yu DW, Ji Y, Emerson BC, Wang X, Ye C, Yang C, Ding Z (2012) Biodiversity soup: metabarcoding of arthropods for rapid biodiversity assessment and biomonitoring. Methods Ecol Evol 3:613–623
Zhu J, Sun Y, Zhao F-Q, Yu J, Craig R, Hu S (2009) Analysis of tarantula skeletal muscle protein sequences and identification of transcriptional isoforms. BMC Genomics 10:e117
Zhu H-F, Wang Z-Y, Wang Z-L, Yu X-P (2019) Complete mitochondrial genome of the crab spider *Ebrechtella tricuspidata* (Araneae: Thomisidae): a novel tRNA rearrangement and phylogenetic implications for Araneae. Genomics 111:1266–1273
Zuckerkandl E, Pauling L (1965) Molecules as documents of evolutionary history. J Theor Biol 8:357–366

Chapter 3
Ischnocolinae and Schismatothelinae

José Paulo Leite Guadanucci

Abstract Theraphosidae systematics has undergone some changes in the last 20 years, with many phylogenetic hypotheses supporting unforeseen relationships and several new genera described. Amongst all 11 subfamilies currently included in Theraphosidae, Ischnocolinae is the most problematic, as it was originally established based on a plesiomorphic feature, the divided tarsal scopula. A recent phylogenetic hypothesis based only on morphology showed that most of Ischnocolinae representatives do not form monophyletic units, and instead, are scattered across the family tree, except for those comprising Schismatothelinae: *Sickius*, *Neoholothele*, *Guyruita*, *Euthycaelus*, and *Schismatothele*. Schismatothelinae is endemic to the Neotropics and its diversity is still underestimated, as many new species are described every year. Although well sampled in phylogeny mentioned above, the status of the remaining Neotropical Ischnocolinae genera (*Catumiri*, *Dolichothele*, *Acanthopelma*, *Holothele*) are far from being resolved, as morphological characters show high degrees of homoplasy. Not until a comprehensive sampling of Ischnocolinae is achieved with molecular data, new nomenclatural acts should be further proposed. The Neotropical spiders included in Ischnocolinae and Schismatothelinae occur in all different environments across Central and South Americas. The full diversity of Neotropical Ischnocolinae, and Schismatothelinae will be considered in this chapter, with diagnoses for genera, identification keys, and pictures.

3.1 Introduction

The subfamily Ischnocolinae has been very taxonomically problematic since its original establishment as a tribe, by Simon (1892). Simon regarded the divided tarsal scopula as the main diagnostic feature for the ischnocolines. However, as later pointed by Pocock (1897), and then later by Pérez-Miles (1994), the tarsal scopula undergoes an ontogenetic change in most of the theraphosids, as it remains divided in juveniles stages and becoming entire as the spider grows adult. Although promptly

J. P. L. Guadanucci (✉)
Departamento de Biodiversidade, Instituto de Biociências, Universidade Estadual Paulista, Rio Claro, SP, Brazil

F. Pérez-Miles (ed.), *New World Tarantulas*, Zoological Monographs 6,
https://doi.org/10.1007/978-3-030-48644-0_3

pointed by Pocock (1897), the divided tarsal scopula remained as the key character for Ischnocolinae for more than a century. Raven (1985) drew attention to the fact that only a phylogenetic hypothesis would show monophyletic units and support a new classification. The New World Ischnocolinae could only be recognized by exclusion, as they lack the synapomorphic characters of Theraphosinae and Aviculariinae, and it comprised a non-monophyletic group. In 2014, Guadanucci included representatives of all Theraphosidae subfamilies in a phylogenetic analysis based on morphological characters and proposed the following changes: (1) Ischnocolinae as defined after Raven (1985) is paraphyletic and should be broken into less inclusive groups; the name Ischnocolinae should be restricted to a monophyletic group the includes the New World genera *Holothele*, *Acanthopelma*, *Reichlingia*, *Trichopelma*, and the African and Middle East genus *Ischnocolus*, and it must be denoted as Ischnocolinae sensu stricto; (2) part of the former Ischnocolinae comprise a monophyletic subfamily of exclusive New World representatives named Schismatothelinae (*Euthycaelus*, *Schismatothele*, *Guyruita*, *Sickius*, and *Neoholothele*); (3) other New World genera (*Catumiri* and *Dolichothele*) without clearly defined relationships, as they do not fit into any of the two other groups. More recently, Lüddecke et al. (2018), using six molecular markers (among nuclear and mitochondrial) did not find support for monophyly of Schismatothelinae. However, sampling across Theraphosidae and Schismatothelinae was very limited and the question regarding subfamilial monophyly and their classification is still doubtful. Further, genera included in Ischnocolinae *sensu stricto*, and other with unresolved affinities (*e.g.*, *Dolichothele*, *Catumiri*, *Trichopelma*, *Holothele*, and also the non-Neotropical *Heterothele*, *Chaetopelma*, *Ischnocolus*) are completely lacking in that analysis, and are in urgent need of a molecular approach. Despite the disagreement between morphological and molecular phylogenies (which is expected considering the high level of homoplasy of morphological features), I present below the main New World Theraphosidae genera currently included in Schismatothelinae and "Ischnocolinae," that includes Ischnocolinae *sensu stricto*, and the two remaining *Catumiri* and *Dolichothele*. All genera included here fit into the following description: small- or medium-sized spider, never reaching the size of some Theraphosinae or Aviculariinae spiders, without urticating or stridulating setae, with male palpal bulb presenting a simple aspect, without keels or great modifications. They are found underneath rock, fallen trunks, where they build their retreat with a lot of silk, with the aspect of sheet web. Some species belonging to either group also make use of natural crevices on rocks or manmade rocky walls, also protected with large amount of silk.

3.2 Subfamily Schismatothelinae

3.2.1 Identification Key (All steps in the key require examination of sexual features, such as male palpal bulb and female spermathecae. Therefore, adult specimens are required for proper generic recognition.)

1. A. Males: palpal tibiae strongly swollen with several short and thick spines on prolateral face; palpal bulb with short embolus and small keels, subtegulum large (Figs. 3.1c, d and 3.2c, d). Females: spermathecae strongly sclerotized, receptacles are stiff (Figs. 3.1e and 3.2e). Spiders from Northern South America (Colombia, Venezuela and Northern Brazil)..2
   B. Males: palpal tibia of similar width of other articles, if spines are present, they are long and thin, similar to those on legs; palpal bulb with long embolus, that can be straight, slightly curved, or sinuous, with thin apex (Figs. 3.1h, m, 3.2h, 3.3c, h and 3.4c, h). Females: spermathecae with flexible receptacles, usually slender (longer then wide) (Note: some species of *Guyruita* have spermathecal receptacles that are wider than long, but with only apical lobes strongly sclerotized; spermatheca is absent in *Sickius*)...3
2. A. Males: palpal tibia with two oblique rows of thick spines (Fig. 3.1c); palpal bulb elongated, with subtegulum large and separated from tegulum (Fig. 3.1c); retrolateral branch of tibial spur very long (Fig. 3.1d). Females: spermathecae composed of a pair of single receptacles heavily sclerotized (Fig. 3.1e)..*Euthycaelus*
   B. Males: palpal tibia with on row, or a small group, of spines with sharp apex (Fig. 3.2c); palpal bulb globose, with subtegulum large and not separated from tegulum (Fig. 3.2c). Females: spermathecae composed of a central bulky receptacle (ventral) strongly sclerotized and two side receptacle that are less sclerotized (dorsal) (Fig. 3.2e)...*Schismatothele*
3. A. Males: palpal tibia slightly wider at proximal half than apical (Fig. 3.2h); palpal bulb with globose tegulum and long, curved and cilindrical embolus (Fig. 3.2h); retrolateral branch of tibial spur with sharp apex, metatarsus I curved and with ventral short spur on apical half (Fig. 3.2i). Females: without spermathecae. Spiders with brown body coloration, with golden setae on dorsal abdomen (Fig. 3.5g)...*Sickius*
   B. Males: palpal tibia slender (Fig. 3.1h, m); palpal bulb with piriform tegulum, embolus long, slightly curved and tapering, with thin apex (Figs. 3.1h, m, 3.3c, h and 3.4c, h). Females: with spermathecae. Spiders with coloration different from described above..4
4. A. Males: retrolateral branch of tibial spur with enlarged apex, with two apical spines inserted on opposite sides (Fig. 3.1n). Females: with long and slender spermathecal receptacles (Fig. 3.1o). Spiders with black carapace covered by bright golden setae on margin, cephalic region, thoracic furrow

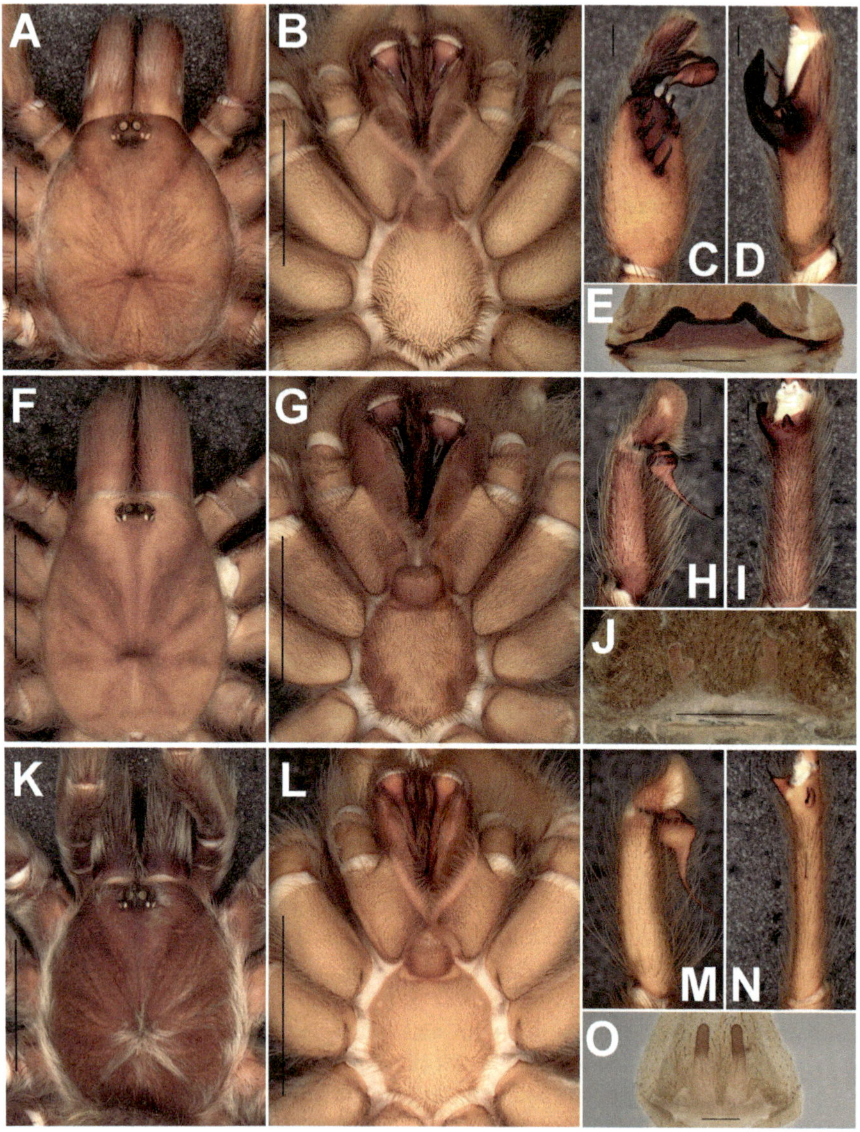

Fig. 3.1 (**a**)–(**d**) *Euthycaelus colonicus*, male. (**a**) Prosoma, dorsal view (scale = 5 mm). (**b**) Prosoma, ventral view (scale = 5 mm). (**c**) Palpal tibia, cymbium and bulb, retrolateral view (scale = 1 mm). (**d**) Tibial apophysis, retrolateral view (scale = 1 mm). (**e**) *Euthycaelus colonicus*, female, spermatheca, dorsal view (scale = 1 mm). (**f**)–(**i**) *Guyruita cerrado*, male. (**f**) Prosoma, dorsal view (scale = 5 mm). (**g**) Prosoma, ventral view (scale = 5 mm). (**h**) Palpal tibia, cymbium and bulb, retrolateral view (scale = 1 mm). (**i**) Tibial apophysis, retrolateral view (scale = 1 mm). (**j**) *Guyruita* sp., female, spermatheca, dorsal view. (**k**)–(**n**) *Neoholothele fasciaaurinigra*, male. (**k**) Prosoma, dorsal view (scale = 5 mm). (**l**) Prosoma, ventral view (scale = 5 mm). (**m**) Palpal tibia, cymbium and bulb, retrolateral view (scale = 1 mm). (**n**) Tibial apophysis, retrolateral view (scale = 1 mm). (**o**) *Neoholothele fasciaaurinigra*, female, spermatheca, dorsal view (scale = 1 mm)

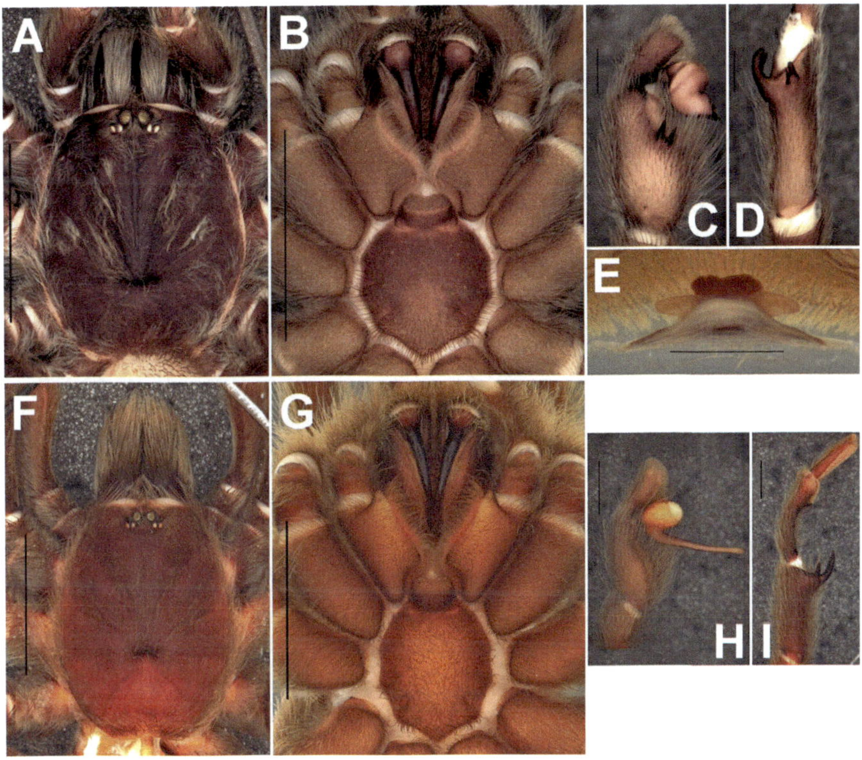

Fig. 3.2 (**a**)–(**d**) *Schismatothele* sp., male. (**a**) Prosoma, dorsal view (scale = 5 mm). (**b**) Prosoma, ventral view (scale = 5 mm). (**c**) Palpal tibia, cymbium and bulb, retrolateral view (scale = 1 mm). (**d**) Tibial apophysis, retrolateral view (scale = 1 mm). (**e**) *Schismatothele* sp., female, spermatheca, dorsal view (scale = 1 mm). (**f**)–(**i**) *Sickius longibulbi*, male. (**f**) Prosoma, dorsal view (scale = 5 mm). (**g**) Prosoma, ventral view (scale = 5 mm). (**h**) Palpal tibia, cymbium, and bulb, retrolateral view (scale = 1 mm). (**i**) Tibial apophysis, retrolateral view (scale = 1 mm)

and striae, and striped abdomen (not very evident in males) (Fig. 3.5d, e)..*Neoholothele*

B. Males: retrolateral branch of tibial spur of constant width, with two short apical spines inserted on opposite sides (Fig. 3.1i). Females: with spermathecal receptacles with lobes, usually more sclerotized than the rest of the receptacle (Fig. 3.1j)..*Guyruita*

Genus *Euthycaelus* Simon, 1889 (Figs. 3.1a–e and 3.5a, b)

Composition. *Euthycaelus amandae* Guadanucci & Weinmann, 2014; *E. colonica* Simon, 1889; *E. guane* Valencia-Cuellar, Perafán & Guadanucci, 2019 and *E. norae* Guadanucci & Weinmann, 2014.

Geographical distribution. From Colombia, the records are from the Eastern Mountain Range (Cordillera Oriental) and also extending into Western and Northern regions of Venezuela (Guadanucci and Weinmann 2014).

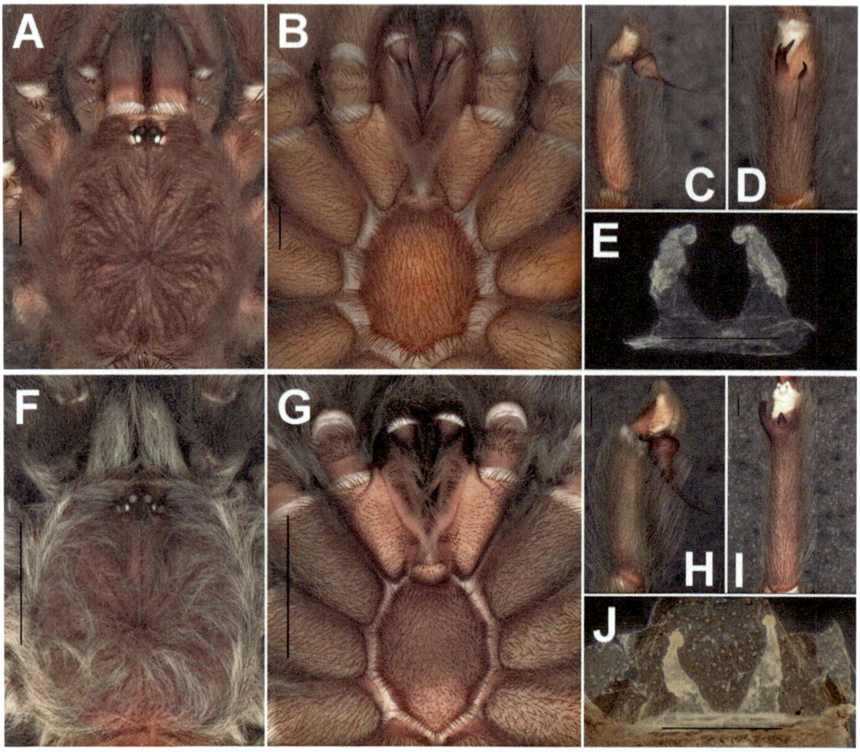

Fig. 3.3 (**a**)–(**d**) *Catumiri petropolium*, male. (**a**) Prosoma, dorsal view (scale = 1 mm). (**b**) Prosoma, ventral view (scale = 1 mm). (**c**) Palpal tibia, cymbium, and bulb, retrolateral view (scale = 1 mm). (**d**) Tibial apophysis, retrolateral view (scale = 1 mm). (**e**) *Catumiri parvum*, female, spermatheca, dorsal view (scale = 1 mm). (**f**)–(**i**) *Dolichothele diamantinensis*, male. (**f**) Prosoma, dorsal view (scale = 5 mm). (**g**) Prosoma, ventral view (scale = 5 mm). (**h**) Palpal tibia, cymbium, and bulb, retrolateral view (scale = 1 mm). (**i**) Tibial apophysis, retrolateral view (scale = 1 mm). (**j**) *Dolichothele exilis*, female, spermatheca, dorsal view (scale = 1 mm)

Remarks. The genus comprises medium-to-small size tarantulas (15–35 mm of body length), varying from pale grey to brown. Males are easily recognized by the swollen palpal tibia bearing thick spines on retrolateral face (Fig. 3.1c); well-developed tibial apophysis with long and thick retrolateral branch (Fig. 3.1d); palpal bulb elongate with separated subtegulum (Fig. 3.1c). Females are distinguished by the heavy sclerotized spermathecae composed of two small receptacula (Fig. 3.1e). The genus was recently removed from the synonymy with *Holothele* by Guadanucci and Weinmann (2014). *Euthycaelus* share with *Schismatothele* many morphological characters, mainly related to male and female dimorphic features (multilobed and strongly sclerotized spermathecae, extended subtegulum, palpal bulb with keels and short embolus, swollen palpal tibia with prolateral megaspines) that support their close evolutionary relationship.

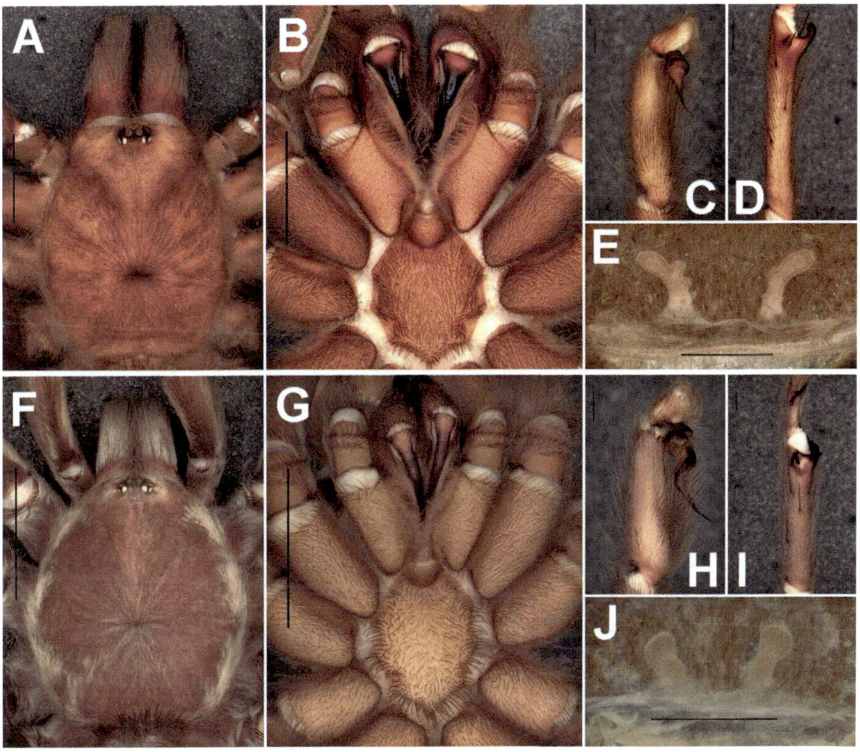

Fig. 3.4 (**a**)–(**d**) *Holothele longipes*, male. (**a**) Prosoma, dorsal view (scale = 1 mm). (**b**) Prosoma, ventral view (scale = 1 mm). (**c**) Palpal tibia, cymbium, and bulb, retrolateral view (scale = 1 mm). (**d**) Tibial apophysis, retrolateral view (scale = 1 mm). (**e**) *Holothele longipes*, female, spermatheca, dorsal view (scale = 1 mm). (**f**)–(**i**) *Holothele* aff. *culebrae*, male. (**f**) Prosoma, dorsal view (scale = 5 mm). (**g**) Prosoma, ventral view (scale = 5 mm). (**h**) Palpal tibia, cymbium, and bulb, retrolateral view (scale = 1 mm). (**i**) Tibial apophysis, retrolateral view (scale = 1 mm). (**j**) *Holothele* aff. *culebrae*, female, spermatheca, dorsal view (scale = 1 mm)

Genus *Guyruita* Guadanucci, Lucas, Indicatti, and Yamamoto, 2007 (Figs. 3.1f–j and 3.5c)

Composition. *Guyruita atlantica* Guadanucci, Lucas Indicatti & Yamamoto, 2007; *G. cerrado* Guadanucci, Lucas Indicatti & Yamamoto, 2007; *G. giupponii* Fukushima and Bertani, 2018; *G. isae* Fukushima and Bertani, 2018; *G. metallophila* Fonseca-Ferreira, Zampaulo & Guadanucci, 2017; *G. waikoshiemi* (Bertani & Araújo, 2005).

Geographical distribution. In Brazil, species can be found along the Atlantic Forest (extending from Southeast into Northeastern region), central Cerrado (states of Distrito Federal, Goiás, Maranhão, Piauí, and Tocantins) and Amazon Forest (Pará, in Brazil, and Southern Venezuela).

Remarks. Small-to-medium size tarantulas (10–30 mm of body length), with large central dark spot on dorsal abdomen (with lateral stripes) or uniform coloration. Males are distinguished by the aspect of male palpal bulb with long and thin embolus (Fig. 3.1h), tibial apophysis with short branches (retrolateral branch

Fig. 3.5 Live specimens, dorsal habitus. (**a**) *Euthycaelus colonicus*, female, from San Luis, Falcón, Venezuela (photo: JPL Guadanucci). (**b**) *Euthycaelus colonicus*, male, from San Luis, Falcón, Venezuela (photo: JPL Guadanucci). (**c**) *Guyruita cerrado*, female, from Miracema do Tocantins, Tocantins, Brazil (photo: RP Indicatti). (**d**) *Neoholothele fasciaaurinigra*, female, from Villavicencio, Meta, Colombia (photo: JPL Guadanucci). (**e**) *Neoholothele fasciaaurinigra*, male, from Villavicencio, Meta, Colombia (photo: JPL Guadanucci). (**f**) *Schismatothele* sp., female, from Santa María, Boyacá, Colombia (photo: JPL Guadanucci). (**g**) *Sickius longibulbi*, female, from Jales, São Paulo, Brazil (photo: VM Ghiroto). (**h**) *Catumiri petropolium*, male, from Petrópolis, Rio de Janeiro, Brazil (photo: RP Indicatti). (**i**) *Dolichothele exilis*, female, from Central, Bahia, Brazil (photo: R Pinto-da-Rocha). (**j**) *Dolichothele mottai*, female, from Catalão, Goiás, Brazil (photo: JPL Guadanucci). (**k**) *Dolichothele bolivianum*, female, from Samaipata, Santa Cruz, Bolivia (photo: JPL Guadanucci). (**l**) *Dolichothele rufoniger*, female, from Central,

slightly longer) (Fig. 3.1i), metatarasus that touches the apophysis when bent. Females with spermathecae with wide base and bearing well sclerotized lobes (Fig. 3.1j). Although recently established (Guadanucci et al. 2007) and three extra species also recently described (Fonseca-Ferreira et al. 2017; Fukushima and Bertani 2018), the examination of collection preserved specimens and other spiders known only from pictures shared among hobbyists, the diversity of the genus should be increased to more than 15 species. These records include Amazonian species from Brazil (states of Pará, Mato Grosso, Amapá, Amazonas) and Guyana. After the formal description of these species, the genus should undergo a revision in order to evaluate its monophyly. A special remark to *G. metallophila* which is found exclusively in iron caves of Carajás region in Pará, Brazil, which was considered to be a troglophile species (Fonseca-Ferreira et al. 2017).

***Neoholothele* Guadanucci and Weinmann, 2015 (Figs. 3.1k–o and 3.5d, e)**
Composition. *Neoholothele fasciaaurinigra* Guadanucci and Weinmann, 2015 and *N. incei* (F. O. Pickard-Cambridge, 1899).

Geographical distribution. Colombia, with a single record from central Eastern Mountain (Cordillera Oriental), and Northern Venezuela, including Isla Margarita, and Trinidad and Tobago.

Remarks. Medium-sized spider, with males reaching up to 30 mm of body length. Species of *Neoholothele* are the largest individuals among Ischnocolinae and Schismatothelinae. They have a dark carapace with golden setae around eye tubercle, thoracic striae, and lateral carapace; legs uniformly beige, and abdomen with clear dark and golden stripes on females, faded striped on males (Fig. 3.5d, e). The genus was established (Guadanucci and Weinmann 2015) to house *N. incei*, removed from *Holothele*, and an extra species with a very particular color pattern (*N. fasciaaurinigra*). Males are easily distinguished from other genera by the long and thin embolus of palpal bulb (Fig. 3.1m), tibial apophysis with two branches, the retrolateral branch with wide apical portion bearing two spines on the sides (Fig. 3.1n). Females have long and slender spermathecal receptacles with longitudinal striae (Fig. 3.1o).

Genus *Schismatothele* Karsch, 1879 (Figs. 3.2a–e and 3.5f)
Composition. *Schismatothele benedettii* Panzera, Perdomo & Pérez-Miles, 2011; *S. hacaritama* Perafán, Valencia-Cuéllar & Guadanucci, 2019; *S. inflata* (Simon, 1889); *S. kastoni* (Caporiacco, 1955); *S. lineata* Karsch, 1879; *S. modesta* (Simon, 1889); *S. olsoni* Guadanucci, Perafán & Valencia-Cuéllar, 2019 and *S. weinmanni* Guadanucci, Perafán & Valencia-Cuéllar, 2019.

Geographical distribution. Few records on Northern Venezuela and a single record in Brazil (Central North Amazonia). Museum specimens show a greater diversity in Colombia, with scattered records in Eastern Mountain and other amazonian localities in Brazil.

Fig. 3.5 (Continue) Bahia, Brazil (photo: R Pinto-da-Rocha). (**m**) *Holothele longipes*, female, from Pamplonita, Norte de Santander, Colombia (photo: JPL Guadanucci). (**n**) *Holothele longipes*, male, from Pamplonita, Norte de Santander, Colombia (photo: JPL Guadanucci). (**o**) *Holothele* aff. *culebrae*, male, Anguilla, Lesser Antiles (photo: K. Questel)

Remarks. Since its original description, by Karsch (1879), the genus remained as monotypic, known only by the female holotype, from Caracas, Venezuela. At the time, it was characterized by the uncommon spermathecal shape, with strongly sclerotized double receptacles (Fig. 3.2e). After few taxonomic changes (synonymized with *Holothele*, by Raven 1985; considered *insertae sedis*, by Pérez-Miles et al. 1996; transferred to Theraphosinae, by Rudloff 1997), the male was described by the first time by Panzera et al. (2011), revealing distinct male characters, such as swollen palpal tibiae with megaspines, palpal bulb with keels, short embolus, and extended subtegulum (Fig. 3.2c), shared with spiders of the genus *Euthycaelus*. Panzera et al. (2011) discussed the inclusion of *Schismatothele*, and consequently *Euthycaelus*, in Theraphosinae, based the presence of keels on bulb and extended subtegulum, and despite the lack or urticating setae. Guadanucci (2014) established the subfamily Schismatothelinae to include, along with *Schismatothele* and *Euthycaelus*, *Guyruita, Neoholothele*, and *Sickius*, which was later questioned by Lüddecke et al. (2018), who found close relation among *Euthycaelus* and *Neoholothele* with the Psalmopoeinae and Aviculariinae.

Genus *Sickius* Soares and Camargo, 1948 (Figs. 3.2f–i and 3.5g)
Composition. *Sickius longibulbi* Soares and Camargo, 1948, monotypic.

Geographical distribution. Records from Central, Western, and Southeastern Brazil, in the states of Distrito Federal, Mato Grosso do Sul, São Paulo and Paraná.

Remarks. Medium-sized spiders with uniform brown coloration with golden setae on all body parts (Fig. 3.5g). They build their silken retreat on burrows dug under rocks or trunks. Males are distinguished by their typical palpal bulb aspect, with long, slender, and cylindrical embolus (Fig. 3.2h), well-developed tibial apophysis, with long retrolateral branch with a sharp end, curved metatarsi I with a short ventral tubercle (Fig. 3.2i). The female lacks spermathecae, the sperm seems to be stored in the oviducts (Bertani and Silva-Junior 2002). They have a very unusual copulatory behavior where the male bites the females before introducing its palpal bulb into the female genital opening.

3.3 Subfamily Ischnocolinae

3.3.1 Identification Key

1. A. Spiders with labium wider than long, with few cuspules (up to 20) (Fig. 3.3b, g)..2
   B. Spiders with labium as wide as long, or trapezoidal, with many cuspules (more than 30) (Fig. 3.4b, g)..3
2. A. Clavate trichobothria on tarsi forming a central and short row on apical portion; anterior tarsal scopula (tarsi I and II) divided by a longitudinal band of thick setae. Males: tarsal claws with teeth in prolateral margin; tibial apophysis (absent in *C. petropolium*) composed of a short retrolateral branch, and very reduced prolateral branch (absent in *C. argentinensis*). Females: spermathecal receptacles with a single termini (no multi-lobes)...................................*Catumiri*

B. Clavate trichobothria on tarsi forming two parallel rows along the tarsus length; anterior tarsal scopula (tarsi I and II) not divided. Males: tarsal claws without teeth; tibial apophysis composed of two branches, retrolateral longer. Females: spermathecael receptacles with many lobes on apical and/or along the receptacles..*Dolichothele*

3. A. Labium trapezoidal (Fig. 3.4b), with many cuspules (more than 80), tarsal claws with median row of small teeth, tarsus IV with pseudosegmentation. Males: palpal bulb with long and slightly curved embolus, well-developed tibal spur, with long branches. Females: spermathecal receptacles long, with lobes on inner margin up to the midlength of each receptacle...*Holothele longipes*

B. Labium trapezoidal (Fig. 3.4g), with many cuspules, tarsal claws without teeth, tarsus IV without psedosegmentation. Males: palpal bulb with long, twisted, and sinuous embolus, with discrete keel along the embolus. Females: spermathecal receptacle single apical termini, without minor lobes along the receptacle...............*Holothele* from the Caribbean (*H. culebrae*, *H. denticulata*, *H. shoemakeri* or *H. sulfurensis*—see further details below)

Genus *Catumiri* Guadanucci, 2004 (Figs. 3.3a–e and 3.5h)
Composition. *Catumiri argentinense* (Mello-Leitão, 1941); *C. chicaoi* Guadanucci, 2004; *C. parvum* (Keyserling, 1878); *C. petropolium* Guadanucci, 2004.

Geographical distribution. Atlantic Forest species (*C. chicaoi* and *C. petropolium*), extreme Southern Brazil and Uruguay (*C. parvum*) and North and Central Argentina (*C. argentinense*).

Remarks. *Catumiri* spiders are the smallest among theraphosids, adult females not reaching 15 mm of body size. They lack urticating setae, wide labium, and few labial and maxillary cuspules (shared with *Dolichothele*) and tarsal clavate trichobothria on central short row are key characters for genus recognition. Males have been found wandering in the leaf litter, under stones and tree bark (pers. obs.).

Genus *Dolichothele* Mello-Leitão, 1923 (Figs. 3.3f–j and 3.5i–l)
Composition. *Dolichothele auratum* (Vellard, 1924); *D. bolivianum* (Vol, 2001); *D. camargorum* Revollo, da Silva & Bertani, 2017; *D. diamantinensis* (Bertani, Santos & Righi, 2009); *D. dominguense* (Guadanucci, 2007); *D. exilis* Mello-Leitão, 1923; *D. mineirum* (Guadanucci, 2011); *D. mottai* Revollo, da Silva & Bertani, 2017; *D. rufoniger* (Guadanucci, 2007); *D. tucuruiense* (Guadanucci, 2007).

Geographical distribution. *Dolichothele* is the most widespread genus among the subfamilies Schismatothelinae and Ischnocolinae, occurring in Central Bolivia and in the Brazilian biomes Cerrado, Caatinga, Atlantic forest, Amazonia.

Remarks. The genus includes mid-sized tarantulas, with setose legs, carapace, and abdomen; some species show colorful body patterns (*e.g.*, *D. diamantinensis*, *D. dominguense*, *D. rufoniger*, *D. mottai*). Many recent studies (Guadanucci 2007, 2011; Bertani et al. 2009; Lucas and Indicatti 2015; Revollo et al. 2017) have clarified many aspects of its diversity, taxonomy, morphology, and biogeography. The lack of urticating setae, wide labium, and few labial and maxillary cuspules are key characters for genus recognition. In general, the species show restricted geographical endemism; three species are sympatric in Espinhaço Mountain Range, in Central Brazilian Cerrado, (*D. mineirum*, *D. diamantinensis* and *D. rufoniger*) and two in the Caatinga (*D. exilis* and *D. rufoniger*). These four species have been studied for courtship and copulatory behavior (Guadanucci et al. 2017b): they share many courtship behavioral characters, but they perform these characters in distinct sequences and intensities, suggesting a strong reproductive isolation of sympatric species.

***Holothele* Karsch, 1879 (Figs. 3.4a–j and 3.5m–o)**
Composition. *Holothele culebrae* (Petrunkevitch, 1929); *H. denticulata* (Franganillo, 1930); *H. longipes* (L. Koch, 1875); *H. shoemakeri* (Petrunkevicth, 1926); *H. sulfurensis* Maréchal, 2005.

Geographical distribution. Caribbean (*H. culebrae*, *H. denticulata*, *H. shoemakeri*, and *H. sulfurensis*) and Northern South America (*H. longipes*).

Remarks. The genus *Holothele* has a very intricate systematic history, which is beyond this chapter to describe (see Guadanucci 2014 for phylogenetics; Guadanucci et al. 2017a for taxonomy). Two morphogical distinct groups within *Holothele* can be recognized as follows: (1) *Holothele longipes* has been recently considered senior synonym of all species in Northern South America (records in Venezuela, Colombia, Suriname, French Guiana, Bolivia, Peru, and northern Brazil). Although they show a wide geographical distribution and variation in color (dark legs with setae varying from beige to red on carapace, abdomen, and chelicerae), their genitalia (palpal bulbs, tibial apophysis, and spermathecae) are very morphologically conservative. It is very likely that a complex of species is housed under *Holothele longipes*; however, only fine morphometric and/or molecular approach would clarify this question. (2) The second group comprises the species in the Caribbean Islands (*H. culebrae*, *H. denticulata*, *H. shoemakeri*, and *H. sulfurensis*). They are smaller spiders (compared to *H. longipes*), varying from golden to dark uniform body coloration; males have a distinct palpal bulb, with twisted and sinuous embolus, with a distinct keel along the length. Although they share characters with *H. longipes* and other genera, as *Trichopelma*, *Reichlingia*, and *Acanthopelma*, comprising the Ischnocolinae *sensu stricto* according to Guadanucci (2014), their support is low and the classification of the "Ischnocolinae" is still a wide open door. It is my opinion that, along with precise methods of morphometric analysis, broad molecular sampling will shed light in the question.

Other less known genera—*Acanthopelma* F. O. Pickard-Cambridge, 1897; *Reichlingia* Rudloff, 2001 and *Trichopelma* Simon, 1888
The excess of homoplastic characters within Theraphosidae has challenged the search for a comprehensive and well-supported classification. Lüddecke et al. (2018) have given an important step toward a general classification across Theraphosidae, although it still lacks important taxa, such as the ones currently included in "Ischnocolinae." The issue with these ischnocolines is that they lack any other synapomorphic characters of the other subfamilies (Guadanucci 2014), and in

general body and sexual characters, they are not very distinct from many Neotropical Barychelidae. Theraphosidae and Barychelidae are two close related families, for which the morphological limits are doubtful in many cases that include Neotropical Sasoninae and Ischnocolinae. Neotropical Barychelidae (except the Barychelinae, that shows a distinct eye pattern) tend to show spinnerets size reduction, fewer labial and maxillary cuspules, maxillary anterior process reduced, reticulated marks, chevrons and/or rings on abdomen and legs. Regardless of the morphological system adopted to define both families (Raven 1985, 1994; Goloboff 1993; Rudloff 1997; Bond et al. 2012; Guadanucci 2014) we do not have a precise distinction (set of morphological synamorphies) to distinguish among spiders of these two families. The three following genera have been moved back and forth Theraphosidae and Barychelidae and we have not reach a consensus regarding their position.

Acanthopelma (*A. beccarii* Caporiacco, 1947, from Guyana and *A. rufescens* F. O. Pickard-Cambridge, 1897, from Guatemala) was deeply commented by Smith (1995), who drew attention to the rigid spines dividing the tarsal scopula as a generic feature, the affinities with ischnocolines, and the doubtful inclusion of *A. beccarii* in the genus. Smith (1995) also remarked that more species could be discovered in Central and South America, but until now, the genus in only know by the male holotype of the two species, and no other recent material have been assigned to *Acanthopelma*.

Trichopelma currently comprises 17 species, according to the World Spider Catalog (2018). Raven (1985) established the subfamily Trichopelmatinae (that included *Trichopelma* and *Psalistops*) in the family Barychelidae. Later, Raven (1994) suggested the transfer of Trichopelmatinae to Theraphosidae, but without further comments. Guadanucci (2014) showed the affinities of *Trichopelma* within Theraphosidae, but did not include *Psalistops* in the analysis. Without a thorough revision, it is not possible to assure the morphological limits of the genus or the diversity of species. *Trichopelma* and *Psalistops* representatives fall into the group where familial distinctiveness are uncertain, and should be dealt with in a broader approach, including more species of Theraphosidae and Barychelidae.

Reichlingia annae (Reichling, 1997), originally included in *Acanthopelma* and later removed to the newly described genus *Reichlingia* Rudloff, 2001, was also included into Trichopelmatinae, Barychelidae. Its sole species, described from Belize, is right in the core of the uncertain limits between Barychelidae and Theraphosidae.

The tropical environments of South and Central America still hold an unknown diversity of spiders belonging to the groups included in this chapter, and whose evolutionary affinities are very far from being solved. They are all small-to-medium size tarantulas, varying from 15 to 35 mm of body length, and they are found underneath rocks, fallen trunks, logs, crevices, among the leaf litter, under tree barks, in between lifted roots, or shallow dug burrows. The number of specimens in scientific collecting has grown abruptly in the last decades, as many groups have been revised and described in the last years.

Acknowledgments I would like to thank the University of São Paulo, Museum of Zoology of the University of São Paulo and the Zoological Museum, University of Copenhagen for in-kind support during the years of my grad course. I am also thankful to CNPq, CAPES, and FAPESP for funding during masters and PhD, when most of the material was gathered and examined.

References

Bertani R, Araújo Y (2005) A new species of *Holothele* Karsch, 1879 (Theraphosidae, Ischnocolinae) from Venezuela. Rev Iber Aracnol 12:13–16
Bertani R, da Silva PI Jr (2002) The first mygalomorph spider without spermathecae: *Sickius longibulbi*, with a revalidation of *Sickius* (Araneae. Theraphosidae, Ischnocolinae). J Arachnol 30:519–526
Bertani R, Santos T, Righi AF (2009) A new species of *Oligoxystre* Vellard 1924 (Araneae, Theraphosidae) from Brazil. ZooKeys 5:41–51
Bond JE, Hendrixson BE, Hamilton CA, Hedin M (2012) A reconsideration of the classification of the spider infraorder Mygalomorphae (Arachnida: Araneae) based on three nuclear genes and morphology. PLoS One 7:1–11
di Caporiacco L (1947) Diagnosi preliminari de specie nuove di aracnidi della Guiana Brittanica raccolte dai professori Beccari e Romiti. Monit Zool Ital 56:20–34
di Caporiacco L (1955) Estudios sobre los aracnidos de Venezuela. 2a parte: Araneae. Acta Biol Venezuel 1:265–448
Fonseca-Ferreira R, Zampaulo RDA, Guadanucci JPL (2017) Diversity of iron cave-dwelling mygalomorph spiders from Pará, Brazil, with description of three new species (Araneae). Trop Zool 30:178–199
Franganillo BP (1930) Arácnidos de Cuba: Mas arácnidos nuevos de la Isla de Cuba. Mem Inst Nac Invest Cient 1:47–99
Fukushima CS, Bertani R (2018) Two new species of *Guyruita* Guadanucci et al., 2007 (Araneae, Theraphosidae) from Brazil. Zootaxa 4370:395–408
Goloboff PA (1993) A reanalysis of mygalomorph spider families (Araneae). Am Mus Novit 3056:1–32
Guadanucci JPL (2004) Description of *Catumiri* n. gen. and three new species (Theraphosidae: Ischnocolinae). Zootaxa 671:1–14
Guadanucci JPL (2007) A revision of the Neotropical spider genus *Oligoxystre* Vellard 1924 (Theraphosidae, Ischnocolinae). Zootaxa 1555:1–20
Guadanucci JPL (2011) Cladistic analysis and biogeography of the genus *Oligoxystre* Vellard 1924 (Araneae: Mygalomorphae: Theraphosidae). J Arachnol 39:320–326
Guadanucci JPL (2014) Theraphosidae phylogeny: relationships of the 'Ischnocolinae'genera (Araneae, Mygalomorphae). Zool Scr 43:508–518
Guadanucci JPL, Weinmann D (2014) The spider genera *Euthycaelus* Simon and *Schismatothele* Karsch (Mygalomorphae, Theraphosidae). Zootaxa 3795:275–288
Guadanucci JPL, Weinmann D (2015) Description of *Neoholothele* gen. nov. (Araneae, Theraphosidae, Schismatothelinae). Stud Neotrop Fauna E 50:221–228
Guadanucci JPL, Lucas SM, Indicatti RP, Yamamoto FU (2007) Description of *Guyruita* gen. nov. and two new species (Ischnocolinae, Theraphosidae). Rev Bras Zool 24:991–996
Guadanucci JPL, Perafán C, Valencia-Cuéllar D (2017a) The genus *Holothele* Karsch, 1879: the identity of the type species (Mygalomorphae, Theraphosidae). Zoosystema 39:263–271
Guadanucci JPL, Edagi HH, Souza JABS, Teixeira JAA (2017b) Evolução do comportamento de corte e cópula no gênero *Dolichothele* (Theraphosidae). In: V Congresso Latinoamericano de Aracnologia, 2017, Caeté, resumos, p 116
Karsch F (1879) Arachnologische Beiträge. Z Gesammten Naturwiss 52:534–562

Keyserling E (1878) Spinnen aus Uruguay und einigen anderen Gegenden Amerikas. Verh Kaiserlich-Königlichen Zool-Bot Ges in Wien 27:571–624

Lucas SM, Indicatti RP (2015) Revalidation of *Dolichothele* Mello-Leitão and notes on *Hapalotremus* Simon (Araneae, Mygalomorphae, Theraphosidae). Stud Neotrop Fauna E 50:204–212

Lüddecke T, Krehenwinkel H, Canning G, Glaw F, Longhorn SJ, Tänzler R, Wendt I, Vences M (2018) Discovering the silk road: Nuclear and mitochondrial sequence data resolve the phylogenetic relationships among theraphosid spider subfamilies. Mol Phylogenet Evol 119:63–70

Maréchal P (2005) Description d'une nouvelle espèce d'*Holothele* (Arachnida, Araneae, Mygalomorphae, Theraphosidae) de Guadeloupe (Antilles françaises) et commentaires sur la répartition du genre. Zoosystema 27:211–218

Mello-Leitão CF (1923) Theraphosideas do Brasil. Ver Mus Paulista 13:1–438

Mello-Leitão CF (1941) Las arañas de Córdoba, La Rioja, Catamarca, Tucumán, Salta y Jujuy colectadas por los Profesores Birabén. Rev Mus La Plata (N.S., Zool.) 2:99–198

Panzera A, Perdomo C, Pérez-Miles F (2011) *Schismatothele benedettii*, a new species of tarantula from amazonic Brazil (Araneae, Theraphosidae). Bull Br Arachnol Soc 15:130–132

Pérez-Miles F (1994) Tarsal scopula division in Theraphosinae (Araneae, Theraphosidae): its systematic significance. J Arachnol 22:46–53

Pérez-Miles F, Lucas SM, da Silva PI Jr, Bertani R (1996) Systematic revision and cladistic analysis of Theraphosinae (Araneae: Theraphosidae). Mygalomorph 1:33–68

Petrunkevitch A (1926) Spiders from the Virgin Islands. T Conn Acad Arts Sci 28:21–78

Petrunkevitch A (1929) The spiders of Porto Rico. Part one. T Conn Acad Arts Sci 30:1–158

Pickard-Cambridge FO (1897) Arachnida – Araneida and Opiliones. Biologia Centrali-Americana, Zoology, London 2:1–40

Pickard-Cambridge FO (1899) On new species of spiders from Trinidad, West Indies. Proc Zool Soc Lond 66:890–900

Pocock RI (1897) On the spiders of the Suborder Mygalomorphs from the Ethiopian region contained in the collection of the British Museum. Proc Zool Soc Lond 65:724–774

Raven RJ (1985) The spider infraorder Mygalomorphae (Araneae): cladistics and systematics. Bull Am Mus Nat Hist 182:1–180

Raven RJ (1994) Mygalomorph spiders of the Barychelidae in Australia and the western Pacific. Mem Queensland Mus 35:291–706

Revollo IS, da Silva PI Jr, Bertani R (2017) Two new *Dolichothele* Mello-Leitão, 1923 species from Brazil and Bolivia (Araneae, Theraphosidae). ZooKeys 724:1–20

Rudloff JP (1997) Revision der Gattung *Holothele* Karsch, 1879 nebst Aufstellung einer neuen Gattung *Stichoplastoris* gen. nov. (Araneae, Theraphosidae) und Wiedereinsetzung einiger weiterer Gattungen der Mygalomorphae. Arachnol Mag 5:1–19

Rudloff JP (2001) Anmerkungen zur systematischen Stellung von *Acanthopelma rufescens* F.O.P.-Cambridge, 1897 und *Acanthopelma annae* Reichling, 1997 (Ischnocolinae: Theraphosidae: Mygalomorphae), sowie die Einrichtung einer neuen Gattung *Reichlingia* gen. nov. (Mygalomorphae: Barychelidae: Trichopelmatinae). Arthropoda 9:14–20

Simon E (1888) Etudes arachnologiques. 21e Mémoire. XXIX. Descriptions d'espèces et de genres nouveaux de l'Amérique centrale et des Antilles. Ann Soc Entomol France 8:203–216

Simon E (1889) Voyage de M. E. Simon au Venezuela (décembre 1887 – avril 1888). 4° memoire. Arachnides. Ann Soc Entomol France 6:169–220

Simon E (1892) Histoire naturelle des araignées (1). Roret, Paris

Smith AM (1995) Tarantula spiders: tarantulas of the U.S.A. and Mexico. Fitzgerald Publishing, London

Soares BAM, Camargo HFA (1948) Aranhas coligidas pela Fundação Brasil-Central (Arachnida-Araneae). Bol Mus Para Emilio Goeldi 10:355–409

Vellard J (1924) Etudes de zoologie. Arch Inst Vital Brazil 2(1–32):121–170

Vol F (2001) Description de *Pseudoligoxystre bolivianus* sp. et gen. n. (Araneae: Theraphosidae: Ischnocolinae), de Bolivie. Arachnides 50:3–10

World Spider Catalog (2018) World Spider Catalog, version 19.5. Natural History Museum, Bern. http://wsc.nmbe.ch. Accessed Nov 2018

Chapter 4
Arboreal Tarantulas and Their Allies: Aviculariinae and Psalmopoeinae

Yeimy Cifuentes and Carlos Perafán

Abstract The Aviculariinae spiders *sensu lato* are known as the American arboreal tarantulas. They are characterized mainly by having legs with few or no spines, laterally extended tarsal and metatarsal scopulae, resulting in a spatulate appearance of the appendices, absence of spiniform setae on the prolateral maxillae, females with completely separated spermathecae, males with palpal bulb with subtegulum not extended, and long and thin embolus without keels (except *Antillena*). Some Aviculariinae, together with all Theraphosinae, are the only spiders that evolutionarily acquired urticating setae as a defense mechanism. The primary mechanism for releasing the urticating setae in Theraphosinae is by the friction of the legs with the abdomen, which throws the urticating setae into the air, in contrast, in most Aviculariinae the releasing mechanism occurs by direct contact. The Aviculariinae tarantulas have received considerable taxonomic and biological attention and the validity as a monophyletic group has been discussed extensively. Some phylogenetic studies suggest at least two subfamilies for the American arboreal tarantulas and their kin: Aviculariinae and Psalmopoeinae. Likewise, the phylogenetic relationships of these groups have been questioned, linking these tarantulas more closely with African or American taxa. Taxonomy, systematics and some aspects of its natural history, behavioral and distribution are addressed in this chapter.

Y. Cifuentes (✉)
Instituto de Biociências, Letras e Ciências Exatas, Universidade Estadual Paulista "Júlio de Mesquita Filho" (IBILCE-UNESP), São José do Rio Preto, São Paulo, Brazil

Laboratório de Ecologia e Evolução (LEEV), Instituto Butantan, São Paulo, Brazil

C. Perafán
Departamento de Ecología y Biología Evolutiva, Instituto de Investigaciones Biológicas Clemente Estable (IIBCE), Montevideo, Uruguay

Entomología, Facultad de Ciencias, UDELAR, Montevideo, Uruguay

F. Pérez-Miles (ed.), *New World Tarantulas*, Zoological Monographs 6,
https://doi.org/10.1007/978-3-030-48644-0_4

4.1 Introduction

In a broad sense, the Aviculariinae spiders are considered as the group of arboreal theraphosids of the Neotropics (Fig. 4.1), represented by species of a variety of sizes and colors; some even have metallic-color patterns. The group is traditionally known as the bird-eating spiders. These curious habits have been documented for centuries. The first documented observation of an aviculariine eating a bird is attributed to the naturalist and scientific illustrator Maria Sibylla Merian (1705), who published in

Fig. 4.1 Aviculariinae arboreal spider *habitus*. (**a**, **b**) *Antillena rickwesti*, (**a**) male, (**b**) female. (**c**, **d**) *Caribena versicolor*, (**c**) juvenile, (**d**) female. (**e**) *Avicularia avicularia*, female. (**f**) *Avicularia minatrix,* female. (Photos: courtesy of Rick West)

Metamorphosis Insectorium Surinamensis the representation of an arboreal tarantula feeding on a hummingbird (Fukushima and Bertani 2017). According to Madame Merian, these spiders surprise small birds that are in their nests, and suck their blood with avidity. The illustration and description initially was controversial and considered as a senseless idea, but it was probably the story that Lamarck (1818) took into account for naming these spiders as *Avicularia*, referring to the fact that these tarantulas prey on birds ("*avis*" in Latin) (Fukushima and Bertani 2017).

Aviculariinae *sensu lato* can be distinguished morphologically from the other theraphosids primarily by having legs with few or no spines, and laterally extended tarsal and metatarsal scopulae, resulting in a spatulate appearance of the appendices (Fig. 4.2). Other additional features presented are: the absence of spiniform setae on the prolateral maxillae, females with completely separated spermathecae, males with palpal bulb with subtegulum not extended, and long and thin embolus without keels (except *Antillena* Bertani, Huff and Fukushima 2017) (Raven 1985; Fukushima and Bertani 2017).

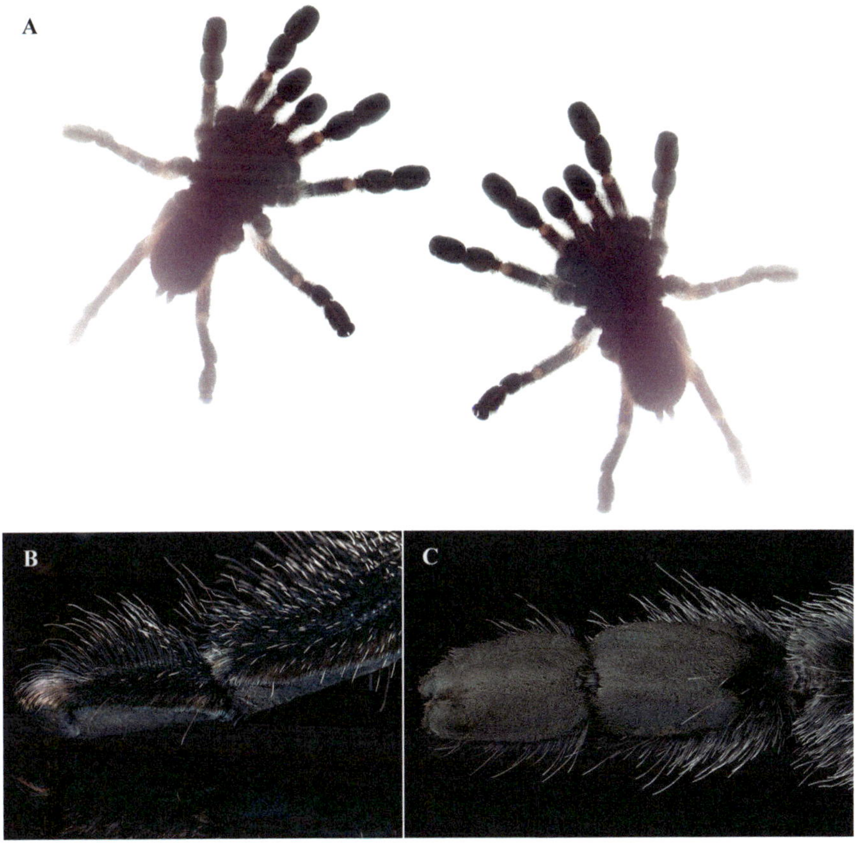

Fig. 4.2 Tarsal and metatarsal scopulae, and claw tufts in Neotropical arboreal tarantulas. (**a**) *Caribena versicolor*, ventral view. (**b**, **c**) Detail of the adhesive pads in *Avicularia* sp., (**b**) Lateral view, (**c**) Ventral view. (Photos: courtesy of Julian Kamzol)

4.1.1 Aviculariinae

The taxonomic history of the aviculariines goes back to the first arachnological studies, beginning with the first system of spider classification proposed by the Baron Walckenaer (1802), who used the term "*Mygale*" (a nonscientific name) to distinguish the "*mineuses*" and "*aviculaires*" spiders (where it included species with two pairs of book lungs and paraxial chelicerae) from the other spiders. The name *Mygale* was subsequently used by Latreille (1802) to designate a spider genus, where he described the most mygalomorph spiders known until then. Later, Olivier (1811) following Walckenaer's system adopted the genus *Mygale* separated from *Aranea*, but he restricted *Mygale* to "*les Araignées mineuses*," excluding all the other *Mygale* species or "*les Araignées aviculaires*," which he referred to *Aranea* (Thorell 1870). Then, Lamarck (1818) followed the Olivier proposal, but for "*les Araignées aviculaires*" proposed a separated genus, *Avicularia* Lamarck, 1818; at this time *Avicularia* included three species, *A. canceridea* (Latreille 1806), *A. blondii* (Latreille 1804), and *A. fasciata* (Latreille 1804) (Fukushima and Bertani 2017). Several years later, Latreille (1802) named *Ctenize* to "*les Araignées mineuses*," and in opposition to Olivier (1811) and Lamarck (1818) gave to "*les Araignées aviculaires*" the name *Mygale*, which was unavailable later (Thorell 1870), because it is a junior homonym of a mammal genus described by Cuvier (1800).

Simon (1892) in his famous work "Histoire naturelle des Araignées" established a classification for the order Araneae where he proposed two suborders "Araneae theraphosae" and "Araneae verae." The first was composed of the Mesothelae and mygalomorph spiders, and divided into Liphistiidae, Aviculariidae, and Atypidae. At the same time, he configured some subfamilies, one of them the Aviculariinae subfamily (which corresponds currently to the family Theraphosidae Thorell, 1869) divided into 10 groups. One of these groups was named Avicularieae, which was composed by two genera, *Avicularia* and *Tapinauchenius* Ausserer, 1871. Additionally, Simon (1892) synonymized *Typhochlaena* C.L. Koch, 1850 with *Avicularia*. Afterwards, Pocock (1901) included *Ephebopus* Simon, 1892, *Psalmopoeus* Pocock, 1895, *Pachistopelma* Pocock, 1901, and *Iridopelma* Pocock, 1901 into Avicularieae, and also suggested the inclusion of the African genera *Scodra* Becker, 1879 (currently *Stromatopelma* Karsch, 1881) and *Heteroscodra* Pocock, 1899 (Pocock 1901). Soon after, Simon (1903) following Pocock (1901) included the African genera mentioned, transferred *Ephebopus* to Phoneyuseae (currently part of Eumenophorinae), and synonymized *Iridopelma* with *Avicularia*. Sometime later, Mello-Leitao (1923) considered only American genera as valid taxa within the "Avicularias" group, added *Ancylochiros* Mello-Leitao, 1920, revalidated *Typhochlaena*, and synonymized *Iridopelma* with this genus. Finally, Roewer (1942) used for his catalog the taxonomic category of subfamily for spiders Avicularieae, additionally transferred *Heteroscodra* and *Scodra* back to this group, included the genera *Avicuscodra* Strand, 1908 (currently *Avicularia*), synonymized *Typhochlaena* with *Avicularia* again, transferred *Ancylochiros* to Ischnocolinae subfamily and *Ephebopus* to Eumenophoriinae subfamily.

It is well known that in 1985 Raven performed the first taxonomic revision and phylogenetic analysis of the infraorder Mygalomorphae based on morphological characters and without cladistics software. Within his results, Raven (1985) established *Avicularia*, *Iridopelma*, *Pachistopelma*, and *Tapinauchenius* genera as the Aviculariinae subfamily. Likewise, he transferred *Ephebopus* to Theraphosinae; *Stramatopelma* and *Heteroscodra* to Eumenophorinae; and *Psalmopoeus* to Selenocosmiinae. Raven (1985) considered the following characteristics as diagnostic of the Aviculariinae subfamily: presence of a spinose process between the lobes of the male palpal tarsus (except *Tapinauchenius*), tarsi broad as or broader than the metatarsi, and legs with few spines or absent. Lucas et al. (1991) described the male of *E. murinus* and, based on the possession of a sinuous embolus without keels, together with the characters used by Raven (1985), transferred *Ephebopus* to Aviculariinae. Additionally, they noted that males of both *Ephebopus* and *Tapinauchenius* lacked the spinose process on the palpal tarsi, but considered the wide tarsal pads, elongated embolus, and absence of spines on the legs as synapomorphies of Aviculariinae, allowing as a consequence the inclusion of these genera within the subfamily (West et al. 2008).

The first approximation to the study of the phylogenetic relationships into the Aviculariinae subfamily was carried out by West et al. (2008), with the taxonomic review and cladistic analysis of *Ephebopus*. Aviculariinae was considered as a monophyletic group which includes various lineages, supported by the presence of a "well-developed scopulae" that extends laterally on tarsi and metatarsi, mainly those of legs I and II (with parallelism in *Poecilotheria*: Selenocosmiinae subfamily). According to West et al. (2008), Aviculariinae includes two West African genera, *Stromatopelma* and *Heteroscodra*, and the New World genera *Avicularia*, *Ephebopus*, *Iridopelma*, *Pachistopelma*, *Psalmopoeus*, and *Tapinauchenius*. This composition is in complete harmony with Pocock's (1901) previous proposal. However, according to their cladogram obtained, the conformation of two subfamilies could be inferred: one of them composed by *Avicularia*, *Heteroscodra*, *Iridopelma*, *Pachistopelma*, and *Stromatopelma*, and the other by *Ephebopus*, *Psalmopoeus*, and *Tapinauchenius* (Fig. 4.3a; West et al. 2008: Fig. 9).

Bertani (2012) performed the taxonomic review of the genera *Iridopelma*, *Pachistopelma*, and *Typhochlaena*, revalidating again this last problematic genus. His results recovered Aviculariinae as a paraphyletic group in the majority of obtained trees, with a clade (*Ephebopus* (*Psalmopoeus* + *Tapinauchenius*)) as sister group of *Pelinobius* (Eumenophorinae) + *Phlogiellus* (Selenocosmiinae); and another clade composed by *Avicularia*, *Heteroscodra*, *Iridopelma*, *Typhochlaena*, *Pachistopelma*, and *Stromatopelma*, as sister group of *Encyocratella* (Stromatopelminae), *Haplopelma* (Ornithoctoninae), and *Poecilotheria* (Selenocosmiinae) (Fig. 4.3b; Bertani 2012: Figs. 177, 178). The author mentioned the possibility of changes in the Aviculariinae composition, which could be supported with the revision of other genera such as *Avicularia* and *Psalmopoeus* and additional information of the outgroups (Bertani 2012).

The genus *Avicularia* was recently revised by Fukushima and Bertani (2017), who also described three new genera in Aviculariinae: *Antillena*, *Caribena*, and

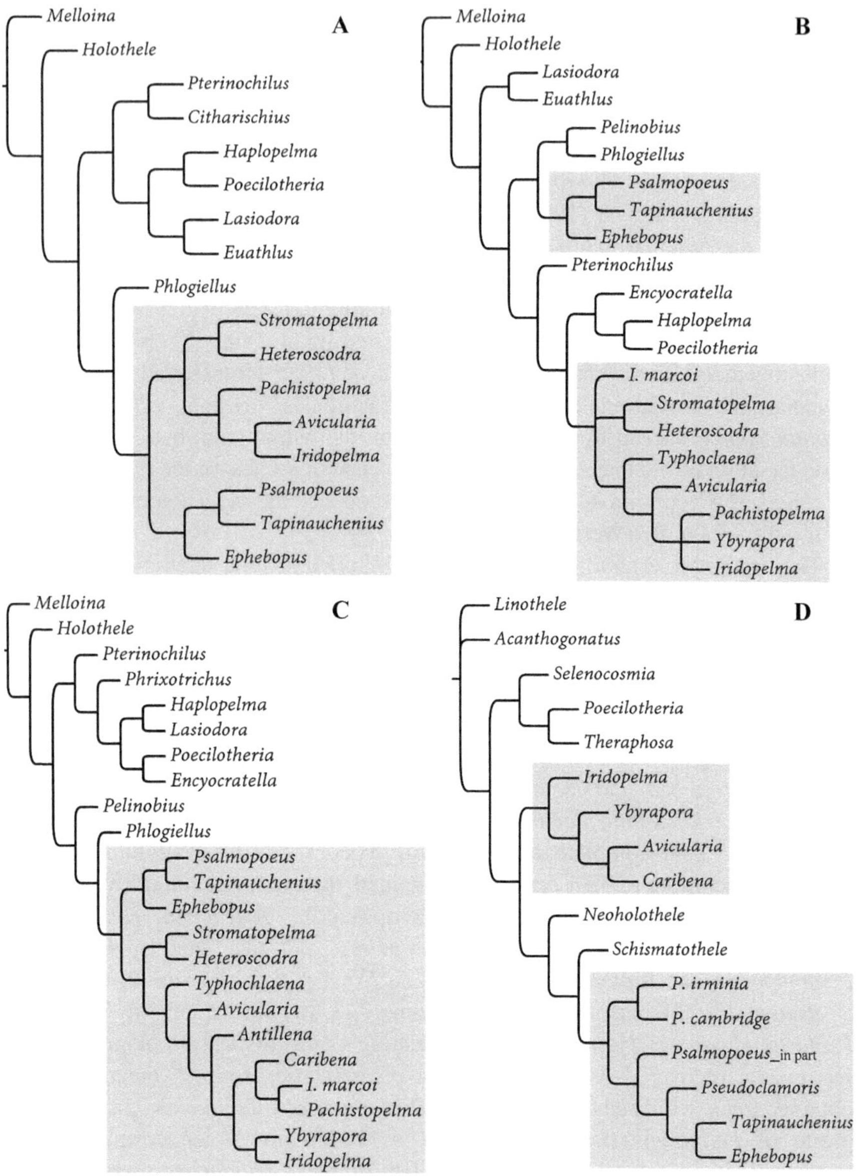

Fig. 4.3 Preferred phylogenetic hypothesis of the arboreal tarantulas and their allies. (**a**) West et al. (2008). (**b**) Bertani (2012). (**c**) Fukushima and Bertani (2017). (**d**) Hüsser (2018)

Ybyrapora. They stated the monophyly of the subfamily, agreeing with West et al. (2008) and based on the following homoplastic characters: presence of well-developed scopulae on tarsi, and metatarsi very extended laterally. According to Fukushima and Bertani (2017), Aviculariinae is composed of: *Antillena, Avicularia,*

Caribena, Ephebopus, Heteroscodra, Iridopelma, Pachistopelma, Psalmopoeus, Stromatopelma, Tapinauchenius, Typhochlaena, and *Ybyrapora.* The clade (*Ephebopus* (*Psalmopoeus* + *Tapinauchenius*)) was retrieved as sister group to all the remaining Aviculariinae (Fig. 4.3c; Fukushima and Bertani 2017: Fig. 325).

The monophyly and the relationships within Aviculariinae have been extensively debated in the last decade, and the classification of the node (*Ephebopus* (*Psalmopoeus* + *Tapinauchenius*)) has been controversial (West et al. 2008; Bertani 2012, Fukushima and Bertani 2017; Foley et al. 2019). Recent phylogenetic analyses with molecular data consider this clade out of Aviculariinae, as another subfamily Psalmopoeinae (Foley et al. 2019; Hüsser 2018; Lüddecke et al. 2018; Turner et al. 2018). The cladograms obtained by Hüsser (2018), Lüddecke et al. (2018), and Turner et al. (2018) recover the clade composed by Aviculariinae taxa as the sister group of a clade composed by Schismatothelinae + Psalmopoeinae (Fig. 4.3d). Additionally, Foley et al. (2019) recover the same evidence from transcriptome data: Aviculariinae as sister group of Schismatothelinae + Psalmopoeinae.

4.1.2 Psalmopoeinae

The genus *Psalmopoeus* Pocock, 1895 has had a controversial history of changes between Theraphosidae subfamilies. Initially, Pocock (1901), followed by other authors, allocated *Psalmopoeus* in Aviculariinae, but later Raven (1985) transferred it to Selenocosmiinae, based on the presence of numerous bacilliform setae in the prolateral surface of the maxillae (lyriform organ) (Fig. 4.4a). Samm and Schmidt (2008) proposed the Sinurticantinae subfamily, composed by *Psalmopoeus* and *Tapinauchenius*, based on the absence of urticating setae, presence of male tibial apophysis bipartite, and differences on the copulatory palpal bulbs regarding Selenocosmiinae males. However, Sinurticantinae was considered an invalid nomenclatural proposal because no type genus was specified and none corresponded with the "subfamily" name. Then, Schmidt (2010) modified the subfamily name to Psalmopoeinae, but did not present a phylogenetic analysis to justify the new group, although he endorsed the nature of the group by the lacking abdominal urticating setae, morphology of the tibial apophysis and palpal bulb, as well as other aspects such the Neotropical distribution. Additionally, Schmidt (2010) proposed that *Psalmopoeus* and *Tapinauchenius* are more closely related phylogenetically with Aviculariinae. This last proposal did not have a place in the academic community, perhaps due to the lack of a robust phylogenetic analysis, the fact that the absence of urticating setae would not be enough to create a subfamily, and in some sense this would be far from solving the group's taxonomic problems.

On the other hand, *Ephebopus* also has been a theraphosid genus that has had a very uncertain taxonomic position over the decades. Summarized, first, *Ephebopus* was included in the Selenocosmieae by Simon (1892), and then it was transferred variously to the Avicularieae (Pocock 1901), Phoneyuseae (Simon 1903), back to Aviculariinae (Mello-Leitão 1923), Eumenophorinae (Roewer 1942), Theraphosinae

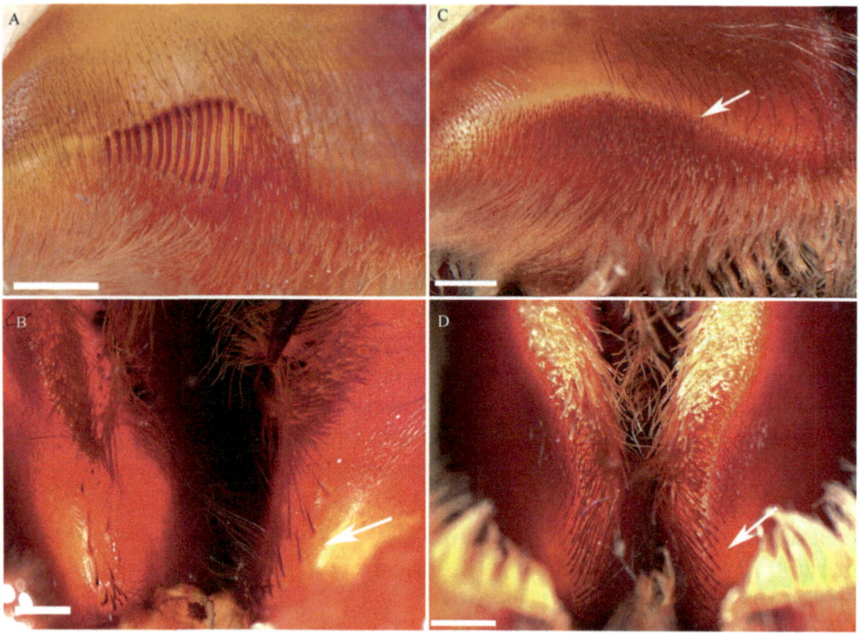

Fig. 4.4 Stridulatory organ in Psalmopoeinae. (**a**, **b**) *Psalmopoeus cambridgei*, (**a**) Maxillary lyra, composed by one row of thick bristles, (**b**) Base of chelicerae, ventral view. (**c**, **d**) *Pseudoclamoris gigas*, (**c**) Maxillary lyra, composed by a group of slender bristles, (**d**) Base of chelicerae, ventral view. White arrows point to the rows of filiform setae strikers

(Raven 1985), again to Aviculariinae (Lucas et al. 1991; West et al. 2008; Fukushima and Bertani 2017), and recently in Psalmopoeinae (Hüsser 2018).

Cladistic analyses carried out by West et al. (2008), Bertani (2012), and Fukushima and Bertani (2017) considered *Ephebopus*, *Tapinauchenius*, and *Psalmopoeus* as Aviculariinae genera. However, other studies validate the inclusion of *Psalmopoeus* and *Tapinauchenius*, and recently *Ephebopus*, into Psalmopoeinae subfamily (Lüddecke et al. 2018; Turner et al. 2018; Hüsser 2018). Recently molecular analyses of Theraphosidae have supported the hypothesis that Psalmopoeinae is closely related to Schismatothelinae, forming the clade sister of Aviculariinae. Lüddecke et al. (2018) and Turner et al. (2018) suggest (*Psalmopoeus* + *Tapinauchenius*) as the sister group of Schismatothelinae, and paraphyletic with respect to Aviculariinae. Lüddecke et al. (2018) also discussed the position of the two African genera *Stromatopelma* and *Heteroscodra*, transferred out of Aviculariinae, and proposed Stromatopelmatinae as the sister group of Harpactirinae, with a phylogenetic analysis strongly supported.

A phylogenetic study focused on *Tapinauchenius* and *Psalmopoeus* based on morphological and molecular data, conducted by Hüsser (2018), suggests Psalmopoeinae as a paraphyletic group (=morphological data) or monophyletic group (=molecular data), including *Ephebopus*, *Tapinauchenius*, *Psalmopoeus*, and a new genus described *Pseudoclamoris* (Fig. 4.3d; Hüsser 2018: Figs. 16–17). The obtained tree with morphological data recovered *Psalmopoeus* as the sister group of (*Tapinauchenius* (*Ephebopus* (*Pseudoclamoris* + Aviculariinae))), while with molecular data recovered the clade (Psalmopoeinae + Schismatothelinae) as sister group of Aviculariinae, as follow (Aviculariinae (Schismatothelinae (*Psalmopoeus* (*Pseudoclamoris* (*Ephebopus* + *Tapinauchenius*))))) (Hüsser 2018). The last tree recovered *Psalmopoeus* as paraphyletic group. Cifuentes (2018) in a taxonomic revision and cladistic analysis of *Psalmopoeus* and *Tapinauchenius,* based on morphological characters, recovered the clade (*Ephebopus* (*Tapinauchenius* + *Psalmopoeus*)) as monophyletic, related as the sister group of the Aviculariinae. Both genera *Tapinauchenius* and *Psalmopoeus* were recovered as monophyletic; nevertheless, no Schismatothelinae taxa were included in the analysis (Cifuentes 2018).

Traditionally, several authors have highlighted the taxonomic difficulties of Theraphosidae and the problems to delimit species, because of its great morphological homogeneity (Raven 1990, Bertani 2000; Perafán and Pérez-Miles 2014); Aviculariinae is no exception. Therefore, deep taxonomic studies and additional molecular analyses are necessary in order to know the identity of the species and attain better resolution of the relationships inside this group. *Tapinauchenius* and *Psalmopoeus* have variety and stability in some characters (sexual characters, setae of the legs, morphology of the maxillae lyra, and strikers) that permit the identification of groups and species. The addition of those characters in cladistic analyses will therefore attempt to better resolve the trees and improve the comparison with earlier hypotheses (Cifuentes 2018).

The validity of Psalmopoeinae as a monophyletic subfamily, as well as the phylogenetic relationships within the group are still questionable. However, we have preferred to maintain the two proposed subfamilies within the group of Neotropical arboreal tarantulas and their kin, according to the recent molecular evidence, for taxonomic treatment within this chapter: Aviculariinae and Psalmopoeinae.

4.2 Taxonomy

4.2.1 Aviculariinae Simon, 1892

Type species: *Avicularia avicularia* (Linnaeus, 1758).

Genera included: *Antillena* Bertani, Huff and Fukushima, 2017, *Avicularia* Lamarck, 1818, *Caribena* Fukushima and Bertani, 2017, *Iridopelma* Pocock, 1901, *Pachistopelma* Pocock, 1901, *Typhochlaena* C.L. Koch, 1850 and *Ybyrapora* Fukushima and Bertani, 2017.

Note: recent molecular study carried out by Lüddecke et al. (2018) considered *Stromatopelma* and *Heteroscodra,* distributed in West and Central Africa, out of Aviculariinae, composing the family Stromatopelminae together with *Encyocratella*. Foley et al. (2019), Hüsser (2018), and Lüddecke et al. (2018) recovered *Ephebopus*,

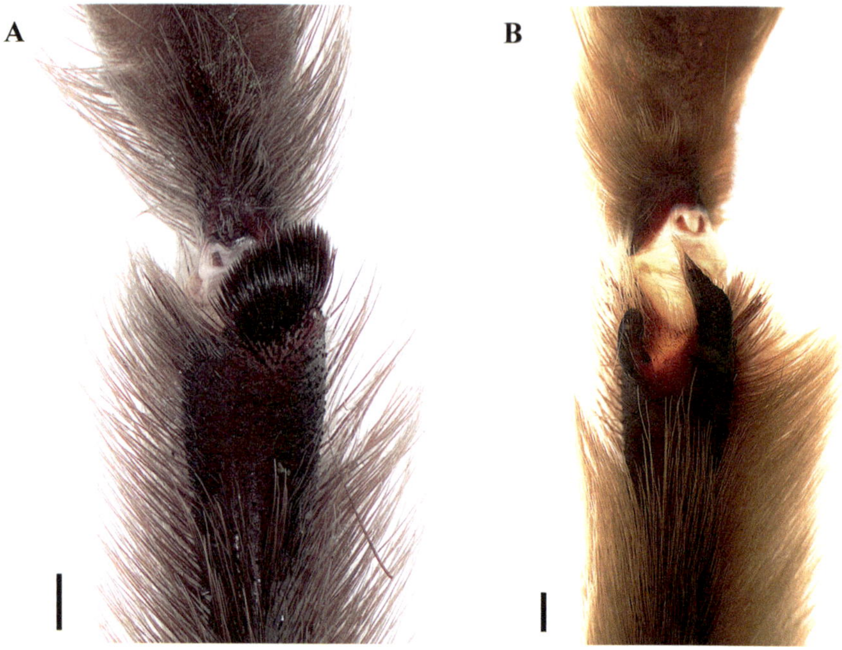

Fig. 4.5 Male tibial apophysis in Aviculariinae and Psalmopoeinae, ventral view. (**a**) *Avicularia* sp. (**b**) *Psalmopoeus* sp.

Psalmopoeus and *Tapinauchenius* in a clade closely related to Schismatothelinae before with the rest of the Aviculariinae taxa.

Diagnosis: Aviculariinae subfamily can be recognized by the conjunction of the following characters: legs without spines; scopulae in tarsus and metatarsus laterally extended, mainly on anterior legs, giving a spatulate appearance (Fig. 4.2); absence of spiniform setae on prolateral maxillae and presence of abdominal urticating setae type II (most species, absent in adult females of both *I. marcoi* Bertani, 2012 and *Pachistopelma*). Females with spermathecae long and completely separated. Males with tibia I with a spinose spur (Fig. 4.5a) (except *Typhochlaena, Ybyrapora* and some *Avicularia,* in which it's lacking), and palpal bulb with subtegulum not extended, and embolus long and thin without keels (except *Antillena*) (Bertani 2012; Fukushima and Bertani 2017).

Distribution: Southern Mexico, Central America, some Caribbean islands, and northern and central South America.

Taxonomic Key for the New World Aviculariinae

(Based on Bertani 2012; Fukushima and Bertani 2017)

1. Adults with sternum as long as wide, truncated behind; posterior lateral spinnerets with domed, short distal article. Males without both tibial spur and spiniform process on the cymbium (Bertani 2012: Figs. 1–28) .... *Typhochlaena*

– Sternum longer than wide; posterior lateral spinnerets with digitiform distal article. Males with tibial spur and spiniform process on the cymbium 2

2. Anterior row of eyes straight or slightly procurved. Females with abdomen dorso-ventrally flattened and without urticating setae (Bertani 2012: Figs. 29–80) ....*Pachistopelma*

– Anterior row of eyes strongly procurved. Females with abdomen not flattened and with urticating setae type II (Fukushima and Bertani 2017: Figs. 13, 15–17) 3

3. Spermathecae very short and broad, with distal half strongly sclerotized (Fukushima and Bertani 2017: Fig. 284). Male palpal bulb with very flattened embolus and well-developed keels (Fukushima and Bertani 2017: Figs. 285–288) .... *Antillena*

– Spermathecae long, not strongly sclerotized. Male palpal bulb lacking keels 4

4. Urticating setae type II very slender on a conspicuous patch on abdomen dorsum (Fukushima and Bertani 2017: Figs. 18, 254) ....*Caribena*

– Urticating setae type II stout spread over most abdominal area (Fukushima and Bertani 2017: Figs. 15–17, 302–304.) 5

5. Spermathecae lacking an accentuated outwards curvature medially, with distal portion far from the base (Fukushima and Bertani 2017: Fig. 14). Male withtibial spur on legs I and II (Bertani 2012: Figs. 85, 86) ....*Iridopelma*

– Spermathecae with an accentuated outwards curvature medially (Fukushima and Bertani 2017: Fig. 21). Males lacking tibial apophysis on leg II 6

6. Spermathecae virtually non-sclerotized (Fukushima and Bertani 2017: Figs. 257–259). Male cymbium with or without a process on retrolateral lobe, tibial spur on leg I absent (Fukushima and Bertani 2017: Fig. 308) (southeastern and part of northeastern Brazil) ....*Ybyrapora*

– Spermathecae with a well-sclerotized area (Fukushima and Bertani 2017: Fig. 21). Male cymbium with or without a process on retrolateral lobe, tibial spur on leg I well-developed or tibia with a discrete elevation covered by a cluster of setae on apical portion (Fukushima and Bertani 2017: Figs. 309–311) (México, Central America and South America) ....*Avicularia*

Antillena **Bertani, Huff and Fukushima, 2017**
Type species: *Antillena rickwesti* (Bertani and Huff, 2013)

Species included: *Antillena rickwesti*

Diagnosis: *Antillena* females can be distinguished by having two very short, broad, and apically sclerotized spermathecae (Fukushima and Bertani 2017:

Fig. 284), and by spiniform setae in prolateral and/or retrolateral side of coxae I–IV. Males differ by the copulatory palpal bulb with embolus very flattened and with keels (Fukushima and Bertani 2017: Figs. 285–288). Additionally, males are characterized by abdominal leaf pattern coloration (Fukushima and Bertani 2017: Fig. 294) (Fukushima and Bertani 2017).

Distribution: Dominican Republic (Bertani and Huff, 2013)

Avicularia **Lamarck, 1818**

Type species: *Avicularia avicularia* (Linnaeus, 1758)

Species included: *Avicularia avicularia*, *Avicularia caei* Fukushima and Bertani, 2017, *Avicularia glauca* Simon, 1891, *Avicularia hirschii* Bullmer, Thierer-Lutz and Schmidt, 2006, *Avicularia juruensis* Mello-Leitão, 1923, *Avicularia lynnae* Fukushima and Bertani, 2017, *Avicularia merianae* Fukushima and Bertani, 2017, *Avicularia minatrix* Pocock, 1903, *Avicularia purpurea* Kirk, 1990, *Avicularia rufa* Schiapelli Gerschman, 1945, and *Avicularia taunayi* (Mello-Leitão, 1920), and *Avicularia variegata* F. O. Pickard-Cambridge, 1896.

Diagnosis: *Avicularia* can be distinguished from some Aviculariinae, except *Caribena*, *Iridopelma*, *Typhochlaena*, and *Ybyrapora*, by having the anterior row of eyes procurved (Fukushima and Bertani 2017). It can be recognized from those groups by the following: from *Caribena* by stout urticating setae on abdomen dorsum and rounded process of cymbium (Fukushima and Bertani 2017: Figs. 15–17, 307); from *Iridopelma* males by lacking tibial apophysis on tibiae II and females by spermathecae long with accentuated medial curvature directed outward (Fukushima and Bertani 2017: Fig. 21); from *Typhochlaena* by digitiform apical article in PLS (Fukushima and Bertani 2017: Fig. 10); and from *Ybyrapora* by the sclerotized spermathecae in females and the presence of developed tibial apophysis in males (Fukushima and Bertani 2017: Figs. 309–311), except *A. minatrix*, *A. lynnae*, *A. caei*, and *A hirschii*. Males of these species can be differentiated from *Y. diversipes* by the embolus less curved in frontal view (Fukushima and Bertani 2017: Fig. 30), from *Y. sooretama* and *Y. gamba* by the presence of a well-developed process on cymbium (Fukushima and Bertani 2017: Fig. 307), except *A. minatrix*; *A. minatrix* can be distinguished from the last *Ybyrapora* species mentioned by the abdomen coloration, dorsum black with orange spots on its side (Fukushima and Bertani 2017: Fig. 89).

Distribution: Mexico; Central America: Costa Rica, Panama; South America: Bolivia, Brazil, Colombia, Ecuador, Guyana, Peru, Suriname, Trinidad and Tobago and Venezuela. In Brazil, it occurs in the states of Roraima, Amapá, Amazonas, Pará, Maranhão, Tocantins, Acre, western Bahia, Rondônia, Mato Grosso, Goiás, Minas Gerais, São Paulo, and The Distrito Federal (Fukushima and Bertani 2017).

Caribena **Fukushima and Bertani, 2017**

Type species: *Caribena laeta* (C. L. Koch, 1842)

Species included: *Caribena laeta*, and *Caribena versicolor* (Walckenaer, 1837).

Diagnosis: *Caribena* species are characterized by the presence of a conspicuous abdominal patch of type II urticating setae, setae long and slender (more than 1 mm long and less than 0.009 mm wide) (Fukushima and Bertani 2017: Figs. 18, 243, 241, 254). Additionally, males have a sharp spiniform process in the retrolateral lobe of cymbium (Fukushima and Bertani 2017: Fig. 306) (Fukushima and Bertani 2017).

Distribution: Puerto Rico, the U. S. Virgin Island, and Martinique (Fukushima and Bertani 2017).

Iridopelma **Pocock, 1901**

Type species: *Iridopelma hirsutum* Pocock, 1901

Species included: *Iridopelma hirsutum*, *Iridopelma katiae* Bertani, 2012, *Iridopelma marcoi* Bertani, 2012, *Iridopelma oliverai* Bertani, 2012, *Iridopelma vanini* Bertani, 2012 and *Iridopelma zorodes* (Mello-Leitao, 1926).

Diagnosis: *Iridopelma* males are characterized by having tibial spurs on leg I and II (Bertani 2012: Figs. 85, 86) and females by having the anterior row of eyes strong procurved (Fukushima and Bertani 2017: Fig. 13). The last characteristic is shared with *Avicularia* and *Typhochlaena*, but it can be recognized from those genera, respectively, by having spermathecae lacking accentuated curvature (Fukushima and Bertani 2017: Fig. 14), and by the distal PLS segment digitiform (Fukushima and Bertani 2017: Fig. 10) (Bertani 2012).

Distribution: Northeastern Brazil, from the state of Bahia, northwards, and some records for states of Para and Tocantins (Bertani 2012).

Pachistopelma **Pocock, 1901**

Type species: *Pachistopelma rufonigrum* Pocock, 1901

Species included: *Pachistopelma bromelicola* Bertani, 2012, and *Pachistopelma rufonigrum*.

Diagnosis: *Pachistopelma* differs from those of other aviculariines by having the anterior row of eyes straight or slightly procurved (Bertani 2012: Fig. 37).

Note: It is worth noting that the characteristic of anterior row of eyes straight or slightly procurved is shared with *Ephebopus*, *Tapinauchenius*, and *Psalmopoeus* genera (Psalmopoeinae). However, *Pachistopelma* can be recognized from those genera by having legs aspinose, abdominal urticating setae (except in mature female), males with a spinose spur on tibia I, and females with abdomen dorsoventral flattened (Bertani 2012: Fig. 34) and paired long spermathecae with slight curvature medially, without any constriction (Bertani 2012: Figs. 58–65) (Bertani 2012). **Distribution:** Northeastern Brazil, from the state of Rio Grande do Norte to state of Bahia, mostly in the coastal region (Bertani 2012).

Typhochlaena **C. L. Koch, 1850**

Type species: *Typhochalena seladonia* (C.L. Koch 1841)

Species included: *Typhochalena amma* Bertani, 2012, *Typhochalena costae* Bertani, 2012, *Typhochalena curumin* Bertani, 2012, *Typhochalena seladonia*, and *Typhochalena paschoali* Bertani, 2012.

Diagnosis: *Typhochlaena* can be differentiated from other aviculariines by the short, domed apical segment of PLS (Bertani 2012: Fig. 21), and adults with a sternum as long as wide, truncated behind (Bertani 2012: Fig. 20). They are characterized by small sizes. Additionally, males lack tibial spurs and spiniform process in the cymbium (Bertani 2012).

Distribution: Brazil: Northeastern, part of Central-West (Tocantins state), and part of southeastern (Espirito Santo state) (Bertani 2012).

Ybyrapora **Fukushima and Bertani, 2017**

Type species: *Ybyrapora sooretama* (Bertani and Fukushima 2009).

Species included: *Ybyrapora diversipes* (C.L. Koch 1842), *Ybyrapora gamba* (Bertani and Fukushima 2009), and *Ybyrapora sooretama*.

Diagnosis: *Ybyrapora* species can be distinguished from some aviculariine by having an anterior row of eyes procurved; a characteristic that is shared with *Avicularia*, *Caribena*, *Iridopelma*, and *Typhochlaena*. It can be distinguished from *Avicularia* by the presence of virtually non-sclerotized spermathecae in females (Fukushima and Bertani 2017: Figs. 257–259), while males differ by lacking developed tibial apophysis (Fukushima and Bertani 2017: Fig. 308), except from *A. minatrix*, *A. lynnae*, *A. caei*, and *A hirschii*. Males of *Y. diversipes* can be differentiated from those four species by the presence of a very long embolus, strongly curved in the frontal view (Fukushima and Bertani 2017: Fig. 278); *Y. sooretama* and *Y. gamba* differ from those of *A. lynnae, A. caei*, and *A. hirschii* by lack a well-developed process on cymbium (Fukushima and Bertani 2017: Fig. 305); *Ybyrapora* males can be distinguished from *A. minatrix* by the presence of a single red or dark longitudinal stripe on the abdomen dorsum (Fukushima and Bertani 2017: Fig. 297). *Ybyrapora* can be distinguished from *Caribena* by the presence of stout urticating setae on the abdomen dorsum in both males and females, and the absence of a sharp process with thin setae on the retrolateral lobe of the cymbium in males; from *Iridopelma* by males lacking a tibial apophysis on tibiae II and females by the presence of a long spermathecae medially curved and directed outwards; and from *Typhochlaena* by the presence of a digitiform apical article in PLS (Fukushima and Bertani 2017).

Distribution: Brazil, from Bahia to southern Rio de Janeiro (Fukushima and Bertani 2017).

4.2.2 Psalmopoeinae Samm and Schmidt (2008)

Type species: *Psalmopoeus cambridgei* Pocock, 1895

Genera included: *Ephebopus* Simon, 1892, *Psalmopoeus* Pocock, 1895, *Pseudoclamoris* Hüsser, 2018, and *Tapinauchenius* Ausserer, 1871 (Fig. 4.6).

Diagnosis: Males and females of Psalmopoeinae can be distinguished from the majority of Theraphosidae subfamilies, except Aviculariinae, by the scopulae in

Fig. 4.6 Psalmopoeinae spiders *habitus*. (**a**) *Ephebopus murinus*, male. (**b**) *Ephepobus cyanognathus*, female. (**c**) *Pseudoclamoris gigas*, female. (**d**) *Tapinauchenius polybtes*, female. (**e**, **f**) *Psalmopoeus ecclesiasticus*, (**e**) juvenil, (**f**) female. (Photos: **a**–**d** courtesy of Rick West; **e**–**f** Carlos Perafán)

tarsus and metatarsus laterally extended, giving a spatulate appearance. They differ from most Aviculariinae species by the absence of abdominal urticating setae; urticating setae only present in palpal femur of *Ephebopus*. Also, by the anterior row of eyes straight (except from *Pachistopelma*). Additionally, *Psalmopoeus*,

Fig. 4.7 (**a**) *Psalmopoeus ecclesiasticus*, showing the long lateral setae in the legs. (**b**) *Psalmopoeus cambridgei*, dorsal view of palp and leg I, showing the group of smooth short setae. Arrows point to the lateral setae and the group of smooth short setae in legs

Tapinauchenius, and *Pseudoclamoris* males can be distinguished by the long covering setae laterally directed in their legs (Fig. 4.7a) and tibial apophysis composed by two branches (Fig. 4.5b) (Hüsser 2018).

Distribution: Mexico; Central America: Belize, Costa Rica and Panama, Caribbean islands: Saint Vincent Island, Saint Lucia island, Union island, The Grenadines, Trinidad and Tobago islands and South America: Brazil, Colombia, Ecuador, French Guiana, Peru, and Venezuela (World Spider Catalog 2020; West 1983).

Taxonomic Key for Psalmopoeinae

1. Presence of a patch of urticating setae (type V) on prolateral distal palpal femora (West et al. 2008: Figs. 19–20; Fukushima and Bertani 2017: Fig. 4) ....*Ephebopus*

   – Absence of a patch of urticating setae on femora 2

2. Stridulatory setae forming a lyriform organ on prolateral maxillae (Fig. 4.4a, c; Fukushima and Bertani 2017: Fig. 11) 3

   – Stridulatory setae absent on maxillae ....*Tapinauchenius*

3. Thick bristles on the maxillary lyra disposed in one row (Fig. 4.4a; Fukushima and Bertani 2017: Fig. 11) ....*Psalmopoeus*

4. Slender bristles on the maxillary lyra arranged in an oval patch, composed by several rows of slender setae (Fig. 4.4c; Hüsser 2018: Fig. 2) ....*Pseudoclamoris*

Ephebopus **Simon, 1892**
Type species: *Ephebopus murinus* (Walckenaer, 1837)

Species included: *Ephebopus cyanognathus* West and Marshall, 2000, *Ephebopus foliatus* West, Marshall, Fukushima and Bertani, 2008, *Ephebopus murinus*, *Ephebopus rufescens* West and Marshall, 2000, and *Ephebopus uatuman* Lucas, Silva and Bertani, 1992.

Diagnosis: *Ephebopus* can be distinguished from all Theraphosidae genera by the presence of a patch of urticating setae (type V) on prolateral distal palpal femora (West et al. 2008: Figs. 19–20; Fukushima and Bertani 2017: Fig. 4).

Distribution: Northeastern and Central Brazilian Amazon, French Guiana, Southern Suriname, and Southwestern Guyana (West et al. 2008).

Psalmopoeus **Pocock, 1895**
Type species: *Psalmopoeus cambridgei* Pocock, 1895

Species included: *Psalmopoeus cambridgei*, *Psalmopoeus ecclesiasticus* Pocock, 1903, *Psalmopoeus emeraldus* Pocock, 1903, *Psalmopoeus intermedius* Chamberlin, 1940, *Psalmopoeus irminia* Saager, 1994, *Psalmopoeus langenbucheri* Schmidt, Bullmer and Thierer-Lutz, 2006, *Psalmopoeus plantaris* Pocock, 1903, *Psalmopoeus pulcher* Petrunkevitch, 1925, *Psalmopoeus reduncus* (Karsch, 1880), and *Psalmopoeus victori* Mendoza, 2014.

Diagnosis: The species of *Psalmopoeus* can be distinguished by the presence of a lyriform organ composed by one row of thick bristles on prolateral maxillae in conjunction with a row of long filiform setae strikers on the base of chelicerae (Fig. 4.4a, b). Additionally, *Psalmopoeus* have a group of short and smooth hair on the retrolateral side of the palpal trochanter and femur, and on prolateral and retrolateral sides of trochanter and femur of leg I (Fig. 4.7b) (Absent in *Tapinauchenius* and *Pseudoclamoris*) (Cifuentes 2018).

Distribution: Mexico; Central America: Belize to Panama; Caribbean: Trinidad and Tobago islands; South America: Colombia, Ecuador, Venezuela, Guyana, and Brazil.

Pseudoclamoris **Hüsser, 2018**
Type species: *Pseudoclamoris gigas* (Caporiacco, 1954)

Species included: *Pseudoclamoris elenae* (Schmidt, 1994), *Pseudoclamoris gigas*, and *Pseudoclamoris burgessi* Hüsser, 2018.

Diagnosis: *Pseudoclamoris* species resemble *Psalmopoeus* genus by the presence of a lyriform organ on prolateral maxillae. However, *Pseudoclamoris* can be distinguished by having an oval maxillary lyra composed by several rows of slender setae (thick bristles on *Psalmopoeus*) (Fig. 4.4c). Additionally, females have

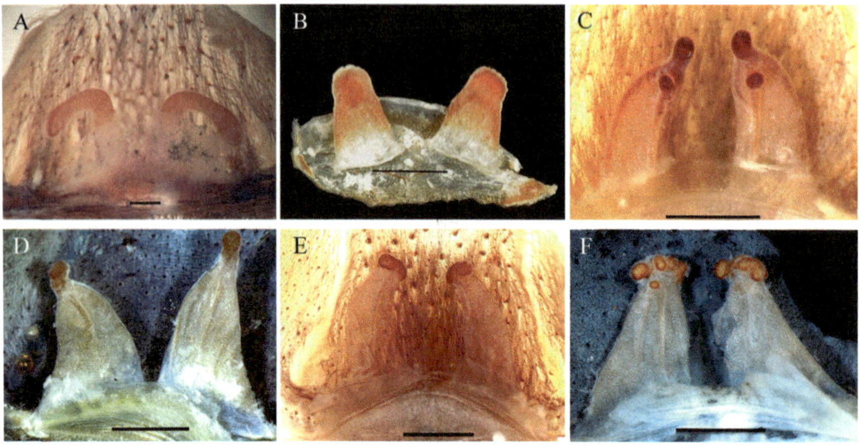

Fig. 4.8 Morphology of the female's spermatheca in *Avicularia* and Psalmopoeinae. (**a**) *Avicularia* sp. (**b**) *Ephebopus rufescens.* (**c**) *Psalmopoeus irminia.* (**d**) *Tapinauchenius polybotes.* (**e**) *Pseudoclamoris gigas.* (**f**) *Pseudoclamoris elenae* (Photos: **a**, **c–f** Yeimy Cifuentes; **b** courtesy of Rick West)

spermathecae with one subsegmented apical lobule (Fig. 4.8e) or with multiple apical lobules (Fig. 4.8f).

Distribution: Colombia, Ecuador, Brazil, Peru, and French Guiana.

Tapinauchenius **Ausserer, 1871**

Type species: *Tapinauchenius plumipes* (C. L. Koch, 1842)

Species included: *Tapinauchenius brunneus* Schmidt, 1995, *Tapinauchenius concolor* (Caporiacco, 1947), *Tapinauchenius cupreus* Schmidt and Bauer, 1996, *Tapinauchenius latipes* L. Koch, 1875, *Tapinauchenius plumipes* (C. L. Koch, 1842), *Tapinauchenius polybotes* Hüsser, 2018, *Tapinauchenius rasti* Hüsser, 2018, *Tapinauchenius sanctivincenti* (Walckenaer, 1837), and *Tapinauchenius violaceus* (Mello-Leitão, 1930).

Diagnosis: *Tapinauchenius* can be distinguished from *Ephebopus* by lack urticating setae, and from *Psalmopoeus* and *Pseudoclamoris* by the absence of a maxillary lyra. Additionally, males have palpal bulb embolus with an apical constriction, small and curved ventrally tip (Fig. 4.9b) (except in *T. cupreus*), and known females have spermathecae with one apical lobule (Fig. 4.8d) (Cifuentes 2018).

Distribution: Caribbean: Saint Vincent island, Saint Lucia island, Union island, Trinidad and Tobago island; South America: Colombia, Ecuador, Venezuela, Guyana, Suriname, French Guiana, and Brazil.

Fig. 4.9 Morphology of copulatory palpal bulb in Aviculariinae and Psalmopoeinae. (**a**) *Avicularia* sp. (**b**) *Tapinauchenius plumipes*

4.3 Distribution and Natural History

One of the most important and conspicuous characteristics in the arboreal tarantulas is the presence of very well-developed claw tufts and scopulae on the tarsi and metatarsi. These structures participate in the adhesion mechanisms, transcendental in predatory and locomotory function, which facilitate the climbing and descending ability on vertical surfaces. Likewise, claw tufts and scopulae played an important role in the arboreal spiders' evolution, maybe because it was involved in the colonization from the ground to vegetation (Pérez-Miles et al. 2015). On the other hand, most aviculariines species possess urticating setae as a defense mechanism, an evolutionary novelty that appeared only in tarantulas of the New World. A detailed analysis of function and morphology of adhesive setae and urticating setae is found in the Chapters 12 and 9, respectively.

The majority of Neotropical arboreal aviculariine species present an outstanding ontogenetic differentiation in the coloration patterns. Usually, juveniles present abdominal patterns of longitudinal and/or transverse stripes, with striking colors, while in adults these patterns tend to disappear and black, bluish, and reddish colors are predominant. There is also sexual dimorphism in the coloration of mature individuals (Figs 4.1a–d). However, contrary to most aviculariines, ontogenetic changes are not known in species of *Typhochlaena* (Bertani 2012). The ontogenetic and sexual differences in color contributed in the past to erroneous descriptions of different species, because the coloration patterns were used before sexual structures as valid characters to separate species. This has brought difficulties for the current taxonomy and the consequent long list of synonymized species.

Aviculariinae species inhabit predominantly tropical lowland wet forest. *Avicularia* is distributed in the tropical areas of the Americas, from Costa Rica to Bolivia and Brazil, although there are also some unofficial records from southern Mexico (Fukushima and Bertani 2017). In addition to their arboreal habits, usually they make their silk retreats on vegetation but they also tend to use the human structures for establishing their retreats. These habits can change throughout their life development, as is known for many tarantulas. Stradling (1994) reported that *A. avicularia,* in a population closely related to human settlements in Trinidad, use leaves of herbaceous plants, tree barks, branches of tree trunks for their silk retreats, and that the juveniles in first instars use leaves of low growing plants.

Avicularia has type II abdominal urticating setae which present a peculiar defense mechanism. These setae are transferred by direct contact when the spider is disturbed, from the abdomen against the target (Bertani and Marques 1996), and are not expelled into the air, as in the majority of Theraphosinae (except *Kankuamo* Perafán, Galvis and Gutiérrez 2016). Urticating setae type II and VII (present in *Kankuamo*) have a different shape and size in relation to other urticating setae; they are thicker and heavier than the slender setae of theraphosines, preventing them from being carried by the air (Cooke et al. 1972; Bertani and Marques 1996; Perafán et al. 2016). A particularity of the urticating setae type II is the basal localization of the penetrating tip being very close to the supporting stalk, which gives it a special release mechanism: In contact, the urticating setae pivots on its stalk, so that the apical end moves away from the target, while the basal penetration tip instead lodges into the target as the stalk releases (see Bertani and Marques 1996: Fig. 3). However, there is a curious exception observed in *C. versicolor*, whereby it has urticating setae type II but its releasing mechanism is similar to theraphosines. According to Bertani et al. (2003), this constitutes a case of convergence between aviculariines and theraphosines. The type II urticating setae of *C. versicolor* are slender, similar to Theraphosinae urticating setae, and this morphology facilitates to said setae to go airborne.

Iridopelma, *Pachistopelma*, and *Typhochalena* are principally distributed in Atlantic forests, Restinga, and Caatinga areas of north and northeastern of Brazil. The majority of *Iridopelma* species live in dry open environments (Bertani 2012). These spiders generally build their retreats with two leaves connected with silk, or in a rolled single leaf, although *I. zorodes* and *I. katiae* use bromeliads as a refuge. This change in the habitat has been considered by Bertani (2012) as a derived condition in the group according to its cladistic analysis, which could have occurred due to climatic changes that carried a shift in the vegetation, forming grasslands on rocky and sandy soils, leading to the use of bromeliads and rocks as the available refuges for those spiders (Bertani 2012). Under this same scenario, *Pachistopelma* spp. are strictly associated with bromeliads (Santos et al. 2002, 2004; Dias et al. 2000; Dias and Brescovit 2003, 2004; Bertani 2012), as well some *Avicularia* juveniles that eventually use these refuge plants (Bertani 2012); nonetheless, other aviculariines are uncommon in bromeliads (Bertani 2012).

Pachistopelma rufonigrum is a species from Restinga vegetation and apparently strictly-dependent of bromeliads as a resource of water and as a refuge (Bertani 2012; Almeida-Silva et al. 2008), as well as *P. bromelicola*, distributed in the Brazilian Atlantic forest, Restinga, and Caatinga areas (Bertani 2012). *Typhochlaena* species are also found in the Brazilian Atlantic forest, but *T. costae* have been reported in drier environments (Bertani 2012). Some of these species construct silk retreats under loose tree bark, or under low tree leaves (Bertani 2012), and there is evidence that *T. seladonia* and *T. costae* close their shelters with a thin lid (Andre and Hüsser 2018).

Ephebopus species are distributed in the Amazonian rainforest, from northeastern and central Brazilian Amazon, French Guiana, southern Suriname, and southwestern Guyana (West et al. 2008). In relation to their shelters, West et al. (2008) described older juvenile and adult *E. murinus* refuges as fossorial channels ending in a blind chamber with a burrow entrance that has a large and elaborate trumpet shaped turret of silk, in contrast to juveniles that build silk refuges inside terrestrial bromeliads. Other species have been found in retreats in hollow logs and stems, both on and above the ground, in holes and hollows of standing trees, in arboreal termite nests, and in mosses on the sides of shaded rock faces; additionally, they can extend its retreat with tubular silk and camouflage it with soil and vegetative debris (West et al. 2008). Marshall and West (2008) made emphasis in the ontogenetic habitat shift behavior presented by *E. murinus* in a study made in a population in French Guiana. They observed that early instars spiderlings constructed tubular silk refuges in the vegetation above the ground (this correlated with the presence of bromeliads), and the older juveniles and adult females were found in retreats in the ground. They theorized that the ontogenetic habitat shift can be considered as a strategy for avoiding intraspecific competition and cannibalism, but they also suggested that it is possible that the early instars retrieve an ancestry habit (Marshall and West 2008). Retreats of *E. uatuman* and *E. cyanognathus* have been found in damp soil covered by litter and were described as flare-mouth burrows, with the entrance surrounded by a collar of leaves attached by silk and the channel ending in a deep chamber (West et al. 2008). *E. rufescens* was found above the ground (1–4 m) with no evidence of retreats on the soil, and *E. foliatus* has been reported as arboreal (West et al. 2008).

Ephebopus represents an exceptional case in the evolution of urticating setae in tarantulas of the New World. It presents an urticating setae patch on the distal prolateral face of palpal femora, composed by setae described as type V (Marshall and Uetz 1990). In contrast to the majority of aviculariines, these setae are released to the air by the movement of the basal segment of the chelicerae against the area of the urticating setae (Lucas et al. 1991; Foelix et al. 2009).

Psalmopoeus has a fairly wide Neotropical distribution, inhabiting Amazonian forest from Ecuador, Colombia, Brazil and Venezuela; Andean and Pacific regions from Ecuador and Colombia; dry forest from Andean and Caribbean regions in Colombia; tropical forest of Panama, Costa Rica, Belize, México, and Trinidad and Tobago islands. In addition, they are the only known arboreal tarantulas found in the Andean forest. *Psalmopoeus* is recognized by its arboreal habits and abilities as an opportunistic taxon. Mendoza-M (2014) reported that *P. victori* construct retreats in tree cavities in rainforest of Veracruz, Mexico, from an area that is under pressure from human activities; *P. pulcher* has been reported as opportunistic, adapting an arboreal lifestyle (Gabriel and Sherwood 2018), and has been observed living in environments close to human settlements (personal observations). Records in Colombia indicate that they construct retreats in palm trees, and are associated with dry forest remnants of Caribbean regions (information of museum specimens).

Valerio (1979) recorded *P. reduncus* as an arboreal species, living up to 4 m above the ground, hidden in trunk cavities and bases of palm trees in forested areas in Costa Rica; *P. irminia* was reported in the Paracaima region of the Amazonian area of Brazil at 900 m.a.s.l., predominantly covered by savanna vegetation, in a tubular silk retreats at 1.60 m above into rocky cavities (Bertani et al. 2016). Saager (1994) described this species from the "Gran Sabana" in Venezuela, an area surrounded by tropical forest galleries inside Mahogany and other tree roots, in cavities, lianas, and under tree barks (Saager 1994). *P. ecclesiasticus*is distributed in Colombia and Ecuador, in the biogeographic Chocó region, it has been found in the Pacific foggy forest at 1400 m.a.s.l. from Nariño, Colombia, living over trunks, and juveniles in retreats inside mosses that cover standing trees. This data constitutes the highest altitudinal limit registered for the genus.

Tapinauchenius has been reported as opportunistic, even constructing retreats in abandoned human constructions (personal communication with R. Bertani). Apparently, *Pseudoclamoris* species are restricted to the Amazonian rainforest, but information about microhabitats has not been reported.

4.4 Conservation

It is widely known that large theraphosids are commonly sold and kept as exotic pets in many countries, due to their exotic beauty, highlighted by their colors, easy maintenance, and long life in captivity. Although, some tarantula dealers act under legal regulations, most of the trade is illegal through animal trafficking (Cifuentes and Perafán personal observations). This activity that is not only used as a lucrative business but also used as scientific traffic. These activities, added to the environmental pressures that all biota of the Earth face due to contamination and reduction of natural habitats, have more than one Aviculariinae species on alert. Unfortunately, we are still far from calculating the full scale of these problems.

Bertani and Fukushima (2009) discussed some traits about the conservation status of the endemic species from Brazilian Atlantic Forest, *Y. diversipes*, *Y. sooretama,* and *Y. gamba*. As expected, the three species are considered as affected, especially by the reduction and fragmentation of their habitat range caused by deforestation in the ecosystems. The Atlantic Forest is one of the most emblematic biogeographic regions of South America, considered as one hotspot of biodiversity and probably the most endangered ecosystem in the planet (Galindo-Leal and Câmara 2003); it is an area of high endemicity, and is considered to house at least 60% of the planet's diversity (Galindo-Leal and Câmara 2003). Bertani and Fukushima (2009) also considered the illegal wildlife trafficking of *Y. diversipes* as one factor that can be affecting the populations; they reported in 2009 that the species was commercialized in Europe, Asia, and United States, with high online sales demand. They consider that, due to the illegal collection, the restricted distributions and the altered habitats, all gravely affecting the populations, the inclusion of these species in the CITES and IUCN threatened list of species should be considered (Bertani and Fukushima 2009).

According to Bertani (2012), some species of *Pachistopelma*, *Iridopelma*, and *Typhochlaena* are very endemic: rare (*Typhochlaena* spp.), specialist (*Pachistopelma* spp. as a result of being strictly bromeliaceous), and scarcely studied (*Iridopelma* are distributed only in threatened areas of Brazilian Atlantic forest, Restinga, and Caatinga). These species can be affected by the reduction of the distribution areas and threatened microhabitats as the bromeliad species of the distribution regions (Bertani 2012).

Wildlife trafficking is one of the most lucrative illegal activities, and it has been reported that only 10% of the captured animals arrive alive to the destination, due to the stress and mistreatment of the animals in the transportation (Caldas et al. 2018). It is considered that the rate of survival of rescued animals, along with the successful reintroduction to their natural habitats, is low or without benefit for the species (Caldas et al. 2018). An extraordinary case was reported by Caldas et al. (2018) where they recorded the study over a case of wildlife trafficking of 52 individuals of *Pachistopelma bromelicola* (Aviculariinae) that were found by the Brazilian Institute of the Environment and Renewable Natural Resources (IBAMA, portuguese acronym) in cruel and poor conditions, transported inside matchboxes and presumably for being sold in Slovakia. Of the 52 individuals found captive, only 22 were alive. During the rehabilitation process 12 of them died and from the remainder, only 4 presented traits of good health (molting). After monitoring the live animals, the researchers realized that some of those spiders presented protuberances "tumor-like" on their opisthosoma, a zoonotic concern for being reintroduced to their natural habitat. Finally, from the total of individuals only 7 (13%) survived (Caldas et al. 2018).

Studies in Aviculariinae and Psalmopoeinae populations are not developed, so the state of the species inside the principal list for animal conservation is generally lacking in information; this information vacuum does not permit the regulation in the use of the natural resources of the region.

4.5 Final Considerations

There is an interesting and deep discussion currently about the monophyly of the Neotropical arboreal theraphosids. As it happens widely in the animal world, the evidence suggests that adaptation to arboreal life to be an evolutionary convergence in tarantulas. The latest phylogenetic analyses also suggest a close relationship of these taxa with South American lineages, and each other, rather than with lineages of African or Asian arboreal theraphosids. The evolutionary evidence theorized that the South American theraphosids make up a unique and ancient lineage. Although the growing interest in the study of Neotropical arboreal tarantulas has improved their taxonomic knowledge, there are still few works related to their natural history and there are still many unsolved questions about the ecology and evolution of the urticating setae, as well as the stridulatory structures. Deepening these fields of research will not only expand the knowledge of this fascinating group but also affect the decision-making related to its conservation and the environments it occupies, which are generally just relicts of natural environments.

Acknowledgments Special thanks to Rick C. West for his recommendations and corrections, and for his collaboration with the images in Fig. 4.1a–f, 4.6a–d, and 4.8b. Thanks to Julian Kamzol for the images in Fig. 4.2, and Laura Montes de Oca for the help with the edition of the cladograms in Fig. 4.3. We are also thankful to M. Hüsser for the revision of the manuscript and his comments, and to Brad Zamft, for his suggestions and corrections to our English grammar. Y.C. wants to thank the Programa Estudiante-Convenio de Posgraduacão (PEC-PG) from the Brazilian Government, and The Conselho Nacional de Desenvolvimento Científico e Tecnológico (CNPq) for the academic fellow process [190543/2015-9].

References

Almeida-Silva LM, Camacho A, Brescovit AD, Lucas SM, Brazil TK (2008) Redescription and notes on the natural history of the arboreal tarantula *Iridopelma seladonium* (Araneae: Theraphosidae: Aviculariinae). Rev Bras Zool 25:728–736

Andre C, Hüsser M (2018) About trapdoors and bridges- new insights in the little-known ecology and lifestyle of the genus *Typhochlaena* C.L. Koch, 1850. Br Tarantula Soc J 32:3–29

Bertani R (2000) Male palpal bulbs and homologous features in Theraphosinae (Araneae, Theraphosidae). J Arachnol 28:29–42

Bertani R (2012) Revision, cladistic analysis and biogeography of *Typhochlaena* CL Koch, 1859, *Pachistopelma* Pocock, 1901 and *Iridopelma* Pocock, 1901 (Araneae, Theraphosidae, Aviculariinae). ZooKeys 230:1–94

Bertani R, Fukushima CS (2009) Description of two new species of *Avicularia* Lamarck, 1818 and redescription of *Avicularia diversipes* (C.L. Koch, 1842) (Araneae, Theraphosidae, Aviculariinae) three possibly threatened Brazilian species. Zootaxa 2223:25–47

Bertani R, Huff J (2013) *Avicularia rickwesti* sp. nov., a remarkable new species of *Avicularia* (Theraphosidae: Aviculariinae) from Dominican Republic. Zoologia (Curitiba) 30:333–337

Bertani R, Marques OAV (1996) Defensive behaviors in mygalomorph spiders: release of urticating hairs by some Aviculariinae. Zool Anz 234:161–165

Bertani R, Boston T, Evenou Y, Guadanucci JPL (2003) Release of urticating hairs by *Avicularia versicolor* (Walckenaer, 1837) (Araneae, Theraphosidae). Bull Br Arachnol Soc 12:395–398

Bertani R, de Sena SA, Diesel Abegg A, Roncolatto Ortiz F, de Freitas MA (2016) First record of the genus *Psalmopoeus* (Araneae: Theraphosidae) in Brazil. CheckList 12:1–4

Caldas ATM, Dias MA, Peres M (2018) Invertebrate (Araneae: Mygalomorphae) Illegal trade: an ignored side of wildlife trafficking. Am J Zool 1:20–23

Cifuentes Y (2018) Revisão taxonômica e analise cladística dos gêneros de tarântulas arborícolas *Psalmopoeus* Pocock, 1985 e *Tapinauchenius* Ausserer, 1871 (Araneae: Theraphosidae: Aviculariinae). (Unpublished Másters thesis). Universidade Estadual Paulista, Instituto de Biociências, Letras e Ciências Exatas, São José do Rio Preto, São Paulo, Brazil

Cooke JA, Roth VD, Miller FH (1972) The urticating hairs of theraphosid spiders. Am Mus Nov 2498:1–48

Cuvier G (1800) Leçons d'anatomie comparée. Tome I, Baudouin, Paris, p 520

de Mello-Leitão CF (1923) Theraphosideas do Brasil. Revista do Museu Paulista 13:1–438

Dias SC, Brescovit AD (2003) Notes on the behavior of *Pachistopelma rufonigrum* Pocock (Araneae, Theraphosidae, Aviculariinae). Rev Bras Zool 20:13–17

Dias SC, Brescovit AD (2004) Microhabitat selection and co-occurrence of *Pachistopelma rufonigrum* Pocock (Araneae, Theraphosidae) and *Nothroctenus fuxico* sp. nov. (Araneae, Ctenidae) in tank bromeliads from Serra de Itabaiana, Sergipe, Brazil. Rev Bras Zool 21:789–796

Dias SC, Brescovit AD, Santos LT, Couto ECG (2000) Aranhas em bromélias de duas restingas do estado de Sergipe, Brasil. Biologia Geral e Experimental, São Cristóvão 1:22–24

Foelix R, Rast B, Erb B (2009) Palpal urticating hairs in the tarantula *Ephebopus*: fine structure and mechanism of release. J Arachnol 37:292–298

Foley S, Lüddecke T, Chen D-Q, Krehenwinkel H, Kuenzel S, Longhorn SJ, Wendt I, von Wirth V, Taenzler R, Vences M, Piel WH (2019) Tarantula phylogenomics: A robust phylogeny of multiple tarantula lineages inferred from transcriptome data sheds light on the prickly issue of urticating setae evolution. bioRxiv 501262. doi:https://doi.org/10.1101/501262

Fukushima CS, Bertani R (2017) Taxonomic revision and cladistic analysis of *Avicularia* Lamarck, 1818 (Araneae, Theraphosidae, Aviculariinae) with description of three new aviculariine genera. ZooKeys 659(Suppl 1–5):1–185

Gabriel G, Sherwood D (2018) Re-description of the male of *Psalmopoeus pulcher* Petrunkevitch, 1925 with the first description of the female (Araneae: Theraphosidae). Br Tarantula Soc J 33:8–18

Galindo-Leal C, Câmara IG (2003) Atlantic Forest hotspot status: an overview. In: Galindo-Leal C, Câmara IG (eds) The Atlantic Forest of South America: biodiversity status, threats and outlook. CABS and Island Press, Washington, pp 3–11

Hüsser M (2018) A first phylogenetic analysis reveals a new arboreal tarantula genus from South America with description of a new species and two new species of *Tapinauchenius* Ausserer, 1871 (Araneae, Mygalomorphae, Theraphosidae). ZooKeys 784:59–93

Lamarck JB (1818) Classe Séptiéme, Les Arachnides, Les Aranéides ou Arachnides Fileuses, 88–108. In: Histoire naturelle des animaux sans vertèbres. Tome 5, Paris, p 612

Latreille PA (1802) Histoire naturelle, générale et particulière des Crustacés et des Insectes. Paris 3:48–59

Lucas S, da Silva Junior PI, Bertani R (1991) The genus *Ephebopus* Simon, 1892. Description of the male of *Ephebopus murinus* (Walckenaer, 1837), (Araneae, Theraphosiidae, Aviculariinae). Spixiana 14:245–248

Lüddecke T, Krehenwinkel H, Canning G, Glaw F, Longhorn SJ, Tänzler R, Wendt I, Vences M (2018) Discovering the silk road: nuclear and mitochondrial sequence data resolve the phylogenetic relationships among theraphosid spider subfamilies. Mol Phylogenet Evol 119:63–70

Marshall SD, Uetz GW (1990) The pedipalpal brush of *Ephebopus* sp. (Araneae, Theraphosidae): evidence of a new site for urticating hairs. Bull Br Arachnol Soc 8:122–124

Marshall SD, West R (2008) An ontogenetic shift in habitat use by the Neotropical tarantula *Ephebopus murinus* (Araneae, Theraphosidae, Aviculariinae). Arachnology 14:280–284

Mendoza-M JI (2014) *Psalmopoeus victori*, the first arboreal theraphosid spider described for Mexico (Araneae: Theraphosidae: Aviculariinae). Rev Mex Biodivers 85:728–735

Merian MS (1705) Metamorphosis insectorum surinamensium, ofte verandering der urinaamsche insecten; Waar in de Surinaamsche rupsen en wormen met alle des zelfs veranderingen na het leven afgebeeld en beschreeven worden, zynde elk geplaast op die gewassen, bloemen en vruchten, daar sy op gevonden zyn; waar in ookde generatie der kikvorschen, wonderbaare padden, hagedissen, slangen, spinnen en mieren werden ver toond en beschreeven, alles in America na het leven en levensgroote geschildert en beschreeven. Selbstverl, Amsterdam, p 60

Olivier GA (1811) Articles: mygale, oxyope. Encyclopédie méthodique, histoire naturelle, insectes. Paris 8:83–87

Perafán C, Pérez-Miles F (2014) The Andean tarantulas *Euathlus* Ausserer, 1875, *Paraphysa* Simon, 1892 and *Phrixotrichus* Simon, 1889 (Araneae: Theraphosidae): phylogenetic analysis, genera redefinition and new species descriptions. J Nat Hist 48:2389–2418

Perafán C, Galvis W, Gutiérrez M, Pérez-Miles F (2016) *Kankuamo*, a new theraphosid genus from Colombia (Araneae, Mygalomorphae), with a new type of urticating setae and divergent male genitalia. ZooKeys 601:89–109

Pérez-Miles F, Perafán C, Santamaría L (2015) Tarantulas (Araneae: Theraphosidae) use different adhesive pads complementarily during climbing on smooth surfaces: experimental approach in eight arboreal and burrower species. Biol Open 4:643–1648

Pocock RI (1901) Some new and old genera of South American Avicularidae. Ann Mag Nat Hist 8:540–555

Raven RJ (1985) The spider infraorder Mygalomorphae (Araneae): cladistics and systematics. Bull Am Mus Nat Hist 182:1–180

Roewer CF (1942) Katalog der Araneae von 1758 bis 1940. Bremen 1:1040

Saager F (1994) *Psalmopoeus irminia* sp. n., Beschreibung einer neuen Aviculariinae (Theraphosidae, Aviculariinae, genus *Psalmopoeus*) inclusive einem Vergleich mit *Psalmopoeus cambridgei*. Arthropoda 2:59–71

Samm R, Schmidt G (2008) Sinurticantinae subfamilia nov.--eine neue Unterfamilie der Theraphosidae (Araneae). Tarantulas World 140:3–14

Santos RL, Almeida MG, Nunes JV (2002) Notes on the association of *Pachistopelma rufonigrum* Pocock, 1901 (Theraphosidae) with phytotelm bromeliads in eastern Rio Grande do Norte State, NE-Brazil. J Bromeliad Soc 52:122–124

Santos RL, Almeida MG, Tinoco LDS, Martins LB, Maia MG (2004) Biogeography and conservation of the bromeliad tarantula *Pachistopelma rufonigrum* (Araneae, Theraphosidae) in Rio Grande do Norte, Brazil. J Bromeliad Soc 54:153–157

Schmidt G (2010) Psalmopoeinae subfamilia nov. -- eine neue Unterfamilie der Theraphosidae (Araneae). Tarantulas World 142:35–41

Simon E (1892) Histoire naturelle des araignées. Paris Tome 1:1084

Simon E (1903) Histoire naturelle des araignées. Paris Tome 2:1080

Stradling DJ (1994) Distribution and behavioral ecology of an arboreal "tarantula" spider in Trinidad. Biotropica 26:84–97

Thorell T (1870) On European spiders, Part. I. Review of the European genera of spiders. Nova Acta Reg Soc Sci Upsal 7:109–242

Turner SP, Longhorn SJ, Hamilton CA, Gabriel R, Pérez-Miles F, Vogler AP (2018) Re-evaluating conservation priorities of New World tarantulas (Araneae: Theraphosidae) in a molecular framework indicates non monophyly of the genera, *Aphonopelma* and *Brachypelma*. Syst Biodivers 16:89–107

Valerio CE (1979) Arañas terafósidas de Costa Rica (Araneae: Theraphosidae). II. *Psalmopoeus reduncus*, redescripción, distribución y el problema de dispersión en terafósidas. Revista de Biología Tropical 27:301–308

Walckenaer CA (1802) Faune parisienne, Insectes, ou Histoire abrégée des insectes de environs de Paris, classes d'aprés le systême de Fabricius; précédée d'um discours sur les insects em general, pour servir d'introduction à l'étude de l'entomologie. Paris 2:187–250

West RC (1983) An introduction to the tarantulas of Trinidad and Tobago. Living World Magazine. J Trin Tob Field Nat Club 1983–1984:54–58

West RC, Marshall SD, Fukushima CS, Bertani R (2008) Review and cladistic analysis of the Neotropical tarantula genus *Ephebopus* Simon, 1892 (Araneae: Theraphosidae) with notes on the Aviculariinae. Zootaxa 1849:35–58

World Spider Catalog (2020) World Spider Catalog. Version 21.5. Natural History Museum Bern, online at http://wsc.nmbe.ch, accessed on 09/11/20. doi: 10.24436/2

Chapter 5
Theraphosinae

Fernando Pérez-Miles and Carlos Perafán

Abstract Theraphosinae is the most diverse subfamily of tarantulas in the world, with more than 500 known species. The group is endemic to the New World and its geographic distribution ranges from Southern North America to temperate zones of South America. The largest and most long-lived spiders in the world belong to this subfamily. They inhabit almost all terrestrial environments from sea level up to 4500 m altitude. Beyond the morphological diagnostic characters of the subfamily, this group is also characterized by a singular mechanism of defense, employing abdominal urticating setae that they can release to the air when they are disturbed. For over a century, this subfamily remained poorly known from both biological and taxonomical aspects. In fact, most genera were established during the last few decades and several papers on behavior, ecology, and reproduction were published. Due to their size, longevity, sex dimorphism, relatively easy conditions for breeding, and other biological characteristics, this group constitutes an interesting model for studies on metabolism, thermoregulation, biomechanics, communication, reproduction, and development. This chapter will present an updated overview of the taxonomy and phylogeny of the group as well as a revision of recent scientific contributions in different fields of general biology.

5.1 Introduction

Theraphosinae is the most speciose subfamily of the spiders known in America as tarantulas or bird-eating spiders, with 67 genera and more than 500 species currently described (World Spider Catalog 2019; Kambas 2019), being highly diverse

F. Pérez-Miles (✉)
Sección Entomología, Facultad de Ciencias, Universidad de La República, Montevideo, Uruguay
e-mail: myga@fcien.edu.uy

C. Perafán
Sección Entomología, Facultad de Ciencias, Universidad de La República, Montevideo, Uruguay

Laboratorio de Etología, Ecología y Evolución, IIBCE, Montevideo, Uruguay

F. Pérez-Miles (ed.), *New World Tarantulas*, Zoological Monographs 6,
https://doi.org/10.1007/978-3-030-48644-0_5

in the Neotropics. Theraphosinae is perhaps the Theraphosidae subfamily that least offers doubts about its monophyly, since Raven's formal proposal in 1985 and its subsequent internal phylogenetic analysis carried out by Pérez-Miles et al. in 1996. The monophyly is based on the combined presence of keels on the copulatory bulb, extended subtegulum, and abdominal urticating setae. The urticating setae are the most distinctive characteristic of the subfamily, and one of the most enigmatic of the tarantulas. These setae are an exclusive evolutionary novelty of New World tarantulas, shared only by Theraphosinae and some Aviculariinae (see Chap. 9).

Theraphosinae spiders are distributed in tropical regions of Central and South America, southern Mexico, and some Caribbean islands, even though some species inhabit temperate zones in the southwestern United States, northern Mexico, Chile, Argentina, and Uruguay (see Chap. 6). They inhabit a great diversity of environments, from tropical wet and montane forests, high Andean environments to sea coastlines, savannas, and deserts. Their retreats typically include cavities in the ground, crevices, and underneath rocks or fallen trunks. They mainly exhibit coloration between brown and black, although some species have patterns of other colors in the opisthosoma or prosoma and legs. The subfamily comprises species of a large variety of sizes, including the largest known spider, the Goliath tarantula *Theraphosa blondi* (Latreille 1804), which can measure up to 30 cm of leg-span (Fig. 5.1).

The taxonomic history of the subfamily has been almost as extensive as the taxonomic history of the spiders, their numerous nomenclatural changes have affected the validity of their scientific names, but also several problems have been detected with vernacular names. The first spider species recognized as tarantula was the wolf spider (Lycosidae) *Lycosa tarantula* (Linnaeus 1758) from Tarento, Italy (hence the specific epithet), whereas in America the first known spiders of Theraphosidae were colloquially called by the European explorers bird-eating spiders or tarantulas.

Fig. 5.1 *Theraphosa blondi*, female habitus (Photo: R. West)

5.2 Taxonomic History

The spider group currently considered as Theraphosidae was initially placed in the Walckenaer's classification (1837) in "Theraphoses," and in the Lucas' (1840) approach within "Tetrapneumones." In 1870, Thorell established the family Theraphosidae (named Theraphosoidae), within the suborder Territelariae which was characterized by the "peculiar direction of the mandibular claw" oriented parallel to the body axis (paraxial), four book lungs, and "generally with two pairs of spinnerets and two tarsal claws." Thorell (1870) drew attention to the following nomenclature aspects: the first one, the synonymy of *Theraphosa* (Walckenaer) with *Mygale* (Walckenaer), and the second one, the differentiation between *Theraphosa* and *Avicularia* (Lamarck).

Thorell (1870) stressed Walckenaer's (1802) nomenclatural inaccuracies which included using the name *Mygale* (a taxonomic name for shrews initially proposed by Cuvier in 1800) to name a separate group of spiders with four book lungs of the "spiders properly so called" (*Aranea* Walckenaer), but also proposing a different genus called *Theraphosa* (Walckenaer), which "according to the original description is absolutely synonymous with *Mygale*" (Thorell 1870). Furthermore, Thorell (1870) emphasized that both Olivier (1811) and de Lamarck (1818) adopted the *Mygale* genus proposed by Walckenaer but restricted it to "*les Araignées mineuses*," excluding the other *Mygale* species (or "*les Araignées aviculaires*"). Nevertheless, de Lamarck (1818) postulated a new genus for the latter group called *Avicularia*. Consequently, Thorell (1870) proposed that *Theraphosa* should include the species "*les Araignées aviculaires*," which in Koch and Simon were named *Mygale* (Thorell 1870), partly because *Theraphosa* is the oldest denomination of the two, and also because according to Thorell the name *Avicularia* should be restricted to a group of species within "*les Mygales aviculaires*," which included *Aranea avicularia* (Lamarck). Thus, Thorell (1870) divided "*les Mygales aviculaires*" in four genera: *Theraphosa* (Walckenaer), *Avicularia* (Lamarck), *Trechona* (C. Koch), and *Diplura* C. Koch.

Ausserer (1871) kept Thorell's (1870) proposal and built a classification mainly based on characteristics of the chelicerae, maxillae, palps, and elevation of the cephalic area for the high ranks, and division of the tarsal scopulae, shape of the fovea, ocular arrangement, and others for the lower ranks. Ausserer (1871) recognized three subfamilies Atypinae (=Atypidae), Eriodontinae (=Actinopodidae), and Theraphosinae (composed by all the other known mygalomorphs). Almost two decades later, Simon (1889) performed the first general classification that included a representative sample of mygalomorphs taxa, with material from Venezuela and the Colombian west. Subsequently, Simon (1892, 1903) added and redescribed several taxa, and published his results in a memorable paper called "*Histoire naturelle des araignées*." In this work, Simon (1892) preserved the classification criterion proposed by Ausserer (1871), but added the character extension of the metatarsal scopulae, and recognized two families within mygalomorphs, Atypidae and Aviculariidae. Aviculariidae included diverse subfamilies (Aviculariinae,

Barychelinae, Actinopodinae, Ctenizinae, Diplurinae, Miginae, and Paratropidinae) each one divided in groups (many of them now currently considered as subfamilies). Aviculariinae (Simon 1892) was divided into the following ten groups, by features related to the legs and fovea: Ischnocoleae, Chaetopelmateae, Crypsidromeae, Phlogieae, Selenocosmieae, Paecilotherieae, Avicularieae, Theraphoseae, Eurypelmateae, and Homoeommateae.

Pocock (1897, 1903) and F.O.P. Cambridge (1896) developed new proposals, reestablished new taxa and taxonomically upgraded some of the ranks proposed by Simon (1892, 1903) at the family and subfamily levels. Likewise, Pocock (1897, 1903) revised the stridulatory organs on the external faces of the chelicerae and maxillae, and the internal trochanters and coxae of the palp and first pair of legs. He also studied relationships between body measurements (patella + tibia/cephalothorax) and spination on the legs as taxonomic characters. Pocock (1897) argued that for practical purposes in the classification of theraphosid spiders it would be sufficient to refer to the classification proposed by Simon in 1892, which was an extension of Ausserer's (1871) proposal. Consequently, Pocock (1897) synonymized Aviculariinae (=Aviculariidae) (Simon 1892) with Theraphosidae (Thorell 1870), and discussed the validity of the scopulae condition as a taxonomic character, demonstrating ontogenetic differences within some species. Concurrently, he proposed the following subfamilies: Ornithoctoninae (Oriental Region), Harpactirinae (South and East Africa), Selenocosminae (Oriental Region and Australia), and Theraphosinae (Tropical and Subtropical America, Madagascar, Ethiopian Region, and the Mediterranean area of Paleartic); and strongly highlighted the imperious need to deeply examine South American species.

In 1928, Petrunkevitch made his first classification based on external morphology following the bases of Simon (1892) and Pocock (1897), with some innovations, and discussed the taxonomic validity of the stridulatory setae proposed by Pocock (1897, 1903). Petrunkevitch (1928) proposed Barychelidae and Theraphosidae as related taxonomically, and the following theraphosid subfamilies: Ornithoctoninae (including Simon's Ornithoctoneae and Thrigmopoeae), Selenocosmiinae (including Simon's Selenocosmiae, Poecilotheriae, and Harpactireae), Eumenophorinae (Simon's Phoneyuseae), Aviculariinae, Ischnocolinae, Theraphosinae (including Simon's Theraphoseae and Lasiodoreae), and Grammostolinae (Simon's Eurypelmateae). Later, he performed a review using internal anatomic features (particularly cardiac ostia) (Petrunkevitch 1933), but proposed Paratropididae related with Theraphosidae, and both related with Ctenizidae and Dipluridae; all of them in a group named Octostiatae (four pairs of cardiac ostia).

One of the first classification works that included an important representation of Neotropical mygalomorph families was carried out by Mello-Leitão in 1923, which divided six Brazilian families based on presence/absence of claw tufts and rastellum, and labium condition (merged or free). According to Mello-Leitão (1923), the family Aviculariidae (Pocock 1897) would correspond to the true "caranguejeiras" (=tarantulas), represented in tropical America only by some Theraphosiinae, which he divided in four groups based on spination of the legs, scopulae condition, and presence/absence of retrolateral scopulae on femur IV: Ischnocoleas (= Ischnocolae) (Simon, 1903), Grammostoleas (= Eurypelmateae) (Simon, 1903), Theraphoseas (= Lasiodoreae + Theraphoseae) (Simon, 1903), and Avicularias (= Aviculariae + Phoneyuseae (only *Ephebopus*) (Simon, 1903). Grammostoleas and Theraphoseas were related by the presence of spines and entire scopulae in all legs, and differentiated from each other by the presence (Theraphoseas) or absence (Grammostoleas) of scopulae in retrolateral face of femur IV; contrary, Ischnocoleas have posterior tarsi with divided scopulae and Avicularias lack spines (Mello-Leitão 1923). The Mello-Leitão (1923) classification related Actinopodidae + Paratropididae as a sister group of Ctenizidae + Dipluridae and Theraphosidae + Barychelidae (Raven 1985).

By the middle of the twentieth century the research developed by Bucherl (e.g., 1947a, b, 1949a, b, 1951, 1957) on Brazilian taxa, and later by Schiapelli and G. de Pikelin (e.g., 1961, 1962, 1964, 1970, 1979) and G. de Pikelin and Schiapelli (e.g., 1970, 1972, 1973) on South America taxa was transcendental for Theraphosinae taxonomy and systematics. These authors revised several genera and established the morphology of reproductive organs as taxonomic characters that allowed the best "natural" systematic delimitation. Bücherl (1957) stressed the importance of comparing and describing the palpal bulb and tibial spur of the males, and judged that the use of spination legs, eye interdistances, and its disposition were subject to individual variations. Meanwhile, the Argentinean researchers Schiapelli and G. de Pikelin made valuable contributions such as detailed description and elaboration of drawings of the taxonomic characters, including palpal bulbs, and highlighted the importance of female genitalia in the mygalomorph system (Raven 1985).

Cooke et al. (1972) examined for the first time the morphology of the urticating setae in detail, comparing the abdominal setae of several theraphosid species. They recognized four urticating setae types (I–IV), studied the effects of these setae on a variety of animals and demonstrated the value of urticating setae in the systematics of Theraphosidae. Later, Marshall and Uetz (1990), Pérez-Miles (1998), and Perafán et al. (2016) discovered and described three other urticating setae types (V–VII). Urticating setae have been widely recognized as an important systematic character (e.g., Raven 1985; Pérez-Miles et al. 1996; Bertani and Guadanucci 2013; Turner et al. 2018; Kaderka et al. 2019). See detailed information in Chap. 9.

Raven's masterpiece (1985) is considered as a starting point in the systematics of Mygalomorphae (Coddington and Levy 1991; Goloboff 1993; Pérez-Miles et al. 1996); admittedly, much of today's classification scheme is still based on this work. Raven (1985) realized the first approach to a detailed examination of the phylogeny of Mygalomorphae, as well as the first revision of genera and species of the infraorder based on the type material comparison. Raven (1985) built phylogenetic trees by hand based on careful examination of morphological characters followed by logic-based arguments of primary and secondary homology. Regarding the family Theraphosidae, Raven (1985) advocated a close phylogenetic relationship between Paratropididae and Theraphosidae, supported by the presence of an anterior lobe on the maxillae and numerous labial cuspules. He considered both to also be related to Barychelidae with the combined three families constituting the Theraphosoidina: based on the shape of the male tibial spur, presence of claw tuft, absence of maxillary serrula, and female with few teeth on the tarsal claws. Raven's Theraphosidae cladogram showed only subfamilial relationships with several unresolved polytomies: Theraphosinae was related in a tetrachotomy with Harpactirinae, Aviculariinae, and Iscnocolinae (in part). These groups share the presence of tarsal scopulae in all legs, absence of inferior tarsal claw (Raven 1985), and only one row of few small teeth on the superior tarsal claw or alternatively, with such teeth absent (Goloboff 1993). Thereby, Theraphosinae was considered in a close phylogenetic relationship with Harpactirinae by sharing unilobed spermathecae or with Aviculariinae due to the presence of urticating setae in the abdomen (Lucas et al. 1991; Pérez-Miles 1992).

Raven (1985) synonimized Grammostoleae Mello-Leitão, 1923, Crypsidromeae Simon, 1889, and Homoeommeae, Simon 1892 with Theraphosinae Thorell, 1870, arguing that the conformation of the first groups was based upon artificial characters and/or characters of only tribal significance. He also expanded the subfamily by incorporating several genera previously located in Ischnocolinae. Thus, based on Schiapelli and Gerschman (1979), Valerio (1980) and the morphological review, Raven (1985) diagnosed the Theraphosinae subfamily as follows: male palpal bulb with modified embolus distally stout and broad or keeled; large subtegulum, extending down the bulb for half of the tegulum; and, additionally, femur IV sometimes with scopulae in retrolateral face. The resulting Theraphosinae group consisted of 31 genera distributed from southern North America to South America.

The first cladistic analysis of Theraphosinae was carried out by Pérez-Miles et al. (1996). They developed a data matrix with all genera known at the time (30) and 27 morphological characters, and analyzed the subfamily under a Maximum Parsimony approach. They suggested the monophyly of Theraphosinae and elucidated much of the phylogenetic relationships of the genera described up to that point (Fig. 5.2), but the phylogenetic relationships between Theraphosinae and other theraphosid subfamilies was not resolved. Pérez-Miles et al. (1996) discarded the presence of retrolateral scopulae of the femur IV as a synapomorphy for Theraphosinae and highlighted for the first time the importance of urticating setae within the systematics of the group. They proposed the presence of urticating setae type III as a new synapomorphy for the subfamily, with reversion in some genera (*Aphonopelma*, *Citharacantus*, *Cyrtopholis*, *Spaerobothria*, *Vitalus*, and *Xenesthis*) (Pérez-Miles et al. 1996). Additionally, their results indicated that types I and IV urticating setae were likely synapomorphies for two clades within Theraphosinae: type I present in most genera with a reasonable degree of phylogenetic resolution and on the contrary, type IV including only a small group of genera related within a large polytomy (Fig. 5.2).

Likewise, Pérez-Miles et al. (1996) emphasized that the unilobed receptacles of the spermathecae cannot be considered a synapomorphy of the group, because this interpretation would require three parallels for its explanation in some other subfamilies (Harpactirinae and some genera of Aviculariinae and Ischnocolinae). Thus,

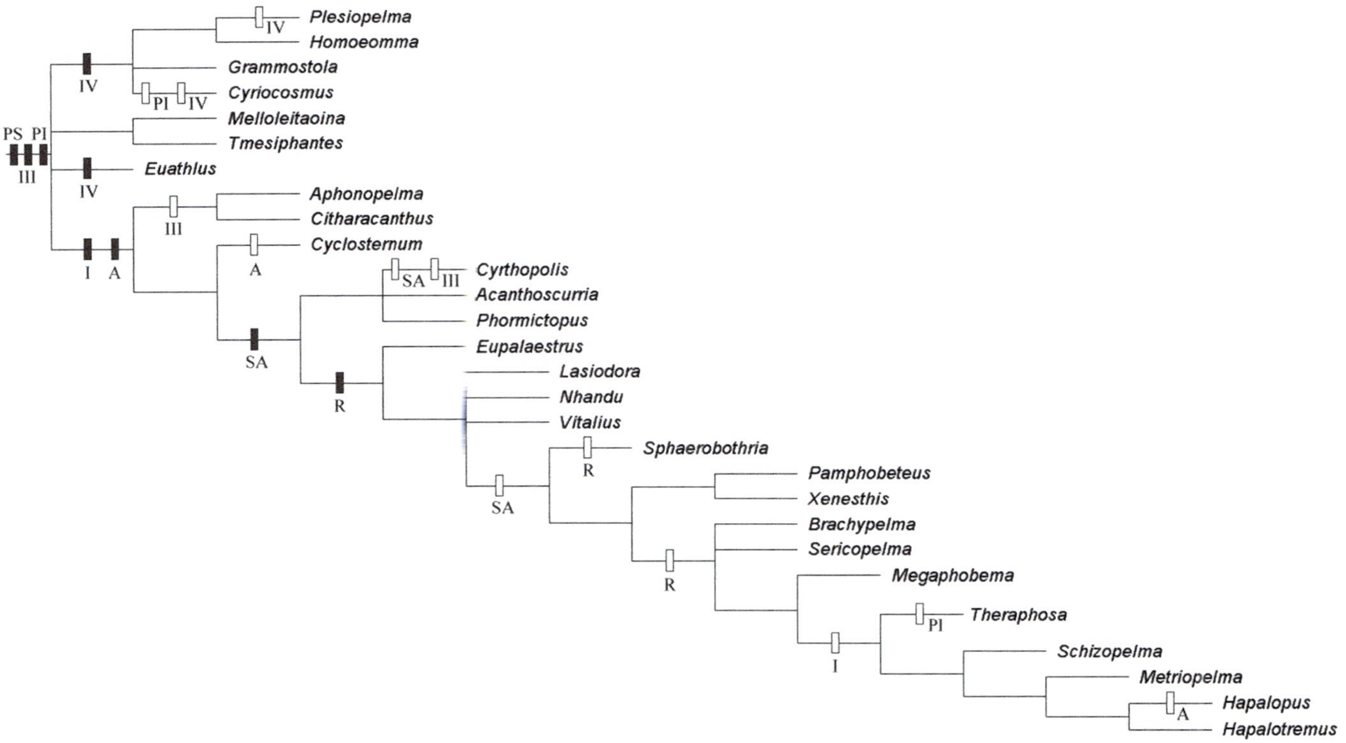

Fig. 5.2 Cladogram of Theraphosinae, modified from Pérez-Miles et al. (1996), keel indications modified from Bertani (2000) and urticating setae modified from Bertani and Guadanucci (2013). In black acquisitions, in white reversions. I = type I urticating setae, III = type III urticating setae, IV = type IV urticating setae, A = apical keel, PI = prolateral inferior keel, PS = prolateral superior keel, R = retrolateral keel, SA = sub-apical keel

the arboreal South American Aviculariinae (excluding Psalmopoeinae, formerly in Aviculariinae) was proposed as the preferred sister group to the Theraphosinae, since both subfamilies share the presence of abdominal urticating setae (Type II in Aviculariinae) and a defensive display with abdominal movements as synapomorphies (Pérez-Miles et al. 1996). Pérez-Miles et al. (1996) removed *Acanthopelma* and *Holothele* from Theraphosinae and transferred them to Ischnocolinae, because they lack urticating setae and keels on palpal bulb. Subsequently, Pérez-Miles (2000) increased taxon sampling in his morphological analysis with the addition of a newly described genus (recently renamed as *Bumba*), but his results remained similar as in previous data.

Bertani (2000) carried out an important contribution to the study of Theraphosinae cladistics and the description of new species. He performed the first approach to homologize male palpal bulb keels of Theraphosinae and proposed a new terminology to name the homologous keels. Thereby, his work increased the available number of characters for taxonomic studies and incorporated elements of the copulatory bulb in the phylogenetic analyzes. Furthermore, he studied the morphology and distribution of palpal bulb keels in over 60 species in 27 genera, resulting in four basic groups of keels (Fig. 5.3): prolateral inferior (PI) and prolateral superior (PS) keels, for the two more or less parallel keels found on the prolateral bulb face; apical (A) keel, for the ventral keel located just before the apex of the embolus; subapical (SA) keel, for a keel located just before the apical keel; and, retrolateral (R) keel, for the keel located in the retrolateral region, originating in the apical region and extending backward. Also, Bertani (2000) discussed that other palpal bulb keels, as well as other structures, were found in some genera and/or species constituting apomorphies for these groups. Bertani (2000) mapped the bulb characters in the cladogram proposed by Pérez-Miles et al. (1996) to explore the evolution of the male palpal keels in Theraphosinae, and as a result he found many concordances with the keel's evolution with few exceptions (Fig. 5.2). Later, Bertani (2001) focused on specific South American lineages (*Vitalius*, *Nhandu*, and *Proshapalopus*) and performed a phylogenetic analysis where he incorporated the recently proposed homology of the keels. In the last few decades, the description of genera and species with new keels is demanding a new revision of Bertani's proposal.

The phylogenies up to this point showed an acceptable resolution for most lineages except for the genera with type IV urticating setae (Pérez-Miles et al. 1996; Pérez-Miles 2000; Bertani 2001; Fukushima et al. 2008), of which the majority of workers previously defined in the Grammostolinae (Mello-Leitão 1923). Perafán (2010) in his master's thesis tested the monophyly of that group, where he constructed a data matrix with 40 species as terminals, belonging to 18 genera, and 78 morphological characters, gathering the characters proposed in previous works. Under Maximum Parsimony analysis, Perafán (2010) determined that genera with type IV urticating setae constitute a monophyletic group, additionally supported by the presence of metatarsal II scopulae occupying the distal half of the segment. In addition, he argued that this group is more closely related to the genera that possess both I + III urticating setae, than to those which possess only type I (Fig. 5.4).

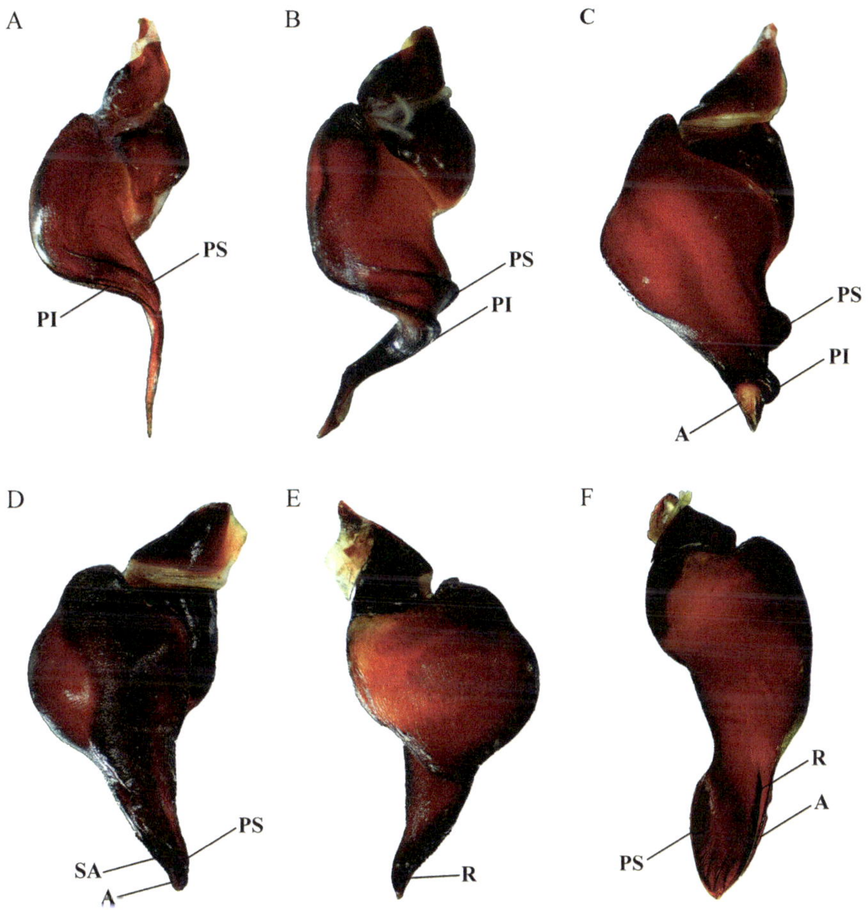

Fig. 5.3 Palpal bulb morphology and keels (**a**) *Plesiopelma longisternale*, (**b**) *Tmesiphantes nubilus*, (**c**) *Acanthoscurria cordubensis* (Ex *A. suina*) (**d**) *Eupalaestrus weijenberghi*. (**e**) Eupalaestrus weijenberghi, (**f**) *Pamphobeteus verdolaga*. (**a–d**) prolateral views, (**e–f**) retrolateral view. *A* apical keel, *PI* prolateral inferior keel, *PS* prolateral superior keel, *R* retrolateral keel, *SA* sub-apical keel

Turner et al. (2018) highlighted that Theraphosidae's phylogenetic approaches are largely reliant of the attributes of a set of morphological characters many of which may be plesiomorphic and/or homoplastic, a problem commonly seen in all mygalomorph families. Also, these authors discussed that the majority studies with molecular data focused in the evolutionary history of Theraphosidae have little understanding beyond the generic limits (e.g., Hamilton et al. 2011, 2014, 2016; Hendrixson et al. 2013, 2015; Longhorn et al. 2007; Montes de Oca et al. 2016; Ortiz and Francke 2015, 2016; Petersen et al. 2007; Wilson et al. 2013), which have been historically inferred solely from morphological data (e.g., Pérez-Miles et al. 1996; Pérez-Miles 2000; Bertani 2001; Bertani 2012; Guadanucci 2014; Gabriel 2016). Turner et al. (2018) explored a mtDNA gene tree based on the

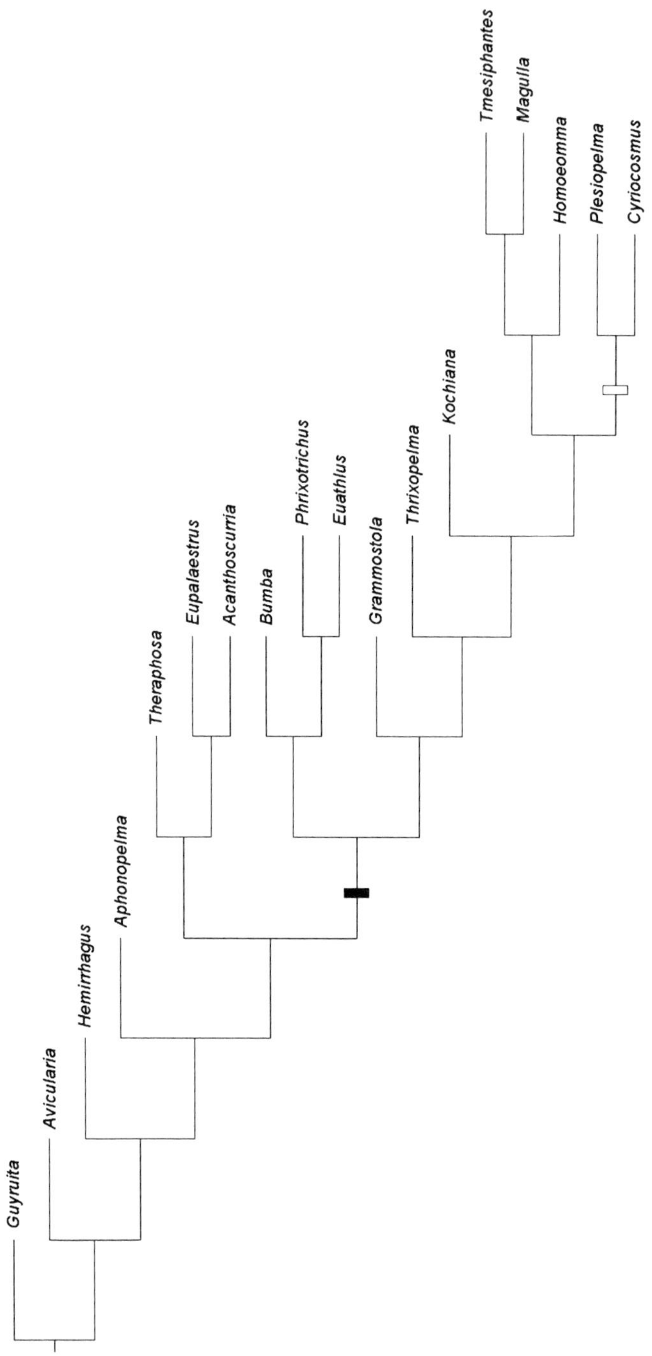

Fig. 5.4 Cladogram of "Grammostolinae" from Perafán (2010) (unpublished M.Sc. thesis). In black type IV urticating setae acquisition, in white reversions

mitochondrial 16S-tRNA (leu)-ND1 gene region, frequently used as a taxonomic "barcode" in spiders, to redefine the taxonomic limits and evolutionary relationships between some genera and species within the Theraphosinae subfamily.

Turner et al. (2018) confirmed the monophyly of Theraphosinae but uncertainty remains regarding relationships between subfamilies due to limited sampling of genes and phylogenetically informative characters, limited taxon sampling from Old World theraphosids, and few New World "Ischnocolinae." The hypotheses of Turner et al. (2018) supported in a single node the American theraphosids, related in two internal clades: one of them composed by Theraphosinae and the other composed by Aviculariinae as a sister group of the New World "Ischnocolinae," Schismatothelinae and Psalmopoeinae. However, according to Turner et al. (2018) the internal nodes relating to the subfamilies in the second node were weakly supported, and Aviculariinae (excluding Psalmopoeinae) may remain as the preferred sister-group of Theraphosinae. On the other hand, Turner et al. (2018) emphasized the presence of deep splits within the Theraphosinae, dividing the subfamily into three tribes within two supergroups (see Chap. 1, Fig. 1.11b). Furthermore, they stressed that the types of urticating setae seem especially congruent with the inferred groupings. One of the lineages named Grammostolini (with *Grammostola* as the type genus) includes several genera that ancestrally possess both type III and type IV urticating setae. The second major lineage was defined as the tribe Hapalopini, considered to contain multiple genera with only type III urticating setae ancestrally (with *Hapalopus* as the type genus). Finally, the major lineage defined as the tribe Theraphosini (with *Theraphosa* as the type genus, and as the sister genus to all others in this tribe), consists of the remaining genera that hypothetically possess urticating setae type I ancestrally.

Lüddecke et al. (2018) provided a preliminary molecular phylogenetic hypothesis of relationships among theraphosid subfamilies, based on 3.5 kbp of three nuclear and three mitochondrial markers. Like in previous morphological analysis, this study strongly validates the prevalent subfamily level classification within Theraphosidae. Besides the problems aroused from phylogenetic inferences obtained with morphological data, Foley et al. (2019) argued that traditional Sanger sequencing of preselected genes does not offer strong support for supra-generic clades. In this sense, they addressed the first robust phylogenetic hypothesis of theraphosid evolution inferred from transcriptome data, where they recovered a monophyletic group that comprises the vast majority of New World theraphosid subfamilies. As a result, the clade has been divided into the subfamily Theraphosinae and three other subfamilies Aviculariinae, Psalmopoeinae, and Schismatothelinae that together form another clade (Foley et al. 2019).

5.3 Taxonomy

5.3.1 Theraphosinae Thorell, 1870

Genera included: *Acanthoscurria, Acentropelma, Aenigmarachne, Agnostopelma, Aguapanela, Aphonopelma, Bistriopelma, Bonnetina, Brachypelma, Bumba, Proshapalopus, Cardiopelma, Catanduba, Chromatopelma, Citharacanthus, Clavopelma, Cotztetlana, Crassicrus, Cubanana, Cyclosternum, Cyriocosmus, Cyrtopholis, Davus, Euathlus, Eupalaestrus, Eurypelmella, Grammostola, Hapalopus, Hapalotremus, Hemirrhagus, Homoeomma, Kankuamo, Kochiana, Lasiodora, Lasiodorides, Magnacarina, Longilyra, Magulla, Megaphobema, Melloleitaoina, Metriopelma, Miaschistopus, Munduruku, Mygalarachne, Neischnocolus, Neostenotarsus, Nesipelma, Nhandu, Ozopactus, Pamphobeteus, Phormictopus, Phrixotrichus, Plesiopelma, Pseudhapalopus, Pterinopelma, Reversopelma, Sandinista, Schizopelma, Sericopelma, Sphaerobothria, Stichoplastoris, Theraphosa, Thrixopelma, Tliltocatl, Tmesiphantes, Umbyquyra, Vitalius, Xenesthis***.**

Diagnosis: Differs from other Theraphosidae in the presence of urticating setae types I, III, IV, V, VI, or VII. Male palpal bulb with subtegulum extended and keels present along the embolus.

Distribution: Theraphosinae are mainly found in tropical regions of Central and South America, southern Mexico, and some Caribbean islands, but many species inhabit temperate climates in the southwestern United States, northern Mexico, Chile, Argentina, and Uruguay (see Chap. 6).

5.4 Taxonomic Key

We present a brief key and diagnosis of generic groups (tribes) inspired by the proposal of Turner et al. (2018), summarize a few characteristics of the genera and give an orientation to selected literature. The presently available information on several of these genera is not enough to recognize them easily, and some of them are monotypic; they are indicated here by an asterisk. The references given after the characteristics of the genera are the main source of information used here. The number of species and geographic distribution are taken from World Spider Catalog (2019).

Urticating setae types III and/or IV present; type I absent .... Group I
Urticating setae type III present, type I absent (Note: in a few genera the additional presence of type IV is questionable, but if present type III is nonetheless predominant) .... Group II
Urticating setae III and/or I present, type IV absent .... Group III
Urticating setae otherwise, types I, III, IV absent .... Group IV

Group I (Fig. 5.5)
Without prominent abdominal pattern in adults (patterning present faintly at rear in some *Euathlus*). Male tibial apophysis with two branches (lost in *Aguapanela*). No nodule on male metatarsi I. Femur IV without retrolateral scopulae. Presence of types III and IV urticating setae. Absence of coxal stridulatory setae (except *Aguapanela* and *Grammostola*). Absence of trochanter and femur stridulatory setae (except *Aguapanela*). Male palpal bulb piriform, usually with cylindrical filiform embolus. Presence of superior and inferior keels (PS and PI). Two spermathecal receptacles with separated bases.

Comprises: *Agnostopelma*, *Aguapanela*, *Bistriopelma*, *Bumba*, *Euathlus*, *Grammostola*, *Homoeomma*, *Magulla*, *Melloleitaoina*, *Phrixotrichus*, and *Tmesiphantes*.

Agnostopelma Pérez-Miles and Weinmann 2010. Differs from all other Theraphosinae in the absence of scopulae on leg tarsi IV. Two species from Colombia, Boyacá (Pérez-Miles and Weinmann 2010).

Aguapanela Perafán et al. 2015. Characterized by the presence of stridulatory plumose setae on coxae, trochantera, and femora of legs I and II. Males lack tibial

Fig. 5.5 Representatives of Theraphosinae Group I. (**a**) *Euathlus* sp. (**b**) *Bistriopelma matuskai*. (**c**) *Grammostola quirogai*. (**d**) *Aguapanela arvi*. (Photos **a**, **b**: R. Kaderka)

apophysis and females have two digitiform spermathecal receptacles. One species from Colombia, Medellin (Perafán et al. 2015).

Bistriopelma Kaderka 2015. Differs from all other Theraphosinae (except *Phrixotrichus* and *Tmesiphantes hypogeus*) by the presence of two patches of type III urticating setae. From *Phrixotrichus* and *Tmesiphantes* in the morphology of female of spemathecae and by the position of urticating patches, respectively. Three species from Peru (Kaderka 2015, 2017).

Bumba Pérez-Miles et al. 2014. Characterized by the presence of type IV urticating setae, a retrolateral process on male palpal tibiae, palpal bulb resting in a ventral distal excavation of the palpal tibia and spiniform setae on prolateral and retrolateral maxillae and coxae I–IV. Four species from Brazil, Ecuador, and Venezuela (Pérez-Miles 2000; Pérez-Miles et al. 2014; Perafán and Pérez-Miles 2014b).

Euathlus Ausserer 1875. Males have a palpal bulb with two prolateral keels and tip directed retrolaterally, the tibial apophyses with retrolateral spines, a subapical spine on retrolateral branch and a basal spine on prolateral branch. Females have two spermathecal receptacles with a lateral spheroid chamber. Differs from *Phrixotrichus* in the presence of only one central urticating setae patch. Ten species from Argentina and Chile (Perafán and Pérez-Miles 2014a, b; Ferretti 2015).

Grammostola Simon 1892. Stridulatory setae present on coxae of palp and leg I. Male palpal bulb with filiform embolus and prolateral superior and inferior keels subequal; tibial apophysis with two separated branches. Females with two spermathecal receptacles subconical with a spheroid chamber on apex (a subdistal constriction). Twenty-two species from Argentina, Brazil, Bolivia, Chile, Paraguay, and Uruguay (Schiapelli and Gerschman 1961; Ferretti et al. 2011, 2013; Montes de Oca et al. 2016).

Homoeomma Ausserer 1871. Both sexes small in size, males are characterized by palpal bulb with digitiform tegular apophysis and embolous twisted. In some species males present a baso-ventral nodule on metatarsus I. Females have two separated spermathecal sacular receptacles of variable morphology. Some species have only type IV urticating setae and lack type III. Fifteen species from Argentina, Brazil, Chile, Colombia, Peru, and Uruguay (Gerschman and Schiapelli 1972, 1973; Pérez-Miles et al. 1996; Montenegro et al. 2018; Sherwood et al. 2018).

Magulla Simon 1892. Both sexes small in size with comparatively short metatarsi. Male palpal bulb with embolous twisted; similar to *Homeomma* but differing by the lack of a digitiform tegular apophysis and differentiated from *Tmesiphantes* by the apex of the bulb not dilated. Female spermathecae formed by two receptacles with long neck, slightly dilated apex, which presents a ventral loop in lateral view. Four species from Brazil (Indicatti et al. 2008).

Melloleitaoina Gerschman and Schiapelli 1960. Extremely reduced number of labial cuspules (6–15) in both sexes. Males with thickened femur III and palpal bulb with long curved embolous (similar to *Tmesiphantes*). Females with two separated spermathecae with a slight subdistal constriction and spiniform setae on promargin of coxae III and IV. Four species from Argentina (Perafán and Pérez-Miles 2014a).

Phrixotrichus Simon 1889. Urticating setae type IV gathered in two dorso-lateral patches in both sexes. Male palpal bulb with two subequal prolateral superior and inferior keels, embolus tip directed retrolaterally. Female with two spermathecal receptacles with a lateral spheroid chamber. Five species from Argentina and Chile (Perafán and Pérez-Miles 2014b).

Tmesiphantes Simon 1892. Both sexes small in size with an absence of scopulae on metatarsi IV. Male with palpal bulb with conspicuous superior and inferior prolateral keels, and a long and curved embolus. Metatarsus I bends to the retrolateral side of the retrolateral tibial apophysis. Females with spermathecae slightly constricted near the apex. Nine species from Brazil (Yamamoto et al. 2007; Bertani et al. 2013; Fabiano-da-Silva et al. 2015).

Group II (Fig. 5.6)

Usually with an abdominal pattern in adults of both sexes (lateral banding and/or spots), reduced in some genera as *Thrixopelma*, *Plesiopelma*, *Schizopelma*, *Hapalopus*, and others. Some genera have species with abdominal patterning and others without it. Male tibial apophysis with two branches often fused at the bases; modified in *Schizopelma* (a megaspine) and in *Bonnetina* (accessory apophysis). Metatarsi I of adult males can have retrolateral nodule with spines (*Bonnetina*, *Magnacarina*) or basal process (*Plesiopelma*, *Catanduba*). Femur IV without retrolateral scopulae. Type III urticating setae present and predominant, type I absent. Stridulating setae absent. Male palpal bulb often short and strong (except *Plesiopelma* and *Catanduba*); can have very modified keels. Spermathecae usually single, fused, and sclerotized or, if separated, have sinuous or spiral morphology.

Fig. 5.6 Representatives of Theraphosinae Group II. (**a**) *Cyriocosmus perezmilesi*. (**b**) *Chromatopelma cyaneopubescens*. (**c**) *Hapalotremus* sp. (**d**) *Hapalopus formosus*. (Photos **a**, **b**, **c**: R. West; **d**: R. Kaderka)

Comprises: *Aenigmarachne*, *Bonnetina*, *Cardiopelma*, *Catanduba*, *Chromatopelma*, *Cyriocosmus*, *Davus*, *Hapalopus*, *Hapalotremus*, *Kochiana*, *Magnacarina*, *Munduruku*, *Plesiopelma*, *Schizopelma*, and *Thrixopelma*.

Aenigmarachne Schmidt 2005. Characterized by the presence of type IV urticating setae distally bent; males lack tibial apophysis. One species from Costa Rica, female unknown (Schmidt 2005).*

Bonnetina Vol 2000. This genus differs from all other theraphosine because males have three tibial apohysis on leg I. Females with widely fused spermathecal receptacles, or if narrow, with wide bases. Seventeen species, endemic to Mexico (Ortiz and Francke 2017).

Cardiopelma Vol 1999. The genus was characterized by the spermathecae morphology, heart shaped with a wide basis. Male unknown. Locality unknown (Vol 1999). The validity of this genus offers serious doubts due to the absence of locality data and because is poorly diagnosed. Heart shaped spermathecae is also present in some *Hapalopus* species. One species without locality data (Vol 1999; Perdomo et al. 2009).*

Catanduba Yamamoto, Lucas and Brescovit 2012. Males of *Catanduba* are characterized by the presence of a prolateral inferior keel with a tooth in the middle of the embolus and a triangular basal nodule on metatarsus I (as in *Homoeomma*). Female spermathecae with a large number of spherical nodules. Seven species from Brazil (Yamamoto et al. 2012).

Chromatopelma Schmidt 1995. Similar to *Aphonopelma* but females differ by the abdominal pattern, presence of a single fused spermathecae, presence of type IV urticating setae, tarsal scopulae of leg III divided and larger posterior median eyes. Male palpal bulb with a triangular prolateral keel. Tibial apophysis with two separated and unequal branches. One species from Venezuela (Schmidt 1995; Schmidt and Herzig 1997).*

Cyriocosmus Simon 1903. Males are characterized by the presence of a paraembolic apophysis on palpal bulb, as in *Hapalopus*, but can be distinguished from *Hapalopus* by the divergent branches of tibial apophysis. Females have a sinuous or spiral spermathecae neck with a convex base and a caliciform fundus. Twenty-one species from Argentina, Bolivia, Brazil, Colombia, Paraguay, Trinidad and Tobago, Venezuela (Pérez-Miles 1998; Fukushima et al. 2005; Kaderka 2010, 2016).

Davus O. Pickard Cambridge 1892. Differ from most Theraphosinae in the laterally striped abdominal pattern. Males with palpal bulb with apical keel and embolus retrolaterally twisted. Palpal tibia with a retrolateral domed process, tibial apophysis with two convergent branches with bases fused. Female with single spermathecae with variable morphology; subquadrate, subtriangular, subspheric. Four species from Costa Rica, Guatemala, Nicaragua, Mexico, and Panama (Gabriel 2016).

Hapalopus Ausserer 1875. Most species have a dotted or striped abdominal pattern. Males with palpal bulb with prolateral inferior keel ring shaped, and retrolateral face of palpal tibia with a process or field of spines. Metatarsus I strongly curved. Females with single spermathecae usually very sclerotized or if two receptacles very fused at the bases. Nine species from Brazil, Bolivia, Colombia, Guyana,

Panamá, and Venezuela (Fukushima et al. 2005; Perdomo et al. 2009; Gabriel 2011; Fonseca-Ferreira et al. 2017).

Hapalotremus Simon 1903. Male palpal bulb with embolus long, strongly curved, but not twisted, prolateral inferior keel curved, long, and not divided. Females with single spermathecae subrectangular, without membranose base. Nine species from Argentina, Bolivia, and Peru (Cavallo and Ferretti 2015; Ferretti et al. 2018).

Kochiana Koch 1842. Males with palpal bulb with long and narrow embolus twisted, pointing downward, prolateral accessory keel present. Females with horn shaped spermathecae with large granules. One species from Brazil (Fukushima et al. 2008)*.

Magnacarina Mendoza, Locht, Kaderka, Medina and Pérez-Miles 2016. Males with tri-branched tibial apophysis on leg I, bifid palpal bulb, and retrolateral nodule on metatarsus I with a field of megaspines. Females with low semicircular single spermathecae. Four species, endemic from Mexico (Mendoza et al. 2016).

Munduruku Miglio, Bonaldo and Pérez-Miles 2013. Abdomen laterally patterned with zig-zag light marks in both sexes. Male palpal bulb with prolateral superior, inferior, and subapical keels. Female spermathecae with two spheroid distal receptacles, with a straight funnel-shaped neck bearing a sclerotized area. One species from Brazil (Miglio et al. 2013)*.

Plesiopelma Pocock 1901. Males of most species have a small subapical tooth on the palpal bulb and a retrolateral basal nodule on metatarsus I. Females are characterized by the spiral-shaped spermathecae. Eleven species from Argentina, Brazil, Paraguay, Uruguay, and Venezuela (Ferretti and Barneche 2013).

Schizopelma F.O. Pickard Cambridge 1897. Males are characterized by the palpal bulb with bifid embolus and extended prolateral superior and subapical keels. Females with single semicircular spermathecae. Two species from Mexico and Central America (Gabriel 2016; Ortiz et al. 2018).

Thrixopelma Schmidt 1994. Teeth present on tarsal claws (usually seven large teeth). Males with palpal bulb subconical. Females with two sacular spermathecae. Some species have type IV urticating setae with few barbs. Five species from Chile and Peru (Schmidt 1994a, b; Peters 2005).

Group III (Fig. 5.7)

Without abdominal pattern in adults (except some *Neischnocolus* and *Neostenotarsus*). Male tibial apophysis variable: absent, with one branch or with two branches. Male metatarsi without retrolateral nodule, nor basal process. Femur IV usually with retrolateral scopulae. Type I urticating setae present (except *Theraphosa*); type III present (with some secondary losses). Type IV urticating setae absent. Some genera have coxal and trochanteral stridulating setae. Male palpal bulb highly variable, usually broad and/or spatulate with numerous keels. Spermathecae also variable, fused or twin receptacles with based fused or not.

Comprises: *Acanthoscurria*, *Acentropelma*, *Aphonopelma*, *Brachypelma*, *Citharacanthus*, *Clavopelma*, *Cotztetlana*, *Crassicrus*, *Cubanana*, *Cyclosternum*, *Cyrtopholis*, *Eupalaestrus*, *Eurypelmella*, *Lasiodora*, *Lasiodorides*, *Longilyra*,

Fig. 5.7 Representatives of Theraphosinae Group III. (**a**) *Aphonopelma mooreae*. (**b**) *Phormictopus* sp. (**c**) *Pamphobeteus* sp. (**d**) *Brachypelma albiceps*. (Photos **a**, **b**, **d**: R. West)

Megaphobema, *Mygalarachne*, *Metriopelma*, *Miaschistopus*, *Neischnocolus*, *Neostenotarsus*, *Nesipelma*, *Nhandu*, *Pamphobeteus*, *Phormictopus*, *Proshapalopus*, *Pseudhapalopus*, *Pterinopelma*, *Reversopelma*, *Sericopelma*, *Sphaerobothria*, *Stichoplastoris*, *Theraphosa*, *Tliltocatl*, *Vitalius*, and *Xenesthis*.

Acanthoscurria Ausserer, 1871. The genus is characterized by the presence of few stridulatory setae on retrolateral face of the palpal trochanter and prolateral face of trochanter I (except *A. sacsayhuaman*). Some species can reach a large size. Males have palpal bulb with wide embolus and in some species very developed keels (PI) and a nodule on retrolateral face of palpal tibia. Tibial apophysis constituted by one branch. Females have two sub-spherical or secular spermathecae. Twenty-seven species from Central and South-America (Schiapelli and G. de Pikelin 1964; Lucas et al. 2010; Rodríguez-Manzanilla and Bertani 2010; Gonzalez-Filho et al. 2012; Paula et al. 2014; Ferretti et al. 2016).

Acentropelma Pocock 1901. Stridulatory setae present on retrolateral face of palpal trochanter and in prolateral face of trochanter I and femur I. Males with palpal bulb not twisted with two prolateral keels. Females with two spermathecae with short neck and sub-spherical fundus. Four species from Belize, Guatemala, and Mexico (Gabriel 2016).

Aphonopelma Pocock 1901. Only type I urticating setae present in both sexes. Hair-like, spiniform, or plumose setae on retrolateral face of palpal trochanter and prolateral face of trochanter I. Absence of stridulatory setae. Setae on the prolateral surface of coxa I hair-like and not basally swollen, spiniform and basally swollen, or distinctly stout and thorn-like. Scopulae divided on tarsi IV. Males with lower process of tibial apophysis curving prolaterodistally and widening apically, with apical or subapical megaspine; upper branch stout and shorter, with an inner megaspine. Fifty-nine species from Mexico, United States, and Central America (Hamilton et al. 2016; Hendrixson 2019; Sherwood 2019). Turner et al. (2018) proposed that *Aphonopelma* is polyphyletic and includes four different lineages.

Brachypelma Simon 1891. Differs from all other known Theraphosinae genera (except the genus *Tliltocatl*) having just claviform stridulating setae on the prolateral face of leg I trochanter/femur and on the retrolateral face trochanter of the palp. Male palpal bulb distally wide and flattened (spoon-shaped) and presents prolateral superior and apical keels and can have a small or reduced prolateral inferior keel. Females can have a divided spermatheca with each lobe as large as wide or a simple undivided/fused spermatheca; spermatheca baseplate divided and sclerotized. Absence of plumose pad of setae on leg IV femur in both sexes. *Brachypelma* differs from *Tliltocatl* by the red/orange coloration pattern on legs and/or carapace (except *B. albiceps*) and in the morphology of genitalia in both sexes. Eight species endemics from Mexico (Longhorn 2001; Mendoza and Francke 2017, 2019; Turner et al. 2018).

Citharacanthus Pocock 1901. Urticating setae I and III (see Chap. 9), type I simple or with longer area b (similar to *Neischnocolus*) in both sexes. Spiniform or clavated setae on palpal and leg coxae. Male palpal bulb piriform with apically serrated prolateral inferior keel, tibial apophysis with two branches. Female with two subspherical spermathecae. Nine species from Mexico and Central America (Reichling and West 2000; Estrada-Alvarez et al. 2013).

Clavopelma Chamberlin 1940. Spatulate lanciform setae on retrolateral face of palpal trochanter and prolateral face of trochanter I. Male palpal bulb with straight narrow embolus. Females have twin spermathecae with wide basis, narrow neck, and subspherical fundus. One species from Mexico (Prentice 1997, Schmidt 1998).*

Cotztetlana Mendoza 2012. The genus is characterized by the incrassate barrel-shaped tibia IV. Type I urticating setae present. Absence of scopulae on metatarsus IV. Metatarsus and tarsus IV wide and strong. Females with two spermathecae with lobular receptacles a constriction at the neck and separated basis. Two species endemic from Mexico (Mendoza 2012; Estrada-Alvarez 2014).

Crassicrus Reichling and West 1996. Only type I urticating setae. Presence of cuneiform thorn-like setae on prolateral face of leg coxae, thicker near the ventral region. Male palpal bulb with proventral face of subapical region of embolus convex. Several keels present: prolateral superior, prolateral inferior, subapical (serrated), apical, retrolateral superior, retrolateral median, and retrolateral inferior. Females have spiniform setae on the ventral and proventral surfaces of leg femora II–IV and two spermathecae partially fused by a heavily sclerotized median region. Tibia IV can be slightly or very incrassate. Five species from Belize and Mexico (Candia-Ramírez and Francke 2017).

Cubanana Ortíz 2008. Retrolateral face of femora IV covered by a pad of ciliate hairs. Absence of stridulatory setae. Male palpal bulb with smooth apical and prolateral inferior keels and serrated subapical keel. Male palpal tibiae with retrolateral nodule. Tibial apophysis with two branches on a common base. Females with twin spermathecae with fundus subspherical, neck marked, and wide separated basis. One species endemic from Cuba (Ortiz 2008)*.

Cyclosternum Ausserer 1871. Tarsal scopulae divided in both sexes. Male palpal bulb not twisted with prolateral superior and inferior keels slightly developed. Female with two spermathecae with a short neck and sub-spherical fundus. Eleven species from Bolivia, Brazil, Colombia, Costa Rica, Ecuador, Mexico, Paraguay, and Venezuela (Pérez-Miles et al. 1996; Gabriel 2016). This genus is probably not monophyletic.*

Cyrtopholis Simon, 1892. Stridulatory setae on retrolateral trochanter of palp and prolateral trochanter of leg I in combination with the presence of retrolateral nodule on male palpal tibia. Female with two spermathecae with base wider than sub-spherical fundus, median-sized neck narrowing apically. Twenty-four species from Antilles and South America (Pérez-Miles et al. 1996).

Eupalaestrus Pocock, 1901. Tibia IV thickened. Male palpal bulb with retrolateral serrated keel and an embolus slightly flattened. Females with two spermathecae, receptacles low, with base and fundus of similar size, fundus sub-spheric. Five species from Argentina, Brazil, Guyana, Paraguay, and Uruguay (Pérez-Miles 1992; Pérez-Miles et al. 1996, Bertani 2001; Ferretti and Barneche 2012).

Eurypelmella Strand 1907. Male palpal bulb piriform with a strongly curved at the subapical region, two tibial apophysis. Female unknown. One species from Guatemala (Gabriel 2016). This genus was restored by Gabriel (2016) based on the differences with *Schizopelma* found in the original description of Strand (1907) but the types were destroyed in the Second World War.*

Lasiodora Koch 1850. Stridulatory setae on prolateral face of coxae I and II. Male palpal bulb with a subapical triangular keel. Female with two sub-spheric spermathecae separated by a weakly sclerotized area. Thirty-three species from Argentina, Bolivia, Brazil, Costa Rica, and Uruguay. Localities out of Brazil are doubtful (Bertani 2001).

Lasiodorides Schmidt and Bischoff 1997. Resembles *Lasiodora* but differs from it in the absence of stridulatory setae and of retrolateral scopulae on femur IV. Four species from Ecuador and Peru (Schmidt and Bischoff 1997; Peters 2000).*

Longilyra Gabriel 2014. Small tarantulas characterized by the presence of few long club-shaped stridulatory setae on retrolateral face of palpal trochanter and prolateral face of trochanter I. Female spermathecae with a sub-rectangular fused base and two apical lateral sub-spherical receptacula (fundus). Male unknown. One species from El Salvador (Gabriel 2014)*.

Megaphobema Pocock 1901. Large-sized tarantulas. Males palpal bulb with very wide concave–convex embolus with prolateral superior and inferior keels, and apical and prolateral accessory keels. Females have one spermathecal receptacle transversely striated. Five species from Brazil, Ecuador, Colombia, and Costa Rica (Pérez-Miles et al. 2006).

Mygalarachne Ausserer 1871. Females with a single sub-rectangular spermathecae medially notched. Types I and III urticating setae present. Absence of stridulatory setae. Male unknown. One species from Honduras (Gabriel and Longhorn 2011).*

Metriopelma Becker 1878. Absence of tibial apophysis. Palpal bulb lacking a tegular heel. Prolateral superior, prolateral inferior, apical, subapical, retrolateral superior, retrolateral inferior and retrolateral accesory keels present and weakly developed. Subapical keel serrate. Female unknown. One species from Mexico (Gabriel 2016)*.

Miachistopus Pocock 1897. Male palpal bulb strongly subquadrate with a very well developed subapical keel placed from the base of the bulb to the apical keel. Female unknown. One species from Venezuela (Gabriel 2016)*.

Neischnocolus Petrunkevitch 1925. The genus is characterized by the presence of modified urticating setae type I with long area b (median zone without barbs, see Chap. 9). Males with one or two conical processes on retrolateral palpal tibia. Females differ from those of all other theraphosines by their spermathecae with paired ventral receptacles attached to a semicircular, sclerotized back-plate or a single notched receptaculum. Eight species from Brazil, Colombia, Costa Rica, Ecuador, Panama, Peru, and Venezuela (Pérez-Miles et al. 2008, 2019; Kaderka 2014; Lapinski et al. 2018).

Neostenotarsus Pribik and Weinmann 2004. Resembles *Aphonopelma* but differs by the presence of types I and III urticating setae, procurved fovea and male palpal bulb distally widened. Basal-ventral nodule on metatarsus I. Female with single spermathecae, widely fused at the base with two sub-spherical lateral apical cameras (fundi). One species from French Guiana (Tesmoingt and Schmidt 2002).*

Nesipelma Schmidt and Kovarik 1996. Resembles *Acanthoscurria* by the presence of stridulatory setae on retrolateral face of palpal trochanter and prolateral trochanter I. Male differs from this genus by the palpal bulb with embolus narrower and acuminate, and tibial apophysis with two branches. Female with two separated saccular spermathecae. One species from St. Kitts and Nevis (Schmidt and Kovařík 1996).*

Nhandu Lucas 1983. Male palpal bulb with triangular subapical keel and absence of accessory keel. Tibial apophysis absent or if present two parallel branches on a common base, the retrolateral branch constricted in the middle. Females with retrolateral scopulae on femur IV, two spermathecal receptacles separated by a hard sclerotized short area. Long curly hairs over the carapace. Five species from Brazil and Paraguay (Bertani 2001).

Pamphobeteus Pocock, 1901. Male palpal bulb with broad embolus and slightly developed apical keel, prolateral inferior keel and retrolateral keel of similar length to apical, and usually a prolateral accessory keel present. Prolateral superior and retrolateral keels well developed. Females have spermathecae with a wide base and very short seminal oval receptacles, usually curved toward the center. Thirteen species from Bolivia, Brazil, Colombia, Ecuador, Panama, and Peru (Bertani et al. 2008; Cifuentes et al. 2016).

Phormictopus Pocock 1901. Plumose bristles on coxae and trochanters in both sexes. Male palpal tibiae with a retrolateral nodule, a tegumentary prominence and ventral subapical zone of tibiae of legs I with two branched spurs. Retrolateral face of legs IV covered by a scopulae of plumose setae. Fifteen species from Argentina, Brazil, Cuba, Hispaniola, Caribbean region (Ortiz and Bertani 2005).

Proshapalopus Mello-Leitao 1923. Male palpal bulb with an accessory keel present under the prolateral inferior keel. Females can be distinguished by having a thickened tibia IV together with absence of type III urticating setae (*Proshapalopus amazonicus* Bertani, 2001); or type I urticating setae with the region "a" shorter than the region "b." Four species from Brazil and Colombia (Bertani 2001; Perafán and Valencia-Cuéllar 2018).

Pseudhapalopus Strand 1907. Male palpal tibia very spinose on prolateral and dorsal faces. Palpal bulb short and wide, subconical, with prolateral superior and inferior keels present. Females with two separated spermathecal receptacles, tubular, sinuous and divergent or sacular. Type I urticating setae present. Five species from Bolivia, Colombia, Ecuador, and Trinidad (Schmidt and Weinmann 1997; Gabriel 2016).

Pterinopelma Pocock 1901. Resembles *Lasiodora*, *Vitalius*, and *Nhandu*, by the absence of accessory prolateral keels and presence of prolateral superior, prolateral inferior, retrolateral and apical keels in palpal bulb. Prolateral inferior keel with denticles and subapical keel weak or absent. Females differ from those genera by the spermathecae short, separated by a heavily sclerotized short area and absence of long setae on carapace. Both sexes are distinguished from those of *Lasiodora* by the absence of stridulatory setae on the prolateral coxae. Three species from Brazil (Bertani et al. 2011).

Reversopelma Schmidt 2001. Differs from other genera in the modification of type I urticating setae having barbs apically directed on the base of the setae, basally directed in the median region and apically directed in the subapically zone. One species from Ecuador or Peru (Schmidt 2001)*.

Sericopelma Ausserer, 1875. Absence of stridulatory setae on trochanter I. Male without tibial apophysis, palpal bulb with spoon shaped distal region, prolateral superior and apical keels fused in the apex of the embolus. Female with single spermathecae swollen at the apex forming a P-shaped cross-section. Dense scopulae of plumose setae present on retrolateral femur of leg IV. Fourteen species from Brazil, Central America (Costa Rica, Panama), and Mexico (Gabriel and Longhorn 2015; Andre and Esche 2011; Gabriel 2016).

Sphaerobothria Karsch 1879. Differs from other genera in the presence of a prominent spheroid tegumentary protuberanceon the fovea. One species from Costa Rica and Panama (Valerio 1980)*.

Stichoplastoris Rudloff 1997. Male palpal bulb with tegulum and subtegulum wide, embolus short, prolateral superior, and inferior keels presents, two tibial apophysis convergent and fused in the base. Female with two sacular/tubular spermathecae weakly fused at their base. Eight species from Costa Rica, El Salvador, and Panama (Rudloff 1997; Valerio 1980).*

Theraphosa Thorell 1870. Very large tarantulas characterized by the presence of stridulatory setae on prolateral coxae I and II. Type III urticating setae short and type I absent. Males with or without tibial apophysis, palpal bulb sub-cylindrical, concave–convex with apical keel completely fused with prolateral superior keel. Females with fused sclerotized spermathecae. Three species from Brazil, Colombia, Guyana, and Venezuela (Bertani 2001; Almeida et al. 2018).

Tliltocatl Mendoza and Francke 2019. Differs from all other Theraphosinae genera (except *Brachypelma*) having claviform stridulating setae on the prolateral face of trochanter/femur I and retrolateral face trochanter of the palp. Male palpal bulb distally wide and flattened (spoon-shaped) with prolateral superior and apical keels united at the apex, prolateral superior, and prolateral inferior keels are joined at their distal end and widely separating toward the embolus base, the apical keel can extend widely to backward just as the prolateral inferior keel. Females have a single fused spermatheca, apically narrowed, spermathecal baseplate absent or slightly developed. Plumose pad of setae on femur of legs IV absent. Differs from *Brachypelma* by the coloration of legs, which are black or have long, whitish setae and by the morphology of genitalia. Seven species from Mexico, Guatemala, Belize, Honduras, El Salvador, Nicaragua, and Costa Rica (Mendoza and Francke 2019).

Vitalius Lucas, Silva and Bertani 1993. Characterized by the absence of stridulatory setae on the superior region of coxae I and II in both sexes. Many long curly scattered setae on carapace. Male palpal bulb with triangular subapical keel present, accessory keel absent. Two tibial apophysis present, metatarsus closing over the tibial apophysis, between the branches. Females with two spermathecae separated by a heavily sclerotized short area. Ten species from Argentina and Brazil (Bertani 2001; Ferretti et al. 2015).

Xenesthis Simon 1891. Large tarantulas without stridulatory setae. Male palpal bulb spoon shaped with prolateral superior and apical keels fused at the apex of the embolus. Two unequal tibial apophysis and metatarsus closing between branches. Females with two spermathecae widely fused in the base with spheric apical cameras. Metatarsal scopulae on leg IV entire. Three species from Colombia, Panama, and Venezuela (Schiapelli and Gerschman 1979; Pérez-Miles et al. 1996).

Group IV: Incertae Sedis (Fig. 5.8)

Hemirrhagus Simon 1903. Characterized by retrolateral projections (heels) on ventral face of coxae of all legs. Type VI urticating setae arranged in one dorsomedian patch, two dorsal paramedian or lateral patches, or two lateral patches (excepting the troglobite species). Male palpal bulb with a slender embolus (except *H. embolulatus*). Subapical keel extended posteriorly and retrolaterally, a ventral grove presents between the tegulum and the embolus. Females with spermathecae paired, unilobular, which can be completely separated (most of the troglobitic species) or fused at their base (epigean species). Twenty-seven species endemic from Mexico (Mendoza 2014; Mendoza and Francke 2018).

Kankuamo Perafán, Galvis and Pérez-Miles 2016. The genus is characterized by having a distinct type of urticating setae type (VII), similar to type II but having a small distal patch of lanceolated barbs arranged in reversed direction, regarding the main barbs, oriented with their tips toward the penetration tip. Males have a curved sub-conical palpal bulb with many conspicuous keels distributed throughout the majority of the subtegulum and embolus, especially developed on the dorsal and

Fig. 5.8 Representatives of Theraphosinae Group IV. (**a**) *Hemirrhagus benzaa*. (**b**) *Kankuamo marquezi*. (Photo **a**: J. Mendoza)

prolateral faces, most of them with serrated edges. Tibial apophysis absent. Females have a spermathecae with a single notched receptacle, with two granulated lobes, and several irregular sclerotized longitudinal striations. One species from Colombia (Perafán et al. 2016)*.

Ozopactus Simon 1889. Resembles *Crypsidromus* but differs by the sternum convex and more spinose legs. Anterior and posterior lateral eyes subequal. One species from Venezuela (Simon 1889, 1892).*

Acknowledgments We thank R. West, J. Mendoza, R. Kaderka, R. Bertani, and D. Valencia-Cuéllar, who generously gave us photos and figures used in this chapter. To Nelson Ferretti and Daniella Sherwood for the critical reading; L.A. Gómez who helped us with English.

References

Almeida MQ, Salvatierra L, Morais JW (2018) First record of *Theraphosa apophysis* (Tinter, 1991) (Araneae, Mygalomorphae, Theraphosidae) in Brazil. Check List 14:647–650. https://doi.org/10.15560/14.4.647

Andre C, Esche S (2011) Die Gattung *Sericopelma*: Habitat, Lebensweise, Terrarienhaltung und Nachzucht von *Sericopelma melanotarsum* Valerio, 1980. Arachne 16:10–31

Ausserer A (1871) Beiträge zur Kenntnis der Arachniden-Familie der Territelariae Thorell (Mygalidae Autor). Verh Zool Bot Ges Wien 21:117–224

Bertani R (2000) Male palpal bulbs and homologous features in Theraphosinae (Araneae, Theraphosidae). J Arachnol 28:29–42

Bertani R (2001) Revision, cladistic analysis, and zoogeography of *Vitalius*, *Nhandu*, and *Proshapalopus*; with notes on other theraphosine genera (Araneae, Theraphosidae). Arquiv Zool 36:265–356

Bertani R (2012) Revision, cladistic analysis and biogeography of *Typhochlaena* C. L. Koch, 1850, *Pachistopelma* Pocock, 1901 and *Iridopelma* Pocock, 1901 (Araneae, Theraphosidae, Aviculariinae). Zookeys 230:1–94

Bertani R Fukushima CS Silva Jr PI da (2008) Two new species of *Pamphobeteus* Pocock 1901 (Araneae: Mygalomorphae: Theraphosidae) from Brazil, with a new type of stridulatory organ. Zootaxa 1826:45–58

Bertani R, Guadanucci JPL (2013) Morphology, evolution and usage of urticating setae by tarantulas (Araneae: Theraphosidae). Zoologia (Curitiba) 30(4):403–418. https://doi.org/10.1590/S1984-46702013000400006

Bertani R, Nagahama RH, Fukushima CS (2011) Revalidation of *Pterinopelma* Pocock 1901 with description of a new species and the female of *Pterinopelma vitiosum* (Keyserling 1891) (Araneae: Theraphosidae: Theraphosinae). Zootaxa 2814:1–18

Bertani R, Bichuette ME, Pedroso DR (2013) *Tmesiphantes hypogeus* sp. nov. (Araneae, Theraphosidae), the first troglobitic tarantula from Brazil. An Acad Bras Ciênc 85:235–243. https://doi.org/10.1590/s0001-37652013005000007

Bücherl W (1947a) Estudo comparativo das espécies brasileiras do gênero *Pamphobeteus* Pocock, 1901 (Mygalomorphae). Mem Inst Butantan 20:233–282

Bücherl W (1947b) Duas novas espécies do gênero Eupalaestrus Pocock, 1901. Mem Inst Butantan 20:297–314

Bücherl W (1949a) Em torno das três espécies insulares e praianas do gênero *Pamphobeteus* Pocock, 1901 (Mygalomorphae). Mem Inst Butantan 21:117–136

Bücherl W (1949b) Estudos sobre o gênero Magulla Simon, 1892 (Theraphosidae, Ischnocolinae). Mem Inst Butantan 21:261–274

Bücherl W (1951) Estudos sobre a biologia e a sistemática do gênero *Grammostola* Simon, 1892. Monog Inst Butantan 1:1–203

Bücherl W (1957) Sobre a importância dos bulbos copuladores e das apófisis tibiais dos machos na sistemática das aranhas caranguejeiras (Orthognatha). Instituto Butantan, São Paulo. An Acad Bras Ciênc 29:377–420

Cambridge FOP (1896) On the Theraphosidae of the lower Amazons: being an account of the new genera and species of this group of spiders discovered during the expedition of the steamship "Faraday" up the river Amazons. Proc Zool Soc Lond 1896:716–766

Candia-Ramírez DT, Francke OF (2017) Taxonomic revision of the genus *Crassicrus* Reichling and West, 1996 (Araneae: Theraphosidae: Theraphosinae), with the description of additional keels on the embolus. J Arachnol 45:67–98. https://doi.org/10.1636/JoA-S-16-005.1

Cavallo P, Ferretti NE (2015) The first *Hapalotremus* Simon, 1903 (Araneae: Theraphosidae) from Argentina: description and natural history of *Hapalotremus martinorum* sp. nov. J Nat Hist 49:873–887. https://doi.org/10.1080/00222933.2014.953226

Cifuentes Y, Estrada-Gomez S, Vargas-Muñoz LJ, Perafán C (2016) Description and molecular characterization of a new species of tarantula, *Pamphobeteus verdolaga*, from Colombia (Araneae: Mygalomorphae: Theraphosidae). Zoologia (Curitiba) 33:1–6. https://doi.org/10.1590/s1984-4689zool-20160113

Coddington JA, Levy HW (1991) Systematics and evolution of spiders (Araneae). Annu Rev Ecol Syst 22:565–592

Cooke JA, Roth VD, Miller FD (1972) The urticating hairs of theraphosidae. Am Mus Novit 2498:1–43

de Lamarck JB (1818) Histoire naturelle des animaux sans vertèbres. Paris 5:1–612 (Araneae, pp. 88–108)

Estrada-Alvarez JC (2014) New data from mygalomorph spiders (Araneae: Mygalomorphae) of Estado de Mexico, with taxonomic comments about the genus *Davus* O. Pickard-Cambridge, 1892. Dugesiana 21:55–66

Estrada-Alvarez JC, Guadarrama RCA, Martínez OM (2013) Nueva especie de *Citharacanthus* Pocock, 1901 (Theraphosidae: Theraphosinae) para México. Dugesiana 20:63–66

Fabiano-da-Silva W, Guadanucci JPL, DaSilva MB (2015) *Tmesiphantes mirim* sp. nov. (Araneae: Theraphosidae) from the Atlantic Forest of Bahia, Brazil, biogeographical notes and identification keys for species of the genus. Zoologia (Curitiba) 32:151–156. [errata: Zoologia (Curitiba)32: 541]. doi:https://doi.org/10.1590/S1984-46702015000200006

Ferretti N (2015) On three new *Euathlus* tarantulas from Argentina and cladistic analysis of the genus. J Arachnol 43:313–326. https://doi.org/10.1636/0161-8202-43.3.313

Ferretti N, Barneche J (2012) New species of *Eupalaestrus* from Argentina (Araneae, Theraphosidae, Theraphosinae). Iheringia Sér Zool 102:327–330

Ferretti N, Barneche J (2013) Description of two new species of *Plesiopelma* (Araneae, Theraphosidae, Theraphosinae) from Argentina. Iheringia Sér Zool 103:374–380. https://doi.org/10.1590/S0073-47212013000400007
Ferretti N, Pompozzi G, Pérez-Miles F (2011) The species of *Grammostola* (Araneae: Theraphosidae) from central Argentina: taxonomy, distribution, and surface ultrastructure of coxal setae. Zootaxa 2828:1–18
Ferretti N, Pompozzi G, González A, Pérez-Miles F (2013) The genus *Grammostola* Simon 1892 (Araneae: Theraphosidae): a new species from western Argentina, new synonymy and distributional data. J Nat Hist 47:2961–2977. https://doi.org/10.1080/00222933.2013.791945
Ferretti N, Copperi S, Schwerdt L, Pompozzi G (2015) First record of *Vitalius longisternalis* Bertani, 2001 (Araneae, Theraphosidae) in Argentina and notes on its natural history in Misiones province. Check List 11:1–5. https://doi.org/10.15560/11.5.1748
Ferretti N, Ochoa JA, Chaparro JC (2016) Una nueva especie de *Acanthoscurria* (Araneae: Theraphosidae: Theraphosinae) sin órgano estridulador, del sur de Perú. Rev Peru Biol 23:89–94. https://doi.org/10.15381/rpb.v23i2.12375
Ferretti N, Cavallo P, Chaparro JC, Ríos-Tamayo D, Seimon TA, West R (2018) The Neotropical genus *Hapalotremus* Simon, 1903 (Araneae: Theraphosidae), with the description of seven new species and the highest altitude record for the family. J Nat Hist 52:1927–1984. https://doi.org/10.1080/00222933.2018.1506521
Foley S, Lüddecke T, Chen DQ, Krehenwinkel H, Künzel S, Longhorn S, Wendt I, von Wirth V, Tänzler R, Vences M, Piel WH (2019) Tarantula phylogenomics: a robust phylogeny of deep theraphosid clades inferred from transcriptome data sheds light on the prickly issue of urticating setae evolution. Mol Phylogenet Evol 140(106573):1–12. https://doi.org/10.1016/j.ympev.2019.106573
Fonseca-Ferreira R, Zampaulo Rde A, Guadanucci JPL (2017) Diversity of iron cave-dwelling mygalomorph spiders from Pará, Brazil, with description of three new species (Araneae). Trop Zool 30:178–199. https://doi.org/10.1080/03946975.2017.1367590
Fukushima CS, Bertani R, da Silva Jr PI (2005) Revision of *Cyriocosmus* Simon, 1903, with notes on the genus *Hapalopus* Ausserer, 1875 (Araneae: Theraphosidae). Zootaxa 846:1–31
Fukushima CS, Nagahama RH, Bertani R (2008) The identity of *Mygale brunnipes* C.L. Koch 1842 (Araneae, Theraphosidae), with a redescription of the species and the description of a new genus. J Arachnol 36:402–410
Gabriel R (2011) A new species of *Hapalopus* Ausserer, 1875 from Guyana (Araneae: Theraphosidae). J Br Tarantula Soc 26:76–80
Gabriel R (2014) A new genus and species of theraphosid spider from El Salvador (Araneae: Theraphosidae). J Br Tarantula Soc 29:146–153
Gabriel R (2016) Revised taxonomic placement of the species in the Central American genera *Davus* O. Pickard-Cambridge, 1892, *Metriopelma* Becker, 1878, and *Schizopelma* F. O. Pickard-Cambridge, 1897, with comments on species in related genera (Araneae: Theraphosidae). Arachnology 17:61–92. https://doi.org/10.13156/arac.2006.17.2.61
Gabriel R, Longhorn SJ (2011) Redescriptions of the holotypes of *Mygalarachnae* [sic] Ausserer 1871 and *Harpaxictis* Simon (1892) (Araneae: Theraphosidae) with rebuttal of their synonymy with *Sericopelma* Ausserer 1875. Rev Ibér Aracnología 19:157–165
Gabriel R, Longhorn SJ (2015) Revised generic placement of *Brachypelma embrithes* (Chamberlin and Ivie, 1936) and *Brachypelma angustum* Valerio, 1980, with definition of the taxonomic features for identification of female *Sericopelma* Ausserer, 1875 (Araneae, Theraphosidae). ZooKeys 526:75–104. https://doi.org/10.3897/zookeys.526.6315
Gerschman de Pikelin BS, Schiapelli RD (1970) Discusión de los caracteres válidos en la sistemática de las arañas Theraphosomorphae. Bulletin Du Museum National D'Histoire Naturelle. 2da. Serie, Tomo 41, Suplemento No. 1, pp 150–154
Gerschman de Pikelin BS, Schiapelli RD (1972) El género *Homoeomma* Ausserer 1871 (Araneae: Theraphosidae). Physis (C) 31:237–258
Gerschman de Pikelin BS, Schiapelli RD (1973) La subfamilia «Ischnocolinae» (Araneae: Theraphosidae). Revista del Museo Argentino de Ciencias Naturales «Bernardino Rivadavia»

e Instituto Nacional de Investigación de las Ciencias Naturales. Entomología Tomo IV, n°2: 43–77

Goloboff PA (1993) A reanalysis of mygalomorph spider families (Araneae). Am Mus Novit 3056:1–32

Gonzalez-Filho HMO, Lucas SM, Paula F, Indicatti RP, Brescovit AD (2012) On the taxonomy of *Acanthoscurria* Ausserer from southeastern Brazil with data on the natural history of *A. gomesiana* Mello-Leitão (Araneae, Mygalomorphae, Theraphosidae). Int J Zool 721793:1–11

Guadanucci JPL (2014) Theraphosidae phylogeny: relationships of the 'Ischnocolinae' genera (Araneae, Mygalomorphae). Zool Scr 43:508–518

Hamilton CA, Formanowicz DR, Bond JE (2011) Species delimitation and phylogeography of *Aphonopelma hentzi* (Araneae, Mygalomorphae, Theraphosidae): Cryptic diversity in North American tarantulas. PLoS One 6:e26207

Hamilton CA, Hendrixson BE, Brewer MS, Bond JE (2014) An evaluation of sampling effects on multiple DNA barcoding methods leads to an integrative approach for delimiting species: a case study of the North American tarantula genus *Aphonopelma* (Araneae, Mygalomorphae, Theraphosidae). Mol Phylogenet Evol 71:79–93

Hamilton CA, Hendrixson BE, Bond JE (2016) Taxonomic revision of the tarantula genus *Aphonopelma* Pocock, 1901 (Araneae, Mygalomorphae, Theraphosidae) within the United States. Zookeys 560:1–340

Hendrixson BE (2019) A new species of *Aphonopelma* (Araneae: Mygalomorphae: Theraphosidae) from the Madrean pine-oak woodlands of northeastern Sonora, Mexico. Zootaxa 4688:519–534

Hendrixson BE, DeRussy BM, Hamilton CA, Bond JE (2013) An exploration of species boundaries in turret-building tarantulas of the Mojave Desert (Araneae, Mygalomorphae, Theraphosidae, *Aphonopelma*). Mol Phylogenet Evol 66:327–340

Hendrixson E, Guice AV, Bond JE (2015) Integrative species delimitation and conservation of tarantulas (Araneae, Mygalomorphae, Theraphosidae) from a North American biodiversity hotspot. Insect Conserv Diver 8:120–131

Indicatti RP, Lucas SM, Guadanucci JPL, Yamamoto FU (2008) Revalidation and revision of the genus *Magulla* Simon 1892 (Araneae, Mygalomorphae, Theraphosidae). Zootaxa 1814:21–36

Kaderka R (2010) *Cyriocosmus venezuelensis* sp. n. from Venezuela (Araneae: Theraphosidae: Theraphosinae). Rev Ibér Aracnología 18:87–96

Kaderka R (2014) *Ami armihuariensis*, a new species from Peru (Araneae: Theraphosidae: Theraphosinae). Rev Per Biol 21:207–212. https://doi.org/10.15381/rpb.v21i3.10906

Kaderka R (2015) *Bistriopelma*, a new genus with two new species from Peru (Araneae: Theraphosidae: Theraphosinae). Rev Per Biol 22:275–288. https://doi.org/10.15381/rpb.v22i3.11432

Kaderka R (2016) The Neotropical genus *Cyriocosmus* Simon, 1903 and new species from Peru, Brazil and Venezuela (Araneae: Theraphosidae: Theraphosinae). J Nat Hist 50:393–465. https://doi.org/10.1080/00222933.2015.1076082

Kaderka R (2017) Description of the male of *Bistriopelma matuskai* Kaderka 2015 and a new species of *Bistriopelma* from Peru (Araneae: Theraphosidae: Theraphosinae). Rev Per Biol 24:231–242. https://doi.org/10.15381/rpb.v24i3.13904

Kaderka R, Bulantová J, Heneberg P, Rezák M (2019) Urticating setae of tarantulas (Araneae: Theraphosidae): morphology, revision of typology and terminology and implications for taxonomy. PLoS One 14(11):e0224384. https://doi.org/10.1371/journal.pone.0224384

Kambas D (2019) Tarantupedia: an online taxonomic database for the worlds largest spiders. www.tarantupedia.com. Accessed on 2019-x-10

Lapinski W, Gabriel R, Pérez-Miles F (2018) Description of the female of *Ami bladesi* Pérez-Miles, Gabriel and Gallon, 2008 (Araneae: Theraphosidae), with notes on its distribution, ecology, and behaviour. Arachnology 17:419–426. https://doi.org/10.13156/arac.2017.17.8.419

Longhorn SJ (2001) Genetic evaluation of CITES II protectedspiders of the genus *Brachypelma* (Theraphosidae); includingan outline of non-lethal tissue sampling. Masters Thesis. Imperial College, University of London, London

Longhorn SJ, Nicholas M, Chuter J, Vogler AP (2007) The utility of molecular markers from non-lethal DNA samples of the CITES II protected "tarantula" *Brachypelma vagans* (Araneae, Theraphosidae). J Arachnol 35:278–292

Lucas H (1840) Histoire naturelle des Animaux articulés, Annélides, Crustacés, Arachnides, Myriapodes et insectes. Paris:1–601 [Aranéides: 328–441]

Lucas S, da Silva PI, Bertani R (1991) The genus *Ephebopus* Simon, 1892. Description of the male of *Ephebopus murinus* (Walckenaer), 1837. Spixiana 14:245–248

Lucas SM, Paula F, Gonzalez HMO, Brescovit AD (2010) Redescription and new distribution records of *Acanthoscurria paulensis* (Araneae: Mygalomorphae: Theraphosidae). Zoologia (Curitiba) 27:563–568

Lüddecke T, Krehenwinkel H, Canning G, Glaw F, Longhorn SJ, Tänzler R, Wendt I, Vences M (2018) Discovering the silk road: nuclear and mitochondrial sequence data resolve the phylogenetic relationships among theraphosid spider subfamilies. Mol Phylogenet Evol 119:63–70. https://doi.org/10.1016/j.ympev.2017.10.015

Marshall SD, Uetz GW (1990) The pedipalpal brush of *Ephebopus* sp. (Araneae, Theraphosidae): evidence of a new site for urticating hairs. Bull Br Arachnol Soc 8:122–124

Mello-Leitão CF (1923) Theraphosoideas do Brasil. Rev Mus Paul 13:1–449

Mendoza JI (2012) *Cotztetlana omiltemi*, a new genus and species of tarantula from Guerrero, México (Araneae, Theraphosidae, Theraphosinae). Arthropoda Scientia 2:2–7

Mendoza JI (2014) Taxonomic revision of *Hemirrhagus* Simon, 1903 (Araneae: Theraphosidae, Theraphosinae), with description of five new species from Mexico. Zool J Linnean Soc 170:634–689. https://doi.org/10.1111/zoj.12112

Mendoza JI, Francke OF (2017) Systematic revision of *Brachypelma* red-kneed tarantulas (Araneae: Theraphosidae), and the use of DNA barcodes to assist in the identification and conservation of CITES-listed species. Inv Syst 31:157–179. https://doi.org/10.1071/IS16023

Mendoza JI, Francke OF (2018) Five new cave-dwelling species of *Hemirrhagus* Simon 1903 (Araneae, Theraphosidae, Theraphosinae), with notes on the generic distribution and novel morphological features. Zootaxa 4407:451–482. https://doi.org/10.11646/zootaxa.4407.4.1

Mendoza JI, Francke OF (2019) Systematic revision of Mexican threatened tarantulas Brachypelma (Araneae: Theraphosidae: Theraphosinae), with a description of a new genus, and implications on the conservation. Zool J Linnean Soc 20:1–66

Mendoza JI, Locht A, Kaderka R, Medina F, Pérez-Miles F (2016) A new genus of Theraphosid spider from Mexico, with a particular palpal bulb structure (Araneae, Theraphosidae, Theraphosinae). Eur J Taxon 232:1–28. https://doi.org/10.5852/ejt.2016.232

Miglio LT, Bonaldo AB, Pérez-Miles F (2013) On *Munduruku*, a new theraphosid genus from Oriental Amazonia (Araneae, Mygalomorphae). Iheringia Sér Zool 103:185–189. https://doi.org/10.1590/s0073-47212013000200013

Montenegro VR, Aguilera MA, Casanueva ME (2018) First record of *Homoeomma* Ausserer, 1871 in Chile and description of two new species (Araneae, Theraphosidae). Spixiana 41:13–25

Montes de Oca L, D'Elía G, Pérez-Miles F (2016) An integrative approach for species delimitation in the spider genus *Grammostola* (Theraphosidae, Mygalomorphae). Zool Scripta 45:322–333. https://doi.org/10.1111/zsc.12152

Olivier GA (1811) Articles: Mygale, Oxyope. Encyclopédie Méthodique, Histoire Naturelle, Insectes, Paris 8:83–87. [601-604]

Ortiz D (2008) Description of *Cubanana cristinae*, a new genus and species of theraphosine tarantula (Araneae: Theraphosidae) from the island of Cuba. Bol Soc Ent Aragonesa 42:107–122

Ortiz D, Bertani R (2005) A new species in the spider genus *Phormictopus* (Theraphosidae: Theraphosinae) from Cuba. Rev Ibér Aracnol 11:29–36

Ortiz D, Francke OF (2015) Two new species of *Bonnetina* tarantulas (Theraphosidae: Theraphosinae) from Mexico: contributions to morphological nomenclature and molecular characterization of types. J Nat Hist 49:1–23

Ortiz D, Francke OF (2016) Two DNA barcodes and morphology for multi-method species delimitation in *Bonnetina* tarantulas (Araneae: Theraphosidae). Mol Phylogenet Evol 101:176–193

Ortiz D, Francke OF (2017) Reconciling morphological and molecular systematics in tarantulas (Araneae: Theraphosidae): revision of the Mexican endemic genus *Bonnetina*. Zool J Linnean Soc 180:819–886. https://doi.org/10.1093/zoolinnean/zlw013

Ortiz D, Francke OK, Bond JA (2018) A tangle of forms and phylogeny: extensive morphological homoplasy and molecular clock heterogeneity in Bonnetina and related tarantulas. Mol Phylogenet Evol 127:55–73

Paula F, Gabriel R, Indicatti RP, Brescovit AD, Lucas SM (2014) On the Brazilian Amazonian species of *Acanthoscurria* (Araneae: Theraphosidae). Zoologia (Curitiba) 31:63–80. https://doi.org/10.1590/S1984-46702014000100008

Perafán C (2010) Revisión taxonómica y análisis filogenético de un grupo basal de Theraphosinae (Araneae, Mygalomorphae, Theraphosidae). MSc Thesis, Facultad de Ciencias, Universidad de la República, Montevideo

Perafán C, Pérez-Miles F (2014a) Three new species of *Melloleitaoina* Gerschman and Schiapelli, 1960 (Araneae, Mygalomorphae, Theraphosidae) from northern Argentina. ZooKeys 404:117–129. https://doi.org/10.3897/zookeys.404.6243

Perafán C, Pérez-Miles F (2014b) The Andean tarantulas *Euathlus* Ausserer, 1875, *Paraphysa* Simon, 1892 and *Phrixotrichus* Simon, 1889 (Araneae: Theraphosidae): phylogenetic analysis, genera redefinition and new species descriptions. J Nat Hist 48:2389–2418. https://doi.org/10.1080/00222933.2014.902142

Perafán C, Valencia-Cuéllar D (2018) *Proshapalopus marimbai*, a new tarantula species (Mygalomorphae, Theraphosidae) and first genus record from Colombia. Trop Zool 31:200–213. https://doi.org/10.1080/03946975.2018.1493181

Perafán C, Cifuentes Y, Estrada-Gomez S (2015) *Aguapanela*, a new tarantula genus from the Colombian Andes (Araneae, Theraphosidae). Zootaxa 4033:529–542. https://doi.org/10.11646/zootaxa.4033.4.4

Perafán C, Galvis W, Gutiérrez M, Pérez-Miles F (2016) *Kankuamo*, a new theraphosid genus from Colombia (Araneae, Mygalomorphae), with a new type of urticating setae and divergent male genitalia. ZooKeys 601:89–109. https://doi.org/10.3897/zookeys.601.7704

Perdomo C, Panzera A, Pérez-Miles F (2009) *Hapalopus aymara* a new species of tarantula from Bolivia (Araneae, Theraphosidae, Theraphosinae). Iheringia Sér Zool 99:53–55. https://doi.org/10.1590/s0073-47212009000100007

Pérez-Miles F (1992) Revisión del género *Eupalaestrus* Pocock 1901 (Araneae, Theraphosidae). Rev Bras Biol 52:27–35

Pérez-Miles F (1998) Notes on the systematics of the little known theraphosid spider *Hemirrhagus cervinus*, with a description of a new type of urticating hair. J Arachnol 26:120–123

Pérez-Miles F (2000) *Iracema cabocla* new genus and new species of a theraphosid spider from amazonic Brazil (Araneae, Theraphosidae). J Arachnol 28:141–148

Pérez-Miles F, Weinmann D (2010) *Agnostopelma*: a new genus of tarantula without a scopula on leg IV (Araneae: Theraphosidae: Theraphosinae). J Arachnol 38:104–112

Pérez-Miles F, Lucas SM, Silva PI, Bertani R (1996) Systematic revision and cladistic analysis of Theraphosinae (Araneae: Theraphosidae). Mygalomorph 1:33–68

Pérez-Miles F, Miglio LT, Bonaldo AB (2006) *Megaphobema teceae* n. sp. (Araneae, Theraphosidae), a new theraphosine spider from Brazilian Amazonia. Zootaxa 1115:61–68

Pérez-Miles F, Gabriel R, Miglio L, Bonaldo A, Gallon R, Jimenez JJ, Bertani R (2008) *Ami*, a new theraphosid genus from Central and South America, with the description of six new species (Araneae: Mygalomorphae). Zootaxa 1915:54–68

Pérez-Miles F, Bonaldo AB, Miglio LT (2014) *Bumba*, a replacement name for *Maraca* Pérez-Miles, 2005 and *Bumba lennoni*, a new tarantula species from western Amazonia (Araneae, Theraphosidae, Theraphosinae). ZooKeys 448:1–8. https://doi.org/10.3897/zookeys.448.7920

Pérez-Miles F, Gabriel R, Sherwood D (2019) *Neischnocolus* Petrunkevitch, 1925, senior synonym of *Ami* Perez-Miles, 2008 and *Barropelma* Chamberlin, 1940 (Araneae: Theraphosidae). Arachnology 18(2):150–155. https://doi.org/10.13156/arac.2018.18.2.150
Peters HJ (2000) Tarantulas of the world: Kleiner Atlas der Vogelspinnen - Band 2. Published by the Author, 162 p
Peters HJ (2005) Tarantulas of the world: Kleiner Atlas der Vogelspinnen - Band 3. Published by the Author, 130 p
Petersen SD, Mason T, Akber S, West R, White B, Wilson P (2007) Species identification of tarantulas using exuviae for international wildlife law enforcement. Conserv Genet 8:497–502
Petrunkevitch A (1928) Systema aranearum. Trans Conn Acad Arts Sci 29:1–270
Petrunkevitch A (1933) An inquiry into the natural classification of spiders, based on a study of their internal anatomy. Trans Conn Acad Arts Sci 31:299–389
Pocock RI (1897) On the spiders of the Suborder Mygalomorphae from the Ethiopian region contained in the collection of the British Museum. Proc Zool Soc Lond, pp 724–774
Pocock RI (1903) On some genera and species of South American Aviculariidae. Ann Mag Nat Hist 11:81–115
Prentice TR (1997) Theraphosidae of the Mojave Desert west and north of the Colorado River (Araneae, Mygalomorphae, Theraphosidae). J Arachnol 25:137–176
Raven RJ (1985) The spider infraorder Mygalomorphae (Araneae): cladistics and systematics. Bull Am Mus Nat Hist 182:1–180
Reichling SB, West RC (2000) A new species of tarantula spider (Araneae, Mygalomorphae, Theraphosidae) from the Cayo District of Belize. Southwest Nat 45:128–132
Rodríguez-Manzanilla C, Bertani R (2010) Descripción de *Acathoscurria turumban* sp. nov. (Araneae: Theraphosidae) y primer registro del género para Venezuela. Bol Soc Ent Aragonesa 46:143–145
Rudloff JP (1997) Revision der Gattung *Holothele Karsch*, 1879 nebst Aufstellung einer neuen Gattung *Stichoplastoris* gen. nov. (Araneae, Theraphosidae) und Wiedereinsetzung einiger weiterer Gattungen der Mygalomorphae. Arachnologisches Magazin 5(2):1–19
Schiapelli RD, Gerschman BS (1961) Las especies del género *Grammostola* Simon 1892, en la República Argentina (Araneae, Theraphosidae). Act Trab Cong Sud Zool, La Plata I (La Plata, 1959) 3:199–208
Schiapelli RD, Gerschman BS (1962) Importancia de las espermatecas en la sistemática de las arañas del Suborden Mygalomorphae (Araneae). Physis 23(64):69–75
Schiapelli RD, Gerschman BS (1964) El género *Acanthoscurria* Ausserer, 1871 (Araneae, Theraphosidae) en la Argentina. Physis 24(68):391–417
Schiapelli RD, Gerschman BS (1970) El género *Ceropelma* Mello-Leitão 1923 (Araneae, Theraphosidae). Physis 30(80):225–239
Schiapelli RD, Gerschman BS (1979) Las arañas de la subfamilia Theraphosinae (Araneae, Theraphosidae). Rev Mus Arg C Nat B Rivadavia (Ent) 5:287–300
Schmidt G (1994a) Eine neue Vogelspinnenart aus Peru, *Thrixopelma ockerti* gen. et sp. n. (Araneida: Theraphosidae: Theraphosinae). Arachnol Mag 2:3–8
Schmidt G (1994b) Das Männchen von *Thrixopelma ockerti* Schmidt 1994 (Araneida: Theraphosidae: Theraphosinae). Arachnol Mag 2:1–4
Schmidt G (1995) *Chromatopelma* gen.n.; eine neue Gattung der Theraphosidae (Arachnida: Araneida: Theraphosidae: Theraphosinae). Arthropoda 3:25–26
Schmidt G (1998) Bestimmungsschlüssel für die Gattungen der Unterfamilie Theraphosinae (Araneae: Theraphosidae). 2. Aktualisierte Auflage. Arachnol Mag 4:1–28
Schmidt G (2001) *Reversopelma petersi* sp. n. (Araneae: Theraphosidae: Theraphosinae), eine neue Spinnenart aus dem Nordesten Südamerikas. Arachnol Mag 9:1–10
Schmidt G (2005) *Aenigmarachne sinapophysis* gen. et sp. n., eine neue Vogelspinnenart aus Costa Rica (Araneae: Theraphosidae: Theraphosinae). Tarantulas World 112:3–9
Schmidt G, Bischoff B (1997) Die Gattung *Lasiodorides* n. gen. und ihre Typus-Art *Lasiodorides polycuspulatus* n. sp. (Arachnida: Araneae: Theraphosidae: Theraphosinae). Ent Zeitschrift 107:153–159

Schmidt G, Herzig V (1997) Das Männchen von *Chromatopelma cyanopubescens* (Strand, 1907) (Araneae: Theraphosidae: Theraphosinae). Arthropoda 5:11–14

Schmidt G, Kovařík F (1996) *Nesipelma insulare* gen. and sp. n. from the Nevis Island, Lesser Antilles (Arachnida; Araneida: Theraphosidae). Arachnol Mag 4:1–9

Schmidt G, Weinmann D (1997) Eine neue *Pseudhapalopus*-Art aus Kolumbien (Arachnida: Araneae: Theraphosidae: Theraphosinae). Ent Zeitschrift 107:69–72

Sherwood D (2019) *Aphonopelma braunshausenii* Tesmoingt, 1996 is a nomen dubium, with review of some historic morphological characters ineffective at species delineation (Araneae: Theraphosidae). Zootaxa 4657:573–580

Sherwood D, Gabriel R, Longhorn SJ (2018) A new species of *Homoeomma* Ausserer, 1871 from Chile (Araneae: Theraphosidae). J Br Tarantula Soc 33(3):19–33

Simon E (1889) Arachnides. In: Voyage de M. E. Simon au Venezuela (décembre 1887–avril 1888). 4e Mémoire. Ann Soc Ent France 6(9):169–220

Simon E (1892) Histoire naturelle des araignées. Deuxième édition, tome premier. Roret, Paris, pp. 1–256

Simon E (1903) Histoire Naturelle des Araignées. Deuxième édition, Tome 2, fascicule 4. Paris: 669–1080

Strand E (1907) Aviculariidae und Atypidae des Kgl. Naturalienkabinetts in Stuttgart. Jahreshefte des Vereins für vaterländische Naturkunde in Württemberg 63:1–100

Tesmoingt M, Schmidt G (2002) *Stenotarsus scissistylus* gen. et sp. n. (Theraphosidae: Theraphosinae), eine Vogelspinne aus Französisch Guyana. Tarantulas World 76:3–12

Thorell T (1870) On European spiders. Nova Acta Regiae Societatis Scientiarum Upsaliensis 7(3):109–242

Turner SP, Longhorn SJ, Hamilton CA, Gabriel R, Pérez-Miles F, Vogler AP (2018) Re-evaluating conservation priorities of New World tarantulas (Araneae: Theraphosidae) in a molecular framework indicates nonmonophyly of the genera, *Aphonopelma* and *Brachypelma*. Syst Biodivers 16:89–107

Valerio CE (1980) Arañas terafósidas de Costa Rica (Araneae: Theraphosidae). III. *Sphaerobothria*, *Aphonopelma*, *Pterinopelma*, *Citharacanthus*, *Crypsidromus* y *Stichoplastus*. Rev Biol Trop 28:271–296

Vol F (1999) A propos d'une spermatheque inhabituelle. Arachnides 42:1–13

Walckenaer CA (1802) Faune parisienne. Insectes Ou histoire abrégée des insectes de environs de Paris. Paris 2:187–250

Walckenaer CA (1837) Histoire naturelle des insectes. Aptères Paris 1:1–682

Wilson JS, Gunnell CF, Wahl DB, Pitts JP (2013) Testing the species limits of the tarantulas (Araneae: Theraphosidae) endemic to California's Southern Coast Ranges, USA. Insect Conserv Diver 6:365–371

World Spider Catalog (2019) World Spider Catalog.Version 20.5. Natural History Museum Bern. http://wsc.nmbe.ch. Accessed 10 Oct 2019. doi: 10.24436/2

Yamamoto FU, Lucas SM, Guadanucci JPL, Indicatti RP (2007) Revision of the genus *Tmesiphantes* Simon (Araneae, Mygalomorphae, Theraphosidae). Rev Bras Zool 24:971–980

Yamamoto FU, Lucas SM, Brescovit AD (2012) *Catanduba*, a new Theraphosinae spider genus from central Brazil (Araneae, Theraphosidae). Zootaxa 3172:1–19

Chapter 6
Biogeography of New World Tarantulas

Carlos Perafán, Nelson Ferretti, and Brent E. Hendrixson

Abstract Biogeography is a multidisciplinary field that is concerned with delimiting and explaining the geographic distributions of organisms in space and time. Due to their distribution patterns and interesting biology (e.g., ancient lineage with a nearly cosmopolitan distribution, sedentary lifestyle with limited dispersal capabilities), tarantulas are an appealing taxonomic group for addressing a variety of biogeographic questions concerning the Earth's history. In this chapter, we discuss some biogeographic basic concepts, delve into the distribution patterns of New World tarantulas, and explore some of the historical explanations that may have led to these distributions. We mostly review and highlight the results of recent studies but also include personal observations and unpublished data. The distributions of higher-level taxonomic groups (subfamilies and tribes) are described and we detail their latitudinal and elevational limits. We also review the distributions of groups with unique insular habitats such as those found on islands surrounded by seas, forested "islands" surrounded by "seas" of deserts, and caves. Furthermore, we discuss the distribution of some unique morphological characters of taxonomic importance such as urticating setae. Finally, we review a handful of studies that have explicitly investigated the biogeography of New World tarantulas using a variety of different analytical methods.

C. Perafán (✉)
Laboratorio de Etología, Ecología y Evolución, IIBCE, Montevideo, Uruguay

Entomología, Facultad de Ciencias, Iguá 4225, Montevideo, Uruguay

N. Ferretti
Instituto de Ciencias Biológicas y Biomédicas del SUR (INBIOSUR), CONICET-UNS, Buenos Aires, Argentina
e-mail: nferretti@conicet.gov.ar

B. E. Hendrixson
Department of Biology, Millsaps College, Jackson, MS, USA
e-mail: brent.hendrixson@millsaps.edu

F. Pérez-Miles (ed.), *New World Tarantulas*, Zoological Monographs 6,
https://doi.org/10.1007/978-3-030-48644-0_6

6.1 Introduction

In its simplest sense, the discipline of biogeography is concerned with delimiting and explaining the geographic distributions of organisms. In other words, biogeographers are engaged in answering three fundamental questions about organisms: What is it? Where is it found? and Why is it found there? The first two questions help identify and describe the *patterns* of organismal distributions whereas the third attempts to analyze and interpret the *processes* responsible for generating these patterns. Two conceptually different approaches have been developed to aid in the interpretation of organismal distributions: ecological biogeography and historical biogeography. The primary difference between both approaches is a matter of scale (i.e., ecological biogeography tends to be concerned with smaller spatial and shorter temporal scales than historical biogeography).

Ecological biogeography examines the role that biotic and abiotic environmental factors play in confining an organism to its *present-day* distribution. This field draws upon information from a variety of *ecological* disciplines including physical geography, climatology, community ecology, and physiology. Historical biogeography, on the other hand, examines the roles that dispersal and vicariance (Fig. 6.1) have played in shaping organismal distributions over *evolutionary time* and utilizes information from *historical* disciplines such as geology, paleontology, and phylogenetics to address its aims.

Despite most biogeographers falling into only one of the two camps, integration of the two fields is necessary in order to best understand *all* the processes that have shaped organismal distributions over ecological *and* evolutionary time (Crisci et al. 2006; Ricklefs and Jenkins 2011; Wiens et al. 2019). Biogeography is an exciting field of study and has experienced a revolution as of late due to many recent technical and analytical advancements (e.g., phylogenomics, molecular clocks, ecological niche modeling, etc.) and has moved from a largely descriptive discipline to one that is hypothesis-driven, complex, and challenging (Crisci 2001; Posadas et al. 2006).

6.1.1 Overview of New World Tarantula Biogeography

The "New World" is composed of three major biogeographical regions, characterized by their distinctive biotas: Nearctic, Neotropical and Andean (Morrone 2015a). The Neotropical region, in a broad sense, encompasses the entirety of Central and South America, southern Mexico, the Caribbean, and South Florida in the United States. The Nearctic region includes northern Mexico and the remainder of North America. Together, these two regions exhibit complex topography with elevations ranging from below sea level to above 6900 m and encompass an incredible diversity of biomes including some of the harshest deserts and most humid forests on Earth. The amazing biodiversity of the Neotropics in particular, as compared with the rest of the planet, is well illustrated by higher plants which have been considered

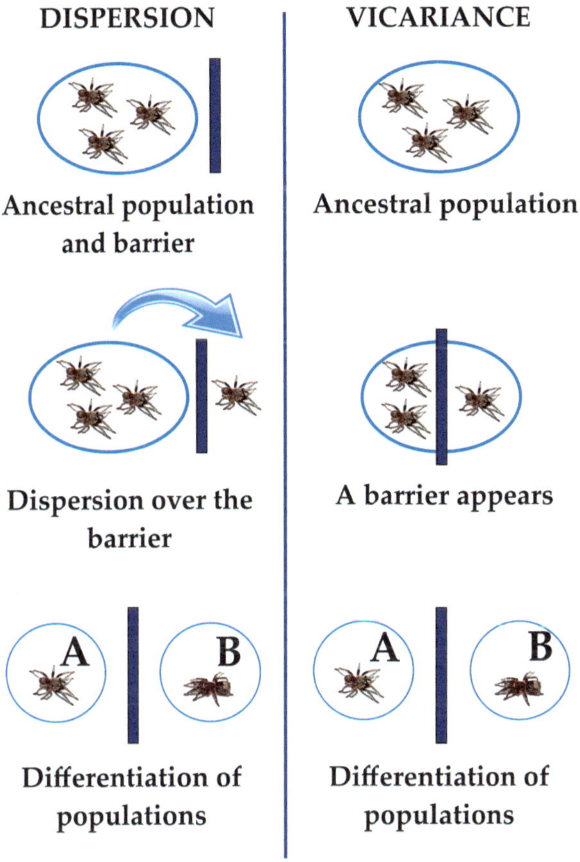

Fig. 6.1 Schematic representation of historical biogeographical explanations for disjunct distributions in closely related taxa. Vicariance describes the process by which a geographic barrier (e.g., a river, mountain range, arid valley) forms and isolates populations. Dispersal is the process by which organisms reach new regions by traversing a preexisting geographic barrier (e.g., trans-oceanic rafting). Vicariance is thought to be the most important explanation for disjunct distributions in tarantulas due to the spiders' sedentary habits and limited dispersal capabilities. Dispersal is probably rare for theraphosids but likely accounted for the distributions for at least some taxa

surrogates for estimating general biological diversity (Usinowicz et al. 2017). This pattern of rich Neotropical diversity is similar for many animal taxa including reptiles, amphibians, mammals, birds, butterflies and many others (Antonelli et al. 2015). Therefore, the Neotropics are generally considered the most biodiverse region of the world and provide a well-suited location to study the drivers, processes, and mechanisms involved in shaping the geographical distribution of organisms.

Tarantulas (family Theraphosidae) are the most diverse lineage of mygalomorph spiders and are particularly well represented in the New World (Table 6.1). They are found in a large variety of environments and inhabit all 10 biodiversity hotspots that

Table 6.1 Diversity and distribution of tarantulas in the New World organized by country based on data obtained from Kambas (2019) and World Spider Catalog (2019)

Region/Country	Subfamilies	Genera	Species
North America			
Mexico	2	15	101
United States	1	1	29
Central America			
Belize	3	5	5
Costa Rica	3	13	35
El Salvador	1	2	4
Guatemala	1	7	9
Honduras	1	2	2
Nicaragua	1	1	2
Panama	4	13	18
South America			
Argentina	2	15	45
Bolivia	3	12	19
Brazil	5	42	188
Caribbean Islands[a]	5	16	63
Chile	2	6	17
Colombia	4	19	32
Ecuador	3	12	19
French Guiana	3	6	7
Guyana	4	8	9
Paraguay	1	7	11
Peru	3	12	36
Suriname	2	2	2
Uruguay	2	7	16
Venezuela	5	20	37

Note: This table should only be used to identify general trends in diversity and distributions. Some values (especially number of species) are undoubtedly erroneous for some countries due to unpublished information (e.g., unpublished taxonomic data, incorrect taxonomic placements, etc.)
[a]For simplicity, all of the Caribbean Islands were grouped into a single category under South America

have been recognized in the New World (https://www.cepf.net/our-work/biodiversity-hotspots), threatened regions known for their exceptional biodiversity and high levels of endemism. But despite their diversity and abundance, potential relevance as top invertebrate predators in ecosystems (Sergio et al. 2008), importance to pharmacological and therapeutic research (Escoubas and Rash 2004; Cardoso et al. 2015), and conservation concern (Bertani and Fukushima 2009; Hendrixson et al. 2015; Mendoza and Francke 2017), theraphosids have been largely neglected in biological research and very little is known about their biogeography (Bertani 2012). But why is the group's biogeography so poorly understood?

The first questions that biogeographers seek to answer about organismal distributions ("What is it?" and "Where is it found?") appear straightforward but have been incredibly complicated for theraphosids. Species and higher-level taxa must be identified and defined to understand patterns of biodiversity. Biogeographical studies assume that the taxonomic entities in question have been accurately delimited and that their distributions have been carefully documented. Neither assumption can be taken lightly in the case of theraphosids. Tarantulas, like most mygalomorph taxa, appear very similar to each other and do not possess the complex and diverse structures found in araneomorph spiders. As a consequence, morphological homogeneity between closely related taxa and high levels of intraspecific variation have posed significant challenges for species delimitation (Prentice 1997; Hamilton et al. 2016) and higher-level classification (Raven 1985). The taxonomic problem is furthered complicated by the proliferation of monotypic genera (e.g., nearly 1 out of every 3 theraphosine genera are known from a single species, World Spider Catalog 2019) whose phylogenetic positions remain unknown.

In addition to inadequately defined taxa (genera and species) and our lack of phylogenetic data, the geographic distributions for most New World tarantulas are poorly known. Distribution records for most taxa are restricted to only a few sites or type localities. Moreover, researchers face a number of challenges when attempting to document the distribution of New World tarantulas. It is uncommon to find large series of theraphosids housed in scientific collections, due in part to the difficulty of finding and collecting these cryptic animals from their retreats (e.g., inside burrows, under rocks, etc.). Only under special conditions it is possible to collect more than a few individuals (Bertani 2001).

Only within the last few decades have New World tarantulas received any significant attention from the scientific community in the form of taxonomic revisions and phylogenetic analyses, constituting a starting point for future biogeographical studies (e.g., Raven 1985; Pérez-Miles et al. 1996; Bertani 2001, 2012; Guadanucci 2014; Perafán and Pérez-Miles 2014; Ferretti 2015; Hamilton et al. 2016; Fukushima and Bertani 2017). The adoption of integrative taxonomy has greatly improved our understanding of species boundaries and distributions in some New World taxa (e.g., Hendrixson et al. 2013, 2015; Hamilton et al. 2011, 2014, 2016; Mendoza and Francke 2017; Montes de Oca et al. 2016; Ortiz and Francke 2016, 2017) and provides a robust methodological approach for future taxonomic revisions. Likewise, advances in molecular systematics and phylogenomics should complement recent phylogenetic treatments of Theraphosidae (Lüddecke et al. 2018; Turner et al. 2018; Foley et al. 2019; see discuss in Chap. 2) and provide the phylogenetic framework required for addressing the family's historical biogeography.

6.2 Biogeography of the Infraorder Mygalomorphae

The spider infraorder Mygalomorphae is fascinating because it possesses numerous characteristics that make it an excellent model for biogeographical research at various spatial and temporal scales. The group is distributed on every continent (except Antarctica), has an ancient origin dating back to the Carboniferous

(Ayoub et al. 2007; Opatova et al. 2019), has a largely sedentary lifestyle, exhibits high habitat fidelity (e.g., Pétillon et al. 2012), and generally has limited dispersal abilities (i.e., most mygalomorphs are restricted to terrestrial locomotion).

Pocock's (1903) seminal paper "On the Geographical Distribution of Spiders of the Order Mygalomorphae" was the first to shed light on some aspects of the biogeography of mygalomorphs. He discussed different dispersal mechanisms in spiders and noted the relevance of limited vagility in Mygalomorphae from a geographical standpoint. He also provided a detailed description of mygalomorph spider distributions (for families, subfamilies, and genera), spearheading subsequent biogeographical research on the group. In the 1970s and 1980s, numerous researchers began investigating the biogeography of Mygalomorphae in more depth. There was also a shift from producing largely descriptive works detailing distributional patterns to testing specific biogeographical hypotheses. Coyle (1971) analyzed the distributions of Antrodiaetidae in the United States and Japan, and proposed a sequence of geological and climatic events that may have shaped their distribution. Raven (1980, 1984a) did the same for Hexathelidae and Migidae in Australia, and for Dipluridae and Barychelidae (Raven 1984b, 1986). Main (1981a, b) presented some biogeographical considerations for these spiders in New Guinea and discussed the eco-evolutionary radiation of mygalomorphs in Australia. These early contributions used a cladistic or phylogenetic biogeographic approach to unveil and explain the geographical patterns of these spiders.

Cladistic or phylogenetic biogeography is an "area biogeography" approach whose aim is to identify patterns in the relationships among areas of endemism that arise repeatedly in the phylogenies for different taxa, which may correspond to shared events in Earth's history (Crisci et al. 2003). Areas of endemism are understood to be biogeographical units that can be described as geographic areas that are defined by the spatial congruence of two or more taxa. Spatial congruence between taxa implies a shared history in space and time (Morrone 1994; Noguera-Urbano 2016) and suggests the same historical process(es) may be responsible for shaping their shared distributions.

More recently, advances in molecular phylogenetics (see Opatova et al. 2019) and the founding of phylogeography (Avise et al. 1987) have had an enormous impact on our understanding of mygalomorph biogeography. Molecular-based phylogenies provide the evolutionary framework for historical biogeographic methods and the incorporation of molecular clocks provide the temporal context for testing specific hypotheses and inferring causal biogeographic events. For example, if a defined tectonic event is hypothesized as the causal explanation for a taxon's disjunct distribution, molecular clock analyses can provide evidence that supports or rejects such a hypothesis (Posadas et al. 2006). If a molecular clock analysis indicates that sister taxa diverged well after the last known connection between two currently isolated regions, a dispersal event—rather than vicariance—can be implicated.

Several phylogenetic and phylogeographic studies of Mygalomorphae (especially non-theraphosid taxa)—with an explicit biogeographic component—have been published during the last two decades. Some of them cover the phylogenetic and geographic breadth of the infraorder (e.g., Bond et al. 2001; Cooper et al. 2011;

Godwin et al. 2018; Hedin et al. 2013, 2015, 2019; Hendrixson and Bond 2007; Opatova and Arnedo 2014a, b; Opatova et al. 2013, 2016, 2019; Rix et al. 2017; Starrett and Hedin 2007; Starrett et al. 2013, 2018). As expected, most studies demonstrate that vicariance has played an important role in shaping the distributions of mygalomorph taxa at all spatial scales (i.e., from local to continental). Phylogeographic research has revealed several important trends consistent with a history of vicariance: (1) long-term persistence of populations; (2) deep phylogenetic divergence between populations; (3) and short-range endemism. Drifting continents, extensional tectonics, submersion of land bridges, climate change, and other phenomena have all been hypothesized to have led to vicariance in mygalomorphs. However, some mygalomorphs are able to passively disperse by ballooning (Coyle et al. 1985; Bell et al. 2005; Ferretti et al. 2013) and other means including short-range (e.g., Hedin et al. 2013; Opatova and Arnedo 2014b) and even trans-oceanic rafting (Harrison et al. 2017).

Non-phylogenetic methods have also been developed for the study of historical biogeography. Such methods only consider the distributions of taxa and are primarily focused on analyzing patterns of endemism. Some of these methods (such as panbiogeography) have been criticized for lacking a phylogenetic framework but they do provide evidence of change in distributions over time and identify hypotheses of primary biogeographical homology, including identifying the geographic origins of entire biotas (Rosen 1988; Nihei 2006; Casazza and Minuto 2009; Morrone 2011, 2014a, 2015b). There are only a few studies that utilize these approaches in mygalomorphs and some concerning theraphosids will be discussed later in the chapter.

6.3 Distribution of New World Tarantulas

The New World is particularly rich in theraphosid spiders [more than half of all described species are reported from there (World Spider Catalog 2019)], comprising representatives of five currently recognized subfamilies: Aviculariinae, Ischnocolinae, Psalmopoeinae, Schismatothelinae, and Theraphosinae (Raven 1985; West et al. 2008; Guadanucci 2014; Lüddecke et al. 2018; Foley et al. 2019). Tarantula diversity is highest in the Neotropics but considerable species-level diversity is also known from the Nearctic region in northern Mexico and the southwestern United States (Table 6.1).

6.3.1 Geographical Distribution of Theraphosidae Subfamilies

The monophyly and composition of Aviculariinae have been widely investigated in recent years (West et al. 2008; Bertani 2012; Fukushima and Bertani 2017; Lüddecke et al. 2018; Turner et al. 2018; Hüsser 2018; Foley et al. 2019). According to the

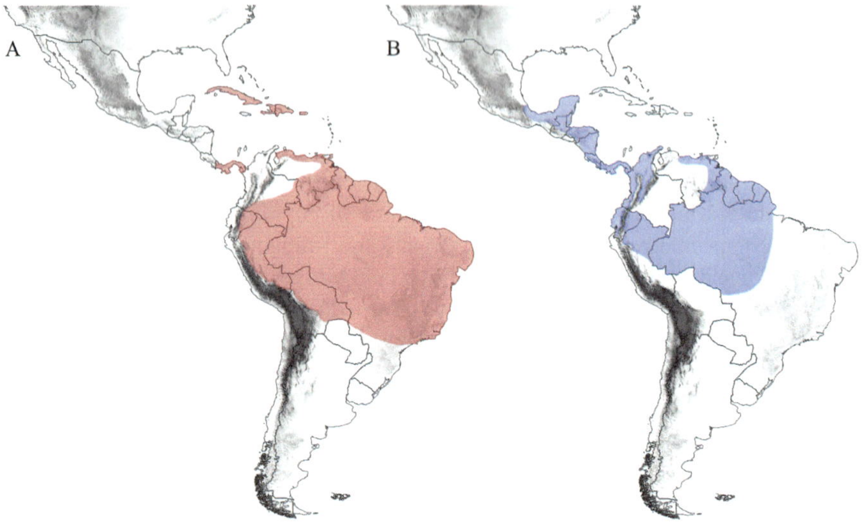

Fig. 6.2 Approximate distribution area of the subfamilies (**a**) Aviculariinae and (**b**) Psalmopoeinae

latest molecular phylogenetic evidence, Aviculariinae is endemic to the Neotropics and comprises the genera *Antillena*, *Avicularia*, *Caribena*, *Iridopelma*, *Pachistopelma*, *Typhochlaena*, and *Ybirapora* (see discuss in Chap. 4). The subfamily is distributed from Central America, north to central South America, and is present on some Caribbean islands (Fig. 6.2a; Bertani 2012; Fukushima and Bertani 2017). Aviculariine species have mainly arboreal habits and predominantly inhabit tropical rainforests, although *Iridopelma*, *Pachistopelma*, and *Typhochalena* are principally distributed in Atlantic forests of northern and northeastern Brazil (Bertani 2012). The genus *Avicularia* has the widest distribution and can be found throughout Costa Rica, Panama, Trinidad and Tobago, Venezuela, Guyana, Suriname, French Guiana, Colombia, Ecuador, Peru, Bolivia, and Brazil. Conversely, the genera *Antillena* and *Caribena* have limited distributions restricted to various Caribbean islands (Fig. 6.2a; Fukushima and Bertani 2017).

The geographic distribution of Aviculariinae in tropical rainforest areas and with few representatives in northern Caatinga and Cerrado provinces (Brazil) could be interpreted as that the group was ancestrally widely distributed before vicariance of the Antilles, Early Jurassic-Early Cretaceous (Morrone 2014b). In this sense, the currently restricted distribution in the open environments of Caatinga and Cerrado would be explained by the later paleoecological changes which turning them to open areas; that is to say, that the presence of only a few species in Caatinga and Cerrado could be a relic (Pérez-Miles and Perafán 2015). In agreement with this hypothesis, vicariance events during the Miocene-Pliocene were used to explain endemism for species of *Iridopelma* and *Pachistopelma* (Bertani 2012). The distribution of aviculariines is also consistent with the Neotropical region proposed by

Morrone (2014b, c) in his biogeographical regionalization, and could constitute additional evidence of the naturalness of this region (Pérez-Miles and Perafán 2015). According to Morrone (2014b, c) the Neotropical region corresponds to the tropical areas of the New World. On the other hand, and interestingly, the subfamily Aviculariinae is not present in the Andes Mountains or other mountain ranges in Central America.

The subfamily Psalmopoeinae (validated by Hüsser 2018) consists of four Neotropical genera including *Ephebopus*, *Psalmopoeus*, *Pseudoclamoris*, and *Tapinauchenius* (see Chap. 4), distributed throughout Central America and northern South America (Fig. 6.2b). Of these, *Ephebopus* is the only fossorial genus (Marshall and West 2008) in an otherwise arboreal subfamily. *Ephebopus* is distributed in northern Brazil, French Guiana, Suriname, and Guyana where they can be found in lowland and upland rainforests and fringing grasslands (West et al. 2008). *Psalmopoeus* is known from northern South America (including Trinidad), Central America, and southern Mexico (Mendoza-Marroquín 2014b); in South America it inhabits mainly in wet Andean forests. *Pseudoclamoris* has an apparent disjunct distribution with some members in Ecuador, Peru, and Colombia and others in French Guiana (Hüsser 2018). *Tapinauchenius* is more broadly distributed throughout northern South America with some species on various Caribbean islands (Hüsser 2018).

The subfamily Ischnocolinae is an enigmatic group with a problematic taxonomic history and there remains considerable confusion regarding its generic composition (Raven 1985; Guadanucci 2014; Lüddecke et al. 2018; Foley et al. 2019). It has the widest distribution among theraphosid subfamilies with species located in the Americas, Asia, Africa, and Europe. A systematic revision of the group is badly needed and recent phylogenetic analyses suggest that the subfamily is not monophyletic (Guadanucci 2014; Foley et al. 2019). Despite this confusion, five ischnocoline genera (which themselves probably do not form a monophyletic group; see Hüsser 2018; Foley et al. 2019) are reported from the New World including *Catumiri*, *Dolichothele*, *Holothele*, *Reichlingia*, and *Trichopelma* (see discuss in Chap. 3). Most species are distributed in tropical areas of central and northern South America, Central America, and some Caribbean islands (Fig. 6.3b). Only *Catumiri* inhabits more temperate zones in South America. A few *Dolichothele*, *Holothele* and *Trichopelma* species inhabit cave environments (Bloom et al. 2014; Fonseca-Ferreira et al. 2017; Guadanucci et al. 2017). Interestingly (and strangely for theraphosids), *Holothele longipes* occupies a very large range in northern South America. It is reported from Bolivia, Peru, Brazil, Colombia, Venezuela, Suriname, and Trinidad and Tobago where it can be found from sea level along the Caribbean and Atlantic coasts to above 2000 m in Colombia's Eastern Range (Guadanucci et al. 2017).

The subfamily Schismatothelinae comprises five Neotropical genera including *Euthycaelus*, *Guyruita*, *Neoholothele*, *Schismatothele*, and *Sickius* (see Chap. 3). This group of taxa formerly placed in Ischnocolinae may not be monophyletic (Guadanucci 2014) but recent phylogenetic analyses suggest a close relationship to

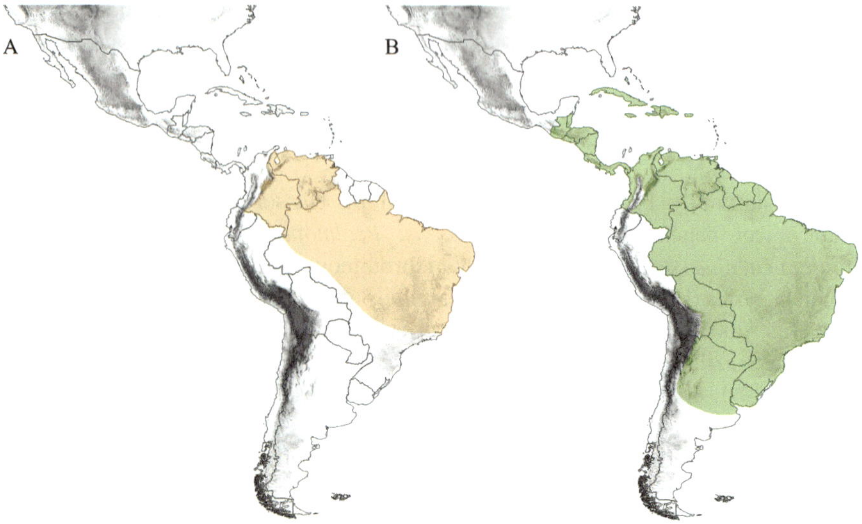

Fig. 6.3 Approximate distribution area of the subfamilies (**a**) Schismatothelinae and (**b**) Ischnocolinae (Neotropical)

Psalmopoeinae (Hüsser 2018; Lüddecke et al. 2018; Foley et al. 2019). With the exception of *Guyruita*, the other genera have very restricted distributions (Fig. 6.3a). *Guyruita* is distributed in Brazilian Cerrado and Atlantic Rainforest areas, and in the Amazon region of Venezuela and northern Brazil (Guadanucci et al. 2007; Fukushima and Bertani 2018). *Neoholothele*, *Euthycaelus*, and *Schismatothele* are distributed in northeastern South America. *Euthycaelus* and *Schismatothele* have a cis-Andean distribution along the Colombia's Eastern Range and Venezuela's Mérida Range and can also be found in northeastern Venezuela, from sea level to about 1700 m (Guadanucci and Weinmann 2014; Valencia-Cuéllar et al. 2019). Likewise, *Neoholothele* is distributed in northeastern Venezuela, Trinidad and Tobago, and central Colombia east of the Eastern Range (Guadanucci and Weinmann 2015). The monotypic genus *Sickius* is distributed in Brazil in the eastern State of Mato Grosso to the Distrito Federal, and south to the States of Mato Grosso do Sul, São Paulo, and western Paraná (Bertani and da Silva 2002).

Theraphosinae is the most diverse theraphosid subfamily and currently comprises 68 genera and more than 500 species (World Spider Catalog 2019; Kambas 2019) including the only species found in the Nearctic. These tarantulas are mainly found in tropical regions of Central and South America, southern Mexico, and some Caribbean islands, but many species inhabit more temperate climates found in the southwestern United States, northern Mexico, Chile, Argentina, and Uruguay (Fig. 6.4). Theraphosines inhabit a great diversity of environments, from tropical wet forests, cloud forests, montane forests, and alpine tundra to sea coastlines, savannas, and deserts.

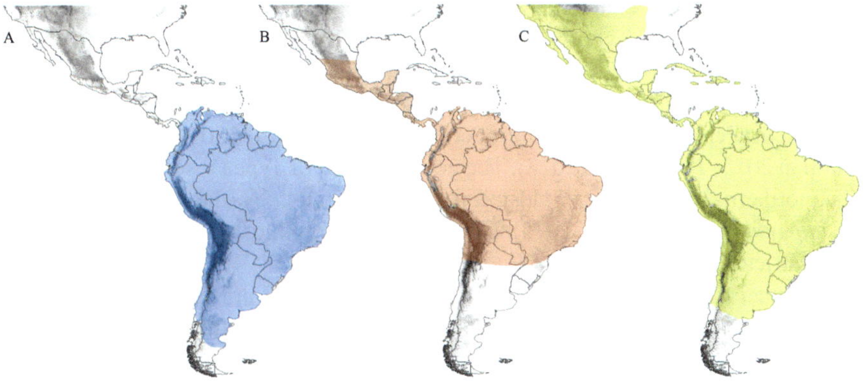

Fig. 6.4 Approximate distribution area of the Theraphosinae tribes (**a**) Gramostolini, (**b**) Hapalopini, and (**c**) Theraphosini

6.3.2 Geographical Distribution of Urticating Setae Types

While not all tarantulas in the New World possess urticating setae (*Psalmopoeus*, *Pseudoclamoris*, *Tapinauchenius*, some troglobitic *Hemirrhagus* species, and the ischnocoline/schismatotheline genera lack them), the presence of these interesting structures is unique to New World tarantulas. Seven different types of urticating setae have been described (reviewed by Bertani and Guadanucci 2013, see also Perafán et al. 2016, and Chap. 9) and some interesting biogeographic patterns have emerged from their distributions. These setae are most diverse in the Theraphosinae (including Types I, III, IV, VI, and VII) but unique types are found in a handful of non-theraphosine taxa including Type II in Aviculariinae and Type V in the psalmopoeine genus *Ephebopus*.

Recently, Turner et al. (2018) investigated the phylogenetic relationships of some theraphosine genera and proposed three tribes for the subfamily including the Grammostolini, Hapalopini, and Theraphosini. The tribes are distinguished primarily by the types and combinations of abdominal urticating setae present in each group, confirming that the different setal conditions are closely linked to deeper-level relationships in the group (Pérez-Miles et al. 1996; Bertani and Guadanucci 2013; Pérez-Miles and Perafán 2015; Ortiz et al. 2018). Turner et al. (2018) also discussed that the tribes are largely congruent with distinct geographic regions, in agreement with Pérez-Miles and Perafán et al. (2015) who studied the geographic distribution of theraphosid abdominal urticating setae types and combinations. It should be noted that some theraphosine genera with unique urticating setae types [Type VI in *Hemirrhagus* (Pérez-Miles 1998) and Type VII in *Kankuamo* (Perafán et al. 2016)] could not be assigned to a tribe (Turner et al. 2018).

Grammostolini is characterized by taxa that possess both Types III and IV urticating setae but never Type I (Turner et al. 2018). The tribe is predominantly distributed in southern South America in temperate countries such as Chile, Argentina, and Uruguay, but also ranges into Paraguay, Bolivia, Amazonian Brazil, and Colombia (Perez-Miles et al. 1996; Perafán and Pérez-Miles 2014; Pérez-Miles and Perafán 2015; Turner et al. 2018). The northern limit of Type IV setae (and therefore the northern limit of Grammostolini) is represented by the Colombian genus *Aguapanela* (Perafán et al. 2015). Consequently, Grammostolini and Type IV setae are restricted to South America (Fig. 6.4a).

Members of the tribe Hapalopini possess Type III urticating setae but never Type I; a few genera also possess Type IV but there is some confusion in the literature regarding how to distinguish among the two types in some genera (Turner et al. 2018). Hapalopini primarily occurs in South America with some genera (e.g., *Cyriocosmus* and *Hapalopus*) widespread throughout the Amazonian and Andean regions, but the tribe also extends northward into Central America and Mexico (Fig. 6.4b). The northern limit of Hapalopini is represented by the Mexican genus *Magnacarina* in the Pacific coast states of Sinaloa and Nayarit (Mendoza et al. 2016).

Finally, members of the tribe Theraphosini uniquely possess Type I urticating setae (except for the genus *Theraphosa*) and most genera have Type I in combination with Type III; Type IV is conspicuously absent in this tribe (Turner et al. 2018). Theraphosini is most diverse in tropical regions of South America but numerous genera are distributed throughout Central America, some Caribbean islands, Mexico, and northward into the southwestern United States (Fig. 6.4c; Turner et al. 2018).

Pérez-Miles and Perafán (2015) concluded that the genera that only possess Type III urticating setae (but without type I or IV) are widely distributed throughout most of the Neotropics with the exception of the Antillean subregion and the Southeastern Amazonian dominion (Morrone 2014c). On the other hand, the co-occurrence of types I and III urticating setae (which is the most common combination in Theraphosinae) is widespread throughout the Neotropics (Morrone 2014c) and into the Neartic (Morrone 2014c).

The widespread distribution of type III setae occurring alone or in combination with other types of urticating setae supports the hypothesis that type III is the ancestral condition for Theraphosinae (Pérez-Miles et al. 1996; Bertani and Guadanucci 2013; Turner et al. 2018). The unique presence of type I (without co-occurrence of type III) is interpreted as a secondary loss of type III setae (Pérez-Miles et al. 1996; Bertani and Guadanucci 2013; Turner et al. 2018) and mainly occurs in northern South America, Central America, and North America. This pattern could reflect differential ecological pressures between these areas given that type I setae are used more in passive defense (Pérez-Miles and Perafán 2015). Moreover, the presence of type III setae in Mexico (e.g., *Brachypelma*) supports the hypothesis that portions of the Neotropical entomofauna migrated into the Nearctic (Halffter 1987; Morrone 2005).

Fig. 6.5 (**a**) High altitude habitat of *Hapalotremus* species at Cordillera Vilcanota in Peru, at about 4000 m high, (**b**) *Hapalotremus carabaya*, habitus. Photos: Tracie Seimon

6.3.3 Elevational and Latitudinal Limits for New World Tarantulas

There are relatively few tarantula species found living at high elevations (most species are distributed below 1600 m) but particularly two genera (*Bistriopelma* and *Hapalotremus*) have been documented from 4000–4500 m high up in the Peruvian Andes (Kaderka 2015; Ferretti et al. 2018), and represent the highest elevations reported for theraphosid spiders (Fig. 6.5). Other genera with at least some high-elevation members include *Euathlus* in Chile and Argentina (up to 3200 m, Perafán and Pérez-Miles 2014; Ferretti 2015), *Cyclosternum* in Ecuador (around 3000 m, Perafán 2017), *Hemirrhagus* in Mexico (around 3000 m, Mendoza-Marroquín 2014a), *Cyriocosmus* in Peru (up to 3000 m, Pérez-Miles and Weinmann 2009), *Aphonopelma* in the United States (up to 2850 m, Hamilton et al. 2016; Hendrixson personal observation), *Bonnetina* in Mexico (around 2700 m, Ortiz and Francke 2017), and from Colombia *Agnostopelma* (up to 3000 m, Pérez-Miles and Weinmann 2010), *Aguapanela* (up to 2400 m, Perafán et al. 2015), and *Neischnocolus* (up to 2400 m, Perafán 2017).

In the United States, the theraphosine *Aphonopelma iodius* appears to reach the northernmost latitudinal limit for tarantulas (near 42°N) in the state of Utah (Hamilton et al. 2016). Interestingly, there are no reliable observations of this species in southern Idaho despite its close proximity (<40 km) to the border. The southern latitudinal limit for theraphosids (near 47°S) is reached by the theraphosine *Euathlus* sp. at the province of Santa Cruz in Argentina (Ferretti personal observation).

6.3.4 Islands

Insularity is one of the most interesting and important concepts for students of biogeography to understand in that it can reveal the evolutionary processes involved in generating and maintaining biodiversity. Islands tend to be less biologically complex than their continental counterparts and therefore provide the ideal geographical and historical settings for studying these phenomena. If the timing of an island's formation is known, such studies can be conducted within a well-defined time frame. Moreover, if an island is part of an archipelago, each island represents a natural repeated experiment and therefore provides increased statistical power for testing ideas related to the patterns and processes of island biogeography (Parent et al. 2008).

MacArthur and Wilson (1967) famously proposed their "Theory of Island Biogeography" and surmised that the number of species on an island is in a dynamic equilibrium between immigration and extinction (Fig. 6.6). They also considered how an island's size and distance from the mainland affect these two parameters (Fig. 6.6). Island size largely affects extinction rate (i.e., small islands have higher extinction rates compared to larger islands) whereas distance from the mainland primarily affects immigration rate (i.e., close islands have higher immigration rates compared to islands further away). In this way, a large island that is close to the mainland is predicted to harbor more species than a small island that is far from the mainland due to a combination of its higher immigration and lower extinction rates. This model remains fundamentally important in biogeography but downplays the importance of in situ speciation and diversification. Species arriving at oceanic islands may undergo speciation and adaptive radiations (Schluter 2000; Ricklefs and Bermingham 2008) such that species richness may not entirely depend on successful immigration and colonization. Doubtless, one reason why the species richness-area relationship has been so well established in different models without consideration of speciation is that the impact of speciation on species richness can be expected to be in the same direction as the impacts of colonization and extinction.

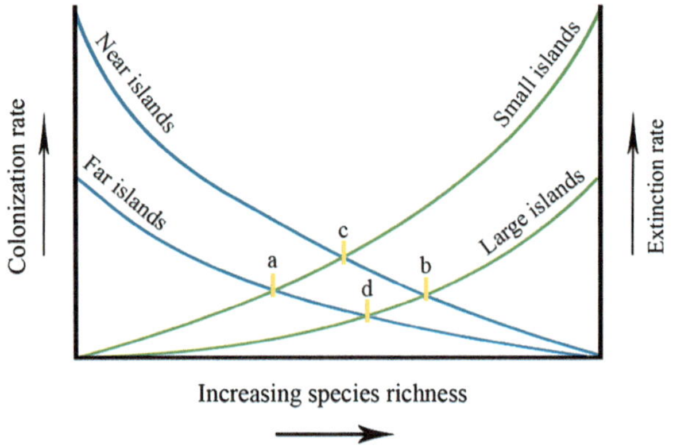

Fig. 6.6 The number of species on an island is set by equilibrium between colonization rates and extinction rates. (**a**) Small far islands have few species in the equilibrium state, (**b**) Large near islands have many species in equilibrium, (**c**) Small and near islands, and (**d**) Large and far islands have an intermediate number of species. Modified version of the island biogeography model proposed by MacArthur and Wilson (1967); taken from: http://www.islandbiogeography.org/

Spiders have figured prominently in studies of island biogeography (e.g., see http://www.islandbiogeography.org, Gillespie 2002; Pugh 2004; Kuntner and Agnarsson 2011; Agnarsson et al. 2018, and many others) but no such studies exist for New World theraphosids despite their occurrence on several Caribbean islands. For those taxa that inhabit these islands, some genera are endemic whereas others include congeners that can also be found on the mainland. The monotypic genera *Antillena*, *Cubanana*, and *Nesipelma* are endemic to specific islands including Hispaniola (Dominican Republic), Cuba, and Nevis Island (Saint Kitts and Nevis), respectively. The genera *Caribena* and *Cyrtopholis* are likewise endemic to various Caribbean islands but are more broadly distributed among them. Genera that occur on Caribbean islands and the mainland include *Acanthoscurria*, *Avicularia*, *Citharacanthus*, *Holothele*, *Neoholothele*, *Phormictopus*, *Psalmopoeus*, *Tapinauchenius*, and *Trichopelma* (World Spider Catalog 2019).

Very little is known about New World theraphosids on islands outside of the Caribbean region. In the United States, *Aphonopelma anax* has been observed on South Padre Island (Hamilton et al. 2016), a narrow barrier island located in the Gulf of Mexico near the main coastline of Texas. Theraphosids are conspicuously absent from California's Channel Islands despite the occurrence of other dispersal-limited arachnids there (e.g., Bryson et al. 2013b). In Mexico, reliable records of tarantulas have been reported from several islands in the Gulf of California, on Isla María Madre off the coast of Nayarit, and on Islas Coronado off the Pacific coast of northern Baja California (see https://www.inaturalist.org/observations?place_id=6793&taxon_id=47424).

There are few islands in the Tropical Eastern Pacific and most of them are located far from the mainland so it is not surprising that there are very few records of theraphosids (and mygalomorphs in general) on these islands. Nevertheless, a species of *Neischnocolus* and an unidentified ischnocoline have been confirmed on Isla Gorgona off the coast of Colombia (Rico-G et al. 2005; Perafán personal observations).

6.3.5 *Caves*

Caves are natural cavities on the Earth's surface that extend to depths where sunlight is unable to penetrate. They are generally formed through mechanical or chemical weathering processes but other types of caves such as lava tubes and talus caves are formed through different geological processes (e.g., volcanic activity and rockfalls, respectively). Most cave systems consist of surface-subsurface hydrological connections within a drainage basin, but lithological and tectonic constraints often limit connectivity of passages across drainage basins. As a consequence, most cave systems are discontinuous and can be considered biogeographic islands (Romero 2009).

Cave-adapted organisms generally have small distributions and exhibit high levels of endemism, with many species restricted to single caves (Christman et al. 2005). These distinctive geographic patterns have led researchers to investigate a variety of questions regarding how organisms colonize, adapt, and persist in subterranean environments (Romero 2009). It is largely recognized that the distributions of cave-adapted organisms can be explained by a combination of both dispersal and vicariance events, reflecting the processes that likely occurred in ancestral surface populations before they colonized the subsurface (Porter 2007).

Although the environmental conditions inside caves remain fairly stable over time (e.g., constant temperature, high humidity), cave-adapted organisms face many unique challenges. In addition to living in a world of total darkness, these organisms must also deal with limited food availability and access to mating partners. Consequently, cave-adapted fauna often present a unique suite of regressive (e.g., reduction or loss of eyes and pigment, thinning of the cuticle) and/or progressive (e.g., appendage elongation, enhanced sensitivity to non-visual stimuli) traits known as troglomorphies (Christiansen 1962). Many studies have thus focused on answering the following question: "Was the ability to live in subsurface habitats derived from preadaptive metabolic, behavioral, or even genetic traits of ancestral individuals, or was it from genetic modification or phenotypic plasticity of individuals over countless generations?" (Romero 2009).

Not all animals associated with cave environments have such adaptations, however. As such, cave fauna can be placed into three categories based on the amount of time they spend in caves. Trogloxenes are temporary cave residents which freely move in and out of caves but never spend their entire life cycle inside them. Troglophiles can spend their entire life cycle in the dark zones of a cave but can also survive outside of caves. Troglobites are the "true cave dwellers" (with troglomorphies) that spend their entire life cycle in the dark zone of caves and cannot survive on the surface (Trajano 2012).

Even though Theraphosidae comprises mostly sedentary spiders (as is the case for mygalomorphs in general), only a few species have been observed associated with caves. *Dolichothele tucuruiense*, *Hapalopus aymara*, *Acanthoscurria geniculata*, *Theraphosa blondi*, *Nhandu coloratovillosus*, *Guyruita metallophila*, *Hapalopus serrapelada*, and *Holothele longipes* have been observed in cave environments but are considered trogloxenes or troglophiles (Fonseca-Ferreira et al. 2017; Guadanucci et al. 2017). The only described troglobitic theraphosids include *Tmesiphantes hypogeus* from two caves in Bahia State, Brazil; *Trichopelma maddeni* (originally placed in Barychelidae but subsequently transferred to Theraphosidae by Guadanucci 2014) from Cueva Seibo in Altagracia Province, Dominican Republic (Bloom et al. 2014); and *Hemirrhagus* spp. in Mexico (see below). An undescribed species that has been discovered from a cave in Colombia is the only theraphosid known that lacks scopulae on all of its tarsi (the presence of claw tuft and Type IV urticating setae confirm its placement in Theraphosidae) (Perafán personal observations). We interpret the loss of scopulae in this specimen as an adaptation for living in caves. The unique presence of claw tufts in the absence of scopulae in Mygalomorphae has only been reported in a species of the paratropidid genus *Melloina* which interestingly also inhabits caves (Bertani 2013; Perafán personal observations).

The genus *Hemirrhagus* is fascinating from a biogeographic and evolutionary standpoint. More than half of the 27 described species have been collected from inside caves (Mendoza-Marroquín 2014a; Mendoza and Francke 2018) and in at least one case, two species (*H. grieta* and *H. billsteelei*) occupy different depths within the same cave (Cueva de la Grieta in Oaxaca). Several cave-dwelling species are bona fide troglobites that possess well-defined troglomorphies (e.g., reduction in size or complete loss of eyes, reduction of ocular pigmentation, elongation of appendages, loss of urticating setae). Future phylogenetic work (also see Pérez-Miles and Locht 2003) should help elucidate a variety of interesting biogeographic phenomena in this captivating genus.

6.3.6 *Exotic Species Introductions*

The native range of tarantulas in the United States (genus *Aphonopelma*) is bound longitudinally by the Pacific Ocean to the west and the Mississippi River Embayment to the east (Hamilton et al. 2016). However, *Tliltocatl vagans* (native to Central America and southern Mexico) has reportedly become established in South Florida (Edwards and Hibbard 1999). The exact origin of the Florida population is unknown but it is suspected to be from exotic pet trade specimens that were released or escaped from captivity. There have been no follow-up studies on this introduced population so it is unclear if these tarantulas have become invasive.

6.4 Historical Events Affecting the Distributional Patterns of New World Theraphosids

The mygalomorph fossil record extends as far back as the Triassic (Selden and Gall 1992; Dalla Vecchia and Selden 2013) but recent phylogenetic and molecular clock analyses suggest that the group is considerably older, placing the root node of Mygalomorphae in the Permian (Garrison et al. 2016), Carboniferous (Opatova et al. 2019), or even the Devonian (Hedin et al. 2019). The oldest mygalomorph known to date is *Rosamygale rauvogeli*, found at the Triassic locality in the Vosges, France (Selden and Gall 1992) and was provisionally assigned to the family Hexathelidae. A Triassic atypoid was described from Italy (Dalla Vecchia and Selden 2013) while other Mesozoic mygalomorphs include members of "Dipluridae" found in Brazil dating from 115 Ma (Selden et al. 2006) and a representative of "Nemesiidae" found in the United Kingdom dating from 90 Ma (Selden 2002). A particular rich mygalomorph fauna was described from the Cretaceous (Eskov and Zonshtein 1990), including representatives of the family Atypidae, Antrodiaetidae and Mecicobothriidae. Those authors proposed with their findings that could be related with an "age of mygalomorphs" during the Mesozoic when mygalomorphs could be the predominant spiders.

But what about the age of Theraphosidae? Despite being the most diverse mygalomorph family with approximately 1000 described extant species, only three fossils have been convincingly assigned to the group [all of Neogene (Miocene) age]: *Ischnocolinopsis acutus* from Dominican amber (Wunderlich 1988) and two species (*Hemirrhagus* sp. and an unidentified genus) from Chiapas amber (García-Villafuerte 2008; Dunlop et al. 2008). Two recent studies (Ortiz et al. 2018; Opatova et al. 2019), however, place the root node of Theraphosidae in the Cretaceous (discussed in more detail in the next section).

Below, we summarize the main paleoenvironmental changes of the New World from the Cretaceous to the present (with some earlier historical context), emphasizing the relationships between biomes and the geological forces that, through different climatic-environmental factors, could have driven the evolution and distribution of the New World tarantula fauna. Because Theraphosidae likely originated in Gondwana (see Opatova et al. 2019), much of the following review concerns the history of South America.

6.4.1 *Origin and Diversification of the New World Theraphosid Fauna*

The most significant events concerning the origin and diversification of the New World tarantula fauna are related to the isolation of South America from Gondwana, dramatic changes in land-sea geographical patterns (e.g., continental drift, changes in sea level, formation of dispersal corridors), uplift and deformation events of the American Cordillera, additional consequences of plate tectonics during the Cenozoic, and climatic variations during the Pleistocene (Gentry 1982; Palminteri and Powell 2001; Rull 2018).

North and South America (as parts of Laurasia and Gondwana, respectively) were joined together as part of the supercontinent Pangea during the Carboniferous up until their initial separation in the early Jurassic ca. 185 Ma (Veevers 2004). From then, Gondwana continued fragmenting and South America began separating from Africa in the early Cretaceous (ca. 130–110 Ma). The South America-Australia connection (via Antarctica) occurred up until at least the late Cretaceous (ca. 80 Ma). During the Oligocene (ca. 30 Ma) South America completed its transformation into an island continent after its definitive separation from Antarctica and the opening of Drake Passage or Sea of Hoces.

The composition of South America's biota was and is very much dependent upon those periods of connection and fragmentation. Until the early Cretaceous, the flora and fauna of South America were connected to the remainder of Gondwana through a system of corridors. The eventual separation and isolation of South America ultimately resulted in the evolution of the continent's unique biota (Smith and Klicka 2010). At the beginning of the Cenozoic, however, occasional long-distance dispersal events and temporary corridors (e.g., land bridges, island arcs) created opportunities for organisms to colonize South America after its split. The organisms that utilized these corridors became an integral part of the South American biota (Simpson 1980; Oliveira et al. 2009; Chávez Hoffmeister 2016).

When the biotic exchanges between South America and the remainder of Gondwana began to decline, they increased between South America and North America. It is generally accepted that North and South America remained separated until the Pliocene but the closure of the Isthmus of Panama may have been completed as early as the Miocene (Coates and Stallard 2013; Montes et al. 2015, but see O'Dea et al. 2016). Regardless, numerous trans-oceanic corridors permitted biotic exchanges between the two continents—albeit sporadically (Woodburne et al. 2006; Woodburne 2010)—prior to the Pliocene (Chávez Hoffmeister 2016). But during the Pliocene, following the *definitive* closure of the Isthmus, there were no longer any major barriers blocking the passage of biotas between North and South America; Central America became a terrestrial bridge linking the two continents. This newly formed corridor permitted mass migration of continental faunas between North and South America, particularly during the late Neogene and early Quaternary, and is known as the Great American Biotic Interchange (GABI) (Woodburne 2010). GABI is considered one of the most important biogeographical events that has shaped the composition of the modern fauna on both continents (Chávez Hoffmeister 2016).

Although GABI is considered to have been a bidirectional process, it appears that movement mostly occurred in the north to south direction, at least initially (Chávez Hoffmeister 2016). The evidence for theraphosids, however, suggests that the migration mainly occurred in the south to north direction. This hypothesis is supported by the presumed Gondwanide origin of the family, the group's high diversity (including endemics) in South America, and recent molecular phylogenetic analyses (Opatova et al. 2019).

As noted above, theraphosids occupy several Caribbean Islands. As is the case for many island biotas, over-water dispersal has been evoked to explain the origin of the Caribbean fauna (e.g., Hedges et al. 1992). However, some researchers think that a land bridge may have also played a role. This idea, known as the GAARlandia hypothesis (Iturralde-Vinent and MacPhee 1999), suggests that a terrestrial connection between South America and the Greater Antilles existed for one- to two million years during the mid-Cenozoic (ca. 34 Ma). GAARlandia has been criticized by some authors (e.g., Ali 2012) but a few recent studies (including some on araneomorph spiders) show evidence in support of it (e.g., Chamberland et al. 2018; Tong et al. 2019). Future biogeographical research on Caribbean theraphosids may provide additional insight on the GAARlandia hypothesis.

The uplift of the Andes during the Mesozoic produced a diverse number of new environments and brought with it a profound impact on the history of the continent's biota (Palminteri and Powell 2001). In the late Eocene, the northern Andes were isolated by a Pacific marine incursion known as the "Western Andean Portal" or "Guayaquil Gap" (Antonelli et al. 2009). Subsequently, with the closure of this marine incursion and the gradual uplift of the Eastern Range, the Andes played an important role that can be summarized by three main points: (1) they have acted as an important dispersal route throughout South America (Antonelli et al. 2009); (2) they represent the most recent separation between the Amazonian and western biotas (Hernández-Camacho et al. 1992); and (3) they have formed a major barrier between the crossing American biota. Their formation has also led to rapid diversification through allopatric fragmentation and ecological specialization in the highlands (Antonelli et al. 2009).

During the Paleocene–Oligocene (ca. 65–55 Ma) in southern South America, the sequences of the Andean dominion showed considerable variation in depositional regimes from north to south, and consequently volcanic rocks were deposited from southwestern Peru to Puna latitudes (Pascual et al. 1996; Ortíz-Jaureguizar and Cladera 2006). The absence of large continental topographic barriers allowed a widespread Atlantic marine incursion during the late Cretaceous and early Paleocene. This seaway covered most of southern South America from Patagonia to Bolivia and Peru. Consequently, the continent was divided into northeastern and southeastern sections (Pascual et al. 1996; Ortíz-Jaureguizar and Cladera 2006). During much of the Paleocene, the Atlantic incursion was transformed into a series of broad alluvial plains and large lake basins (Ortíz-Jaureguizar and Cladera 2006).

During much of the Eocene (ca. 55–33 Ma), the geological characteristics were largely similar to that of the Paleocene. Toward the end of the Eocene, however, tectonic deformation in the Andean basin of Bolivia, Peru, and southern Argentina/Chile occurred (Salfity and Marquillas 1999). This reactivation and uplift of the southern Andes correlated with a sharp marine regression (Haq et al. 1987). By the late Oligocene and early Miocene (ca. 26–20 Ma), the present structural configuration of the Andes began to develop and as a consequence, a new Atlantic marine incursion occurred as well as a Pacific incursion along the west margin of South America (Malumián 1999).

During the mid-Miocene (ca. 17 Ma), landscapes in the southern Neotropics were dominated by the processes that led to the present configuration of the Andes (Uliana and Biddle 1988). Successive Atlantic incursions separated terrestrial environments and spread over virtually all of eastern Argentina, western Uruguay, southern Paraguay, and southeastern Bolivia (Pascual et al. 1996). The northwestern part of this incursion covered widespread areas among the Andean chain and the Guayanian and Brazilian bedrocks (see Rässänen et al. 1995; Webb 1995; Ortíz-Jaureguizar and Cladera 2006). At this time, a significantly higher equator-to-pole thermal gradient developed (Ortíz-Jaureguizar and Cladera 2006). During the late Miocene and early Pliocene (ca. 11–3 Ma), the widespread marine incursions were succeeded by the spread of vast plains, extending from northern Patagonia to central and northern Argentina, Uruguay, along the eastern slopes of the Andes in northern Bolivia, southern Peru, Venezuela, and into the upper Amazon Basin (Pascual et al. 1996; Ortíz-Jaureguizar and Cladera 2006). This event was correlated with a new uplift in the main Andean chain ca. 14–10 Ma that progressively produced a major barrier to moisture-laden south Pacific winds (Riccardi and Rolleri 1980; Ortíz-Jaureguizar and Cladera 2006). This produced a rain shadow in the eastern

Patagonian landscapes that led to the first stages on the differentiation of the present biogeographic subregions in southern South America (Pascual et al. 1996).

The modern Amazon Basin was formed during the Miocene when the Andean uplift caused a major shift in the drainage patterns of northern South America. Up until the early Miocene, most drainages of the western Amazon region were directed to the north. By the late Miocene, the Andean uplift drastically changed this pattern and formed the huge Amazon and Orinoco River systems that drain eastward into the Atlantic Ocean (Hoorn et al. 1995; Rull 2018). During the Pliocene (ca. 5–2 Ma), a new orogenic phase uplifted the Southern Andes in Argentina and Chile, the eastern orographic systems in Argentina, and the Mesopotamia region (Ortíz-Jaureguizar and Cladera 2006). This event resulted in a series of drastic climatic and ecological changes to the region including the extreme aridity and climatic conditions of the Puna (e.g., Atacama Desert) we see today.

The "Tertiary" (Paleogene and Neogene) and Quaternary periods can largely be differentiated on the basis of the amplitude and frequency of environmental changes that occurred during those times. As Ortíz-Jaureguizar and Cladera (2006) stated: "*During the Tertiary, the geologic and biotic records apparently show long periods of uniform environmental conditions, sporadically interrupted by geologically short but highly disruptive periods of change. Contrarily, the climatic changes that occurred during the Quaternary had larger amplitude and frequency.*" As a consequence, the biotic composition (both taxonomic and ecological) largely changed by means of cladogenesis, extinction, and migration during Paleogene and Neogene whereas distributional shifts (e.g., contractions and expansions) occurred more during the Quaternary. During the Pleistocene (2.5–0.01 Ma), the South American fauna was primarily affected by periods of glaciation and the arrival of North American immigrants (Pascual et al. 1996; Ortíz-Jaureguizar and Cladera 2006). Glacial cycles caused the expansion and contraction of arid (e.g., savannas, steppes) and humid (e.g., tropical and subtropical forests) biomes. During cold and dry climates, humid biomes contracted and a savanna corridor formed along the eastern Andes, linking the steppes and grasslands of southern Argentina with those of Colombia. These biomes continued northward across the Panamanian land bridge into southeastern North America (Webb and Rancy 1996). Conversely, when open areas retreated during interglacial periods, rainforest biomes were uniformly distributed throughout the North and South American tropics (Webb and Rancy 1996).

Orogenic and tectonic activity, marine incursions, and climatic fluctuations during Cenozoic Era played similarly important roles in shaping North America's biota. Many of the continent's mountain ranges were uplifted during this time providing the impetus for the origin and spread of several unique biomes (e.g., North American deserts, Madrean pine-oak woodlands) that harbor tarantulas today. The separation of Baja California from mainland Mexico during the Miocene-Pliocene along with its significant latitudinal coverage and diversity of habitats has resulted in the evolution of numerous endemic taxa along the peninsula, especially arthropods (e.g., Johnson and Ward 2002; Williams 1980). In addition, extensional tectonics in the Basin and Range Physiographic Province of northern Mexico and the western United States has presumably isolated some montane biotas through the creation of low-elevation barriers (e.g., Bryson et al. 2013a, b; Graham et al. 2015).

The Pleistocene glaciations had a tremendous impact on the North American fauna, especially those taxa found in more northern latitudes (e.g., the southwestern United States). The rapidly changing climate during glacial periods affected organisms by shifting their distributions, habitats, and population sizes, causing frequent and repeated extinctions. As a consequence, large numbers of organisms are thought to have experienced a general shift to more southern latitudes and lower elevations during glacial maxima, and often occupying a distributional range very different than the one in which they are found today (Hamilton et al. 2011). During interglacial periods, the opposite trends are expected and these predictions have been supported by numerous phylogeographic studies. Southern populations often exhibit relatively high amounts of genetic variation whereas their northern counterparts tend to possess much lower amounts of variation. It is hypothesized that the greater diversity of alleles found in southern populations is the result of their persistence in refugia which allowed them to accumulate genetic variation during the glacial-interglacial cycles. Conversely, the lack of diversity in northern populations is thought to be due to "founder effects" caused by rapid post-glacial expansion into more northern latitudes (Hewitt 1999, 2004).

6.5 Published and Ongoing Research on the Biogeography of New World Tarantulas

Despite a growing body of literature on the biogeography of mygalomorphs, only a handful of studies have explicitly examined or discussed the biogeography of New World theraphosids. Below, we review some of these studies and also include a brief discussion of ongoing research on the biogeography of North American tarantulas.

6.5.1 Divergence Dating

Very few researchers have investigated phylogenetic relationships involving Theraphosidae but even fewer have incorporated molecular divergence dating analyses into their studies. Recent higher-level research on Mygalomorphae places the divergence between Theraphosidae and Barychelidae in the Cretaceous (Ayoub et al. 2007; Bond et al. 2014; Garrison et al. 2016; Ortiz et al. 2018; Opatova et al. 2019) but only two studies have investigated the timing of divergence between multiple New World genera (Note: Hamilton et al. (2011) and Graham et al. (2015) also employed molecular clocks in their studies but focused on the North American genus *Aphonopelma*, see below for details). Ortiz et al. (2018) included *Hemirrhagus eros* and several theraphosine genera in their investigation of *Bonnetina*. They noted that Theraphosinae exhibits a relatively deep history and rapidly diversified into its three main lineages at a time when South America was isolated. The subfamily remained confined in South America for millions of years before expanding to Central and North America and the Caribbean islands. Interestingly, Ortiz et al. (2018) also suggested that the colonization of Central America and Mexico by Hapalopini may have occurred as a single event during the Oligocene (ca. 30 Ma), long before the rise of the Isthmus of Panama. Dispersal across such a formidable barrier seems unlikely for tarantulas but similar patterns have been observed in other mygalomorphs with limited dispersal capabilities (Opatova and Arnedo 2014b; Harrison et al. 2017). Opatova et al. (2019) included nine tarantulas in their study of mygalomorph phylogeny, six of which were from the New World. The ischnocoline *Catumiri parvum* diverged from the remaining theraphosids during the Cretaceous and several theraphosine genera from Central and North America diverged from each other during the Neogene.

The molecular phylogenies presented by Lüddecke et al. (2018) and Foley et al. (2019) have improved higher-level sampling for Theraphosidae but neither study included divergence dating analyses.

6.5.2 South America

Bertani (2001) was among the first to explicitly investigate the biogeography of New World theraphosids. He used cladistic biogeography to investigate patterns of endemism and identified a few biogeographical patterns, including a disjunction between a larger northwestern track and smaller southeastern track (Bertani 2001). The proposed northwestern track included Central America, the northwestern part of South America, and the eastern part of the Amazon. This is supported by the geographical distribution of the genera *Theraphosa*, *Pamphobeteus*, *Xenesthis*, *Brachypelma*, *Megaphobema*, and *Sericopelma*. A probable vicariant event related to the division of the Amazon by a lake along the Amazonas, Madeira and Mamoré Rivers during the Late Cretaceous (Amorim and Pires 1996) was considered responsible for the separation (Bertani 2001).

A southeastern track was found to be correlated with the geographical distributions of *Nhandu* and *Vitalius* which are primarily found in the Atlantic Forest of southeastern Brazil (Bertani 2001). It was presumed that their distributions could be related to historical geological events that took place in the region. For example, the formation of the Serra do Mar and Serra da Mantiqueira mountain systems (Oligocene to Pleistocene) and the peripherial depression along the eastern boundary of Paraná Basin (Cenozoic Era) are congruent with the observed distributions (Ross 1996; Bertani 2001).

Bertani (2001) also found that *Proshapalopus amazonicus* was separated from its sister group (formed by *P. anomalus* and *P. multicuspidatus*) by a region characterized by more open vegetation. A possible explanation for the absence of *Proshapalopus* in this area could be due to an extinction event that resulted from environmental changes associated with the Quaternary glacial cycles (Vivo 1997).

Recently, *P. marimbai* was described from the Chocó Region of western Colombia, greatly expanding the known distribution of the genus (Perafán and Valencia-Cuellar 2018). Even though the authors warn about the inclusion of this species within *Proshapalopus* due to a lack of taxonomic studies in the region, they argued that if the species does indeed belong to genus, it suggests a highly fragmented evolutionary history between the Amazon Forest, Atlantic Forest, and Chocó Region. The environments where these species are currently distributed have similar characteristics: dense vegetation, high temperature, and high humidity with significant amounts of rain. This provides additional evidence of the connections between the eastern and western biota of northern South America, since separated by the uplift of the northern Andes (Antonelli et al. 2009; Perafán and Valencia-Cuellar 2018).

Guadanucci (2011) constructed an area cladogram from a cladistic analysis of the genus *Oligoxystre* (since synonymized with *Dolichothele*) and provided a discussion related to the origin of the group. Based on the cladogram, he concluded that the monophyly of species distributed in the northeastern Brazilian Cerrado (*D. bolivianum*, *D. dominguense*, and *D. diamantinensis*) suggests a single diversification event within the region as well as a historical relationship between the Cerrado and Caatinga plus Amazonia. On the other hand, observed sympatry between two distantly related species in Caatinga (*D. caatinga* and *D. rufoniger*) suggested the lineages diverged allopatrically but one may have dispersed after its original location. Finally, Guadanucci (2011) highlighted the importance of the Serra do Espinhaco and its role in shaping distribution patterns of *Dolichothele*. It represents the longest pre-Cambrian orogenic belt of the Brazilian territory (ca. 1200 km) and is composed of a series of hills and tablelands that vary between 900 and 2000 m in elevation. The range separates three major river basins (Bacia do Rio São Francisco, Rio Jequitinhonha, and Rio Doce) and represents the limit for the three major biomes in Brazil (Caatinga, Cerrado, and Atlantic Forest) (Almeida-Abreu and Renger 1998; Guadanucci 2011). Not surprisingly, the mountain range has been recognized for its high levels of endemism.

Ferretti et al. (2012) took a formal approach to understanding the historical biogeography of the genus *Cyriocosmus* in the Neotropics. The authors proposed that the genus could have originated in the areas of Amazonian and Paramo-Punan. During the late Oligocene and early Miocene, those areas could be considered a single unit due to their homogeneous climates (Donato 2006; Ortíz-Jaureguizar and Cladera 2006), and in the pre-Quaternary, the South American Neotropical biota expanded northward to Central America and Mexico and southward to Paramo-Punan and the Andean region (Ferretti et al. 2012). Some of the vicariant events identified that could have shaped the present geographical distribution of *Cyriocosmus* species appear related to geological changes during the Miocene. For example, the separation of an Amazonian clade (formed by *C. fasciatus* and *C. fernandoi*) from the Caribbean clade (*C. elegans* and *C. leetzi*) could be the result of the Atlantic marine transgression and a new phase of Andean orogenesis during the Middle and Late Miocene (Ferretti et al. 2012). In addition, the equator-to-pole thermal gradient registered during the Middle Miocene could have acted as a barrier that split *Cyriocosmus bertae* in the north from *C. versicolor* in the south (Ferretti et al. 2012).

Bertani (2012) discussed the biogeography of the aviculariine genera *Typhochlaena*, *Pachistopelma*, and *Iridopelma* in northeastern Brazil. He noted that the distributions of congeners never overlapped and most of the identified areas of endemism were concordant with river systems as has been proposed for Atlantic rainforest in northeastern and southeastern Brazil (Pellegrino et al. 2005; Bertani 2012). One of these endemic regions (occupied by *Iridopelma hirsutum*, *Pachistopelma rufonigrum*, and *Typhochlaena curumim*) is located from the State of Ceará–Río Grande do Norte southward to the State of Alagoas (Bertani 2012). To the south, a second area of endemism (occupied by *Iridopelma zorodes*, *Pachistopelma bromelicola*, and *Typhochlaena seladonia*) was identified between Rio São Francisco and Rio Paraguaçú. An additional area of endemism (occupied by *Typhochlaena pschoali*, *Avicularia diversipes* and *A. gamba*) from Rio Paraguaçú south to Rio Jequitinhonha was proposed. Finally, the southernmost area of endemism for aviculariines was recognized between Rio Doce and Rio Paraíba do Sul with *Typhochlaena amma* as the species endemic to this region (Bertani 2012).

At least two hypotheses have been proposed to explain the existence of regional endemism within the Atlantic forest of northeastern and southeastern Brazil (Pellegrino et al. 2005). The first states that the region experienced a semiarid climate through the late Miocene and early Pliocene. Later in the Pliocene, however, the region experienced higher humidity and rainfall resulting in the formation of the major river systems that are observed today. These rivers bisected deposits of the Barreiras Formation (Suguio and Nogueira 1999) and ultimately isolated the interriverine regions from each other. Other authors (e.g., Carnaval and Mortiz 2008) provide evidence for the existence of forest refugia during the Quaternary that are congruent with areas of endemism known for Atlantic Forest in northeastern Brazil.

In the same region of the Brazilian Atlantic Forest, Fabiano-da-Silva et al. (2015) provided notes on the biogeography of some species of *Tmesiphantes*. They argued that the occurrence of four sympatric species (three of which are endemic to the southern Bahia) supports the hypothesis that the forests remained relatively stable and served as refugia throughout the Quaternary (Carnaval and Mortiz 2008). The region has also been recognized as relevant area of endemism in the Atlantic Forest (DaSilva and Pinto-da-Rocha 2011; Fabiano-da-Silva et al. 2015).

Perafán (2017) used panbiogeographic analyses to study the distribution and patterns of endemicity of mygalomorph spiders in the Northern Andes, one of the most biodiverse regions of the world (Cuesta et al. 2009). Results from this research suggest that the Northern Andes do not represent a single biogeographical unit, but instead, comprises two distinct units (i.e., a northern and southern section) that were historically separated and isolated during gradual uplift of the range. The northern unit corresponds to the northern portion of the Central and Western Ranges (Colombia), the Eastern Range (Colombia), and the Mérida Range (Venezuela) whereas the southern unit includes the Ecuadorian Andes and southern portion of the Colombian Andes. The northern unit is supported by the distributions of *Aguapanela arvi*, *Cyriocosmus leetzi*, *Euthycaelus amandae*, *E. norae*, *Hapalopus triseriatus*, *Holothele longipes*, *Neischnocolus weinmanni*, *Pamphobeteus fortis*, *P. nigricolor*, *P. verdolaga*, *Psalmopoeus emeraldus*, *Pseudhapalopus spinulopalpus*, *Schismatothele olsoni*, *S. weinmanni*, *Xenesthis immanis*, and *X. intermedia*. The southern unit is supported by *Avicularia purpurea*, *Cyclosternum schmardae*, *Megaphobema velvetosoma*, *Pamphobeteus augusti*, *P. insignis*, *P. vespertinus* and *Psalmopoeus ecclesiasticus*.

The northern section of the Northern Andes also suggests a historical connection between the Magdalena Valley and northern part of the Cauca valley (both in Colombia) (Perafán 2017). Additionally, there is evidence for a historical connection between the Mérida Range and Cordillera de la Costa (Venezuela). The latter region is a major center of endemism for mygalomorphs, harboring approximately one-third of the species known from the country. The distribution patterns of the mygalomorphs also allowed Perafán (2017) to infer different areas of endemism and other biogeographical units of unique co-occurrence of species (Crother and Murray 2011), which are related to historical and current biogeographic processes. These areas were mainly found in the foothills and the inter-Andean valleys, to the north of Ecuador and in the center of Colombia.

6.5.3 Mexico

The theraphosid fauna of Mexico is quite diverse (Table 6.1) and interest in the group has been revitalized in the past decade. Even though few studies have explicitly discussed biogeographical patterns or processes in Mexico, several recent taxonomic revisions and species delimitation accounts have been published that provide the foundation for future biogeographic research (e.g., Mendoza-Marroquín 2014a; Ortiz and Francke 2016, 2017; Candia-Ramírez and Francke 2017; Mendoza et al. 2016; Mendoza and Francke 2017, 2018; Ortiz et al. 2018). In their taxonomic revision of the red-kneed species of *Brachypelma* in Mexico, Mendoza and Francke (2017) noted that the Balsas River Basin acts as a geographic barrier between the morphologically similar species *B. smithi* and *B. hamorii*. Ortiz and Francke (2017) suggested that the surprisingly high diversity of *Bonnetina* in central and southern Mexico can be attributed to the region's geological and ecological complexity as well as its proximity to five biotic provinces. Diversification within *Bonnetina* appears to have largely taken place during the Miocene (Ortiz et al. 2018), a period of intense geological activity in Mexico (e.g., Gans 1997; Molina-Garza et al. 2015).

6.5.4 United States

Although there are some questions regarding the generic placement of tarantulas in the United States (see Turner et al. 2018), the diversity and distribution of these spiders is fairly well understood (Hamilton et al. 2011, 2014, 2016; Hendrixson et al. 2013, 2015; Graham et al. 2015). Biogeographic research on the group has proceeded more slowly, however, because we know so little about related tarantulas in northern Mexico. The origin of the US fauna (1 genus with 29 species) is certainly due to northward dispersal from Mexico but it is unclear *when* or *how many times* this happened. Despite our meager knowledge regarding the biogeography of US theraphosids, two recent phylogeographic studies have helped set the foundation.

Hamilton et al. (2011) investigated phylogeographical patterns in the widely distributed species *Aphonopelma hentzi*. Divergence dating analyses and patterns of mitochondrial haplotype diversity led the authors to suggest that *A. hentzi* likely persisted in southern refugia during the Last Glacial Maximum (LGM) but rapidly expanded northward following the LGM. They concluded that climatic fluctuations in the Pleistocene likely led to habitat retraction and fragmentation during glacial periods and to habitat expansion during interglacial periods for a number of species in the US. This prediction has been confirmed for widespread desert-dwelling species such as *A. gabeli*, and *A. vorhiesi* (Hendrixson unpublished data), can be inferred for some members of the *A. iodius* species group (see Hamilton et al. 2016), and has been documented within various lineages of the *Aphonopelma mojave* group (Graham et al. 2015).

Graham et al. (2015) investigated patterns of diversification in turret-building tarantulas belonging to the *Aphonopelma mojave* group (Hendrixson et al. 2013) in the Mojave and Sonoran Deserts. They identified six major clades within the group that originated during the Miocene with subsequent lineage diversification during the Pleistocene (see above paragraph). Interestingly, the patterns of lineage formation during the Miocene were concordant with the time frame when tectonic activity was known to have displaced regional mountain ranges in the southwestern US (McQuarrie and Wernicke 2005). As a result, Graham et al. (2015) hypothesized that the six clades were the products of vicariance as populations became isolated when displaced mountain ranges formed low-elevation valleys that operated as barriers to dispersal. The authors argued that Miocene extensional tectonics played an important role in the assembly of desert biota in the southwestern US, a hypothesis that had not received much serious attention prior.

Much work remains to be done on the biogeography of US tarantulas. Current research is focused on the assembly and diversification of *Aphonopelma* in the Madrean Archipelago Ecoregion near the US-Mexico border in southeastern Arizona and the "bootheel" of New Mexico. This region is of great biogeographical interest due to its location at the intersection of five major biotic provinces and is renowned for its exceptional biodiversity. Many taxa (including tarantulas) reach their distributional limits here (Hamilton et al. 2016), adding to the region's biogeographic intrigue. Of particular interest are tarantulas distributed in the Madrean woodlands (Hendrixson et al. 2015; Hamilton et al. 2016), a unique habitat consisting of various pines and/or oaks that is only found at higher elevations or in cool/moist canyons associated with the numerous mountain ranges (commonly referred to as the "Madrean Sky Islands") that span the "cordilleran gap" between the Rocky Mountains/Colorado Plateau and Sierra Madre Occidental (Mexico). These mountain ranges are separated from each other by valleys consisting of arid grassland and desert scrub habitat.

As a consequence, the Madrean woodlands where tarantulas are found are also isolated (hence the term "sky island"). Previous research indicates that the Madrean tarantula fauna is monophyletic and that individual mountain ranges are also monophyletic (Hendrixson et al. 2015, Hamilton et al. 2016). These phylogeographic patterns are consistent with a history of vicariance (rather than dispersal) in the group but neither study proposed a biogeographic explanation for these patterns. Based on information that is available for co-distributed organisms, two hypotheses that may account for this pattern of lineage diversification should be tested: climate change (e.g., Wiens et al. 2019) or extensional tectonics (e.g., Bryson et al. 2013a, b). Under the first scenario, Madrean tarantulas may have been widespread at lower elevations when climates were cooler and subsequently became isolated through local extinction when temperatures warmed. Alternatively, Madrean tarantulas may have persisted in contiguous highlands and became isolated when the highlands fragmented and rifted due to extensional tectonics. Additional sampling throughout the Madrean Sky Islands along with divergence dating analyses should resolve this matter.

6.6 Final Comments

Tarantulas are a diverse and ancient lineage that likely originated during the Cretaceous. They have cosmopolitan distribution and have adapted to living in nearly every conceivable terrestrial ecosystem, except Antarctica, from the most humid rainforests and most arid deserts to the harshest andine environments and even deep inside caves. They are long-lived, sedentary animals with limited dispersal capabilities and high levels of endemism, and have embraced fossorial, arboreal, and other lifestyles. It is no coincidence that these remarkable animals are so appealing to biogeographers; they make excellent models for understanding the ecological, evolutionary, and biogeographical processes that have helped shape the planet's terrestrial biota. Insights gained from such studies will also facilitate our understanding of tarantula conservation, especially for imperiled groups that are at risk of extinction due to habitat destruction, climate change, and illegal trade.

Our knowledge of tarantula biogeography is still in its infancy and we are far from understanding these spiders' full potential for biogeographical research. The most daunting challenge is simply being able to answer the questions "What is it?" and "Where is it?" The family's problematic taxonomy and lack of collections have slowed progress in being able to define the fundamental units for biogeographical studies. Despite these issues, recent advances in molecular systematics (including species delimitation), improved analytical methods, and new international collaborations have helped accelerate progress in this field. There are so many interesting questions and it is an incredibly exciting time to be a tarantula biogeographer!

References

Agnarsson I, Van Patten C, Sargeant L, Chomitz B, Dziki A, Binford G (2018) A radiation of the ornate Caribbean 'smiley-faced spiders' – with descriptions of 15 new species (Araneae, Theridiidae, *Spintharus*). Zool J Linnean Soc 182:758–790

Ali JR (2012) Colonizing the Caribbean: is the GAARlandia land-bridge hypothesis gaining a foothold? J Biogeogr 39:431–433

Almeida-Abreu PA, Renger FE (1998) Effects of the West-Congolian belt orogeny on the tectonic evolution of the eastern São Francisco craton during Neoproterozoic times. In: Abstracts of the 14th international conference on basement tectonics. Ouro Preto, Brazil, pp 38–40

Amorim DS, Pires MRS (1996) Neotropical biogeography and a method for maximum biodiversity estimation. In: Bicudo CEM, Menezes MA (eds) Biodiversity in Brazil. A first approach. Conselho Nacional de Desenvolvimento Científico e Tecnológico, Sao Paulo, p 326

Antonelli A, Nylander JAA, Persson C, Sanmartín I (2009) Tracing the impact of the Andean uplift on Neotropical plant evolution. Proc Natl Acad Sci U S A 106:9749–9754

Antonelli A, Zizka A, Silvestro D, Scharn R, Cascales-Miñana B, Bacon CD (2015) An engine for global plant diversity: highest evolutionary turnover and emigration in the American tropics. Front Genet 6:130

Avise JC, Arnold J, Ball RM, Bermingham E, Lamb T, Neigel JE, Reeb CA, Saunders NC (1987) Intraespecific phylogeography: the mitochondrial DNA bridge between population genetics and systematics. Annu Rev Ecol Evol Syst 18:489–522

Ayoub NA, Garb JE, Hedin M, Hayashi CY (2007) Utility of the nuclear protein-coding gene, elongation factor-1 gamma (EF-1γ), for spider systematics, emphasizing family level relationships of tarantulas and their kin (Araneae: Mygalomorphae). Mol Phylogenet Evol 42:394–409

Bell J, Bohan D, Shaw E, Weyman G (2005) Ballooning dispersal using silk: world fauna, phylogenies, genetics and models. Bull Entomol Res 95:69–114

Bertani R (2001) Revision, cladistics analysis, and zoogeography of *Vitalius*, *Nhandu* and *Proshapalopus*; with notes on other theraphosine genera (Araneae: Theraphosidae). Arq Zool 36:265–356

Bertani R (2012) Revision, cladistic analysis and biogeography of *Typhochlaena* C. L. Koch, 1850, *Pachistopelma* Pocock, 1901 and *Iridopelma* Pocock, 1901 (Araneae, Theraphosidae, Aviculariinae). Zookeys 230:1–94

Bertani R (2013) A new species of *Melloina* (Araneae: Paratropididae) from Venezuela. Fortschr Zool 30:101–106

Bertani R, da Silva PI (2002) The first mygalomorph spider without spermathecae: *Sickius longibulbi*, with a revalidation of *Sickius* (Araneae, Theraphosidae, Ischnocolinae). J Arachnol 30:519–526

Bertani R, Fukushima CS (2009) Description of two new species of *Avicularia* Lamarck, 1818 and redescription of *Avicularia diversipes* (C. L. Koch, 1842) (Araneae, Theraphosidae, Aviculariinae)— three possibly threatened Brazilian species. Zootaxa 2223:25–47

Bertani R, Guadanucci JPL (2013) Morphology, evolution and usage of urticating setae by tarantulas (Araneae: Theraphosidae). Fortschr Zool 30:403–418

Bloom T, Binford G, Esposito LA, Alayón Garcia G, Peterson I, Nishida A, Loubet-Senear K, Agnarsson I (2014) Discovery of two new species of eyeless spiders within a single Hispaniola cave. J Arachnol 42:148–154

Bond JE, Hedin MC, Ramirez MG, Opell BD (2001) Deep molecular divergence in the absence of morphological and ecological change in the Californian coastal dune endemic trapdoor spider *Aptostichus simus*. Mol Ecol 10:899–910

Bond JE, Garrison NL, Hamilton CA, Godwin RL, Hedin M, Agnarsson I (2014) Phylogenomics resolves a spider backbone phylogeny and rejects a prevailing paradigm for orb web evolution. Curr Biol 24:1765–1771

Bryson RW, Riddle BR, Graham MR, Smith BT, Prendini L (2013a) As old as the hills: Montane scorpions in southwestern North America reveal ancient associations between biotic diversification and landscape history. PLoS One 8:e52822

Bryson RW, Savary WE, Prendini L (2013b) Biogeography of scorpions in the *Pseudouroctonus minimus* complex (Vaejovidae) from south-western North America: implications of ecological specialization for pre-Quaternary diversification. J Biogeogr 40:1850–1860

Candia-Ramírez DT, Francke OF (2017) Taxonomic revision of the genus *Crassicrus* Reichling & West, 1996 (Araneae: Theraphosidae: Theraphosinae), with the description of additional keels on the embolus. J Arachnol 45:67–98

Cardoso FC, Dekan Z, Rosengren KJ, Erickson A, Vetter I, Deuis JR, Herzig V, Alewood PF, King GF, Lewis RJ (2015) Identification and characterization of ProTx-III [μ-TRTX-Tp1a], a new voltage-gated sodium channel inhibitor from venom of the tarantula *Thrixopelma pruriens*. Mol Pharmacol 88:291–303

Carnaval AC, Mortiz C (2008) Historical climate modeling predicts patterns of current biodiversity in the Brazilian Atlantic forest. J Biogeogr 35:1187–1201

Casazza G, Minuto L (2009) A critical evaluation of different methods for the determination of areas of endemism and biotic elements: an Alpine study. J Biogeogr 36:2056–2065

Chamberland L, McHugh A, Kechejian S, Binford GJ, Bond JE, Coddington J, Dolman CA, Hamilton CA, Harvey MS, Kuntner M, Agnarsson I (2018) From Gondwana to GAARlandia: evolutionary history and biogeography of ogre-faced spiders (*Deinopis*). J Biogeogr 45:2442–2457

Chávez Hoffmeister MF (2016) El origen de la fauna Sudamericana moderna: de Gondwana al Gran Intercambio Americano. En: Pino M. El Sitio Pilauco Osorno, Patagonia Noroccidental de Chile. Universidad Austral de Chile, pp 47–74

Christiansen KA (1962) Proposition pour la classification des animaux cavernicoles. Spelunca 2:76–78

Christman MC, Culver DC, Madden MK, White D (2005) Patterns of endemism of the eastern North American cave fauna. J Biogeogr 32:1441–1452

Coates AG, Stallard RF (2013) How old is the Isthmus of Panama? Bull Mar Sci 89(4):801–813

Cooper SJB, Harvey MS, Saint KM, Main BY (2011) Deep phylogeographic structuring of populations of the trapdoor spider *Moggridgea tingle* (Migidae) from southwestern Australia: evidence for long-term refugia within refugia. Mol Ecol 20:3219–3236

Coyle FA (1971) Systematics and natural history of the mygalomorph spider genus *Antrodiaetus* and related genera (Araneae: Antrodiaetidae). Bull Mus Comp Zool 141:269–402

Coyle FA, Greenstone MH, Hultsch AL, Morgan CE (1985) Ballooning mygalomorphs: estimates of the masses of *Sphodros* and *Ummidia* ballooners (Araneae: Atypidae, Ctenizidae). J Arachnol 13:291–296

Crisci JV (2001) The voice of historical biogeography. J Biogeogr 28:157–168

Crisci JV, Katinas L, Posadas P (2003) Historical biogeography: an introduction. Harvard University Press, Cambridge, p 262

Crisci JV, Sala OE, Katinas L, Posadas P (2006) Bridging historical and ecological approaches in biogeography. Aust Syst Bot 19:1–10

Crother BI, Murray CM (2011) Ontology of areas of endemism. J Biogeogr 38:1009–1015

Cuesta F, Peralvo M, Valarezo N (2009) Los bosques montanos de los Andes tropicales: una evaluación regional de su estado de conservación y de su vulnerabilidad a efectos del cambio climático. Serie Investigación y Sistematización #5. Programa Regional ECOBONA – INTERCOOPERATION. Quito

Dalla Vecchia FM, Selden PA (2013) A Triassic spider from Italy. Acta Palaeontol Pol 58:325–330

DaSilva MB, Pinto-da-Rocha R (2011) História biogeográfica da Mata Atlântica: opiliões (Arachnida) como modelo para suainferência. In: De Carvalho CJB, Almeida EAB (eds) Biogeografia da América do Sul – Padrões e Processos. São Paulo, Roca, pp 221–238

Donato M (2006) Historical biogeography of the family Tristiridae (Orthoptera: Acridomorpha) applying dispersal vicariance analysis. J Arid Environ 66:421–434

Dunlop JA, Harms D, Penney D (2008) A fossil tarantula (Araneae: Theraphosidae) from Miocene Chiapas amber, Mexico. Rev Iber Aracnol 15:9–17

Edwards GB, Hibbard KL (1999) The Mexican redrump, *Brachypelma vagans* (Araneae: Theraphosidae), an exotic tarantula established in Florida. Florida Department of Agriculture Entomology Circular 394:1–2

Escoubas P, Rash L (2004) Tarantulas: eight-legged pharmacists and combinatorial chemists. Toxicon 43:555–574

Eskov K, Zonshtein S (1990) First Mesozoic mygalomorph spiders from the Lower Cretaceous of Siberia and Mongolia, with notes on the system and evolution of the infraorder Mygalomorphae (Chelicerata: Araneae). N Jb Geol Paläont Abh 178:325–368

Fabiano-da-Silva W, Guadanucci JPL, DaSilva MB (2015) *Tmesiphantes mirim* sp. nov. (Araneae: Theraphosidae) from the Atlantic Forest of Bahia, Brazil, biogeographical notes and identification keys for species of the genus. Fortschr Zool 32:151–156

Ferretti N (2015) On three new *Euathlus* tarantulas from Argentina and cladistic analysis of the genus. J Arachnol 43:313–326

Ferretti N, González A, Pérez-Miles F (2012) Historical biogeography of the Genus *Cyriocosmus* (Araneae: Theraphosidae) in the neotropics according to an event-based method and spatial analysis of vicariance. Zool Stud 51:526–535

Ferretti N, Pompozzi G, Copperi S, Schwerdt L (2013) Aerial dispersal by *Actinopus* spiderlings. J Arachnol 41:407–408

Ferretti N, Cavallo P, Chaparro JC, Ríos-Tamayo D, Seimon TA, West R (2018) The Neotropical genus *Hapalotremus* Simon, 1903 (Araneae: Theraphosidae), with the description of seven new species and the highest altitude record for the family. J Nat Hist 52:1927–1984

Foley S, Lüddecke T, Chen D-Q, Krehenwinkel H, Kuenzel S, Longhorn SJ, Wendt I, von Wirth V, Taenzler R, Vences M, Piel WH (2019) Tarantula phylogenomics: a robust phylogeny of multiple tarantula lineages inferred from transcriptome data sheds light on the prickly issue of urticating setae evolution. bioRxiv 501262. doi:https://doi.org/10.1101/501262

Fonseca-Ferreira R, de Almeida ZR, Guadanucci JPL (2017) Diversity of iron cave-dwelling mygalomorph spiders from Pará, Brazil, with description of three new species (Araneae). Trop Zool 3:178–199

Fukushima CS, Bertani R (2017) Taxonomic revision and cladistic analysis of *Avicularia* Lamarck, 1818 (Araneae, Theraphosidae, Aviculariinae) with description of three new aviculariine genera. Zookeys 659:1–185

Fukushima CS, Bertani R (2018) Two new species of *Guyruita* Guadanucci et al., 2007 (Araneae, Theraphosidae) from Brazil. Zootaxa 4370:395–408

Gans PB (1997) Large-magnitude Oligo-Miocene extension in southern Sonora: implications for the tectonic evolution of northwest Mexico. Tectonics 16:388–408

García-Villafuerte MA (2008) Primer registro fósil del género *Hemirrhagus* (Araneae, Theraphosidae) en ámbar del Terciario, Chiapas, México. Rev Iber Aracnol 16:43–47

Garrison NL, Rodriguez J, Agnarsson I, Coddington JA, Griswold CE, Hamilton CA, Hedin M, Kocot KM, Ledford JM, Bond JE (2016) Spider phylogenomics: untangling the Spider Tree of Life. PeerJ 4:e1719

Gentry AH (1982) Neotropical floristic diversity: phytogeographical connections between Central and South America, pleistocene climatic fluctuations, or an accident of the Andean orogeny? Ann Missouri Bot Gard 69:557–593

Gillespie R (2002) Biogeography of spiders on remote oceanic islands of the Pacific: archipelagoes as stepping stones? J Biogeogr 29:655–662

Godwin RL, Opatova V, Garrison NL, Hamilton CA, Bond JE (2018) Phylogeny of a cosmopolitan family of morphologically conserved trapdoor spiders (Mygalomorphae, Ctenizidae) using Anchored Hybrid Enrichment, with a description of the family, Halonoproctidae Pocock 1901. Mol Phylogenet Evol 126:303–313

Graham MR, Hendrixson BE, Hamilton CA, Bond JE (2015) Miocene extensional tectonics explain ancient patterns of diversification among turret-building tarantulas (*Aphonopelma mojave* group) in the Mojave and Sonoran deserts. J Biogeogr 42:1052–1065

Guadanucci JPL (2011) Cladistic analysis and biogeography of the genus *Oligoxystre* Vellard 1924 (Araneae: Mygalomorphae: Theraphosidae). J Arachnol 39:320–326
Guadanucci JPL (2014) Theraphosidae phylogeny: relationships of the 'Ischnocolinae' genera (Araneae, Mygalomorphae). Zool Scripta 43:508–518
Guadanucci JPL, Weinmann D (2014) The spider genera *Euthycaelus* Simon and *Schismatothele* Karsch (Mygalomorphae, Theraphosidae). Zootaxa 3795:275–288
Guadanucci JPL, Weinmann D (2015) Description of *Neoholothele* gen. nov. (Araneae, Theraphosidae, Schismatothelinae). Stud Neotrop Fauna Environ 50:221–228
Guadanucci JPL, Lucas SM, Indicatti RP, Yamamoto FU (2007) Description of *Guyruita* gen. nov. and two new species (Ischnocolinae, Theraphosidae). Revista Brasileira de Zoologia 24:991–996
Guadanucci JPL, Perafán C, Valencia-Cuéllar D (2017) The genus *Holothele* Karsch, 1879: the identity of the type species (Mygalomorphae, Theraphosidae). Zoosystema 39:263–271
Halffter G (1987) Biogeography of the montane entomofauna of Mexico and Central America. Annu Rev Entomol 32:95–114
Hamilton CA, Formanowicz DR, Bond JE (2011) Species delimitation and phylogeography of *Aphonopelma hentzi* (Araneae, Mygalomorphae, Theraphosidae): cryptic diversity in North American tarantulas. PLoS One 6(10):e26207
Hamilton CA, Hendrixson BE, Brewer MS, Bond JE (2014) An evaluation of sampling effects on multiple DNA barcoding methods leads to an integrative approach for delimiting species: a case study of the North American tarantula genus *Aphonopelma* (Araneae, Mygalomorphae, Theraphosidae). Mol Phylogenet Evol 71:79–93
Hamilton CA, Hendrixson BE, Bond JE (2016) Taxonomic revision of the tarantula genus *Aphonopelma* Pocock, 1901 (Araneae, Mygalomorphae, Theraphosidae) within the United States. ZooKeys 560:1–340
Haq BU, Hardenbol J, Vail PR (1987) Chronology of fluctuating sea levels since the Triassic. Science 235:1156–1167
Harrison SE, Harvey MS, Cooper SJB, Austin AD, Rix MG (2017) Across the Indian Ocean: a remarkable example of trans-oceanic dispersal in an austral mygalomorph spider. PLoS One 12:e0180139
Hedges SB, Hass CA, Maxson LR (1992) Caribbean biogeography: molecular evidence for dispersal in West Indian terrestrial vertebrates. Proc Natl Acad Sci U S A 89:1909–1913
Hedin M, Starrett J, Hayashi C (2013) Crossing the uncrossable: novel trans-valley biogeographic patterns revealed in the genetic history of low-dispersal mygalomorph spiders (Antrodiaetidae, *Antrodiaetus*) from California. Mol Ecol 22:508–526
Hedin H, Carlson D, Coyle F (2015) Sky island diversification meets the multispecies coalescent – divergence in the spruce-fir moss spider (*Microhexura montivaga*, Araneae, Mygalomorphae) on the highest peaks of southern Appalachia. Mol Ecol 24:3467–3484
Hedin M, Derkarabetian S, Alfaro A, Ramírez MJ, Bond JE (2019) Phylogenomic analysis and revised classification of atypoid mygalomorph spiders (Araneae, Mygalomorphae), with notes on arachnid ultraconserved element loci. PeerJ 7:e6864
Hendrixson BE, Bond JE (2007) Molecular phylogeny and biogeography of an ancient Holarctic lineage of mygalomorph spiders (Araneae: Antrodiaetidae: Antrodiaetus). Mol Phylogenet Evol 42:738–755
Hendrixson BE, DeRussy BM, Hamilton CA, Bond JE (2013) An exploration of species boundaries in turret-building tarantulas of the Mojave Desert (Araneae, Mygalomorphae, Theraphosidae, *Aphonopelma*). Mol Phylogenet Evol 66:327–340
Hendrixson BE, Guice AV, Bond JE (2015) Integrative species delimitation and conservation of tarantulas (Araneae, Mygalomorphae, Theraphosidae) from a North American biodiversity hotspot. Insect Conserv Diver 8:120–131
Hernández-Camacho J, Walschburger T, Ortíz-Quijano R, Hurtado-Guerra A (1992) Origen y distribución de la biota suramericana y colombiana. In: Halffter G (ed) La Diversidad Biológica de Iberoamérica I, Acta Zoológica Mexicana (nueva serie), Volumen Especial. Xalapa, México, Instituto de Ecología, pp 55–104

Hewitt G (1999) Post-glacial recolonization of European biota. Biol J Linn Soc 68:87–112
Hewitt GM (2004) Genetic consequences of climatic oscillations in the Quaternary. Philos Trans R Soc Lond Ser B Biol Sci 359:183–195
Hoorn C, Guerrero J, Sarmiento GA, Lorente MA (1995) Andean tectonics as a cause for changing drainage patterns in Miocene northern South America. Geology 23:237–241
Hüsser M (2018) A first phylogenetic analysis reveals a new arboreal tarantula genus from South America with description of a new species and two new species of *Tapinauchenius* Ausserer, 1871 (Araneae, Mygalomorphae, Theraphosidae). ZooKeys 784:59–93
Iturralde-Vinent MA, MacPhee RDE (1999) Paleogeography of the Caribbean region: implications for Cenozoic biogeography. Bull Am Mus Nat Hist 238:1–95
Johnson RA, Ward PS (2002) Biogeography and endemism of ants (Hymenoptera: Formicidae) in Baja California, Mexico: a first overview. J Biogeogr 29:1009–1026
Kaderka R (2015) *Bistriopelma*, a new genus with two new species from Peru (Araneae: Theraphosidae: Theraphosinae). Rev Peru Biol 22:275–288
Kambas D (2019) Tarantupedia: an online taxonomic database for the worlds largest spiders. https://www.tarantupedia.com. Last accessed 7 July 2019
Kuntner M, Agnarsson I (2011) Biogeography and diversification of hermit spiders on Indian Ocean islands (Nephilidae: Nephilengys). Mol Phylogenet Evol 59:477–488
Lüddecke T, Krehenwinkel H, Canning G, Glaw F, Longhorn SJ, Tänzler R, Wendt I, Vences M (2018) Discovering the silk road: Nuclear and mitochondrial sequence data resolve the phylogenetic relationships among theraphosid spider subfamilies. Mol Phylogenet Evol 119:63–70
MacArthur RH, Wilson EO (1967) The theory of Island biogeography. Princeton University Press
Main BY (1981a) Eco-evolutionary radiation of mygalomorph spiders in Australia. In: Keast A (ed) Ecological biogeography of Australia. Dr. W. Junk, The Hague, pp 853–872
Main BY (1981b) Some zoogeographic considerations of families of spiders occurring in New Guinea. In: Gressit JL (ed) Biogeography and ecology in New Guinea. Dr. W Junk, TheHague, pp 583–602
Malumián N (1999) La sedimentación y el volcanismo terciarios en la Patagonia extra andina. La sedimentación en la Patagonia extra andina. In: Caminos R (ed) Geología Argentina. Anales del Instituto de Geología y Recursos Minerales, Buenos Aires, pp 557–612
Marshall SD, West R (2008) An ontogenetic shift in habitat use by the Neotropical tarantula *Ephebopus murinus* (Araneae, Theraphosidae, Aviculariinae). Arachnology 14:280–284
McQuarrie N, Wernicke BP (2005) An animated tectonic reconstruction of southwestern North America since 36 Ma. Geosphere 1:147–172
Mendoza J, Francke O (2017) Systematic revision of *Brachypelma* red-kneed tarantulas (Araneae: Theraphosidae), and the use of DNA barcodes to assist in the identification and conservation of CITES-listed species. Invertebr Syst 31:157–179
Mendoza JI, Francke OF (2018) Five new cave-dwelling species of *Hemirrhagus* Simon 1903 (Araneae, Theraphosidae, Theraphosinae), with notes on the generic distribution and novel morphological features. Zootaxa 4407:451–482
Mendoza JI, Locht A, Kaderka R, Medina F, Pérez-Miles F (2016) A new genus of theraphosid spider from Mexico, with a particular palpal bulb structure (Araneae, Theraphosidae, Theraphosinae). Eur J Taxon 232:1–28
Mendoza-Marroquín JI (2014a) Taxonomic revision of *Hemirrhagus* Simon, 1903 (Araneae: Theraphosidae, Theraphosinae), with description of five new species from Mexico. Zool J Linnean Soc 170:634–689
Mendoza-Marroquín JI (2014b) *Psalmopoeus victori*, the first arboreal theraphosid spider described for Mexico (Araneae: Theraphosidae: Aviculariinae). Rev Mex Biodivers 85:728–735
Molina-Garza RS, Geissman JW, Wawrzyniec TF, Alonso TAP, Iriondo A, Weber B, Aranda-Gómez J (2015) Geology of the coastal Chiapas (Mexico) Miocene plutons and the Tonalá shear zone: syntectonic emplacement and rapid exhumation during sinistral transpression. Lithosphere 7:257–274
Montes de Oca L, D'Elía G, Pérez-Miles F (2016) An integrative approach for species delimitation in the spider genus *Grammostola* (Theraphosidae, Mygalomorphae). Zool Scr 45(3):322–333

Montes C, Cardona A, Jaramillo C, Pardo A, Silva JC, Valencia V, Ayala C, Pérez-Ángel LC, Rodríguez-Parra LA, Ramírez V, Niño H (2015) Middle Miocene closure of the Central American Seaway. Science 348:226–229

Morrone JJ (1994) On the identification of areas of endemism. Syst Biol 43:438–441

Morrone JJ (2005) Hacia una síntesis biogeográfica de México. Rev Mex Biodivers 76:207–252

Morrone JJ (2011) Sistemática, biogeografía, evolución: los patrones de la biodiversidad en tiempo-espacio. Facultad de Ciencias, UNAM, Textos de Biología

Morrone JJ (2014a) Parsimony analysis of endemicity (PAE) revisited. J Biogeogr 41:842–854

Morrone JJ (2014b) Cladistic biogeography of the Neotropical region: identifying the main events in the diversification of the terrestrial biota. Cladistics 30:202–214

Morrone JJ (2014c) Biogeographical regionalisation of the Neotropical region. Zootaxa 3782:001–110

Morrone JJ (2015a) Biogeographical regionalization of the world: a reappraisal. Aust Syst Bot 28:81–90

Morrone JJ (2015b) Track analysis beyond panbiogeography. J Biogeogr 42:413–425

Nihei SS (2006) Misconceptions about parsimony analysis of endemicity. J Biogeogr 33:2099–2106

Noguera-Urbano EA (2016) Areas of endemism: travelling through space and the unexplored dimension. Syst Biodivers 14:131–139

O'Dea A, Lessios HA, Coates AG, Eytan RI, Restrepo-Moreno SA et al (2016) Formation of the Isthmus of Panama. Sci Adv 2:e1600883

Oliveira F, Cassola E, Marroig G (2009) Paleogeography of the South Atlantic: a route for primates and rodents into the new world? In: Garber PA et al (eds) South American primates, developments in primatology: progress and prospects. Springer, pp 55–68

Opatova V, Arnedo MA (2014a) From Gondwana to Europe: inferring the origins of Mediterranean *Macrothele* spiders (Araneae: Hexathelidae) and the limits of the family Hexathelidae. Invertebr Syst 28:361–374

Opatova V, Arnedo MA (2014b) Spiders on a hot volcanic roof: colonization pathways and phylogeography of the Canary Islands endemic trap-door spider *Titanidiops canariensis* (Araneae, Idiopidae). PLoS One 9:e115078

Opatova V, Bond JE, Arnedo MA (2013) Ancient origins of the Mediterranean trap-door spiders of the family Ctenizidae (Araneae, Mygalomorphae). Mol Phylogenet Evol 69:1135–1145

Opatova V, Bond JE, Arnedo MA (2016) Uncovering the role of the Western Mediterranean tectonics in shaping the diversity and distribution of the trap-door spider genus *Ummidia* (Araneae, Ctenizidae). J Biogeogr 43:1955–1966

Opatova V, Hamilton CA, Hedin M, Montes de Oca L, Král J, Bond JE (2019) Phylogenetic systematics and evolution of the spider infraorder Mygalomorphae using genomic scale data. bioRxiv 531756. doi:https://doi.org/10.1101/531756

Ortiz D, Francke OF (2016) Two DNA barcodes and morphology for multi-method species delimitation in *Bonnetina* tarantulas (Araneae: Theraphosidae). Mol Phylogenet Evol 101:176–193

Ortiz D, Francke OF (2017) Reconciling morphological and molecular systematics in tarantulas (Araneae: Theraphosidae): revision of the Mexican endemic genus *Bonnetina*. Zool J Linnean Soc 180:819–886

Ortiz D, Francke OF, Bond JE (2018) A tangle of forms and phylogeny: extensive morphological homoplasy and molecular clock heterogeneity in *Bonnetina* and related tarantulas. Mol Phylogenet Evol 127:55–73

Ortíz-Jaureguizar E, Cladera GA (2006) Paleoenvironmental evolution of southern South America during the Cenozoic. J Arid Environ 66:498–532

Palminteri S, Powell G (eds) (2001) Visión de la biodiversidad de los Andes del norte. World Wilelife Fund - WWF, Santiago de Cali

Parent CE, Caccone A, Petren K (2008) Colonization and diversification of Galápagos terrestrial fauna: a phylogenetic and biogeographical synthesis. Philos Trans R Soc Lond Ser B Biol Sci 363:3347–3361

Pascual R, Ortiz-Jaureguizar E, Prado JL (1996) Land mammals: paradigm of Cenozoic South American geobiotic evolution. In: Arratia G (ed) Contribution of Southern South America to Vertebrate Paleontology. Müncher Geowissenschaftliche Abhandlungen, Münich, pp 265–319

Pellegrino KCM, Rodrigues MT, Waite AN, Morando M, Yassuda YY, Sites JRJW (2005) Phylogeography and species limits in the *Gymnodactylus darwinii* complex (Gekkonidae, Squamata): genetic structure coincides with river systems in the Brazilian Atlantic Forest. Biol J Linn Soc 85:13–26

Perafán C (2017) Distribución actual e histórica del infraorden Mygalomorphae (Araneae) en los Andes del norte. PhD thesis (umpublished), Facultad de Ciencias, Universidad de la República, Montevideo, Uruguay

Perafán C, Pérez-Miles F (2014) The Andean tarantulas *Euathlus* Ausserer, 1875, *Paraphysa* Simon, 1892 and *Phrixotrichus* Simon, 1889 (Araneae: Theraphosidae): phylogenetic analysis, genera redefinition and new species descriptions. J Nat Hist 48:2389–2418

Perafán C, Valencia-Cuellar D (2018) *Proshapalopus marimbai*, a new tarantula species (Mygalomorphae, Theraphosidae) and first genus record from Colombia. Trop Zool 31:200–213

Perafán C, Cifuentes Y, Estrada-Gómez S (2015) *Aguapanela*, a new tarantula genus from the Colombian Andes (Araneae, Theraphosidae). Zootaxa 4033:529–542

Perafán C, Galvis W, Gutiérrez M, Pérez-Miles F (2016) *Kankuamo*, a new theraphosid genus from Colombia (Araneae, Mygalomorphae), with a new type of urticating setae and divergent male genitalia. Zookeys 601:89–109

Pérez-Miles F (1998) Notes on the systematics of the little known theraphosid spider *Hemirrhagus cervinus*, with a description of a new type of urticating hair. J Arachnol 26:120–123

Pérez-Miles F, Locht A (2003) Revision and cladistic analysis of the genus *Hemirrhagus* Simon, 1903 (Araneae, Theraphosidae, Theraphinae). Bull Br Arachnol Soc 12:365–375

Pérez-Miles F, Perafán C (2015) Geographic patterns of abdominal urticating setae types in Neotropical tarantulas (Araneae, Theraphosidae). Bol Soc Zool, Uruguay (2ª época) 24: 103–116

Pérez-Miles F, Weinmann D (2009) Two new species of *Cyriocosmus* Simon, 1903 from Peru and the highest altitude record for the genus (Araneae, Theraphosidae, Theraphosinae). Rev Iber Aracnol 17:29–35

Pérez-Miles F, Weinmann D (2010) *Agnostopelma*: a new genus of tarantula without a scopula on leg IV (Araneae: Theraphosidae: Theraphosinae). J Arachnol 38:104–112

Pérez-Miles F, Lucas SM, da Silva PI, Bertani R (1996) Systematic revision and cladistic analysis of Theraphosinae (Araneae: Theraphosidae). Mygalomorph 1:33–68

Pétillon J, Deruytter D, Decae A, Renault D, Bonte D (2012) Habitat use, but not dispersal limitation, as the mechanism behind the aggregated population structure of the mygalomorph species *Atypus affinis*. Anim Biol 62:181–192

Pocock RI (1903) On the geographical distribution of spiders of the order Mygalomorphae. Proc Roy Soc Lond 1:340–368

Porter ML (2007) Subterranean biogeography: what have we learned from molecular techniques? J Cave Karst Stud 69:179–186

Posadas P, Crisci JV, Katinas L (2006) Historical biogeography: a review of its basic concepts and critical issues. J Arid Environ 66:389–403

Prentice TR (1997) Theraphosidae of the Mojave Desert west and north of the Colorado River (Araneae, Mygalomorphae, Theraphosidae). J Arachnol 25:137–176

Pugh PJA (2004) Biogeography of spiders (Araneae: Arachnida) on the islands of the Southern Ocean. J Nat Hist 38:1461–1487

Rässänen ME, Linna AM, Santos JCR, Negri FR (1995) Late Miocene tidal deposits in the Amazonian foreland basin. Science 269:386–390

Raven RJ (1980) The evolution and biogeography of the mygalomorph spider family Hexathelidae (Araneae, Chelicerata). J Arachnol 8:251–266

Raven RJ (1984a) Systematics and biogeography of the mygalomorph spider family Migidae in Australia. Aust J Zool 32:379–390

Raven RJ (1984b) A revision of the *Aname maculata* species group (Araneae, Dipluridae) with notes on biogeography. J Arachnol 12:177–193

Raven RJ (1985) The spider infraorder Mygalomorphae (Araneae): cladistics and systematics. Bull Am Mus Nat Hist 182:1–180

Raven RJ (1986) Revision of the spider genus *Sason* Simon (Sasoninae, Barychelidae, Mygalomorphae) and its historical biogeography. J Arachnol 14:47–70
Riccardi A, Rolleri E (1980) Cordillera Patagónica Austral. Acad Nac Ci 2:1173–1306
Ricklefs R, Bermingham E (2008) The West Indies as a laboratory of biogeography and evolution. Philos Trans R Soc Lond Ser B Biol Sci 363(1502):2393–2413
Ricklefs RE, Jenkins DG (2011) Biogeography and ecology: towards the integration of two disciplines. Philos Trans R Soc Lond Ser B Biol Sci 366(1576):2438–2448
Rico-G A, Beltrán JP, Álvarez A, Flórez-D E (2005) Diversidad de arañas (Arachnida: Araneae) en el Parque Nacional Natural Isla Gorgona, pacífico colombiano. Biota Neotrop 5:99–110
Rix MG, Cooper SJB, Meusemann K, Klopfstein S, Harrison SE, Hrvey MS, Austin AD (2017) Post-Eocene climate change across continental Australia and the diversification of Australian spiny trapdoor spiders (Idiopidae: Arbanitinae). Mol Phylogenet Evol 109:302–320
Romero A (2009) Cave biology: life in darkness. Cambridge University Press
Rosen BR (1988) From fossils to earth history: applied historical biogeography. In: Myers AA, Giller PS (eds) Analytical biogeography. Springer, New York, pp 437–481
Ross JLS (1996) Geografia do Brasil. Editora da Universidade de São Paulo, São Paulo
Rull V (2018) Neotropical diversification: historical overview and conceptual insights. PeerJ Preprints 6: e27294v1. doi:https://doi.org/10.7287/peerj.preprints.27294v1
Salfity JA, Marquillas RA (1999) La cuenca Cretácico-Terciaria del norte argentino. In: Caminos R (ed) Geología Argentina. Anales del Instituto de Geología y Recursos Minerales, Buenos Aires, pp 613–626
Schluter D (2000) The ecology of adaptive radiation. Oxford University Press
Selden PA (2002) First Mesozoic spider from Cretaceous amber of the Isle of Wight, Southern England. Palaeontology 45:973–983
Selden PA, Gall JC (1992) A Triassic mygalomorph spider from the northern Vosges, France. Palaeontology 35:211–235
Selden PA, Casado FDC, Mesquita V (2006) Mygalomorph spiders (Araneae: Dipluridae) from the Lower Cretaceous Crato Lagerstätte, Araripe Basin, north-east Brazil. Palaeontology 49:817–826
Sergio F, Caro T, Brown D, Clucas B, Hunter J, Ketchum J, McHugh K, Hiraldo F (2008) Top predators as conservation tools: ecological rationale, assumptions, and efficacy. Annu Rev Ecol Evol Syst 39:1–19
Simpson GG (1980) Splendid isolation: The curious history of South American mammals. Yale University Press, New Haven, CT, p 266
Smith BT, Klicka J (2010) The profound influence of the Late Pleistocene Panamanian uplift on the exchange, diversification, and distribution of New World birds. Ecography 33:333–342
Starrett J, Hedin M (2007) Multilocus genealogies reveal multiple cryptic species and biogeographical complexity in the California turret spider *Antrodiaetus riversi* (Mygalomorphae, Antrodiaetidae). Mol Ecol 16:583–604
Starrett J, Hedin M, Ayoub N, Hayashi CY (2013) Hemocyanin gene family evolution in spiders (Araneae), with implications for phylogenetic relationships and divergence times in the infraorder Mygalomorphae. Gene 524:175–186
Starrett J, Hayashi CY, Derkarabetian S, Hedin M (2018) Cryptic elevational zonation in trapdoor spiders (Araneae, Antrodiaetidae, *Aliatypus janus* complex) from the California southern Sierra Nevada. Mol Phylogenet Evol 118:403–413
Suguio K, Nogueira ACR (1999) Revisão crítica dos conhecimentos geológicos sobre a formação (ou grupo?) Barreiras do Neógeno e o seupossível significado como testemunho de alguns eventos geológicos mundiais. Geociências, São Paulo 18:461–479
Tong Y, Binford G, Rheims CA, Kuntner M, Liu J, Agnarsson I (2019) Huntsmen of the Caribbean: multiples tests of the GAARlandia hypothesis. Mol Phylogenet Evol 130:259–268
Trajano E (2012) Ecological classification of subterranean organisms. In: White WB, Culver DC (eds) Encyclopedia of caves. Academic, Waltham
Turner SP, Longhorn SJ, Hamilton CA, Gabriel R, Pérez-Miles F, Vogler AP (2018) Re-evaluating conservation priorities of New World tarantulas (Araneae: Theraphosidae) in a molecular

framework indicates non monophyly of the genera, *Aphonopelma* and *Brachypelma*. Syst Biodivers 16:89–107

Uliana MA, Biddle KT (1988) Mesozoic–Cenozoic paleogeographic and geodynamic evolution of southern South America. Rev Bras Geoc 18:172–190

Usinowicz J, Chang-Yang CH, Chen YY, Clark JS, Fletcher C, Garwood NC, Hao Z, Johnstone J, Lin Y, Metz MR, Masaki T, Nakashizuka T, Sun IF, Valencia R, Wang Y, Zimmerman JK, Ives AR, Wright SJ (2017) Temporal coexistence mechanisms contribute to the latitudinal gradient in forest diversity. Nature 550:105–108

Valencia-Cuéllar D, Perafán C, Guerrero RJ, Guadanucci JPL (2019) Schismatothelinae spiders (Araneae, Mygalomorphae, Theraphosidae) from Colombia: four new species and an approach to their diversity. Zootaxa 4545:548–562

Veevers JJ (2004) Gondwanaland from 650–500 Ma assembly through 320 Ma merger in Pangea to 185–100 Ma breakup: supercontinental tectonics via stratigraphy and radiometric dating. Earth Sci Rev 68:1–13

Vivo M (1997) Mammalian evidence of historical ecological change in the Caatinga semiarid vegetation of northeastern Brazil. J Comp Biol 2:65–73

Webb SD (1995) Biological implications of the middle Miocene Amazon seaway. Science 269(5222):361–362

Webb SD, Rancy A (1996) Late Cenozoic evolution of the neotropical mammal fauna. In: Jackson JBC, Budd AF, Coates AG (eds) Evolution and environments in tropical Americas. The University of Chicago Press, Chicago, pp 335–358

West RC, Marshall SD, Fukushima CS, Bertani R (2008) Review and cladistic analysis of the Neotropical tarantula genus *Ephebopus* Simon 1892 (Araneae: Theraphosidae) with notes on the Aviculariinae. Zootaxa 1849:35–58

Wiens JJ, Camacho A, Goldberg A, Jezkova T, Kaplan ME, Lambert SM, Miller EC, Streicher JW, Walls RL (2019) Climate change, extinction, and Sky Island biogeography in a montane lizard. Mol Ecol 28:2610–2624

Williams SC (1980) Scorpions of Baja California, Mexico, and adjacent islands. Occasional Papers of the California Academy of Sciences 135:1–127

Woodburne MO (2010) The Great American biotic interchange: dispersals, tectonics, climate, sea level and holding pens. J Mamm Evol 17:245–264

Woodburne MO, Cione AL, Tonni EP (2006) Central American provincialism and the Great American Biotic Interchange. In: Carranza-Castañeda Ó, Lindsay EH (eds) Advances in late Tertiary vertebrate paleontology in Mexico and the Great American Biotic Interchange, vol 4. Universidad Nacional Autónoma de México, Instituto de Geología and Centro de Geociencias, Publicación Especial, pp 73–101

World Spider Catalog (2019) Version 20.0. Natural History Museum Bern. http://wsc.nmbe.ch. Last accessed 7 July 2019

Wunderlich J (1988) Die fossilen Spinnenim Dominikanischen Bernstein. Beitr Araneol 2:1–378

Chapter 7
Tarantulas and Their Habitats

Witold Lapinski

Abstract New World Theraphosidae inhabit a great variety of habitats from deserts and semi-deserts, grasslands to forests of all types. Many species thrive in man-made habitats such as soccer fields and backyards, others prefer pristine regions. High Andean species occur even above 4000 m a.s.l., others dwell in Mexican and Brazilian caves. Based on their lifestyles, they may be assigned to two main sub-guilds: terrestrial and vegetation dwelling. Many terrestrial species make burrows in the soil (fossorial species), others hide under various objects on the ground like stones or logs without constructing burrows. The vegetation dwellers may use different plant types, mainly trees and their epiphytes up to more than 35 m above ground (arboreal species) or are restricted to certain plants like terrestrial tank bromeliads. Many arboreal species thrive also on buildings. In some species small juveniles use different microhabitats than later instars (ontogenetic habitat shift). Local assemblages may contain several species which either use slightly different areas or heights of the habitat, differ in their basic life style or use the same patch of habitat at different times. However, habitat use by Theraphosidae at various scales and their community ecology have been largely neglected by arachnologists being focused mainly on tarantula taxonomy and systematics. Knowledge of habitat use by tarantulas, both at local assemblage level and at regional level may contribute greatly to a better understanding of the biology of these fascinating and beautiful animals as well as to their conservation.

7.1 Introduction

Everything in the universe is interconnected. Imagine visiting a square kilometer of any habitat; let us say a semi-desert in Mexico or a tropical lowland rainforest along the Amazon. What tarantula assemblage will you find there, consisting of how many species? Which species? And now it becomes even more interesting, now it becomes beautiful. How do they share the habitat? Are there intraspecific differences in life

W. Lapinski (✉)
Munich, Germany

F. Pérez-Miles (ed.), *New World Tarantulas*, Zoological Monographs 6,
https://doi.org/10.1007/978-3-030-48644-0_7

style among instars? What traits are linked to their specific lifestyles? Why are some of the species strongly associated with undisturbed habitats while others may also thrive on football fields or buildings? To what extent are arboreal tarantulas really arboreal and am I afraid of heights preventing me from investigating this?

Scanning the existing literature for basic life history information of Theraphosid species results all too often in rather minimalistic statements, even in the recent millennium. Basically we are stuck in the naming and counting process. Why are we not three or four steps further than that? Most studies on tarantulas deal with their taxonomy and systematics but it seems the chaos is still unresolved. Some authors have included more or less detailed natural history paragraphs to descriptions of species or taxonomic revisions (e.g., Indicatti et al. 2008; West et al. 2008; Mendoza 2014a, b; Kaderka 2015; Fukushima and Bertani 2017; Ferretti et al. 2018), sometimes even mentioning other tarantula species living in sympatry with the described species (e.g., Marshall and West 2008; Ortiz 2008; Ortiz and Bertani 2014; Lapinski et al. 2018). On rare occasions taxonomy and ecology go hand in hand exploring possible segregation mechanisms (Prentice 1997; Bertani 2012). In genus revisions sometimes data concerning habitat or other natural history traits are included (e.g., Fukushima and Bertani 2017), but often species from the same genus are not sympatric (Bertani 2012).

Due to the paucity of ecological studies on local tarantula assemblages it is still impossible to infer patterns of their community ecology and test hypotheses regarding the driving mechanisms behind syntopy and sympatry that may apply to tarantulas. In general, for relatively mobile animals habitat selection may be an important factor (Morin 2011; Rao 2017). They may move among areas and actively choose where they forage, reproduce, or live. Thus, habitat choice may affect community patterns. Factors influencing habitat choice are the avoidance of physiological stress, competitors and enemies, as well as resource availability. For example, the avoidance of predators may result in the cost of reduced opportunities for foraging. Thus, when trying to understand and explain community patterns it is imperative to keep in mind that the patterns may result from animals making choices *before* becoming part of a community and from interactions at biotic and abiotic level *after* they became part of the community (Morin 2011). This calls for a multi-focus approach surveying local communities over longer periods of time instead of rapid assessments.

With respect to habitat use, Rao (2017) introduced the term topophilia arguing that although site selection in spiders depends on multiple factors, substrate characteristics are of primary importance followed by subsequent site evaluation (but see Lapinski and Tschapka 2014). Topophilia contains both behavioral and structural preferences of spiders. It is not static yet constrained and may change with changing conditions of the environment or ontogeny the individual. The author emphasizes the role of microhabitat structure and recognizes three levels of topophilia: (1) Macro level: habitat types and environmental characteristics; (2) Micro level: microhabitats, e.g., certain species or types of plants or soils; and (3) Nano level: certain structural features within microhabitats. Of course there exists some overlap between those levels. For example, *Eustala perfida* (Araneidae) occurred only inside of the forest on tree trunks of large trees with rough bark (Messas et al. 2014). Or *Avicularia* species (Theraphosidae) living in the Amazonian rainforest on tree trunks and in epiphytic bromeliads as well as on palm thatched huts (this chapter). If recent papers on tarantulas have any sort of habitat information they refer mostly to topophilia at macro level.

In the following, I summarize information on habitat use by New World Theraphosidae (Fig. 7.1). Most information was obtained from species descriptions and genus revisions and some from the few existing ecological studies. Additionally, I present a case study comparing the tarantula assemblages of the Sarapiquí region, Costa Rica and Río Napo, Ecuador with special focus on the arboreal lifestyle. Maps were created with QGIS 3.6 Noosa (available at www.qgis.org) using ESRI

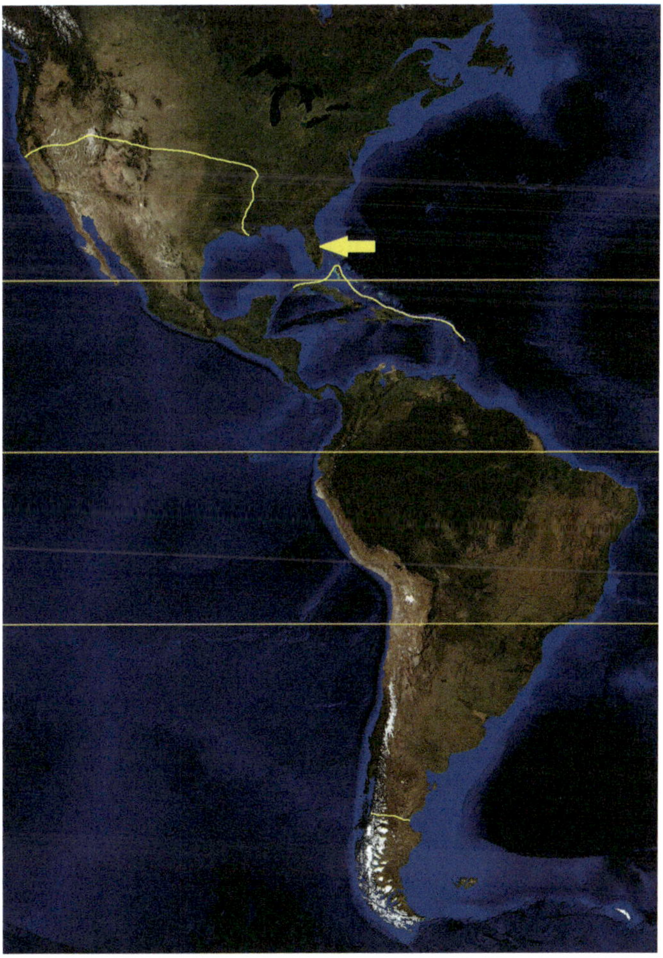

Fig. 7.1 Approximate northern and southern distribution limits of New World Theraphosidae. Arrow indicates the introduced *Tliltocatl vagans* to Florida, USA

raster layer from OGS eb Map Service. Distributional information was added manually according to the information given in the respective publications dealing with the taxon in question.

7.2 USA: *Aphonopelma*

According to Hamilton et al. (2016), in the United States, 29 species of the genus *Aphonopelma* are distributed from west of the Mississippi River to California (but see Turner et al. 2018) (Fig. 7.1, Table 7.6). Additionally, *Aphonopelma* species occur in Mexico and Central America. The North American species of *Aphonopelma* live across a wide range of physical and climatic conditions from hot and arid regions near Death Valley (California, USA) to more temperate high-elevation forests along the Mogollon Rim and Madrean Archipelago (Arizona, USA). *Aphonopelma* species construct mostly silk-lined burrows in the ground and inhabit nearly every habitat throughout their distribution range where trees are usually scarce or completely lacking. Being fossorial, *Aphonopelma* species construct four burrows types: (1) "scrape" or burrow under rock in rocky habitats, (2) burrow with a turreted mound around the entrance, (3) burrow with a distinct crescent mound around the entrance, and (4) typical freestanding burrow without distinct mound around the entrance in desert, grassland, or rocky habitats.

The lineages within the *Paloma* species group are miniaturized species and are usually found in syntopy with a larger species (generally from the *Iodius* species group). Syntopy among members of the *Paloma* species group is rare. The larger *Aphonopelma* species usually are distributed beyond the distribution range of individual members of the *Paloma* species group. Body size frequently affects the niche of an organism (Hanken and Wake 1993) so that this miniaturization and the associated changes in niche utilization may have evolved as a consequence of ancestral character displacement (Hamilton et al. 2016). Thus members of the *Paloma* species group were enabled to use available microhabitats where the larger species were less abundant (Prentice 1997), allowing persistence in syntopy with larger taxa. As various species still are "problematic" in terms of identity, ecological differences among syntopic/sympatric species are difficult to assess. Syntopy among some *Aphonopelma* species has been documented with burrows sometimes in very close proximity (50 cm); in some cases sympatry is only assumed but not documented (Hamilton et al. 2016) but syntopy or sympatry analyses of the assemblages with respect to habitat use are lacking. So besides taxonomy there is still much to be learned about the ecology of North American tarantulas.

Prentice (1997) showed very nicely that the combination of taxonomy and the integration of quite detailed distributional and different life history data is well possible. He studied three sympatric *Aphonopelma* species from the Mojave Desert (USA): *A. iodius* and two significantly smaller species from the *Paloma* species group (*A. joshua* and *A. mojave*) (Fig. 7.2). All three species co-occurred only in the north-western part of the Joshua Tree National Monument. Since *A. joshua* and

Fig. 7.2 Approximate distributions of some selected tarantula taxa. Yellow area: *Aphonopelma mojave*, *A. joshua*, and *A. iodius*, yellow diamond: study site *A. hentzi*. Red line: northern distribution limit of *Brachypelma* spp. in Mexico. Light blue areas: *Hemirrhagus* spp. Yellow dots: Las Vegas and Los Angeles (USA), Mexico City (Mexico). References: see the respective chapter paragraph

A. mojave had similar ecological requirements they were found in syntopy only in the before mentioned habitat. Both species placed their burrows in similar sandy soil of various particle size with similar vegetation communities. They occurred between 550 and 1600 m a.s.l. on flat or gently sloping terrain. *Aphonopelma iodius* in contrast was found at elevations between 300 and 1700 m a.s.l. and had higher tolerance to various soil types (rocky, clayey), and thrived in xeric conditions as well as in cooler and mildly hydric climates. Overall, Prentice (1997) showed that sympatric species may show differences in respective distribution ranges and specific tolerances of substrate quality and climate. Besides that syntopic species of similar body size and habitat use may differ in their hunting and breeding activity patterns. The study indicates the importance of integrated research reaching far beyond taxonomy. However, from that some questions arise: Why is *A. joshua* restricted to such a small area? How does habitat use of different instars differ intra- and interspecifically? At least in *A. mojave* the author found second and third year juveniles living together with the mother tarantula in its burrow.

Punzo and Henderson (1999) examined a population of *Aphonopelma hentzi* in the northern region of the Chihuahuan Desert (USA) (Fig. 7.2). The species occurred throughout this area. Adult females either excavated their own burrows or occupied burrows abandoned by rodents. Adult males sometimes occupied a burrow but were more frequently found within rock crevices or beneath rocks. The early instars were rarely observed and were typically found under rocks or dead vegetation in this area but were never found within a burrow. Larger juveniles rarely occupied burrows but seemed to prefer rock crevices and were sometimes found within crevices along the sides of rocky canyon walls up to 4 m off the ground. Temperature at the deepest point of the burrows was constant during the course of the day (ranging 26–29 °C), being significantly lower than surface temperature at early afternoon (59 °C) and slightly higher during the night. This shows the adaptive advantage of burrows in areas with extreme climate conditions. However, since juveniles apparently do not excavate burrows one might ask: How is the relationship between microclimate of typical shelters of juveniles and the ambient microclimate? Furthermore, as juveniles were found on canyon walls up to 4 m: How high were the walls and how is the vertical distribution of the juveniles? According to Hamilton et al. (2016) *Aphonopelma hentzi* is the most widely distributed tarantula species in the United States. The species inhabits a great variety of elevations and habitats including metropolitan areas and population density can be quite high. Depending on the habitat the tarantulas excavate free standing burrows or scrapes (i.e., burrows under rocks or wood). Although *A. hentzi* may be syntopic to several other *Aphonopelma* species throughout its distribution range not much is known with respect to interspecific differences in habitat use or other interactions. An interesting behavioral shift in association with change of habitat structure was reported by Smith (1994). In an area previously heavily grazed by cattle the grazing was reduced considerably. Consequently the grass height increased significantly and during mating season balloons of silk appeared above the entrances of the burrows of adult females of *A. hentzi*. The balloons were held in position by the grass strands. The function of these balloons was assumed to guide searching males to the females by their pheromones. This example shows how grazing animals may modify certain habitat features which in turn may affect the lives of small ground dwellers such as tarantulas. How does this apply to American bison habitats or the African savannas? Unfortunately this phenomenon has not been examined in greater detail *A. hentzi*.

7.3 México: *Brachypelma* and *Tliltocatl*

Brachypelma species have received considerable attention both from taxonomical and autecological perspective. Currently, 17 *Brachypelma* species are recognized as valid (World Spider Catalog 2020, but see Turner et al. 2018). However, the most recent revision on the genus splits it into *Brachypelma* (8 "red knee" species) and *Tliltocatl* (7 species, former "red rump" *Brachypelma*) (Mendoza and Francke 2020). Certainly the attractive appearance and the resulting popularity of

Fig. 7.3 The "red-knee" tarantula *Brachypelma hamorii*, female, dry forest, foothills of Sierra Madre del Sur, Colima, Mexico. Photo: Eddy Hijmensen

Brachypelma species in pet trade converted them into flagship species among tarantulas for spider enthusiasts and biologists, and even non-biologists. Popularity as pets and habitat destruction make *Brachypelma* and *Tliltocatl* species good model organisms for arthropod conservation. In Mexico two lineages of *Brachypelma* occur in a great variety of habitats (Fig. 7.2, Table 7.6). Along the Pacific coast several "red-knee" species of *Brachypelma* sensu stricto (after Turner et al. 2018) inhabit relatively small neighboring, slightly overlapping areas (Locht et al. 1999; Mendoza and Francke 2017) (Fig. 7.3). The "red-rump" *Tliltocatl* species ranges from Veracruz, Oaxaca, Chiapas, and the entire Yucatán peninsula southward to Costa Rica (Valerio 1980; Arisqueta-Chablé et al. 2010; Locht et al. 1999; Rudloff 2008). *Tliltocatl vagans* (Fig. 7.4) is the only species of the former *Brachypelma*, if not the only theraphosid species, of which several aspects of natural history traits were intensively studied by various researchers (see below). *Brachypelma* and *Tliltocatl* species are usually fossorial, but in *T. vagans* smaller juveniles live rather in rotten logs or beneath some sort of debris (Reichling 2003). However, since almost all of the studies on tarantula natural history were conducted on late instars it is unknown if it also is true for other species of the genus.

7.3.1 "Red-knee" *Brachypelma*

These species are distributed along the Pacific coast of Mexico with *B. emilia* being the northernmost species of the genus and *B. smithi* inhabiting the southernmost range of "red-knee" *Brachypelma* (Locht et al. 1999; West 2005a; Mendoza and Francke 2017). The most detailed account on habitat types and burrow placement in Mexican species of *Brachypelma* was given by West (2005a). According to West (2005a), dry thorn forest is inhabited by all *Brachypelma* species with the exception

Fig. 7.4 The "red-rump" tarantula *Tliltocatl vagans*, male left, female right, tropical lowland evergreen forest, Biosfera de Calakmul, Campeche, Mexico. Photo: Eddy Hijmensen

of *B. auratum* which occurs at higher elevations in topical deciduous forest and pine-oak forest. These forest types are inhabited also by *B. emilia* and *B. klaasi*. Most species are not restricted to pure forests but occur also in transition habitats and some live in cattle pastures. Usually the burrows are found beside some larger objects on the ground like logs, boulders or tree roots and the burrows are excavated underneath those objects. Some species may occur in sympatry where distribution range limits overlap (West 2005a; Mendoza and Francke 2017) but no data are available by what means the respective sympatric species share the habitat. Even less is known about other sympatric tarantula taxa and the ecology of local assemblages. At the Biological Station Chamela (Jalisco), for example, *Brachypelma klaasi* co-occurs with *Bonnetina cyanifemur*, *Magnacarina cancer*, and a species of *Aphonopelma* (Mendoza et al. 2016). Yáñez and Floater (2000) studied the spatial distribution and habitat preference of *B. klaasi* in more detail at Chamela Biological Station. The species occurs at elevations between 300 and 1400 m a.s.l. along the Pacific coast from Tepic, Nayarit in the north to Chamela, Jalisco in the south. The habitat at Chamela was tropical lowland deciduous forest on neutral sandy soils. Spiderlings were found in small holes in the ground, intermediate juveniles occupied temporary burrows and if later instars had encountered suitable sites they inhabited a permanent burrow for many years. Interestingly *B. klaasi* appeared to be a diurnal species being most active in the morning and in the early evening. The spiders adopted either the sit-and-wait strategy or searched actively for prey even climbing up to 2 m in the vegetation. Yáñez and Floater (2000) found that at a spatial scale all specimens combined were aggregated but not at lower scale. Burrow distance ranged between 3 and 257 m (mean = 103 m, n = 13) and many burrows were in south-western areas with low tree and canopy density. No specimens were found in the north-eastern areas of the reserve. Sometimes the spiders needed several square meters of soil for the construction of their elaborate burrows. The authors found high evening humidity (and the associated low evening temperature) to be the best predictor for burrow presence. This may be associated with the ecophysiology of *B. klaasi* allowing the spiders to be diurnal, at least at that site. When the sky remained overcast for several consecutive days, I have sometimes found females of the fossorial tarantula of *Stichoplastoris obelix* (Fig. 7.10d) in a sit-and-wait posture at their burrow entrances in a Costa Rican lowland rainforest (see below). The vegetation dwelling ctenids *Cupiennius coccineus* and *C. getazi* were also found quite often during such days outside of their shelters (pers. obs.). At that Costa Rican site habitat use of large wandering spiders was linked to desiccation resistance (Lapinski and Tschapka 2014). Thus, microclimate may affect both habitat use and activity patterns of spiders. Additionally, Yáñez and Floater (2000) found that the number of prey items was significantly positively related to abundance and species richness of plants around the burrow indicating that plants may indirectly affect habitat choice by this species. Some questions arise from the findings of Yáñez and Floater (2000). Are all instars of *B. klaasi* diurnal or is there a difference among instars? Is *B. klaasi* diurnal both in the dry and in the rainy season? Do the other sympatric tarantula species occur in the north-eastern areas at Chamela, where *B. klaasi* is absent, resulting in a "checkerboard" pattern of species occurrences (Morin 2011)? Does the desiccation resistance of *B. klaasi* differ from the other sympatric species resulting in a physiologically mediated segregation by activity time or by area and its microclimate? As mentioned above, nocturnal sympatric *Aphonopelma* tarantulas may differ in their activity patterns throughout the year (Prentice 1997). Medina Soriano (2006) found in Chiapas at exactly the same localities adults of *Cupiennius salei* (Ctenidae) only during the wet season while adults of *C. chiapanensis* occurred only in the dry season. In the present case of sympatric tarantulas at Chamela much work is still to be done in order to describe the tarantula assemblage and to be able to formulate any hypotheses explaining the mechanisms driving the coexistence of those species.

7.3.2 "Red-Rump" *Tliltocatl*

The species of that lineage occur around the southern and south-eastern distribution limits of the "red-knee" *Tliltocatl* (Locht et al. 1999; West 2005a). *Tliltocatl verdezi* inhabits dry thorn forests in sympatry with *smithi* (West 2005a). *Tliltocatl schroederi* was described from the Acapulco area being sympatric with *T. verdezi* (Rudloff 2003) and hence probably with *B. smithi*, too. *Tliltocatl kahlenbergi* has been described from the town area of Veracruz on the Caribbean coast (Rudloff 2008).

Since the arboreal *Psalmopoeus victori* was described from the rainforest in San Andrés Tuxtla (Veracruz) (Mendoza 2014b) it is possible that in Veracruz it co-occurs with either *T. kahlenbergi* or *T. vagans*, forming an assemblage consisting of at least one fossorial and one arboreal species.

According to Locht et al. (1999), *Tliltocatl vagans* occupies the entire southern part of Mexico. Furthermore the species is distributed all over Belize even on the dry cays offshore (Reichling 2003) and northern Guatemala (Hijmensen 2009). The species was introduced to Florida (USA) around 1986 (Edwards and Hibbard 2003). On the Yucatán peninsula *T. vagans* is sympatric with *T. epicureanum* (Arisqueta-Chablé et al. 2010) although their respective ecological differences have not been assessed yet. Similar to the above mentioned *Aphonopelma* species from the Mojave Desert (USA) (Prentice 1997) in addition to slight differences in habitat use one segregating mechanism of both *Tliltocatl* species might be the mating season. Most males of *T. vagans* were found in July and August whereas most males of *T. epicureanum* were collected in October (Arisqueta-Chablé et al. 2010). Hijmensen (2009) found in Calotmul on a recently burnt corn field three sympatric fossorial tarantula species: *Tliltocatl epicureanum*, *Crassicrus lamanai*, and *Aphonopelma* cf. *stoicum*. Various aspects of life history and ecology of the widely distributed *Tliltocatl vagans* were examined by several researchers during the last two decades. It is ecologically by far the most adaptive species of the genus thriving in a great variety of different habitats from tropical deciduous and tropical evergreen forest to cattle pastures, plantations, the respective transitional habitats, residential lawns, garbage dumps, and soccer fields (Reichling 2003; West 2005a; Machkour-M'rabet et al. 2005; Hijmensen 2009). In Campeche (Mexico) the species was not found in the interior of primary and secondary forests but preferred clearly rather disturbed sites, forming there very dense populations. This suggests that originally *T. vagans* inhabited open early succession areas being thus preadapted to inhabit anthropogenically modified habitats. However, within those anthropogenically disturbed areas, densities of adult females were more negatively affected by intensity of human activity than juveniles (Machkour-M'rabet et al. 2005). In Belize *T. vagans* was also strongly associated with highly disturbed areas living in dense populations. Shaw et al. (2011) report that at the clearing of the research station, which was surrounded by lowland tropical rainforest, more than 100 burrows of juveniles and adults were located over the course of two consecutive years. Individual specimens changed their burrow placement quite often. Regularly flooded areas had almost no burrows. In case of permanent burrows existing in such areas, the burrows were placed closely to objects (rocks, building edges, large roots) that stabilized the soil preventing the burrows from collapsing when inundated. The high adaptability of *T. vagans* to human activity was also reported by Hijmensen (2009). In Orange Walk Town (Belize) a whole colony was found in a hotel backyard. Another dense colony of *T. vagans* and *Crassicrus lamanai* was found on the lawn around the Mayan ruins at Lamanai. In the town of San Ignacio (Belize) close to the border with Guatemala every lawn was full of *T. vagans* of all instars, with spiderlings living in wall crevices and the larger instars in burrows (Hijmensen 2009).

Given the high adaptability of *Tliltocatl vagans* the species appears not to be in danger of extinction due to habitat loss. Au contraire, as anthropogenically modified habitats expand and connect (and pristine forests shrink getting isolated) *T. vagans* may benefit through both gaining suitable habitats and increasing gene flow among formerly isolated disturbed areas (Machkour-M'rabet et al. 2012). Thus *T. vagans* may function as a bioindicator of habitat degradation as proposed for the Argentine tarantula *Grammostola vachoni* by Schwerdt et al. (2018). However, other taxa, not being legally protected by CITES or otherwise, may face negative effects of habitat fragmentation and destruction. Clearly, much more work has still to be done in order to understand habitat use among syntopic and sympatric tarantulas and to being able to implement any protective measures in Mexico and elsewhere.

7.4 Exit Light, Enter Night: Cave Dwellers of Mexico, and Beyond

Caves are very special habitats. Apart from the entrance area the various effects of sunlight in deep caves are nil and climate inside the caves fluctuates little throughout the year. Caves may contain water bodies from small pools to subterranean rivers and in deeper regions of caves the chemical composition water may deviate quite strongly from the "normal" state outside caves. Besides that, caves are mostly very confined and isolated habitats. All this may lead to high endemism due to mostly impeded gene flow among populations. Every cave may have its particular conditions of abiotic factors resulting in unique cave-specific evolution of organisms living there (Culver and White 2005).

Beside species accounts there is virtually nothing known about any biological aspect of cave dwelling Theraphosidae. Most species were collected in extremely low numbers so that the few natural history data are very preliminary. Sixteen out of 27 *Hemirrhagus* species were collected only in caves 20–800 m from the entrance (Mendoza 2014a; Mendoza and Francke 2018, Fig. 7.5). Thus this is the only genus with epigean, troglophile, and troglobitic species. The known distribution range of *Hemirrhagus* is central to southern Mexico (Mendoza and Francke 2018) living thus in sympatry with other Mexican tarantulas (Fig. 7.2, Table 7.6). Cave-living *Hemirrhagus* species are segregated from other sympatric tarantulas by the obligatory cave-dwelling lifestyle. Within caves the spiders mainly occur on the floor and on walls, using rocks and crevices as shelters. While Mendoza (2014a) assumes *H. pernix* to occur in pine forest, Horstkotte et al. (2010) reports the species from the cave Cueva del Azufre. It lived in syntopy with large-sized araneomorphs (Ctenidae, Pisauridae) and Amblypygi. In a feeding experiment the *H. pernix* female consumed all four presented cave mollies that usually live in the toxic water of the cave.

Another new-world tarantula genus that contains both epigean and at least one troglobitic species is *Tmesiphantes*. Bertani et al. (2013) described *T. hypogeus* from two sandstone/quarzitic caves in the state of Bahia (Brazil), found on humid rocky substrates (Table 7.6). Dias et al. (2015) report on a relatively high density of theraphosids in a 150-m long "hot cave" in the state of Sergipe, also north-eastern Brazil. "Hot caves" are characterized by high temperatures and humidity maintained by dense populations of bats. In one case a *Lasiodora* species was observed consuming a bat on the cave floor and at least seven other tarantulas were in its vicinity (Table 7.6). Judging from the photograph the *Lasiodora* specimen belongs to an epigean species suggesting that epigean tarantula species may enter caves and stay there if conditions are favorable. This is also suggested by the encounter of

Fig. 7.5 *Hemirrhagus* sp., female, Mexico. Photo taken in captivity in the lab at UNAM, Mexico. Photo: Eddy Hijmensen

three unidentified tarantulas which lived in shelters among rocks just at the entrance of the "Río Frío" cave in the Mountain Pine Ridge, Belize (Hijmensen 2009). The presented cases show how little we know about caves as habitats for organisms in general and cave-dwelling tarantulas in particular. Caves certainly may become fruitful playgrounds for researchers that are eager to express their precious adventurer genes.

7.5 Brazil: Mata Atlântica

The habitats of north-eastern Brazil are a mixture of different vegetation zones: caatinga = seasonally dry tropical forest (Cardoso et al. 2017), restinga = strings of beaches and sandy dunes covered by herbaceous and arbustive-arboreal vegetation (Rocha et al. 2007), cerrado = Brazilian savanna encompassing open grassland to forest with a discontinuous grass layer (Oliveira and Marquis 2002), and of course the rainforest fragments along the Brazilian Atlantic coast. Due to this mosaic of heterogeneous habitats, which are connected by transitional areas or separated by habitats modified by man, the biodiversity is high and the region shows several zones of endemism with respect to Theraphosids (Bertani 2012). In those various biomes at least five vegetation-dwelling Aviculariinae genera have been recorded: *Avicularia*, *Ybyrapora*, *Iridopelma*, *Pachistopelma*, and *Typhochlaena* (Fig. 7.6, Table 7.6).

With an adult body length between 10 and 20 mm the currently five known *Typhochlaena* species belong to the smallest arboreal theraphosids and show the most deviating lifestyle among this subguild. The species inhabit the lowland Atlantic rainforest, the nearby restinga, and adjacent drier habitats caatinga and cerrado from Paraiba in the northeast to Espirito Santo in the southeast, and Maranhão and Tocantins further to the west (Bertani 2012; Andre and Hüsser 2016). All species were found on vegetation, under loose tree bark or under low leaves of bushes. In areas close to the forest, wandering males may occur also in houses searching for

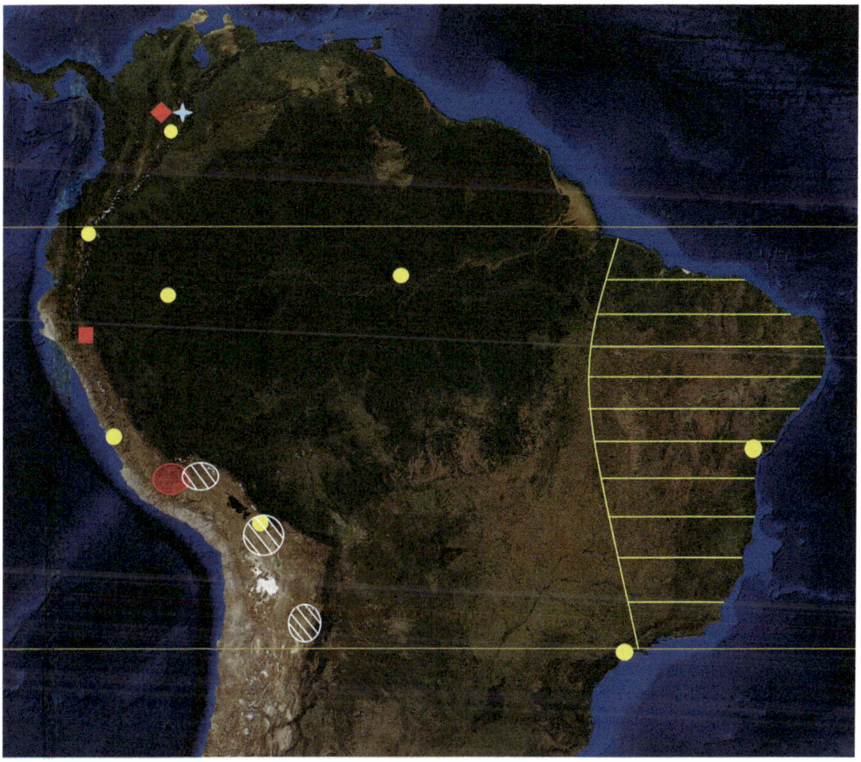

Fig. 7.6 Approximate distributions of some selected South American tarantula taxa. Red diamond: *Aguapanela arvi*, blue cross star: *Agnostopelma* spp., red square: highland *Cyriocosmus* spp., red circle area: *Bistriopelma* spp., white areas: *Hapalotremus* spp., yellow area: *Dolichothele* spp., *Ybyrapora* spp., *Iridopelma* spp., *Pachistopelma* spp., *Typhochlaena* spp., and a few *Avicularia* spp. Yellow dots anticlockwise: Bogotá (Colombia), Quito (Ecuador), Iquitos and Lima (Perú), La Paz (Bolivia), São Paulo, Salvador da Bahia, and Manaus (Brazil)

females. Unlike in other arboreal species the shelters of *Typhochlaena* species are very cryptic refuges that are rather short, located in a small hole or crevice in the bark covered by a trap door. Often the spiders rasp pieces of bark from inside of the shelter and include them into the silk of the trap door thus enhancing the camouflage effect (Almeida-Silva et al. 2008; Bertani 2012; Andre and Hüsser 2016). The miniaturization may have allowed those tarantulas to occupy such microhabitats (Hanken and Wake 1993). In a lowland rainforest area Andre and Hüsser (2016) found some shelters of *T. seladonia* with and without occupant in heights between 1.6 and 5.0 m above ground, in one case even three shelters on the same tree. The authors found *T. seladonia* in syntopy with *Ybyrapora diversipes* and *Iridopelma zorodes*. Due to the small body size and very secretive lifestyle of *Typhochlaena* species thorough ecological field research is quite difficult and it takes much patience and dedication to obtain useful data. Here surely the rope climbing technique (see below and info box) would help to assess more completely the three dimensional habitat use by these beautiful tarantulas.

Canopy Arachnology
Forest canopies are challenging yet fascinating habitats to explore (literature see below). Most studies use fogging to sample tree fauna but they do not allow to actually observe the animals and record their heights above ground, their microhabitats or any type of interaction or behavior. Arthropods living in bromeliads, tree holes or silken tubes may get stuck and not even be detected. Thus scientists have to access the canopy by some means. Canopy cranes, suspension bridges and rafts are mostly very expensive and cannot be used flexibly at any site. The rope climbing technique is inexpensive and allows surveying trees from ground to high crowns; however, it requires good physical condition and certain climbing skills. Here the adventurer and scientist genes are expressed simultaneously. Depending on the training level outer crown regions may be accessed or not.
The basic procedure is to shoot or throw a weight of approx. 200–300 g with a thin line over a strong branch close to the tree trunk. When trees are not more than 30 m high and the understory not dense it is possible to throw a throw bag with a throw line over the branch. As soon as the weight descends to the ground at the other side one attaches the climbing rope to the line and pulls it over the branch. One end of the climbing rope may be either attached to any healthy tree trunk (base anchor) and the loose end of rope to be used for climbing, or attaching the climbing rope directly to the branch using a butterfly knot or a carabiner (crown anchor). When the climbing rope has been installed there are several ways to climb. The most effective is the single rope technique with a Rope Wrench by ISC or ZigZag/Chicane by Petzl. The devices allow the climber to efficiently ascend, to descend without changing from ascending to descending system and to move along the branches into the outer canopy. Equipment can be purchased from various arboriculture suppliers. It is necessary to get proper climbing training and become familiar with the gear before climbing in the rainforest. Bullet ants, bees, wasps, heavy rain and other surprises may challenge you sufficiently so knowing almost blindly your equipment may help to avoid unnecessary trouble when hanging 30 m above ground.
Basset et al. (2003), Jepson (2000), Lowman & Nadkarni (1995), Lowman & Rinker (2004) Lowman et al. (2012), Mitchell et al. (2002), Stork et al. (1997) (see References)
Avicularia rufa, 35 m
Avicularia purpurea, shelter, 23 m
Psalmopoeus reduncus, 1.7 m
Pseudoclamoris sp., 3.6 m
Avicularia hirschii, 1.6 m

Pachistopelma bromelicola and *P. rufonigrum* occur more or less along the Atlantic coast of eastern Brazil from Rio Grande do Norte to Alagoas (*P. rufonigrum*) and from Sergipe to Bahia (*P. bromelicola*). Both species inhabit mostly terrestrial tank bromeliads, preferably exposed to direct sunlight. The habitats are mostly restinga, caatinga, and rainforest areas. In the latter case *P. rufonigrum* were found in terrestrial bromeliads on rocky outcrops, a rather hot and dry habitat compared to the surrounding rainforest. Such patches of bromeliads are quite separated from each other resulting in isolated populations. *Pachistopelma bromelicola* was also found in tank bromeliads used as ornamental plants in gardens. Bertani (2012) did not find any *P. bromelicola* in a rainforest area. However, do tank bromeliads occur there on trees in the canopy and could it be that the spiders just track the bromeliads up into the canopy? In general, large tank bromeliads may store water for a long time and organic matter may accumulate among the leaf axils; this is a good base for many arthropods, reptiles, and amphibians. They use the bromeliads for shelter, as water source and are feeding either on the accumulated organic matter or forage for prey living in the bromeliads. Even more than tank bromeliads of the rainforest canopy these plants are key for survival of many small animals in hot and otherwise dry habitats lacking the temperature and humidity gradient of a forest (Santos et al. 2003, 2004; Dias and Brescovit 2003, 2004; Bertani 2012; Lapinski and Tschapka 2014).

The genus *Iridopelma* is distributed from Pará in the west to Rio Grande do Norte and Bahia in the northeast. *Iridopelma hirsutum* and *I. zorodes* inhabit the rainforest fragments along the Atlantic coast, while the other four species occur farther west to Pará in drier habitats like cerrado and caatinga. Interestingly, *I. hirsutum* and *I. zorodes* make their retreats in a similar way connecting leaves with silk or inhabit bromeliads or rolled *Heliconia* leaves. The four species of the drier habitats were found occasionally similar to *Pachistopelma* spp. in terrestrial tank bromeliads but also under loose bark and, rather untypical for Aviculariinae, under fallen tree trunks and rocks. The greatest height at which *Iridopelma* spp. were found was 1 m above ground, which was certainly a bias of the survey method (Bertani 2012).

Ybyrapora species occur along the Atlantic coast of Brazil from Bahia southward to the state of Rio de Janeiro (Bertani and Fukushima 2009). *Ybyrapora gamba* is only known from Elísio Medrado near Salvador (Bahia) where it has been found in syntopy with *Y. diversipes* (Fig. 7.7); however, the latter species ranges farther southward along the Atlantic coast of Bahia. *Ybyrapora sooretama* occurs from the southernmost Bahia to southern Rio de Janeiro state. All three species are associated with the Atlantic rain forest, although the densest population of *Y. diversipes* was found in a secondary forest recovering for about 30 years from cocoa plantation. In a nearby experimental cocoa plantation the population was less dense. All three species constructed their shelters on tree trunks, large leaves and occasionally in tank bromeliads, similar to *Iridopelma zorodes* from the same region. The height of *Ybyrapora* specimens ranged from several centimeters to 2 meters above ground. According to the authors, *Y. gamba* constructed retreats also on grass (Poaceae), which also was reported for a juvenile specimen of *Avicularia hirschii* in Ecuador (Hüsser 2017).

Avicularia taunayi is the southernmost species of the genus in Brazil (Fukushima and Bertani 2017) and unlike its rainforest congeners it inhabits the cerrado located between Amazonia to its west and north, and Mata Atlântica to its east and south-east

Fig. 7.7 Juvenile of *Ybyrapora diversipes*, occurring in the Atlantic tropical rainforest region of north-eastern Brazil. Photo taken in captivity

(Bertani and Motta 2013) where *Typhochlaena*, *Pachistopelma*, *Iridopelma*, and *Ybyrapora* species occur (see above). The pristine cerrado and anthropogenically modified areas form a mosaic of habitats in that region. In this savanna-like region, *A. taunayi* live mainly on small trees which have relatively many holes in their trunks in which the spiders have their silken shelter tubes. The tarantulas have been found between 0.5 and 2 m above ground on the scattered trees. Similarly, Bullmer et al. (2006) found *Avicularia hirschii* on a pasture in Ecuador. Like *A. taunayi* the spiders inhabited scattered small trees with grass in between; however, I have found *A. hirschii* also in the Amazonian rainforest (see below).

Another theraphosid genus from north-eastern Brazil is *Dolichothele* (Guadanucci 2011). Judging from the distribution map presented by Guadanucci (2011), these Ischnocolinae are sympatric with some of the above mentioned Aviculariinae. *Dolichothele* species are terrestrial and at least one species, *D. bolivianum*, construct burrows strongly lined with camouflaged silk beneath stones and logs (Guadanucci 2007).

Sympatry: Syntopy in the Mata Atlântica

The complex habitat mosaic of the Mata Atlântica of north-eastern Brazil is inhabited by several tarantula species of which only the vegetation dwelling aviculariines have been examined in more detail. Bertani (2012) suggested that distribution ranges of most congeneric species are separated by major rivers in that region and that the Mata Atlântica shows various centers of endemism. As summarized above, those genera mainly use available plants in order to forage for prey and construct shelters, while species in forested areas live on trees and understory herbs, species of caatinga and restinga use the terrestrial tank bromeliads. Where species live in syntopy they are segregated by their lifestyle. For example, in one area of endemism between Rio São Francisco and Rio Paraguaçú the species *Iridopelma zorodes*, *Pachistopelma bromelicola*, and *Typhochlaena seladonia* are found in sympatry.

While *I. zorodes* lives in forested areas on trees and herbs, sometimes also in bromeliads, *T. seladonia* also lives on trees but constructs the trap-door retreats and uses small crevices in or under the bark. *Pachistopelma bromelicola* lives outside of forested habitats in tank bromeliads. For a more detailed analysis of biogeography of those genera see Bertani (2012). The question arises why *Pachistopelma* species specialized on terrestrial tank bromeliad in open habitats while sympatric *Iridopelma* species inhabit the vegetation of forested habitats Bertani (2012). To me the question is: are *Pachistopelma* species really restricted to open areas? In an attempt to explain the assumed segregation pattern Bertani (2012) suggests that large wet rainforest areas of the region were converted into xeric habitats due to climate fluctuations of the Neogene and the Quaternary periods. Because of their adaptations to hot and dry conditions of rainforest canopies bromeliads were preadapted to thrive on the ground of those xeric habitats providing valuable shelter and food resources for local fauna. Later the climate became humid again and some areas restored the forest character, but some remained xeric, forming a mosaic of forested and open habitats, all being inhabited by their specific faunas. However, the comparison of canopy with terrestrial bromeliads suggests that conditions among these regions may differ strongly (Gonçalves-Souza et al. 2010).

Exploring vertical transects from the forest ground to the canopy in the forested areas should reveal whether *Pachistopelma* species actually are restricted to tank bromeliads in open areas. Due to their assumed physiological adaptations to hot and dry environments, they might also thrive in the nearby forest canopy, not necessarily in epiphytic tank bromeliads, while *Iridopelma zorodes* and *I. hirsutum* be restricted to lower parts of the forests.

7.6 High Up on the Mountain

Tarantulas are of tropical origin and hence their basic physiology is adapted to temperatures between 20 and 30 °C and relatively high humidity. But there are exceptions to that. High up on the mountains, where clouds stroke their slopes, where calderas and glaciers almost meet tarantulas, the spiders face other challenges. The climate may be arid, semi-arid, or very humid with a wide range of temperatures. At an elevation of ca. 2000 m a.s.l. in the central Chilean Andes, for example, the air temperature can range from 0 to 40 °C, with surface temperature exceeding 60 °C (Alfaro et al. 2013). Snow may fall even in summer and during winter time the snow cover me be up to 3 meters (Cubillos et al. 2018). Records of tarantulas occurring at elevations above 2000 m a.s.l. are accumulating, yet not much is known about the habitat use of highland tarantulas. Physiological adaptations of South American tarantulas to cold were studied by Veloso et al. (2012), Alfaro et al. (2013), and Cubillos et al. (2018).

In the United States a few *Aphonopelma* species from the *Marxi* species group inhabit mountain ranges between 2000 and 2700 m a.s.l. mainly in oak woodland, pine-oak woodland, and mixed conifer communities of northern Arizona and New

Mexico (Table 7.6). The biology of highland tarantulas in the United States is poorly known; the burrows or shelters of *A. chiricahua*, for example, have not even been observed yet. In *A. madera* only one scrape burrow has been observed, all other males and females were observed walking around during day time while all juveniles occurred under stones without any burrows (Hamilton et al. 2016). In Mexico, epigean species of the genus *Hemirrhagus* live in cold climate zones between 1500 and 3100 m a.s.l. and can be found mainly in pine, oak, or pine-oak forest (Mendoza 2014a) (Table 7.6).

The highest elevation records for New World theraphosids are from the Andes (Fig. 7.6, Table 7.6). In Colombia *Agnostopelma tota* and *A. gardel* occur at approximately 3000 m a.s.l. in shelters under stones, nothing more has been reported about those species (Pérez-Miles and Weinmann 2010). Another Colombian highlander is *Aguapanela arvi* from the Piedras Blancas Ecological Park between 2100 and 2400 m a.s l., which take shelter in rock crevices and inhabit high Andean cloud forests (Perafán et al. 2015). According to the authors, the park with recovering cloud forest is an area of high diversity of Mygalomorphae; other sympatric theraphosids are *Hapalopus* sp. and *Ami* sp. The authors report also, based on museum material, that in Colombia there are also highland species from the genera *Catumiri* (2004 m a.s.l.) and *Holothele* (3000 m a.s.l.). Even less is known about highland tarantulas from Ecuador; all known species were described from the mountain ranges around Quito: *Cyclosternum gaujoni* from elevations between 2600 and 3600 m a.s.l. (Berland 1913), *C. schmardae* from 4000 to 5000 m a.sl. (Peters 2005 citing Ausserer 1871), *Pamphobeteus augusti* from forests on Pichincha volcano and *P. vespertinus*, both occur at approximately 2800 m a.s.l. (Pocock 1903; Peters 2003). In Peru species from at least three genera can be termed highlanders. Pérez-Miles and Weinmann (2009) described from northern Peru *Cyriocosmus pribiki* and *C. rogerioi* inhabiting cloud forests at elevations between 2200 and 3000 m a.s.l. In the southern Andes of Peru, two species of *Bistriopelma* and six species of *Hapalotremus* were discovered at even greater elevations. *Bistriopelma lamasi* and *B. matuskai* live at elevations between 3749 and 4398 m a.s.l. on grass slopes in burrows under stones and rocks. The average annual temperature ranges from 9.5 to 11.2 °C and may reach 0 °C (Kaderka 2015). Slightly to the south-east of their distribution range six species of *Hapalotremus* inhabit the high Peruvian Andes at elevations between 2500 and 4524 m a.s.l.; this is to date the highest elevation for any theraphosid and the record goes to *H. vilcanota*—if the above mentioned elevation range of *Cyclosternum schmardae* in the Ecuadorian Andes not be verified. Peruvian *Hapalotremus* species were found both in burrows and in crevices under rocks while ambient temperature may fall well below 0 °C (Ferretti et al. 2018). The distribution of *Hapalotremus* continues southward to Bolivia and north-western Argentina. In the Bolivian Andes two *Hapalotremus* species inhabit the edges of Yungas at elevations between 3200 and 4252 m a.s.l. In north-western Argentina *H. martinorum* occurs in high Andean cloud forest at around 1500 m a.s.l. and *H. chasqui* lives in the grass land with tree groves at 1300–2000 m a.s.l. Here the temperatures vary between as high as 26 °C in summer and 16 °C in winter, often reaching −10 °C (Ferretti et al. 2018).

Four highland *Euathlus* species occur from northern to central Chile at elevations from 1900 to 3200 m a.s.l. living in semi-arid habitats where temperatures may drop well below 0 °C (Veloso et al. 2012; Alfaro et al. 2013; Perafán and Pérez-Miles 2014; Cubillos et al. 2018). In northwestern Argentina *E. diamante* inhabits semi-arid grass and shrub land at 2200 m a.s.l. Two more species live even further south at lower elevations between 1042 and 1400 m a.s.l. in small crevices and short burrows under stones in the Patagonic steppe (*E. sagei*) and in quite humid Andean-Patagonic forests (*E. tenebrarum*) (Ferretti 2015). Veloso et al. (2012) found that *Euathlus parvulus* (central Chilean Andes) may select shelter sites (rocks) that have certain climatic regimes, which in combination with temperature in the deeper burrow allow the spiders to thermoregulate behaviorally.

7.7 Costa Rica and Ecuador: Local Assemblages Compared

7.7.1 *Methods*

In the light of a tremendous paucity of comparative studies at assemblage level, often even lacking basic natural history data I present a case study with ecological data collected mainly in the Sarapiquí region, Costa Rica and compare this assemblage to that found in the Oriente of Ecuador (Fig. 7.8, Table 7.6).

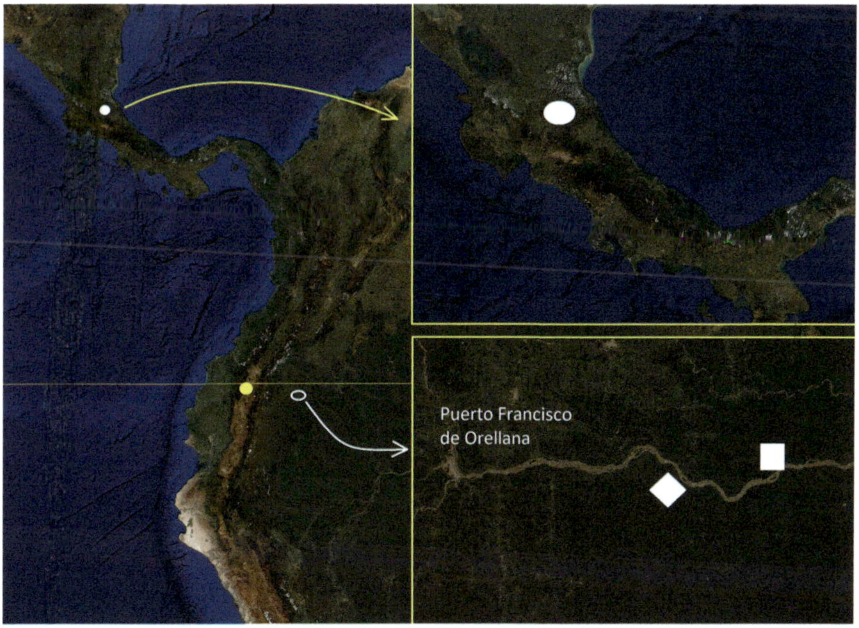

Fig. 7.8 Study sites in Costa Rica (white dot: Sarapiquí region) and Ecuador (white circle: along Río Napo). Yellow dot: Quito (Ecuador). Detailed location of the study sites along Río Napo (Ecuador): Centro de Interpretación Yaku Kawsay (white diamond) and Sani Lodge (white square)

I used the World Spider Catalog (version 21.0) to obtain information about total number of described Theraphosidae species from both Costa Rica and Ecuador. Then I assigned the known species to two subguilds: (1) terrestrial and (2) arboreal in order to compare the proportion of both lifestyles in Costa Rica and Ecuador.

Methods Costa Rica

On several short trips and long-term stays I collected data on basic tarantula ecology in Puerto Viejo de Sarapiquí (March 2001), at Reserva Biológica Tirimbina (August 2006 to June 2007, April to July 2008, and March 2010 to March 2012), the banana plantation of Chiquita Finca Nogal and a nearby forest fragment Agrícola Sofía were surveyed monthly from November 2010 to March 2012. The banana plantation of Finca Guayacán and the forest of Refugio Nogal were visited sporadically when visiting Finca Nogal and Agrícola Sofía. Tarantula species were identified using Valerio (1979–1980), Pérez-Miles et al. (2008) and comparing them to the specimens deposited at the Museo de Zoologia, Universidad de Costa Rica. Alcohol and still living specimens are in the private collection of the author. *Neischnocolus panamanus* specimens are deposited in the Staatliches Museum für Naturkunde Karlsruhe, Germany (Lapinski et al. 2018) and in the private collection of the author.

Study Areas The sites were located in the Sarapiquí region at elevations between 40 and 220 m asl, Heredia Province, Costa Rica (Fig. 7.8). Mean annual temperature is 25.3 °C and mean annual precipitation is 3777 mm. (1) Puerto Viejo de Sarapiquí is a small town at approximately 50 m a.s.l. surrounded by a mosaic of fragments of lowland rainforest, banana plantations, pastures, and and other agroecosystems. To its south there is La Selva Biological Station and the northernmost part of Braulio Carrillo National Park. (2) Refugio Nogal is a small temporarily inundated rainforest fragment at Río Sucio with its understory formed by many *Heliconia* plants. South to it there were (3) the banana plantations of Chiquita: Finca Guayacán and Finca Nogal. The plantations formed together an area of approximately 390 ha and according to Chiquita staff agrochemicals were used only moderately. The banana plants were 3–4 m tall, interspersed with few herbs. The ground was covered by cut banana plants and banana leaves. (4) To their west there is the rainforest fragment Agrícola Sofía of approximately 176 ha with infrequent logging activity. (5) Reserva Biológica Tirimbina (RBT) has an area of approximately 345 ha and is located approximately 16 km south-west of Puerto Viejo de Sarapiquí at an elevation between 180 and 220 m a.s.l. About 85% of the reserve's forest is classified as primary forest, encompassing also areas of secondary forest of various age classes. Canopy height at the forest study sites ranged between 30 and 40 m, with emergent trees up to *ca* 50 m. For a more detailed description of RBT see Lapinski and Tschapka (2013, 2018).

Fieldwork Most fieldwork at all sites was conducted by walking trails during day and night turning objects on the ground, searching for tarantula burrows in the ground and for shelters on all plants up to approximately 4 m, and for active tarantulas. At RBT there are two suspension bridges allowing canopy surveys up to approximately 26 m. However, closer inspection of tree trunks, epiphytes, and trees holes was very restricted. Therefore I used single rope technique (Jepson 2000;

Barker and Standridge 2002; see info-box) between September 2010 and February 2012 to repeatedly survey 22 trees belonging to 12 different species once per month (hereafter called vertical transects, details see Lapinski and Tschapka 2018). As spiders usually are not host-tree specific I surveyed suitable trees from different species (Sørensen 2003 and references therein). Selection criteria of suitable trees were: accessibility by the arrow shot from the ground in order to install the climbing rope and height at least that of the surrounding canopy. Spiders were searched at night on trunks and central portions of the crown of each tree individual and its surrounding vegetation from ground to approximately 3 meters above the branch where the rope was attached to. Data were taken within a survey radius of approximately 5 meters around the rope-climbed tree and only during the ascension to prevent repeated sampling. Prior to ascension spiders were also searched for 15 min on the forest ground, on the lower tree trunk and on understory vegetation within a 5-meter radius around the rope-climbed trees to collect also data on spiders roaming the lower strata of the forest. Spiders of all but the smallest instars were searched during 152 surveys of varying duration, from 1830 to 0415 h, the exact starting hour depended on weather conditions. Surveys on tall and structurally more complex trees usually took longer than on smaller and less complex trees, also the number of actual encounters prolonged the survey duration. Spiders at greater distance from the climbing observer were identified using binoculars (Pentax 8x36 DCF HS). It was never observed that spiders fled when illuminated by head lamp; sometimes they retreated into their shelters when vibrations caused by the climber's movements were too strong.

Methods Ecuador

Study Areas Data on basic tarantula ecology were collected between November 13 and 24, 2017 at two sites along the Río Napo at an elevation of approximately 200 m a.s.l. (Fig. 7.8). The landscape has slightly undulating hills (terra firme), lower parts of rainforest close to Río Napo and its larger tributaries are flooded seasonally (igapó). Mean temperature in the region ranges between 24 and 28 °C and annual rainfall is between 2800 and 3000 mm. The study sites were approximately 30 km apart: (1) Centro de Interpretación Yaku Kawsay run by the Kichwa community Nueva Providencia located along Río Shipati, close to its mouth into Río Napo. The community has an area of 6426 ha and lies in the most north-western part of Yasuní National Park, Orellana Province. The Yasuní National Park has a total area of approximately 1,022,736 ha. The site has secondary and primary lowland rainforest and swamps inside the forest and along the river. On the clearing there were five traditional huts with palm thatched roofs, no insecticides were used. The huts were connected by elevated wooden walkways, in between there were some small fruit trees and the ground was covered by short grass. (2) Sani Lodge is run by the Sani Isla Kichwa community and is located north of Río Napo at the lake Challuacocha. In the cabins and other facilities of the lodge insecticides were applied, the soil of clearing was covered by short-cut grass. Around of and in the vicinity to Challuacocha there are several swamps. The rainforest is classified mainly as primary. An approximately 35 m high observation tower at a Kapok tree

(*Ceiba pentandra*) of approximately 45 m total height allows access to its crown and the epiphytes on its branches. The mainly uninterrupted rainforest north of Río Napo continues northward into Reserva de Producción de Fauna Cuyabeno, resulting in a vast continuous lowland rainforest. Canopy height was approximately 30 m with emergent trees up to 50 m.

Fieldwork The procedure was similar to that applied in Costa Rica. Due to the short stay two trees of unknown identity were climbed at Yaku Kawsay. At Sani Lodge one still living tree clasped by a well-developed strangler-fig was climbed and the canopy observation tower was also used to search for tarantulas in the crown of a Kapok tree (*Ceiba pentandra*) of more than 45 m total height. *Avicularia* species were identified using Fukushima and Bertani (2017) by habitus and two exuviae, *Neischnocolus yupanquii* by habitus using Pérez-Miles et al. (2008), and *Megaphobema velvetosoma* by habitus by the dedicated tarantula aficionado A. Stirm who knows this species well. Three other tarantula species were assigned to morphospecies based on completely different habitus, i.e., physical appearance and coloration. Besides two *Avicularia* exuviae no specimens were collected.

Variables and Statistical Tests During the surveys I collected the following data: date, hour, individual number, species, instar, sex, site, habitat type, microhabitat—which I later subdivided into topophilia at micro- and nano-scale (Rao 2017)—shelter type, height of spider above ground, and diameter of plants, trunks, twigs, etc. where a spider was perching on, and tree species when possible. At RBT I also assessed the following structural variables of the rope-climbed trees: diameter (usually in breast height), total tree height, relative height of lowest branch (=height of lowest branch/total tree height). The amount of epiphytes on the trunk, amount of epiphytes within the crown, number of holes in the trunk, and number of holes within the crown were estimated and subdivided into classes. Epiphyte classes were: 1 = few (≤20% cover), 2 = many (ca 20–50%), 3 = almost full (≥50%), hole classes were: 0 = none, 1 = some, 2 = few, 3 = many. I used SigmaStat (Version 3.5) for all analyses. Continuous non-normally distributed data were analyzed with Kruskal–Wallis one way ANOVA on ranks using Dunn's *post-hoc* test for unequal sample size. This test was used to examine whether (a) sympatric/syntopic tarantula species, (b) different instars of the Costa Rican arboreal *Psalmopoeus reduncus* differed in their vertical distribution, and (c) whether the diameter of plants/parts of plants used by *P. reduncus* differed among instars. To simplify the analysis of differences among instars I assigned them into the following instar classes according to their body length: large ≥3 cm, intermediate =1.6–2.9 cm, and small ≤1.5 cm. I used the Mann–Whitney Rank Sum Test to compare surveys from ground only with surveys along vertical transects in order to examine whether the survey method may affect the picture of vertical distribution in arboreal tarantulas. To do so, I analyzed height data of the arboreal *Psalmopoeus reduncus*, for which sample size is biggest. I created to groups (1) ground-only with height data up to 4 m which may be obtained easily from the ground (trails and ground collected data at vertical transects), and (2) vertical transects only with height data obtained only at the rope-climbed trees from ground to the highest point possible (survey details are described above).

In order to explore which structural tree variables may affect vertical distribution of *Psalmopoeus reduncus* I calculated relative height of spider = height of spider/ total height of the tree on which it was encountered. I used only the data obtained on the actually rope-climbed trees because structural characteristics of neighboring vegetation were not assessed. The data were $\log_{10}(x + 1)$—transformed, and then additionally standardized to omit mathematical problems caused by different units of different variables (Leyer and Wesche 2007). I used only the standardized data for further analyses. Relative height of spider was the dependent variable and the above mentioned tree variables were independent variables which first were entered into stepwise forward regression to identify those which may predict relative height of spider. With the identified predictor variables I ran a multiple linear regression.

7.7.2 Results

Comparison of Tarantula Lifestyle in Costa Rica and Ecuador
Ecuador has a surface area approximately 5.5 times that of Costa Rica but from the latter approximately 1.8 times more tarantula species were described. However, in Ecuador 8 out of 21 known species—or 38%—from four genera are arboreal, in Costa Rica there is only one well-documented arboreal species out of 37 reported species (3%) (Fig. 7.9). In Ecuador only one arboreal species (*Psalmopoeus ecclesiasticus* Pocock 1903) is known from the forests west of the Andes, the other genera and species occur east of the Andes ranging eastward well into the lowlands of Amazonia. Additionally, during the survey in Ecuador I found another arboreal, previously not reported species, resulting in a total of 9 arboreal out of 22 reported species (41%).

Costa Rica During the surveys from 2001 to 2012 the total of 104 individual tarantulas from five genera and five species were found (Table 7.1, Fig. 7.10). Most species were strongly associated with forest (forest interior and gaps in the forest) (Table 7.2). *Lasiodora icecu* occurred also in anthropogenically modified habitats, but both habitats were very close to the surrounding rainforest at RBT. Only one tarantula specimen, an adult male of *Stichoplastoris obelix*, was found in a banana plantation approx. 300 m from the forest fragment Refugio Nogal where a conspecific adult female was encountered.

Psalmopoeus reduncus occurred significantly higher than *Neischnocolus panamanus*, *Lasiodora icecu*, *Stichoplastoris obelix*, and *Sericopelma silvicola* (Fig. 7.11, Kruskal-Wallis ANOVA, $H_4 = 77.72$, $P \leq 0.001$, post hoc test Dunn's method $P < 0.05$). The greatest height recorded for *P. reduncus* was 28 m above ground (relative height = 0.72). The adult female had its shelter in a big tree hole among epiphytes. Median relative height in *P. reduncus* was 0.13 (vertical transects only). Exceptional heights in the terrestrial tarantula species were measured in a

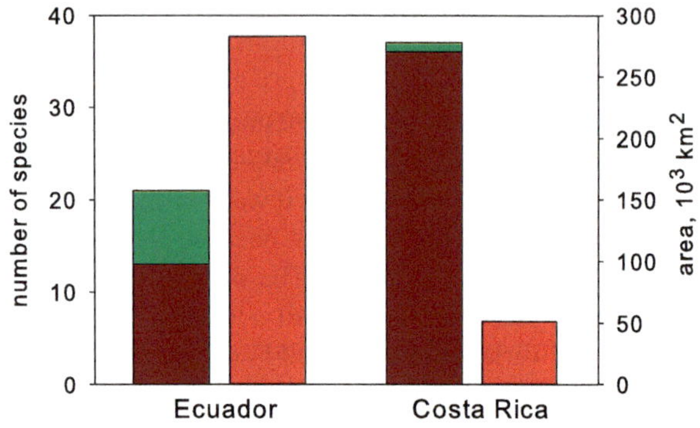

Fig. 7.9 Comparison of the total number of Theraphosidae described from Ecuador and Costa Rica and the proportion of arboreal (green) to terrestrial (dark red) tarantula species recorded from those countries. Red bars represent country area in 10^3 km^2

Table 7.1 Theraphosid species and sample size encountered at the different sites in the Sarapiquí region, Costa Rica

Species	N (RBT)	N (other)
Neischnocolus panamanus (Petrunkevitch 1925)	12	0
Lasiodora icecu (Valerio 1980)	11	0
Psalmopoeus reduncus (Karsch 1880)	54	1 AS
Sericopelma silvicola (Valerio 1980)	13	1 PV
Stichoplastoris obelix (Valerio 1980)	10	1 FG, 1 RN
N (total)	100	4

RBT Reserva Biológica Tirimbina, other include, *AS* Agrícola Sofía, *FG* banana plantation Finca Guayacán, *RN* Refugio Nogal, *PV* sec. Forest in Puerto Viejo de Sarapiquí

female *N. panamanus* found on a trunk of *Vochysia guatemalensis* in 1.4 m, where it retreated under loose bark during capture attempt. Two females of *L. icecu* had their burrows in 1.4 and 1.6 m on a 2-m high and almost vertical embankment of an unpaved road. The burrows of those two females were 20 cm apart.

In order to examine whether the survey method may influence the picture of vertical distribution of arboreal *P. reduncus*, I compared data on “height of spiders above ground” obtained at vertical transects (i.e., searching from ground around the rope-climbed trees and during actual climbing those trees) with height data obtained only from the ground by walking trails and by searching on and around rope-climbed trees from ground only (data from both canopy bridges were omitted because such bridges are, like ropes, a type of technical canopy access system). The height above ground of *P. reduncus* was significantly greater at fully surveyed vertical transects than in ground-only surveys, resulting in documented occurrence up into the canopy region of the rainforest and higher median of height at which *P. reduncus* may occur (Fig. 7.12, Mann–Whitney Test, $T = 949.00$, $n(small) = 22$, $n(big) = 40$, $P \leq 0.001$).

To examine whether there exist ontogenetic habitat shift in the arboreal *P. reduncus* I tested for differences among three instar classes with regard to (1) height above ground and (2) diameter of plant parts on which specimens were found. The different instars of *P. reduncus* did not differ significantly in their height above

Fig. 7.10 Females of tarantula species from Reserva Biológica Tirimbina, Sarapiquí region, Costa Rica. (**a**) *Neischnocolus panamanus*, (**b**) *Lasiodora icecu*, (**c**) *Sericopelma silvicola*, (**d**) *Stichoplastoris obelix*, (**e**) *Psalmopoeus reduncus* with silken tube on a *Philodendron* leaf, (**f**) *Psalmopoeus reduncus* as mostly found on tree trunks, here with a recently captured *Copiphora* sp. (Tettigoniidae)

Table 7.2 Habitat types used by the theraphosid species in the Sarapiquí region, Costa Rica

Species	Forest interior	Gap	Open area	Cacaotal	Banana pl.
Neischnocolus panamanus	11	1	0	0	0
Lasiodora icecu	8	0	2	1	0
Psalmopoeus reduncus	44	10	0	1	0
Sericopelma silvicola	13	0	0	0	0
Stichoplastoris obelix	9	2	0	0	1

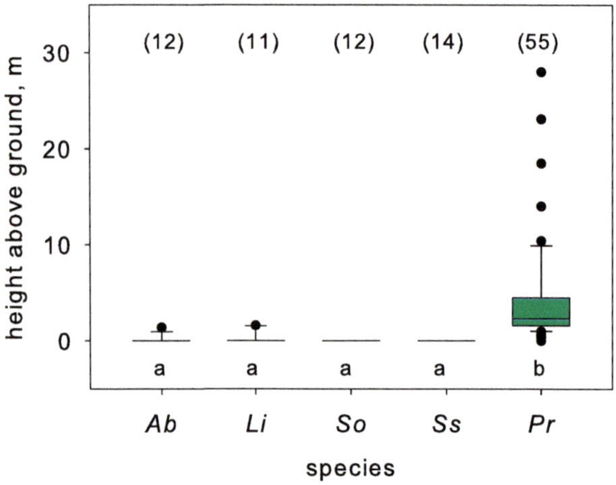

Fig. 7.11 Vertical distribution of the sympatric theraphosid species encountered at several study sites in the Sarapiquí region, Costa Rica. Data from both ground-only and vertical transect surveys pooled. Different letters below the boxes indicate significant differences, figures in parentheses above the boxes show sample size. Horizontal lines in the boxes represent the median, boxes are from Q_{25} to Q_{75}, error bars from Q_{10} to Q_{90}, full circles are outliers. Species abbreviations: Np = *Neischnocolus panamanus*, Li = *Lasiodora icecu*, So = *Stichoplastoris obelix*, Ss = *Sericopelma silvicola*, Pr = *Psalmopoeus reduncus*

ground (Fig. 7.13a, Kruskal-Wallis ANOVA, $H_2 = 2.28$, $P = 0.32$). Although a significant difference in diameter of plants or their parts used by *P. reduncus* was found only between large instars and spiderlings (Fig. 7.13b, Kruskal-Wallis ANOVA, $H_2 = 12.92$, $P = 0.002$, post hoc test Dunn's method $P < 0.05$), overall large and intermediate instars tended to use plants with greater diameter than spiderlings.

Topophilia In the Costa Rican assemblage there was a conspicuous segregation of the species by topophilia at micro scale (Fig. 7.14). Ground and trail banks were used almost exclusively by *L. icecu*, *S. obelix* and *S. silvicola*; *N. panamanus* was not found on banks. Vertical objects, mostly trees, were occupied by *P. reduncus*. There was almost no overlap among the two subguilds at that scale, with the exception of a wandering male of *P. reduncus* on the ground and the above mentioned female of *N. panamanus* on a tree trunk. Figure 7.15 shows segregation of topophilia at nano scale. *Psalmopoeus reduncus* mainly used structures associated with plants. But there was also segregation among the terrestrial species. While *L. icecu*,

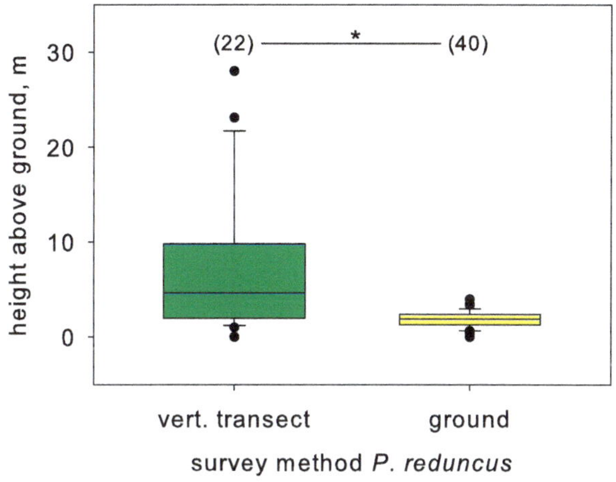

Fig. 7.12 Comparison of the effect of survey method on resulting picture of vertical distribution of the arboreal *Psalmopoeus reduncus* in Costa Rica. The asterisk indicates significant difference, figures in parentheses above the boxes show sample size. Horizontal lines in the boxes represent the median, boxes are from Q_{25} to Q_{75}, error bars from Q_{10} to Q_{90}, full circles are outliers

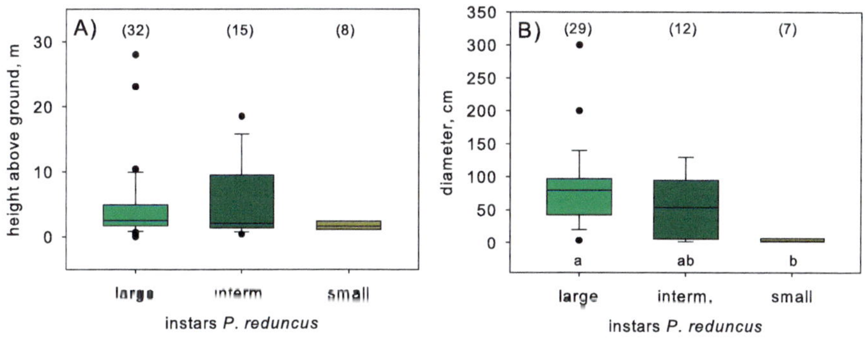

Fig. 7.13 Comparison of three different instar classes of the arboreal *Psalmopoeus reduncus*. (**a**) Height above ground, and (**b**) diameter of plant parts the specimens were found perching on. Instar classes of the tarantulas: large ≥3 cm, intermediate = 1.6–2.9 cm, and small ≤1.5 cm body length. Different letters below the boxes indicate significant differences, figures in parentheses above the boxes show sample size. Horizontal lines in the boxes represent the median, boxes are from Q_{25} to Q_{75}, error bars from Q_{10} to Q_{90}, full circles are outliers

S. obelix and *S. silvicola* were found mostly in or at burrows in the ground, *N. panamanus* occurred predominantly under wood or wandering around among leaf litter. Besides that a few wandering males of all but *S. silvicola* were found without being associated tightly with a certain topophilia.

During the vertical-transect surveys the arboreal *Psalmopoeus reduncus* occurred on seven tree species with different structural characteristics. The tree species were: *Alchorneopsis floribunda* (Euphorbiaceae), *Ceiba pentandra* (Malvaceae-Bombacoideae), *Pentaclethra macroloba* (Fabaceae-Mimosoideae), *Terminalia*

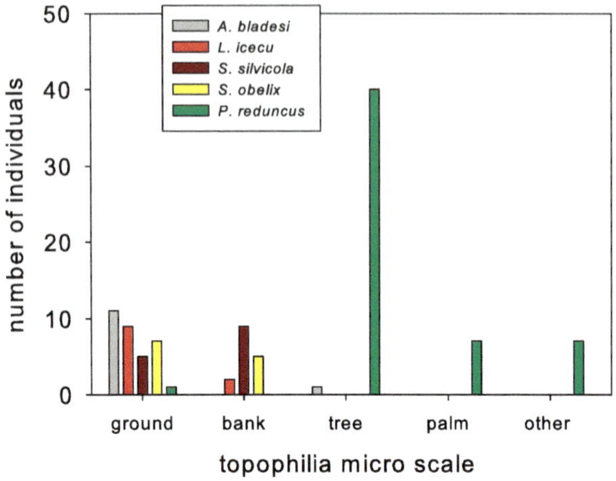

Fig. 7.14 Topophilia at micro scale of tarantulas found in the Sarapiquí region, Costa Rica. Other include: shack, hollow rail of stairs of a suspension bridge, log on ground, liana

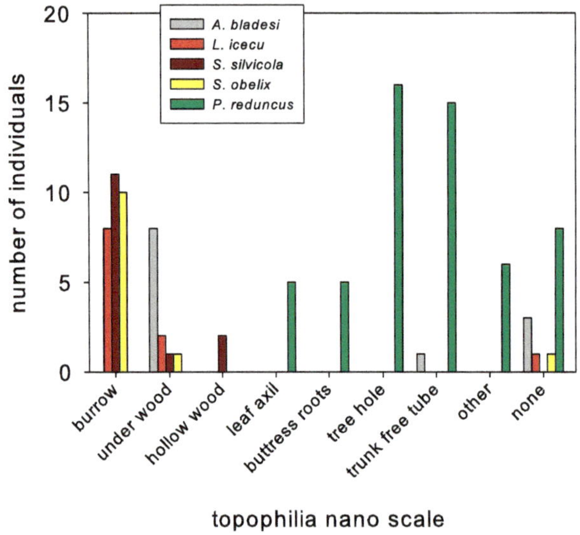

Fig. 7.15 Topophilia at nano scale of tarantulas found in the Sarapiquí region, Costa Rica, showing the placement of shelters. Other include: under roof of a shack, root of *Socratea*, leaf epiphyte. * *Neischnocolus panamanus* female on tree trunk, which retreated under loose bark when disturbed, but no silken structures were found

oblonga (Combretaceae), *Vitex cooperi* (Lamiaceae), *Vochysia guatemalensis* (Vochysiaceae), *Virola koschnyi* (Myristicaceae), and *Socratea* sp. (Arecaceae).

Relative height of the arboreal *Psalmopoeus reduncus* increased significantly with both the amount of epiphytes and amount of holes in the tree crowns (multiple linear regression, Table 7.3). The other assessed structural variables DBH, total tree

Table 7.3 Results of multiple linear regression analysis of relative height of n = 19 *Psalmopoeus reduncus* as dependent variable on n = 10 trees from 7 species and the independent variables trunk diameter, total tree height, relative height of lowest branch, amount of epiphytes on the trunk, amount of epiphytes within the crown, number of holes in the trunk, and number of holes within the crown

	Coefficient	Std. error	t	P	VIF
Constant	9.17E-16	0.13	7.11E-15	1	
Epiphytes in crown	0.72	0.14	5.27	<0.001	1.07
Holes in crown	0.29	0.14	2.15	0.047	1.07
Analysis of variance					
	DF	SS	MS	F	P
Regression	2	12.94	6.47	20.45	<0.001
Residual	16	5.06	0.32		
Total	18	18	1		
R = 0.85	$R^2 = 0.72$	Adj $R^2 = 0.68$			
SE of estimate = 0.56					

For definitions and details of variables see methods

height, relative height of lowest branch, and both amount of epiphytes on tree trunks and amount of holes in tree trunks did not add to the ability of the equation to predict relative height in *P. reduncus*. Greatest heights of *P. reduncus* were measured on *Vitex cooperi* (HS = 14–28 m, three spiders) and on *Ceiba pentandra* (HS = 18.5 m, one spider). Both trees were very tall and densely covered with epiphytes of different types from ground to crown. *Vitex cooperi*, however, had in addition to its epiphyte abundance more holes in the wood in the crown than *C. pentandra*. And exactly deep in a large hole an adult female of *P. reduncus* was found at the greatest height (28 m) in that *V. cooperi* tree. Although many of the surveyed trees in Costa Rica were heavily loaded with tank bromeliads up to 1 m in diameter on the trunks and on branches I never found any arboreal tarantula in a tank bromeliad.

Ecuador

I have found a total of 30 tarantula individuals at both sites in the Oriente of Ecuador belonging to nine genera and species (Table 7.4, Fig. 7.16). Apart from one intermediate juvenile of *Neischnocolus yupanquii* and one spiderling of *Megaphobema velvetosoma* and *Pseudoclamoris* sp., respectively, no small juveniles were found. Hence no data on possible ontogenetic habitat shift are available. Although I was not able to assign the three unidentified Theraphosinae species to any genus, their coloration and habitus differed sufficiently to treat them as different morphospecies belonging to different genera. Theraphosinae 3 is already known in the internet as Yasuní Tiger Tarantula and I use this name hereafter. The species is allegedly in the process of description but no further information is available to me. Yaku Kawsay and Sani Lodge shared only two species: *Megaphobema velvetosoma* and *Avicularia rufa*. However, due to the short time spent at both sites and hence small sample size these results have to be regarded only as preliminary.

Most species were restricted to the forest interior (both primary and secondary) (Table 7.5). Theraphosinae 1 and Theraphosinae 2 were only found on the clearing with huts at Yaku Kawsay. *Avicularia rufa* was found in the forest interior at Sani

Table 7.4 Theraphosid species and sample size encountered at the different sites along Río Napo, Ecuador

Species	N (YK)	N (SL)
Neischnocolus yupanquii (Pérez-Miles et al. 2008)	0	1
Megaphobema velvetosoma (Schmidt 1995)	1	4
Theraphosinae 1	4	0
Theraphosinae 2	1	0
Theraphosinae 3 (Yasuní Tiger Tarantula)	1	0
Avicularia hirschii (Bullmer et al. 2006)	0	1
Avicularia purpurea (Kirk 1990)	0	4
Avicularia rufa (Schiapelli and Gerschman 1945)	8	2
Pseudoclamoris sp. (cf. *burgessi*, Hüsser 2018)	0	3
N (total)	15	15

YK Centro de interpretación cultural Yaku Kawsay, *SL* Sani Lodge

Lodge, at Yaku Kawsay it occurred at forest edge along the bank of Río Shipati and on palm thatched roofs of the huts.

At both sites in Ecuador there were conspicuous differences in vertical distribution among the encountered tarantula species (Fig. 7.17). However, the sample size per species was too small to run a sensible ANOVA in order to examine interspecific differences in height of the spiders. Therefore, I subdivided all typically terrestrial (*Neischnocolus yupanquii* to Yasuní Tiger Tarantula) and all typically arboreal (*Avicularia hirschii* to *Pseudoclamoris* sp.) species into two corresponding groups: terrestrial versus arboreal. Mann–Whitney Test corroborated that the terrestrial species occurred at significantly lower heights than the arboreal ($T = 79.00$, $n(small) = 12$, $n(big) = 18$, $P \leq 0.001$). On the same tree/strangler fig-complex at Sani Lodge one adult female of *Pseudoclamoris* sp. (at 1.0 m) and three subadult/adult females of *A. purpurea* (at 10.1 m, 10.7 m, and at 23 m) were found.

The tarantula species were segregated with respect to topophilia at micro level (Fig. 7.18). The terrestrial species were obviously associated with ground; however, while *N. yupanquii* and *M. velvetosoma* occurred on the ground of primary and secondary forest, Theraphosinae 1 and Theraphosinae 2 were found on ground that was strongly modified by human activity (law and bare soil among and under the huts at Yaku Kawsay). The arboreal species were mainly associated with trees in primary and secondary forest, only *A. rufa* occurred also on manmade structures at Yaku Kawsay. The only specimen of Yasuní Tiger Tarantula, a female, was found in the rainforest understory.

Nano scale topophilia of the theraphosids found at both sites was quite diverse (Fig. 7.19). Among the terrestrial species *M. velvetosoma* and Theraphosinae 1 lived mainly in burrows, *N. yupanquii* was wandering among leaf litter, Theraphosinae 2 was wandering on bare soil and Yasuní Tiger Tarantula was perching on a leaf of an understory Marantaceae, but no shelters of the latter three specimens were discernible. The arboreal species showed an even more diverse topophilia at nano scale. *Avicularia hirschii* lived in a silken tube on a horizontal leaf of a sapling and a dead leaf was attached to the upper side of the tube. *Avicularia purpurea* lived in mainly exposed silken tubes on trees, usually ending in a tree hole or crevice between tree

Fig. 7.16 Tarantula species from Yaku Kawsay and Sani Lodge, Río Napo, Ecuador. (**a**) *Pseudoclamoris* sp., (**b**) *Avicularia purpurea*, 23 m above ground, (**c**) *Avicularia rufa*, 35 m above ground, (**d**) *Avicularia hirschii*, (**e**) Theraphosinae 3 (Yasuní Tiger Tarantula), (**f**) Theraphosinae 1, (**g**) Theraphosinae 2, (**h**) *Megaphobema velvetosoma*, (**i**) *Neischnocolus yupanquii*, juvenile

Table 7.5 Habitat types used by theraphosids at Yaku Kawsay and Sani Lodge, Ecuador

Species	Forest interior	Forest edge	Clearing huts
Neischnocolus yupanquii	1	0	0
Megaphobema velvetosoma	5	0	0
Theraphosinae 1	0	0	4
Theraphosinae 2	0	0	1
Theraphosinae 3 (Yasuní Tiger Tarantula)	1	0	0
Avicularia hirschii	1	0	0
Avicularia purpurea	4	0	0
Avicularia rufa	2	2	6
Pseudoclamoris cf. *burgessi*	3	0	0

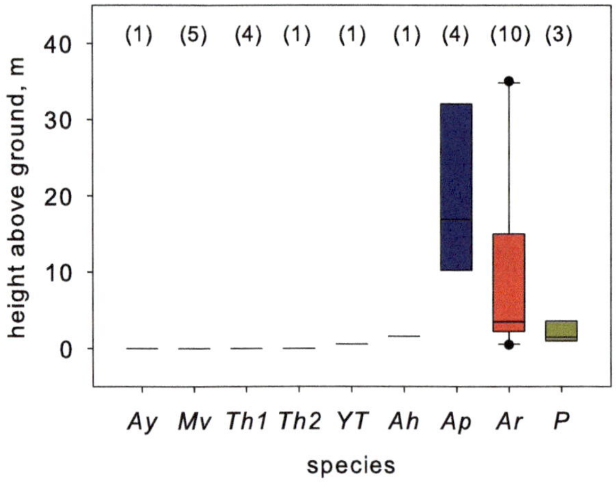

Fig. 7.17 Vertical distribution of the sympatric theraphosid species at Yaku Kawsay and Sani Lodge, Ecuador. Figures in parentheses above the boxes show sample size. Horizontal lines in the boxes represent the median, boxes are from Q_{25} to Q_{75}, error bars from Q_{10} to Q_{90}, full circles are outliers. Species abbreviations: Ny = *Neischnocolus yupanquii*, Mv = *Megaphobema velvetosoma*, Th1, Th2 = Theraphosinae 1 and 2, YT = Yasuní Tiger Tarantula (Theraphosinae 3), Ah = *Avicularia hirschii*, Ap = *Avicularia purpurea*, Ar = *Avicularia rufa*, P = *Pseudoclamoris* sp.

trunk and a strangler fig. However, an exuvia of an intermediate juvenile *A. purpurea* was found in a silken tube between the lateral leaves of a huge tank bromeliad 35 m above ground on a branch of a *Ceiba pentandra* of more than 45 m total height; but the occupant was absent. In the same tree crown two *A. rufa* females were found in those tank bromeliads without constructing a significant silken tube. Additionally at Yaku Kawsay, permanent silken tubes were inhabited by a female in a tree hole and by an intermediate juvenile on the board underside of a wooden walkway connecting the huts. The other specimens of *A. rufa* were found on the roofs of palm thatched huts. Interestingly, conspicuous and firm silken tubes were usually not built by *A. rufa* that occurred on those roofs and in the above mentioned bromeliads. Two rather loose silken sacks with fresh exuviae were found on the undersides of the palm thatched roofs, but the recently molted females roosted

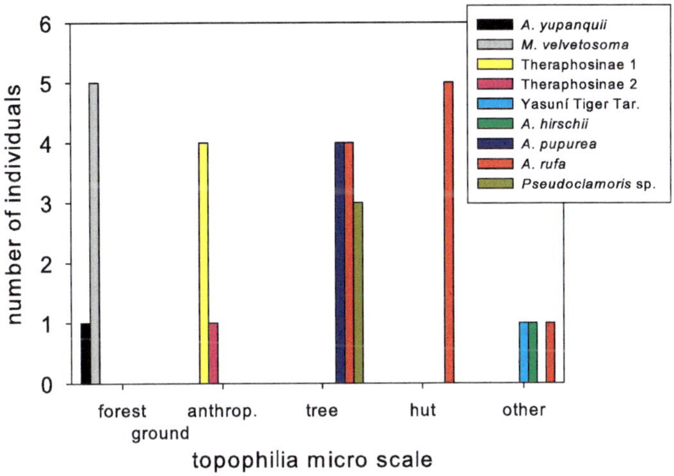

Fig. 7.18 Topophilia at micro scale, showing the microhabitat types used by the sympatric theraphosids at Yaku Kawsay and Sani Lodge, Ecuador. Other include: sapling (*A. hirschii*), wooden walkway (*A. rufa*, juv.), understory herb (Yasuni Tiger Tarantula)

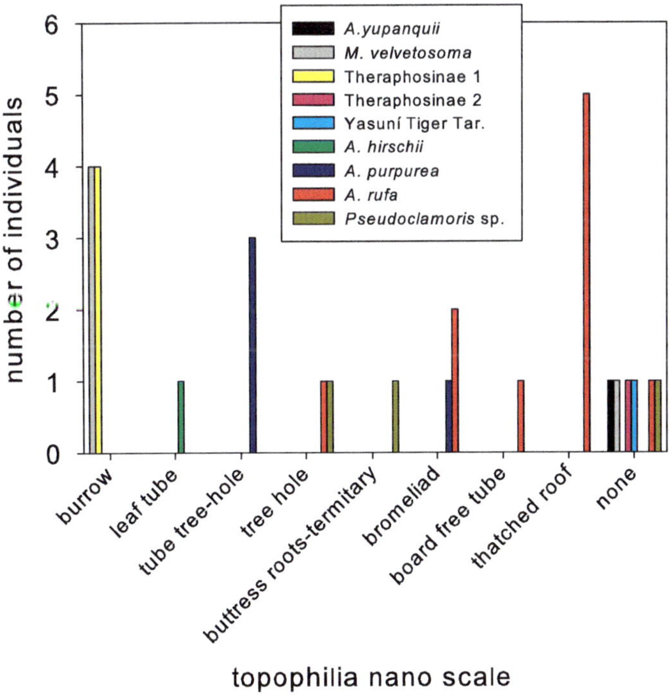

Fig. 7.19 Topophilia at nano scale of sympatric theraphosids at Yaku Kawsay and Sani Lodge, Ecuador, showing the placement and type of shelters

among nearby the palm leaves each day at a different location of the same roof. *Pseudoclamoris* sp. was found only on tree trunks. The spiderling was wandering freely at day upward and both adult females had their silken tubes in a tree hole and in a hole between a buttress root and an active termite nest (tree/strangler fig-complex).

7.7.3 Discussion

Basic Lifestyles of Tarantulas in Costa Rica and Ecuador

When comparing the known Costa Rican and Ecuadorian tarantula species with respect to the basic life styles terrestrial and arboreal, a difference or gradient in the proportion of arboreal to terrestrial habit between both countries and within Ecuador became apparent. The proportion in arboreality in Costa Rica is much lower than in Ecuador. In Ecuador itself the highest proportion of arboreality is found in the Amazon lowlands ranging into the eastern mid-elevations of the Andes, dropping dramatically within Ecuador when the Andes are crossed. This gradient is too conspicuous to be a random phenomenon. Seasonal flooding of vast areas affects many aspects of life in the Amazon basin (Adis 2002; Junk 1997; Kelly et al. 2017). Canopy arthropods are more sensitive to forest type than to biogeography (Basset 2001). The rainforest areas that were originally covering Central America and the Pacific lowlands of Ecuador with only small to intermediate-sized water bodies without pronounced inundations seem to not strongly promote arboreality. In Pantanal regular large-scale flooding influenced changes in both structure and composition of arthropod communities on trees (Marques et al. 2006). In contrast to Costa Rica and the Pacific lowlands of Ecuador, the Amazon basin with its seasonal interaction of forest with large-scale flooding and a mosaic of different terra firme and inundation forest types may drive and sustain high proportion of arboreality in arthropods in general and tarantulas in particular. From there arboreality might spread into surrounding biomes, decreasing in proportion with decreasing degree of seasonal large-scale floodings.

Tarantula Assemblages

Here I present a case study on the habitat use of two tarantula assemblages in the Sarapiquí region of Costa Rica and the middle Río Napo of Ecuador. Overall there was a general similarity in assemblage structure. In both study areas the respective assemblage consisted of several terrestrial and one (Costa Rica) to four (Ecuador) arboreal tarantula species, of which *Pseudoclamoris* sp. seems to be the first record for Ecuador. In both assemblages the terrestrial subguild contained one small species of the genus *Ami* and several intermediate to large species. While species of *Ami* usually take shelter under objects on the forest floor (mostly wood) instead of constructing permanent burrows (Auer 2011; Lapinski et al. 2018) the other terrestrial Theraphosinae mostly occupied burrows in the ground, only few individuals were found wandering on the ground or low in the understory. In Ecuador one arboreal and one fossorial species appear to tolerate or even prefer anthropogenically modified habitats. The facility area at Yaku Kawsay where three species occurred was small, still closely surrounded by rainforest and no insecticides were used. At Sani Lodge no tarantulas were found at the clearing and on the cabins despite proximity to the rainforest. Insecticides are used there and may explain the absence of tarantulas on the main facility area. However, just few meters into the surrounding secondary forest I found four tarantula species. Flexibility in habitat use, including the use of buildings, was also reported for other arboreal tarantulas, e.g., *Avicularia purpurea* from Tena, Ecuador (Kirk 1990) and *A. avicularia* from Trinidad (Stradling 1994). Despite frequent short-term and one long-term stay in Costa Rica, only two females of *Lasiodora icecu* were found along an unpaved road close to rainforest at RBT and only one adult male of *Stichoplastoris obelix* was found in a banana plantation. Weinmann (2003) examined a plantation with mixed crops in Colombia. He found a population of the fossorial *Megaphobema robustum* with an exceptional high density of 225 individuals/ha. A fossorial *Pamphobeteus* sp. shared this habitat, but no further data were given. In Martinique the arboreal *Caribena versicolor* was associated with forest areas and also occurred on nearby buildings but was usually absent from cultivated areas including banana and pineapple plantations (Maréchal et al. 2009). Thus, all five species found during the surveys in Costa Rica appear to be strongly associated with rainforest, rendering them vulnerable to forest destruction by man. In the studied Ecuadorian species at least two of them showed a sufficient degree of adaptability to anthropogenic habitat modification but certainly more thorough studies are needed. The difference in the proportion of arboreal to terrestrial species between the two studied Ecuadorian sites (4:2 at Sani Lodge, 1:4 at Yaku Kawsay) appears to be an artifact of the short survey time at each site. The only study to my knowledge that examines an assemblage of New World theraphosids was conducted in the Loreto region, Peru by Reátegui-Suárez et al. (2014). Like in Ecuador the region around Loreto is seasonally inundated (Auer 2011). At two sites five terrestrial species were encountered. Similar to the Ecuadorian assemblage presented here the authors found *M. velvetosoma* and *N. yupanquii*. In addition the Peruvian assemblage contained *Cyriocosmus sellatus*, *Pamphobeteus petersi*, and *Acanthoscurria ferina*. In Loreto again *N. yupanquii* was the only not fossorial and together with *C. sellatus* the smallest species. Strangely no arboreal species were found, or at least not reported. Only the fossorial species were analyzed in more detail, omitting *N. yupanquii*. The fossorial species placed their burrows in shaded places of the forest interior on slightly inclined relief with not dense understory vegetation and leaf litter of intermediate thickness. The burrows of different species were often only few meters apart.

Vertical Distribution

To my knowledge this is the first study that explores the ecology of theraphosids from forest ground to high rainforest canopy and at the same time includes data on syntopic species. The surveys in Costa Rica and Ecuador revealed a clear segregation of sympatric tarantula species by height above ground, forming a terrestrial and a vegetation-dwelling or arboreal subguild. The results show that arboreal tarantula species may use most of the available height range from low understory to high canopy as high as 35 m above ground as long as sufficient shelter possibilities exist. These results are in accordance with the vertical distribution of an assemblage of large wandering spiders sympatric with the tarantula assemblage at Reserva Biológica Tirimbina, Costa Rica (Lapinski and Tschapka 2013, 2018). During the tree surveys using rope climbing data on vertical distribution of both wandering spiders and tarantulas were collected simultaneously. Although overlapping in vertical distribution, the arboreal theraphosid *Psalmopoeus reduncus* occurred somewhat lower (median = 4.7 m) than the arboreal ctenid *Cupiennius coccineus* (median = 12.9 m). Knowledge of arboreal arthropods is strongly affected by survey methods (Basset 2001). Höfer et al. (1994) used fogging to sample the forest canopy in central Amazonia but not even one specimen of the frequently observed *Avicularia* sp. or other arboreal mygalomorph was collected. According to Valerio (1979) the Costa Rican arboreal *Psalmopoeus reduncus* lives in tree holes up to 4 m above ground. As shown here the species occurs frequently also in greater heights up to 28 m constructing silken tubes with different exposure degree. The comparison of ground-only surveys with surveys along vertical transects showed nicely that the survey method may affect the picture of the vertical distribution and topophilia of arboreal arthropods. The use of a canopy access system (observation tower and rope climbing) proved also to enhance the detection probability of arboreal species, since during the short stay at Sani Lodge I was able to find *Avicularia rufa* (tower) and *A. purpurea* (using both rope climbing and tower) only above 10 m, i.e., not visible from the ground. In addition, those methods facilitated direct observations of the animals, study of their shelters and all this without any necessity to kill them. The fun factor of studying the rainforest canopy in general and using the rope climbing technique in a beautiful tropical rainforest in particular cannot be neglected, either.

Ontogenetic Habitat Shift

Intraspecific differences in habitat use among different instar stages are common in various animal taxa and may reduce intraspecific competition and/or predation of juveniles by older, hence bigger, instars (reviewed by Morin 2011). Marshall and West (2008) found such ontogenetic habitat shift in the tarantula *Ephebopus murinus* in French Guiana, with small juveniles living in terrestrial bromeliads and larger to adult instars being fossorial. Due to the usually low population densities of tarantulas in rainforests in particular (pers. obs.), but also of many other spiders in general (Wise 1993), both intra- and interspecific competition and intraguild predation appear to be of little importance among rainforest theraphosids. In fossorial species small juveniles should be well protected from larger individuals by (a) low population density, (b) the predominantly sessile lifestyle, so that encounters are quite unlikely, and (c) by the small diameter of their burrows, into which a larger conspecific cannot follow easily. However, in the fossorial colonial *Tliltocatl vagans* females may enter burrows of conspecific females resulting in aggressive interactions among similarly sized individuals which are actually able to enter the burrows (Dor et al. 2008). Thus following *homo homini lupus* similar-sized instars in sessile tarantulas should be of greater danger to each other than distinctly bigger for smaller instars. Consequently, Marshall and West (2008) propose for *E. murinus* that the observed ontogenetic habitat shift might have evolved due to the evolutionary history where the fossorial habit may have evolved secondarily from arboreal life style. Within the genus *Ephebopus* a range from predominantly fossorial to rather arboreal species is documented (Marshall and West 2008; West et al. 2008). With respect to the "typically" arboreal tarantulas, West (2005b) reports a reversed ontogenetic habitat shift where size of plants and height of spiders may increase with instar stage. Data from the relatively large sample size of the Costa Rican arboreal *Psalmopoeus reduncus* suggest that this species does not exhibit any ontogenetic habitat shift in vertical distribution and only a slight in plant size expressed by diameter of object the spider was found on. However, the syntopic large ctenid *Cupiennius coccineus* showed a significant difference vertical distribution among three instar classes, with smaller instars occurring at greater heights (Lapinski and Tschapka 2018).

Topophilia at Micro and Nano Levels

At micro level tarantulas from the studied assemblages were associated either with ground (terrestrial species) or with vertical objects, mostly plants (arboreal species), both inside and outside of the rainforest. At nano level the encountered terrestrial species may be subdivided into fossorial species and species living under various objects on the ground (both *Ami* spp.). Microclimate inside of tarantula burrows in the ground stays constantly favorable compared to the outside microclimate (Punzo and Henderson 1999; Striffler 2005). This protective feature of burrows explains why only fossorial species were found on both ground and on banks of trails and roads both inside and outside of the surrounding forest. *Ami* species do not construct burrows (Auer 2011; Lapinski et al. 2018) and hence both *Ami* spp. were found on forest ground covered with leaf litter and wooden debris, avoiding clearings and the mostly bare soil of trail banks. The ecology of the Yasuní Tiger Tarantula (Theraphosinae 3) is completely unknown but inferring from its habitus it is most probably a ground dwelling species. In Costa Rica and in Ecuador the studied arboreal species showed high variability in topophilia. At micro level topophilia ranged from plants of different types to buildings. At nano level the arboreal species not only used almost the entire available height range (see vertical distribution) but even within same species different parts of plants and buildings were used ranging from free silken tubes on trunks, silken tubes ending in tree holes or being placed almost completely to completely in tree holes. The data suggest that vertical distribution of the arboreal tarantulas may be positively correlated by structural complexity of host trees. In the large ctenid *Cupiennius coccineus* a similar effect of structural complexity was found (Lapinski and Tschapka 2018). In Ecuador the arboreal *Avicularia rufa* was particularly flexible in shelter construction. In microhabitats with a superabundance of hiding possibilities like are offered by palm thatched roofs and by branches full of large tank bromeliads the spiders did not construct significant silken tubes (with the exception of thin molting sacks) rendering them wandering spiders. In microhabitats with scarce hiding possibilities *A. rufa* constructed firm and thick silken tubes and the spiders were found each night in their direct proximity. Thus, high complexity of vegetation may not only enhance densities of arthropods through providing shelter and food to both predators and their prey (Halaj et al. 2000; Stunz et al. 2002; Teixeira da Souza and Martins 2004), but also affect in some species shelter building behavior and probably site fidelity. The high flexibility may have evolved as a response to the microclimate range from understory to the rainforest canopy, the highly variable microclimate in the canopy (Lapinski and Tschapka 2014) and the ephemeral characteristics of canopy microhabitats with branches or epiphytes breaking off and falling down to the ground. Thus, many arboreal species may face rather instable and hence more variable conditions than inhabitants of the rainforest ground.

7.8 Concluding Remarks

Tarantulas (Theraphosidae) from the New World are widely distributed from northern Utah and California, USA in the north to northern Patagonia, Argentina in the south. Table 7.6 summarizes the information on the tarantula taxa depicted here according to the sequence of the paragraphs in this chapter. However, the ecology of tarantulas has been largely unappreciated. Scarce habitat information comes mainly from taxonomic works and the observations of dedicated amateurs. Tarantulas occupy a variety of different habitats from remote deserts and Caribbean islands, high Andean mountains, dark caves and deep tropical rainforests to plantations, gardens and soccer fields. Some species prefer pristine habitats, other thrive in anthropogenically modified areas, and some are quite flexible and are found in primary rainforest, on houses and road banks in rural areas. The present case study conducted in Costa Rica and Ecuador shows nicely that neotropical rainforests may harbor assemblages of several syntopic tarantula species with two main lifestyles: terrestrial and arboreal. The proportion of arboreal to terrestrial species in the Neotropics may be enhanced by a combination of forests, tree complexity, the presence of water bodies from small to gigantic accompanied by seasonal inundations of huge areas rendering the ground uninhabitable for long periods of time. Within both subguilds habitat segregation may occur at finer scales and include also seasonality patterns. While most terrestrial species live in burrows few take shelter under debris on the forest ground. Some species are found only in the forest interior, others prefer the open ground of small clearings. Arboreal species show high variability in microhabitat choice as long as it is more or less elevated and presents vertical structures: plants of various types and shapes. Buildings are used by several arboreal species as well, especially in the genus *Avicularia*. Here many species seem to be preadapted due to the highly variable conditions they encounter between the understory and the high canopy.

It is evident how underdeveloped our current knowledge and understanding of almost any ecological aspect of even "well-known" and conspicuous tarantula species is. Habitat use and community ecology of tarantulas is even deeper down the

Table 7.6 Summary of ecology of the tarantula taxa depicted in this chapter according to the sequence of the respective paragraphs

Genus/species	Region	Topophilia macro	Topophilia micro
Aphonopelma spp.	USA, Central America	(semi)deserts to temperate high-elevation forests	Silk-lined burrows in the ground
Brachypelma spp. "red-knee"	Mexico, Pacific slope	Dry thorn to high elevation forests, pastures	Silk-lined burrows in the ground
Tliltocatl spp. "red-rump"	Mexico, S & Carribean slope	Dry thorn to high elev. Forests, trop. Evergreen forests,	Silk-lined burrows in the ground
	Central America, Florida (USA)[a]	Other anthropogenically modified habitats[a]	Silk-lined burrows in the ground
Hemirrhagus spp.	Mexico, central to south	Caves, high elevation forests	On floor and walls
Tmesiphantes hypogeus	Brazil, Bahia	Caves	On humid rocky substrates
Lasiodora sp.	Brazil, Bahia	Cave	On cave floor
Avicularia spp.	Trop. South America, Panama	Trop. Rainforest, anthropogenically modified habitats	On vegetation and buildings
Avicularia taunayi	Brazil, NE	Cerrado	Silk tubes in tree holes
Ybyrapora spp.	Brazil, NE	Lowland Atlantic rainforest, primary and secondary	Silk tubes on tree trunks, among leaves,
			In terrestrial bromeliads
Iridopelma spp.	Brazil, NE	Lowland Atlantic rainforest, Restinga, caatinga, cerrado	Silk tubes on leaves, in terrestrial bromeliads,
			Under logs
Pachistopelma spp.	Brazil, NE	Lowland Atlantic rainforest, Restinga, caatinga	Silk tubes in terrestrial bromeliads
Typhochlaena spp.	Brazil, NE	Lowland Atlantic rainforest, Restinga, caatinga, cerrado	Silk tubes with trap door on trees
Dolichothele spp.	Brazil, NE	Lowland Atlantic rainforest, Restinga, caatinga, cerrado	Silk-lined burrows under stones and logs
Agnostopelma spp.	Colombia, Andes	No information available	Under stones
Aguapanela Arvi,	Colombia, Andes	High Andean cloud forest	In rock crevices
Hapalopus sp., *Ami* sp.	Colombia, Andes	High Andean cloud forest	No information available
Catumiri spp., *Holothele* spp.	Colombia, Andes	No information available	No information available
Cyclosternum gaujoni, C. schmardae	Ecuador, Andes	No information available	No information available
Pamphobeteus augusti, P. vespertinus	Ecuador, Andes	No information available	No information available

(continued)

Table 7.6 (continued)

Genus/species	Region	Topophilia macro	Topophilia micro
Cyriocosmus pribiki, C. rogerioi	Peru, Andes	High Andean cloud forest	No information available
Bistriopelma lamasi, B. matuskai	Peru, Andes	Grass slopes	Under stones and rocks
Hapalotremus spp.	Peru, Andes	Grass slopes	Crevices and burrows under rocks
Hapalotremus spp.	Bolivia, Andes	Yungas	No information available
Hapalotremus martinorum	Argentina, NW, Andes	High Andean cloud forest	No information available
Hapalotremus chasqui	Argentina, NW, Andes	Grass land with tree groves	Under rocks
Euathlus spp.	Chile, Andes	Semi-arid habitats,	No information available
Euathlus diamante	Argentina, NW, Andes	Semi-arid grass and shrubland	No information available
Euathlus sagei	Argentina, NW, Patagonia	Steppe	Crevices and burrows under rocks
Euathlus tenebrarum	Argentina, NW, Patagonia	Humid Andean-Patagonic forests	Crevices and burrows under rocks
Psalmopoeus reduncus	Costa Rica, Sarapiquí region	Tropical rainforest	On trees and other vegetation
Lasiodora icecu	Costa Rica, Sarapiquí region	Tropical rainforest	Burrows in forest ground
Sericoplema silvicola	Costa Rica, Sarapiquí region	Tropical rainforest	Burrows in forest ground
Stichoplastoris obelix	Costa Rica, Sarapiquí region	Tropical rainforest	Burrows in forest ground
Neischnocolus panamanus	Costa Rica, Sarapiquí region	Tropical rainforest	Under wood on forest ground
Neischnocolus yupanquii	Ecuador, Rio Napo	Tropical rainforest	Leaf litter forest ground
Megaphobema velvetosoma	Ecuador, Rio Napo	Tropical rainforest	Burrows in forest ground
Theraphosinae 1	Ecuador, Rio Napo	Tropical rainforest, clearing	Burrows in soil of clearing
Theraphosinae 2	Ecuador, Rio Napo	Tropical rainforest, clearing	On ground
Theraphosinae 3 (Yasuní Tiger Tarant.)	Ecuador, Rio Napo	Tropical rainforest	On leaf understory Maranthaceae
Avicularia hirschii, A.purpurea, A. rufa	Ecuador, Rio Napo	Tropical rainforest	On trees and huts
Pseudoclamoris sp. (cf. *burgessi*)	Ecuador, Rio Napo	Tropical rainforest	On trees

NE north-east, *NW* north-west, *S* south
[a]Refers to *Tliltocatl vagans*

Rabbit Hole—or further up the Tree of Souls—and I hope researchers all over the world start to investigate this overlooked aspect of tarantula biology, daring to enter the higher spheres of the rainforest, to go underground exploring the deeper realms of the earth and observing patiently both the living animals and their surroundings to which they are connected so tightly. Why? Because it is fascinating, important, and fun at the same time!

Acknowledgments I thank the unknown reviewer for his helpful comments on the manuscript. The Ministerio de Ambiente, Energía y Telecomunicaciones (MINAET) kindly granted us the permission to conduct the field work in Costa Rica. The friendly assistance of Javier Guevara (MINAET) is greatly appreciated. I am very grateful to all staff members at the Reserva Biológica Tirimbina and Chiquita (Costa Rica) as well as to the staff of Centro de Interpretación Yaku Kawsay and Sani Lodge (Ecuador) for all their friendliness, support, and for allowing me to realize this project at their sites. Many thanks to José Angel González Ramírez (Organization for Tropical Studies, Costa Rica) for identification of some trees of Tirimbina. I thank Eddy Hijmensen, Leonela Schwerdt, and Gabriel Emilio Vargas Arana for providing some scientific articles. Many thanks to Eddy Hijmensen for providing photographs of *Hemirrhagus* sp., *Tliltocatl hamorii*, and *T. vagans*.

References

Adis J (ed) (2002) Amazonian Arachnida and Myriapoda, Pensoft series faunistica no. 24, Sofia, p 590

Alfaro C, Figueroa DP, Torres H, Veloso C, Venegas F, Canals L, Canals M (2013) Effect of thermal acclimation on preferred temperatures in two mygalomorph spiders inhabiting contrasting habitats. Physiol Entomol 38:20–25

Almeida-Silva LM, Camacho A, Brescovit AD, Lucas SM, Brazil TK (2008) Redescription and notes on natural history of the arboreal tarantula *Iridopelma seladonium* (Aranea: Theraphosidae: Aviculariinae). Rev Bras Zootec 25:728–736

Andre C, Hüsser M (2016) Die Gattung *Typhochlaena* – Historie und Systematik sowie Habitat und Lebensweise von *T. seladonia*, *T. costae* und *T. curumim*. Arachne 3:4–33

Arisqueta-Chablé C, Manrique-Saide P, Pinkus Rendón MA, Meléndez Ramírez V (2010) Noteworthy records of *Brachypelma* (Araneae: Theraphosidae) from peninsula of Yucatan, Mexico. Entomol News 120:566–569

Auer HW (2011) Zwergvogelspinnen. Lebensraum, Haltung, Nachzucht. Natur und Tier, Münster, p 159

Barker M, Standridge N (2002) Ropes as a mechanism for canopy access. In: Mitchell AW, Secoy K, Jackson T (eds) The global canopy handbook. Techniques of access and study in the forest roof. Global Canopy Programme, Oxford, UK, pp 13–23

Basset Y (2001) Invertebrates in the canopy of tropical rain forests. How much do we really know? Plant Ecol 153:87–107

Basset Y, Novotny V, Miller SE, Kitching RE (eds) (2003) Arthropods of tropical forests: spatio-temporal dynamics and resource use in the canopy. Cambridge University Press, Cambridge, UK, p 474

Berland L (1913) Araignées. In: Mission du Service géographique de l'armée pour la mesure d'un arc du méridien équatorial en Amérique du Sud (1899–1906), vol 10, Paris, pp 78–119

Bertani R (2012) Revision, cladistic analysis and biogeography of *Typhochlaena* C. L. Koch, 1850, *Pachistopelma* Pocock, 1901 and *Iridopelma* Pocock, 1901 (Araneae, Theraphosidae, Aviculariinae). ZooKeys 230:1–94

Bertani R, Fukushima CS (2009) Description of two new species of *Avicularia* Lamarck 1818 and redescription of *Avicularia diversipes* (C.L. Koch 1842) (Araneae, Theraphosidae, Aviculariinae) – three possibly threatened Brazilian species. Zootaxa 2223:25–47
Bertani R, Motta PC (2013) Redescription of *Avicularia taunayi* and notes on its habitat and geographical distribution (Araneae: Theraphosidae: Aviculariinae). Zoologia 30:107–114
Bertani R, Bichuette ME, Pedroso DR (2013) *Tmesiphantes hypogeus* sp. nov. (Araneae, Theraphosidae), the first troglobitic tarantula from Brazil. An Acad Bras Ciên 85:235–243
Bullmer M, Thierer-Lutz M, Schmidt G (2006) *Avicularia hirschii* sp. n. (Araneae: Theraphosidae: Aviculariinae), eine neue Vogelspinnenart aus Ekuador. Tarantulas World 124:3–17
Cardoso da Silva JM, Leal IR, Tabarelli M (eds) (2017) Caatinga. The largest tropical dry forest region in South America. Springer, Cham, p 482
Cubillos C, Cáceres JC, Villablanca C, Villarreal P, Baeza M, Cabrera R, Graether SP, Veloso C (2018) Cold tolerance mechanisms of two arthropods from the Andean Range of Central Chile: *Agathemera crassa* (Insecta: Agathemeridae) and *Euathlus condorito* (Arachnida: Theraphosidae). J Therm Biol 74:133–139
Culver DC, White WB (2005) Encyclopedia of caves. Elsevier, Burlington, p 654
Dias SC, Brescovit AD (2003) Note on the behavior of *Pachistopelma rufonigrum* Pocock (Araneae, Theraphosidae, Aviculariinae). Rev Bras Zool 20:13–17
Dias SC, Brescovit AD (2004) Microhabitat selection and co-ocurrence of *Pachistopelma rufonigrum* Pocock (Araneae, Theraphosidae) and *Nothroctenus fuxico* sp.nov. (Araneae, Ctenidae) in tank bromeliads from Serra de Itabaiana, Sergipe, Brazil. Rev Bras Zool 21:789–796
Dias SC, Rocha PA, Silva Bomfim L, Ferrari SF (2015) Predation of the bat *Pteronotus personatus* (Mormoopidae), by a tarantula *Lasiodora* sp. (Theraphosidae, Araneae) in cave in northeastern Brazil. Biotemas 28:173–175
Dor A, Machkour-M'Rabet S, Legal L, Williams T, Hénaut Y (2008) Chemically mediated burrow recognition in the Mexican tarantula *Brachypelma vagans* female. Naturwissenschaften 95:1189–1193
Edwards GB, Hibbard KL (2003) Mexican Redrump Tarantula, *Brachypelma vagans* (Ausserer) (Arachnida: Araneae: Theraphosidae). EENY-287. DPI Entomol Circ 394:1–3
Ferretti N (2015) On three new *Euathlus* tarantulas from Argentina and cladistic analysis of the genus. J Arachnol 43:313–326
Ferretti N, Cavallo P, Chaparro JC, Ríos-Tamayo D, Seimon TA, West R (2018) The Neotropical genus *Hapalotremus* Simon, 1903 (Araneae: Theraphosidae), with the description of seven new species and the highest altitude record for the family. J Nat Hist 52:1927–1984
Fukushima CS, Bertani R (2017) Taxonomic revision and cladistic analysis of *Avicularia* Lamarck, 1818 (Araneae, Theraphosidae, Aviculariinae) with description of three new aviculariine genera. ZooKeys 659:1–185
Gonçalves-Souza T, Almeida-Neto M, Romero GQ (2010) Bromeliad architectural complexity and vertical distribution predict spider abundance and richness. Aust Ecol 36:476–484
Guadanucci JPL (2007) A revision of the Neotropical spider genus *Oligoxystre* Vellard 1924 (Theraphosidae, Ischnocolinae). Zootaxa 1555:1–20
Guadanucci JPL (2011) Cladistic analysis and biogeography of the genus *Oligoxystre* Vellard 1924 (Araneae: Mygalomorphae: Theraphosidae). J Arachnol 39:320–326
Halaj J, Ross DW, Moldenke AR (2000) Importance of habitat structure to the arthropod food-web in Douglas-fir canopies. Oikos 90:139–152
Hamilton CA, Hendrixson BE, Bond JE (2016) Taxonomic revision of the tarantula genus *Aphonopelma* Pocock, 1901 (Araneae, Mygalomorphae, Theraphosidae) within the United States. Zookeys 560:1–340
Hanken J, Wake DB (1993) Miniaturization of body size: organismal consequences and evolutionary significance. Ann Rev Ecol Evol Syst 24:501–519
Hijmensen E (2009) Vogelspinnen der Maya-Welt – eine Vogelspinnen-Expedition in die zentralamerikanischen Länder Mexiko (Yucatan), Belize, Guatemala und Honduras. Arachne 14:4–39

Höfer H, Brescovit AD, Adis J, Paarmann W (1994) The Spider Fauna of neotropical tree canopies in Central Amazonia: first results. Stud Neotrop Fauna E 29:23–32

Horstkotte J, Riesch R, Plath M, Jäger P (2010) Predation by three species of spiders on a cave fish in a Mexican sulphur cave. Bull Br Arachnol Soc 15:55–58

Hüsser M (2017) Ecuador – *Avicularia hirschii* found in a special retreat. https://birdspiders.ch/avicularia-hirschii/. Accessed 6 April 2019

Indicatti RP, Lucas SM, Guadanucci JPL, Yamamoto FU (2008) Revalidation and revision of the genus *Magulla* Simon 1892 (Araneae, Mygalomorphae, Theraphosidae). Zootaxa 1814:21–36

Jepson J (2000) The tree climber's companion. A reference and training manual for professional tree climbers, 2nd edn. Beaver Tree, Longville, p 104

Junk WJ (ed) (1997) The central amazon floodplain: ecology of a pulsing system, Ecological studies book 126. Springer, New York, p 528

Kaderka R (2015) *Bistriopelma*, a new genus with two new species from Peru (Araneae: Theraphosidae: Theraphosinae). Rev Peruana Biol 22:275–288

Kelly M, Truscott H, Andersen TR, Shibuya A (2017) Amazon wildlife ecology. In: Eisemberg CC, Reynolds SJ (eds) An introduction to wildlife conservation in the Brazilian Amazon: a view from northern Australia. Brazilian Amazon Field Intensive (BAFI). Charles Darwin University, Darwin, pp 13–24

Kirk P (1990) *Avicularia purpurea*, a new species of Theraphosid spider from Ecuador. Br Tarantula Soc J 6:15–19

Lapinski W, Tschapka M (2013) Habitat use in an assemblage of Central American wandering spiders. (Araneae: Ctenidae, Trechaleidae). J Arachnol 41:151–159

Lapinski W, Tschapka M (2014) Desiccation resistance reflects patterns of microhabitat choice in a Central American assemblage of wandering spiders. J Exp Biol 217:2789–2795

Lapinski W, Tschapka M (2018) Vertical distribution of wandering spiders in Central America. J Arachnol 46:13–20

Lapinski W, Gabriel R, Pérez-Miles F (2018) *Ami bladesi* Pérez-Miles et al. 2008 (Araneae, Theraphosidae), description of the female with notes on the species' ecology and behavior. Arachnology 17:419–426

Leyer I, Wesche K (2007) Multivariate Statistik in der Ökologie. Eine Einführung. Corrected edition 2008. Springer, Berlin, p 221

Locht A, Yáñez M, Vázquez I (1999) Distribution and natural history of Mexican species of *Brachypelma* and *Brachypelmides* (Theraphosidae, Theraphosinae) with morphological evidence for their synonymy. J Arachnol 27:196–200

Lowman MD, Nadkarni NM (eds) (1995) Forest canopies, 1st edn. Academic Press, San Diego, p 624

Lowman MD, Rinker HB (eds) (2004) Forest canopies, 2nd edn. Elsevier, Burlington, p 517

Lowman MD, Schowalter TD, Franklin JF (eds) (2012) Methods in forest canopy research. University of California Press, Berkeley, p 221

Machkour-M'rabet S, Hénaut Y, Rojo R, Calmé S (2005) A not so natural history of the tarantula *Brachypelma vagans*: interaction with human activity. J Nat Hist 39:2515–2523

Machkour-M'Rabet S, Hénaut Y, Calmé S, Legal L (2012) When landscape modification is advantageous for protected species. The case of synanthropic tarantula, *Brachypelma vagans*. J Ins Cons 16:479–488

Maréchal P, Rolland C, Leguin E-A, Monpiou L (2009) Étude sur l'état des populations de *Avicularia versicolor*, Mygale endémique de la Martinique, Rapport de Synthése, Missions 2007–2008. Muséum national d'Histoire naturelle, Paris. http://www.side.developpement-durable.gouv.fr/EXPLOITATION/DEFAULT/Infodoc/ged/viewportalpublished.ashx?eid=IFD_FICJOINT_0006180&search

Marques MI, Adis J, dos Santos GB, Battirola LD (2006) Terrestrial arthropods from tree canopies in the Pantanal of Mato Grosso, Brazil. Rev Bras Ent 50:257–267

Marshall SD, West R (2008) An ontogenetic shift in habitat use by the Neotropical tarantula *Ephebopus murinus* (Araneae, Theraphosidae, Aviculariinae). Bull Br Arachnol Soc 14:280–284
Medina Soriano FJ (2006) A new species of *Cupiennius* (Araneae, Ctenidae) coexisting with *Cupiennius salei* in a Mexican mangrove forest. J Arachnol 34:135–141
Mendoza JI (2014a) Taxonomic revision of *Hemirrhagus* Simon, 1903 (Araneae: Theraphosidae, Theraphosinae), with description of five new species from Mexico. Zool J Linn Soc 170:634–689
Mendoza JI (2014b) *Psalmopoeus victori*, the first arboreal theraphosid spider described for Mexico (Araneae: Theraphosidae: Aviculariinae). Rev Mex Biod 85:728–735
Mendoza JI, Francke OF (2017) Systematic revision of *Brachypelma* red-kneed tarantulas (Araneae: Theraphosidae), and the use of DNA barcodes to assist in the identification and conservation of CITES-listed species. Inv Syst 31:157–179
Mendoza JI, Francke OF (2018) Five new cave-dwelling species of *Hemirrhagus* Simon 1903 (Araneae, Theraphosidae, Theraphosinae), with notes on the generic distribution and novel morphological features. Zootaxa 4407:451–482
Mendoza JI, Francke OF (2020) Systematic revision of Mexican threatened tarantulas *Brachypelma* (Araneae: Theraphosidae: Theraphosinae), with a description of a new genus, and implications on the conservation. Zool J Lin Soc 188:82–147
Mendoza JI, Locht A, Kaderka R, Medina F, Pérez-Miles F (2016) A new genus of Theraphosid spider from Mexico, with a particular palpal bulb structure (Araneae, Theraphosidae, Theraphosinae). Eur J Tax 232:1–28
Messas YF, Souza HS, Gonzaga MO, Vasconcellos-Neto J (2014) Spatial distribution and substrate selection by the orb-weaver spider *Eustala perfida* Mello-Leitão, 1947 (Araneae: Araneidae). J Nat Hist 48:2645–2660
Mitchell AW, Secoy K, Jackson T (eds) (2002) The global canopy handbook. Techniques of access and study in the forest roof. Global Canopy Programme, Oxford, UK, p 248
Morin PJ (2011) Community ecology. Wiley-Blackwell, Chichester, UK, p 407
Oliveira PS, Marquis RJ (2002) Introduction: development of research in the cerrados. In: Oliveira PS, Marquis RJ (eds) The cerrados of Brazil: ecology and natural history of a Neotropical savanna. Columbia University Press, New York, pp 1–10
Ortiz D (2008) Description of *Cubanana cristinae*, a new genus and species of theraphosine tarantula (Araneae: Theraphosidae) from the island of Cuba. Bol Soc Ent Aragonesa 42:107–122
Ortiz D, Bertani R (2014) A new species in the spider genus *Phormictopus* (Theraphosidae: Theraphosinae) from Cuba. Rev. Ibér Aracnol 11:29–36
Perafán C, Pérez-Miles F (2014) The Andean tarantulas *Euathlus* Ausserer, 1875, *Paraphysa* Simon, 1892 and *Phrixotrichus* Simon, 1889 (Araneae: Theraphosidae): phylogenetic analysis, genera redefinition and new species descriptions. J Nat Hist 48:2389–2418
Perafán C, Cifuentes Y, Estrada-Gomez S (2015) *Aguapanela*, a new tarantula genus from the Colombian Andes (Araneae, Theraphosidae). Zootaxa 4033:529–542
Pérez-Miles F, Weinmann D (2009) Two new species of *Cyriocosmus* Simon, 1903 from Peru and the highest altitude record for the genus (Araneae, Theraphosidae,Theraphosinae). Rev Ibér Aracnol 17:29–35
Pérez-Miles F, Gabriel R, Miglio L, Bonaldo A, Gallon R, Jiménez JJ, Bertani R (2008) Ami, a new theraphosid genus from Central and South America, with the description of six new species (Araneae: Mygalomorphae). Zootaxa 1915:54–68
Pérez-Miles F, Weinmann D (2010) *Agnostopelma*: a new genus of tarantula without a scopula on leg IV (Araneae: Theraphosidae: Theraphosinae). J Arachnol 38:104–112
Peters H-J (2003) Tarantulas of the world. Amerika's Vogelspinnen. Tarantulas of the World, Wegberg, p 329
Peters H-J (2005) Tarantulas of the world: Kleiner Atlas der Vogelspinnen - Band 3. Tarantulas of the World, Wegberg, p 130
Pocock RI (1903) On some genera and species of South-American Aviculariidae. Ann Mag Nat Hist 11:81–115
Prentice TR (1997) Theraphosidae of the Mojave Desert west and north of the Colorado River (Araneae, Mygalomorphae, Theraphosidae). J Arachnol 25:137–176

Punzo F, Henderson L (1999) Aspects of the natural history and behavioural ecology of the tarantula spider *Aphonopelma hentzi* (Girard, 1854) (Orthognatha, Theraphosidae). Bull Br Arachnol Soc 11:121–128

Rao D (2017) Habitat selection and dispersal. In: Viera C, Gonzaga MO (eds) Behavior and ecology of spiders: contributions from the neotropical region. Springer, Cham, pp 85–108

Reátegui-Suárez P, Vásquez-Bardales J, Patiño-Patroni J, Tirado-Herrera E (2014) Algunos aspectos ecológicos de "tarántulas" (Araneae: Theraphosidae) en dos tipos de bosques en San Juan Bautista, Loreto, Perú. Ciencia Amazónica (Iquitos) 4:109–116

Reichling SB (2003) Tarantulas of Belize. Krieger, Malabar, p 127

Rocha CFD, Bergallo HG, Van Sluys M, Alves MAS, Jamel CE (2007) The remnants of restinga habitats in the Brazilian Atlantic Forest of Rio de Janeiro state, Brazil: habitat loss and risk of disappearance. Brazil J Biol 67:263–273

Rudloff J-P (2003) Eine neue *Brachypelma*-Art aus Mexiko, *Brachypelma schroederi* sp. n. (Araneae: Mygalomorphae: Theraphosidae: Theraphosinae). Arthropoda 11:2–15

Rudloff J-P (2008) Eine neue *Brachypelma*-Art aus Mexiko (Araneae: Mygalomorphae: Theraphosidae: Theraphosinae). Arthropoda 16:26–30

Santos RL, Almeida MG, Nunes JV, Tinoco LDS, Martins LB (2003) Bromeliads as a keystone resource for the scorpion *Tityus neglectus* in eastern Rio Grande do Norte State, Brazil. J Bromeliad Soc 53:241–288

Santos RL, Almeida MG, Tinoco LDS, Martins LB, Maia MG (2004) Biogeography and conservation of the bromeliad tarantula *Pachistopelma rufonigrum* (Aranae, Theraphosidae) in Rio Grande do Norte, Brazil. J Bromeliad Soc 54:145–192

Schwerdt L, Villalobos AE, Pérez-Miles F (2018) Spiders as potential bioindicators of mountain grasslands health: the Argentine tarantula *Grammostola vachoni* (Araneae, Theraphosidae). Wild Res 45:64–71

Shaw EM, Bennett SP, Wheater CP (2011) Distribution of *Brachypelma vagans* (Theraphosidae) burrows and their characteristics in Belize over two years. J Arachnol 39:515–518

Smith AM (1994) Tarantula spiders: tarantulas of the U.S.A. and Mexico. Fitzgerald, London, p 200

Sørensen LL (2003) Stratification of the spider fauna in a Tanzanian forest. In: Basset Y, Novotny V, Miller SE, Kitching RL (eds) Arthropods of tropical forests: spatio-temporal dynamics and resource use in the canopy. Cambridge University Press, Cambridge, pp 92–101

Stork NE, Adis J, Didham RK (eds) (1997) Canopy arthropods. Chapman and Hall, London, p 567

Stradling DJ (1994) Distribution and behavioral ecology of an arboreal "tarantula" spider in Trinidad. Biotropica 26:84–97

Stuffler BF (2005) Life history of Goliath Birdeaters – *Theraphosa apophysis* and *Theraphosa blondi* (Araneae, Theraphosidae, Theraphosinae). J Br Tarantula Soc 21:26–33

Stunz S, Ziegler C, Simon U, Zotz G (2002) Diversity and structure of the arthropod fauna within three canopy epiphyte species in central Panama. J Trop Ecol 18:161–176

Teixeira da Souza AL, Martins RP (2004) Distribution of plant dwelling spiders: inflorescences versus vegetative branches. Aust Ecol 29:342–349

Turner SP, Longhorn SJ, Hamilton CA, Gabriel R, Pérez-Miles F, Vogler AP (2018) Re-evaluating conservation priorities of New World tarantulas (Araneae: Theraphosidae) in a molecular framework indicates nonmonophyly of the genera, *Aphonopelma* and *Brachypelma*. Syst Biod 16:89–107

Valerio CE (1979) Arañas terafósidas de Costa Rica (Araneae: Theraphosidae). II. *Psalmopoeus reduncus*, redescripción, distribución y el problema de dispersión en terafósidas. Revi Biol Trop 27:301–308

Valerio CE (1980) Arañas terafósidas de Costa Rica (Araneae, Theraphosidae) I. *Sericopelma* y *Brachypelma*. Brenesia 18:259–288

Veloso V, Luhr D, Marfull R, Torres-Contreras H, Figueroa Pérez D, Sabat P, Canals M (2012) Characterization of the thermal micro-environment of *Paraphysa parvula* Pocock 1903 (Araneae: Theraphosidae), a spider from the Chilean Andes. J Arachnol 40:34–38

Weinmann D (2003) Populationsuntersuchungen an einer Kolonie der Vogelspinne *Megaphobema robustum* (Ausserer, 1873) in Kolumbien (Araneae, Theraphosidae, Theraphosinae). Arthropoda 11:23–30
West RC (2005a) The *Brachypelma* of Mexico. J Br Tarantula Soc 20:108–119
West RC (2005b) Die Vogelspinnen aus Französisch Guayana. Arachne 10:4–13
West R, Marshall SD, Fukushima CS, Bertani R (2008) Review and cladistic analysis of the Neotropical tarantula genus *Ephebopus* Simon 1892 (Araneae: Theraphosidae) with notes on the Aviculariinae. Zootaxa 1849:35–58
Wise DH (1993) Spiders in ecological webs. Cambridge University Press, Cambridge, UK, p 328
World Spider Catalog (2020) World spider catalog, version 21.0. Natural History Museum Bern. http://wsc.nmbe.ch. Accessed 20 Feb 2020
Yáñez M, Floater G (2000) Spatial distribution and habitat preference of the endangered tarantula *Brachypelma klaasi* (Araneae: Theraphosidae) in Mexico. Biodiv Cons 9:795–810

Chapter 8
Predation and Other Interactions

Yann Hénaut and Salima Machkour-M'Rabet

Abstract Theraphosids interact with numerous species from invertebrates to humans. Their diet predominantly consists of insects, spiders, and worms; however, their prey cover a wide range of taxa, including mammals, birds, reptiles, fish, and even the very toxic poison dart frogs. Practically blind, they detect prey by vibrations and may also use odors. Their predatory activity may affect the distribution of other predatory species such as other spiders or scorpions. Furthermore, predation may include cannibalism between females. Despite a variety of antipredatory strategies—which include urticating hairs, burrows, or special coloration—theraphosids are preyed on by vertebrates such as coatis and also by some invertebrates. Parasitoid hawk wasps hunt tarantulas, and they are also parasitized by flies. Tarantulas may be found in commensalistic associations with toads or bromeliads. In their natural range, they also interact with humans and native people in different regions of the Americas, who use them for food or as part of traditional medicine. The diversity of these interactions and adaptations may be considered a result of a long evolutionary history.

8.1 Introduction

Tarantulas are one of the oldest groups of spiders in terms of evolution and characteristics. As Mygalomorphae, they diverged with other spiders 300 million years ago (Ayoub and Hayashi 2009). In this context, their interactions with prey and predators, as well as their associations with other organisms, are of particular importance in relation to other spiders. Tarantulas are also very large and heavy spiders; for example, the largest species of tarantula, the Goliath tarantula (*Theraphosa blondi* (Latreille, 1804)), may reach a leg span of up to 30 centimeters and a weight of 170 grams, and in the field, an individual of the species *Tliltocatl vagans* Ausserer, 1875 may easily reach a leg span of 13.5 centimeters and a weight of around 10–15

Y. Hénaut (✉) · S. Machkour-M'Rabet
El Colegio de la Frontera Sur, Unidad Chetumal, Chetumal, Quintana Roo, Mexico
e-mail: yhenaut@ecosur.mx

F. Pérez-Miles (ed.), *New World Tarantulas*, Zoological Monographs 6,
https://doi.org/10.1007/978-3-030-48644-0_8

grams (Machkour-M'Rabet et al. 2007; Hénaut et al. 2015). In comparison, an adult female of one of the larger orb web spiders such as *Nephila clavipes* (Linnaeus, 1767) (Araneae, Araneidae) weighs around 4 grams with a leg span of approximately 5 centimeters. The size and weight of tarantulas are certainly factors that influence their interactions with other organisms.

Like practically all species of spider, tarantulas are predators, with the notable and unique exception of the jumping spider *Bagheera kiplingi* Peckham & Peckham, 1896 (Araneae, Salticidae), found in Central America (Jackson 2009), which inhabits acacia trees and mainly feeds on the Beltain bodies of the plant (Meehan et al. 2009). Generally, spiders prey on insects (Foelix 2011), using a great variety of strategies such as active pursuit, ambush, or use of silk in a wide variety of ways, principally by weaving webs with different architectures adapted for different kinds of prey (Zschokke et al. 2006; Herberstein and Tso 2011; Nelson and Jackson 2011). In comparison with their consumption of insects, spiders rarely prey on vertebrates. However, semiaquatic spiders from the Pisauridae, Trechaleidae, Lycosidae, and Ctenidae families occasionally prey on small fish (Nyffeler and Pusey 2014). Arachnids, with the exception of large web-building spiders, rarely catch mammals or birds; however, *Nephila* species have been observed preying on bats (Nyffeler and Knörnschild 2013) and on birds, principally hummingbirds (Peloso and Pereira de Souza 2007; Brooks 2012). The generally small size of spiders ensures that predation on vertebrates remains uncommon. In contrast, the larger size and weight of the tarantula allows them to prey on bigger prey, particularly vertebrates. Despite their size, tarantulas are generally limited to a "sit-and-wait" predation strategy. While other spiders use a variety of perception traits to detect prey, such as the excellent vision of jumping spiders (Su et al. 2007), tarantulas are much more limited in how they perceive their prey and environment (Blein et al. 1996). Furthermore, although tarantulas detect prey well through vibrations, they have difficulty in detecting smaller organisms; thus, potential prey fall within a specific size and weight range (Dor and Hénaut 2013).

Spiders are susceptible to being preyed on by a large variety of organisms. According to Foelix (2011), citing Jackson (1992), the principal enemies of spiders are spiders themselves. Spiders may prey on other species of spider, and they are also commonly cannibalistic. The well-known case of *Portia* spp. (Karsch, 1878; Araneae, Salticidae) and its behavioral plasticity to catch other spiders is an example of spider-eating specialization (see Jackson works cited in Foelix 2011), as are spiders such as Mimetidae feeding exclusively on spiders (Foelix 2011). Examples of cannibalism in spiders include wolf spiders preying on smaller juvenile spiders (Samu et al. 1999), the classical example of sexual cannibalism (Persons and Uetz 2005), or the spectacular case of matryphagy in *Amaurobius ferox* (Walckenaer, 1830) (Araneae, Amaurobiidae) (Kim and Horel 1998). Other invertebrates such as parasitoids or parasites are spider enemies—in particular, Pompilidae wasps and nematodes (Foelix 2011). As vertebrates, birds are considered a major threat to spiders, particularly in forest ecosystems (Gunnarson 2007). Spiders are also eaten by lizards, fish, amphibians, and some mammals (Foelix 2011); however, those vertebrates appear not to be a major threat. Focusing on tarantulas, once again their size prevents them from being preyed on by the same predators that usually prey on smaller spiders. However, tarantulas may be an attractive host for parasites, considering their weight as well as their capacity to live for a long time period under fasting conditions, even for more than 2 years without water and food access in the most extreme known cases (Baerg 1938). Also, a tarantula's body contains about 63% protein, in comparison with vertebrates, which contain only around 20% (see Taylor 1975; West 1997), making tarantulas particularly interesting prey for predators, including humans.

Spiders are both predators and prey, but they also may live in association with other organisms. As kleptoparasites, some spiders such as *Argyrodes* live together with other spider called hosts, particularly *Nephila* (Whitehouse et al. 2002; Hénaut et al. 2005), and steal their prey. Furthermore, spiders are frequently found to be associated with plants such as bromeliads; the tank bromeliad is an attractive species for several spiders, including jumping spiders (Salticidae; De Omena and Romero 2010) and curtain web spiders (Dipluridae; Hénaut et al. 2014). One spectacular association between spiders and other organisms is provided by myrmecophiles spiders, which demonstrate a close evolution with ant species, living alongside or within ant colonies (see review in Cushing 2012). According to Cushing (2012), the ant guest spiders are also mirmecomorphs and mimic the host colony odor; they may be kleptoparasitic (eating host prey), eat other insects found in the ant nest, or prey on their hosts. While other species of spider have frequently been observed in association with ants, this has rarely been observed with tarantulas. However, despite the perception we have of tarantulas as ferocious predators, the interactions between these spiders and other organisms is not restricted to a predator–prey relationship with tarantulas taking the role of top predators. These large spiders are also found in association with other organisms such as bromeliads and have a particular association with frogs across the Americas.

Human activities affect spider communities (Shochat et al. 2004), and human perception of spiders is often negative, as spiders provoke fear or are considered disgusting by humans (Gerdes et al. 2009). As with other spiders and organisms, human activities may affect tarantula populations (Machkour-M'Rabet et al. 2005, 2007). However, the size and colors of tarantulas make them attractive for some humans, giving them a particular status and provoking more attraction than fear in comparison with other spiders. Both attraction and fear associated with tarantulas make them the target of illegal wildlife trafficking for the pet market, or they are killed merely because of human fear and ignorance. In addition, spiders are also used in medicine and sometimes as food; tarantulas are also part of the diet of certain ethnic groups and an ingredient in human traditional and modern medicine (Machkour-M'Rabet et al. 2011). Finally, people are not indifferent with regard to especially large spiders; thus, tarantulas have a special place in human cultures.

8.2 Predation

Tarantulas generally hunt during the night, given that the large majority are burrowing spiders, which inhabit underground burrows but may also live behind stones and in other crevices, as do species belonging to the *Tliltocatl* genus (Locht et al. 1999; Yañez and Floater 2000; Machkour-M'Rabet et al. 2005, 2007). A few tarantulas are arboreal, such as *Avicularia* species, which live in tree trunks, between leaves, or in bromeliads (Bertani and Fukushima 2009; Bertani and Motta 2013). Theraphosids are considered sit-and-wait predators, waiting for prey at the entrance to their burrow or retreat. Few studies have focused on the sensory channels used by those spiders to detect their prey; however, the first account was given by Blein et al. (1996). The Brazilian salmon pink bird-eating tarantula *Lasiodora parahybana* Mello-Leitão, 1917 (a burrowing species from Brazil) uses mainly vibrations to detect prey, but this predator also catches prey by using sound or sound with vision, although not odors or odors with vision (Blein et al. 1996). In this context, the silk present in the entrance to a burrow or retreat, together with the quality of the substrate, may play an important role, as the vibrations provoke and facilitate prey detection (Blein et al. 1996). However, other tarantulas such as *Tliltocatl vagans* were observed leaving their burrow by night and attacking congeners in their own burrow, taking their refuge, and eating the other tarantula (Hénaut and Machkour-M'Rabet 2005; Hénaut et al. 2015). In this particular context of cannibalism, Dor et al. (2008) demonstrated that the active tarantula uses odors originating from the silk produced by its victim. This phenomenon of attraction by the silk of a conspecific is called sericophily and is common in spiders (Hodge and Storfer-Isser 1997). Several studies suggest that tarantulas are limited by the size of the prey; as the use of vibration is their principal channel of perception, they cannot detect small prey (Dor and Hénaut 2011). For example, during observation of interactions between *T. vagans* and wolf spiders, smaller lycosids are neither preyed on nor detected by the tarantula (Dor and Hénaut 2011). Similarly, very active and larger prey such as cockroaches are preyed on by *T. vagans* more often than other types of prey (Dor et al. 2011). However, whereas smaller organisms are not detected by tarantulas, as occurs with other spiders (Hénaut 2000; Hénaut et al. 2005), theraphosids are able to detect and catch bigger prey than practically any other arachnid in the world of spiders. In fact, a global characteristic of tarantulas around the world is their ability and capacity to prey on vertebrates such as bats (see Nyffeler and Knörnschild 2013), with the examples of *Poecilotheria rufilata* Pocock, 1899 feeding on the pipistrelle bat *Pipistrellus ceylonicus* Kelaart, 1852 (Chiroptera, Vespertilionidae) in India (Das et al. 2012) and South American tarantulas also consuming these mammals (Fig. 8.1a), as observed by West (in Nyffeler and Knörnschild 2013) (Table 8.1). It is also remarkable that tarantulas observed eating bats are arboreal or wandering predators whereas tarantulas observed in the wild eating other kinds of vertebrates are burrowing spiders (Table 8.1). Bats are not the only mammals that theraphosids may capture and eat; there have been many observations of tarantulas kept in laboratories or as pets eating other mammals, principally mice (Table 8.1). These observations have principally involved large tarantulas such as the Goliath birdeater

Fig. 8.1 Tarantulas eating mammals and birds. (**a**) Female *Lasiodora parahybana* found eating a bat at night near a forest creek in João Pessoa, Brazil. (**b**) Female *Xenesthis immanis* eating a hummingbird by a forest trail near Maracaibo Lake, Venezuela. (Photos: Rick C. West)

(*T. blondi*) or *L. parahybana* but also smaller tarantulas such as several *Brachypelma* species (Table 8.1). There are only a few observations of mammals being preyed on: bats, a kangaroo rat, and a mouse opossum are the only mammals reportedly preyed on by tarantulas in the wild. This lack of observations may be associated with a low frequency of predation of mammals and an absence of reports. Tarantulas—in particular, large American tarantulas such as *T. blondi*–are clearly able to catch rodents, and it is reasonable to consider that in the wild, tarantulas sometimes prey on small mammals, even though we cannot be certain that these types of prey are common for tarantulas.

Large American tarantulas are commonly called bird-eating spiders; in fact, these spiders were described as bird eaters by European settlers who arrived in the Americas. The illustration shown in Fig. 8.2, created in 1705 by the naturalist and artist Anna Maria Sibylla Merian (1647–1717), contributed to this reputation of large American tarantulas, inspiring the name *Avicularia*, which originates from the

Table 8.1 Vertebrates preyed on by Theraphosidae

Prey type/species/ family	Taxa	Tarantula species	Location	Source
Mammals				
Marmosops noctivagus	Opossum (Didelphidae)	*Pamphobeteus* sp.	Peru	von May et al. (2019)
Mouse, gerbil	Rodent	*Theraphosa blondi*	Breeders (induced)	Internet source; Gallon (2000)
Mouse	Rodent	*Lasiodora parahybana*[a]	Breeders (induced)	Internet source; Gallon (2000)
Mouse	Rodent	*Brachypelma smithi, B. boehmei, T. vagans, T. verdezi*	Breeders (induced)	Internet source; Gallon (2000)
Kangaroo rat	Rodent	*Aphonopelma chalcodes*	Arizona (USA)	Skinner (2009)
Saccopteryx bilineata	Bat	*Avicularia juruensis*[b]	Peru	West, in Nyffeler and Knörnschild (2013)
Myotis nigricans	Bat	*Avicularia* sp.[b]	Ecuador	Schmid, in Nyffeler and Knörnschild (2013)
No information	Bat	*Lasiodora parahybana*[a]	Brazil	West, in Nyffeler and Knörnschild (2013)
Birds				
Willisornis poecilinotus	Bird	*Theraphosa blondi*	Brazil (induced)	Carvalho et al. (2016)
Hummingbird	Bird	*Xenesthis immanis*	Venezuela	West (pers. comm. 2019)
Reptiles				
Bothrops atrox	Snake[c]	*Theraphosa blondi*	Venezuela (induced)	West (1992)
Erythrolamprus almadensis	Snake	*Grammostola quirogai*	Brazil	Borges et al. (2016)
Leptophis ahaetulla	Snake	*Pamphobeteus* sp.	Peru (induced)	Cocroft and Hambler (1989)
Anolis ortonii	Lizard (Squamata)	*Pseudoclamoris elenae*	Peru	West (pers. comm. 2019)
Ptychoglossus bicolor	Lizard (Squamata)	*Pamphobeteus ferox*	Colombia	Castillo Rodriguez and Méndez Galeano (2017)
Scincidae	Lizard (Squamata)	*Theraphosa blondi*	Brazil	Azevedo and Smith (2004), in Menin et al. (2005)

(continued)

Table 8.1 (continued)

Prey type/species/ family	Taxa	Tarantula species	Location	Source
Tropidurus hispidus	Lizard (Squamata)	*Lasiodora klugi*	Brazil	Vieira et al. (2012)
Tropidurus semitaeniatus	Lizard (Squamata)	*Acanthoscurria natalensis*	Brazil (induced)	Ferreira et al. (2014)
Micrablepharus maximiliani	Lizard (Squamata)	Unidentified		Sousa and Freire (2010)
Amphibians				
Hypsiboas sp.	Frog	*Theraphosa blondi*	Brazil	Menin et al. (2005)
Leptodactylus knudseni	Frog	*Theraphosa blondi*	French Guyana	Boistel and Pauwels (2002b)
Phyllomedusa vallantii	Frog	*Theraphosa blondi*	Brazil	Menin et al. (2005)
Dendrobates auratus	Frog[d]	*Sericopelma rubronitens*	Central America (Panama)	Summers (1999)
Physalaemus pustulosos	Frog	*Sericopelma rubronitens*	Panama (induced)	Gray et al. (2010)
Hamptophryne boliviana	Frog	*Pamphobeteus* sp.	Peru	von May et al. (2019)
Hyla fasciata	Frog	*Pamphobeteus* sp.	Peru (induced)	Cocroft and Hambler (1989)
Phyllomedusa palliata	Frog	*Pamphobeteus* sp.	Peru (induced)	Cocroft and Hambler (1989)
Eleutherodactylus peruvianus	Frog	*Pamphobeteus* sp.	Peru (induced)	Cocroft and Hambler (1989)
Leptodactylus wagneri	Frog	*Pamphobeteus* sp.	Peru (induced)	Cocroft and Hambler (1989)
Scarthyla goinorum	Frog	*Avicularia juruensis*	Peru	West (pers. comm. 2019)
Leptodactylus pentadactylus	Frog	*Theraphosa blondi*	French Guyana	West (pers. comm. 2019)
Unknown species	Frog	*Lasiodorides striatus*	Peru	West (pers. comm. 2019)
Bufo marinus	Toad	*Theraphosa blondi*	Brazil	Menin et al. (2005)
Bufo marinus	Toad	*Xenesthis immanis*	Peru (induced)	Cocroft and Hambler (1989)
Oscaecilia zweifeli	Caecilian	*Theraphosa blondi*	French Guyana	Boistel and Pauwels (2002b)

(continued)

Table 8.1 (continued)

Prey type/species/ family	Taxa	Tarantula species	Location	Source
Fish				
Poecilia mexicana	Fish	*Hemirrhagus pernix*	Mexico	Horstkotte et al. (2010)

[a]Wandering close to the burrow species
[b]Arboreal species
[c]Venom
[d]Poison

Fig. 8.2 *Spiders, Ants and Hummingbird on a Branch of a Guava* (illustrated by Anna Maria Sibylla Merian in 1705), showing pink-toed tarantulas (*Avicularia avicularia*), a ruby-topaz hummingbird (*Chrysolampis mosquitus*), hunstman spiders (*Heteropoda venatoria*), and army ants (*Eciton* sp.) on a guava tree (*Psidium guineense*)

Latin term *avicula*, meaning "little bird" (Teixeira et al. 1991). However, although birds are frequently observed in the webs of *Nephila* species and being preyed on by these spiders, there is no strong evidence related to bird capture and consumption by theraphosids. However, West (2019, personal communication) once observed a hummingbird being eaten by a tarantula (Fig. 8.1b) although he did not observe whether the bird was caught by the spider. A report was published by Carvalho et al. (2016), regarding a common scale-backed antbird (*Willisornis poecilinotus* Cabanis, 1847; Passeriformes, Thamnophilidae) entangled in a mist net installed close to the ground, thus allowing an individual of *T. blondi* to take the prey. As for mammals, large tarantulas are clearly able to catch and prey on birds but do not do so as commonly as we imagine. The only observed natural instances of tarantulas preying on birds or mammals were practically all induced directly or indirectly by humans, with the notable exceptions of the hummingbird observed by West and bats caught in the wild by arboreal tarantulas. Furthermore, under natural conditions, burrowing tarantulas have been observed preying more frequently on reptiles (see Table 8.1). In the wild, Borges et al. (2016) observed a snake (*Erythrolamprus almadensis*) being preyed on by a tarantula (*Grammostola quirogai*) when entering a spider retreat described as a crevice between rocks (Fig. 8.3a). This observation is the sole report of a snake being preyed on by a tarantula in the wild without being induced. Tarantulas have also been observed preying on snakes, such as the venomous species *Bothrops atrox* caught by *T. blondi* (West 1992), but these have all been induced events (Fig. 8.3b). Lizards (skinks and *Tropidurus species*) were also observed as tarantula prey in different events in Brazil (Table 8.1), sometimes induced as in the case of *Tropidurus semitaeniatus* (Ferreira et al. 2014) but also in a natural environment as with *Tropidurus hispidus* (Vieira et al. 2012).

In comparison with other vertebrates, frogs and toads appear to be much more frequently preyed on by arthropods, including spiders, because of their smaller size (Menin et al. 2005). Theraphosids in America were observed capturing different species of anurans (see Table 8.1 and Fig. 8.4). There is also a report of a serpentine amphibian, *Oscaecilia zweifeli* (a caecilian), being preyed on by *T. blondi* (Boistel

Fig. 8.3 Tarantulas eating snakes. (**a**) *Grammostola quirogai* preying on the snake *Erythrolamprus almadensis* in the state of Rio Grande do Sul, southern Brazil (photo: Gabriela Franzoi Dri (Borges et al. 2016)) (**b**) *Theraphosa blondi* eating a 56 cm fer-de-lance viper (*Bothrops atrox*) in its burrow in Rio Siapa, Amazonas, Venezuela (photo: Rick C. West)

Fig. 8.4 Tarantulas eating frogs. (**a**) Subadult female *Avicularia juruensis* eating a *Scarthyla goinorum* frog at night in Rio Momon, Loreto, Peru. (**b**) Female *Lasiodorides striatus* found eating a frog at night on a forest trail in the Tumbes region of Peru. (Photos: Rick C. West)

and Pauwels 2002a, b). *Sericopelma rubronitens* preys on frogs, including a poisonous dendrobatid frog (Summers 1999). Although fish are frequently preyed on by spiders throughout the world, only two tarantulas have been observed catching fish and this occurred in captivity and under experimental conditions (Nyffeler and Pusey 2014). One of these spiders, *Hemirrhagus pernix*, was observed in a Mexican sulfur cave, preying on a fish (*Poecilia mexicana*) in an experimental setup inside the cave, demonstrating the capability of this theraphosid to catch this prey. However, the researchers (Horstkotte et al. 2010) observed that the spiders were generally found on the edge of the pool where *P. mexicana* individuals gathered; therefore, they assumed that this predation may also occur naturally. Predation on vertebrates by tarantulas is remarkable and demonstrates their abilities to capture such prey. The species generally observed are *Avicularia* species (which prey on bats), large arboreal spiders, and the larger fossorial tarantulas *T. blondi* and *Lasiodora* species; their size allows them to prey on larger prey than other spiders, but it is also possible that their size makes them attractive species for humans to induce impressive predation scenarios.

We generally consider insects classical prey for spiders, including tarantulas (Gallon 2000). However, other invertebrates may also be preyed on (Table 8.2 and Fig. 8.5). Surprisingly, it is easier to find information about tarantula predation on vertebrates under natural conditions than information about their predation on

Table 8.2 Invertebrates preyed on by Theraphosidae

Prey type/species/family	Taxa	Tarantula species	Location	Source
Giant earthworm/*Rhinodrilus* sp.	Worm	*Theraphosa blondi*	French Guyana, Guyana,	Nyffeler et al. (2017)
Giant earthworm	Worm	*Theraphosa apophysis*	Venezuela, Ecuador, Peru	West, in Nyffeler et al. (2017)
Giant earthworm	Worms	*Megaphobema velvetosoma*	Venezuela, Ecuador, Peru	West, in Nyffeler et al. (2017)
Tityus bahiensis	Scorpion[a]	*Vitalius* sp.	Brazil	West (pers. comm. 2019)
Centruroides ochraceus/Buthidae	Scorpion[a]	*Tliltocatl vagans*	Mexico (induced)	Dor et al. (2011)
Centruroides gracilis/Buthidae	Scorpion[a]	*Tliltocatl vagans*	Mexico (induced)	Dor et al. (2011)
Lycosa subfuscal/Lycosidae	Spider	*Tliltocatl vagans*	Mexico	Dor and Henaut (2011, 2013)
Tliltocatl vagans/Theraphosidae	Spider	*Tliltocatl vagans*	Mexico	Henaut and Machkour-M'Rabet (2005); Henaut et al. (2015)
Unknown species 2	Millipede[b]	*Megaphobema mesomelas*	Costa Rica (induced)	Klem (2017)
Unknown species 1	Millipede	*Megaphobema mesomelas*	Costa Rica (induced)	Klem (2017)
Spirostreptid millipede	Millipede[b]	*Aphonopelma chalcodes*	Arizona (USA)	Minch (1978)
Gryllus integer/Gryllidae	Insect (Orthoptera)	*Aphonopelma* spp.	California, Arizona (USA)	Hedrick and Kortet (2006)

(continued)

Table 8.2 (continued)

Prey type/species/family	Taxa	Tarantula species	Location	Source
Tenebrio molitor/ Tenebrionidae	Insect (Coleoptera)	*Paraphysa* sp.	Central Andes (induced)	Canals et al. (2012)
Passalidae	Insect (Coleoptera)	*Tliltocatl vagans*	Mexico	Machkour-M'Rabet et al. (2007)
Formicidae	Insect (Hymenoptera)	*Tliltocatl vagans*	Mexico	Machkour-M'Rabet et al. (2007)
Belostomatidae	Insect (Hymenoptera)	*Tliltocatl vagans*	Mexico	Machkour-M'Rabet et al. (2007)
Zophobas morio/ Tenebrionidae	Insect (Coleoptera)	*Tliltocatl vagans*	Mexico (induced)	Dor et al. (2011)
Periplaneta Americana/ Blattidae	Insect (Blatyodea)	*Tliltocatl vagans*	Mexico (induced)	Dor et al. (2011)
Acheta domesticus/ Gryllidae	Insect (Orthoptera)	*Tliltocatl vagans*	Mexico (induced)	Dor et al. (2011)
Diloboderus abderus	Insect (Coleoptera)	*Eupalaestrus weijenberghi*	Uruguay	Pérez-Miles et al. 2005
Diloboderus abderus	Insect (Coleoptera)	*Acanthoscurria cordubensis*	Uruguay	Pérez-Miles et al. 2005
Periplaneta Americana, Acheta domesticus, Carabus sp., *Drosophila mojavensis, D. melanogaster*	Insect	*Aphonopelma steindachneri*	USA (laboratory, induced)	Punzo (2005)
Acheta domestica	Insect (Orthoptera)	*Aphonopelma chalcodes*	Arizona (USA)	Minch (1978)
Eleodes obscura, E. armata, E. longicollis	Insect (Coleoptera)[b]	*Aphonopelma chalcodes*	Arizona (USA)	Minch (1978)
Slug (species not mentioned)	Mollusk	*Lasiodora parahybana*	Captivity (induced)	Nyffeler and Symondson (2001)
Snail (species not mentioned)	Mollusk	*Selenocosmia crassipes*	Captivity (induced)	Nyffeler and Symondson (2001)

[a]Venom
[b]Toxic volatile compound

insects or other invertebrates. Field observations made by Pérez-Miles et al. (2005) show that two tarantulas—*Eupalaestrus weijenberghi* and *Acanthoscurria cordubensis* (Ex *A. suina*)—prey on *Diloboderus abderus*, a beetle whose larvae are considered pests for several crops. This beetle has few natural predators, making tarantulas an efficient

Fig. 8.5 Tarantulas eating invertebrates. (**a**) *Theraphosa blondi* pulling a captured giant earthworm (presumably *Rhinodrilus* sp.) into its burrow in a rainforest in French Guyana (photo: C.E. Timothy Paine (in Nyffeler et al. 2017)). (**b**) *Brachypelma auratum* eating a cockroach in breeding condition in Mexico (photo: Aarón Zúñiga Sánchez). (**c**) *Aphonopelma seemanni* eating a cricket in breeding condition in Mexico (photo: Aarón Zúñiga Sánchez)

biological agent for use against this particular pest (Pérez-Miles et al. 2005). Considering that the majority of tarantulas leave prey parts in the burrows they inhabit (Machkour-M'Rabet et al. 2007), it is easy to acquire information on tarantula prey; however, there have been practically no studies focused on prey remains. Only insects were found (see Table 8.2 for details) in a study of the prey remains of the Mexican redrump tarantula (Machkour-M'Rabet et al. 2007), but softer prey are certainly eaten entirely, making this information incomplete. Nevertheless, in the same location, these tarantulas were observed preying on wolf spiders (Lycosidae) and on congeners, but no spider remains were found in the burrows (Dor and Hénaut 2011, 2013; Hénaut and Machkour-M'Rabet 2005; Hénaut et al. 2015) (see Table 8.2). Tarantulas may prey less frequently on other spiders than on insects because larger wolf spiders avoid tarantula territory (Dor and Hénaut 2011). Other chelicerates may also be preyed on by tarantulas, as observed with *Centruroides* scorpions; however, these observations have mostly been made in the laboratory and only rarely in the natural environment (see Table 8.2), because of the low presence of scorpions in areas inhabited by tarantulas (Dor et al. 2011). Millipedes are other arthropods preyed on by tarantulas in experimental studies (Klem 2017). Other invertebrates preyed on by tarantulas in the Americas include giant earthworms belonging to the genus *Rhinodrilus*. In French Guyana, *T. blondi* tarantulas were observed capturing giant earthworms (Fig. 8.5a)—one approximately 60 centimeters long and another approximately 1 meter long (Nyffeler et al. 2017). According to West (mentioned in Nyffeler et al. 2017), tarantulas use the same technique for pulling giant earthworms into their burrows as they do for snake species. Furthermore, West mentioned (in Nyffeler et al. 2017) that in different events and areas in South America, theraphosids were observed eating oligochaetes (Table 8.2). In captivity, tarantulas (see Table 8.2) were also observed preying on mollusk species, including snails and slugs (Nyffeler and Symondson 2001).

Taking into consideration that in the Americas there are few reports and observations of tarantulas preying on vertebrates in the wild, we cannot state that tarantulas in the Americas are a common predator of vertebrates and that this may influence their ecological assemblages in terms of antipredator strategies and evolution. However, the observations we report in this chapter are certainly not complete, or this information is not accessible. We clearly need further formal studies to analyze with precision the natural diet of tarantulas, focusing on larger prey to bring to light their real impact on fauna. Nevertheless, the capacity these spiders have to predate on a large range of prey, including several vertebrates and invertebrates, merits discussion. As for vertebrates, the lack of accurate information about insects and other invertebrates as natural prey of tarantulas prevents us establishing predator–prey relationships and understanding how these may influence the evolution of prey and tarantula biological traits. However, it is interesting to observe the capacity tarantulas have to predate on a very large range of prey from mammals to earthworms, including large prey.

Several prey caught by tarantulas, even if the predation is induced by humans, have poisonous or chemical defenses, as observed in millipedes, scorpions, dendrobatid frogs, and other prey with a real potential for defense, such as snakes (including the fer-de-lance), large lizards, and other theraphosids (see Tables 8.1 and 8.2 for details). In the case of dendrobatid frogs, *Sericopelma rubronitens* is the only predator observed preying on this very poisonous frog (Gray et al. 2010). This tarantula may represent a strong selective force for the toxicity of this organism. As a sit-and-wait predator, and being almost blind but very reactive to vibrations when preying (Blein et al. 1996), tarantulas (particularly those living in burrows) cannot distinguish prey colors, shapes, or forms, and are in this sense immune to aposematic warning or to any signal that their prey may be dangerous in nature. In these situations, the spiders attack every kind of potential prey without discrimination. The surprise and velocity of the attack, and the weight of the tarantula, together with its venom, are able to knock down its prey (such as dendrobatid frogs, which secrete skin-toxic fluid when faced with a predator (Gray et al. 2010)) before it can even detect the tarantula and release its poison, or simply prevent the prey from reacting and defending itself. For example, when *Megaphobema mesomelas* preys on a millipede, its attack lasts for less than 1 second while the millipede needs at least 1.4 seconds to release cyanide (Klem 2017). This strategy of no discrimination in tarantulas, which presents a high degree of risk, especially when considering the dangerousness of its potential prey, is counterbalanced by the surprise, violence, and velocity of the attack, allowing tarantulas to prey on a huge variety of prey. In the same way, as species that rarely forage out of their retreats, tarantulas are sit-and-wait predators that depend on prey not only to be available locally but also to actually pass by their burrows. This strategy is associated with the capacity these spiders have to fast for a long time (Baerg 1938) by taking advantage of every kind of prey that passes close to the burrow entrance. This life strategy certainly drives the evolution of capacities for being fast and trapping large prey, including dangerous ones, and so during an attack, the tarantula's actions are quicker than those of the prey it has caught (Gray et al. 2010; Klem 2017). Some prey (such as millipedes) with chemical defenses may be captured and eaten but consist of a very small proportion of their diet; in one study, coleopterans from the *Eleodes* genus were captured less frequently than inoffensive prey such as crickets (Minch 1978). In addition, Minch (1978) mentioned that cricket capture decreased dramatically when a chemical defense was provided or the prey was presented after a rejection of noxious prey. According to Minch (1978), the rejection of prey with defensive secretions is not totally effective and the technique the tarantula uses to grab the prey determines whether the spider comes into contact with the secretions.

Tarantulas are generally solitary and dispersed, but sometimes they are observed in high densities, as demonstrated by *Tliltocatl vagans* (Machkour-M'Rabet et al. 2005; Hénaut et al. 2015). In this context, they are able to impact potential prey distribution, as observed with scorpions and wolf spiders. For example, *T. vagans* preys on the wolf spider *Lycosa subfusca*, catching larger wolf spiders but never attacking smaller ones. At the same time, larger wolf spiders prey on smaller lycosids and are also able to catch spiderlings of *T. vagans* (Dor and Hénaut 2013). The presence of tarantulas in high densities is linked with a higher presence of small wolf spiders and a lower presence of larger lycosids, thus contributing to a spatial size-dependent segregation of wolf spiders (Dor and Hénaut 2011). In the same way, a lower presence of *Centruroides* scorpions (Buthidae) appears to be associated with the presence of *T. vagans*, which preys on them, in the Yucatan Peninsula (Dor et al. 2011).

Cannibalism in spiders is commonly observed and is generally size dependent, as observed in wolf spiders (Dor and Hénaut 2011), and is linked with territorial defense (Moya-Laraño et al. 2002). Cannibalism can also be sexual, as observed in the Mediterranean tarantula, or tarantula wolf spider (*Lycosa tarantula* (Lycosidae)), allowing females to produce up to 30% more offspring with better body condition (Rabaneda-Bueno et al. 2008). One spectacular case of cannibalism in spiders is matriphagy, observed in *Amaurobius ferox* (Kim and Horel 1998). In New World tarantulas, sexual cannibalism is not usual. According to Costa and Pérez-Miles (2002), the absence of sexual cannibalism seems to be widespread in theraphosids despite some observations during male approaches in *Acanthoscurria cordubensis* and *Stromatopelma griseipes* (Celerier 1981 in Costa and Pérez-Miles 2002) and some postcopulatory cases, as observed in *Aphonopelma iodium* (Prentice 1997), *Aphonopelma* sp. (Shillington and Verrell 1997), *Aphonopelma anax* (Shillington 2002), and *Brachypelma klaasi* (Yañez et al. 1999). However, in *Acanthoscurria atrox*, females appear to be very aggressive toward their males and cannibalize them (Lourenço 1978). Sexual cannibalism was also observed in *Aphonopelma hentzi* in larger proportions than occurs accidentally in other tarantulas such as those previously mentioned (Punzo and Henderson 1999). Ferretti and Ferrero (2008) also observed cannibalism of males by females in *Grammostola schulzei* but in the absence of previous courtship behavior. Marshall and West (2008) propose that to reduce intraspecific competition and cannibalism, *Ephebopus murinus* exhibits an ontogenetic habitat shift. However, with the exception of some cases of sexual cannibalism, few cases of cannibalism have been noted in New World tarantulas, which may show agonistic behavior but not intraspecific predation, as observed by Trabalon (2011) in *Tliltocatl albopilosus*. However, both in the field and under laboratory conditions, cannibalism between females was observed in *Tliltocatl vagans* (Hénaut and Machkour-M'Rabet 2005) and in populations with a large number of individuals living in a restricted area and during months with food scarcity due to dry weather conditions (Machkour-M'Rabet et al. 2005; Hénaut et al. 2015). During this unusual predation on conspecifics in tarantulas, females leave their own burrows, walk toward the retreat of another, which is detected by emitted volatile compounds (Dor et al. 2008), and then the intruder attacks, kills, and eats the female resident tarantula before taking control of her burrow. The intruders demonstrate a success rate of approximately 80% (Hénaut and Machkour-M'Rabet 2005). When approaching a burrow with a male resident, the female takes the burrow but does not attack and kill the male; cannibalism occurs only between females (Hénaut and Machkour-M'Rabet 2005). However, observations of both sexual and nonsexual cannibalism in tarantulas are rare. As predators, tarantulas interact mostly with other organisms.

8.3 Predators and Parasites

8.3.1 Predators

Tarantulas in the Americas are generally nocturnal; they inhabit burrows or, in the case of arboreal species, retreats. They rarely wander, with the exception of some species, particularly arboreal spiders. Consequently, predators have limited access to spot and catch tarantulas, with the notable exception of wandering males looking for females (Shillington 2002; Shillington and Verrell 1997) or wandering spiderlings leaving their mother, as observed in *T. vagans* (Reichling 2000; Shillington and McEwen 2005; Dor and Hénaut 2012). Furthermore, flooding may force individuals to leave their burrows, as observed in *T. vagans* (Shaw et al. 2011), and, in doing so, tarantulas are momentarily vulnerable to predators. Living in a burrow or using a retreat is the first defense of tarantulas; however, in the New World, most tarantulas also have urticating hairs, with the exception of New World Ischnocolinae and some Aviculariinae. Urticating hairs are absent in all Old World theraphosids (Marshall and Uetz 1990a). According to their habitats, tarantulas have morphological and behavioural differences in the type of hairs and in the way they are used. Those that burrow employ a hair-flicking strategy, while arboreal spiders use contact urticating hairs (Bertani and Marques 1996). *Theraphosa blondi* and *Megaphobema* sp. in Ecuador were observed incorporating setae from their abdomen into the silk of the egg sac and into silk mats as protection from the parasitoid *Megaselia scalaris* (Diptera: Phoridae) (Marshall and Uetz 1990b). Some tarantulas may be active and avoid predator cues, as observed in wandering juveniles of *Aphonopelma steindachneri*, avoiding substrates containing chemosensory cues associated with one of their natural predators, *Hogna carolinensis* (Araneae: Lycosidae) (Punzo 2005).

According to Edwards and Hibbart (1999), juvenile individuals of *T. vagans* are preyed on by invertebrates, while adults are eaten by small mammals. In fact, the most well-known predators of theraphosids in the New World are the coatis *Nasua nasua* (in South America) and *Nasua narica* (in Central America). *Nasua nasua* is described as eating tarantulas after intense prey manipulations, rolling them on the ground to rid them of their defense (Alves-Costa et al. 2004), evidently the urticant hairs. In Central America, *N. narica* was also observed eating tarantulas, such as *T. vagans* in Southern Mexico, becoming very excited when finding a tarantula and also cleaning them by rolling them before consumption (Hénaut, personal observation). However, even if the consumption of tarantulas by coatis is remarkable, this feeding behavior appears to be largely opportunistic and exceptional, as consumption of invertebrates (spiders and millipedes) in this mostly frugivorous mammal is relatively rare (Hirsch 2009). Chickens (*Gallus gallus*) have also been observed preying on tarantulas (*T. vagans*), prudently cutting the legs to eat this part of the spider, but without directly killing and consuming the entire tarantula (Hénaut, personal observation), indicating that some birds may eat tarantulas. An adult female of *T. vagans* was also observed being entirely eaten by a skink, *Mesoscincus schwartzei* (Reptilia, Scincidae), during the day after the spider left the burrow for unknown reasons (Hénaut, personal observation). Amphibians are generally considered prey of spiders; however, the large mountain chicken frog, *Leptodactylus fallax*, was twice observed feeding on the theraphosid *Cyrtopholis femoralis* in the Lesser Antilles (Rosa et al. 2012, Fig. 8.6). All of these observations of vertebrates eating adult or subadult tarantulas are isolated observations of rare or relatively rare cases and appear to be linked to opportunistic behavior. However, the blood snake, *Stenorrhina freminvillei*, which inhabits Central America (Mexico to Panama), is believed to feed principally on scorpions and tarantulas. In laboratory observations of this snake, it actively sought and precisely followed the trail of *T. vagans* (Hénaut, personal observations). Moreover, out of 12 blood snakes, seven had *T. vagans* remains in their gastrointestinal tract, making tarantulas the principal prey of this reptile, followed by whip spiders (Amblypygi) and scorpions

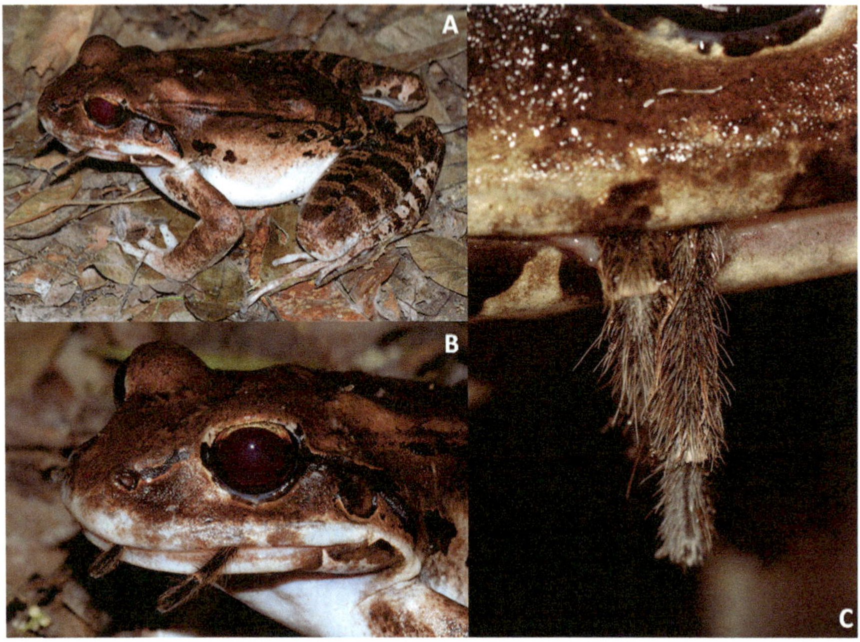

Fig. 8.6 A frog eating a tarantula. The mountain chicken frog (*Leptodacylus fallax*) eating a *Cyrtopholis femoralis* in the Centre Hills, Monserrat, Lesser Antilles. (**a**) Complete view of the frog eating the tarantula. (**b**) and (**c**) Details of the tarantula legs. (Photos: Gonçalo M. Rosa)

(Köhler et al. 2017). In contrast to the opportunistic predators presented above, the blood snake could be considered an active predator of tarantulas and indeed the only known specialist one; however, this relationship requires further research.

Predation on tarantulas by invertebrates is rare. In the Amazonian forest, Dias and Lo-Man-Hung (2009) observed a velvet worm (Onychophora, Peripatidae, *Peripatus* sp.) feeding on an adult female theraphosid spider (*Hapalopus butantan*); however, they did not observe whether the tarantula was caught and killed by the onychophoran. The velvet worm normally preys on small invertebrates, including small spiders. In the laboratory, wolf spiders have been observed preying on tarantula spiderlings (Dor and Hénaut 2013), which leave the mother in a column "ant-like" behavior, as observed previously (Dor and Hénaut 2012). In field and laboratory studies, we also observed female *T. vagans* preying on female congeners (Hénaut and Machkour-M'Rabet 2005; Hénaut et al. 2015), as mentioned in the preceding paragraph. The lack of observations of invertebrates preying on tarantulas is partly due to the fact that they are particularly well protected by their burrow and defensive behavior. Moreover, their large size protects them against attacks from smaller invertebrates, with the exception of spiderlings, which are much more susceptible to being preyed on by other invertebrates.

8.3.2 Parasites and Parasitoids

A parasite can be defined as an organism that lives at the expense of another (the host). It may live on the host (e.g., lice), live in the host (e.g., tapeworms), or feed occasionally on a host (e.g., mosquitoes), but generally parasites do not kill their hosts. A parasitoid can be defined as an organism that, during the first instars of its cycle, develops on or in the host and always kills it (Godfray and Shimada 1999). Parasitoid organisms can be assimilated as predators with a specific foraging strategy to find their prey (Korenko et al. 2011), which they always kill (Godfray and Shimada 1999). Parasitoids can attack different development stages of their host, the most frequent being the eggs, larvae, and pupae; less commonly they may attack the adult stage (Korenko et al. 2011). Parasitoid species most commonly belong to the order Hymenoptera, followed by Diptera (principally the Tachinidae family), plus a few species from other orders (Korenko et al. 2011).

Many species of spider are parasitized by a variety of insects and a few nematodes (Eason et al. 1967; Penney and Bennett 2006; van Helsdingen 2011), as well as other spiders such as those belonging to the genus *Argyrodes*, which are considered kleptoparasites because they steal food from their host spiders (Hénaut 2000; Whitehouse et al. 2002; Hénaut et al. 2005; van Helsdingen 2011). Reports of parasites or parasitoids of Mygalomorphae spiders are limited, and if we consider only the true tarantula in the New World, the information is quite scarce. The best known parasitoids of Theraphosidae are *Pepsis* spp. (Hymenoptera, Pompilidae), also known as tarantula hawk wasps (Fig. 8.7), with some laboratory observations and a few reports in the field (Punzo and Garman 1989; Vardy 2000; Cambra et al. 2004; Costa et al. 2004; Rego et al. 2004; Punzo 2007), but none of this information concerns the Mexican *Pepsis*–Theraphosidae interaction. In the Hymenoptera order, *Pepsis* is the main genus reported to be parasitoids of Theraphosidae. Pizzi (2009)

Fig. 8.7 (**a**) Predation of a juvenile *Avicularia* sp. by *Pepsis frivaldszkyi* in the municipality of Itacoatiara, Amazonas, Brazil (photo: Felipe Rego (in Rego et al. 2004)). (**b**) Pompilidae wasp (*Pepsis pallidolimbata*) dragging a paralyzed female *Aphonopelma gabeli* tarantula in Pinos Altos, Grant County, New Mexico, USA (photo: Jeremy Huff)

mentions the possibility that ichneumonid ectoparasitoids (Hymenoptera) prey on captive tarantulas, but no such information has been reported for tarantulas in the field. Pérez-Miles et al. (2005) formulated the hypothesis that the burrow entrance–closing behaviour of tarantulas using silk and/or soil can have a protective function against attack from both parasites and parasitoids. Two Diptera families are reported to be parasitoids of tarantulas: Phoridae, which were described as parasitoids in laboratory and field studies (Marshall and Uetz 1990a; Weinman and Disney 1997; Pizzi 2009; Machkour-M'Rabet et al. 2015; Cuéllar-Balleza et al. 2017), and Acroceridae, which were reported to be endoparasitoids (Cady et al. 1993; Pizzi 2009; Barneche et al. 2013; Gillung and Borkent 2017), particularly species of the genus *Exetasis* Erichson (Barneche et al. 2013), of which *E. eickstedtae* Schlinger, 1972 was described as an internal parasitoid of *Lasiodora klugi* (C. L. Koch, 1841) in Brazil (Schlinger 1972; Cady et al. 1993), and *E. jujuyensis* Gillung sp. nov. was reported to be a parasitoid of *Acanthoscurria sternalis* Pocock, 1903 in Argentina (Barneche et al. 2013).

Paré et al. (2001) report an infestation of a captive *Brachypelma smithi* (Araneae, Theraphosidae) by a parasitic larva of Dermestidae (Coleoptera). Pizzi (2009) mentions two families of parasitic nematodes, one more commonly associated with wild tarantulas (Mermithidae) and the other more commonly associated with captive tarantulas (Panagrolaimidae). More recently, Cuéllar-Balleza et al. (2017) reported an infestation of captive Theraphosidae in Queretaro (Mexico) by Rhabditoidae. Acarian parasites of Theraphosidae species have also been described. In particular, species of the genus *Ljunghia* Oudemans, 1932 (Acari, Mesostigmata, Laelapidae) are obligate parasites of Theraphosidae spiders in Australia, Africa, and possibly Central America, and all life stages have been reported as parasitic (Paré and Dowling 2012). Very few studies have described spiders parasitized/infected by bacteria or fungi. Cuéllar-Balleza et al. (2017) mentioned the presence of Gram-positive bacteria in nine Theraphosidae species in laboratory research, and Barbosa et al. (2016) reported a parasitic association between spiders (Theraphosidae) and arthropod-pathogenic fungi. In an urban forest fragment in southern Brazil, these authors observed four specimens belonging to the genus *Grammostola* (Simon, 1892) and one *Pterinopelma vitiosum* (Keyserling, 1891) infested by a fungus of the species *Ophiocordyceps caloceroides* (Berk. & M.A. Curtis) Petch. In Argentina, the fungus *Cordyceps ignota* March was identified on a "large spider" targeted as Theraphosidae (Mains 1954).

Parasites and parasitoids of Theraphosidae in the New World, even if they are few, come from a great variety of taxa and, as invertebrates, appear to be more diverse than tarantula predators. In fact, these large spiders may be sensitive to parasites and parasitoids, because of their size, and are useful hosts for a variety of them. In addition, the parasites and parasitoids may have an impact on tarantula breeders and pet owners, and on rearing of tarantulas for research. Parasitism in tarantulas may be considered a veterinary concern.

8.4 Associations and Other Interactions

8.4.1 Frogs and Tarantulas

The most notorious association in New World tarantulas is certainly the frog–theraphosid one. Moreover, this association appears to be a global pattern and has been observed throughout the world (Cocroft and Hambler 1989; Siliwal and Ravichandran 2008; Suranjan Karunarathna and Thasun Amarasinghe 2009). Although tarantulas are great predators of amphibians—particularly frogs and toads, including poisonous ones, as previously mentioned—some anurans live in the burrows of theraphosids. This unexpected fact has been observed on a large geographic scale in the Americas, from the southern USA to South America and with different species of anurans and tarantulas.

The first description of a frog living together with a tarantula was written by Blair (1936), observing the western narrow-mouthed toad, which is in fact the frog *Gastrophryne olivacea*, living generally alone or with two or three congeners in 75% of burrows of the species *Dugesiella hentzi* (now known as *Aphonopelma hentzi*). Dundee et al. (2012) observed the same species association, finding up to 22 frogs sharing a burrow with a tarantula. *Gastrophryne olivacea* and *A. hentzi* are commonly observed together in the southern USA, particularly in Oklahoma and Texas. Dundee et al. (2012) also observed that the presence of an associated anuran does not impact the predatory behavior of the tarantula toward other prey.

In Mexico, the túngara frog (*Physalaemus pustulosus*) has been observed inhabiting a burrow of an *Aphonopelma* sp. together with the tarantula (Powell et al. 1984), and, as mentioned by West (2005), the endemic frog *Eleutherodactylus occidentalis* inhabits the burrow of *Brachypelma auratum*. In Bolivia, two other anurans—*Leptodactylus bufonius* (a frog) and *Rhinella major* (a toad) were observed in the burrows of theraphosids (*Acanthoscurria* sp.) (Schalk and Sezano 2014). These burrows are used especially by conspicuous calling males, but it seems that in this case, the burrows were not inhabited by spiders. The frog and the toad used those abandoned burrows only as refuges against predators (Schalk and Sezano 2014). In southeastern Peru, Cocroft and Hambler (1989) studied an association between the dotted humming frog (*Chiasmocleis ventrimaculata*) and the tarantula *Xenesthis immanis*, which was subsequently confirmed to be an undescribed *Pamphobeteus* sp. (Fig. 8.8). These authors observed that the tarantula emerged first from the burrow and stayed close to the refuge while waiting for prey. Up to four frogs emerged around 1 hour after the tarantula, also staying close to the burrow. However, the frogs looked more active in areas with some vegetation, while the tarantulas used the open area. Around half of the frogs moved under the tarantula after emerging or hid there when detecting potential danger. No attempts by the tarantula to prey on the frogs were observed, although the tarantulas did capture and eat five species of anurans from the families Bufonidae, Hylidae, and Leptodactylidae (see Table 8.1) and rejected only a sixth one, *Physaleamus petersi*, presumably because of their pustular skin (Cocroft and Hambler 1989).

Fig. 8.8 Association between dotted humming frogs (*Chiasmocleis ventrimaculata*) and *Pamphobeteus* sp. tarantulas in southeastern Peru. Each *white arrow* indicates a frog. (**a**) Dotted humming frog emerging from the burrow. (**b**) and (**c**) *C. ventrimaculata* under a tarantula. (**d**) *Pamphobeteus* sp. adult female and three juveniles outside the burrow, with a frog under a juvenile. (Photos: Reginald B. Cocroft)

The possible toxicity of anurans does not explain why tarantulas do not attack and consume frogs that cohabit with them; in fact, Cocroft and Hambler (1989) demonstrated that a toxic frog, *P. petersi*, was attacked by *A. hentzi* and then released. According to Dundee et al. (2012), the *Aphonopelma* reaction is not associated with learning from previous contact with toxic anurans, as naive tarantulas react in the same way as non-naive individuals with associated frogs. Furthermore, although Dor and Hénaut (2011, 2013) observed that tarantulas do not attack small prey they cannot detect, as mentioned by Cocroft and Hambler (1989), the frog falls within the size range of prey captured by tarantulas. The same authors suggested that chemosensory cues are involved in the recognition of the frogs by the tarantulas. In fact, the frogs involved in these associations present noxious skin secretions, which may protect them from an eventual tarantula attack. Rödel et al. (2013) demonstrated that the frog *Phrynomantis microps* cohabits with ponerine ants (*Paltothyreus tarsatus*), thanks to chemical compounds secreted from the skin of the frog that act as an appeasement allomone. A similar process may occur between frogs and tarantulas, involving frog skin secretion and chemosensory hairs on tarantulas' feet (Foelix et al. 2012).

The association between frogs and tarantulas was first considered commensal because the tarantulas protect the amphibians from predators and the frogs do not negatively affect the spiders. In fact, in observations of *A. hentzi* and *G. olivacea*

(Hunt 1980), an introduced snake (*Thamnophis proximus*) ate frogs around the burrow but was repelled by the tarantula, which protected the frogs inside the burrow. In addition, the frog possibly takes advantage of the remains of the tarantula prey. Cocroft and Hambler (1989) mentioned that tarantulas' discarded prey parts may attract small invertebrates as potential prey for the frogs. Tarantula burrows provide protection against climatic variations and may protect frogs from desiccation and temperature changes (Powell et al. 1984). However, some authors consider that this association may be mutualistic because tarantulas have no defenses against ant invasion (Baerg 1958); therefore, they may benefit from the capability of the frog to clean ants from the burrow and thereby help protect the tarantulas' eggs or spiderlings (Dundee et al. 2012). In addition, the frog never preys on the young of the tarantula (Hunt 1980). According to Dundee et al. (2012), the association between frogs and tarantulas is not critical for the survival of these organisms but appears to be advantageous for the anurans and the theraphosids, and deserves to be studied and observed further in more species in the Americas and worldwide.

8.4.2 Bromeliads and Tarantulas

Bromeliads are attractive plants for a large number of organisms. For example, the tank bromeliad provides a refuge and food for microbes or algae and a large variety of animals such as crustaceans, insects, mollusks, amphibians, reptiles, and mammals (Dejan and Olmsted 1997; Frank and Lounibos 2009; Jabiol et al. 2009; McCracken and Forstner 2014; Brouard et al. 2011). Spiders are also associated with the tank bromeliad, particularly Salticidae (Romero and Vasconcellos-Neto 2004; Santos and Romero 2004), Ctenidae (Barth et al. 1988; Schuster et al. 1994), Dipluridae (Hénaut et al. 2014), and Theraphosidae (Calaça-Dias and Domingos-Brescovit 2004).

The theraphosid *Pachistopelma rufonigrum*, within the Aviculariinae subfamily, is a strictly arboreal tarantula that inhabits bromeliads (*Aechmea* sp. and *Hohenbergia* sp.), with strictly one female spider per bromeliad (Calaça-Dias and Domingos-Brescovit 2004). The tarantulas were observed to be inactive most of the time inside the leaves of the bromeliad and during the night left the retreat to position themselves in the external base of the plant, waiting for prey (Calaça-Dias and Domingos-Brescovit 2004). Silk retreats were also observed, with tarantulas closing space between leaves and females with egg sacs using a silk retreat in the central tank of the bromeliad. According to the same authors, *P. rufonigrum* is one of a few cases of organisms that present morphological adaptations (a flattened body and low eye tubercle) that allow this theraphosid to inhabit bromeliads. This very specific and rare association of a tarantula and a bromeliad is certainly advantageous for the tarantula, as it is protected by the bromeliad against predators and climatic variations, as observed in other spiders, and also because the tank bromeliad attracts several organisms that are potential prey for the tarantula (Hénaut et al. 2014).

8.4.3 Interactions with Humans

As tarantulas are large and remarkable spiders, human beings are never indifferent toward them. They may be attractive for some people or a source of fear for others. Generally, spiders are considered dangerous, and tarantulas, particularly the larger ones, provoke fear and disgust in western culture (Kleinknecht 1982; Smits et al. 2002). Costa-Neto (2006) observed in Brazil that the traditional zoological knowledge of the inhabitants of Pedra Branca in Brazil includes perceptions of tarantulas as particularly venomous and able to cause health problems. Álvarez del Toro (1992) noted in Chiapas (Mexico) and in other parts of Central and South America that traditionally people have named tarantulas *hierba* (grass), *mala hierba* (weed), *mata caballo* (horse killer), or *pica caballo* (horse stinger). People have often mentioned that tarantulas bite horses' hooves, resulting in the hoof falling off, as well as a rotten leg and maybe the death of the animal. This affirmation is not true but is a popular belief (personal observations). Some medicinal studies mention that the urticarious hairs of tarantulas kept as pets can cause eye diseases (Blaikie et al. 1997; Spraul et al. 2003).

Despite their bad reputation, tarantulas are used as a resource in several traditional cultures across the New World. The fangs of *Theraphosa blondi* are used as medicine to treat *erysipelas* (or "Holy fire"), asthma, and toothache in the state of Alagoas in Brazil (see Costa-Neto 2006). In the Amazonian basin, Yanomamo and Piaroa Indians eat roasted tarantulas (*T. blondi*) and other large tarantulas (see West 1997 and Costa-Neto 2006). According to observations by West (West 1997), Piaroa call tarantulas *a-ou-kay*, and the shaman performs a ritual ceremony before collecting tarantulas in the forest (Fig. 8.9). Only large heavy females with eggs from the species *T. blondi* and *Teraphosa apophysis* are collected. The abdominal contents, including eggs, are cooked in a rolled leaf, while the tarantula body is grilled directly on the fire before consumption (see West 1997 for details).

Enríquez Vázquez et al. (2006) mentioned that tarantulas are used also for fortification of teeth and treatment of asthma in Chiapas (Mexico) by the Tzotzil and Tzetzal ethnic groups. To treat tumors, the same ethnic groups in southern Mexico use the bites of tarantulas around the affected zone (Hunn 1977). Machkour-M'Rabet et al. (2011) observed a ritual performed by a medicine man using a beverage made with a tarantula (*Tliltocatl vagans*) to treat *aire de tarantula* ("tarantula wind"), which has symptoms including asthma, chest pain, and cough. In addition, because of the similarity in the recipes in different villages, the authors suggested that the use of the tarantula in traditional medicine is ancestral (Fig. 8.10). In more recent observations, Hénaut et al. (2016) mentioned that in the same areas, people use tarantula juice to wash their eyes and improve eyesight. In these villages, tarantulas are kept and protected in the backyards of the houses, together with medicinal plants, to have medicine on hand (Hénaut et al. 2016). In fact, the perception of the tarantula is positive in those places where people do not consider them dangerous animals (Machkour-M'Rabet et al. 2011) and tarantulas are found in abundance

Fig. 8.9 (**a**) Piaroa Indian shaman preparing half a dozen *Theraphosa apophysis* to be cooked and eaten as a nutritional food source in Gavilan, Amazonas, Venezuela. (**b**) Piaroa shaman's tarantula mask in Coromoto, Amazonas, Venezuela. (Photos: Rick C. West)

within villages (Machkour-M'Rabet et al. 2005, 2007). In the Yucatan peninsula, when Maya people are coming home and a tarantula crosses their path (*T. vagans* generally), this is considered a presage that somebody is damaging their house (personal observation). West (2003) also observed divination with tarantulas in Cameroon.

In the medical context, tarantulas have been well studied for the benefits they provide in human medicine, particularly through pharmacology research on the venom of the Chilean rose tarantula (*Grammostola rosea* (Walckenaer, 1837) (synonym: *G. spatulata*)), which is used to inhibit atrial fibrillation (Bode et al. 2001). Another modern use of tarantulas is the pet market. This activity is becoming more common by the day and could be very problematic for tarantula conservation, considering the extraction of individuals from the wild by poachers. The absence of knowledge, the fear reaction, and the fascination all contribute to the trafficking of tarantulas, which, together with the current degradation of habitats throughout the Americas, due to human activities, represent a combination of threats that may be very problematic for the future of tarantulas in the New World.

Fig. 8.10 Chol ritual using *Tliltocatl vagans* to prepare a tarantula-based beverage in southern Mexico. (**a**) Crushing the *T. vagans*. (**b**) Mixing the tarantula with spirit alcohol and some additional herbs. (**c**) Filtering to eliminate fragments and urticating hairs. (Photos: Yann Henaut)

8.5 Conclusion and Perspectives

In comparison with other groups of spiders, the interactions between tarantulas and other organisms is diverse in terms of the nature of the tarantulas' interactions as predators, prey, hosts, commensals, and eventually symbionts, but also in terms of the diversity of organisms interacting. Particularly in the Americas, the use of tarantulas by humans as pets and for traditional or modern human medicine is extraordinary for this spider group. The diversity of interactions may be related to the age of this taxon, considering its evolutionary time, which has allowed these spiders to develop particular and subtle interactions, as observed, for example, with frogs. The anatomical characteristics of tarantulas, considered the largest spiders in the world, clearly modulate their interactions, opening up possibilities particularly for prey taxa that are not common for spiders. The development of these advantages has certainly been driven by evolution, such as their superior size, which allows them, as spiders, to occupy a distinct ecological niche, preying on organisms that are not preyed on by other arthropods. On the other hand, tarantulas have needed to develop special defenses such as urticant hairs and nocturnal habits, as their size makes tarantulas attractive prey or hosts for predators, parasites, and parasitoids. In the New World, the interactions between tarantulas and toxic prey, and the associations of those spiders with amphibians, also merit further study. Such interactions involve behavioral traits and adaptations against venoms and chemical defenses that have rarely been explored.

Although tarantulas are popular animals, studies have mainly focused on their natural history and behavior, and there is still very little information related to tarantula interactions, without a doubt because it is not always easy to observe them in the wild. The available information about their prey is largely based on mostly induced interactions during field observations or in laboratories, providing us with information about their great capacities as predators but no precise information about their principal prey in the natural environment. Are these spiders notable predators of vertebrates to such an extent that they depend on them, or is vertebrate predation occasional and opportunistic? Furthermore, we could pose the same question about other invertebrate taxa, and more research is required to help us improve our understanding of the role, the necessity, and the impact of these species of spider in the ecosystem. Moreover, the physiological consequences of eating certain types of prey versus with others are not known; significantly, this information would contribute to the understanding of how the taxonomic origin of the prey may influence tarantulas' growth and further reproduction capacities. This question may be important, considering the development of tarantula breeding for diverse uses such as pets or medicine, improving breeding and also protection strategies. In addition, it could help us to understand how changes in their environment may affect their survival and how tarantulas may survive and adapt to environmental changes, as they have for millions of years.

References

Álvarez del Toro M (1992) Arañas de Chiapas. Universidad Autónoma de Chiapas, Tuxtla Gutiérrez

Alves-Costa CP, Da Fonseca GAB, Christófaro C (2004) Variation in the diet of the brown-nosed coati (*Nasua nasua*) in southern Brazil. J Mammal 85:478–482

Ayoub NA, Hayashi CY (2009) Spiders (Araneae). In: Hedges SB, Kumar S (eds) The timetree of life. Oxford University Press, New York, pp 255–259

Baerg WJ (1938) Tarantula studies. J N Y Entomol Sci 46:31–43

Baerg WJ (1958) The tarantula. University of Kansas Press, Lawrence

Barbosa BC, Maciel TT, Abegg AD, Borges LM, Rosa CM, Vargas-Peixoto D (2016) Arachnids infected by arthropod-pathogenic fungi in an urban fragment of Atlantic Forest in southern Brazil. Nat Online 14:11–14

Barneche JA, Gillung JP, Gonzalez A (2013) Description and host interactions of a new species of *Exetasis* Walker (Diptera: Acroceridae), with a key to species of the genus. Zootaxa 3664:525–536

Barth FG, Seyfarth EA, Bleckmann H, Schüch W (1988) Spiders of the genus *Cupiennius* Simon 1891 (Araneae, Ctenidae). I. Range distribution, dwelling plants, and climatic characteristics of the habitats. Oecologia 77:187–193

Bertani R, Fukushima SC (2009) Description of two new species of *Avicularia* Lamarck 1818 and redescription of *Avicularia diversipes* (C. L. Koch 1842) (Araneae, Theraphosidae, Aviculariinae)—three possibly threatened Brazilian species. Zootaxa 2223:25–47

Bertani R, Marques OAV (1996) Defensive behaviors in mygalomorph spiders: release of urticating hairs by some Aviculariinae (Araneae, Theraphosidae). Zool Anz 234:161–165

Bertani R, Motta PC (2013) Redescription of *Avicularia taunayi* and notes on its habitat and geographical distribution (Araneae: Theraphosidae: Aviculariinae). Fortschr Zool 30:107–114

Blaikie AJ, Ellis J, Sanders R, MacEwen CJ (1997) Eye disease associated with handling pet tarantulas: three case reports. Br Med J 314:1524–1525

Blair WF (1936) A note on the ecology of *Microhyla olivacea*. Copeia 1936:115

Blein W, Fauria K, Hénaut Y (1996) How does Tarantula *Lasiodora parahybena* Mello Leitto (Aranea, Theraphosidae) detect its prey. Rev Suisse Zool Hors Série:1–8

Bode F, Sachs F, Franz MR (2001) Tarantula peptide inhibits atrial fibrillation. Nature 409:35–36

Boistel R, Pauwels OSG (2002a) *Oscaecilia zweifeli* (Zweifel's caecilian). Predation. Herpetol Rev 33:120–121

Boistel R, Pauwels OSG (2002b) *Leptodactylus knudseni* (Knudsen's bullfrog). Predation. Herpetol Rev 33:303

Borges LM, da Rosa CM, Dri GF, Bertani R (2016) Predation of the snake *Erythrolamprus almadensis* (Wagler, 1824) by the tarantula *Grammostola quirogai* Montes De Oca, D'Elía & Pérez-Miles, 2016. Herpetol Notes 9:321–322

Brooks DM (2012) Birds caught in spider webs: a synthesis of patterns. Wilson J Ornithol 124:345–353

Brouard O, Le Jeune A-H, Leroy C, Céréghino R, Roux O, Pelozuelo L, Dejean A, Corbara B, Carrias J-F (2011) Are algae relevant to the detritus-based food web in tank-bromeliads? PLoS One 6:e20129. https://doi.org/10.1371/journal.pone.0020129

Cady A, Leech R, Sorkin L, Stratton G, Caldwell M (1993) Acrocerid (Insecta: Diptera) life histories, behaviors, host spiders (Arachnida: Araneida), and distribution records. Can Entomol 125:931–944

Calaça-Dias S, Domingos-Brescovit A (2004) Microhabitat selection and co-occurrence of *Pachistopelma rufonigrum* Pocock (Araneae, Theraphosidae) and *Nothroctenus fuxico* sp. nov. (Araneae, Ctenidae) in tank bromeliads from Serra de Itabaiana, Sergipe, Brazil. Rev Bras Zool 21:789–796

Cambra RA, Quintero T, Miranda RJ (2004) Presas, comportamiento de anidación y nuevos registros de distribución en pompílidos neotropicales (Hymenoptera: Pompilidae). Tecnociencia 6:95–109

Canals L, Figueroa D, Torres-Contreras H, Veloso C, Canals M (2012) Mealworm (*Tenebrio molitor*) diets relative to the energy requirements of small mygalomorph spiders (*Paraphysa* sp). J Exotic Pet Med 21:203–206

Carvalho WD, Norris D, Michalski F (2016) Opportunistic predation of a common scale-backed antbird (*Willisornis poecilinotus*) by a Goliath bird-eating spider (*Theraphosa blondi*) in the eastern Brazilian Amazon. Stud Neotrop Fauna E 51:239–241

Castillo Rodriguez N, Méndez Galeano MA (2017) *Ptychoglossus bicolor* (Werner's large scaled lizard) predation. Herpetol Rev 48:658–659

Cocroft RB, Hambler K (1989) Observations on a commensal relationship of the microhylid frog *Chiasmocleis ventrimaculata* and the burrowing theraphosid spider *Xenesthis immanis* in Southeastern Peru. Biotropica 21:2–8

Costa FG, Pérez-Miles F (2002) Reproductive biology of Uruguayan theraphosids (Araneae, Mygalomorphae). J Arachnol 30:571–587

Costa FG, Pérez-Miles F, Mignone A (2004) Pompilid wasp interactions with burrowing tarantulas: *Pepsis cupripennis* versus *Eupalaestrus weijenberghi* and *Acanthoscurria suina* (Araneae, Theraphosidae). Stud Neotrop Fauna E 39:37–43

Costa-Neto EM (2006) Bird-spiders (Arachnida, Mygalomorphae) as perceived by the inhabitants of the village of Pedra Branca, Bahia State, Brazil. J Ethnobiol Ethnomed 2:50

Cuéllar-Balleza L, Mosqueda J, Vergara-Pineda S (2017) Parásitos de tarántulas (Araneae: Theraphosidae): estudio de caso en Querétaro. Bol Soc Mex Entomol (nueva serie) 3:77–82

Cushing PE (2012) Spider–ant associations: an updated review of myrmecomorphy, myrmecophily, and myrmecophagy in spiders. Psyche 23:151989

Das KSA, Sreekala LK, Abdurahiman O (2012) Predation on the Kelaart's pipistrelle bat, *Pipistrellus ceylonicus* Kelaart (Chiroptera: Vespertilionidae), by the reddish parachute tarantula, *Poecilotheria rufilata* Pocock (Araneae: Theraphosidae), in Chinnar Wildlife Sanctuary, Kerala, India. Trop Nat Hist 12:257–260

De Omena PM, Romero GQ (2010) Using visual cues of microhabitat traits to find home: the case study of a bromeliad-living jumping spider (Salticidae). Behav Ecol 21:690–695

Dejean A, Olmsted I (1997) Ecological studies on *Aechmea bracteata* (Swartz) (Bromeliaceae). J Nat Hist 31:1313–1334

Dias SC, Lo-Man-Hung NF (2009) First record of an onychophoran (Onychophora, Peripatidae) feeding on a theraphosid spider (Araneae, Theraphosidae). J Arachnol 37:116–118

Dor A, Calmé S, Hénaut Y (2011) Predatory interactions between *Centruroides* scorpions and the tarantula *Brachypelma vagans*. J Arachnol 39:201–204

Dor A, Hénaut Y (2011) Are cannibalism and tarantula predation structuring factors of the spatial distribution of the wolf spider *Lycosa subfusca* (Araneae, Lycosidae)? Ethol Ecol Evol 23:375–387

Dor A, Hénaut Y (2012) Silk use and spiderling behavior in the tarantula *Brachypelma vagans* (Araneae: Theraphosidae). Acta Zool Mex 28:1–12

Dor A, Hénaut Y (2013) Importance of body size and hunting strategy during interactions between the Mexican red-rump tarantula (*Brachypelma vagans*) and the wolf spider *Lycosa subfusca*. Can J Zool 91:545–553

Dor A, Machkour-M'Rabet S, Legal L, Williams T, Hénaut Y (2008) Chemically mediated burrow recognition in the Mexican tarantula *Brachypelma vagans* female. Naturwissenschaften 95:1189–1193

Dundee HA, Shillington C, Yeary CM (2012) Interactions between tarantulas (*Aphonopelma hentzi*) and narrow-mouthed toads (*Gastrophyrne olivacea*): support for a symbiotic relationship. Tulane Stud Zool Bot 32:31–38

Eason RR, Peck WB, Whitcomb WH (1967) Notes on spider parasites including reference list. J Kansas Entomol Soc 40:422–434

Edwards GB, Hibbard LH (1999) The Mexican Redrump, *Brachypelma vagans* (Araneae: Theraphosidae) an exotic tarantula established in Florida. Entomol circular no 394, Florida Department of Agriculture & Consumer Services, Division of Plant Industry. http://www.freshfromflorida.com/content/download/10780/140971/ent394.pd

Enríquez Vázquez P, Mariaca Méndez R, Retana Guiascón ÓG, Naranjo Piñera EJ (2006) Uso medicinal de la fauna silvestre en los altos de Chiapas, México. Interciencia 31:491–499

Ferreira AS, De Oliveira SA, Moura Da Conceiçao B, De Souza CD, Santos P, Gomes Faria R (2014) Predation on *Tropidurus semitaeniatus* (Squamata: Tropiduridae) by *Acanthoscurria natalensis* (Aranea: Theraphosidae) in the semiarid Caatinga region of northeastern Brazil. Herpetol Notes 7:501–503

Ferretti NE, Ferrero AA (2008) Courtship and mating behavior of *Grammostola schulzei* (Schmidt 1994) (Araneae, Theraphosidae), a burrowing tarantula from Argentina. J Arachnol 36:480–483

Foelix RF (2011) Biology of spiders, 3rd edn. Oxford University Press, New York

Foelix RF, Rast B, Peattie AM (2012) Silk secretion from tarantula feet revisited: alleged spigots are probably chemoreceptors. J Exp Biol 215:1084–1089

Frank JH, Lounibos LP (2009) Insects and allies associated with bromeliads: a review. Terrestrial Arthropod Rev 1:125–153

Gallon RC (2000) The natural history of tarantula spiders. British Tarantula Society, London. http://www.thebts.co.uk/old_articles/natural.htm

Gerdes ABM, Uhl G, Alpers GW (2009) Spiders are special: fear and disgust evoked by pictures of arthropods. Evol Hum Behav 30:66–73

Gillung JP, Borkent CJ (2017) Death comes on two wings: a review of dipteran natural enemies of arachnids. J Arachnol 45:1–19

Godfray HCJ, Shimada M (1999) Parasitoids as model organism for ecologists. Res Popul Ecol 41:3–10

Gray HM, Kaiser H, Green DM (2010) Does alkaloid sequestration protect the green poison frog, *Dendrobates auratus*, from predator attacks? Salamandra 46:235–238

Gunnarson B (2007) Bird predation on spiders: ecological mechanisms and evolutionary consequences. J Arachnol 35:509–530

Hedrick AV, Kortet R (2006) Hiding behavior in two cricket populations that differ in predation pressure. Anim Behav 72:1111–1118

Hénaut Y (2000) Host selection by a kleptoparasitic spider. J Nat Hist 34(5):747–753

Hénaut Y, Machkour-M'Rabet S (2005) Cannibalism and cleptobiosis in *Brachypelma vagans*. Entomol Mex 4:30–32

Hénaut Y, Delme J, Legal L, Williams T (2005) Host selection by a cleptobiotic spider. Naturwissenschaften 92:95–99

Hénaut Y, Corbara B, Pelozuelo L, Azemar F, Cereghino R, Herault DA (2014) A tank bromeliad favors spider presence in a Neotropical inundated forest. PLoS One 9:e114592. https://doi.org/10.1371/journal.pone.0114592

Hénaut Y, Machkour-M'Rabet S, Weissenberger H, Rojo R (2015) Dimorphism and population size of the Mexican redrump tarantula, *Brachypelma vagans* (Araneae: Theraphosidae), in southeast Mexico. Rev Mex Biodivers 86:737–743

Hénaut Y, Tchibozo S, Machkour-M'Rabet S (2016) The use of tarantulas in traditional medicine on three continents. In: Cushing PE (ed) Program and abstracts 20th International Congress of Arachnology July 2–9, 2016, Colorado School of Mines, Golden, Colorado. Denver Museum of Nature and Science Reports, vol 3, p 106

Herberstein ME, Tso IM (2011) Spiders webs: evolution, diversity and plasticity. In: Herberstein ME (ed) Spider behaviour flexibility and versatility. Cambridge University Press, New York, pp 31–56

Hirsch BT (2009) Seasonal variation in the diet of ring-tailed coatis (*Nasua nasua*) in Iguazu, Argentina. J Mammal 90:136–143

Hodge MA, Storfer-Isser A (1997) Conspecific and heterospecific attraction: a mechanism of web-site selection leading to aggregation formation by web-building spiders. Ethology 103:815–826

Horstkotte J, Riesch R, Plath M, Jäger P (2010) Predation by three species of spiders on a cave fish in a Mexican sulphur cave. Bull Br Arachnol Soc 15:55–58

Hunn ES (1977) Tzeltal folk zoology: the classification of discontinuities in nature. Academic, New York

Hunt RH (1980) Toad sanctuary in a tarantula burrow. Nat Hist 89:49–53

Jabiol J, Corbara B, Dejean A, Céréghino R (2009) Structure of aquatic insect communities in tank bromeliads in an East-Amazonian rainforest in French Guiana. For Ecol Manag 257:351–360

Jackson RR (1992) Eight-legged trickster. Spiders that specialize in catching other spiders. BioScience 42:591–598

Jackson DE (2009) Nutritional ecology: a first vegetarian spider. Curr Biol 19:R894–R895

Kim KW, Horel A (1998) Matriphagy in the spider *Amaurobius ferox* (Araneidae, Amaurobiidae): an example of mother–offspring interactions. Ethology 104:1021–1037

Kleinknecht RA (1982) The origins and remission of fear in a group of tarantula enthusiasts. Behav Res Ther 20:437–443

Klem ML (2017) Effectiveness of chemical defense in a tropical millipede species on a potential predator, the tarantula *Megaphobema mesomelas*. Occam's Razor 7:1–10

Köhler G, Cedeño-Vázquez JR, Kraus ED, Beutelspacher-García PM, Domíguez-Lepe JA (2017) The Chetumal snake census: generating biological data from road-killed snakes. Part 5. *Imantodes tenuissimus*, *Lampropeltis triangulum*, and *Stenorrhina freminvillii*. Mesoam Herpetol 4:774–789

Korenko S, Michalková V, Zwakhals K, Pekár S (2011) Host specificity and temporal and seasonal shifts in host preference of a web-spider parasitoid *Zatypota percontatoria*. J Insect Sci 11:101–112. https://doi.org/10.1673/031.011.10101

Locht A, Yáñez M, Vásquez L (1999) Distribution and natural history of Mexican species of *Brachypelma* and *Brachypelmides* (Theraphosidae, Theraphosinae) with morphological evidence for their synonymy. J Arachnol 27:196–200
Lourenço WR (1978) Notas sobre a biologia de *Acanthoscurria atrox* Vellard, 1924 (Araneae: Theraphosidae). Rev Bras Biol 38:161–164
Machkour-M'Rabet S, Hénaut Y, Rojo R, Calmé S (2005) A not so natural history of the tarantula *Brachypelma vagans*: interaction with human activity. J Nat Hist 39:2515–2523
Machkour-M'Rabet S, Hénaut Y, Sepulveda A, Rojo R, Calmé S, Geissen V (2007) Soil preference and burrow structure of an endangered tarantula *Brachypelma vagans* (Mygalomorphae: Theraphosidae). J Nat Hist 41:1025–1033
Machkour-M'Rabet S, Hénaut Y, Rojo R, Winterton P (2011) A case of zootherapy with the tarantula *Brachypelma vagans* Ausserer, 1875 in traditional medicine of the Chol Mayan ethnic group in Mexico. J Ethnobiol Ethnomed 7:12. https://doi.org/10.1186/1746-4269-7-12
Machkour-M'Rabet S, Dor A, Hénaut Y (2015) *Megaselia scalaris* (Diptera: Phoridae): an opportunistic endoparasitoid of the endangered Mexican redrump tarantula, *Brachypelma vagans* (Araneae: Theraphosidae). J Arachnol 43:115–119
Mains EB (1954) Species of cordyceps on spiders. Bull Torrey Bot Club 81:492–500
Marshall SD, Uetz G (1990a) The pedipalpal brush of *Ephebopus* sp. (Araneae, Theraphosidae): evidence for a new site for urticating hairs. Bull Br Arachnol Soc 8:122–124
Marshall SD, Uetz G (1990b) Incorporation of urticating hairs into silk: a novel defense mechanism in two Neotropical tarantulas (Araneae, Theraphosidae). J Arachnol 18:143–149
Marshall SD, West R (2008) An ontogenetic shift in habitat use by the Neotropical tarantula *Ephebopus murinus* (Araneae, Theraphosidae, Aviculariinae). Bull Br Arachnol Soc 14:280–284
McCracken SF, Forstner MRJ (2014) Oil road effects on the anuran community of a high canopy tank bromeliad (*Aechmea zebrina*) in the upper Amazon basin, Ecuador. PLoS One 9:e85470. https://doi.org/10.1371/journal.pone.0085470
Meehan CJ, Olson EJ, Reudink MW, Kyser TK, Curry RL (2009) Herbivory in a spider through exploitation of an ant–plant mutualism. Curr Biol 19:R892–R893
Menin M, Rodrigues DJ, Azevedo CS (2005) Predation on amphibians by spiders (Arachnida, Araneae) in the Neotropical region. Phyllomedusa 4:39–47
Minch EW (1978) The effectiveness of chemical defences against predation by the tarantula *Aphonopelma chalcodes* Chamberlin. Bull Br Arachnol Soc 4:275–278
Moya-Laraño J, Orta-Ocan JM, Barrientos JA, Bach C, Wise DH (2002) Territoriality in a cannibalistic burrowing wolf spider. Ecology 83:356–361
Nelson XJ, Jackson RR (2011) Flexibility in the foraging strategies of spiders. In: Herberstein ME (ed) Spider behaviour flexibility and versatility. Cambridge University Press, New York, pp 31–56
Nyffeler M, Symondson WOC (2001) Spiders and harvestmen as gastropod predators. Ecol Entomol 26:617–628
Nyffeler M, Knörnschild M (2013) Bat predation by spiders. PLoS One 8:e58120. https://doi.org/10.1371/journal.pone.0058120
Nyffeler M, Pusey BJ (2014) Fish predation by semi-aquatic spiders: a global pattern. PLoS One 9(6):e99459. https://doi.org/10.1371/journal.pone.0099459
Nyffeler M, Lapinski W, Snyde A, Birkhofer K (2017) Spiders feeding on earthworms revisited: consumption of giant earthworms in the tropics. J Arachnol 45:242–247
Paré JA, Dowling APG (2012) An overview of acariasis in captive invertebrates. J Zoo Wildlife Med 43:703–714
Paré JA, Pellitteri PJ, Pinckney RD (2001) *Trogoderma ornatum* pseudoectoparasitism in a Mexican red-knee tarantula (*Brachypelma smithi*). J Zoo Wildlife Med 32:274–277
Peloso PL, Pereira de Sousa V (2007) Predation on *Todirostrum cinereum* (Tyrannidae) by the orb-web spider *Nephilengys cruentata* (Aranae, Nephilidae). Rev Bras Ornitol 15:461–463
Penney D, Bennett SP (2006) First unequivocal mermithid–linyphiid (Araneae) parasite–host association. J Arachnol 34:273–278

Pérez-Miles F, Costa FG, Toscano-Gadea C, Mignone A (2005) Ecology and behaviour of the 'road tarantulas' *Eupalaestrus weijenberghi* and *Acanthoscurria suina* (Araneae, Theraphosidae) from Uruguay. J Nat Hist 39:483–498

Persons MH, Uetz GW (2005) Sexual cannibalism and mate choice decisions in wolf spiders: influence of male size and secondary sexual characters. Anim Behav 69:83–94

Pizzi R (2009) Parasites of tarantulas (Theraphosidae). J Exotic Pets Med 18:283–288

Powell R, Little LW, Smith DD (1984) Eine Wohngemeinschaft von *Physalaemus pustulosus* (Cope, 1864) (Salientia: Leptodactylidae) mit einer bodenbewohnenden Vogelspinne. Salamandra 20:273–274

Prentice TR (1997) Theraphosidae of the Mojave Desert west and north of the Colorado River (Araneae, Mygalomorphae, Theraphosidae). J Arachnol 25:137–176

Punzo F (2005) The ability of *Aphonopelma steindachneri* (Ausserer) (Araneae: Theraphosidae) to detect and respond to chemosensory cues associated with a predator, *Hogna carolinensis* (Araneae: Lycosidae). Bull Br Arachnol Soc 13:169–171

Punzo F (2007) Interspecific variation in hunting behavior of *Pepsis grossa* (Fabricius) and *Pepsis thisbe* Lucas (Hymenoptera: Pompilidae): a field study. J Hymenopt Res 16:297–310

Punzo F, Garman B (1989) Effects of encounter experience on the hunting behavior of the spider wasp, *Pepsis formosa* (Say) (Hymenoptera: Pompilidae). Southwest Nat 34:513–518

Punzo F, Henderson L (1999) Aspects of the natural history and behavioural ecology of the tarantula spider *Aphonopelma hentzi* (Girard, 1854) (Orthognatha, Theraphosidae). Bull Br Arachnol Soc 11:121–128

Rabaneda-Bueno R, Rodríguez-Gironés MÁ, Aguado-de-la-Paz S, Fernández-Montraveta C, De Mas E, Wise DH, Moya-Laraño J (2008) Sexual cannibalism: high incidence in a natural population with benefits to females. PLoS One 3:e3484. https://doi.org/10.1371/journal.pone.0003484

Rego FNAA, Rheims CA, Venticinque EM (2004) Notes on the predation of an aviculariine spider (Araneae: Theraphosidae, *Avicularia* sp.) by *Pepsis frivaldszkyi* (Hymenoptera: Pompilidae) in Brazilian Amazonia. J Br Tarantula Soc 20:18–25

Reichling SB (2000) Group dispersal in juvenile *Brachypelma vagans* (Araneae, Theraphosidae). J Arachnol 28:248–250

Rödel MO, Brede C, Hirschfeld M, Schmitt T, Favreau P, Stöcklin R, Wunder C, Mebs D (2013) Chemical camouflage—a frog's strategy to co-exist with aggressive ants. PLoS One 8:e81950. https://doi.org/10.1371/journal.pone.0081950

Romero GQ, Vasconcellos-Neto J (2004) Spatial distribution patterns of jumping spiders associated with terrestrial bromeliads. Biotropica 36:596–601

Rosa GM, Bradfield K, Fernández-Loras A, Garcia G, Tapley B (2012) Two remarkable prey items for a chicken: *Leptodactylus fallax* Müller, 1926 predation upon the theraphosid spider *Cyrtopholis femoralis* Pocock, 1903 and the colubrid snake *Liophis juliae* (Cope, 1879). Trop Zool 25:135–140

Samu F, Toft S, Kiss B (1999) Factors influencing cannibalism in the wolf spider *Pardosa agrestis* (Araneae, Lycosidae). Behav Ecol Sociobiol 45:349–354

Santos JA, Romero GQ (2004) A new bromeliad-dwelling jumping spider (Araneae, Salticidae) from Brazil. J Arachnol 32:188–190

Schalk CM, Sezano M (2014) Observations on the use of tarantula burrows by the anurans *Leptodactylus bufonius* (Leptodactylidae) and *Rhinella major* (Bufonidae) in the Dry Chaco ecoregion of Bolivia. Acta Herpetol 9:99–102

Schlinger EI (1972) A new Brazilian panopine species, *Exetasis eickstedtae*, reared from the theraphosid spider, *Lasiodora klugi* (Koch), with a description of its immature larval stages (Diptera, Acroceridae). Pap Avul Zool 267:73–82

Schuster M, Baurecht D, Mitter E, Schmitt A, Barth FG (1994) Field observations on the population structure of three ctenid spiders (*Cupiennius*, Araneae, Ctenidae). J Arachnol 22:32–38

Shaw EM, Bennett SP, Wheater CP (2011) Distribution of *Brachypelma vagans* (Theraphosidae) burrows and their characteristics in Belize over two years. J Arachnol 39:515–519

Shillington C (2002) Thermal ecology of male tarantulas (*Aphonopelma anax*) during the mating season. Can J Zool 80:251–259

Shillington C, Verrell P (1997) Sexual strategies of a North American 'tarantula' (Araneae: Theraphosidae). Ethology 103:588–598

Shillington C, McEwen B (2005) Activity of juvenile tarantulas in and around the maternal burrow. J Arachnol 34:261–265

Shochat E, Stefanov WL, Whitehouse MEA, Faeth SH (2004) Urbanization and spider diversity: influences of human modification of habitats structure and productivity. Ecol Appl 14:268–280

Siliwal M, Ravichandran B (2008) Commensalism in microhylid frogs and mygalomorph spiders. Zoos' Print Mag 23:13

Skinner M (2009). Flickr photos. https://www.flickr.com/photos/12921146@N04/3862145569

Smits JAJ, Telch MJ, Randall PK (2002) An examination of the decline in fear and disgust during exposure-based treatment. Behav Res Ther 40:1243–1253

Sousa PAG, Freire EMX (2010) *Micrablepharus maximiliani* (blue-tailed lizard). Predation. Herpetol Rev 41:82–83

Spraul CW, Wagner P, Lang GE, Lang GK (2003) *Ophthalmia nodosa* caused by the hairs of the bird spider (family Theraphosidae) or hairy megalomorph (known in the US as tarantula): case report and review of the literature. Klin Monatsbl Augenheilkd 220:20–23

Su KF, Meier R, Jackson RR, Harland DP, Li D (2007) Convergent evolution of eye ultrastructure and divergent evolution of vision-mediated predatory behaviour in jumping spiders. J Evol Biol 20:1478–1489

Summers K (1999) *Dendrobates auratus*. Predation. Herpetol Rev 30:91

Suranjan Karunarathna DMS, Thasun Amarasinghe AA (2009) Mutualism in *Ramanella nagaoi* Manamendra-Arachchi and Pethiyagodi, 2001 (Amphibia: Microhylidae) and *Poecilotheria* species (Aracnida: Tereposidae) from Sri Lanka. Taprobanica 1:6–18

Taylor RL (1975) Butterflies in my stomach: insects in human nutrition. Woodbridge, Santa Barbara

Teixeira DM, Luigi G, Schloemp IM (1991) Aves brasileiras como presas de artrópodes. Ararajuba Rev Bras Ornitol 2:69–74

Trabalon M (2011) Agonistic interactions, cuticular and hemolymphatic lipid variation during the foraging period in spider females *Brachypelma albopilosa* (Theraphosidae). J Insect Physiol 57:735–743

van Helsdingen PJ (2011) Spiders in a hostile world (Arachnoidea, Araneae). Arachnologische Mitteilungen 40:55–64

Vardy CR (2000) The New World tarantula-hawk wasp genus *Pepsis* Fabricius (Hymenoptera: Pompilidae). Part I. Introduction and the *P. rubra* species-group. Zool Verhand 332:1–86

Vieira WLS, Gonçalves MBR, Nógrega RP (2012) Predation on *Tropidurus hispidus* (Squamata: Tropiduridae) by *Lasiodora klugi* (Aranea: Theraphosidae) in the semiarid Caatinga region of northeastern Brazil. Biota Neotrop 12:263–265

von May R, Biggi E, Cárdenas H, Diaz MI, Alarcón C, Herrera V, Santa-Cruz R, Tomasinelli F, Westeen EP, Sánchez-Paredes CM, Larson JG, Title PO, Grundler MR, Grundler MC, Davis Rabosky AR, Rabosky DL (2019) Ecological interactions between arthropods and small vertebrates in a lowland Amazon rainforest. Amph Reptile Cons 13:65–77

Weinman D, Disney RHL (1997) Two news species of Phoridae (Diptera) whose larvae associate with large spiders (Araneae: Theraphosidae). J Zool 243:319–328

West RC (1992) Beware blond snake-eaters. Forum Am Tarantula Soc 1:76–77

West RC (1997) Eating bird-eating spiders. Fauna 1:46–53

West RC (2003) Cameroon adventure and "spider divination". J Br Tarantula Soc 18:76–84

West RC (2005) The *Brachypelma* of Mexico. J Br Tarantula Soc 20:108–119

Whitehouse M, Agnarsson I, Miyashita T, Smith D, Cangialosi K, Masumoto T, Li D, Hénaut Y (2002) *Argyrodes*: phylogeny, sociality and interspecific interactions. J Arachnol 30:238–245

Yañez M, Floater G (2000) Spatial distribution and habitat preference of the endangered tarantula, *Brachypelma klaasi* (Araneae: Theraphosidae) in Mexico. Biodivers Conserv 9:795–810

Yañez M, Locht A, Macías-Ordóñez R (1999) Courtship and mating behavior of *Brachypelma klaasi* (Araneae, Theraphosidae). J Arachnol 27:165–170

Zschokke S, Hénaut Y, Benjamin SP, García-Ballinas JA (2006) Prey-capture strategies in sympatric web-building spiders. Can J Zool 84:964–973

Chapter 9
Enemies and Defences: Urticating Setae of Theraphosidae

Radan Kaderka

Abstract Tarantula urticating setae are modified setae located on the abdomen or pedipalps, which represent an effective defensive mechanism against vertebrate and invertebrate predators and intruders. They are also useful taxonomic tools as morphological characters facilitating the classification of New World theraphosid spiders. Up to now, seven morphological types of urticating setae, types I, II, III, IV, V, VI, and VII, have been described. They differ in location, shape, size, and orientation of barbs along the shaft and the length/width ratio. The urticating setae, which were originally among type III in spiders with type I setae, or were considered setae of intermediate morphology between types I and III, are now considered to be ontogenetic derivatives of type I. Recently published papers focused on the molecular phylogenetic reconstruction of Theraphosidae revealed the monophyly of theraphosine taxa with type I setae and the reclassification is also supported by the specific development of type I setae during ontogeny and by the morphological differences between type III setae in taxa with types I + III and types III or III + IV. The hypothesis of urticating setae evolution proposed and based on published molecular analyses gave proof that theraphosids with abdominal urticating setae of type I (Theraphosinae, part) and type II (Aviculariinae) had evolved independently and represent two non-sister but monophyletic groups. Although the taxa with urticating setae of types III and III + IV represent two monophyletic sister groups, the systematic position of *Hemirrhagus* with type VI setae and *Kankuamo* with type VII setae is still unclear, similarly to the evolution of both types of urticating setae.

9.1 Urticating Setae

Urticating setae are modified setae located on the abdomen or pedipalps of New World theraphosid spiders. The release of these setae represents an effective defensive mechanism against vertebrate and invertebrate predators and intruders (Cooke

R. Kaderka (✉)
Roztoky, Česká republika, Czechia

F. Pérez-Miles (ed.), *New World Tarantulas*, Zoological Monographs 6,
https://doi.org/10.1007/978-3-030-48644-0_9

et al. 1972). The urticating setae penetrate into the attacker's skin or mucous membranes, whereby inducing a physical irritation. There is no evidence of any chemical irritant being involved (Cooke et al. 1972). Some Theraphosinae also use urticating setae as an additional component for making egg sacs and the silk mat for moulting as a passive defensive strategy against ants or the larvae of phorid flies (Bertani et al. 2003; Marshall and Uetz 1990b; Bertani and Guadanucci 2013). This defensive mechanism is found in roughly 540 of the 600 Neotropical theraphosid species (Bertani and Guadanucci 2013) and is absent in theraphosids from other regions.

Representatives of all known species of the subfamily Theraphosinae (except for some cave-dwelling species of *Hemirrhagus*), and species of the Aviculariinae genera *Avicularia* Lamarck, 1818, *Iridopelma* Pocock, 1901, *Pachistopelma* Pocock, 1901, and *Typhochlaena* C.L. Koch, 1850 have abdominal urticating setae. *Ephebopus* Simon, 1892 (Psalmopoeinae) is the only genus to have palpal urticating setae. The arboreal tarantulas *Psalmopoeus* Pocock, 1895 and *Tapinauchenius* Ausserer, 1871, as well as the Ischnocolinae and Schismathotelinae genera represent New World theraphosids that lack urticating setae (Bertani and Guadanucci 2013).

The morphological types of urticating setae have been used in the taxonomy and systematics of Theraphosidae, and are useful as a set of characters for the differentiation of subfamilies and genera, as shown in previously published phylogenetic analyses (Cooke et al. 1972; Pérez-Miles et al. 1996; Bertani 2001, 2012; Pérez-Miles 2002; West et al. 2008; Bertani and Guadanucci 2013; Perafán et al. 2016; Fukushima and Bertani 2017). Bertani and Marques (1996) considered the absence of urticating setae a plesiomorphic state within theraphosid spiders.

9.1.1 Morphology and Typology of Urticating Setae

Seven different types of urticating setae have been described (Figs. 9.1, 9.2, 9.3, 9.4, 9.5, 9.6a, 9.7, 9.8, 9.9, 9.10 and 9.11). Setae types were based on their location and morphology differing in the shape of shaft, the size, and the orientation of barbs scattered along the shaft, the number of penetrating tips, and the form of insertion on the body surface through stalks or insertion sockets. One type is known in Aviculariinae (type II), one type in Psalmopoeinae (type V), and another five types in Theraphosinae (types I, III, IV, VI, and VII) (Cooke et al. 1972; Marshall and Uetz 1990a; Pérez-Miles 1998; Perafán et al. 2016). Except for type V in *Ephebopus* which occurs on the distal prolateral surface of palpal femora (Marshall and Uetz 1990a; Foelix et al. 2009), all other types are found on the dorsum of the abdomen. They can be located in one central area (in most genera) or in two separate areas (*Bistriopelma* Kaderka, 2015, *Phrixotrichus* Simon, 1889, or *Tmesiphantes hypogeus* Bertani, Bichuette and Pedroso, 2013) (Kaderka 2015a; Perafán and Pérez-Miles 2014; Bertani et al. 2013). The abdominal urticating setae of types I, II, III, IV, VI, and VII are connected with body surface through a supporting stalk whose apical part represents a break-off zone. In palpal setae of type V, the attachment is through specialised insertion sockets. The lengths of the individual types of

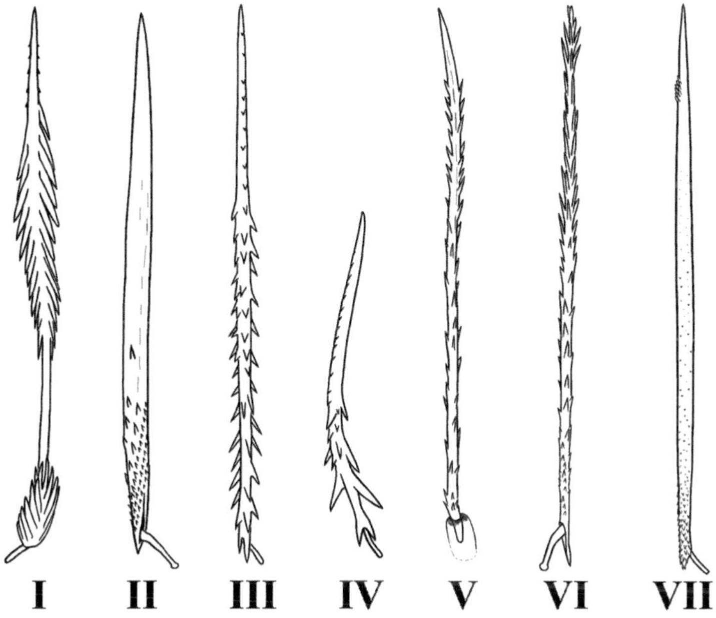

Fig. 9.1 Urticating setae types of theraphosid spiders (Kaderka et al. 2019)

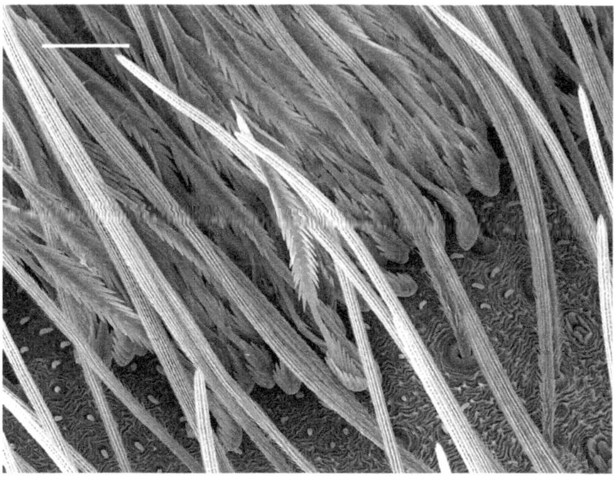

Fig. 9.2 Densely packed type I urticating setae of *Brachypelma albopilosum* Valerio, 1980, intermixed with covering setae, supporting stalks without urticating setae are located in the lower half of the figure. Scale bar: 50 micrometres. (Photo: Jana Bulantová)

urticating setae are shown in Table 9.1. The distribution of urticating setae on genera of Theraphosinae, Aviculariinae and Psalmopoeinae is shown in Table 9.2.

Cooke et al. (1972) observed that in spiders possessing more than one type of urticating setae it appears that, in some species at least, the different types are

Fig. 9.3 (**a**) Basal part of modified type I urticating setae of *Eupalaestrus weijenberghi* (Thorel, 1894) with reduced basal barbs, (**b**) apical part of type I setae of *Brachypelma smithi* (Cambridge, 1897) with reversed denticles, (**c**) basal parts of modified type I urticating setae of *Brachypelma smithi* (Cambridge, 1897) with reduced basal barbs, (**d**) basal part of type I urticating setae of *Brachypelma smithi* (Cambridge, 1897). Scale bar: 10 micrometres. (Photo: Jana Bulantová)

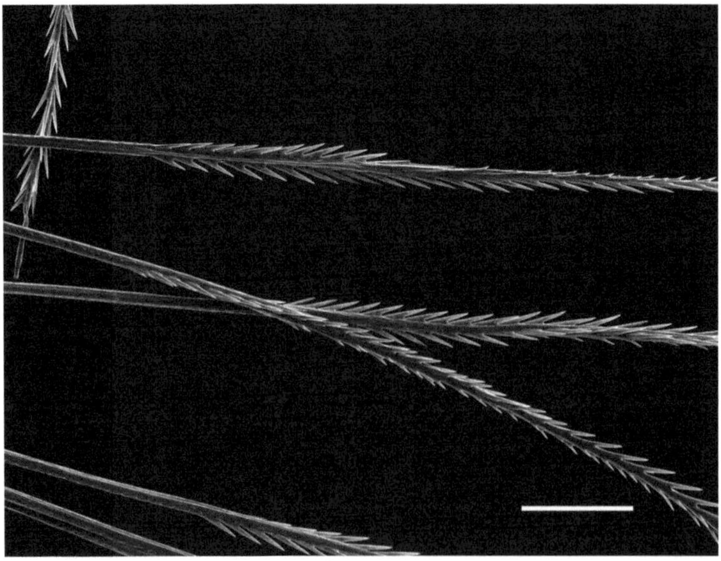

Fig. 9.4 Modified type I urticating setae of *Nhandu vulpinus* (Dresco, 1984). Scale bar: 50 micrometres. (Photo: Jana Bulantová)

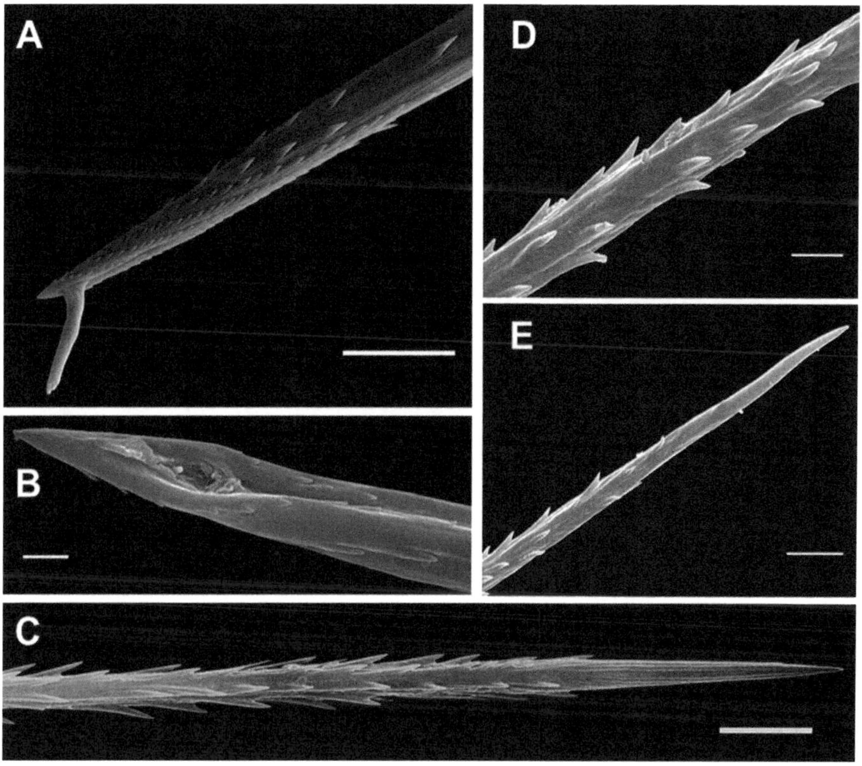

Fig. 9.5 (**a**) Type II setae of *Avicularia* sp., female from Bolivia, Beni province, basal part with a supporting stalk, scale bar = 20 micrometres, (**b**) basal part of type II setae of *Caribena versicolor* (Walckenaer, 1837), juvenile specimen (car. 3.5), scale bar = 2 micrometres, (**c**) apical part of type II setae of *C. versicolor*, adult male, scale bar = 10 micrometres, (**d**) central part of type II setae of *C. versicolor*, juvenile specimen (car. 2.5), scale bar = 5 micrometres, (**e**) apical part of *C. versicolor*, juvenile specimen (car. 3.5), basic morphology of urticating setae in juveniles is identical with mature specimens, but the barbs are less developed and less sharpened, scale bar = 10 micrometres. Abbreviations: car. = length of carapace. (Photo: Jana Bulantová)

segregated into distinct areas of the patch of urticating setae. However, it was not clear whether all species have the setae arranged in this way or whether in some species the setal types are mixed randomly over the entire patch. Examination of the abdomen under a binocular microscope presented a confused picture because the setae are so densely packed. However, when the abdominal wall was sectioned, the arrangement of the setae on their supporting stalks was clearly visible and the question of the distribution of different setal types could probably be resolved by a program of serial sectioning (Cooke et al. 1972). Bertani and Guadanucci (2013) sampled six different areas of the dorsal abdomen, six areas in Aviculariinae with type II and in Theraphosinae with type I, seven areas were sampled in taxa with types III + IV.

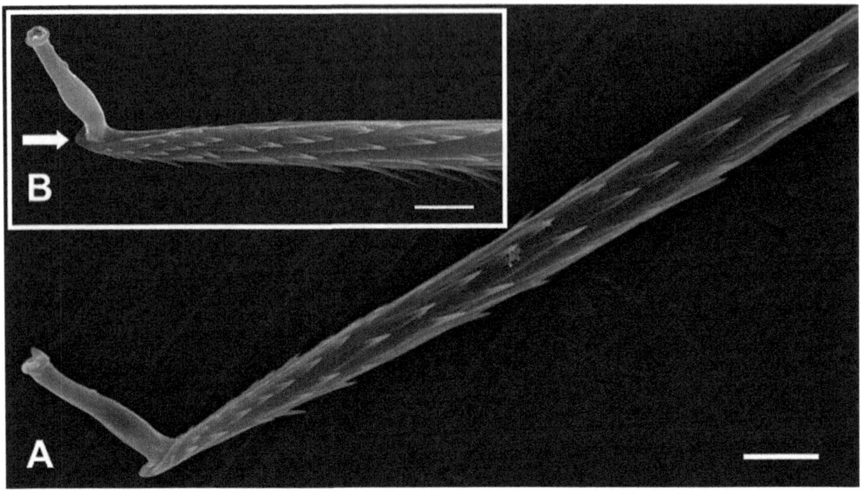

Fig. 9.6 (**a**) Type II setae of a female *Antillena rickwesti* Bertani and Huff, 2013, basal section with a supporting stalk, the apical section is bare, (**b**) a detail of seta of intermediate morphology between body setae and type II in *A. rickwesti*, basal section with a supporting stalk, the arrow shows a precursor of the basal tip, scale bar = 10 micrometres. (Photo: Jana Bulantová)

Bertani and Guadanucci (2013) figured nine distribution patterns of abdominal urticating setae in Theraphosinae (Bertani and Guadanucci 2013: Figs. 5–13). There is a question as to whether this method of sampling is sufficient for a description of the individual distribution patterns of abdominal urticating setae presented by the authors. Theraphosinae with type I setae may have only one to two patterns and differ from the taxa with types III, IV or III + IV in which another seven patterns were recognised. Another type may be found in the theraphosine genus *Magulla* Simon, 1892. Type IV setae are placed in the central dorsal patch, while type III setae are in two separate anterior patches connected by the central patch (Indicatti et al. 2008).

Studying the urticating setae of the Arizona theraphosid *Aphonopelma chalcodes* Chamberlin, 1939, Cooke et al. (1972) revealed that the density of type I setae is approximately 9800–11,400 setae per square mm, which means that they occur in a distinctly higher density than body setae (Cooke et al. 1972). Foelix et al. (2017) recorded the density of type I setae in *Brachypelma kahlenbergi* Rudloff, 2008 as being up to 10,000 setae per square mm. In *Avicularia*, the density of type II setae is substantially less (Cooke et al. 1972), being approximately 6600 setae per square mm (Foelix et al. 2017: Fig. 7a). The density of palpal type V setae in *Ephebopus cyanognathus* West and Marshall, 2000 is approximately 5500 setae per square mm (Foelix et al. 2009).

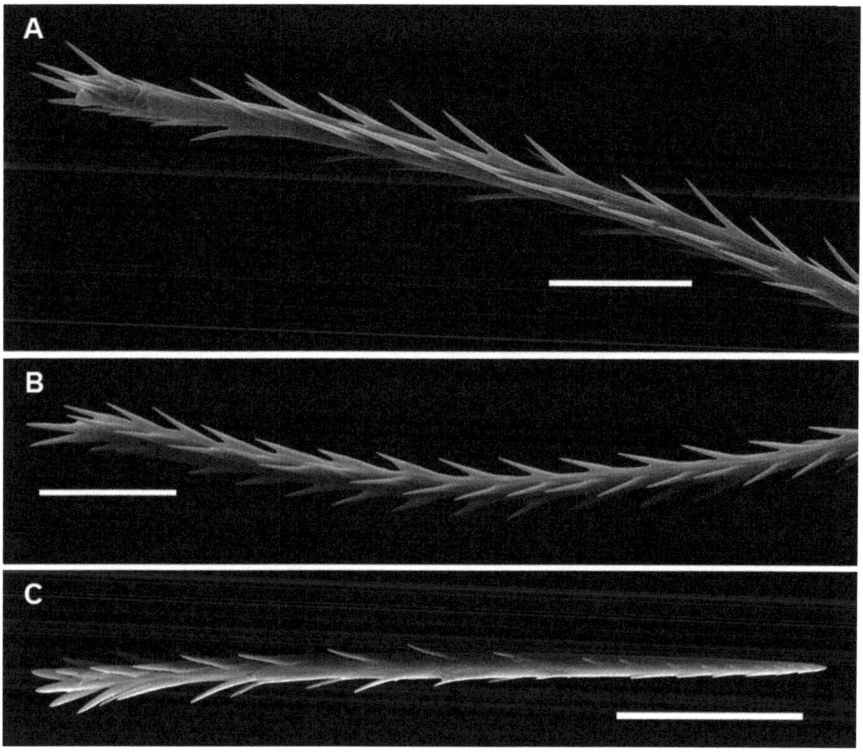

Fig. 9.7 (**a**) Basal part of type III urticating setae of *Hapalotremus* sp. from Peru, (**b**) basal part of long type III urticating setae of *Theraphosa blondi* (Latreille, 1804), scale bar: 20 micrometres, (**c**) short type III urticating setae of *Theraphosa blondi* (Latreille, 1804), scale bar: 10 micrometres. (Photo: Jana Bulantová)

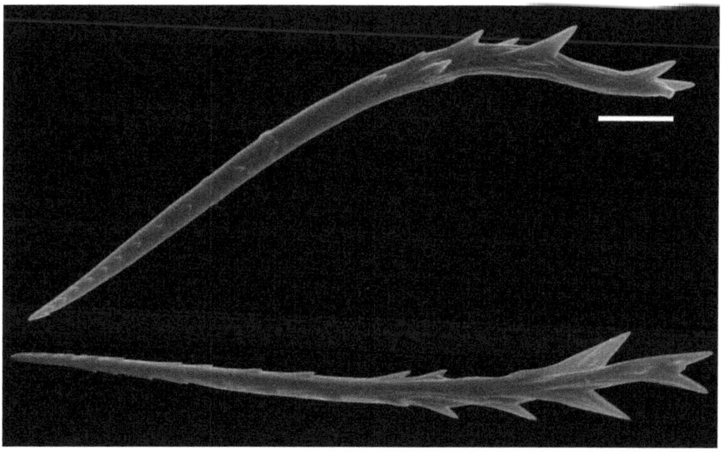

Fig. 9.8 Type IV urticating setae of *Chromatopelma cyanopubescens* (Strand, 1907). Scale bar: 10 micrometres. (Photo: Jana Bulantová)

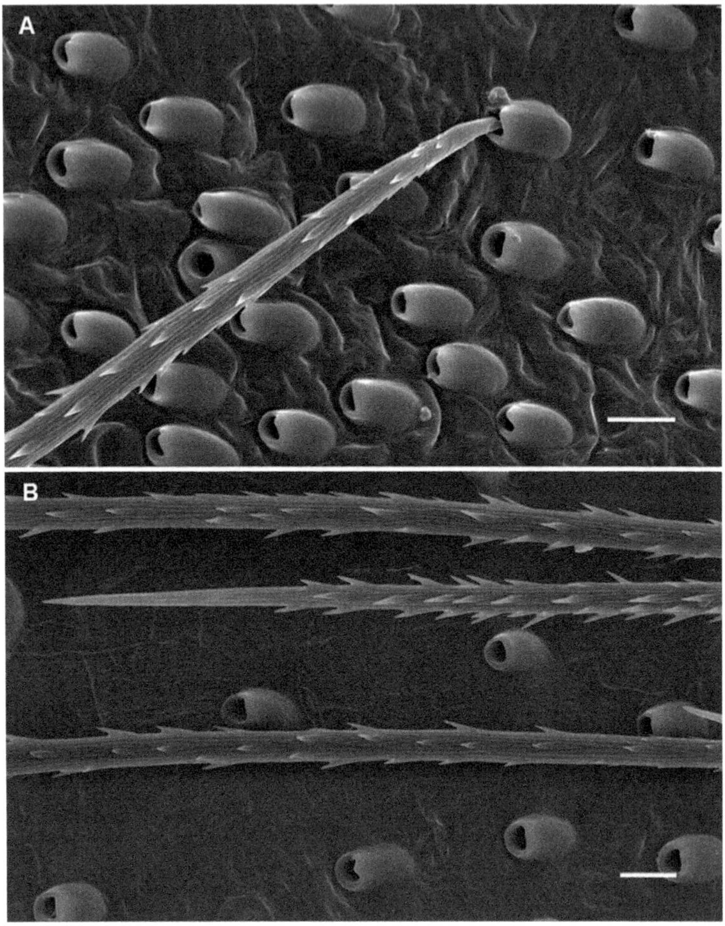

Fig. 9.9 Palpal type V urticating setae of *Ephebopus cyanognathus* West & Marshall, 2000, (**a**) basal part of type V seta connected with the palpal surface by an insertion socket, (**b**) apical (middle) and two central parts of type V setae with a few empty insertion sockets. Scale bar: 10 micrometres. (Photo: Jana Bulantová)

9.1.2 Active Defence of Theraphosids Using Urticating Setae

Some theraphosine taxa may have more than one morphological type or subtype of abdominal uticating setae. Bertani and Guadanucci (2013) proposed that spiders use the various types of urticating setae differently and according to the different targets: type I setae, when incorporated either into the moulting web or egg sac (passive defence), are more effective against invertebrates (ants and phorid flies) than the modified type I setae and type III setae, which are used against vertebrate predators (active defence). When disturbed, spiders frequently disperse their abdominal urticating setae by moving the posterior leg against the abdomen (Figs. 9.12 and

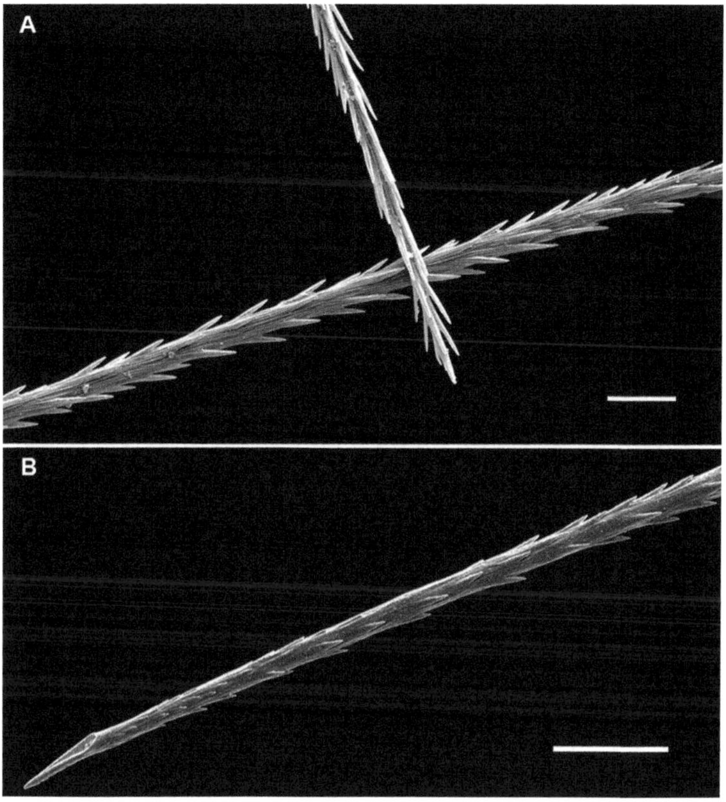

Fig. 9.10 (**a**) Type VI urticating setae of *Hemirrhagus papalotl* Perez-Miles & Locht, 2003, apical and central part, scale bar: 10 micrometres, (**b**) Type VI urticating setae of *Hemirrhagus eros* Perez-Miles & Locht, 2003, basal part, scale bar: 20 micrometres. (Photo: Jana Bulantová)

9.13), or, in the case of palpal setae, by moving the palps along the basal segments of chelicerae and directing these airborne setae toward the intruder (Cooke et al. 1972; Marshall and Uetz 1990a; Foelix et al. 2009). Bertani and Marques (1996) mentioned that Theraphosinae with spines concentrated on the ventral metatarsus IV use this leg segment to comb the urticating setae out from the dorsum of the abdomen. These spines become erect at the moment of release and many setae remain attached to the metatarsal spines (Bertani and Marques 1996). Most representatives of Aviculariinae (*Avicularia*, *Iridopelma*, *Pachistopelma*, and *Typhochlaena*) and probably also *Kankuamo* (Theraphosinae) require direct contact with the intruder to release their urticating setae (Bertani and Marques 1996; Bertani et al. 2003; Perafán et al. 2016). These tarantulas direct the abdomen towards the stimulus and transfer the urticating setae when the abdomen touches a target. In contrast to the airborne urticating setae, which are longer and slender, these setae have a significantly lower length/width ratio.

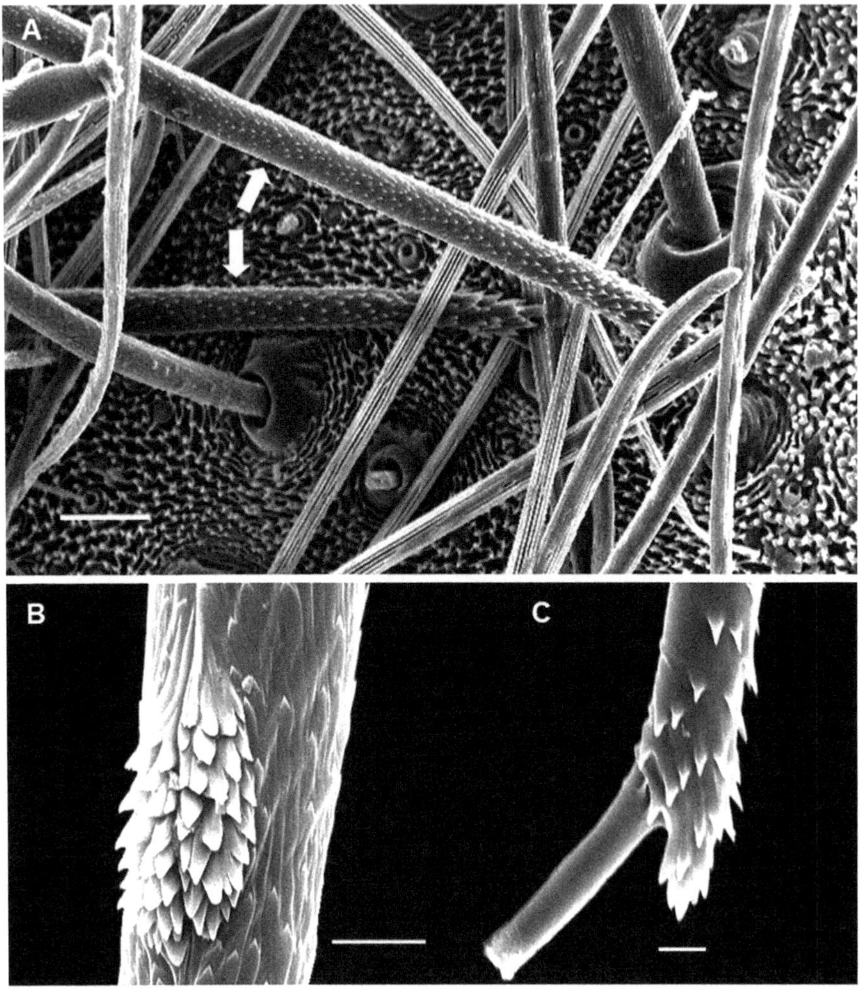

Fig. 9.11 Type VII setae of *Kankuamo marquezi* Perafán et al., 2016, (**a**) two basal ends (marked by the arrows), (**b**) distal part of seta with an area of lanceolate barbs, (**c**) detail of the basal part with a supporting stalk. Scale bar: 50 micrometres (**a**) and 10 micrometres (**b** and **c**). (Photos: Carlos Perafán)

Concerning the length of setae in Theraphosinae, Bertani and Guadanucci (2013) observed that towards the central and posterior areas of the abdomen the length of urticating setae mostly increases. These long setae are released by the spider as a part of its active defensive behaviour against vertebrate predators. Bertani and Guadanucci (2013) proposed that the length of setae is directly proportional to their efficacy towards a predator, i.e. long setae cause greater reactions than short ones.

Bertani and Marques (1996) proposed that the differences in shape, size and thickness between Theraphosinae and Aviculariinae urticating setae could explain

Table 9.1 Lengths of single urticating setae types (according to Kaderka et al. 2019)

Urticating setae types	Length ranges in micrometres	References
Type I	210–600	Kaderka et al. (2019)
Modified type I	270–1640	Kaderka et al. (2019)
Type II	450–1660	Kaderka et al. (2019)
Type III	70–1250	Kaderka et al. (2019)
Type IV	80–210	Kaderka et al. (2019)
Type V	550–670	Kaderka et al. (2019)
Type VI	640–1210	Kaderka et al. (2019)
Type VII	1082–1162	Perafán et al. (2016)

the two releasing mechanisms. The morphological characteristics of all known urticating setae types in Theraphosinae, particularly their light weight and aerodynamics, would allow them to be airborne. They suggest that short and/or thin setae with a length/width ratio of 100:1 or 200:1 are able to float through the air. Bertani et al. (2003) compared the length/width ratio between urticating setae of types II and III and concluded that all airborne setae (including the airborne version of type II setae of *A. versicolor*) are narrow, with an average width ranging from 6 to 7 μm, and a length/width ratio ranging from 98.3 to 208.3. In contrast, for urticating setae released by direct contact with the target, the width was greater, ranging from 15 to 22 μm, with a length/width ratio of 37.3–46.8 (Bertani et al. 2003, Table 9.1). Urticating setae of type VII is 33 ± 3 μm wide with a length/width ratio of 34 and represent the thickest setae known to-date, and with the lowest known length/width ratio, aspects that together reinforce the hypothesis that their releasing mechanism is by direct contact (Perafán et al. 2016). Conversely to the number and size of barbs scattered along the shaft (types I, III, IV, VI), in both urticating setae types II and VII, which are not presumably airborne, the barbs represent only residual denticles, being much smaller than those of other types with known air dispersal (Perafán et al. 2016).

Cooke et al. (1972) questioned whether the urticating setae provide any protection against such insect parasites as acrocerid flies or pompilid wasps, and indeed attacks by pompilids, although quite often observed (e.g. Wiliams 1956; Cazier and Mortenson 1964), have never been reported to induce a flicking of urticating setae. *Aphonopelma* sometime releases setae half-heartedly if surrounded by harvester ants (*Pogonomyrmex* sp.) under experimental conditions, but this is probably a generalised response indicating irritability and does not appear to have the least effect on the ants (Cooke et al. 1972). Bertani and Guadanucci (2013) performed an experiment on ants *Camponotus rufipes* (Fabricius, 1775) and described the effects of urticating setae of *Lasiodora* sp. on them.

Table 9.2 Distribution of urticating setae types according to the previously published papers or unpublished observations, respecting the taxa with type I setae as monophyletic group (Turner et al. 2017; Lüddecke et al. 2018; Ortiz et al. 2018), with a specific development of type I setae during ontogeny[a]

Genus	Urticating setae types							References
	I[a]	II	III	IV	V	VI	VII	
AVICULARIINAE								
Antillena Bertani, Huff and Fukushima, 2017	–	+	–	–	–	–	–	Fukushima and Bertani (2017)
Avicularia Lamarck, 1818	–	+	–	–	–	–	–	Bertani (2012)
Caribena Fukushima and Bertani, 2017	–	+	–	–	–	–	–	Fukushima and Bertani (2017)
Iridopelma Pocock, 1901	–	+	–	–	–	–	–	Bertani (2012)
Pachistopelma Pocock, 1901	–	+	–	–	–	–	–	Bertani (2012)
Typhochlaena C. L. Koch, 1850	–	+	–	–	–	–	–	Bertani (2012)
Ybyrapora Fukushima and Bertani, 2017	–	+	–	–	–	–	–	Fukushima and Bertani (2017)
PSALMOPOEINAE								
Ephebopus Simon, 1892	–	–	–	–	+	–	–	Marshall and Uetz (1990a)
Psalmopoeus Pocock, 1895	–	–	–	–	–	–	–	Schmidt (2003)
Pseudoclamoris Hüsser, 2018	–	–	–	–	–	–	–	Hüsser (2018)
Tapinauchenius Ausserer, 1871	–	–	–	–	–	–	–	Cooke et al. (1972), Schmidt (2003)
THERAPHOSINAE								
Acanthoscurria Ausserer, 1871	+	–	+[b]	–	–	–	–	Pérez-Miles et al. (1996), Bertani and Guadanucci (2013)
Acentropelma Pocock, 1901	+	–	–	–	–	–	–	Gabriel (2016), Turner et al. (2017)
Aenigmarachne Schmidt, 2005[c]	+	–	–	–	–	–	–	Kaderka, pers. obs.
Agnostopelma Pérez-Miles and Weinmann, 2010	–	–	+	+	–	–	–	Pérez-Miles and Weinmann (2010)
Aguapanela Perafán and Cifuentes, 2015	–	–	+	+	–	–	–	Perafán et al. (2015)
Aphonopelma Pocock, 1901	+	–	–	–	–	–	–	Cooke et al. (1972), Prentice (1992), Pérez-Miles et al. (1996), Bertani and Guadanucci (2013)
Bistriopelma Kaderka, 2015	–	–	+	–	–	–	–	Kaderka (2015a)

Genus	Urticating setae types							References
	I[a]	II	III	IV	V	VI	VII	
Bonnetina Vol, 2000	–	–	+	–	–	–	–	Locht and Medina (2008)
Brachypelma Simon, 1891	+	–	+[b]	–	–	–	–	Pérez-Miles et al. (1996), Bertani and Guadanucci (2013)
Bumba Pérez-Miles, Bonaldo and Miglio, 2014	–	–	+	+	–	–	–	Bertani and Da Silva (2003), Bertani and Guadanucci (2013)
Cardiopelma Vol, 1999	–	–	+	–	–	–	–	Kaderka, pers. obs.
Catanduba Yamamoto, Lucas and Brescovit, 2012	–	–	+	–	–	–	–	Yamamoto et al. (2012)
Chromatopelma Schmidt, 1995	–	–	+	+	–	–	–	Schmidt (2003), Kaderka, pers. obs.
Citharacanthus Pocock, 1901	+	–	–	–	–	–	–	Pérez-Miles et al. (1996), Bertani and Guadanucci (2013)
Clavopelma Chamberlin, 1940	+	–	–	–	–	–	–	Schmidt (2003)
Cotztetlana Mendoza Marroquín, 2012	+	–	–	–	–	–	–	Mendoza-Marroquín (2012)
Crassicrus Reichling and West, 1996	+	–	–	–	–	–	–	Schmidt (2003)
Cubanana Ortiz, 2008	+	–	–	–	–	–	–	Ortiz (2008)
Cyclosternum Ausserer, 1871	+	–	–	–	–	–	–	Kaderka, pers. obs. (type)
Cyriocosmus Simon, 1903	–	–	+	–	–	–	–	Fukushima et al. (2005), Bertani and Guadanucci (2013), Kaderka (2015b)
Cyrtopholis Simon, 1892	+	–	–	–	–	–	–	Cooke et al. (1972), Pérez-Miles et al. (1996), Schmidt (2003)
Davus Pickard-Cambridge, 1892	–	–	+	–	–	–	–	Schmidt (2003), Gabriel (2016), Kaderka, pers. obs.
Euathlus Ausserer, 1875	–	–	+	+	–	–	–	Bertani and Guadanucci (2013), Perafán and Pérez-Miles (2014)
Eupalaestrus Pocock, 1901	+	–	+[b]	–	–	–	–	Pérez-Miles et al. (1996), Bertani and Guadanucci (2013)
Eurypelmella Strand, 1907	+	–	–	–	–	–	–	Gabriel (2016), Turner et al. (2017)

(continued)

Table 9.2 (continued)

Genus	Urticating setae types							References
	I[a]	II	III	IV	V	VI	VII	
Grammostola Simon, 1892	–	–	+	+	–	–	–	Cooke et al. (1972), Pérez-Miles et al. (1996): as *Phrixotrichus*
Hapalopus Ausserer, 1875[d]	–	–	+	–	–	–	–	Pérez-Miles et al. (1996), Bertani and Guadanucci (2013)
Hapalotremus Simon, 1903	–	–	+	–	–	–	–	Pérez-Miles et al. (1996)
Hemirrhagus Simon, 1903	–	–	–	–	–	+	–	Pérez-Miles (1998), Pérez-Miles and Locht (2003), Mendoza Marroquin (2014)
Homoeomma Ausserer, 1871	–	–	+	+	–	–	–	Pérez-Miles et al. (1996), Bertani and Guadanucci (2013)
Kankuamo Perafán, Galvis and Pérez-Miles, 2016	–	–	–	–	–	–	+	Perafán et al. (2016)
Kochiana Fukushima, Nagahama and Bertani, 2008	–	–	+	–	–	–	–	Fukushima et al. (2008)
Lasiodora C. L. Koch, 1850	+	–	+[b]	–	–	–	–	Pérez-Miles et al. (1996), Bertani and Guadanucci (2013)
Lasiodorides Schmidt and Bischoff, 1997	+	–	+[b]	–	–	–	–	Schmidt (2003), Bertani and Guadanucci (2013)
Longilyra Gabriel, 2014	+	–	–	–	–	–	–	Gabriel (2014)
Magnacarina Mendoza et al., 2016	–	–	+	–	–	–	–	Mendoza et al. (2016)
Magulla Simon, 1892	–	–	+	+	–	–	–	Indicatti et al. (2008)
Megaphobema Pocock, 1901	+	–	+[b]	–	–	–	–	Pérez-Miles et al. (1996), Bertani and Guadanucci (2013)
Melloleitaoina Gerschman and Schiapelli, 1960	–	–	+	+	–	–	–	Pérez-Miles et al. (1996), Perafán and Pérez-Miles (2014)
Metriopelma Becker, 1878[e]	+	–	–	–	–	–	–	Locht, pers. comm., Gabriel, pers. comm.
Miaschistopus Pocock, 1897	+	–	–	–	–	–	–	Gabriel (2016), Turner et al. (2017)

Genus	Urticating setae types							References
	I[a]	II	III	IV	V	VI	VII	
Munduruku Miglio et al., 2013	–	–	+	+	–	–	–	Miglio et al. (2013)
Mygalarachne Ausserer, 1871	+	–	–	–	–	–	–	Gabriel and Longhorn (2011)
Neostenotarsus Pribik and Weinmann, 2004	+	–	–	–	–	–	–	Tesmoingt and Schmidt (2002)
Neischnocolus Petrunkevitch 1925	+	–	–	–	–	–	–	Pérez-Miles et al. (2008, 2019)
Nesipelma Schmidt and Kovařík, 1996	+	–	–	–	–	–	–	Schmidt (2001a, 2003)
Nhandu Lucas, 1981	+	–	+[b]	–	–	–	–	Pérez-Miles et al. (1996), Bertani and Guadanucci (2013)
Pamphobeteus Pocock, 1901	+	–	+[b]	–	–	–	–	Pérez-Miles et al. (1996), Bertani and Guadanucci (2013)
Phormictopus Pocock, 1901	+	–	+[b]	–	–	–	–	Pérez-Miles et al. (1996), Bertani and Guadanucci (2013)
Phrixotrichus Simon, 1889	–	–	+	+	–	–	–	Perafán and Pérez-Miles (2014)
Plesiopelma Pocock, 1901	–	–	+	+	–	–	–	Pérez-Miles et al. (1996)
Proshapalopus Mello-Leitão, 1923	+	–	+[b]	–	–	–	–	Bertani (2001), Bertani and Guadanucci (2013)
Pseudhapalopus Strand, 1907	+	–	+[b]	–	–	–	–	Schmidt (2003), Bertani and Guadanucci (2013)
Pterinopelma Pocock, 1901	+	–	–	–	–	–	–	Bertani et al. (2011)
Reversopelma Schmidt, 2001	+	–	–	–	–	–	–	Schmidt (2001b)
Sandinista Longhorn and Gabriel, 2019	+	–	–	–	–	–	–	Longhorn and Gabriel (2019)
Schizopelma Pickard-Cambridge, 1897	–	–	+	–	–	–	–	Pérez-Miles et al. (1996)
Scopelobates Simon, 1903	+	–	–	–	–	–	–	Guadanucci et al. (2017)
Sericopelma Ausserer, 1875	+	–	–	–	–	–	–	Pérez-Miles et al. (1996)
Sphaerobothria Karsch, 1879	+	–	–	–	–	–	–	Pérez-Miles et al. (1996)
Stichoplastoris Rudloff, 1997	+	–	–	–	–	–	–	Schmidt (2003), Bertani and Guadanucci (2013)
Theraphosa Thorell, 1870	–	–	+	–	–	–	–	Pérez-Miles et al. (1996), Bertani and Guadanucci (2013)

(continued)

Table 9.2 (continued)

Genus	Urticating setae types							References
	I[a]	II	III	IV	V	VI	VII	
Thrixopelma Schmidt, 1994	–	–	+	+	–	–	–	Schmidt (2003)
Tliltocatl Mendoza and Francke, 2020	+	–	+[b]	–	–	–	–	Mendoza and Francke (2020)
Tmesiphantes Simon, 1892	–	–	+	+	–	–	–	Pérez-Miles et al. (1996), Yamamoto et al. (2007), Bertani and Guadanucci (2013)
Umbyquyra Gargiulo, Brescovit and Lucas, 2018	+	–	+[b]	–	–	–	–	Gargiulo et al. (2018)
Vitalius Lucas, Silva and Bertani, 1993	+	–	+[b]	–	–	–	–	Cooke et al. (1972), Schmidt (2003); Pérez-Miles et al. (1996), Bertani and Guadanucci (2013)
Xenesthis Simon, 1891	+	–	+[b]	–	–	–	–	Pérez-Miles et al. (1996), Bertani and Guadanucci (2013)

Legend: "+" = present, "–" = absent. Classification of urticating setae types follows Cooke et al. (1972), Marshall and Uetz (1990a), Pérez-Miles (1998) and Perafán et al. (2016)

[a]The type I setae including ontogenetic derivatives (so-called modified type I setae). Type I setae are always present in the first nymphal stages, modified type I setae occur later during ontogeny

[b]Urticating setae in genera *Acanthoscurria*, *Brachypelma*, *Eupalaestrus*, *Lasiodora*, *Lasiodorides*, *Megaphobema*, *Nhandu*, *Pamphobeteus*, *Phormictopus*, *Proshapalopus*, *Pseudhapalopus*, *Tliltocatl*, *Umbyquyra*, *Vitalius*, and *Xenesthis* are misinterpreted as type III. The urticating setae of type III in taxa with type I setae differ from type III setae in taxa with type III or III + IV in having a broad shaft at the basal end, at high magnification, this region shows flattened barbs (Pérez-Miles 2002). As they appear later as a product of ontogenetic development and they are derived from type I seta morphology, they may be called modified type I setae or ontogenetic derivatives of type I setae. Pérez-Miles (2002) hypothesised that type III setae represent two different types of setae masked by morphological similarity (homoplasy). According to the published phylogenetic analyses based on molecular data (Turner et al. 2017; Lüddecke et al. 2018; Ortiz et al. 2018), the taxa with type I setae and the taxa with setae of type III and III + IV evolved independently

[c]The modified type I setae found in the holotype of *Aenigmarachne sinapophysis* Schmidt, 2005 were erroneously interpreted as type VI in the original description

[d]In *Hapalopus* Ausserer, 1875, the type IV setae are present only in *Hapalopus butantan* (Pérez-Miles 1998; Fukushima et al. 2005). The revision of *H. butantan* and *H. lesleyae* Gabriel, 2011, both with similar male palpal bulb morphology, is needed

[e]Based on the topotype of *Metriopelma breyeri* Becker, 1878 from Guanajuato, Mexico (in BMNH). The holotype is lost

Fig. 9.12 *Grammostola quirogai* Montes de Oca et al., 2015 releasing abdominal urticating setae by friction of the hind legs, (**a**) a release with the right leg, lateral view, (**b**) a release with the leftt leg, front view. (Photos: A. Laborda)

Fig. 9.13 Defensive behaviour of *Thrixopelma ockerti* Schmidt, 1994 consisting in a lifting of the abdomen as the first warning to a potential intruder before releasing abdominal urticating setae by friction of the hind leg. (Photo: R. Kaderka)

9.1.3 Passive Defence of Theraphosids Using Urticating Setae

Some Theraphosinae spiders use the urticating setae as an additional component for making egg sacs and the silk mat for moulting (Figs. 9.14 and 9.15) as a passive defensive strategy against ants (Hymenoptera) or the larvae of phorid flies (Diptera) (Bertani et al. 2003; Marshall and Uetz 1990b; Bertani and Guadanucci 2013). Cooke et al. (1972) also mentioned that type II setae had been incorporated into the silk walls of the retreat in arboreal *Avicularia surinamensis* (nomen dubium sensu Fukushima and Bertani 2017). It was not known whether the spider deliberately sheds setae for incorporation into the retreat, as seems likely, or whether the presence of setae is fortuitous (Cooke et al. 1972), e.g. during the period of moulting when lying on the dorsal side, the spider's abdomen is in direct contact with the silk mat for moulting. Studying the silk mats of *Phormictopus* sp. from the Dominican Republic with type I urticating setae and their ontogenetic derivatives and *Chromatopelma cyanopubescens* (Strand, 1907) from Venezuela with types III and IV, it is apparent that the use of urticating setae during the construction of silk mats is not fortuitous. The first layer of scattered urticating setae (Fig. 9.15a) is covered with the silk layer (Fig. 9.15b). It means that during the act of ecdysis, the spider's soft cuticle does not come into contact with the urticating setae. Bertani and Marques (1996) did not find any type II setae incorporated into the webs of the tested Aviculariinae (three *Avicularia* spp. and *Pachistopelma rufonigrum* Pocock, 1901).

When incorporating setae into the moulting web, the spider dislodges them by distal metatarsal friction produced by downward movements of the fourth leg against the abdomen, which is different from the upward movements used to throw urticating setae (Pérez-Miles and Costa 1994). However, in egg sac construction the setae are scraped from specific areas of the abdominal setal patch, from the centre and the edge of the anterior area and from the lateral posterior area, which also contrasts with the shedding of setae from posterior areas during active defensive

Fig. 9.14 *Theraphosa blondi* (Latreille, 1804) sitting on a moulting web with incorporated type III setae of two different lengths. (Photo: R. Kaderka)

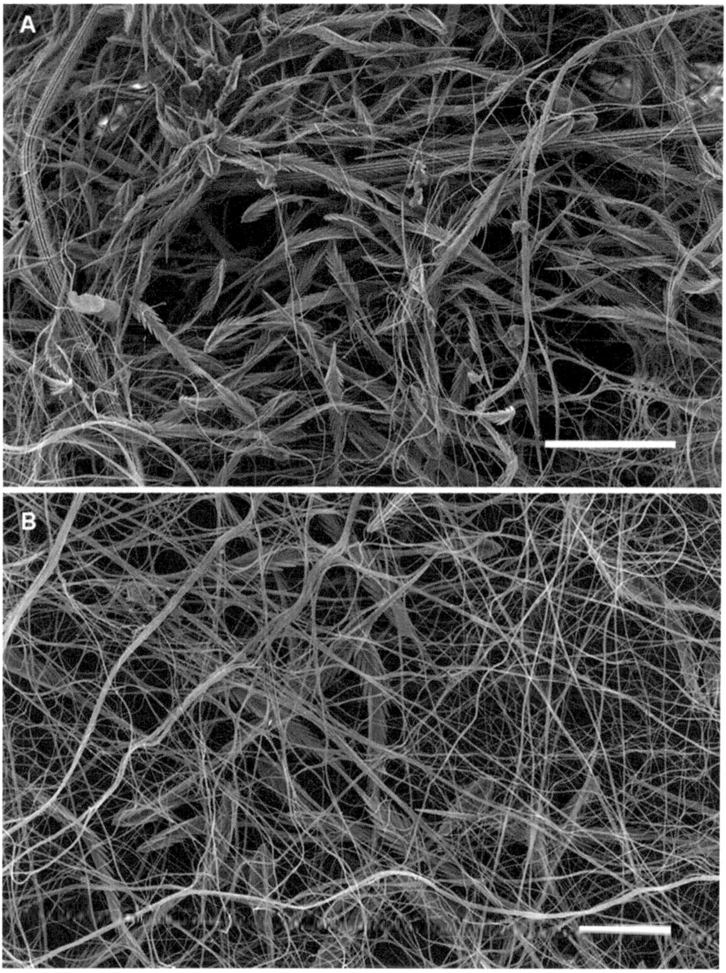

Fig. 9.15 Moulting web of *Phormictopus* sp. from the Dominican Republic with incorporated type I setae, (**a**) bottom, scale bar: 100 micrometres, (**b**) upper side, scale bar: 50 micrometres. (Photo: Jana Bulantová)

behaviour (Marshall and Uetz 1990b). Interestingly, apart from *Theraphosa blondi* Latreille, 1804, all species known to readily incorporate urticating setae in the web mat or egg sac bear type I setae, which are very effective against phorid larvae, stopping the larval movements (Bertani and Guadanucci 2013).

9.1.4 Development of Urticating Setae During Ontogeny

The abdominal urticating setae of Theraphosinae emerge approximately in the first or second nymphal stage, in the stage when the nymphs leave the female's burrow to lead independent lives. Living together in a egg sac and later in the security of the female's burrow, entirely dependent on yolk reserves, the larval stages lack urticating setae and so do not use the defensive strategy based on the release of urticating setae (nomenclature of instars determined after Foelix et al. 2009).

In species with only type III, the urticating setae morphology is relatively constant during ontogeny. In species with types III and IV in the terminal instars, the missing type, type III or IV, as well as the setae of intermediate morphology, appear later during ontogeny (Pérez-Miles 2002; Kaderka et al. 2019). Type III is usually located in central and posterior regions of the abdomen, whereas type IV setae have been found in marginal regions surrounding the patch of type III setae (Bertani and Guadanucci 2013). In species with both types III and IV, a high degree of variability in the total length and curvature of the setae has been recorded, just as in the length and diameter of the basal barbs. Both types III and IV represent the extremes of a morphological continuum. Some Theraphosinae species may have type IV setae only, at least one of both sexes, but mostly females.

In taxa with type I setae, the first nymphal stages possess only type I setae and modified type I setae, together with intermediate forms, appear in the later instars and simultaneously with type I (Pérez-Miles 2002; Kaderka et al. 2019). The modified type I setae are usually located in central and posterior regions of the abdomen, whereas type I setae have been found in marginal regions of the patch of the urticating setae (Bertani and Guadanucci 2013). Some Theraphosinae have type I setae only.

In the majority of spiders, moulting ceases at maturity. However, females of mygalomorph spiders continue to moult approximately every year throughout their lives, acquiring a fresh vestment of setae, including defensive urticating setae at each molt (Cooke et al. 1972).

Studying the ontogeny of five Uruguayan species, Peréz-Miles (2002) found that type III setae occurred later than the other types present (I or IV). He hypothesised that type III setae represent two different types of setae masked by morphological similarity (homoplasy) and derived from type I and IV, respectively. Pérez-Miles (2002) also found some differences in the morphology of type III setae in species with the co-occurrence of types I + III and III + IV: the basal end of the type III seta in specimens having types I + III has a broad shaft, and at high magnification, this region shows flattened barbs. In specimens with types III + IV, the basal end of the type III seta has no flattened barbs and the shaft is not extended to the tip.

Later, in congruence with published phylogenetic analyses based on molecular data (Turner et al. 2017; Lüddecke et al. 2018; Ortiz et al. 2018), Kaderka et al. (2019) recognised four different forms of development of urticating setae during ontogeny in Theraphosinae (two of them were studied in detail) and suggested a new terminology for the ontogenetic derivatives of the type I, so-called modified type I setae, which were described as subtypes, derived from the basic morphology of type I setae (Fig. 9.16). The modified type I setae, which were originally among type III or were considered setae of intermediate morphology between types I and III, are now considered to be ontogenetic derivatives of type I. Type I setae, including their subtypes as morphological repetitions, are generally characterised by a broad basal end. The basal barbs are present and may be developed (Figs. 9.3d and

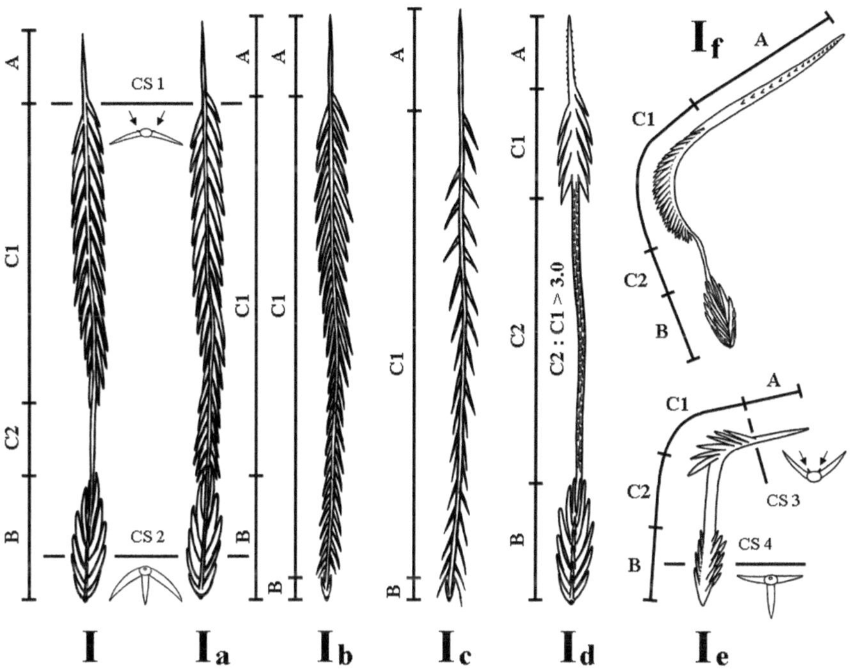

Fig. 9.16 Type I setae and their modifications called subtypes (Kaderka et al. 2019), including the arrangement of barbs in cross-sections 1–4. Abbreviations: B = basal section with barbs, C1 = central section with reversed barbs, C2 = central section without barbs or with two longitudinal rows of barbs confluent with a shaft, A = apical section with denticles arranged in two opposite rows, CS = cross-section

9.16), reduced (Figs. 9.3c and 9.16) or strongly reduced (Figs. 9.3c and 9.16). The connection of setae with a supporting stalk, which represents a so-called break-off zone, is beneath the basal section of non-reversed barbs; the one exception is in subtype I_f (sensu Kaderka et al. 2019) with an additional break-off zone between the basal section and the central section with reversed barbs. In types III and IV, the connection of the seta with a supporting stalk is between the tips and basal ends of the basalmost reversed barbs.

9.1.5 Evolution of Urticating Setae

Bertani and Guadanucci (2013) hypothesised that urticating setae evolved from the body setae and noted (1) the identical manner in which they are inserted into the spider tegument, (2) the resemblance between the truncated basal part of body setae and stalks in urticating setae and (3) the morphological similarity of basal barbs in some variants of body setae and urticating setae. The existence of intermediates between short body setae and urticating setae of type III strengthens this idea (Bertani and Guadanucci 2013). Bertani and Guadanucci (2013) supposed that both types I and IV probably evolved from an ancestor type III setae as they found intermediates between types I and III and types III and IV but no intermediates between type I and IV were found. Unfortunately, this hypothesis does not offer an explanation of how the morphologically more complex type I setae evolved from the ancestor type III setae and why the ancestor type III setae evolved into two different types of urticating setae. Kaderka et al. (2019) studied body setae in some Aviculariinae with type II setae (*Avicularia rickwesti*, *Iridopelma hirsutum*) and hypothesised that type II setae evolved from body setae by the fusion of barbs with a shaft. They found intermediates between body setae and type II setae and specific arrangements of barbs along the shaft of body setae. Considering the hypothesis that type II, III and IV evolved from body setae because the relevant intermediates were found in contemporary taxa, it was still unclear whether all three types of urticating setae evolved from different type of body setae or they evolved from the same type of body setae but in different ways. In that latter case, the urticating setae should be considered homologous.

Later, in 2017 and 2018, molecular analyses combining the nuclear and mitochondrial genes (Turner et al. 2017; Lüddecke et al. 2018; Ortiz et al. 2018) revealed for the first time that the taxa with type I setae represent a monophyletic group on the contrary to previously published phylogenetic analyses based on morphological characters including the set of urticating setae types. This new phylogenetic hypothesis is in accordance with the specific development of type I setae during ontogeny, which is described above and which differs from that found in species with urticating setae of type III or III + IV, and offers an explanation for the morphological differences found in type III in taxa with or without type I setae, which were pointed out by Pérez-Miles (2002). According to the phylogenetic hypotheses proposed by Turner et al. (2017), Lüddecke et al. (2018) and Hüsser (2018), the subfamilies Aviculariinae associated with type II setae and Psalmopoeinae represent monophyletic but non-sister groups. According to Lüddecke et al. (2018), it seems likely that urticating setae of type II in Aviculariinae evolved independently from types I, III and IV found in Theraphosinae. Both subfamilies, Aviculariinae and Theraphosinae, evolved independently and represent two non-sister but monophyletic groups. The co-occurrence of type II with types I, III or IV, or their intermediates, has not been recorded (Bertani and Guadanucci 2013). Although Perafán et al. (2016) proposed that type VII setae in *Kankuamo* evolved from type I setae, the evolution of abdominal urticating setae within Theraphosinae, similarly to the systematic position of *Hemirrhagus* with type VI setae and *Kankuamo* with type VII, are still unclear.

According to Cooke et al. (1972), a basic pelage consists of strongly pilose setae, so-called "short body setae" sensu Bertani and Guadanucci (2013), with a length of approximately 0.5 mm and a density of only 300–400 per square mm. In a comparison to the density of 9800–11,400 in type I setae of *Aphonopelma chalcodes*, it would mean an enormous increase in setae per square mm during evolution of up to 30 times.

In species with urticating setae of type III and IV (Theraphosinae, part), and type II (some Aviculariinae), body setae were found in the areas where urticating setae were lacking, on the boundary of the patch of urticating setae, or intermixed (Bertani and Guadanucci 2013). In species having type I setae and in *Theraphosa blondi* (Latreille, 1804) with only type III setae, body setae were found mainly on the ventral side of the abdomen (Bertani and Guadanucci 2013). In this group, the authors found another non-urticating setae type on the dorsal abdominal area. They are highly plumose and longer than typical body setae (0.7–2.0 mm). These setae bear a close morphological resemblance to the body setae because of the strong plumosity and the similar insertion into the abdomen but they differ in the helicoidal basal part. They are released along with type I urticating setae (Bertani and Guadanucci 2013).

The occurrence of palpal urticating setae of type V is unique, representing an autapomorphy. There is no reason to consider the type V urticating setae being homologous with abdominal urticating setae of Aviculariinae or Theraphosinae, because of its distinct position (Bertani and Marques 1996).

To better understand the evolutionary process consisting of morphological specialisations of body setae and resulting in different types of defensive urticating setae in Theraphosinae and Aviculariinae, it seems crucial to recognise from which morphological types of body setae different types of urticating setae evolved and to understand how it probably occurred. The existence of such types of body setae is presumed based on the existence of already documented intermediates between a particular type of urticating setae and corresponding body setae.

References

Bertani R (2001) Revision, cladistic analysis, and zoogeography of *Vitalius*, *Nhandu*, and *Proshapalopus*; with notes on other theraphosine genera (Araneae, Theraphosidae). Arq Zool 36:265–356

Bertani R (2012) Revision, cladistic analysis and biogeography of *Typhochlaena* C. L. Koch, 1859, *Pachistopelma* Pocock, 1901 and *Iridopelma* Pocock, 1901 (Araneae, Theraphosidae, Aviculariinae). Zoo Keys 230:1–94

Bertani R, Da Silva SC (2003) Notes on the genus *Iracema* Pérez-Miles, 2000 with the first description of the male of *I. horrida* (Schmidt, 1994) (Araneae: Theraphosidae). Zootaxa 362:1–8

Bertani R, Guadanucci JPL (2013) Morphology, evolution and usage of urticating setae by tarantula (Araneae: Theraphosidae). Fortschr Zool 30:403–418

Bertani R, Marques OAV (1996) Defensive behaviors in mygalomorph spiders: release of urticating hairs by some Aviculariinae (Araneae, Theraphosidae). Zool Anz 234:161–165

Bertani R, Boston T, Evenou Y, Guadanucci JPL (2003) Release of urticating hairs by *Avicularia versicolor* (Walckenaer, 1837) (Araneae, Theraphosidae). Bull Br Arachnol Soc 12:395–398

Bertani R, Nagahama RH, Fukushima CS (2011) Revalidation of *Pterinopelma* Pocock, 1901 with description of a new species and the female of *Pterinopelma vitiosum* (Keyserling, 1891) (Araneae: Theraphosidae: Theraphosinae). Zootaxa 2814:1–18

Bertani R, Bichuette ME, Pedroso DR (2013) *Tmesiphantes hypogeus* sp. nov. (Araneae, Theraphosidae), the first troglobitic tarantula from Brazil. An Acad Bras Cięnc 85:235–243

Cazier MA, Mortenson MA (1964) Bionomical observations on tarantula-hawk and thein prey (Hymenoptera: Pompilidae: *Pepsis*). Ann Entomol Soc Am 57:533–541
Cooke JA, Roth VD, Miller F (1972) The urticating hairs of theraphosidae. Am Mus Novit 2498:1–43
Foelix RF, Bastian R, Erb B (2009) Palpal urticating hairs in the tarantula *Ephebopus*: fine structure and mechanism of release. J Arachnol 37:292–298
Foelix RF, Bastian R, Erb B (2017) Die Feinstructur der Brennhaare einiger Vogelspinnen. Arachne 6:4–12
Fukushima CS, Bertani R (2017) Taxonomic revision and cladistic analysis of *Avicularia* Lamarck, 1818 (Araneae, Theraphosidae, Aviculariinae) with description of three new aviculariine genera. ZooKeys 659:1–185
Fukushima CS, Bertani R, Da Silva PI Jr (2005) Revision of *Cyriocosmus* Simon, 1903, with notes on the genus *Hapalopus* Ausserer, 1875 (Araneae: Theraphosidae). Zootaxa 846:1–31
Fukushima CS, Nagahama RH, Bertani R (2008) The identity of *Mygale brunnipes* C. L. Koch, 1842 (Araneae, Theraphosidae), with redescription of the species and the description of a new genus. J Arachnol 36:402–410
Gabriel R (2014) A new genus and species of theraphosid spider from El Salvador (Araneae: Theraphosidae). J Br Tarantula Soc 29:146–153
Gabriel R (2016) Revise taxonomic placement of the species in the Central American genera *Davus* O. Pickard-Cambridge, 1892, *Metriopelma* Becker, 1878, and *Schizopelma* F. O. Pickard-Cambridge, 1897, with comments on species in related genera (Araneae: Theraphosidae). Arachnology 17:61–92
Gabriel R, Longhorn SJ (2011) Redescription of the holotypes of *Mygalarachne* Ausserer 1871 and *Harpaxictis* Simon (1892) (Araneae: Theraphosidae) with rebuttal of their synonymy with *Sericopelma* Ausserer, 1875. Rev Iber Aracnol 19:157–165
Gargiulo F, Brescovit AD, Lucas SM (2018) *Umbyquyra* gen. nov., a new tarantula spider genus from the Neotropical region (Araneae, Mygalomorphae, Theraphosidae), with a description of eight new species. Eur J Taxon 457:1–50
Guadanucci JPL, Perafán C, Valencia-Cuéllar D (2017) The genus *Holothele* Karsch, 1879: the identity of the type species (Mygalomorphae, Theraphosidae). Zoosystema 39:263–271
Hüsser M (2018) A first phylogenetic analysis reveals a new arboreal tarantula genus from South America with description of a new species and two new species of *Tapinauchenius* Ausserer, 1871 (Araneae, Mygalomorphae, Theraphosidae). ZooKeys 784:59–93
Indicatti RP, Lucas SM, Guadanucci JPL, Yamamoto FU (2008) Revalidation and revision of the genus *Magulla* Simon, 1892 (Araneae, Mygalomorphae, Theraphosidae). Zootaxa 1814:21–36
Kaderka R (2015a) *Bistriopelma*, a new genus with two new species from Peru (Araneae: Theraphosidae: Theraphosinae). Rev Peru Boil 22:275–288
Kaderka R (2015b) The neotropical genus *Cyriocosmus* Simon, 1903 and new species from Peru, Brazil and Venezuela (Araneae: Theraphosidae: Theraphosinae). J Nat Hist 50:393–465. https://doi.org/10.1080/00222933.2015.1076082
Kaderka R, Bulantová J, Heneberg P, Řezáč M (2019) Urticating setae of tarantulas (Araneae: Theraphosidae): morphology, revision of typology and terminology and implications for taxonomy. PLoS One 14(11):e0224384. https://doi.org/10.1371/journal.pone.0224384
Locht A, Medina F (2008) Una nueva especie de tarántula del género *Bonnetina* Vol, 2000 (Araneae, Theraphosidae, Theraphosinae), de México. Bol Soc Entomol Aragon 43:45–48
Longhorn S, Gabriel R (2019) Revised taxonomic status of some Mexican and Central American tarantulas (Araneae, Theraphosidae), with transfers from *Aphonopelma* Pocock, 1901, and a new genus from the Pacific lowlands of Nicaragua and Costa Rica. Arachnology 18(2):101–120
Lüddecke T, Krehenwinkel H, Canning G, Glaw F, Longhorn SJ, Tänzler R, Wendt I, Vences M (2018) Discovering the silk road: nuclear and mitochondrial sequence data resolve the phylogenetic relationships among theraphosid spider subfamilies. Mol Phylogenet Evol 119:63–70
Marshall SD, Uetz GW (1990a) The pedipalpal brush of *Ephebopus* sp. (Araneae, Theraphosidae): evidence of a new site for urticating hairs. Bull Br Arachnol Soc 8:122–124

Marshall SD, Uetz GW (1990b) Incorporation of urticating hairs into silk: a novel defense mechanism in two neotropical tarantulas (Araneae, Theraphosidae). J Arachnol 18:143–149
Mendoza JI, Francke O (2020) Systematic revision of Mexican threatened tarantulas *Brachypelma* (Araneae, Theraphosidae, Theraphosinae), with a description of a new genus, and implication on conservation. Zool J Linnean Soc 188(1):82–147. https://doi.org/10.1093/zoolinnean/zlz046
Mendoza Marroquin JI (2014) Taxonomic revision of *Hemirrhagus* Simon, 1903 (Araneae: Theraphosidae, Theraphosinae), with description of five new species from Mexico. Zool J Linn Soc Lond 170:634–689
Mendoza JI, Locht A, Kaderka R, Medina F, Pérez-Miles F (2016) *Magnacarina,* a new genus of Theraphosid spiders from Mexico with a particular palpal bulb structure (Araneae, Theraphosidae, Theraphosinae). Eur J Taxon 232:1–28
Mendoza-Marroquín JI (2012) *Cotztetlana omiltemi,* a new genus and species of tarantula from Guerrero, Mexico (Araneae, Theraphosidae, Theraphosinae). Arthropoda Scientia 2:2–7
Miglio LT, Bonaldo AB, Pérez-Miles F (2013) On *Munduruku*, a new theraphosid genus from Oriental Amazonia (Araneae, Mygalomorphae). Inheringia Sér Zool Porto Alegre 103:185–189
Ortiz D (2008) Description of *Cubanana cristinae*, a new genus and species of theraphosine tarantula (Araneae: Theraphosidae) from the Island of Cuba. Bol Soc Entomol Aragon 42:107–122
Ortiz D, Francke OF, Bond JE (2018) A tangle of forms and phylogeny: extensive morphological homoplasy and molecular clock heterogeneity in Bonnetina and related tarantulas. Mol Phylogenet Evol 120:55–73
Perafán C, Pérez-Miles F (2014) The Andean tarantulas *Euathlus* Ausserer, 1875, *Paraphysa* Simon, 1892 and *Phrixotrichus* Simon, 1889 (Araneae, Theraphosidae). J Nat Hist 48:1–30
Perafán C, Cifuentes-Gil Y, Estrada-Gómez S (2015) *Aguapanela*, a new tarantula genus from the Colombian Andes (Araneae, Theraphosidae). Zootaxa 4033:529–542
Perafán C, Galvis W, Gutiérrez M, Pérez-Miles F (2016) *Kankuamo*, a new Theraphosid genus from Colombia (Araneae, Mygalomorphae), with a new type of urticating setae and divergent male genitalia. ZooKeys 601:89–109
Pérez-Miles F (1998) Notes on the systematics of the little known theraphosid spider *Hemirrhagus cervinus*, with a description of a new type of urticating hair. J Arachnol 26:120–123
Pérez-Miles F (2002) The occurrence of abdominal urticating hairs during development in Theraphosinae (Araneae, Thraphosidae): phylogenetic implications. J Arachnol 30:316–320
Pérez-Miles F, Costa FG (1994) *Acanthoscurria atrox* incorporates urticating hairs into its shedding mat. Forum Am Tarantula Soc 3:63–64
Pérez-Miles F, Locht A (2003) Revision and cladistic analysis of the genus *Hemirrhagus* Simon, 1903 (Araneae, Theraphosidae, Theraphosinae). Bull Br Arachnol Soc 12:365–375
Pérez-Miles F, Weinmann D (2010) *Agnostopelma*: a new genus of tarantula without a scopula on leg IV (Araneae, Theraphosidae, Theraphosinae). J Arachnol 38:104–112
Pérez-Miles F, Lucas SM, da Silva PI Jr, Bertani R (1996) Systematic revision and cladistic analysis of Theraphosinae (Araneae: Theraphosidae). Mygalomorph 1:33–68
Pérez-Miles F, Gabriel R, Miglio L, Bonaldo A, Gallon R, Jimenez JJ et al (2008) *Ami*, a new Theraphosid genus from Central and South America, with the description of six new species (Araneae: Mygalomorphae). Zootaxa 1915:54–68
Pérez-Miles F, Gabriel R, Sherwood D (2019) *Neischnocolus* Petrunkevitch, 1925, senior synonym of *Ami* Pérez Miles, 2008 and *Barropelma* Chamberlin, 1940 (Araneae, Theraphosinae). Arachnology 18(2):150–155
Prentice TR (1992) A new species of North American tarantula, *Aphonopelma paloma* (Araneae, Mygalomorphae, Theraphosidae). J Arachnol 20:189–199
Schmidt G (2001a) *Nesipelma* Schmidt & Kovařík, 1996 ein Synonym von *Phormictopus* Pocock, 1901? Arachnol Mag 9:1–7
Schmidt G (2001b) *Reversopelma petersi* sp. n. (Aranea: Theraphosidae: Theraphosinae), eine neue Spinnenart aus dem Norswesten Südamerikas. Arachnologische Magazin 9:1–10
Schmidt G (2003) Vogelspinnen, 1st edn. Die Neue Brehm-Bücherei, Westarp Wiessenshaften, Hohenwarsleben

Tesmoingt M, Schmidt G (2002) *Stenotarsus scissistylus* gen. et sp. n. (Theraphosidae: Theraphosinae), eine Vogelspinne aus Französisch Guyana. Tarantulas World 76:3–12

Turner SP, Longhorn SJ, Hamilton CA, Gabriel R, Pérez-Miles F, Vogler AP (2017) Re-evaluating conservation priorities of New World tarantula (Araneae: Theraphosidae) in a molecular framework indicates non-monophyly of the genera, *Aphonopelma* and *Brachypelma*. Syst Biodivers 16:89–107. https://doi.org/10.1080/14772000.2017.1346719

West RC, Marshall SD, Fukushima CS, Bertani R (2008) Review and cladistic analysis of the Neotropical tarantula genus *Ephebopus* Simon, 1892 (Araneae: Theraphosidae). Zootaxa 1849:35–58

Wiliams F (1956) Life history studies of *Pepsis* and *Hemipepsis* wasps in California (Hymenoptera, Pompilidae). Ann Entomol Soc Am 49:447–466

Yamamoto FU, Lucas SM, Guadanucci JPL, Indicatti RP (2007) Revision of the genus *Tmesiphantes* Simon (Araneae, Mygalomorphae, Theraphosidae). Rev Bras Zool 24:971–980

Yamamoto FU, Lucas SM, Brescovit AD (2012) *Catanduba*, a new Theraphosinae spider genus from Central Brazil (Araneae: Theraphosidae). Zootaxa 3172:1–19

Chapter 10
Phsiological Ecology of Tarantulas: Thermoregulation, Metabolism, and Performance Traits

Cara Shillington

Abstract Tarantulas represent interesting models for examining metabolism and performance traits (e.g., sprint speed) because of their large size and sex dimorphism at maturity that leads to different selective pressures. As ectotherms, these traits are also dependent on body temperature and their ability to behaviorally thermoregulate (dealing with both extreme heat and cold) can in turn, have significant impacts on growth, survival, and reproduction. After reaching sexual maturity, most males change their sedentary habits and leave their retreats or webs to search actively for females. Males are relatively short-lived and costs of locomotion may be high. Exposure to large fluctuations in environmental conditions makes males more vulnerable to heat stress, desiccation, and predation. In comparison, females tend to remain in close proximity to their burrows and maintain a larger body size over a longer life span. Because of their sit-and-wait predatory strategy, they may undergo long periods with limited food availability and tend to have very low resting metabolic rates that enables survival with low and unpredictable food resources. This chapter will review various aspects of thermoregulation and coldhardiness, metabolism and performance traits in tarantulas with particular emphasis on differences between males and females and examining these traits in the context of selection pressures due the animal's natural history and life history strategies.

10.1 Thermoregulation

Biological processes depend either directly or indirectly on body temperatures (T_b). Among ectothermic animals these T_bs, in turn, depend on environmental temperatures and, because most animals live in thermally variable environments, their T_bs are then also subject to variation. This resulting variation in T_b impacts most physiological and developmental process such as feeding, digestion, growth, locomotion,

---

C. Shillington (✉)
Department of Biology, Eastern Michigan University, Ypsilanti, MI, USA
e-mail: cshilling@emich.edu

F. Pérez-Miles (ed.), *New World Tarantulas*, Zoological Monographs 6,
https://doi.org/10.1007/978-3-030-48644-0_10

mating, and reproduction and egg development (e.g., Kingsolver 1989; Huey 1991; Casey 1992; Peterson et al. 1993). These processes are all directly linked to survival and fitness. Thermoregulation is an active regulatory process in which animals attempt to maintain their T_bs as close as possible to some preferred T_b range where biological processes occur most efficiently. The extent to which ectotherms thermoregulate is reflected by a combination of costs and benefits associated with these behaviors.

Although thermoregulation is well-documented in a wide variety of both vertebrate and invertebrate organisms, spiders as a group, and tarantulas more specifically, have received comparatively little study. Spiders regulate via behavioral mechanisms that include web orientation and site selection (e.g., Carrel 1978; Casey 1981; Cloudsley-Thompson 1991, 1993; Higgins and Ezcurra 1996; Ramirez et al. 2003; Shillington 2002; Alfaro et al. 2013), posture (Robinson and Robinson 1978), changing their position within a burrow (Baerg 1958; Humphreys 1974, 1978; Minch 1978; Seymour and Vinegar 1973; Lubin and Henschel 1990; Veloso et al. 2012), restricting activities to cooler parts of the day (Minch 1978; Punzo and Henderson 1999; Shillington 2002), and evaporative cooling (Davies and Edney 1952; Chew 1961; Pulz 1987; Punzo and Jellies 1983). In addition, spiders will select retreat sites based on the thermal properties of the microenvironment (Reichert and Tracy 1975; Reichert 1981, 1985; Hammerstein and Reichert 1988) and because the thermal environment influences activities times (Shillington 2002), this can affect behavior and ecology of the animals. The following sections will focus specifically on studies of thermoregulation in tarantulas.

10.1.1 Measuring Thermoregulation

Hertz et al. (1993) suggest a conceptual framework for describing thermoregulation. This includes measuring (1) T_bs of animals in their environments, (2) available thermal options within that same environment (i.e., operative temperatures; see Porter and Gates 1969; Bakken 1981, 1992), and (3) the preferred T_b range selected by animals within a laboratory thermal gradient (often called the preferred or select T_b). These data can then be used to calculate indices that indicate the extent of active thermoregulation. Specifically, these indices indicate the thermal quality of the habitat (for additional information on tarantula habitats see Chap. 7) as well as the extent to which animals experience T_bs within their preferred range (see Hertz et al. 1993). While this framework provides insight on an animal's thermal environment as well as the extent and effectiveness of thermoregulation, there is only one tarantula study that describes and details all these data and indices (*Aphonopelma anax* in Texas, Shillington 2002). A study of *Paraphysa parvula* (Note: this species was recently described as a new species: *Euathlus condorito* (Parafán & Pérez-Miles 2014)); where the species is incorrectly identified in the original study, I will refer to it as *E. condorito** in this chapter. No asterix indicates the correct species name was used in the original paper) in the Chilean Andes, provides extensive details on the thermal microclimate and preferred T_bs (measured in a laboratory thermal environment) as well as T_bs measured in the field (Veloso et al. 2012) but did not calculate the indices suggested by Hertz et al. (1993).

10.1.2 Activity Patterns and Body Temperature

Because tarantulas are predominantly crepuscular/nocturnal, temperature may not strongly impact night-time activity although it may be important in determining circadian rhythms/patterns such as the emergence times and duration of activity. As sit-and-wait predators (see Chap. 8), females typically remain fossorial throughout their life cycle and may inhabit the same burrow for many years and maybe even their entire life span (Baerg 1958). This same life history pattern likely applies to arboreal species (see Chap. 4) as well that can be found in webbed retreats (e.g., Stradling 1994). Female activity around their burrows/retreats as part of this predation strategy includes emergence of the female (typically at or after sunset). A web covering that is often across the top of the burrow is removed and females will sit in the burrow entrance or in close proximity (Fig. 10.1). Light intensity is thought to influence emergences times (Minch 1978; Punzo and Henderson 1999) and, in dry environments (deserts), humidity may also play a role due to low resistance to temperature stress (Punzo 1991). In a study of arboreal tarantulas in Trinidad where temperatures were moderate during an extended rainy season, female tarantulas were still predominantly nocturnal (Stradling 1994).

Male tarantulas perhaps represent more interesting thermoregulatory behaviors because of their active, locomotory lifestyle once they reach sexual maturity. They

Fig. 10.1 *Tliltocatl vagans* from Puebla, Mexico. Photo was taken at night and shows a female sitting at the entrance to her burrow on a silk mat (Photo Brian McEwen)

no longer have the protection and relatively stable thermal environment associated with a permanent retreat. Their ability to actively search for females directly influences their mating opportunities and thus overall fitness. Emergence from temporary daytime retreats is correlated with increasing humidity along with both decreasing ambient temperatures and light (Shillington 2002). Increased cloud cover and intermittent rain, which lead to both cooler temperatures and increased humidity, led to earlier emergence of males (Shillington 2002). Critical thermal maxima (CTM) determined for two species of tarantulas (Punzo 1991; Seymour and Vinegar 1973; approximately 43 °C) suggests that daytime temperatures may limit activity for some species particularly in hot environments. Radio-tagged male tarantulas that were followed during the mating season in Texas, typically remained in temporary burrows throughout the day until environmental temperatures closely reflected their preferred T_bs. In the early morning, activity typically again ceased when temperatures approached the upper limit of their preferred T_bs and before CTM was reached. Calculations using the indices and conceptual framework from Hertz et al. (1993) along with behavioral observations of activity at the burrow entrance in the evening, indicated that males actively thermoregulate in the mornings and evenings but have no need to thermoregulate at night when environmental temperatures are within their preferred T_b range (Shillington 2002).

10.1.3 Preferred Body Temperatures (T_{pref})

T_{pref} is typically measured in a laboratory thermal gradient where animals are not impacted by ecological constraints that may limit their temperature choices. Several studies have measured T_{pref} of various tarantulas. Regardless of habitat or species, mean preferred T_bs are within an 8 °C range (22.64–31.7 °C) suggesting similar homeostatic and thermoregulatory requirements (Table 10.1). In studies that include only females and juveniles, T_{pref} range is typically around 10 °C (Alfaro et al. 2013;

Table 10.1 Comparisons of preferred or selected temperatures (T_{pref}) of tarantulas in a laboratory thermal gradient (from various studies)

Species	T_{pref} range in °C	Mean T_{pref} in °C	Reference
Grammostola rosea (acclimated at 25 °C), (acclimated at 15 °C)	18.75–32.25 19.00–27.25	22.82 22.64	Alfaro et al. (2013)
*Euathlus condorito** (acclimated at 25 °C)	26.25–37.8	27.58	Alfaro et al. (2013)
*Euathlus condorito** (acclimated at 15 °C)	25.75–35.25	29.94	Alfaro et al. (2013)
*Euathlus condorito**	26.2–36.8	31.7	Veloso et al. (2012)
Aphonopelma anax	24.7–35.1	24.7	Shillington (2002)

Veloso et al. 2012). However in one study of males only (Shillington 2002) and another that combined T_{pref} for males and females (Alfaro et al. 2013; *G. rosea* acclimated at 25 °C), the range was >10 °C (Table 10.1). Shillington (2002) suggests that males may have a large range due to their higher activity levels in the thermal gradient and the importance of such activity to mate location (rather than careful thermoregulation). Although Alfaro et al. (2013) indicated there was no difference in mean T_{pref} between males and females, it would be interesting to know more about the inter-sexual ranges of selected temperatures. Alfaro et al. (2013) also showed differences in T_{pref} with time of day which was not the case in the study with just male tarantulas (Shillington 2002).

In their comparison of T_{pref} between tarantulas of differences sizes and habitats, Alfaro et al. (2013), report some interesting variation. T_{pref} was measured at two different acclimation temperatures: warm = 25 °C and cold = 15 °C. In the smaller *E. condorito** used in this study, T_{pref}s are higher at both acclimation temperatures. Additionally, at the warmer acclimation temperature, individuals showed a decreasing T_{pref} with increasing size/mass. The larger *G. rosea* showed no correlation between body mass and T_{pref} at either acclimation temperature. The authors suggest the smaller species likely have higher thermal sensitivity due to larger surface area-to-volume ration (and higher energetic demand, Canals et al. 2008, 2011) and greater plasticity in their thermal responses due to a more variable thermal habitat high in the Andes.

10.1.4 Coldhardiness

Although we typically associate tarantulas with more tropical and subtropical environments (see Chaps. 6 and 7), there are many species found in much colder environments. As with most arthropods, spiders that are exposed to temperatures below which water will freeze, are themselves faced with the threat of freezing. Because spiders are found in environments where freezing is possible (e.g., temperate, polar and high mountain regions), they have developed physiological, behavioral, and biochemical mechanisms to survive as temperatures drop below freezing (see Lee 2010). Cold temperatures can result in damage to tissues and organs and even death as a result of changing metabolism, membrane fluidity, and reduced protein and enzyme function leading to mechanical damage and/or physiological impairment of function (see Ramløv 2000; Lee 2010). Moreover, ice formation can cause direct mechanical damage or produce cell osmotic stress, leading to unregulated water outflow and cell shrinking which is typically associated with cell death (Lee 2010; Storey and Storey 2013). Two major strategies are typically associated with arthropods that are regularly exposed to freezing temperatures and these are (1) freeze tolerance or (2) preventing/avoiding freezing. In freeze-tolerant animals, the extracellular fluid freezes, but individual cells do not while freeze-intolerant organisms survive only as long as ice does not form in their bodies. Animals typically increase their cold tolerance by supercooling mechanisms (see Lee 2010). Cubillos et al.

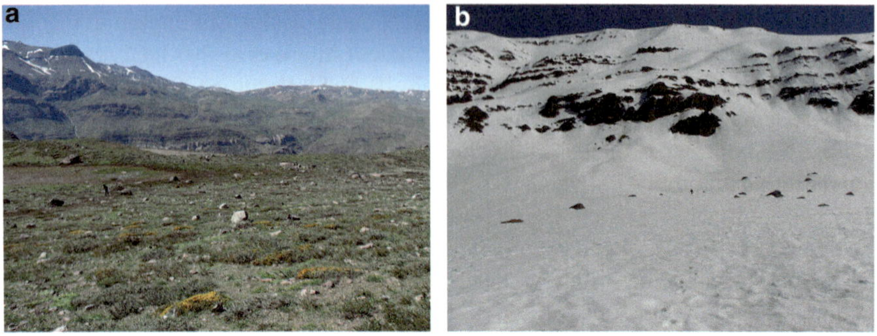

Fig. 10.2 Photo of the summer (**a**) and winter (**b**) habitat for *Euathlus condorito* (Photos Claudio Patricio Veloso)

(2018) worked with *Euathlus condorito* in a population from 2000 meters above sea level in Andes range (see Fig. 10.2a, b showing photos of the summer and winter habitats of this species). They found that the concentration of glucose (~12 mM) in the hemolymph is considerably higher than the concentration measured in lowland tarantulas, e.g., *Aphonopelma californicum* (0.7 mM; Schartau and Leidescher 1983); *Grammostola rosea*, (0.97 mM; Zachariah et al. 2007) and *Theraphosa blondi* (1 mM; Zachariah et al. 2007). These high glucose levels could colligatively depress the melting point and thus, in part, explain the supercooling ability of *E. condorito* (−6.1 °C). Unfortunately, strategies associated with cold-hardiness have not been studied for any of these spiders, and so a proper comparison cannot be made at this time. Hemolymph glucose concentrations in *E. condorito* were higher in the summer compared to winter and but the authors suggest that the differences are not a cold-exposure response because the cold tolerance strategy and supercooling point were constant throughout the year. One explanation could be related to higher energetic and thus metabolic demands during the summer months. In two species of freeze-avoiding spiders, Duman (1979) detected glycerol concentrations that were higher than those found in *E. condorito* thus it is unlikely that this solute alone explains the supercooling capacity in the tarantulas. In addition, Cubillos et al. (2018) detected low levels of potential antifreeze protein activity. Antifreeze proteins typically bind to ice crystals and prevent their growth. These proteins in conjunction with glucose may explain the supercooling ability of the tarantulas.

10.1.5 Retreat Site Selection

Because female and juvenile tarantulas typically inhabit permanent burrows or retreats, selecting the location for these sites, within appropriate thermal limits, is very important. These animals often live in environments that can have large daily and seasonal temperature fluctuations and, as discussed previously, ectotherms rely directly on the external environment to regulate their body temperatures that affect metabolic rates. In addition, as relatively small animals, they have large surface areas that makes them susceptible to rapid fluctuations in heat and water gain/loss across their book lungs (Figueroa et al. 2010). For females with developing eggs within an egg sac, temperature may also affect development of offspring (e.g., Rittschof 2012).

In a field study on *E. condorito** in the Chilean Andes, tarantulas were found in their retreats under rocks and stones (Fig. 10.3) and T_bs and the thermal characteristics of the surrounding microclimate were measured (Veloso et al. 2012) during the day. Specifically, they measured temperatures of the ground, rocks and air in close proximity to the animal's shelter. Among non-reproductives, T_bs were close to preferred T_bs determined in the lab and these T_bs were lower than temperatures recorded from rocks and the soil but higher than air temperatures suggesting active thermoregulation. Interestingly, reproductive females tended to have T_bs that were 3 °C lower than those of the non-reproductives (Veloso et al. 2012).

Mature males are typically active throughout the night and select temporary retreats as temperatures start to increase after sunrise. In a study in southern Texas, temperatures measured within these temporary burrows were stable throughout the day (<10 °C range) compared to more variable ranges in other microhabitats above ground (>20 °C). In addition, these burrow temperatures were well within the animal's preferred temperature range. Anecdotal observations of males in the morning suggest that they are selective in their burrow choice and do not always remain in the first burrow that they find and enter (Shillington, pers. obs.). This may be due to unfavorable thermal environments, but temperatures of "rejected" burrows were not recorded.

On the whole, our understanding of thermoregulation in tarantulas is restricted to just a handful of species and only one study on coldhardiness and freeze-intolerance. These studies provide interesting insights and also suggest areas of future research. To date, there is minimal information regarding Old World or arboreal species. Additional comparisons between age classes/sizes and sexes would also be interesting. There are

Fig. 10.3 Female *Euathlus condorito* (with a cocoon) in a web-lined retreat after the covering stone has been removed. (Photo Claudio Patricio Veloso)

also several groups of sub-social tarantulas and it would be fascinating to examine how social interactions influence retreat site choice or selected temperature ranges.

10.2 Metabolism

Energy expenditure related to internal processes is collectively called metabolism and, when measured as a function of time, is called metabolic rate (MR). Variations in energy expenditure are strongly influenced by both environmental factors (e.g., temperature) and intrinsic factors such as body mass and sex (see Bennett and Dawson 1976; Beaupre 1993; Beaupre et al. 1993). Differences in energy allocations (to growth, activity, reproduction, etc.) within and among individuals may affect behavior, life history, and population dynamics. Understanding variation in metabolism and energetic requirements can provide insights into functional relationships that define large-scale ecological and evolutionary patterns.

As a group, spiders exhibit variable rates of energy expenditure that, similar to most organisms, is strongly affected by body size (Anderson 1970; Greenstone and Bennett 1980). In addition, most spiders and other arachnids demonstrate low aerobic metabolic rates compared to arthropods of similar size (Anderson 1970, 1996; Anderson and Prestwich 1985; Lighton et al. 2001; Nespolo et al. 2011). Various explanations have been suggested to explain this pattern. In an unstable environment where there may be extended periods of reduced prey availability, lower rates of energy expenditure may be adaptive (Lighton et al. 2001). Sit-and-wait strategists and/or animals with a sedentary life style exhibit lower metabolic rates than active predators (Lighton and Fielden 1995; Terblanch et al. 2004) and Lighton and Fielden (1995) suggested that this may be due to a low ratio of actively respiring tissue to body mass. In contrast, based on their study of tarantula book lung structure and function, Canals et al. (2008) suggest diffusion capacities and oxygen delivery that could support higher energetic demands. Energetic requirements associated with different lifestyles (and related physiology) may have led to evolution of differential metabolic rates; or this low there may be phylogenetic constraints. More studies are needed in this area.

To compare MRs among different organisms, different types of metabolic measurements have been defined. Resting metabolic rates are often used as a general measure of the minimal requirements for an organism to maintain physiological homeostasis which is required to sustain life. This is measured when the animal is at rest and post-absorptive (and at ecologically relevant temperatures for ectotherms) and is a measure of aerobic metabolism. Because activities such as locomotion and digestion lead to significant increases above RMR, maximal rates of metabolism and aerobic scope (difference between peak MR and RMR) can also be measured. Finally, the increase in metabolism following feeding is another variable that is often measured. This increase in metabolic rate, as a result of ingesting food, is referred to as specific dynamic action (SDA) and is the increase in energy due to processing of a meal. The following sections will discuss each of these measures as they have been studied in tarantulas.

10.3 Resting Metabolic Rates

Tarantulas represent an excellent model for addressing questions in evolutionary physiology because of differences in life history of adult animals. Soon after reaching sexual maturity, male tarantulas engage in mate searching activity (Prentice 1992; Punzo and Henderson 1999; Shillington 2002, Stoltey and Shillington 2009; Perez-Miles et al. 2005) associated with a scramble competition mating system whereby mating success depends largely are rapid locomotion (Thornhill and Alcock 1983). Compared to females, males are short lived and the energy associated with gamete production is small although locomotory costs are high. Over their longer life span, females typically maintain a larger body size and have higher energetic costs associated with gamete production. Females can also reproduce for multiple years, while males typically die within or after one mating season (Baerg 1928, 1963; Minch 1978). Thus, over a short period, males have a higher energy life style compared to females. Several studies have shown that animals with high energy demands have high RMRs to support this lifestyle (e.g. Daan et al. 1990; Koteja 1991; Ricklefs and Miles 1994; Reinhold 1999; Rogowitz and Chappell 2000).

Table 10.2 Comparison of RMRs from various tarantula studies

Species	Mass (g)	μl CO_2/g^{-1} min^{-1}	Reference
Three unidentified from Mexico	M series: 21.75 (N = 4) H series: 9.77 (N = 1) G series: 14.29 (N = 1)	0.314at 20 °C (RQ = 0.92) 0.410 at 20 °C 0.337 at 20 °C	Anderson (1970)
Brachypelma smithi	26.9 (N = 5)	0.337 at 23 °C (RQ = 0.92)	Anderson and Prestwich (1985)
Aphonopelma anax	12.73 (N = 12) 13.19 (N = 10) 13.88 (N = 13) 13.92 (N = 12)	0.144 at 20 °C 0.202 at 25 °C 0.396 at 30 °C 0.443 at 35 °C	Shillington (2005)
Aphonopelma anax	15.72 (N = 6)	0.233 at 24–26 °C	Shillington and Peterson (2002)
Grammastola rosea	15.21 (N = 2)	0.790 (RQ = 0.92)	Canals et al. (2008)
*Euathlus condorito**	No individual weights provided (6–10 g range)	0.435 at 30 °C 0.623 at 35 °C 0.982 at 40 °C	Canals et al. (2011) (Females without egg sacs measured in the dry season)
Euathlus truncultentus	7.32 (N = 32)	0.490 at 25 ° C	Nespolo et al. (2011)
Grammostola rosea	16.7 (N = 25)	0.362 at 30 °C	Grossi et al. (2016a, b)

Metabolic rates have been measured in female tarantulas from several species all from the New World and all but one from the Americas. Where possible, based on published data, the RMRs are presented in Table 10.2. For ease of comparison among the different studies, all values were converted to μl CO_2g^{-1} min^{-1}. Also, where data were only recorded as rates of O_2use, an RQ = 0.92 (Paul et al. 1989; Lighton and Fielden 1995; Shillington and Peterson 2002) was used to convert the data to rates of CO_2 consumption. Using ratios (e.g., mass-specific metabolism) for variables that scale with body size (such as metabolic rates) does not remove the confounding effects due to body size and may lead to misleading interpretations of the data (Packard and Boardman 1988). However, this measure is commonly found in metabolic studies and I will use it here to easily summarize and compare the various tarantula studies.

In general, RMRs are similar within the various tarantulas species and are significantly lower than those reported for other spiders (e.g., see Lighton and Fielden 1995). In two separate studies of RMRs in *G. rosea*, there is a surprisingly higher RMR in females measured at 20 compared to 30 °C (Canals et al. 2008; Grossi et al. 2016a). However, samples sizes in the 20 °C studies are very small (N = 2) (Canals et al. 2008).

In three of the studies comparing RMRs at different temperatures, MRs increased with increasing temperatures (Canals et al. 2011; Figueroa et al. 2010; Shillington 2005). The specific temperatures chosen for the MR measurements reflect temperatures that the tarantulas would likely experience in their natural environments. In addition, Canals et al. (2011) compared RMRs between animals collected in the wet and dry season, females with and without egg sacs, as well as RMRs associated with different diet regimes. Season by itself did not significantly impact RMRs but interestingly, dry season RMRs were always higher than those measured in the wet season. There was also a complex interaction between diet regime and RMRs. Specifically, diets reflected availability of food (beetle larvae) and/or water (e.g., animals had access to (a) food and water, (b) food and no water, (c) no food and water, and (d) no food and no water, all over a 3 week period). The most important factor affecting RMRs was food availability and this had the biggest impact in the dry season where tarantulas that lacked food (regardless of water availability) always showed decreased RMRs and this was particularly true at the highest temperature. Higher RMRs in the groups with food available ad libitum may be due to SDA (addressed later), but the large differences in groups without food in the dry and wet season suggest a metabolic adjustment to periods when food maybe be reduced or less predictable (Canals et al. 2011). In the comparison of RMRs between females with and without egg sacs, animals with egg sacs had lower RMRs across all temperatures. In addition, there were similar trends with higher MRs in the dry season compared to the wet season. Although the authors reported no differences in weight among all of the groups across season, no individual weights are provided (a weight range of 6–10 grams is reported) (Canals et al. 2011).

RMRs of mature males have only been reported in two studies. In comparison to females, RMRs of males are higher at all temperatures (Shillington 2005; Grossi et al. 2016a). Intra-specific comparisons among other spider groups show mixed results (Humphreys 1977; Tanaka and Ito 1982; Watson and Lighton 1994; Kotiaho 1998; Walker and Irwin 2006). Across animal groups, suggested proximate causes for inter-sexual differences include differences in body composition (Brian et al. 1972; Carrel 1990; Foelix 1996; Cullum 1998). Adaptive explanations for higher RMRs in males is, perhaps, best supported by the aerobic capacity hypothesis (Reinhold 1999) which links RMR and the maximal metabolic rate that an individual can attain. In mature males, a high RMR can sustain high levels of activity which may improve a male's ability to come into contact with well-dispersed females. However, in a 2-year study of male RMRs over the mating season, there was no correlation between RMR and locomotory activity or search area (Stoltey and Shillington 2009). This is consistent with body condition measurements which also showed no change over a season despite visual observations suggesting deterioration (e.g., shrunken abdomen and loss of urticating hairs (see Chap. 9)). The authors suggest that factors other than maximal MRs (e.g., endurance, predator avoidance) need to be examined (Stoltey and Shillington 2009).

In an early study, Anderson (1970) measured RMRs in both adult females and juveniles of the same three unknown species. Unfortunately, sample sizes are very small and the age of the juveniles is unknown (although weights of both adults and juveniles are provided). Juveniles range from threefold smaller than the females (two species) to sevenfold smaller in the third species. Weight-specific RMRs were very similar between juveniles and adults; in two cases RMRs were slightly higher and in the third case slightly lower. With such limited data it is not possible to make any accurate predictions about RMRs and age. There does seem to be a trend with smaller species having higher metabolic rates (Table 10.2) which may be consistent with smaller animals have a larger surface area-to-volume ratio, but again more comparisons across multiple species are needed to support this trend.

Repeatability of whole-animal metabolism has been supported in many animal groups (see Nespolo and Franco 2007) and refers to inter-individual differences that are consistent over time and reflects inter-individual variation. Nespolo et al. (2011) found very low repeatability in RMR in the one lab study with theraphosids. Low repeatability can be interpreted as selection for low RMR (and thus reduced genetic variability) which may be an adaptation for sit-and-wait predators such as tarantulas to minimize maintenance costs (Nespolo et al. 2011). Additional comparisons across multiple tarantula species are needed to determine the consistency of these findings.

10.4 Specific Dynamic Action (*SDA*)

Despite the extensive study of SDA (also called post-prandial response) in animals (see Secor 2009), there is only one study that specifically addresses this phenomenon in tarantulas (Nespolo et al. 2011). SDA is an increase in metabolism after a meal due to digestion and absorption. SDA can range from 20 to 300% above resting rates and has a duration of hours, days or weeks (Lighton and Fielden 1995;

Secor 2009). In addition, SDA can be affected by meal type and size, temperature, and duration since last meal (extensively reviewed by Secor 2009). Several variables associated with SDA can be measured including peak or maximal MR, metabolic scope (maximal MR/RMR), time to peak and duration (time for MR to return to resting levels).

One of the differences in measuring SDA in some arachnids is the fact that digestion starts externally. Thus, for these measurements, both predator and prey are initially live in the metabolic chamber and measurement of SDA variables starts as soon as the prey item is captured (Jensen et al. 2010; Nespolo et al. 2011). In the study with tarantulas, MR measurements were also taken for trials where the cricket was not eaten as a comparison (Nespolo et al. 2011). Because there is only one study on tarantula SDA, it is difficult to make broad comparisons. In a study on an Arctic mite fed different types of food, SDA varied with food type, but it is not clear whether the differences were due to the variation in the amount of food consumed (Young and Block 1980). In comparison, Jensen et al. (2010) found no difference in SDAs of wolf spiders fed diets where lipid and protein content were varied. However, differences in the respiratory exchange ratio suggest that the spiders may have regulated nutrient extraction (Jenson et al. 2010). Tarantulas exhibited a metabolic scope above 600% compared with a twofold increases found in scorpions (Bradley 1982) and fourfold increase in wolf spiders (Jensen et al. 2010). SDA scope typically ranges between 150 and 200% for most invertebrates (Secor 2009), but higher values in tarantulas may be attributable to the large meal size (Nespolo et al. 2011). Duration of SDA in many invertebrate carnivores varies from 3 to 300 h (Secor 2009) and was recorded as 6 h in scorpions (Bradley 1982) and 8 h in tarantulas (Nespolo et al. 2011). In a study that measured RMRs of tarantulas fed different amounts of prey over a 4-month period, animals fed every 5 days had significantly higher RMRs after 1 and 4 months whereas those fed only once a month saw no change in the RMRs over this same period (Philip and Shillington 2010). Although the study did not specifically measure SDA, the authors suggest that the elevated response may be due to SDA even 5 days post-feeding (Philip and Shillington 2010).

10.5 Evaporative Water Loss

Water is a major component of spider body fluids (~60–85%, Pulz 1987) and these body fluids are essential in the hydrostatic pressure required for locomotion (Foelix 1996) (see Chap. 13). Evaporative water loss in arthropods occurs via transpiration across the cuticle and evaporation from respiratory surfaces (Pulz 1987; Cloudsley-Thompson 1991). For xeric adapted arthropods at low temperatures, water loss occurs mostly by evaporation across the respiratory surfaces because the cuticle has a thick and waxy layer that limits water permeability (Davies and Edney 1952; Punzo and Jellies 1983; Pulz 1987). Among mygalomorphs, a low metabolic rate is suggested as an adaptation to unpredictable food availability with limited aerobic activity (site-and-wait predation) and a long life span. Figueroa et al. (2010) suggest that low metabolism may also reduce water loss. In a study that measured weight loss as a measure of water, Vinegar and Seymour (Seymour and Vinegar 1973) found an approximately threefold difference in water loss in *Aphonopelma* sp. acclimated at 10 and 30 °C. In this study, animals were maintained in a dry environment and were weighed regularly. The majority of tarantulas in this study died (100% of those acclimated at 30 °C) and at the higher temperature, animals lost a mean of 18.32% of their initial body weight prior to death.

In comparison, Figueroa et al. (2010) examined evaporative water loss (EWL) during metabolism at 25, 30, 35, and 40 °C in the tarantula *E. condorito** using a hygrometer. In addition to the range of temperatures, they also manipulated the openings to the booklungs and animals were measured at the same temperatures but with two (randomly chosen) openings to the booklungs sealed. At the three lower temperatures, MRs and EWL increased incrementally. RMR values were similar to other tarantulas (Table 10.2) and there were no significant differences in EWL among the three temperatures and experimental groups. However, at 40 °C, there was an approximate tenfold increase in both MR and EWL that was similar between the two experimental groups with either intact or occluded booklung openings (Figueroa et al. 2010). Canals et al. (2011) showed a similar large increase in MR and EWL at 40 °C with animals collected during the dry and wet season and even when animals were provided with *ad libitum* water. *Eusathlus condorito* is a small mygalomorph from the Andes (Perafan and Perez-Miles 2014) and high MRs and EWL at 40 °C (which is typical during the summer months) may limit activity and retreat site choice (Canals et al. 2011; Figueroa et al. 2010; Veloso et al. 2012). The same is likely true for other tarantulas that are found in environments where animals could experiences high daily temperatures during the summer months (see Shillington 2002) but it may be interesting to also compare EWL among animals from xeric versus high humidity environments.

10.6 Performance Traits

There are likely strong selective pressures on locomotor capacities of animals within their natural environment and the physiological components required to support such activities. This is particularly true for ecological relevant activities such as sprint speed and endurance that can be impacted by environment and may be an essential component of fitness (e.g., Careau and Garland Jr 2012; Le Galliard et al. 2013). Differences in physiological performance may be expected in organisms such as tarantulas that exhibit differences in life history, morphology, and size. Females tend to maintain larger body sizes over a significantly longer life span and have lower RMRs compared to males (discussed previously). However, after reaching sexual maturity, male tarantulas change their sedentary habits, abandoning their retreats and searching actively for well-dispersed females. These differences in lifestyle are associated with higher energetic demands for sexually mature males over a shorter life span and males are presumably under greater selective pressure for locomotor efficiency because of the importance of this trait in their mate-finding strategies (Grossi et al. 2016a; Janowski-Bell and Horner 1999; Punzo and Henderson 1999; Shillington and Verrell 1997; Stoltey and Shillington 2009).

Several studies have addressed aspects of locomotor performance in tarantulas (see also Chaps. 12 and 13) and, two of these have focused specifically on differences between males and females (Grossi et al. 2016a; Shillington and Peterson 2002). There is also one study comparing treadmill running by male and female lycosids and salticids (Schmitz 2005). Both Grossi et al. (2016a) and Shillington and Peterson (2002) compared performance traits of male and female tarantulas running on a treadmill; variables measured included maximal aerobic speed (MAS), maximal CO_2 production, and minimum cost of transport (C_{min}). Maximal CO_2 rates occur at the MAS which is the maximum speed that can be sustained aerobically. C_{min} is the slope of the line that can be determined from the regression equation of speed and maximal CO_2 rates (Bennett 1982; Taylor et al. 1970; Gatten Jr et al. 1992; Full 1987). In their study, Shillington and Peterson (2002) reported on behavioral differences between males and females that may have influenced results during trials. Females were typically more resistant to movement and thus their maximal physiological capabilities may not have been reflected in the reported results. Although, males were much more willing to move, their measurements were recorded towards the end of the mating season and the authors suggest that the results did not accurately reflect locomotory abilities of males at their peak performance early in the mating season.

10.6.1 Metabolic Rates During Locomotion

Metabolic rates increase with increases in speed which is typical of most invertebrates (e.g., Full 1987, 1997; Herreid II 1981; Herreid II and Full 1984; Rogowitz and Chappell 2000) including spiders (Grossi et al. 2016a; Schmitz 2005; Shillington and Peterson 2002). To minimize anaerobic activity, recordings of speed and MRs were only recorded at speeds that were considered sustainable, but among the several tarantula studies, anaerobic contributions may explain some of the results. Two studies have specifically addressed differences between male and female tarantulas. Shillington and Peterson (2002) reported no intersexual differences in either maximal CO_2 rates or MAS between males and females *Aphonopelma anax*, but suggest these results may not accurately reflect performance because they were measured late in the mating season. In comparison, Grossi et al. (2016a) found that males had significantly higher MRs at all speeds; higher maximal CO_2 rates and higher MAS in *Grammostola rosea* (Table 10.2). Anderson and Prestwich (1985) measured MRs at set speeds of 72, 90 and 180 m/h (see Table 10.2 for comparison) which correspond to approximately 0.3, 0.4, and 0.8 body lengths/s. These speeds were chosen based on sustained running capabilities observed by the authors prior to the study. However, it is likely that the higher velocities resulted in increased anaerobic contributions and are not sustainable (Anderson and Prestwich 1985; Herreid II 1981;

Fig. 10.4 Female tarantula (*Aphonopelma anax*) running on a treadmill in the 2002 study by Shillington & Peterson

Shillington and Peterson 2002). During running on a treadmill (Fig. 10.4), several studies have reported that tarantulas most likely increase stride frequency (instead of stride length) to maintain the higher velocities (Anderson and Prestwich 1985; Biancardi et al. 2011; Booster et al. 2015) and recovery times after activity (i.e., time to return to resting levels of metabolism) may be long (Anderson and Prestwich 1985).Speed also increased approximately 2.5-fold from 15 to 40 °C. Interestingly, tarantulas do not show clear patterns in the spatiotemporal coordination of the legs with increasing velocities whereas studies of insect gait patterns show clear regularities (see Biancardi et al. 2011).

Factorial scopes (which represent an individual's ability to increase MR above RMR) have been measured in a variety of animals and are usually within a 5–15-fold range (mammals: Taylor et al. 1970; reptiles: Bennett 1982; crustaceans: Full 1987). Among spiders, the range is from 2–10 fold (Miyashita 1969; Peakall and Witt 1976; Ford 1977; Humphreys 1977; Prestwich 1983a; McQueen 1980; Anderson and Prestwich 1985). Within tarantulas specifically, there is a lot of variation (Seymour and Vinegar 1973; Shillington and Peterson 2002). Anderson and Prestwich (1982) suggest that the range for factorial scope in spiders may be limited in part by several physiological factors such as circulation, heart rate, ventilation, and respiratory surfaces. Additional studies are needed to elucidate factorial scope in tarantulas.

10.6.2 Minimum Cost of Transport (C_{min})

C_{min} is the slope of the line that can be determined from the regression equation relating maximal CO_2 production in the aerobic range (dependent variable) to speed (independent variable) and can be thought of as a measure of locomotory efficiency. C_{min} is often used in interspecies comparisons and is typically reported in mass-specific units of oxygen consumption (see Altmann 1987 and Shillington and Peterson 2002 for further discussion about the pros and cons of this convention). This variable is interesting for comparing the metabolic costs associated among animals with different body mass, as well as different numbers of legs (e.g., mammals and birds, Taylor and Heglund 1982; reptiles, Bennett 1982; insects Herreid II and Full 1984; Gatten Jr et al. 1992; crustaceans, Full 1987; spiders, Schmitz 2005). Among these diverse taxa, mass-specific C_{min} typically decreases with increasing body mass. Schmitz (2005) also compared C_{min} with intact and restricted respiratory organs (either lungs or tracheae) in species of lycosid and salticid. This study focused on active metabolism in well-tracheated (salticid) and poorly tracheated (lycosid) spiders and examined the role of the respiratory systems in aerobic and anaerobic activity. Specifically within tarantulas, there is some variation in resulting C_{min} calculations. Both Herreid (Herreid II 1981) and Anderson and Prestwich (1985) reported running tarantulas at high velocities and suggest that the low C_{min} (in mass-specific units) results may be due to contributions from anaerobic respiration. Shillington and Peterson (2002) recorded C_{min} for both male and female tarantulas that was very similar to values predicated using a C_{min}/mass equation determined for several insect taxa (Lighton 1985). This C_{min} was much lower than the previous two studies, but likely is a better representation of C_{min} at submaximal speeds with minimal anaerobic contributions (Shillington and Peterson 2002; Schmitz 2005). Shillington and Peterson (2002) reported no differences in C_{min} between male and female tarantulas but in a similar comparison, Grossi et al. (2016a) indicate that females use almost 40% more energy compared to males when traveling similar distances. In addition, in a comparison of leg length and body size (Grossi et al. 2016b) showed that male tarantulas have a significantly shorter opisthosma and longer legs compared to females of the same species (*Grammostola rosea*) and their results support the prediction of increased locomotory efficiency in male tarantulas (Grossi et al. 2016a, b; Shillington and Peterson 2002).

Separating the contributions of aerobic and anaerobic metabolism requires more study. While spiders in general tend to have low aerobic capabilities they may have well-developed anaerobic abilities (Linzen and Gallowitz 1975; Prestwich 1983a, b, 1988, 2006). Patterns of true CO_2 production during activity is complicated because some CO_2 is retained in the hemolymph for metabolic acidosis (Fincke and Paul 1989). Although Anderson and Prestwich (1985) measured lactate concentration from hemolymph samples before, during and after running activity, the results were very variable.

Within the realm of metabolic studies, RMR have received the most attention among tarantula groups. Once again, the diversity of species covered in these studies is very sparse and is limited to predominantly new world and terrestrial species. With the dearth of studies in this area it is difficult to make comparisons or find trends and many potentially interesting avenues of research remain to be explored. There has been little work done on the links between performance and morphology (see also Chaps. 12 and 13). With the sexual dimorphism evident in tarantulas as well as a range of sizes and life styles (e.g., arboreal versus terrestrial), understanding the links and constraints provides further insight into the evolution of behavior and physical capabilities (see McGinley et al. 2013). There has also been a rising interest in personality studies and integrating personality with performance traits and energetics (see Careau and Garland Jr 2012) but little has been done with tarantulas. Among the tarantulas there is also a wide range of life spans (e.g., Iber et al. 2013) and it is typically expected that individuals that grow faster and mature early also die younger and have higher rates of energy expenditure throughout their lives; tarantulas could be model organisms for such studies. Links between performance and morphology, or trade-offs amongst different performance dimensions.

10.7 Concluding Remarks

The literature related to studies of tarantula physiological ecology is surprisingly sparse considering that tarantulas are not uncommon and are large, dominant predators in a wide variety of habitats. In addition, there is a lack of diversity of species included across current publications. This chapter summarizes what we currently know about thermoregulation (including coldhardiness and freeze intolerance) and metabolism (including SDA and performance traits in tarantulas) and I hope that this stimulates further interest and research as there is still much to learn in relation to these topics. As a phylogenetically basal group of spiders, tarantulas are a successful group that can provide a strong evolutionary perspective for comparative studies with other spiders and arachnids. As ectothermic animals, temperature is exceedingly important in all aspects of their life history, physiology, and behavior and climate change is leading to sometimes large transformations in the thermal quality of available habitats across the world. This may in turn alter activity patterns and performance capabilities as well as availability of prey. To understand how these changes might impact tarantulas, we need a stronger understanding of their current capabilities and adaptations to various microclimates. Finally, tarantulas are found in incredibly diverse habitat providing an opportunity to study adaptations to both extreme heat and extreme cold.

Acknowledgments Thanks to Claudio Patricio Veloso Iriarte for sharing photos of *E. condorito* and for his comments on a draft of this chapter. Also to Fernando Pérez-Miles for the opportunity to contribute to a book on tarantulas. Finally, to Brian McEwen, for his help and support on many field trips.

References

Alfaro C, Figueroa DP, Torres H, Veloso C, Venegas F, Canals L, Canals M (2013) Effect of thermal acclimation on preferred temperatures in two mygalomorph spiders inhabiting contrasting habitats. Physiol Entomol 38:20–25

Altmann SA (1987) The impact of locomotor energetics on mammalian foraging. J Zool Soc Lond 211:215–225

Anderson JF (1970) Metabolic rates of spiders. Comp Biochem Physiol 33:51–72

Anderson JF (1996) Metabolic rates of resting salticid and thomosid spiders. J Arachnol 24:129–140

Anderson JF, Prestwich KN (1982) Respiratory gas exchange in spiders. Physiol Zool 55:72–90

Anderson JF, Prestwich KN (1985) The physiology of exercise at and above maximal aerobic capacity in a theraphosid (tarantula) spider, *Brachypelma smithi* (F.O. Pickard-Cambridge). J Comp Physiol B 155:529–539

Baerg WJ (1928) The life cycle and mating habits of the male tarantula. Q Rev Biol 3:109–116

Baerg WJ (1958) The tarantula. University of Kansas Press, Lawrence, KS

Baerg WJ (1963) Tarantula life history records. J N Y Entomol Soc 71:233–238

Bakken GS (1981) A two-dimensional operative-temperature model for thermal energy management by animals. J Therm Biol 6:23–30

Bakken GS (1992) Measurement and application of operative and standard operative temperatures in ecology. Am Zool 32:194–216

Beaupre SJ (1993) An ecological study of oxygen consumption in the mottled rock rattlesnake, *Crotalus lepidus lepidus*, and the black-tailed rattlesnake, *Crotalus molossus molossus*, from two populations. Physiol Zool 66:437–545

Beaupre SJ, Dunham AE, Overall KL (1993) Metabolism of a desert lizard: the effects of mass, sex, population of origin, temperature, time of day, and feeding on oxygen consumption of *Sceloporus merriami*. Physiol Zool 66:128–147

Bennett AF (1982) The energetics of reptilian activity. In: Gans C, Pough FH (eds) Biology of the reptilian. Academic, New York, pp 155–199

Bennett AF, Dawson WR (1976) Metabolism. In: Gans C, Pough FH (eds) Biology of the reptilian. Academic, New York, pp 127–224

Biancardi CM, Fabrica CG, Polero P, Fegundes Loss J, Minetti AE (2011) Biomechanics of octopedal locomotion: kinematic and kinetic analysis of the spider *Grammostola mollicoma*. J Exp Biol 214:3433–3442

Booster NA, Su FY, Adolph SC, Ahn AN (2015) Effect of temperature on leg kinematics in sprinting tarantulas (*Aphonopelma hentzi*): high speed may limit hydraulic joint actuation. J Exp Biol 218:977–982

Bradley R (1982) Digestion time and reemergence in the desert grassland scorpion *Paruroctonus utahensis*. Oecologia 55:316–318

Brian BL, Gaffney FG, Fitzpatrick LC, Scholes VE (1972) Fatty acid distribution of lipids from carcass, liver and fat bodies in the lizard *Cnemidophorus tigris*, before hibernation. Comp Biochem Physiol B 41:661–664

Canals M, Salazar MJ, Dúran C, Figueroa D, Veloso C (2008) Respiratory refinements in the mygalomorph spider *Grammostola rosea* Walckenaer 1837 (Araneae, Theraphosidae). J Arachnol 35:481–486

Canals M, Figueroa D, Alfaro C, Kawamoto T, Torres-Conteraras H, Sabat P, Veloso C (2011) Effects of diet and water supply on energy intake and water loss in a mygalomorph spider in a fluctuating environment of the Central Andes. J Insect Physiol 57:1489–1494

Careau V, Garland T Jr (2012) Performance, personality, and energetics: correlation, causation, and mechanism. Physiol Biochem Zool B 85:543–571

Carrel JE (1978) Behavioral thermoregulation during winter in an orb-weaving spider. Symp Zool Soc Lond 42:41–50

Casey TM (1981) Behavioral mechanisms of thermoregulation. In: Heinrich B (ed) Insect thermoregulation. Wiley, New York, pp 79–109

Carrel JE (1990) Water and hemolymph content in the wolf spider Lycosa ceratiola (Aranea, Lycosidae). Journal of Arachnology 18:35–40

Casey TM (1992) Biophysical ecology and heat exchange in insects. Am Zool 32:225–237

Chew RM (1961) Ecology of the spiders of a desert community. J N Y Entomol Soc 69:5–41

Cloudsley-Thompson JL (1991) Ecophysiology of desert arthropods and reptiles. Springer, Berlin

Cloudsley-Thompson JL (1993) Ecophysiology of desert Arachnida. Boll. Accad. Gioenia Sci. Nat. 26:53–63

Cubillos C, Cáceres JC, Villablanca C, Villarreal P, Baeza M, Cabrera R, Graether SP, Veloso C (2018) Cold tolerance mechanisms of two arthropods from the Andean range of Central Chile: *Agathemera crassa* (Insecta: Agathemeridae) and *Euathlus condorito* (Arachnida: Theraphosidae). J Therm Biol 74:133–139

Cullum AJ (1998) Sexual dimorphism in physiological performance of whiptail lizards (genus *Cnemidophorus*). Physiol Zool 71:54–552

Daan S, Masman D, Groenewold A (1990) Avian metabolic rates: their association with body composition in nature. Am J Phys 259:R333–R340

Davies ME, Edney EB (1952) The evaporation of water from spiders. J Exp Biol 29:571–582

Duman JG (1979) Subzero temperature tolerance in spiders: the role of thermal-hysteresis-factors. J Comp Physiol B 352:347–352

Figueroa DP, Sabat P, Torres-Contreras H, Veloso C, Canals M (2010) Participation of book lungs in evaporative water loss in *Paraphysa parvula,* a mygalopmorph spider from Chilean Andes. J Insect Physiol 56:731–735

Fincke T, Paul R (1989) Book lung function in arachnids III. The function and control of spiracles. Comp Biochem Physiol B 159:433–441

Foelix RF (1996) Biology of spiders. Oxford University Press, Oxford

Ford MJ (1977) Metabolic costs of the predation strategy of the spider *Pardosa amentata* (Clerck) (Lycosidae). Oecologia 28:333–340

Full RJ (1987) Locomotion energetics of the ghost crab I. Metabolic cost and endurance. J Exp Biol 130:137–153

Full RJ (1997) Invertebrate locomotor systems. In Handbook of Physiology, section 13: Comparative Physiology (ed. W. H. Dantzler) New York: Oxford University Press, pp. 853–930

Gatten RE Jr, Miller K, Full RJ (1992) Energetics at rest and during locomotion. In: Feder ME, Burggren WW (eds) Environmental physiology of the amphibians. University of Chicago Press, Chicago, pp 314–377

Greenstone MH, Bennett AF (1980) Foraging strategy and metabolic rate in spiders. Ecology 61:1255–1259

Grossi B, Solis R, Veloso C, Canals M (2016a) Consequences of sexual size dimorphism on energetics and locomotor performance of *Grammostola rosea* (Araneae; Theraphosidae). Physiol Entomol 41:281–288

Grossi B, Veloso C, Taucare-Ríos A, Canals M (2016b) Allometry of locomotor organs and sexual size dimorphism in the mygalomorph spider *Grammostola rosea* (Walckenaer, 1837) (Araneae, Theraphosidae). J Arachnol 44:99–102

Hammerstein P, Reichert SE (1988) Payoffs and strategies in territorial contests: ESS analyses of two ecotypes of spider *Agelenopsis aperta*. Evol Ecol 2:115–138

Herreid CF II (1981) Energetics of pedestrian arthropods. In: Herreid CF II, Fourtner CR (eds) Locomotion and energetics in arthropods. Plenum Press, New York, pp 491–526

Herreid CF II, Full RJ (1984) Cockroaches on a treadmill: aerobic running. J Insect Physiol 30:395–403

Hertz PE, Huey RB, Stevenson RD (1993) Evaluating temperature regulation by field-active ectotherms: the fallacy of the inappropriate question. Am Nat 145:796–818

Higgins LE, Ezcurra E (1996) A mathematical simulation of thermoregulatory behaviour in an orb-weaving spider. Funct Ecol 10:322–327

Huey RB (1991) Physiological consequences of habitat selection. Am Nat 137:91–115

Humphreys WF (1974) Behavioural thermoregulation in a wolf spider. Nature 251:502–503

Humphreys WF (1977) Respiration studies on *Geolycosa godeffroyi* (Araneae: Lycosidae) and their relationship to field estimates of metabolic heat loss. Comp Biochem Physiol A 57:555–563
Humphreys WF (1978) The thermal biology of *Geolycosa godeffroyi* and other burrow inhabiting Lycosidae (Araneae) in Australia. Oecologia 31:319–347
Iber B, Michalik P, Fischer K (2013) Factors affecting lifespan in bird-eating spiders (Arachnida: Mygalmorphae, Theraphosidae) – a multi-species approach. Zool Anz 253:126–136
Janowski-Bell ME, Horner NV (1999) Movement of the male brown tarantula, *Aphonopelma hentzi* (Araneae, Theraphosidae), using radio telemetry. J Arachnol 27:503–512
Jensen KD, Mayntz T, Wang SJ, Simpson J, Overgaard J (2010) Metabolic consequences of feeding and fasting on nutritionally different diets in the wolf spider *Pardosa prativaga*. J Insect Physiol 56:1095–1100
Kingsolver JG (1989) Weather and the population dynamics of insects: integrating physiological and population ecology. Physiol Zool 62:314–334
Koteja P (1991) On the relation between basal and field metabolic rates in birds and mammals. Funct Ecol 5:56–64
Kotiaho JS (1998) Sexual differences in metabolic rates of spiders. J Arachnol 26:401–402
Le Galliard J, Paquet M, Cisel M, Montes-Poloni L (2013) Personality and the pace-of-life syndrome: variation and selection on exploration, metabolism and locomotor performances. Funct Ecol 27:136–144
Lee RE (2010) A primer on insect cold-tolerance. In: Lee RE, Denlinger DL (eds) Low temperature biology of insects. Cambridge University Press, New York, pp 3–34
Lighton JR (1985) Minimum cost of transport and ventilatory patterns in three African beetles. Physiol Zool 58:390–399
Lighton J, Fielden LJ (1995) Mass scaling of standard metabolism in ticks: a valid case of low metabolic rates in sit-and wait strategists. Physiol Zool 68:43–62
Lighton JRB, Brownell PH, Joos B, Turner RJ (2001) Low metabolic rate in scorpions: implications for population biomass and cannibalism. J Exp Biol 204:607–613
Linzen B, Gallowitz P (1975) Enzyme activity patterns in the muscles of the lycosid spider *Cupennius salei*. J Comp Physiol 96:101–109
Lubin YD, Henshcel JR (1990) Foraging at the thermal limit: burrowing spiders (*Seothyra Eresidae*) in the Namib desert dunes. Oecologia 84:461–467
McGinley RH, Prenter J, Taylor PW (2013) Whole-organism performance in a jumping spider, *Servaea incana* (Araneae: Salticidae): links with morphology and between performance traits. Biol J Linn Soc:110644–110657
McQueen DJ (1980) Active respiration rates for the burrowing wolf spider *Geolycosa domifex* (Hancock). Can J Zool 58:1066–10794
Minch EW (1978) Daily activity patterns in the tarantula *Aphonopelma chalcodes* Chamberlin. Bull Br Arachnol Soc 4:231–237
Miyashita K (1969) Effects of locomotory activity, temperature and hunger on respiratory rates of Lycosa T-insignita (Araneaa: Lycoslidae). Appl Entomol Zool 4:105–113
Nespolo RF, Franco M (2007) Whole-animal metabolic rate is a repeatable trait: a meta-analysis. J Exp Biol 210:2000–2005
Nespolo RF, Correa L, Pérez-Apablaza CX, Cortés P, Bartheld JL (2011) Energy metabolism and the postprandial response of the Chilean tarantuals, *Euathlus truculentus* (Araneae: Theraphosidae). Comp Biochem Physiol A 159:379–382
Packard GC, Boardman TJ (1988) The misuse of ratios, indices, and percentages in ecophysiological research. Physiol Zool 61:1–9
Paul R, Fincke T, Linzen B (1989) Book lung function in arachnids. J Comp Physiol 159:409–418
Peakall DB, Witt PN (1976) The energy budget of an orb web-building spider. Comp Biochem Physiol 54A:187–190
Perafan C, Perez-Miles F (2014) The Andea tarantulas Euathlus Auuserer, 1875, Paraphysa Simon, 1892 and Phrixothrichus Sio, 1889 (Araneae: Theraphosidae): phylogenetic analysis, genera redefinition and new species descriptions. J Nat Hist 48: 39–40

Perez-Miles F, Costa FG, Toscano-Gadea C, Mignone A (2005) Ecology and behavior of the "road tarantulas" *Eupalaestrus weijenberghi* and *Acanthoscurria suina* (Araneae, Therphosidae) from Uruguay. J Nat Hist 39:483–498

Peterson CR, Gibson AR, Dorcas ME (1993) Snake thermal ecology: the causes and consequences of body-temperature variation. In: Seigel RA, Collins JT (eds) Snake ecology and behavior. McGraw-Hill, New York, pp 241–314

Philip BN, Shillington C (2010) The effect of prey availability on metabolism and activity in the tarantula *Phormictopus cancerides*. Can J Zool 88:90–98

Porter WP, Gates DM (1969) Thermodynamic equilibria of animals with environment. Ecol Monogr 29:227–244

Prentice TR (1992) A new species of North American tarantula, *Aphonopelma paloma* (Araneae, Mygalomorphae, Theraphosidae). J Arachnol 20:189–199

Prestwich KN (1983a) The roles of aerobic and anaerobic metabolism in active spiders. Physiol Zool 56:122–132

Prestwich KN (1983b) Anaerobic metabolism in spiders. Physiol Zool 56:112–121

Prestwich KN (1988) The constraints on maximal activity in spiders. J Comp Physiol B 158:437–447

Prestwich KN (2006) Aerobic metabolism and maximal running in the scorpion *Centruroides henti* (Banks) (Scorpiones, Buthidae). J Arachnol 34:351–356

Pulz R (1987) Thermal and water regulations. In: Nentwig W (ed) Ecophysiology of spiders. Springer, Berlin, pp 26–55

Punzo F (1991) Intraspecific variation in responses to thermal stress in the tarantula, *Dugesiella echina* Chamberlin (Orthognatha: Theraphosidae). Bull Br Arachnol Soc 8:277–283

Punzo F, Henderson L (1999) Aspects of the natural history and behavioral ecology of the tarantula, *Aphonopelma hentzi* (Girard 1854) (Orthognatha: Theraphosidae). Bull Br Arachnol Soc 11:121–128

Punzo F, Jellies J (1983) Comparative water relations of araneid and lycosid spiderlings. Comp Biochem Physiol 74A:981–988

Ramirez MG, Wall EA, Medina M (2003) Web orientations of the banded garden spider *Argiope trifasciata* (Araneae, Araneidae) in a California costal population. J Arachnol 31:405–411

Ramløv H (2000) Aspects of natural cold tolerance in ectothermic animals. Hum Reprod 5:26–46

Reichert SE (1981) Games spiders play. II. Resource assessment strategies. Behav Ecol Sociobiol 6:121–128

Reichert SE (1985) Decisions in multiple goal contexts: habitat selection of the spider, *Agelenopsis aperata* (Gertsch). Zietschr Tierpsychol 70:53–69

Reichert SE, Tracy CR (1975) Thermal balance and prey availability: bases for a model relating web-site characteristics to spider reproductive success. Ecology 56:265–284

Reinhold K (1999) Energetically costly behavior and the evolution of resting metabolic rates in insects. Funct Ecol 13:217–224

Ricklefs RE, Miles DB (1994) Ecological and evolutionary inferences from morphology: an ecological perspective. In: Wainwright PC, Reilly SM (eds) Ecological morphology: integrative organismal biology. University of Chicago Press, Chicago, pp 13–41

Rittschof CC (2012) The effects of temperature on egg development and web site selection in *Nephila clavipes*. J Arachnol 40:141–145

Robinson MH, Robinson B (1978) Thermoregulation in orb-web spiders: new descriptions of thermoregulatory postures and experiments on the effect of posture and coloration. Zool J Linnean Soc 64:87–102

Rogowitz GL, Chappell MA (2000) Energy metabolism of eucalyptus-boring beetles at rest and during locomotion: gender makes a difference. J Exp Biol 203:1131–1139

Schartau W, Leidescher T (1983) Composition of the hemolymph of the tarantula *Eurypelma californicum*. J Comp Physiol Biochem 152B:72–77

Schmitz A (2005) Spiders on a treadmill: influence of running activity on metabolic rates in *Pardosa lugubris* (Araneae, Lycosidae) and *Marpissa muscosa* (Araneae, Saticidae). J Exp Biol 208:1401–1411

Secor SM (2009) Specific dynamic action: a review of the postprandial metabolic response. J Comp Physiol B 179:1–56

Seymour RS, Vinegar A (1973) Thermal regulations, water loss and oxygen consumption of a North American tarantula. Comp Biochem Physiol A 44:83–96

Shillington C (2002) Thermal ecology of male tarantulas (*Aphonopelma anax*) during the mating season. Can J Zool 80:251–259

Shillington C (2005) Inter-sexual differences in resting metabolic rates in the Texas tarantula, *Aphonopelma anax*. Comp Biochem Physiol A 142:439–445

Shillington C, Peterson CC (2002) Energy metabolism of male and female tarantulas (*Aphonopelma anax*) during locomotion. J Exp Biol 205:2909–2914

Shillington C, Verrell P (1997) Sexual strategies of a North American 'tarantula' (Araneae: Theraphosidae). Ethology 103:588–598

Stoltey T, Shillington C (2009) Metabolic rates and movements of the male tarantula *Aphonopelma anax* during the mating season. Can J Zool 87:1210–1220

Storey KB, Storey JM (2013) Molecular biology of freezing tolerance. Compr Physiol 3:1283–1308

Stradling DJ (1994) Distribution and behavioral ecology of an arboreal "tarantula" spider of Trinidad. Biotropica 26:84–97

Tanaka K, Ito Y (1982) Decrease in respiratory rate in a wolf spider, *Pardosa astigera* (L. Koch), under starvation. Res Popul Ecol 24:360–374

Taylor CR, Heglund NC (1982) Energetics and mechanics of terrestrial locomotion. Annu Rev Physiol 44:97–107

Taylor CR, Schmidt-Nielsen K, Raab J (1970) Scaling of energetic cost of running to body size in mammals. Am J Phys 219:1104–1107

Terblanch JS, Klock CJ, Marais E, Chown SL (2004) Metabolic rate in the whip-spider, *Damon annulatipes* (Arachnida: Amblypygi). J Insect Physiol 50:637–645

Thornhill R, Alcock J (1983) The evolution of insect mating systems. Harvard University Press, Cambridge

Veloso C, Luhr D, Marfull R, Torres-Contreras H, Pérez DF, Sabat P, Canals M (2012) Characterization of the thermal micro-environment of *Paraphysa parvula* Pocock 1903 (Araneae: Theraphosidae), a spider from the Chilean Andes. J Arachnolog 40:34–38

Walker SE, Irwin JT (2006) Sexual dimorphism in metabolic rate of two species of wolf spider (Araneae, Lycosidae). J Arachnol 34:368–373

Watson PJ, Lighton JR (1994) Sexual selection and the energetics of copulatory courtship in the Sierra dome spider, *Linyphia litigiosa*. Anim Behav 48:615–626

Young SR, Block W (1980) Some factors affecting metabolic rate in an Antarctic mite. Oikos 74:178–185

Zachariah TT, Mitchell MA, Guichard CM, Singh RS (2007) Hemolymph biochemistry reference ranges for wild-caught Goliath birdeater spiders (*Theraphosa blondi*) and Chilean rose spiders (*Grammostola rosea*). J Zoo Wildl Med 38:245–251

Chapter 11
Cuticular Structures of New World Tarantulas: Ultramorphology of Setae and Other Features

José Paulo Leite Guadanucci, Arthur Galleti-Lima, and Rafael Prezzi Indicatti

Abstract Studying morphology of Theraphosidae spiders can be very challenging, especially if the main objective is assembling characters for systematics. Such spiders present a homogeneous morphology, which, according to some specialists, has driven the attention of systematists to other groups of Araneae. Nevertheless, a great diversity of cuticular structures has been overlooked until the widespread use of scanning electron microscopy (SEM) in the last years for theraphosids. Among all mygalomorphs, Theraphosidae spiders possess the greatest variety of cuticular features. Data regarding cuticular features are still incipient, but we have been gathering massive quantity of SEM images of all parts of the spider body, revealing interesting structures to be used in systematics and investigated for functional morphology. In addition to the well-known tarsal adhesive setae of theraphosids and the urticating setae of Theraphosinae, we found putative chemosensitive setae, a great variety of stridulating setae, distinct morphologies of leg and palpal structures, including cuticular projections, labial and maxillary cuspules, trichobothria, as well as other enigmatic features. In this chapter, we aim to present a comprehensive revision of cuticular features of New World Theraphosidae spiders, with descriptions and micrographs.

11.1 Introduction

Spiders, as a member of the Phylum Arthropoda, have their whole body embraced by a hard body shell, named exoskeleton. The exoskeleton is made of the cuticle, a stiff material composed of lipids and waxes, glycosylated and unglycosylates proteins, the polysaccharide chitin and catecholamines. The cuticle provides protection against both desiccation and predators, maintains the shape and structure of the

J. P. L. Guadanucci (✉) · A. Galleti-Lima · R. P. Indicatti
Departamento de Biodiversidade, Instituto de Biociências, Universidade Estadual Paulista, Rio Claro, SP, Brazil

F. Pérez-Miles (ed.), *New World Tarantulas*, Zoological Monographs 6,
https://doi.org/10.1007/978-3-030-48644-0_11

body, allowing locomotion (Moussian 2013). Spiders and insects have cuticles of similar composition (Foelix 1996), and differ from the crustaceans as they may incorporate calcite (Moussian 2013). Spiders also have a mesocuticle, which is absent in insects (Foelix 1996). Rather than a smooth layer, although its lamellate organization, the cuticle bears a great diversity of outgrowths and textures with many distinct functions, e.g., sensitive, adherent and covering setae, bristles, scales, spines, claws, stridulatory ridges and lyrae, among many others.

Tarantulas (Theraphosidae) are notoriously "hairy" spiders, especially when compared to trapdoor or sheet-web mygalomorphs (Fig. 11.1a–d). Not only do they possess greater quantity of setae on their bodies, but they also bear greater diversity of setal forms (Pérez-Miles et al. 2017), providing them with adhesive bristles (adhesive scopula, claw tufts), structural color patterns (Hsiung et al. 2015), urticating setae (Bertani and Guadanucci 2013), stridulating apparatus (Galleti-Lima and Guadanucci 2018), mixed (adhesive and chemosensitive) scopula of males (Pérez-Miles et al. 2017), clavate trichobothria (Guadanucci 2012), tibial apophysis, labial cuspules (Pérez-Miles and Montes de Oca 2005), spigots and epiandrous glands (Ferretti et al. 2017) and other unknown and/or unnamed structures. Recent advances, as in the references cited above, have shown a great diversity of structures, which could only be accessed with scanning electron microscopy, revealing designs and textures not visible under optical microscope.

Fig. 11.1 Habitus, live specimens. Tarantulas and trapdoor spider habitus. (**a**) *Actinopus rufipes*, female. (**b**) *Idiops* sp., male. (**c**) *Homoeomma brasilianum*, female. (**d**) *Dolichothele mottai*, female

Theraphosidae systematics has been greatly influenced by the presence of a stridulatory apparatus since the first records of these setae by Simon (1892) and Pocock (1895). These apparatuses are usually composed by groups of setae (tuft or bristle) inserted on opposite faces (prolateral/retrolateral) of proximal leg and palp articles, such as coxae, trochanters, and femora. The current classification of Theraphosidae at subfamily level greatly relies on the presence, position, and ultrastructure of these setae (e.g., plumose setae, lyra, pegs, thorns). Regarding the Neotropical tarantulas, especially the Theraphosinae, their stridulatory setae have been extensively used as taxonomic characters at generic level, e.g., Ferretti et al. (2011), Bertani et al. (2008), Galleti-Lima and Guadanucci (2018), Gargiulo et al. (2018). To date, different types of stridulating setae have already been found in several theraphosine genera (Figs. 11.2, 11.3, and 11.4) (Pérez-Miles et al. 1996; Galleti-Lima and Guadanucci 2018; pers. obs.). Beyond Theraphosinae, the New World tarantulas belonging to the aviculariinae genus *Psalmopoeus* Pocock, 1895 possess a lyra on prolateral maxillae (West et al. 2008; Fukushima and Bertani 2017), which can be rubbed against a few stiff setae on prolateral chelicerae, acting as a stridulating apparatus.

Theraphosinae and Avicularinae spiders are also notorious for their urticating setae (Chap. 9), a specific detachable kind of seta from which the urticarious reaction has been reported since the naturalists of the nineteenth century (Cooke et al. 1972; Pérez-Miles 2002; Bertani and Guadanucci 2013). Starting with the morphological comprehensive survey by Cooke et al. (1972), a few studies were carried out regarding their morphology and systematics (Marshall and Uetz 1990a; Pérez-Miles 1998; Perafán and Perez-Miles 2010; Bertani and Guadanucci 2013) and usage and behavior (Marshall and Uetz 1990b; Pérez-Miles and Prandi 1991; Pérez-Miles and Costa 1994; Bertani and Marques 1995; Bertani et al. 2003). The two Neotropical endemics Theraphosinae and Aviculariinae are the only spiders that possess these urticating setae, which are lacking in the remaining theraphosids. While in the abdomen of theraphosines one may find types I, III, IV, and VI, Aviculariinae spiders possess type II (abdomen) and type V (palpal femora of *Ephebopus* Simon, 1892). Spiders of these two subfamilies also differ on the way they use the urticating setae: the theraphosines frequently release them by rubbing their hind legs against the abdominal setae patch when disturbed. Their spinose tibia and metatarsi III and IV may have an important role in the mechanism of setae releasing. Type II urticating setae are transferred by contact (except for *Avicularia versicolor,* see Bertani et al. 2003) or by scratching the palpal femur against the chelicera, in *Ephebopus*. The leg spines are numerous in Theraphosinae spiders, and limited to just a few (up to three) apical ones on posterior tibia and metatarsi or completely absent in Aviculariinae. The remaining New World tarantulas, belonging to the subfamilies Schismatothelinae and Ischnocolinae sensu strictu (Guadanucci 2014) lack any kind of urticating setae and have moderate number of leg spines, especially on legs III and IV.

Apart from stridulatory and urticating setae of New World tarantulas, few other cuticular structures received attention and were studied in a comparative framework at the ultramorphological level, and will be dealt with in more detail further below. Trichobothria of Theraphosidae and Barychelidae were described in their

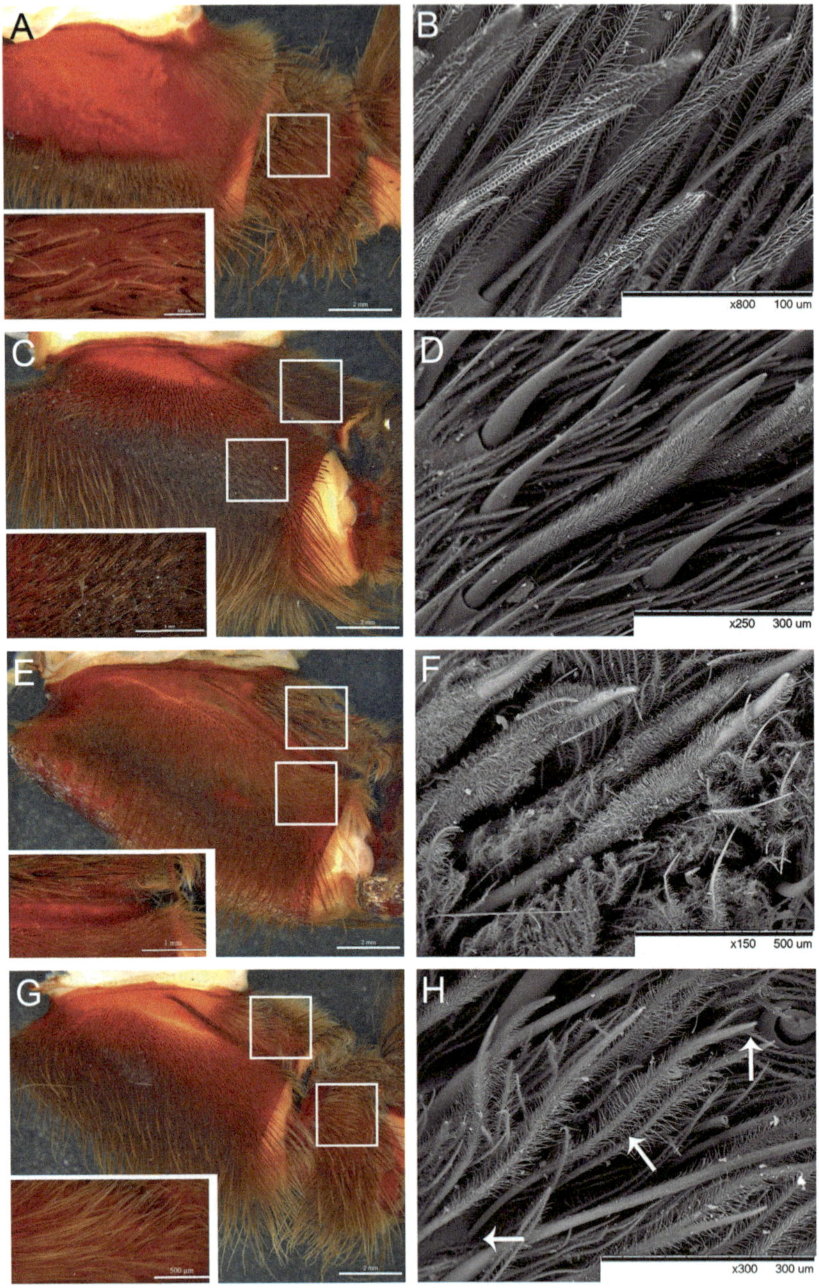

Fig. 11.2 Stridulating setae. (**a**) Retrolateral view of coxae and trochanter of palp in *Acanthoscurria gomesiana*, detail showing claviform stridulating setae. (**b**) Claviform stridulating setae, SEM micrography. (**c**) Prolateral view of coxae I in *Grammostola* sp., detail showing rough stridulating setae. (**d**) Rough stridulating setae, SEM micrography. (**e**) Prolateral view of coxae I in *Lasiodora* sp., detail showing velvet stridulating setae. (**f**) Velvet stridulating setae, SEM micrography. (**g**) Prolateral view of coxae and trochanter I in *Vitalius* sp., detail showing plumose stridulating setae. (**h**) Plumose stridulating setae, SEM micrography

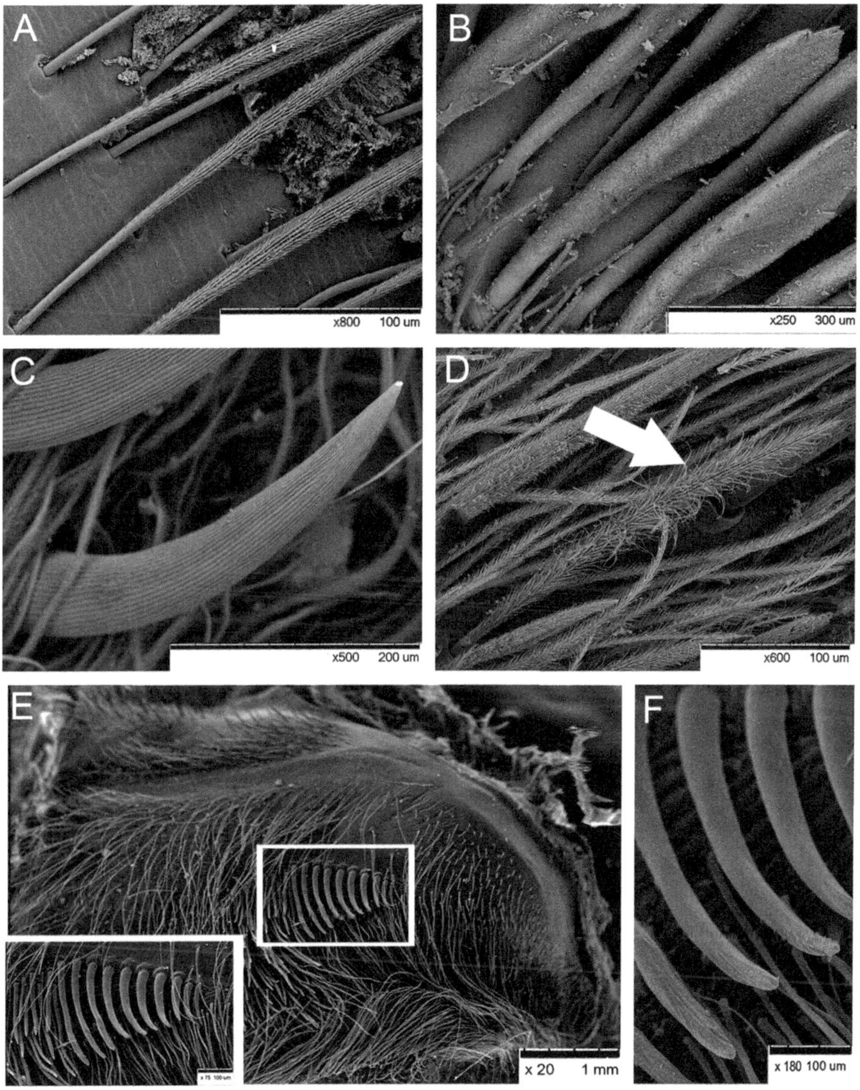

Fig. 11.3 Stridulating setae (cont.). (**a**–**f**) SEM micrographs of distinct types of stridulating setae. (**a**) Sectioned stridulating setae. (**b**) Spatuliform stridulating setae. (**c**) Spiniform stridulating setae. (**d**) Pilose stridulating setae (arrow). (**e**) Prolateral view of maxilla in *Psalmopoeus* sp., showing the maxillary lyra in the detailed square. (**f**) Maxillary lyra, detail of setae, in *Psalmopoeus* sp.

ultrastructure and disposition by Guadanucci (2012); Pérez-Miles et al. (2017) presented data on adhesive tarsal setae (scopula and claw tufts) and recognized the mixed scopula of Theraphosidae males, composed of chemosensitive and adhesive setae; labial and maxillary cuspules are conspicuous cuticular structures shared by Theraphosidae and many other mygalomorph families (except Mecicobothriidae,

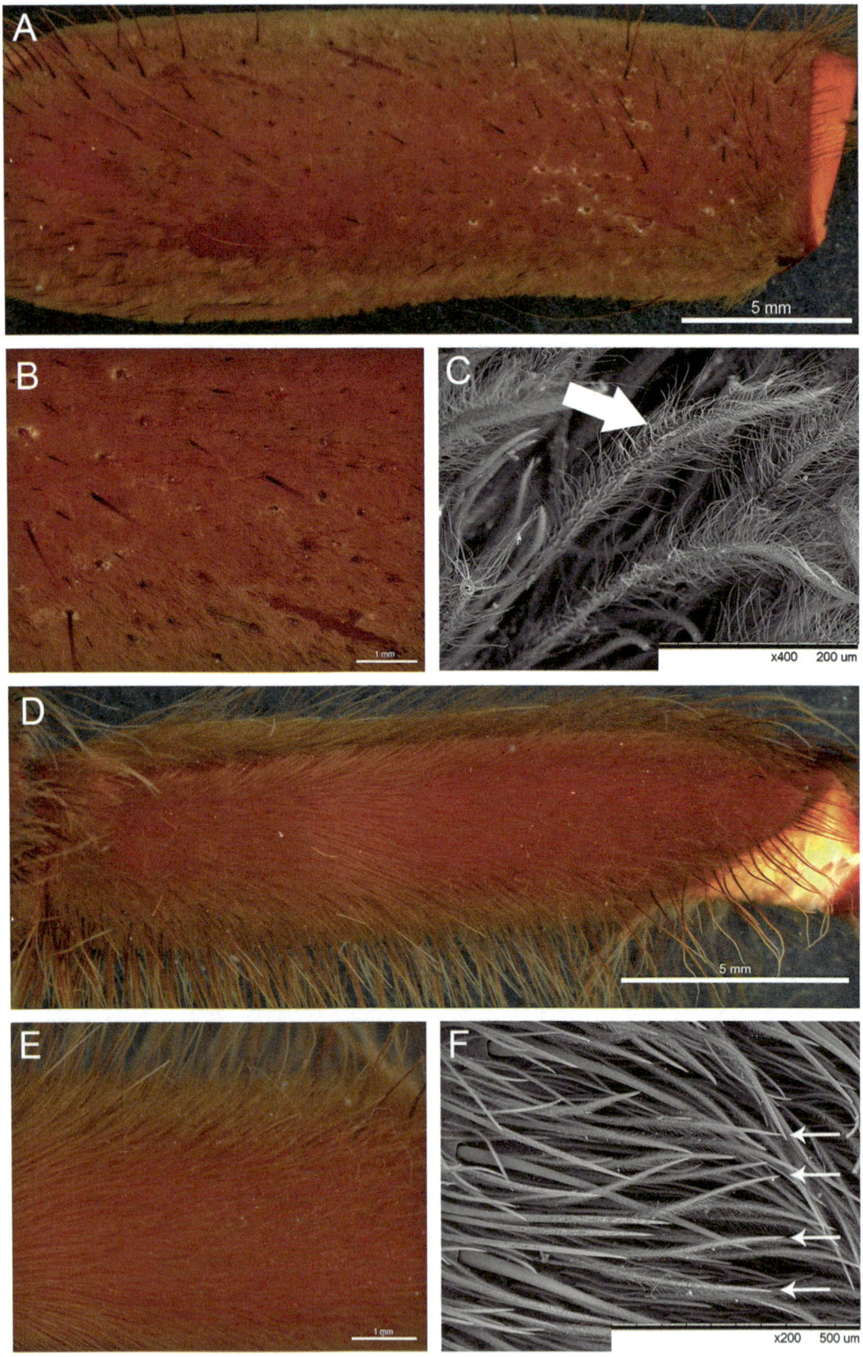

Fig. 11.4 Femoral scopulae. (**a**) Retrolateral view of femur IV in *Theraphosa blondi*. (**b**) Femoral scopula of leg IV in *Theraphosa blondi*. (**c**) Claviform stridulating setae (arrow) of femoral scopulae of leg IV, SEM micrography. (**d**) Retrolateral view of femur IV in *Vitalius* sp. (**e**) Femoral scopulae of leg IV in *Vitalius* sp. (**f**) Plumose stridulating setae (arrows) of femoral scopulae of leg IV, SEM micrography

Hexurellidae, Megahexuridae and Antrodietidae), and they were studied in a few number of theraphosine and aviculariine genera by Pérez-Miles and Montes de Oca (2005). Ferretti et al. (2017) published the first comparative study of the epiandrous apparatus for mygalomorph spiders, which comprises a group of additional silk glands (epiandrous spigots) located at the anterior margin of the genital furrow of males, and included a few New World tarantula spiders; Hsiung et al. (2015) recognized a diversity of nanostructural mechanisms that produces blue reflectance in tarantulas. Except from the ones cited above, no other cuticular feature has received much attention from arachnologists in the last few years, except by some particular taxonomic works for the purpose of morphological description. Recent publications and unpublished surveys have yielded great diversity of cuticular structures that were examined under the SEM, where magnifications above 500× can reveal conspicuous differences in well-known structures (e.g., stridulating setae, tufts, leg scopula, chemosensory setae) and other newly discovered structures for which, at the moment, we can only speculate about their functions.

11.1.1 Stridulating Setae*: Coxae, Trochanters, Femora (Figs. 11.2, 11.3, and 11.4)*

Stridulating apparatus are fairly well known in spiders. A single spider individual (Fam. Zodariidae) has been reported with six stridulating apparatus, according to Jocqué (2005), who also listed many types of stridulatory organs described at least for 22 families of spiders. The production of a hissing sound by some Theraphosidae spiders is well reported (Legendre 1963; Uetz and Stratton 1982; Marshall et al. 1995) and very well known among tarantula hobbyists. Although the detailed mechanism of stridulating has never been elucidated, different authors (Pérez-Miles et al. 2005; Bertani et al. 2008; Galleti-Lima and Guadanucci 2018) agree that the possession of stiff setae, spines, ridges, or pegs on opposite surfaces comprises the apparatus behind the production of sound, caused by the attrition between such structures (Uetz and Stratton 1982). In the case of New World Tarantulas, the stridulating apparatus can be located on distinct proximal palp and legs articles, such as coxae, trochanters and femora, and they are usually composed by a group of setae, with long or short barbs (Figs. 11.2a–h and 11.3a, b) (most genera), short spines (Fig. 11.3c) (*Theraphosa* and *Pamphobeteus*), or peg-like setae forming the lyra (*Psalmopoeus*) (Fig. 11.3e, f).

The few authors that proposed relationship hypothesis among Theraphosinae spider genera in the last years have used the presence of a stridulating apparatus as a systematic character (Pérez-Miles et al. 1996; Bertani 2001; Perafán and Perez-Miles 2014; Bertani et al. 2011). The presence of such a stridulatory apparatus appears as apomorphic for some genera (e.g., *Lasiodora*, *Grammostola*, *Acanthoscurria*, *Cyrtopholis*, *Theraphosa*) and they were all considered of the same morphological type and treated as same character state by the authors above, differing only on where they are located (coxae and/or trochanters). More recent papers have verified the presence of new types of stridulatory setae, some of these setae types are present on spiders for which they were unreported before (although a stridulating apparatus was already known): *Hemirrhagus* (Marroquín 2014), *Pamphobeteus* (Bertani et al. 2008), *Longilyra* (Gabriel 2014), *Aguapanela* (Perafán et al. 2015). Not until the comparative survey on the ultrastructure of the setae on the coxae and trochanters of Theraphosinae (Galleti-Lima and Guadanucci 2018) did we know about the diversity of forms of stridulatory setae. Interestingly, Bücherl (1957) anticipated this diversity, recognized and illustrated differences in the setae of *Acanthocurria*, *Grammostola*, and *Lasiodora*. According to Galleti-Lima and Guadanucci (2018), and also our unpublished data, the diversity of stridulatory setae found on coxae and trochanters of Theraphosidae spiders fits into the following nine types: claviform (Fig. 11.2b), rough (Fig. 11.2d), velvet (Fig. 11.2f), plumose (Fig. 11.2h), sectioned (Fig. 11.3a), spatuliform (Fig. 11.3b), spiniform (Fig. 11.3c), pilose (Fig. 11.3d), and liriform (see Gabriel 2014: Figs. 1–3). Among the Theraphosinae, spiders that have been reported with stridulatory setae on coxae and/or trochanters are listed on Table 11.1.

Table 11.1 Types and location of stridulating setae found on coxa and/or trochanter on Theraphosinae genera

Genera	Type of stridulating setae	Location
Acanthoscurria	Claviform (Fig. 11.2b)	Trochanter
Aguapanela	Plumose (Fig. 11.2h)	Coxa + trochanter
Aphonopelma	Sectioned (Fig. 11.3a)	Coxa + trochanter
Brachypelma	Claviform	Coxa + trochanter
Citharacanthus	Sectioned	Trochanter
Cotztetlana	Spiniform (Fig. 11.3c)	Coxa + trochanter
Crassicrus	Plumose	Trochanter
Cyrtopholis	Claviform	Trochanter
Grammostola	Rough (Fig. 11.2d)	Coxa
Hemirrhagus[a]	Sectioned/spatuliform (Fig. 11.3b)/pilose (Fig. 11.3d)	Coxa + trochanter
Lasiodora	Plumose + velvet (Fig. 11.2f)	Coxa + trochanter
Longilyra	liriform (Gabriel 2014: Figs. 1–3)	Trochanter
Megaphobema	Plumose	Coxa + trochanter
Nesipelma	Claviform	Trochanter
Nhandu	Plumose	Coxa + trochanter
Pamphobeteus	Spiniform	Coxa
Phormictopus	Claviform	Coxa + trochanter
Proshapalopus	Plumose	Coxa + trochanter
Pterinopelma	Plumose	Coxa + trochanter
Theraphosa	Claviform + spiniform	Coxa + trochanter
Umbyquyra	Claviform	Trochanter
Vitalius	Plumose	Coxa + trochanter

[a]Spiders of the genus *Hemirrhagus* have been recorded with distinct types in different species

Nevertheless, Theraphosinae is the most diverse Neotropical group of mygalomorphs, and there is still a great number of spiders to have their coxae and trochanter examined under the SEM, thus reaching enough magnifications to detect these morphological attributes. We are far from a consensus about Theraphosinae relationships and, consequently, far from understanding the evolution of these setae. Few papers suggested the use of stridulation in sexual and defensive behaviors (Legendre 1963; Uetz and Stratton 1982; Marshall et al. 1995). In fact, sexual courtship of theraphosids (and of mygalomorphs in general, see Ferretti et al. 2013) is strongly based on mechanical signals, with legs and body vibrations performed by both males and females. These vibrations, which sometimes are associated with a hissing audible sound, together with those of defensive behaviors, have an important adaptive value, and thus supporting the hypothesis of multiple acquisitions across Theraphosinae (Pérez-Miles et al. 1996; Bertani 2001; Bertani et al. 2008; Galleti-Lima and Guadanucci 2018). The various types of stridulating setae support the hypothesis of multiple acquisitions, following the proposal of Galleti-Lima and Guadanucci (2018) that stridulating setae are modifications of regular body seta. Apart from the Theraphosinae, the only New World theraphosid bearing stridulatory apparatus are spiders of the genus *Psalmopoeus*, showing a conspicuous lyra on the prolateral surface of the maxillae (Fig. 11.3e, f). These peg-like structures that form the lyra are curved, with very short barbs on the apex (Fig. 11.3f—detail), and disposed as a series of parallel setae of increasing length (from basal to apical). For many years, *Psalmopoeus* was included in the Selenocosmiinae, due to the presence of these clavate setae (bacillae) on the prolateral maxillary surface (Raven 1985; West et al. 2012), and thus comprising the sole representative of this subfamily in the Neotropics. *Psalmopoeus* is currently placed within Aviculariinae, and its stridulatory apparatus differ from that of selenocosmiines by having fewer setae (from 9 to 14), while the lyra of selenocosmiines is an oval patch of many setae [obs: the whole organ, composed by the group of setae on prolateral maxillae, has been named lyra by some authors (Raven 1985; West et al. 2012; Mendoza-Marroquín 2014), contrary to Gabriel and Sherwood (2018) that named each setae as lyra.].

The scopula on the retrolateral surface of femur IV (Fig. 11.4a–f) has also been given great weight in Theraphosinae phylogenetics (Raven 1985; Pérez-Miles et al. 1996; Bertani 2001) and its presence is a good taxonomic character for genera diagnosing (pers. obs.). These authors cited above did not draw any attention concerning the presence of a scopula on femora I–III. The femoral scopula is very similar to that on tarsi: it has a rug-like appearance, occupying the whole lateral (pro and/or retrolateral) surface, and it can be intermixed with long setae. Perafán et al. (2016) detected the femoral scopula in species of the following genera: *Acanthoscurria*, *Cyrtopholis*, *Eupalaestrus* Pocock, 1901, *Lasiodora*, *Megaphobema* Pocock, 1901, *Metriopelma* Becker, 1878, *Nhandu*, *Pamphobeteus*, *Phormictopus*, *Schizopelma* F. O. Pickard-Cambridge, 1897, *Sericopelma* Ausserer, 1875, *Theraphosa*, *Vitalius* and *Xenesthis* Simon, 1891, and we report herein the presence of the same structure in *Proshapalopus* and *Pterinopelma*. The femoral scopula is composed of claviform setae (Fig. 11.4a–c) in *Theraphosa* and *Brachypelma*, or plumose setae in the remaining genera cited above. Spiders of the genera *Lasiodora*, *Brachypelma*, and

Table 11.2 Types and location of femoral scopula setae on Theraphosinae genera

Genera	Type of setae	Location (femur)
Acanthoscurria	Plumose (Fig. 11.4f)	I (prolateral) + IV (retrolateral)
Brachypelma	Claviform (Fig. 11.4c)	All legs
Cyrtopholis	Plumose	I (prolateral) + IV (retrolateral)
Eupalaestrus	Plumose	I (prolateral) + IV (retrolateral)
Lasiodora	Plumose	All legs
Megaphobema	Plumose	I (prolateral) + IV (retrolateral)
Metriopelma	Plumose	I (prolateral) + IV (retrolateral)
Nhandu	Plumose	I (prolateral) + IV (retrolateral)
Pamphobeteus	Plumose	I (prolateral) + IV (retrolateral)
Phormictopus	Plumose	I (prolateral) + IV (retrolateral)
Proshapalopus	Plumose	I (prolateral) + IV (retrolateral)
Pterinopelma	Plumose	I (prolateral) + IV (retrolateral)
Schizopelma	Plumose	I (prolateral) + IV (retrolateral)
Sericopelma	Plumose	I (prolateral) + IV (retrolateral)
Theraphosa	Claviform	All legs
Vitalius	Plumose	I (prolateral) + IV (retrolateral)
Xenesthis	Plumose	I (prolateral) + IV (retrolateral)

Theraphosa are the only ones that possess femoral scopula on all legs. Otherwise, in the remaining, these setae are present on femora I (prolateral) and IV (retrolateral) (Table 11.2).

11.1.2 Adhesive Setae*: Scopula and Claw Tufts (Fig. 11.5a, c–e, g, h)*

Many published papers investigated various aspects of these remarkable adhesive setae, providing climbing and prey-capturing skills [see Pérez-Miles et al. (2017) and references cited therein and also Chap. 12], mainly focused on araneomorph spiders. Theraphosids in general are characterized by the possession of claw tufts and dense tarsal scopulae (Raven 1985) (Fig. 11.5a, c–e, g, h), those of which are composed by the same type of adhesive setae, and are similar to those of araneomorphs (Ramírez 2014). They are microtriched setae (spatulated microtrichia), with the difference that those of claw tufts are longer and inserted into tarsal pads located below the claws (Fig. 11.5e). Pérez-Miles et al. (2017) identified two variants of adhesive setae in theraphosids, lamellate and lamellate-crested. Large spiders, as the big theraphosines (e. g., *Theraphosa*, *Pamphobeteus*, *Xenesthis*) bear very dense tarsal scopula, while small theraphosids have a less dense scopula, which can be even less dense in the hind legs. Simon (1892) used the condition of the tarsal scopula (entire or divided by a band of setae) in Theraphosidae to distinguish his tribe Ischnocolea (current subfamily Ischnocolinae) (divided scopula) from the

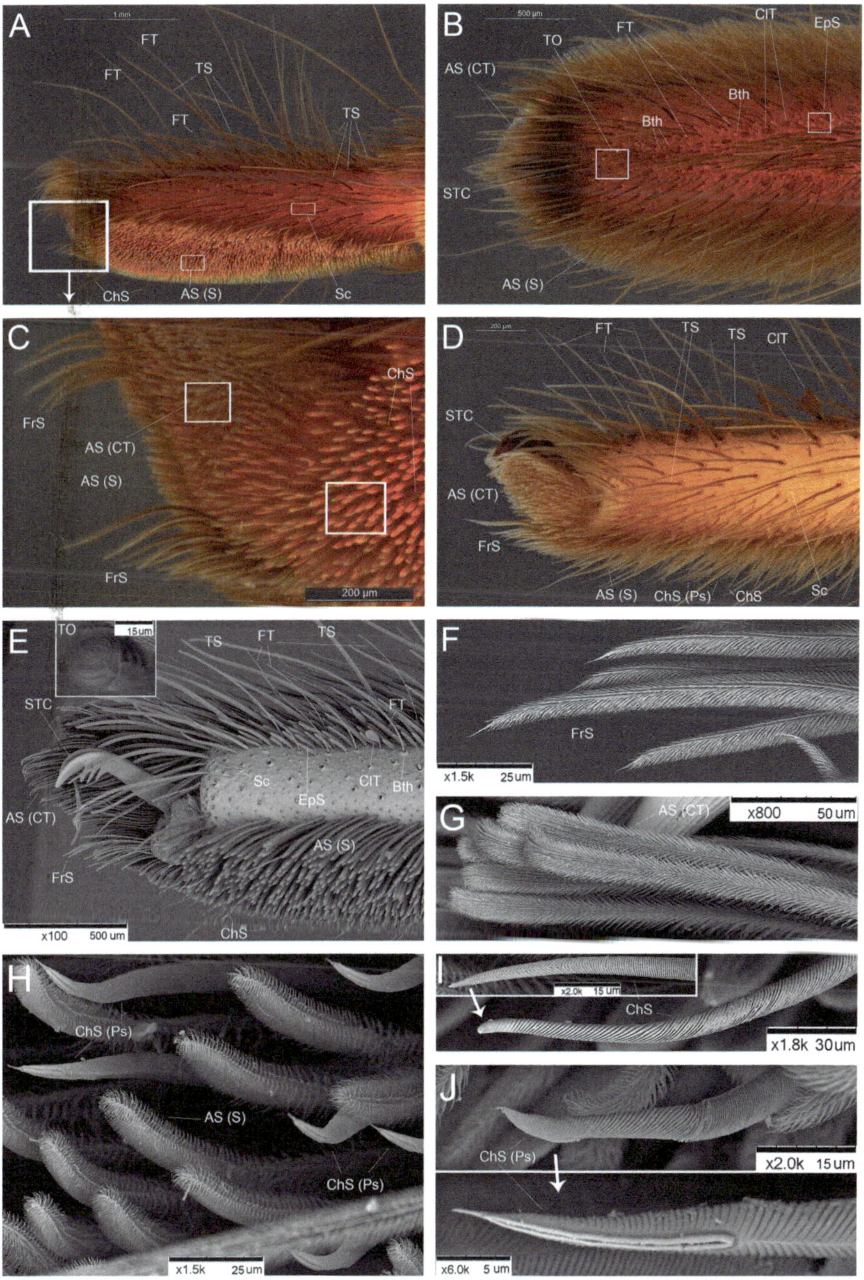

Fig. 11.5 Scopula, pseudoscopula, claw tufts. (**a–c**). Female tarsus I, *Guyruita cerrado*. (**d**) Male tarsus I, *Tmesiphantes riopretano*. (**e**) Female tarsus I, detail of the tarsal organ, *Guyruita* sp. (**f**) Frictional setae, female tarsus I, *Sickius longibulbi*. (**g**) Adhesive setae (claw tufts), female tarsus IV, *Heterothele* sp. (**h**) Scopulae on male tarsi I, *Typhochlaena paschoali*. (**i**) Chemosensory seta, lateral view, male tarsi IV, *Guyruita* sp.; detail, dorsal view. (**j**) Chemosensory seta (male pseudoscopula), lateral view, *Heterothele* sp.; detail, ventral view, *Guyruita* sp. Acronyms: *AS (CT)* adhesive setae (claw tufts), *AS (S)* adhesive setae (scopulae), *Bth* bothrium, *ClT* clavate trichobothrium, *ChS* chemosensory setae, *ChS (Ps)* chemosensory setae (male pseudoscopula), *Sc* scales, *EpS* epitrichobothrial setae, *FT* filiform trichobothrium, *FrS* frictional setae, *STC* superior tarsal claws, *TS* tactile setae (striated setae), *TO* tarsal organ

remaining (entire scopula). Pérez-Miles (1994) verified the ontogenetic differentiation (from divided to entire) and its relation to spider size (divided condition in small spiders) for Theraphosinae. In fact, in theraphosids with divided scopula, there is an anterior-posterior gradation of the width of the band dividing the scopula, being wider on hind legs (Guadanucci 2005; Pérez-Miles et al. 2017). Pérez-Miles et al. (2015) found no differences in friction between species with divided and entire tarsal scopula. However, their results on testing friction of tarantulas' adhesive pads suggested that claw tufts produce adhesion when the legs pull, while scopula functions when legs push, which is in agreement with previous studies with *Cupiennius salei* (Ctenidae) and *Aphonopelma seemani* (Theraphosidae) (Niederegger and Gorb 2006). Many authors have suggested differential usages of these adhesive setae, but it is clear that they are involved both in adhesion for locomotion and in prey capture (Pérez-Miles et al. 2017). Within Mygalomophae, the scopula seems to have been acquired independently by Crassitarsae and Euctenizoidina (Pérez-Miles et al. 2017).

Another important feature concerning the scopula is the condition found in arboreal theraphosids, which for the Neotropical ones are included within Aviculariinae: dense scopula that extends laterally, giving the tarsi a spatulate appearance (Bertani 2012). This condition is also found in those aviculariines that do not have arboreal behavior (*Ephebopus*) and its function if not very well understood (West et al. 2008).

The adhesive setae forming a scopula are also present on ventral metatarsi, varying in density and extension. Except for those few large spiders (e.g., *Xenesthis*, *Theraphosa*, *Pamphobeteus*) with dense metatarsal scopula, the widespread condition is that scopula on metatarsi of hind legs are limited to the distal end (sometimes limited to few sparse setae), whereas the scopula on fore legs can extend from half to the entire article. The condition (extension and density) of the metatarsal scopula has never been closely studied, but it seems plausible to consider this character as an adaptation to prey capture based in two assumptions: (1) the scopula is concentrated on fore legs and (2) the metatarsi do not touch the substrate during locomotion. Wolff et al. (2013), who found a close association between adhesive setae and free hunter lifestyle in spiders, suggesting that the acquisition of a scopula was as an adaption to controlling prey, evolving as a substitute for silk.

11.1.3 Frictional Setae *(Fig. 11.5a, c–f)*

They are conical setae, with no-spatulate microtrichia, inserted on tarsus tip below the claws, and may extend in a median line to the basal portion of the tarsus, dividing the scopula in longitudinal halves (Pérez-Miles 1994; Guadanucci 2005; Pérez-Miles et al. 2017). From lateral view, these setae are easily distinguished from the adhesive ones on apical tarsus, because they stick out from below the dense claw tufts (Fig. 11.5a–d). From ventral view, a patch of plumose set is distinguishable from the dense scopula (Fig. 11.6a). Pérez-Miles (1994) adopted the same terminology proposed by Rovner (1978) for *Lycosa* spp. and named the thick setae that divides the scopula as type B (adhesive setae are type A). Pérez-Miles et al. (2017) considered these as conical setae, and highlighted its presence in the transitional zone between scopula and claws tufts (Fig. 11.6a). They also showed different conditions of the length and width of this band of conical setae, varying from narrow to wide, restricted to either apical or basal portion, or extended along the whole scopula length. Furthermore, Ramírez (2014) recognized in dionychan spiders a morphological intermediate between tenent (adhesive) and plumose setae, and named them pseudotenent setae. The pseudotenent share with frictional setae the position of insertion (below the claws) and the strong morphological similarities, with acute setal shaft and loosely organized expanded barbs (non-spatulate). Ramírez (2014) found these pseudotenent setae in several araneomorph families, as it is present in all Mygalomorphae families (pers. obs.) with the difference that, in Theraphosidae, they have more barbs (plumose aspect) compared to those of other families (except for the barychelids, which whom they share close affinities) with an acute tip (Fig. 11.5f).

A few studies have investigated the adhesive and frictional forces performed by the scopulate spider tarsi (Niederegger and Gorb 2006; Wolff and Gorb 2012, 2013; Wolff et al. 2013; Lapinski et al. 2015), all focused on the typical adhesive setae (claws tufts and scopula setae with spatulate microtrichia). Wolff and Gorb (2012) suggested that these frictional setae play a role in the claw-interlocking mechanism with rough substrates, representing a further adaptation for locomotion on non-horizontal surfaces. However, considering the variation of these setae (disposition and quantity) in tarantulas, it seems that we are still far from a full understanding of the function of these frictional setae.

11.1.4 Chemosensors

Structures that function as chemosensors must have pores or slits, through which any specific substance penetrates and reaches the sensory cells (Barth 2002), that extends inside the setal shaft up to the opening near the tip (Foelix 1996). In spiders, chemoreceptors are concentrated on distal segments of palps and forelegs (Gaskett 2007; Foelix 1996). The most common are chemosensory setae, named hair sensilla by Foelix (1996). In Theraphosidae spiders, as in the remaining mygalomorphs, these chemosensors are more common on tibiae, metatarsus, and tarsus, but they were also found also in proximal palp/leg segments (coxae and trochanters) and spinnerets (pers. obs.). They can be very long setae, slightly curved and usually overpassing in length the surrounding setae, located on all faces of palp and legs (see Fig. 11.5e, few chemosensory setae—ChS, mixed with adhesive scopula). The distal half of these chemosensory setae are ornamented with thin diagonal ridges, resembling the texture of a rope, and the opening is a very small rounded pore (Fig. 11.5i).

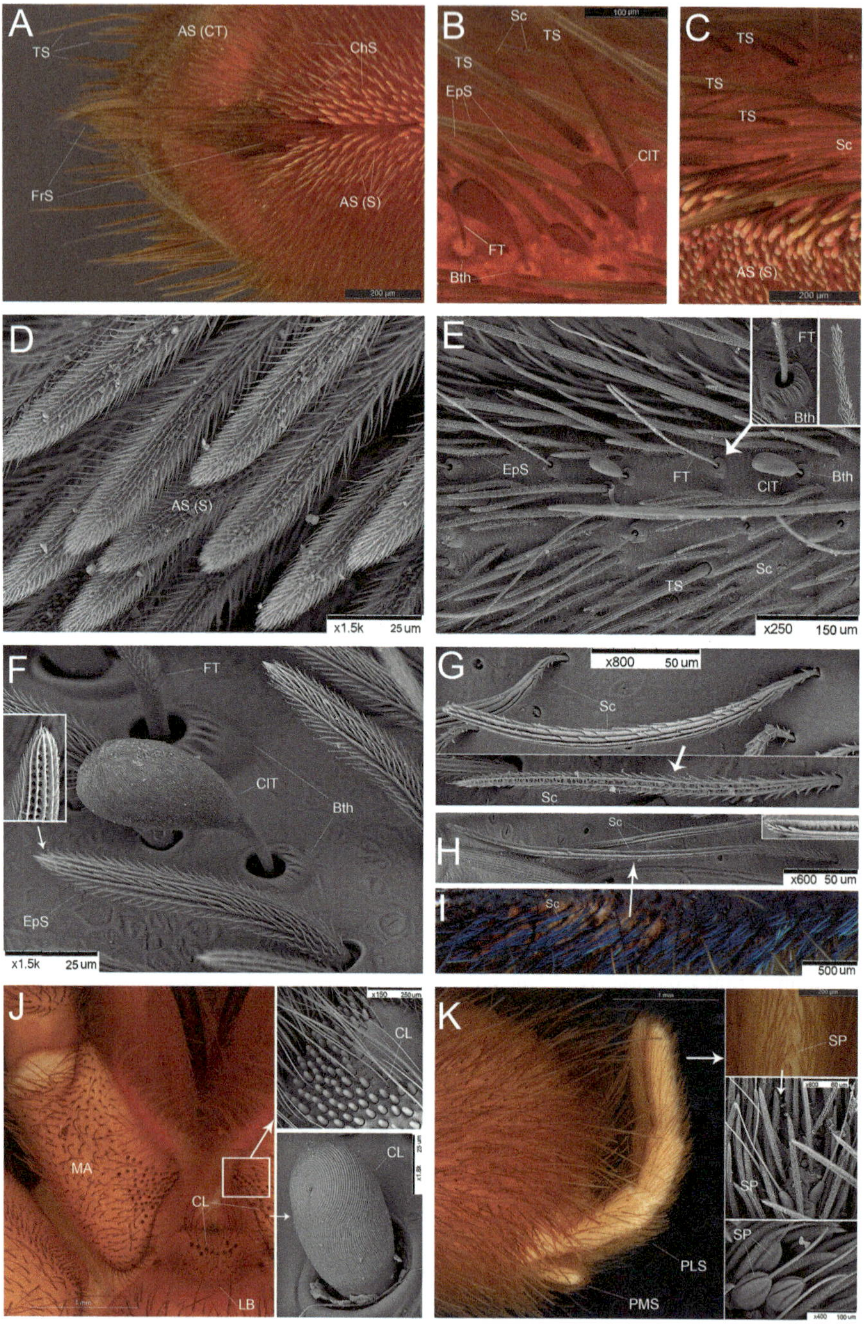

Fig. 11.6 Trichobothria, scales, cuspules, spigots. (**a–c**) Female tarsus I, *Guyruita cerrado*: (**a**) Claw tufts and scopulae, ventral view; (**d**) Female tarsus I, *Guyruita* sp. (**e**) Male tarsus I, dorsal view, *Tmesiphantes riopretano*. (**f**) Female tarsus I, lateral view, *Sickius longibulbi*; detail, dorsal

Recently, Foelix et al. (2010) found a scopula-like concentration of putative chemosensory setae on ventral tarsi of males *Liphistius* and *Idiops*. More recently, Pérez-Miles et al. (2017) reported the same type of chemosensory setae only in males of all mygalomorphs, located ventrally on tarsi and, in a few cases, apical metatarsi (more numerous in forelegs). In mygalomorphs lacking adhesive scopula, these setae form the pseudoscopula, named by Pérez-Miles et al. (2017) to differ from the adhesive scopula, present both in males and females. In Theraphosidae (Fig. 11.5h), and other spiders with adhesive scopula (e.g., diplurines, barychelids, nemesiids and some cyrtaucheniids), these putative chemosensory setae are interspersed with adhesive setae to form the mixed scopula (Pérez-Miles et al. 2017) and are characterized by a similar texture of the surface, similar to the long chemosensory setae described above, and with a distinct subapical opening. In Theraphosidae, the apical portion is slightly swollen with a thin and long opening slit, and a well-marked rim (Fig. 11.5j), condition shared with Barychelidae (pers. obs.).

Foelix et al. (2010) reported a scopula-like structure on ventral tarsi of *Liphistius* (Mesothelae) and *Idiops pylorus* Schwendinger, 1991 (Mygalomorphae), for which they suggested to be probable contact chemoreceptors. Years before Foelix et al. (2010), Gorb et al. (2006) reported to have found tarsal spigots in a female *Aphonopelma seemanni*, and also suggested that the tarsal silk production by these spigots would help the spiders climb vertical surfaces providing extra adhesion. The pictures provided by the author for the alleged spigots are in fact the long chemosensory setae that are present in both sexes, as they share the exact same ultrastructure (blunt tip and thin diagonal ridges). After a few replies (Pérez-Miles et al. 2009; Gorb et al. 2009) and extra morphological and experimental studies (Rind et al. 2011; Pérez-Miles and Ortíz-Villatoro 2012; Foelix et al. 2010), Foelix et al. (2012) showed that they probably comprise chemoreceptors.

The tarsal organ has also been interpreted as a chemoreceptor, although never experimentally tested for this function, and received little attention throughout the years. Raven (1985) described the overall morphology of tarsal organ in mygalomorphs as low or slightly raised, domed, with concentric ridges, located at the median dorsal line close to tarsal anterior margin. Due to the great amount of setae on their body, especially on the legs, the tarsal organ of theraphosids is of difficult visualization, unless some apical setae are removed. The first work regarding the tarsal organ was done by Blumenthal (1935), and the ultrastructural description was performed by Foelix and Chu-Wang (1973). The tarsal organ of theraphosids, which

view. (**g–i**) Scales: (**g**) Male tarsus I, lateral view, *Holothele longipes*; dorsal view, *T. riopretano*. (**h**, **i**) Male tarsus I, dorsal view, *Dolichothele diamantinensis*. (**i**) Live specimen. (**j**) Labial and maxillary cuspules, *T. riopretano*, male; detail of the maxilla, *Typhochlaena paschoali*, male; cuspule detail, *D. diamantinensis*, male. (**k**) Spinnerets, lateral view, *T. riopretano*, male; spigots details, ventral view: *T. riopretano*, male; *S. longibulbi*, female; *T. costae*, male. Acronyms: *AS (CT)* adhesive setae (claw tufts), *AS (S)* adhesive seta (scopulae), *Bth* bothrium, *CL* cuspule, *ClT* clavate trichobothrium, *ChS* chemosensory seta, *Sc* scale, *EpS* epitrichobothrial seta, *FT* filiform trichobothrium, *LB* labium, *MA* maxilla (palpal coxae), *PLS* posterior lateral spinnerets, *PMS* posterior median spinnerets, *FrT* frictional setae, *SP* spigot, *TS* tactile seta (striated setae)

fits in the description above, seems not to be of great value in systematics since it does not show much variation.

11.1.5 Trichobothria*: Clavate, Thickened, Filiform*

Various terrestrial arthropods are equipped with mechanosensitive sensilla named trichobothria, characterized by a cup-shaped cuticular structure (bothrium), in which a seta is inserted (Reissland and Görner 1985). The trichobothria in theraphosid spiders vary from filiform, thickened, and clavate (club-shaped) (Guadanucci 2012). As in other mygalomorph spiders, these trichobothria are present on dorsal tarsi, retrolateral (basal to mid length) and dorsal (distal) metatarsus, and lateral (basal retro and prolateral) and dorsal (from basal extending to apical in two parallel rows) (Raven 1985). Simon (1892) was the first to note the presence of clavate trichobothria in mygalomorph spiders. All known theraphosids bear clavate trichobothria (Fig. 11.6b, e, f) on all tarsi (including the adult male cymbium), and they vary in its ultramorphology, as plumose and non-plumose (i.e., barbed and unbarbed) (Guadanucci 2012). Other mygalomorphs (*Conothele* spp. and *Ummidia* spp., family Halonoproctidae) and *Liphistius* (Mesothelae, Liphistiomorphae) include representatives with clavate trichobothria, but this shared presence with Theraphosidae is considered a convergence (Murphy and Platnick 1981; Raven 1985; Goloboff 1993). Some Barychlidae also bear clavate trichobothria on tarsi, differing in size (smaller than in Theraphosidae) and disposition (few trichobothria in a compact group, rather than in a row as in theraphosids). This shared presence of clavate trichobothria on tarsi is considered synapomorphic, due to the close phylogenetic relationship between Theraphosidae and Barychelidae (Raven 1985; Goloboff 1993; Bond et al. 2012; Guadanucci 2014). Guadanucci (2012) described the following disposition patterns for Neotropical Theraphosidae: two parallel rows in Theraphosinae, Schismatothelinae, *Dolichothele*, *Acanthopelma*, and *Holothele*; two diverging rows (U-shaped row) in Aviculariinae; short apical row in *Catumiri*. The sensitivity to air currents is well known for filiform trichobothria (Foelix 1996; Barth 2002), while the clavate trichobothria have been suggested to be adapted to the detection of vibrations of the soil (Den Otter 1973). Quirici and Costa (2005) demonstrated that seismic communication plays an important role during courtship of two tarantulas from Uruguay (*Eupalaestrus weijenberghi* and *Acanthoscurria cordubensis).* The clavate trichobothria on tarsi of *Sericopelma rubronitens* Ausserer, 1875 (Theraphosinae) showed response to vibrations on a table during electrophysiological experiments, and the removal of these trichobothria resulted in the absence of any stimulus (Den Otter 1973).

It is remarkable the presence of clavate trichobothria on basal lateral tibia of the great majority of Theraphosidae, which was originally reported by Guadanucci (2012). The aviculariine, the Neotropical arboreal tarantulas (except for *Ephebopus*, which according to West et al. 2008 has lost the arboreal behavior) present a distinctive disposition of tibial clavate trichobothria, which are in a compact group.

Considering their arboreal behavior and the presence of similar trichobothria in cerci of crickets, thought to be gravity receptors (Horn and Bishof 1983), it is possible that this configuration is an adaptation to sense in a 3D environment.

On both sides of all trichobothria row (on tarsi, metatarsi and tibiae), forming two parallel rows, tarantulas have the epitrichobothrial setae (Fig. 11.6b, e, f). They are similar to tactile setae, but shorter than the clavate trichobothria. In the aviculariines, as well as in other subfamilies (e.g., Selenocosmiinae, Ornithoctoninae), the epitrichobothrial setae form a dense patch intermixed with the filiform and clavate trichobothria (Guadanucci 2012).

11.1.6 Scales *(Fig. 11.6c, g–i)*

Townsend and Felgenhauer (1998) and Ramírez (2014) defined scales as setae inserted in small sockets and bent immediately after its insertion, so that the setae overlay close to the surface of the cuticle. According to Townsend and Felgenhauer (1999), scales comprise a specific type of setae because they lack innervation. There has been a great diversity of spider cuticular scales described in araneomorphs (Hill 1979; Townsend and Felgenhauer 1998, 1999, 2001; Ramírez 2014), but just few Theraphosidae taxa were included in these studies. In the case of tarantulas, these setae can be seen, under the microscope, as a whitish sub-layer below the numerous tactile setae. SEM micrographs revealed a unique ultramorphology with longitudinal furrows (named lamellar structure by Foelix et al. 2013) and spike-like ornaments (Fig. 11.6g–i), and they are present in all body parts, including appendages as chelicerae, palps, legs, and spinnerets. Many Neotropical tarantulas are brightly colored, ranging from violet to green, most of them have structural colors (iridescent) (Foelix et al. 2013). The intense blue coloration, depicted in Fig. 11.6i of the tarsus of a female *Dolichothele diamantinensis*, is due to light reflectance of these scales. Hsiung et al. (2015) found that the blue color is the most common in tarantulas and is the result of light reflectance form diverse structures in different spiders. According to their results, the setae showing the highest reflectance are produced by setae with longitudinal furrows (named therein as “bladelike” and “lobe-like” protruding hairs).

11.1.7 Spigots

Theraphosidae spiders have spigots on ventral face of both posterior median spinnerets and posterior lateral spinnerets (all three articles) spinnerets (Raven 1985; Palmer 1990; Murphy and Roberts 2015). In Mygalomorphae, as well as within Theraphosidae, the spigots show a uniform morphology, having a short base and a long articulated shaft ending in a terminal pore (Fig. 11.6k). Along the length of the shaft may be ornamented scale-like ornaments, which is also morphologically uniform across mygalomorphs. These spigots are usually characterized as aciniform gland spigots (AC) (Fig. 11.6k, first two small images), and other trap-door mygalomorph also have MAC (Modified Aciniform Spigot), which are similar to AC, but with the base slightly thicker (Murphy and Roberts 2015). We know of at least two *Typhochlaena* species (*T. seladonia* and *T. paschoali*, Fig. 11.6k, last image) with modified aciniform gland spigots (MAC). These MACs are similar to pumpkniform-shape spigots present in *Stenoterommata* (Pycnothelidae), as they both have an enlarged base with longitudinal wrinkle-like marks. These *Typhochlaena* species have a singular behavior of building their web-retreats on tree trunks enclosed by a thin trapdoor (Andre and Hüsser 2016). In the Neotropics, spiders belonging to different families build trapdoors on their retreats (Nemesiidae, Barychelidae, Cyrtaucheniidae), and some other familes are exclusive trapdoor spiders. However, this is very unique among theraphosids.

11.1.8 Cuspules

Cuspules are globular to conical cuticular outgrowths inserted in distinct sockets on labium and/or maxilla. Under the stereomicroscope they appear as small dark dots, but at higher magnifications (above 1000×—SEM) it is possible to make out their ultrastructural surface with several parallel ridges, similar to fingerprints (Fig. 11.6j). Anterior and posterior faces have distinct ornamentations: anterior face with transverse ridges on apical half, basal half with longitudinal ridges; posterior apical half with concentric ridges that fuse with longitudinal ridges on basal half. This general ultramorphological pattern has very little variation within Theraphosidae (Pérez-Miles and Montes de Oca 2005). The number and distribution of these cuspules, and also their disposition are useful diagnostic characters for some genera and suprageneric groups (e.g., *Dolichothele*, *Catumiri*, *Tmesiphantes*, *Hapalopus*, with few cuspules; the Schismatothelinae with extreme numbers and high density of labial cuspules). No clear function has been experimentally demonstrated for these cuspules, but mechanical, glandular, and sensorial usage have been discussed by Pérez-Miles and Montes de Oca (2005). These authors further suggested that due to their position, ornamentations, and unique presence in mygalomorphs (paraxial fangs), the cuspules could help in prey retention by the opposing force from the chelicerae when the spider holds the food item.

11.1.9 Epiandrous Spigots

Recently, Ferretti et al. (2017) published the first comparative study of the epiandrous apparatus of mygalomorph spiders, which comprises a group of additional silk glands associated with epiandrous spigots located at the anterior margin of the genital furrow of males. They are present in most spiders (Marples 1967) and are a possible synapomorphy for Araneae (Ferretti et al. 2017). Their function is still not clear, but it should act during sperm web building and sperm induction. In theraphosids, the number of spigots may be over 100 in large individuals, and are intermixed with distinct types of plumose setae (Ferretti et al. 2017). These plumose setae are well documented in various micrographs by Ferretti et al. (2017) and are, in fact, the ones referred here as scales and tactile setae. This high density of setae and their conspicuous distribution in a distinct patch (half-moon shaped zone) help distinguishing the sex of live spiders, as it appears as a clear small mark on the epigastric furrow of males (Ferretti et al. 2017). In mygalomorphs these spigots are elongated setae, widened at the base and inserted in sockets. Different from the spigots on spinnerets (Fig. 11.6k), the epiandrous spigots lack a globular base.

Acknowledgments We would like to thank CNPq (479377/2012-0—Conselho Nacional de Pesquisa e Tecnologia), CAPES (Coordenação de Aperfeiçoamento de Pessoal de Nível Superior), and FAPESP (2015/06406-4, 2017/11985-9—Fundação de Amparo à Pesquisa do Estado de São Paulo). We are also thankful to all technical staff at the Laboratory of Electronic Microscopy (Mônika Iamonte, Antônio Yabuki and PhD. Odair Correa Bueno).

References

Andre C, Hüsser M (2016) Die Gattung *Typhoclhaena*—Historie und Systematik sowie Habitat und Lebensweise von *T. seladonia*, *T. costae* und *T. curumim*. Arachne 21:33–34

Barth FG (2002) A spider's world: senses and behavior. Springer, Berlin, p 393

Bertani R (2001) Revision, cladistic analysis, and zoogeography of *Vitalius*, *Nhandu*, and *Proshapalopus*; with notes on other theraphosine genera (Araneae, Theraphosidae). Arq Zool 36:265–356

Bertani R (2012) Revision, cladistic analysis and biogeography of *Typhochlaena* CL Koch, 1850, *Pachistopelma* Pocock, 1901 and *Iridopelma* Pocock, 1901 (Araneae, Theraphosidae, Aviculariinae). ZooKeys 230:1–94

Bertani R, Guadanucci JPL (2013) Morphology, evolution and usage of urticating setae by tarantulas (Araneae: Theraphosidae). Fortschr Zool 30:403–418

Bertani R, Marques OAV (1995) Release of urticating hairs by some Aviculariinae. Zool Anz 234:161–165

Bertani R, Boston T, Evenou Y, Guadanucci JPL (2003) Release of urticating hairs by *Avicularia versicolor* (Walckenaer, 1837) (Araneae, Theraphosidae). Bull Br Arachnol Soc 12:395–398

Bertani R, Fukushima CS, Da Silva PI (2008) Two new species of *Pamphobeteus* Pocock 1901 (Araneae: Mygalomorphae: Theraphosidae) from Brazil, with a new type of stridulatory organ. Zootaxa 1826:45–58

Bertani R, Nagahama RH, Fukushima CS (2011) Revalidation of *Pterinopelma* Pocock 1901 with description of a new species and the female of *Pterinopelma vitiosum* (Keyserling 1891) (Araneae: Theraphosidae: Theraphosinae). Zootaxa 2814:1–18

Blumenthal H (1935) Untersuchungen über das 'Tarsalorgan' der Spinnen. Z Morphol Ökol Tiere 29:667–719

Bond JE, Hendrixson BE, Hamilton CA, Hedin M (2012) A reconsideration of the classification of the spider infraorder Mygalomorphae (Arachnida: Araneae) based on three nuclear genes and morphology. PLoS One 7:1–11

Bücherl W (1957) Sobre a importância dos bulbos copuladores e das apófises tibiais dos machos na sistemática das aranhas caranguejeiras (Orthognatha). An Acad Bras Ciênc 29:377–416

Cooke JA, Roth VD, Miller FH (1972) The urticating hairs of theraphosid spiders. Am Mus Novit 2498:1–43

Den Otter CJ (1973) Setiform sensilla and prey detection in the bird-spider *Sericopelma rubronitens* Ausserer (Araneae, Theraphosidae). Neth J Zool 24:219–235

Ferretti N, Pompozzi G, Pérez-Miles F (2011) The species of *Grammostola* (Araneae: Theraphosidae) from Central Argentina: taxonomy, distribution, and surface ultrastructure of coxal setae. Zootaxa 2828:1–18

Ferretti N, Pompozzi G, Copperi S, González A, Pérez-Miles F (2013) Sexual behaviour of mygalomorph spiders: when simplicity becomes complex; an update of the last 21 years. Arachnology 16:85–93

Ferretti N, Pompozzi G, Copperi S, Wehitt A, Galíndez E, González A, Pérez-Miles F (2017) A comparative morphological study of the epiandrous apparatus in mygalomorph spiders (Araneae, Mygalomorphae). Micron 93:9–19

Foelix R (1996) Biology of spiders. Oxford University Press, New York

Foelix RF, Chu-Wang IW (1973) The morphology of spider sensilla II. Chemoreceptors. Tissue Cell 5:461–478

Foelix R, Erb B, Michalik P (2010) Scopulate hairs in male *Liphistius* spiders: probable contact chemoreceptors. J Arachnol 38:599–603

Foelix RF, Rast B, Peattie AM (2012) Silk secretion from tarantula feet revisited: alleged spigots are probably chemoreceptors. J Exp Biol 215:1084–1089

Foelix RF, Erb B, Hill DE (2013) Structural colors in spiders. In: Nentwig W (ed) Spider ecophysiology. Springer, Berlin, pp 333–347

Fukushima CS, Bertani R (2017) Taxonomic revision and cladistic analysis of *Avicularia* Lamarck, 1818 (Araneae, Theraphosidae, Aviculariinae) with description of three new aviculariine genera. ZooKeys 659:1–185

Gabriel R (2014) A new genus and species of theraphosid spider from El Salvador (Araneae: Theraphosidae). J Br Tarantula Soc 29:146–153

Gabriel G, Sherwood D (2018) Re-description of the male of *Psalmopoeus pulcher* Petrunkevitch, 1925 with the first description of the female (Araneae: Theraphosidae). J Br Tarantula Soc 33:8–18

Galleti-Lima AG, Guadanucci JPL (2018) Morphology of setae on the coxae and trochanters of theraphosine spiders (Mygalomorphae: Theraphosidae). J Arachnol 46(2):214–225

Gargiulo FDF, Brescovit AD, Lucas SM (2018) *Umbyquyra* gen. nov., a new tarantula spider genus from the Neotropical region (Araneae, Mygalomorphae, Theraphosidae), with a description of eight new species. Eur J Taxon 457:1–50

Gaskett AC (2007) Spider sex pheromones: emission, reception, structures, and functions. Biol Rev 82:27–48

Goloboff PA (1993) A reanalysis of mygalomorph spider families (Araneae). Am Mus Novit 3056:1–32

Gorb SN, Niederegger S, Hayashi CY, Summers AP, Vötsch W, Walther P (2006) Biomaterials: silk-like secretion from tarantula feet. Nature 443:407

Gorb SN, Niederegger S, Hayashi CY, Summers AP, Vötsch W, Walther P (2009) Gorb et al. reply. Nature 461:E9–E10

Guadanucci JPL (2005) Tarsal scopula significance in Ischnocolinae phylogenetics (Araneae, Mygalomorphae, Theraphosidae). J Arachnol 33:456–467

Guadanucci JPL (2012) Trichobothrial morphology of Theraphosidae and Barychelidae spiders (Araneae, Mygalomorphae). Zootaxa 3439:1–42

Guadanucci JPL (2014) Theraphosidae phylogeny: relationships of the 'Ischnocolinae' genera (Araneae, Mygalomorphae). Zool Scr 43:508–518

Hill DE (1979) The scales of salticid spiders. Zool J Linn Soc 65(3):193–218

Horn E, Bishof HJ (1983) Gravity reception in crickets: the influence of cercal and antennal afferences on the head position. J Comp Physiol 150:93–98

Hsiung BK, Deheyn DD, Shawkey MD, Blackledge TA (2015) Blue reflectance in tarantulas is evolutionarily conserved despite nanostructural diversity. Sci Adv 1:e1500709

Jocqué R (2005) Six stridulating organs on one spider (Araneae, Zodariidae): is this the limit? J Arachnol 33:597–603

Lapinski W, Walther P, Tschapka M (2015) Morphology reflects microhabitat preferences in an assemblage of Neotropical wandering spiders. Zoomorphology 134:219–236

Legendre R (1963) L'audition et l'émission de sons chez les Aranéides. Ann Biol 2:371–390

Marples BJ (1967) The spinnerets and epiandrous glands of spiders. Zool J Linnean Soc 46:209–222

Marroquín JIM (2014) Taxonomic revision of *Hemirrhagus* Simon, 1903 (Araneae: Theraphosidae, Theraphosinae), with description of five new species from Mexico. Zool J Linnean Soc 170:634–689

Marshall SD, Uetz GW (1990a) Pedipalpal brush of *Ephebopus* sp. (Araneae, Theraphosidae): evidence of a new site for urticating hairs. Bull Br Arachnol Soc 8:122–124

Marshall SD, Uetz GW (1990b) Incorporation of urticating hairs into silk: a novel defense mechanism in two Neotropical tarantulas (Araneae, Theraphosidae). J Arachnol 18:143–149

Marshall SD, Thoms EM, Uetz GW (1995) Setal entanglement: an undescribed method of stridulation by a neotropical tarantula (Araneae: Theraphosidae). J Zool 235:587–595

Mendoza-Marroquín JI (2014) *Psalmopoeus victori*, primera araña terafósida arborícola descrita para México (Araneae: Theraphosidae: Aviculariinae). Rev Mex Biodivers 85:728–735

Moussian B (2013) The arthropod cuticle. In: Minelli A, Boxshall G, Fusco G (eds) Arthropod biology and evolution. Springer, Berlin, pp 171–196

Murphy JA, Platnick NJ (1981) On *Liphistius desultor* (Araneae, Liphistiidae). Bull Am Mus Nat Hist 170:46–56

Murphy JA, Roberts MJ (2015) Spider families of the world and their spinnerets. British Arachnological Society, York

Niederegger S, Gorb SN (2006) Friction and adhesion in the tarsal and metatarsal scopulae of spiders. J Comp Physiol A 192:1223–1232

Palmer JM (1990) Comparative morphology of the external silk production apparatus of "primitive" spiders. Thesis (unpublished), Department of Organismic and Evolutionary Biology Harvard University, Harvard

Perafán C, Pérez-Miles F (2010) An unusual setule on type IV urticating setae of *Homoeomma uruguayense* (Araneae: Theraphosidae). J Arachnol 38:153–154

Perafán C, Perez-Miles F (2014) Three new species of *Melloleitaoina* Gerschman & Schiapelli, 1960 (Araneae, Mygalomorphae, Theraphosidae) from northern Argentina. ZooKeys 404:117–129

Perafán C, Cifuentes Y, Estrada-Gomez S (2015) *Aguapanela*, a new tarantula genus from the Colombian Andes (Araneae, Theraphosidae). Zootaxa 4033:529–542

Perafán C, Galvis W, Gutiérrez M, Perez-Miles F (2016) *Kankuamo*, a new theraphosid genus from Colombia (Araneae, Mygalomorphae), with a new type of urticating setae and divergent male genitalia. ZooKeys 601:89–109

Pérez-Miles F (1994) Tarsal scopula division in Theraphosinae (Araneae, Theraphosidae): its systematic significance. J Arachnol 22:46–53

Pérez-Miles F (1998) Notes on the systematics of the little known theraphosid spider *Hemirrhagus cervinus*, with a description of a new type of urticating hair. J Arachnol 26:120–123

Pérez-Miles F (2002) The occurrence of abdominal urticating hairs during development in Theraphosinae (Araneae, Theraphosidae): phylogenetic implications. J Arachnol 30:316–320

Pérez-Miles F, Costa FG (1994) *Acanthoscurria atrox* incorporates urticating hairs into its shedding mat. Forum Am Tarantula Soc 3:63–64

Pérez-Miles F, Montes de Oca L (2005) Surface ultrastructure of labial and maxillary cuspules in eight species of Theraphosidae (Araneae). J Arachnol 33:43–49

Pérez-Miles F, Ortíz-Villatoro D (2012) Tarantulas do not shoot silk from their legs: experimental evidence in four species of New World tarantulas. J Exp Biol 215:1749–1752

Pérez-Miles F, Prandi L (1991) El comportamiento de emision de pelos urticantes en *Grammostola mollicoma* (Aranae Theraphosidae) un analisis experimental. Bol Soc Zool Uruguay 2a época 6:47–53

Pérez-Miles F, Lucas SM, da Silva PI Jr, Bertani R (1996) Systematic revision and cladistic analysis of Theraphosinae (Araneae: Theraphosidae). Mygalomorph 1:33–68

Pérez-Miles F, Costa FG, Toscano-Gadea C, Mignone A (2005) Ecology and behaviour of the 'road tarantulas' *Eupalaestrus weijenberghi* and *Acanthoscurria suina* (Araneae, Theraphosidae) from Uruguay. J Nat Hist 39:483–498

Pérez-Miles F, Panzera A, Ortiz-Villatoro D, Perdomo C (2009) Silk production from tarantula feet questioned. Nature 461:E9–E10

Pérez-Miles F, Perafán C, Santamaría L (2015) Tarantulas (Araneae: Theraphosidae) use different adhesive pads complementarily during climbing on smooth surfaces: experimental approach in eight arboreal and burrower species. Biol Open 00:1–6

Pérez-Miles F, Guadanucci JPL, Jurgilas JP, Becco R, Perafán C (2017) Morphology and evolution of scopula, pseudoscopula and claw tufts in Mygalomorphae (Araneae). Zoomorphology 136:435–459

Pocock RI (1895) Musical boxes in spiders, vol 6. Natural Science, London, pp 44–50

Quirici V, Costa FG (2005) Seismic communication during courtship in two burrowing tarantula spiders: an experimental study on *Eupalaestrus weijenberghi* and *Acanthoscurria suina*. J Arachnol 33:159–166

Ramírez MJ (2014) The morphology and phylogeny of dionychan spiders (Araneae: Araneomorphae). Bull Am Mus Nat Hist 390:1–374

Raven RJ (1985) The spider infraorder Mygalomorphae (Araneae): cladistics and systematics. Bull Am Mus Nat Hist 182:1–180

Reissland A, Görner P (1985) Trichobothria. In: Barth FG (ed) Neurobiology of arachnids. Springer, Berlin, pp 138–161

Rind FC, Birkett CL, Duncan BJA, Ranken AJ (2011) Tarantulas cling to smooth vertical surfaces by secreting silk from their feet. J Exp Biol 214:1874–1879

Rovner JS (1978) Adhesive hairs in spiders: behavioral functions and hydraulically mediated movement. Symp Zool Soc Lond 42:99–108

Simon E (1892) Histoire naturelle des araignées, vol 1. Roret, Paris

Townsend VR Jr, Felgenhauer BE (1998) Cuticular scales of spiders. Invertebr Biol 117:318–330

Townsend VR Jr, Felgenhauer BE (1999) Ultrastructure of the cuticular scales of lynx spiders (Araneae, Oxyopidae) and jumping spiders (Araneae, Salticidae). J Morphol 240:77–92

Townsend VR Jr, Felgenhauer BE (2001) Phylogenetic significance of the morphology of the cuticular scales of the lynx spiders (Araneae: Oxyopidae). J Zool 253:309–332

Uetz GW, Stratton GE (1982) Acoustic communication and reproductive isolation in spiders. In: Witt PN, Rovner JS (eds) Spider communication: mechanisms and ecological significance. Princeton University Press, Princeton, NJ, pp 123–129

West R, Marshall SD, Fukushima CS, Bertani R (2008) Review and cladistic analysis of the Neotropical tarantula genus *Ephebopus* Simon 1892 (Araneae: Theraphosidae) with notes on the Aviculariinae. Zootaxa 1849:35–58

West RC, Nunn SC, Hogg S (2012) A new tarantula genus, *Psednocnemis,* from West Malaysia (Araneae: Theraphosidae), with cladistic analyses and biogeography of Selenocosmiinae Simon 1889. Zootaxa 3299:1–43

Wolff JO, Gorb SN (2012) Comparative morphology of pretarsal scopulae in eleven spider families. Arthropod Struct Dev 41:419–433

Wolff JO, Gorb SN (2013) Radial arrangement of Janus-like setae permits friction control in spiders. Sci Rep 3:1–7

Wolff JO, Nentwig W, Gorb SN (2013) The great silk alternative: multiple co-evolution of web loss and sticky hairs in spiders. PLoS One 8:1–13

Chapter 12
Adhesive Features of the Theraphosid Tarantulas

Fernando Pérez-Miles, Carlos Perafán, and David Ortiz-Villatoro

Abstract Tarantulas are large spiders with adhesive setae on their legs, which enable them to climb on smooth vertical surfaces. The mechanism proposed to explain adhesion in tarantulas is anisotropic friction, where friction is higher when the leg pushes compared to when it pulls. The static friction of live theraphosid spiders on different surfaces and at different inclines was measured and compared between burrowing and arboreal species to test the hypothesis of higher friction in arboreal tarantulas. We analyzed the complementary participation of claw tufts and scopulae of anterior and posterior legs when the tarantula climbs. We also considered the morphology of scopulae and claw tufts setae and compared with similar structures in other families. Adhesive setae, as well as some other setae types found on ventral tarsi are described and characterized. The adhesive face of setae varied in the orientation in different parts of the tarsi, and this variation is more conspicuous in the spiders that have only claw tufts or scopulae. The mechanics of climbing in association with the biological characteristics of the species are analyzed. We discuss the association of adhesive scopulae and claw tufts with burrowing/cursorial mygalomorphs as within Theraphosidae, as was suggested for free-hunter spiders. The morphology, functions, and evolution of scopula and claw tufts are discussed in this chapter.

F. Pérez-Miles (✉) · D. Ortiz-Villatoro
Entomología, Facultad de Ciencias, UDELAR, Montevideo, Uruguay
e-mail: myga@fcien.edu.uy

C. Perafán
Entomología, Facultad de Ciencias, UDELAR, Montevideo, Uruguay

Departamento de Ecología y Biología Evolutiva, Instituto de Investigaciones Biológicas Clemente Estable, Montevideo, Uruguay

F. Pérez-Miles (ed.), *New World Tarantulas*, Zoological Monographs 6,
https://doi.org/10.1007/978-3-030-48644-0_12

12.1 Introduction

Several groups of animals such as arachnids, insects, amphibians, reptiles and mammals, have independently evolved specialized organs on their feet for adhesion (Federle 2006). Adhesive organs are rapidly controllable, can be used repeatedly without any loss of performance, and function on smooth, rough, dirty, and flooded surfaces (Autumn et al. 2014). These organs evolved allowing access to vertical and even inverted surfaces, opening up new habitats (Labonte et al. 2016). The performance of these organs has inspired a considerable amount of work on technical adhesives as they still outperform most artificial adhesives with respect to rapid controllability (Jagota and Hui 2011; Dirks and Federle 2011a). Some researchers also suggest using arthropod-based adhesive mechanisms for more effective tape and binding tools (von Byern and Grunwald 2010; Gorb et al. 2007).

There are two main types of adhesion used by animals: wet adhesion and dry adhesion.

Wet adhesion comprises adhesion mediated by specialized secretions and it is used by insects, amphibians and mammals (Barnes and Jon 2011; Dirks and Federle 2011a; Labonte et al. 2016). Natural chemicals and micromechanical adhesives are often composed of high-molecular compounds containing proteins, polyphenols, lipids, resins, mixtures of long-chain hydrocarbons and mucopolysaccharides, or waxes (von Byern and Grunwald 2010). However, the deep mechanisms are not well understood (von Byern and Grunwald 2010). They are employed for adhesion across many functions such as defense, locomotion, and cocoon building (von Byern and Grunwald 2010).

Dry adhesion relies on van der Waals forces which are intermolecular forces that act between different molecules making them attract or repel. These forces are short-ranging and weak, but if numerous points of close contact are achieved, considerable adhesion forces can be generated (Kesel et al. 2004; IUPAC 2006). Dry adhesive organs have evolved independently at least three times in lizards (Irschick et al. 1996; Williams and Peterson 1982), at least three times in insects (Beutel and Gorb 2001), and occur in some phylogenetically distant groups of spiders and mites (Federle 2006; Wolff et al. 2013; Wolff and Gorb 2016).

Structural differences were found between organs used for wet adhesion in comparison with those used in dry adhesion. As insects employ adhesive substances, the setae in these organs may have relatively blunt tips. On the other hand, lizards and spiders (using only dry adhesion) require setae with extremely fine-spatulated endings (Federle 2006; Wolff and Gorb 2016).

Despite the wide diversity of animals using adhesion (wet or dry), adhesive pads come in only two basic designs: smooth adhesion pads and dense setae adhesion pads (Federle 2006).

12.1.1 Smooth Adhesion Pads

These smooth pads use specialized adhesive substances and have evolved independently in many animal groups such as insects, amphibians, and mammals (Barnes and Jon 2011; Dirks and Federle 2011a; Labonte et al. 2016). Adhesion here is mediated by thin fluid films secreted into the contact zone. As the amount of fluid affects adhesive forces, a control of secretion appears probable (Dirks and Federle 2011b). Some functional principles of smooth pads (adaptability, viscoelasticity, pressure sensitivity) are similar to those known from industrial pressure-sensitive adhesion (Gorb et al. 2007).

Smooth adhesive pads have an internal fibrous structure, are soft and fluid-filled; this way the pad molds to the surface increasing the contact area on rough surfaces (Dirks and Federle 2011a). Adhesive fluid in smooth adhesive systems mainly serves to maximize contact on rough substrates (Bullock et al. 2008).

12.1.2 Setae Adhesion Pads

These pads are composed of specialized setae densely arranged; these setae are covered with specialized setulae or microtrichia. Dense setae adhesion pads are sometimes called "hairy" pads and may use either specialized adhesive secretions or van der Waals forces in order to adhere to a surface (Federle 2006).

In insects, the adhesion of hairy pads is mediated by a liquid secretion, which is released from gland pores at the base of the setae or, in some cases, from an opening under the end-plate at the tip of the hollow adhesive hair (Gorb 1998).

Hairy pads represent an optimized design for surface attachment with increased adhesion, controllable detachment, and can also help to achieve self-cleaning properties (Federle 2006). This could explain why such structures are found among numerous and diverse groups of animals.

12.2 Animal Groups with Adhesive Pads

Amblypygids This group combines structural and functional principles of both smooth and hairy adhesive pads (Beutel and Gorb 2006). Wolff et al. (2015) found a fibrous inner architecture resembling hexagonal structures. These hexagons are rather roof-like with a spatula-like keel. Such thin, spatulate structures are a universal feature of hairy adhesive pads (Wolff et al. 2015). Another interesting feature is the appearance of a viscous fluid (probably lipidic) secretion (Wolff et al. 2015).

Insects Among insects multiple acquisitions of both, dense setae pads and smooth pads can be found (Beutel and Gorb 2006; Gorb 2001; von Byern and Grunwald

2010; Wolff and Gorb 2016). In insects smooth pads or dense setae pads, the adhesion is mediated by a liquid secretion, which is released from gland pores at the base of the setae or, in some cases, from an opening under the end-plate at the tip of the hollow adhesive hair (Gorb 1998).

Amphibia Adhesive pads in frogs are smooth, and like in insects the pads also secrete a fluid. Although the adhesive pads in tree-frogs are generally very similar, it is clear that they have evolved several times independently (Hanna and Barnes 1991). Tree frog toe-pads are made of columnar epithelial cells that are separated from each other at the apices (Barnes and Jon 2011). Pores for mucous glands open into the channels that are between the cells which create a toe pad epithelium that has an array of flat topped cells with mucous filled grooves between them (Barnes and Jon 2011). The purpose of having cells separated at the tip is to allow the toe to conform to the structure it will adhere to (Barnes and Jon 2011). The hexagonal design around the outside of the cells (similar to the crickets) is likely to allow for the mucous to spread evenly over the cell (Barnes and Jon 2011; Hanna and Barnes 1991).

Reptilia Dense setae pads systems of lizards, mainly *Anolis* and Gekkonidae do not produce fluids. Anolids and geckos are very similar and in each case depend upon a profusion of hair-like extensions that can tightly bond onto almost any surface by van der Waals forces (Bullock et al. 2008; Gorb et al. 2007). Geckos have an edge in terms of grip and agility, as they have the ability to fold and unfold their foot-pads allowing a shift from grip to release.

Mammalia Smooth adhesive pads are found in arboreal possums, which are marsupials that glide between trees (Barnes and Jon 2011). The possum is also capable of using smooth adhesive pads to climb vertically, making use of large toe pads (Barnes and Jon 2011). The pads consist of an epidermal layer of stratified squamous epithelium with the outer most layer's cells being flattened (Barnes and Jon 2011). The pad has alternating ridges and grooves with sweat glands emptying into the grooves providing fluid for wet adhesion (Barnes and Jon 2011). Bats have also evolved adhesive pads as an independent acquisition. Some bats make use of an adhesive appendage, which uses wet adhesion (Riskin and Racey 2010).

Araneae Multiple lineages of spiders have independently evolved dense setae adhesive pads (Wolff et al. 2013) that enable them to climb on vertical surfaces (Fig. 12.1). In Araneae adhesive setae are arranged covering ventral surface of tarsi and distal metatarsi (scopulae) and/or in the tip of the tarsi under the claws (claw tufts) (Figs. 12.2 and 12.3). Their pads are much more similar to many lizards and are not like the dense setae pads in insects (Bullock et al. 2008).

On rough surfaces spiders may use their claws as attachment devices; on a smooth surface adhesion is achieved by the adhesive setae on the distal leg segments (Kesel et al. 2004). This is possible due to the miniaturization and multiplication of contact elements (setules) which rely in van der Waal forces (Kesel et al. 2004). Studies on *Evarcha arcuata* (Araneae, Salticidae) have calculated that a single setule can produce an adhesive force of 41 nN perpendicular to a surface, and this

Fig. 12.1 *Psalmopoeus ecclesiasticus* female resting head down on a tree in its natural habitat, Colombian foggy forest

Fig. 12.2 Adhesive pads on the ventral surface of spiders legs, leg IV of *Grammostola anthracina*. *Sc* scopula, *CT* claw tufts (lateral view)

species possesses a safety factor of 173 (meaning the force can support 173 times the weight of the spider) (Kesel et al. 2004). This has been highly overestimated by the authors, as they assume that all visible setules are in contact with the substrate surface (Wolff and Gorb 2013, 2016).

Specialization in Mygalomorphs Morphology and arrangement of scopulae has been widely used in the taxonomy of Mygalomorphae to diagnose families, subfamilies, and even genera (Simon 1892; Pérez-Miles 1994; Guadanucci 2005). Pérez-Miles et al. (2017) described five morphological types of setae forming scopulae and claw tufts in Mygalomorphae. Adhesive setae are present in claw tufts, lateral bands of scopulae in species with divided scopula, and throughout the ventral face of tarsi in species with entire scopula, mostly in burrower/cursorial families (Pérez-Miles et al. 2017).

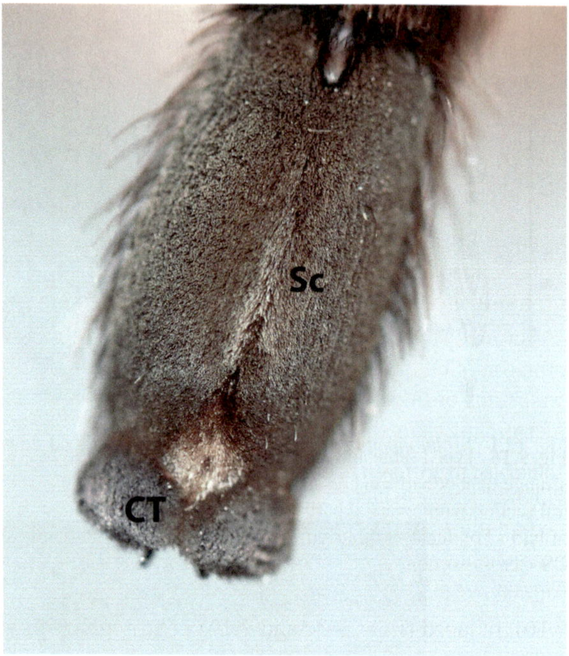

Fig. 12.3 Adhesive pads of tarsus I of *Grammostola anthracina*. *Sc* scopula, *CT* claw tufts (ventral view)

Adhesive setae are present in the following Mygalomorph families: Barychelidae, Cyrtauchenidae, Dipluridae, Euctenizidae, Idiopidae, Microstigmatidae, Nemesidae, Paratropididae, Theraphosidae (Pérez-Miles et al. 2017). At least three types of adhesive setae have been described by Pérez-Miles et al. (2017), for Mygalomorphs:

- Lamellate setae: subcylindrical setae with the distal third widened and apically curved. Apical third is densely covered by spatulate microtrichia mainly on adhesive face. This type of setae appears on the Barychelidae, Theraphosidae, and Paratropididae families.
- Lamellate crested setae: similar to the lamellate type but differs by an apical longer conspicuous microtrichia. Only found on some genera of Theraphosidae.
- Lance-shaped setae: These setae are subcylindrical with the distal third curved and not so widened as in the lamellate type, the apex is pointed. The distal half is densely covered by microtrichia on adhesive face. Present in Nemesidae, Idiopidae, Microstigmatidae, Cyrtauchenidae, and Dipluridae.

12.3 Adhesion Structures in Tarantulas

The spider leg is subdivided in sclerotized podomeres (=articles), which are articulated to each other via movable joints. These joints are operated by muscles inserting on the rim of the distal podomere, immediately after the joint (eudesmatic joints) (Shultz 1989; Ferreti et al. 2017). However, in spiders as well as in other arachnids, there is an additional articulation without muscle insertions (adesmatic joints), dividing the arachnid tip of the leg into two pseudopodomeres, the metatarsus and tarsus (= basitarsus and telotarsus, respectively) (Labarque et al. 2017). The adhesion capacity in tarantulas allows the spiders to climb on vertical surfaces (Fig. 12.1) or even inverted surfaces. This high adhesion is due to the presence of adhesive pads on the ventral surface along the tarsus and metatarsus (scopulae) and under the claws (claw tufts) (Figs. 12.2 and 12.3). The adhesive setae under the claws are organized in two plates; each claw tuft is inserted on an articulate claw tuft plate separated from the tarsus by a movable suture or membrane, thus allowing the claw tuft to move according to changes in haemolymph pressure (Wolff et al. 2013).

The adhesive pads of the scopula and claw tufts consist of a dense array of numerous flexible adhesive setae (Fig. 12.4), composed of thousands of hair-like extensions of the cuticle with spatula-like tips, called setules (Kesel et al. 2003) or microtrichia (Richards and Richards 1979).

The adhesive setae are lamelliform, subcylindrical shape with the distal third widened and apically curved. The basal third of the setae are glabrous, while the median third has sparse microtrichia and apical third is densely covered by spatulate microtrichia on one side (adhesive face) (Fig. 12.4) (Rovner 1978; Wolff et al. 2013; Lapinski et al. 2015; Pérez-Miles et al. 2017). This means these setae exhibit a bi-hierarchical spatula structure, conferring a high capacity of adhesion. Spatula-like microstructures recruit adhesive forces by generating a close contact with the substrate due to elastic deformation (Wolff and Gorb 2016). The curved apical region facilitates movements for the contact of microtrichia and due to only one side being covered with the spatulate microtrichia, friction is highly anisotropic (Niederegger and Gorb 2006; Wolff and Gorb 2013, 2016). These lamelliform setae have a similar morphology to those found in some burrower and weaver mygalomorphs and some cursorial araneomorphs (Wolff et al. 2013; Lapinski et al. 2015; Pérez-Miles et al. 2017). Likewise, among arachnids, spatulate setae occur in the distal tarsomeres of some harvestmen of the sub-order Laniatores and in hooded tickspiders (Ricinulei) (Wolff and Gorb 2016).

The microtrichia of the adhesive setae are lateral extensions of the setae; they have a subcylindrical shape with the basis slightly widened and the apex widened and laterally flattened (Pérez-Miles et al. 2017). The shape of the microtrichia apex varies taxonomically; it can be sub-circular, sub-triangular, or sub-rhomboid (Pérez-Miles et al. 2017). Contrary, the nonadhesive distal face in the adhesive setae has longer and strongly curved microtrichia, not distally widened (not spatulated) and arranged in approximately longitudinal bands (Fig. 12.5) (Pérez-Miles et al. 2017).

The orientation of the adhesive face varies in different areas of the scopula and claw tufts, favoring the adhesion in distinct directions. The adhesive face is oriented on the ventral side of the claw tuft setae (the side normally directed toward the body) and on the dorsal side of the scopula setae (the side directed distally). Additionally, on the tip of the legs (scopula apex or claw tufts base) occurs a slight rotation of the setae with the adhesive face oriented to the axis of the leg (Fig. 12.6) (Pérez-Miles et al. 2017). This anisotropic difference in the orientation of the adhesive setae plays an important role for friction control, a central feature for hunting

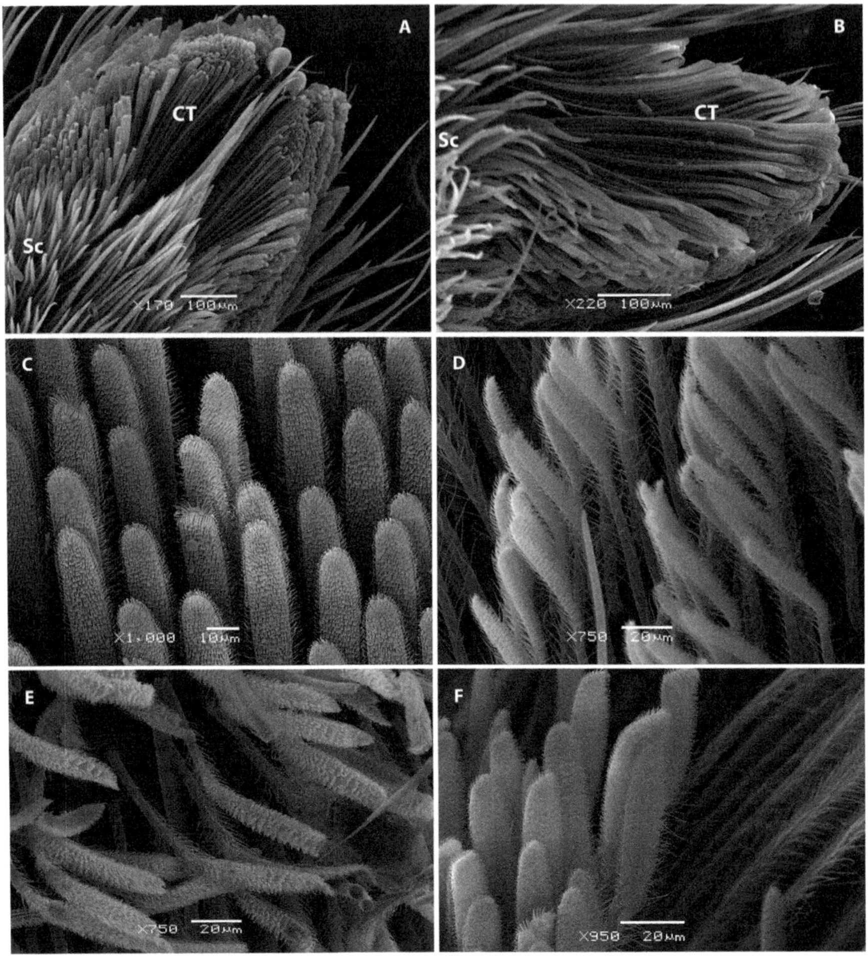

Fig. 12.4 Adhesive lamellate setae of adhesive pads on Theraphosidae. (**a**, **b**) detail of scopula (Sc) and claw tuft (CT) (**a**) *Hapalopus formosus* (**b**) *Holothele longipes*. (**c**–**f**) Close up of adhesive face on the lamellate setae (**c**) *Eupalaestrus weijenberghi* (**d**) *H. formosus* (**e**) *Plesiopelma longisternale* (**f**) *Aphonopelma seemanni* (After Pérez-Miles et al. 2017)

spiders (Wolff and Gorb 2013; Pérez-Miles et al. 2017). This particularity allows the theraphosids to climb, rest head down, and manipulate agile prey.

The scopula is present on ventral surface of all the length tarsi and can be also present on part or whole metatarsi. The morphology and arrangement of the scopula has been widely used in taxonomy and phylogeny of Mygalomorphae, being especially important in Theraphosidae (Simon 1892; Raven 1985; Pérez-Miles 1994; Guadanucci 2005; Bond et al. 2012; Wolff et al. 2013). The condition of scopula and claw tuft on all legs is a distinctive characteristic of Theraphosidae, with few exceptions, and its characteristics has been considered a good taxonomic tool and

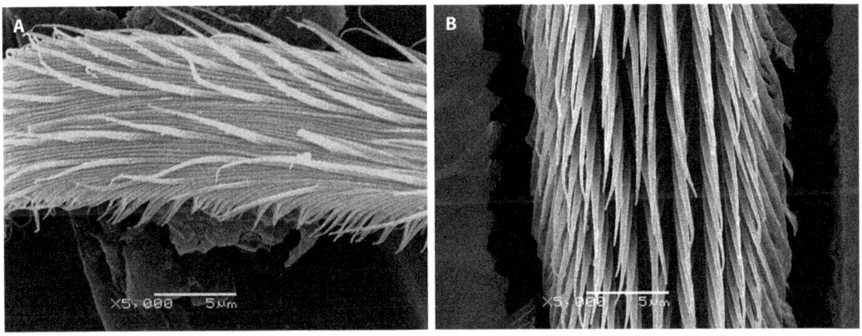

Fig. 12.5 Nonadhesive face of lamellate setae on Theraphosidae. (**a**) *Aphonopelma seemanni* (**b**) *Avicularia* sp. (After Pérez-Miles et al. 2017)

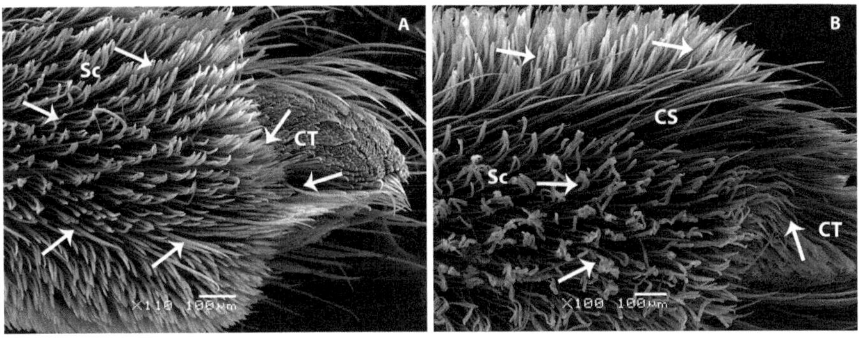

Fig. 12.6 Tips of tarsi showing the orientation of adhesive setae on leg I. (**a**) *Hapalopus formosus* (**b**) *Holothele longipes*. Diagram of the distal part of the tarsus illustrating different orientation of adhesive face of lamellate setae (modified of Wolff and Gorb 2013). Arrows indicate the orientation of adhesive face. *Sc* scopula, *CS* conical setae, *CT* claw tuft

has already been used to diagnose genera and species groups in Theraphosidae. *Agnostopelma* (Pérez-Miles and Weinmann 2010) has an unusual tarsal scopula condition. It has been the only Theraphosinae taxa described with absence of scopula in tarsus IV (Pérez-Miles and Weinmann 2010).

The density, extension, and arrangement of scopula generally vary gradually from front to back, being the front legs denser and more extensive. Additionally, the tarsal scopula can be entire or divided. Entire tarsal scopulae have homogeneous spatulate setae while divided scopulae have a longitudinal band of conical setae (Rovner 1978; Lapinski et al. 2015; Pérez-Miles et al. 2017). These conical setae are longer and thicker than spatulate types (Fig. 12.7). Microtrichia on conical setae are arranged in longitudinal rows, along the entire setae; they are dense, short, and filiform (not spatulated) (Fig. 12.7). Conical setae seem to be more related to traction or propulsion during locomotion (Rovner 1978; Pérez-Miles 1994; Wolff et al. 2013; Pérez-Miles et al. 2017).

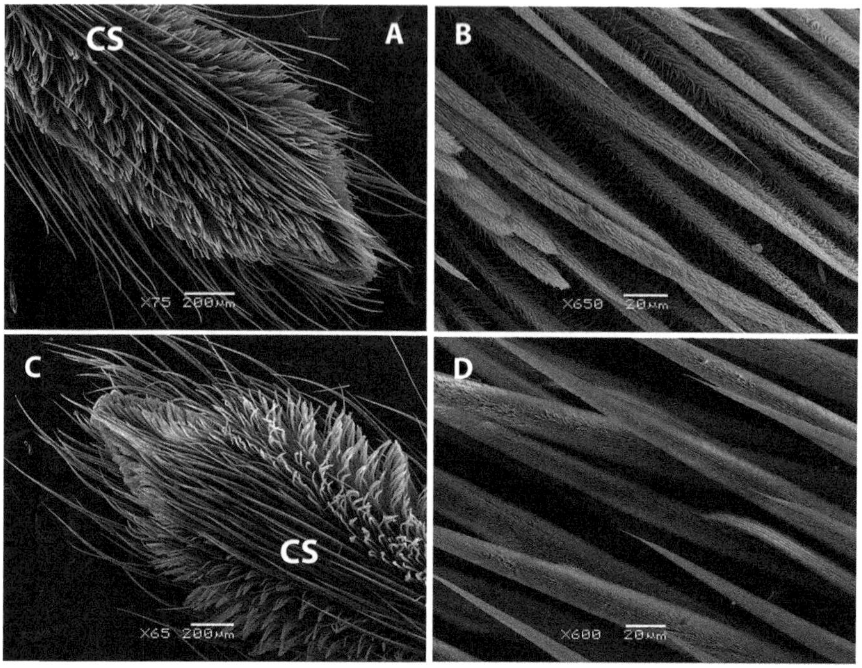

Fig. 12.7 Conical setae (CS). (**a**, **b**) *Hapalopus formosus* (**a**) tarsus IV (**b**) close up of conical setae (**c**, **d**) *Holothele longipes* (**c**) tarsus IV (**d**) close up of conical setae (After Pérez-Miles et al. 2017)

The band of conical setae increases in width from anterior to posterior legs. Consequently, the lateral bands of adhesive setae are reduced toward hind legs. Additionally, other proximal–distal gradations occur. In several theraphosids the median band of conical setae is progressively widening toward the distal part of the tarsi from forelegs to hind legs, while in some theraphosids (e.g., *Ami, Kankuamo*) the anterior–posterior gradation is inverse, on leg I the conical setae are distal in a rhomboid field, and toward the hind legs this field is progressively widening toward the proximal part of tarsi (Fig. 12.8) (Pérez-Miles et al. 2017).

The differentiation of the scopula pattern can be related with ontogenetic status or with spider size. The scopula is divided in most of juvenile theraphosids and becoming entire in adults of some groups (Pocock 1897; Gerschman de Pikelin and Schiapelli 1973; Pérez-Miles 1994), and the condition of scopula divided is related frequently with taxa of small size in Theraphosinae (Pérez-Miles 1994). On the contrary, Guadanucci (2005) found that in the Ischnocolinae, scopula condition appears to have no relation with spider size and should provide useful information for phylogenetic analysis.

Other chemosensory setae appear mixed sparsely with adhesive scopula setae, and present morphological variations between taxa (Fig. 12.9) (Pérez-Miles et al.

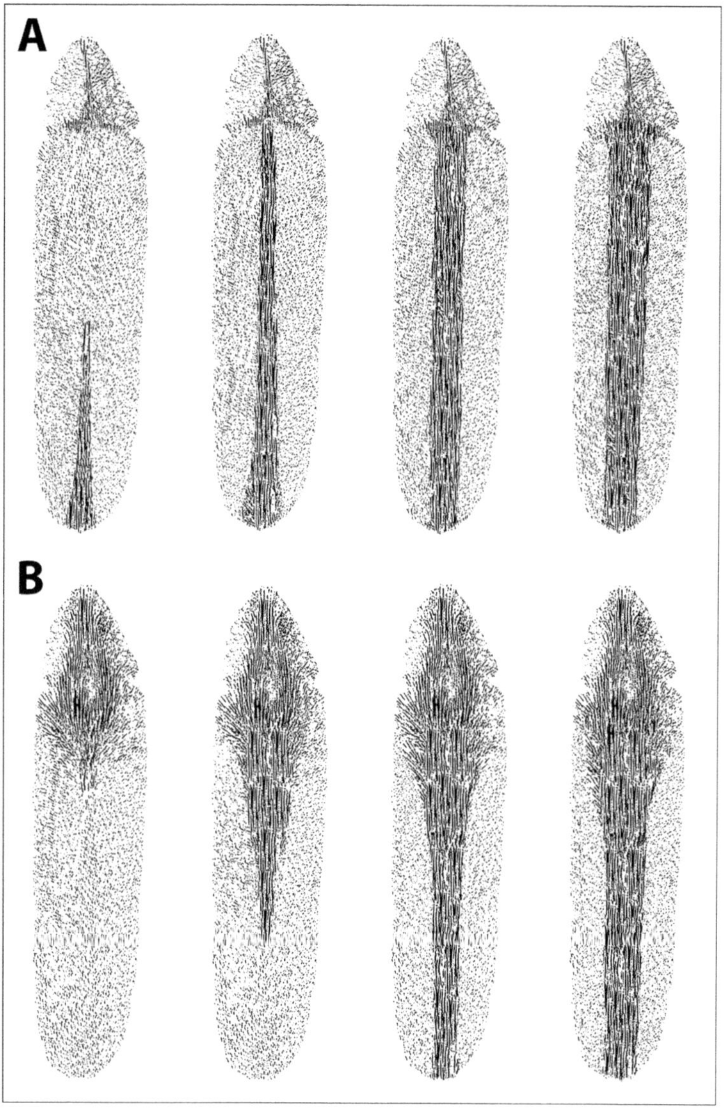

Fig. 12.8 Schematic representation of scopula division: proximal/distal and anterior (left)–posterior (right) gradation (**a**) the median band of conical setae is progressively widening toward the distal part of the tarsi from forelegs to hind legs (**b**) the conical setae are distal in a rhomboid field, and toward the hind legs this field is progressively widening toward the proximal part of tarsi (After Pérez-Miles et al. 2017)

2017). Chemoreception in spiders has been clearly established (Foelix 1970, 2011; Foelix and Chu-Wang 1975; Harris and Mill 1973). Under natural conditions chemoreception is used for testing the quality of food and in recognizing the opposite sex. Chemosensory setae are distinctly curved, blunt-tipped and possess a double

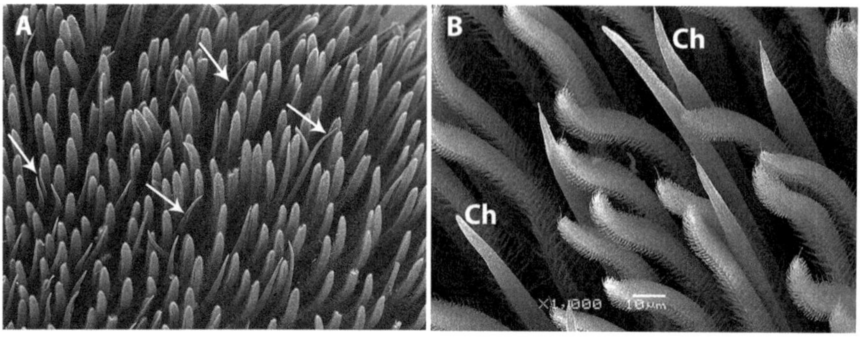

Fig. 12.9 Chemosensory setae (Ch). (**a**) Scopula setae of *Vitalius* sp. Scale bar = 100 μm (**b**) detail of chemosensory setae on scopula of *Plesiopelma longisternale*. Arrows show the chemosensory setae (After Pérez-Miles et al. 2017)

lumen. The smaller circular lumen is entered by dendrites which run up to the open tip (Foelix 1970, Foelix et al. 2013). The outer larger hair lumen does not contain any cellular elements but is filled with some fluid that was called "receptor lymph" (Foelix et al. 2012a).

12.3.1 Function of Scopula and Claw Tuft

In Theraphosidae the dense scopulae and claw tufts occur together (Raven 1985). Scopulae and claw tufts have two important functions in theraphosids: prey capture and locomotion (Homann 1957; Rovner 1978, 1980; Dunlop 1994; Foelix 2011; Pekar et al. 2011; Bond et al. 2012; Foelix et al. 2012b; Niederegger 2013; Wolff and Gorb 2012a, 2015; Wolff et al. 2013; Lapinski et al. 2015; Eggs et al. 2015; Pérez-Miles et al. 2017).

Scopulae and claw tufts consist of thousands of specialized setae that are apically broadened and cover the ventral surfaces of tarsi metatarsi and the tips of the legs under the paired claws (Pérez-Miles et al. 2015). These setae are oriented at a greater angle to the leg axis than covering setae (Pérez-Miles et al. 2015). Setae of scopulae and claw tufts are covered on their distal part by setules with spatula-shaped endings, which increase adhesion. They have a very similar structure but differ in length and density (Foelix et al. 2012b; Wolff et al. 2013; Pérez-Miles et al. 2015).

When the scopula is splayed and pressed against the surface, adhesion comes as the product of each individual setae interacting with the substrate (Peattie et al. 2011). This enables spiders with scopulae to climb even sheer, smooth surfaces such as glass.

12.3.2 Mechanism of Adhesion of the Scopula and Claw Tuft

The original mechanism proposed to explain adhesion in tarantulas is anisotropic friction (Fig. 12.10), where friction is higher when the leg pushes than when it pulls (Niederegger and Gorb 2006). However, Wohlfart et al. (2014) found complementary adhesion in *Cupiennius salei* on glass disabling anterior and posterior legs. In *Cupiennius salei* the direction of highest friction is opposite in claw tufts when compared to scopula (Wolff and Gorb 2013). The specifics of how different pads (scopula and claw tufts) are used by spiders remained unclear for long time (Pérez-Miles et al. 2015).

When a tarantula captures prey the anterior legs are pulled toward the body, so adhesion would be necessary. However, the predator also needs to be able to get rid of the prey very quickly if it turns out to be too large or too dangerous to be handled. In these cases, the ability to push is important (Pérez-Miles et al. 2015).

Pérez-Miles et al. (2015) studied static friction in live theraphosids, to test their climbing abilities on different surfaces and at different inclines and comparing burrower with arboreal species. They found a complementary participation of claw tufts and scopula of anterior and posterior legs when the tarantula climbs. They also tested differences in friction by mixing higher adhesion substrate with a lower one, and comparing friction (Pérez-Miles et al. 2015). This was to test whether an anisotropic friction was responsible of the adhesion as proposed by Niederegger and Gorb (2006) or the alternatives indicated by Wolff and Gorb (2013) and Wohlfart et al. (2014).

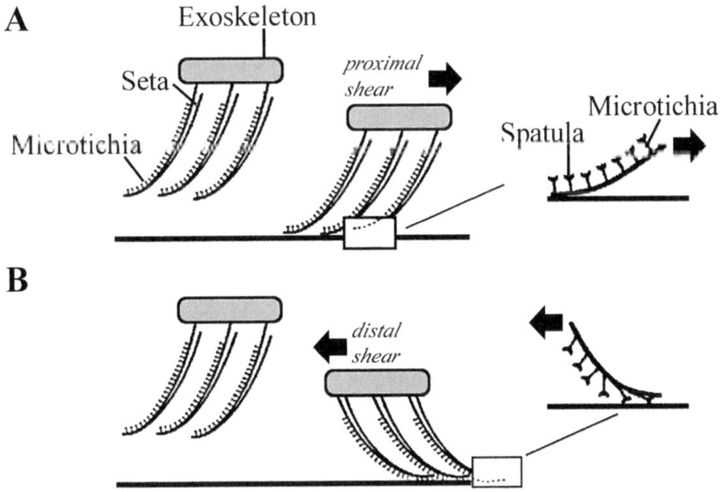

Fig. 12.10 Schematic representation of scopula adhesive setae on tarsus with the microtrichia. (**a**) when the leg pulls toward the body, the nonadhesive face of the setae contacts the substrate. (**b**) when the leg pushes distally, the scopula setae bends backward and the microtrichia enter in contact with the substrate (After Niederegger and Gorb 2006, with permission of the authors)

Pérez-Miles et al. (2015) observed no differences between the main friction of forelegs or hind legs. These results partially agree with the morpho-functional explanation given for both scopulae and claw tufts by Niederegger and Gorb (2006). These authors propose that the scopulae or claw tuft setae are curved in the proximal direction when pushed. Scopulae made contact with the substrate when the leg pushed, while claw tufts made contact when the leg pulled or during locomotion on a horizontal plane (Pérez-Miles et al. 2015).

Considering this observation together with the morphology of scopulae and claw tufts, they may function in a different way: scopulae produce adhesion when the leg pushes while claw tufts produce adhesion when the leg pulls. This observation is congruent with the opposite direction in friction forces of scopula and claw tufts found in *Cupiennius salei* and *Aphonopelma seemanni* (Niederegger and Gorb 2006; Wolff and Gorb 2013). These results could be explained by the opposite arrangement of adhesive setules, which are present on the dorsal (facing substrate) part of scopula setae, as reported by Foelix and Chu-Wang (1975) and Niederegger and Gorb (2006), and on the ventral (facing body) part (Hill 2010; Wolff and Gorb 2012a, b, 2013) of claw tuft setae.

Foelix et al. (2012b) suggested that adhesive setules are on opposite faces on the claw tufts and tarsal scopula setae, but with a different arrangement (ventral for scopulae and dorsal for claw tufts). Wolff and Gorb (2013) also found that the orientation of setae gradually changes in distal and lateral directions of claw tufts and scopula, which could influence the direction of friction force in relation to the part of the leg in contact with the substrate. This factor could optimize the complementary friction of claw tufts and scopulae including lateral components of leg movement (Pérez-Miles et al. 2015, 2017).

In adhesive scopula and claw tufts, the adhesive face is on the ventral side of claw tuft setae (the side normally directed toward the body) and on the dorsal side of the setae on the scopula (the side directed distally). On the tip of the legs (apical scopula or claw tufts) occurs a slight rotation of the setae with the adhesive face oriented to the axis of the leg (Pérez-Miles et al. 2017).

The orientation of the anisotropic adhesive setae differs between proximal tarsus in comparison with the tip of the tarsus (apical scopula or claw tuft), due to different anisotropy in the location of the microtrichia (Pérez-Miles et al. 2017). This is similar to what has been reported for large araneomorph hunting spiders (Wolff and Gorb 2013). This specific arrangement has been hypothesized to play an important role for friction control (Pérez-Miles et al. 2017). Considering the morphology, it is expected that distal setae produce adhesion when the leg pulls while proximal setae produce adhesion when the leg pushes. This hypothetical mechanism is congruent with the observations of Pérez-Miles et al. (2015), but differs from results by Niederegger and Gorb (2006) who found higher adhesion of the scopula when the leg pushes in the theraphosid *Aphonopelma seemanni*.

Pushing adhesion mechanisms are in conflict with the use of adhesive setae for prey capture, at least in the first steps of prey grabbing, because for prey attraction the adhesion must be produced when the forelegs pull the prey toward the chelicerae. However, once capture is advanced and the spider is above the prey and with legs at the sides and below, pushing adhesion may be effective.

Another interpretation is related to the full control of prey by the predator due to the mechanics of moveable setae (Rovner 1978; Eggs et al. 2015; Pérez-Miles et al. 2017). This is also possible in theraphosids since claw tufts setae are moveable (Dunlop 1994; Labarque et al. 2017). Additionally, Wolff and Gorb (2012a) stressed the importance for the predator to release the prey if it turns out to be hazardous, which should be facilitated by the control of adhesion.

Pulling adhesion is expected in claw tufts or distal tarsal scopula, so these features could be involved in both prey capture and locomotion. When the spider climbs vertically upward, adhesion may be produced by distal adhesive setae of forelegs pulling and proximal scopula of hind legs pushing (Pérez-Miles et al. 2015) with a similar dynamics as proposed by Wolff and Gorb (2013) and Wohlfart et al. (2014).

When locomotion is oriented downward, anterior scopulae push and posterior claw tufts (or distal scopulae) pull. Resting position in arboreal theraphosids is usually head down. This behavior could explain the anterior–posterior gradation of scopula indicated by Raven (1985) who observed an increasing development of such features in anterior direction. Obviously front legs are mainly involved in prey capture, as the first explanation for anterior–posterior gradations (Wolff and Gorb 2012a; Eggs et al. 2015). Perhaps mainly tarsal tips are involved in initial stages of prey capture while both tip and most part of tarsi are involved in climbing and late stages of prey capture.

12.3.3 Silk as an Adhesive Fluid?

As far as we know, all spiders secret silk through spigots located on specialized abdominal appendages, the spinnerets. Males also have spigots near the ventral genital opening to build the sperm web (Marples 1967; Ferretti et al. 2017). Singularly, spitting spiders of the genus *Scytodes* could eject a mixture of silk, glue, and venom through their fangs, from cheliceral-prosomatic glands (Monterroso 1928; Millot 1930).

Surprisingly, some years ago Gorb et al. (2006) reported that tarantulas can also secrete silk from their feet and argued that this would provide additional adhesion during climbing locomotion, and avert catastrophic falls, suggesting a third attachment mechanism. They reported that during climbing on vertical smooth surfaces if the tarantula started to slip, silk was secreted as a viscous fluid that solidifies, gluing the thread to the substrate (Gorb et al. 2006; Rind et al. 2011). According to Gorb et al. (2006), the tarsal silk resembles the silk used to cement draglines to substrates described by Vollrath and Knight (2001).

These events were later questioned by Pérez-Miles et al. (2009), who observed tarantulas climbing with free and experimentally sealed spinnerets. When the spinnerets were sealed, they did not observe silk threads on any surface, discarding leg secretion of silk.

Rind et al. (2011) fueled the controversy of tarsal silk by inducing a slight slipping of the tarsi and observing fine silk threads that emerge from ribbed "spigots" on the ventral tarsi. These structures were also reported by Gorb et al. (2006), but Pérez-Miles et al. (2009) found no structures interpretable as silk glands or silk conduits with transverse cuts.

Pérez-Miles and Ortiz-Villatoro (2012) revisited the tarsal silk controversy, testing several species of tarantulas (including an arboreal one). These authors used tarantulas with both free and sealed spinnerets on larger vertical surfaces and induced them to slip, shaking gently the surfaces. They confirmed the results by Pérez-Miles et al. (2009), denying the presence of a silk trail on the legs (when the spinnerets are sealed) and suggested that silk is a light sticky fiber that can easily adhere to a surface. Consequently, passive contamination with spinneret silk is the most likely explanation to the traces found in the studies of Gorb et al. (2006) and Rind et al. (2011).

Peattie et al. (2011) found fluid footprints when arachnids climbed on vertical surfaces, apparently originating from tarsal setae, but their silken nature was not confirmed. These fluids were examined by Pérez-Miles and Ortíz-Villatoro (2012) who determined that this substance clearly differs from silk because they are made up of groups of droplets.

The data presented by Peattie et al. (2011) for spiders show an extreme wetting of the setae, such that the spatulae are submerged in the fluid. Adhesion models show that in hairy pads, secretion enhances adhesion, only if it wets the space between the spatula and the substrate (Huber et al. 2005). If the fluid film is thicker than the spatula, slipping occurs (Bullock et al. 2008; Wolff and Gorb 2012b). In the cases of fluid appearance, the setal tips were completely wetted and the pad slid along the surface. Hence, adhesion was obviously reduced.

The presence of a thin fluid film between the scopula and a smooth substrate may play an important role in the adhesion mechanism (Homann 1957; Wolff and Gorb 2012b; Pérez-Miles and Ortiz-Villatoro 2012). Thus, the secretions observed by Peattie et al. (2011), Wolff and Gorb (2013) and Pérez-Miles and Ortíz-Villatoro (2012) unlikely support adhesion, but may have the function to clean the setae, control (reduce) adhesion, or may even result from unusual behavior of the animals, for example, due to the strong illumination used in such visualization methods (Wolff and Gorb 2013).

According to Foelix et al. (2012b), these footprints may originate from chemosensory setae and can be composed of receptor lymph. This substance can probably appear as fluid and sometimes as dryer filamentous lymph, depending on relative humidity, temperature, and the hydration state of the spider, which could explain the confusion with fibers. Likewise, Foelix et al. (2012b) found that the "silk spigots" reported by Gorb et al. (2006) and Rind et al. (2011) have all diagnostic features of chemosensitive hairs (contact chemoreceptor) in spiders, rather than a silk spigot.

Additionally, Niederegger and Gorb (2006) demonstrated that artificially dried spider scopulae may have diminished adhesion, although this may be due to desiccation of the cuticle (Wolff and Gorb 2013). Experiments with dead specimens of spider legs show that their hairy adhesive pads can generate high adhesion and friction on smooth surfaces even in a dry state (Kesel et al. 2003; Niederegger and Gorb 2006; Wolff and Gorb 2013).

Although the adhesion of tenent setae on the tarsi and claw tufts of legs and pedipalps almost certainly outweighs any potential contribution from the sparsely distributed trails secreted by the ribbed hairs (Foelix et al. 2012b).

12.3.4 *Climbing*

Pérez-Miles et al. (2015) observed that when climbing upward, palps, legs I and II touched the surfaces only with a part of the claw tufts, while leg pair IV touched the surface with the distal portion of tarsal scopulae and only in rare cases with claw tufts. It was also observed that leg IV was more extended than the resting legs.

When climbing, legs III usually touched the surface with a part of the claw tufts but in some cases, when this leg was extended backwards, it also touched the surface with the distal portion of the tarsal scopulae (Pérez-Miles et al. 2015). During increased inclination the contact area of anterior claw tuft and posterior scopula was increased, while the contact area of posterior claw tufts slightly decreased. Usually, just before a leg was elevated for a step, we observed movements in the tarsal claw tufts (Pérez-Miles et al. 2015).

12.4 Ecological and Evolutionary Aspects of the Adhesion Structures in Tarantulas

The spatulate setae as a mechanism in dry adhesion has evolved multiple times independently within arthropods and reptiles (e.g., in geckoes and anoles, mites of the order Trombidiformes, and hunting spiders) (Peattie and Full 2007; Wolff et al. 2013; Ramírez 2014). The widespread occurrence of powerful adhesive setae suggests their importance in locomotion, mainly to climb on inclined or vertical surfaces, and in prey capture.

A hypothesis has emphasized the locomotory function of the adhesive setae as an evolutionary driver of adhesive structures in spiders, due the impressive climbing ability of the spiders, although this predominantly includes only the distal end of the legs, principally the claw tufts (Foelix and Chu-Wang 1975; Roscoe and Walker 1991; Dunlop 1994; Kesel et al. 2003; Niederegger and Gorb 2006; Wolff and Gorb 2012a, c). An the alternative hypothesis has been proposed: prey capture could be the original evolutionary driving force in the evolution of attachment organs in spiders, and that their use in locomotion was a secondary benefit that led to the occupation of new habitats (Rovner 1978; Miller et al. 1988; Wolff et al. 2013). According to Wolff et al. (2013), this can be regarded as a key innovation that promoted the great radiations of the Dionycha in the Eocene and of the Theraphosidae in the Miocene (Penney and Selden 2011).

In spiders, the use of silk for capturing and immobilizing prey is widespread. However, some lineages never evolved a silk-dependent prey capture mode or have independently abandoned web building and explored alternative hunting strategies. Thus, free hunting lifestyles implicate alternative mechanisms for capturing, securing, and handling of prey. The adhesive pads located on spider ventral apical legs have been hypothesized as an evolutionary alternative to prey capture webs. Wolff et al. (2013) found a close association between the adhesive setae and the free hunter lifestyle in spiders. They suggested that scopulae evolved as a substitute for silk in prey control and that the claw tufts are, in most cases, a secondary development. This transition from webs to active hunting occurred independently in many spider taxa and resulted in multiple origins of remarkably similar sticky hairs (Wolff et al. 2013; Ramírez 2014).

The hypothesis that the claw tufts were derived as a distal specialization from scopula (Wolff et al. 2013) was also supported by Pérez-Miles et al. (2017), when describing the rotation of adhesive face in distal scopula in non-claw-tufted mygalomorph families (some nemesiids, idiopids, microstigmatids, cyrtaucheniids, euctenizids, and diplurids). In these scopulae the adhesive face of setae is oriented dorsally in the proximal tarsus (like a normal scopula) and ventrally in the apical tarsus (like a normal claw tuft).

Rovner (1978), Foelix et al. (1984), Pekar et al. (2011), Wolff et al. (2013), and Eggs et al. (2015) proposed the participation of scopulae in prey manipulation in Araneomorphae. The following observations support the hypothesis that scopulae in spiders are generally an adaptation for prey capture (Wolff et al. 2013): (1) Scopulae are often restricted to or more developed in the anterior legs; (2) Scopulate setae are mainly distributed in the pro- and retro-lateral parts of the tarsus, metatarsus and tibia, whereas at the ventral side they are lacking; (3) Most spiders walk on their pretarsal tips; and (4) The adhesive sides of the scopula setae are often facing away from the ground in resting animals and become erect by increased hemolymph pressure.

On the other hand, the evolution of claw tufts is associated with the formation of a highly sclerotized basal plate in the pretarsal region, thus being articulated with the tarsus; it permits spreading and mobility of the pads and may facilitate control of both attachment and detachment (Speck and Barth 1982; Dunlop 1994; Hill 2010; Labarque et al. 2017). It occurs in the derived lineages of the Dionycha (except gnaphosids, most thomisids, and some salticids), Dysderoidea, Theraphosidae, and Ctenidae (Wolff et al. 2013). Thus, although obviously claw tufts also are used for prey capturing, mating, and grooming, there are some arguments supporting the primary locomotory function of the claw tufts differing from the function of the scopulae (Wolff et al. 2013): (1) Claw tufts typically contact the substrate in walking and climbing spiders; (2) In contrast to the scopulae, claw tufts are always well developed in all the legs of those species which feature them; (3) Whereas scopulae are more developed in the anterior legs, claw tufts are often larger in the posterior legs presumably because these produce the highest forward thrust; and (4) Among free hunting spiders those having claw tufts make up a significantly higher proportion found in above-ground and highly structured habitats such as broad leaf litter, than found at ground level and on even substrates. The last argument interprets the claw tufts as an adaptation to enhance climbing ability.

The orientation of the anisotropic adhesive setae that differs between the proximal tarsus in comparison with the tip of the tarsus (apical scopula or claw tuft), and consequently the different orientation of the microtrichia (Wolff and Gorb 2013; Pérez-Miles et al. 2017) suggest that the explanation of adhesive mechanism and biological function of scopula and claw tuft could be more complex. Considering the morphology, it is expected that distal setae produce adhesion when the leg pulls while proximal setae produce adhesion when the leg pushes (Pérez-Miles et al. 2015, 2017). Theraphosids walk on horizontal surfaces using their tarsal tips, but when climbing on vertical surfaces a close contact exists between most part of tarsi which push on the surface (Pérez-Miles et al. 2015). It should be noted that arboreal species of Theraphosidae show more developed scopulae than terrestrial species.

In this sense, Pérez-Miles et al. (2017) questioned the function of scopulae in prey manipulation in Mygalomorphae and argued that this is only applicable to apical scopula or claw tufts, considering that the adhesive faces of most part of scopula setae are oriented dorsally (distally) facilitating pushing adhesion. Pushing adhesion may be effective once capture is advanced and the spider is above the prey and with legs at the sides and below (Pérez-Miles et al. 2017). Another interpretation is related with the full control of prey by the predator due to the mechanics of moveable setae, as happens in mygalomorphs since claw tufts setae are moveable (Dunlop 1994; Labarque et al. 2017). Rovner (1978) proposed that erectile scopular setae could change of orientation during attack which was confirmed in lycosid spiders by Eggs et al. (2015). This could be an alternative explanation for adhesion during prey capture.

Pulling adhesion is expected in claw tufts or in the distal part of tarsal scopula, so these features could be involved in both prey capture and adhesion for locomotion. When the spider climbs vertically upward, adhesion may be produced by apical adhesive setae of forelegs pulling and proximal scopula of hind legs pushing (Pérez-Miles et al. 2015). Inversely, when the locomotion is oriented downward, anterior scopulae push and posterior claw tufts (or distal scopulae) pull. Resting position in arboreal theraphosids is usually head down. This behavior could be explained by the anterior–posterior gradation of scopula indicated above. Although, obviously front legs are mainly involved in prey capture, and is the first explanation for anterior–posterior gradations.

Among the Mygalomorphae the adhesive setae have been also reported in Barychelidae, Cyrtaucheniidae, Idiopidae, Paratropididae, Nemesiidae (Wolff et al. 2013), and in females of the idiopids *Misgolas*, *Euoplos* and *Aganippe*, diplurids *Diplura*, *Trechona* and *Linothele*, and some euctenizids (Pérez-Miles et al. 2017). Some of them use a sheet-web for prey capture or inhabit burrows with wafer-lid trapdoors; therefore, these groups may be exceptions to the hypothesis that suggest that scopulae evolved as a substitute for silk (Wolff et al. 2013). Mygalomorph foraging webs employ silk either in a sheet web or at a burrow entrance to detect, localize, and manipulate prey (Coddington et al. 2019). However, in mygalomorphs silk is not adhesive enough for prey capture as in most araneomorphs, so the participation of adhesive setae would be necessary.

According to Pérez-Miles et al. (2017), adhesive scopulae were acquired in the Crassitarsae, in the Euctenizidae and in some idiopids (*Misgolas*, *Euoplos*, and *Aganippe*) (based on the phylogenetic hypothesis of Bond et al. (2012)), and the presence of adhesive scopula has a relationship with cursorial/burrower taxa, despite of their exceptional presence in some trapdoors and Diplurinae.

Pérez-Miles et al. (2017) hypothesize two scenarios for the origin of adhesive scopulae in Mygalomorphae: (1) two independent acquisitions of adhesive scopula in Crassitarsae and Euctenizoidina with few generic reversions or (2) its acquisition in the Bipectina with a reversion in the Ctenizoidina. According to first hypothesis, the adhesive scopula of Crassitarsae and Euctenizoidina is not homologous (Bond and Opell 2002; Pérez-Miles et al. 2017), and its association with the burrower or cursorial lifestyle indicates that it probably evolved together with ecological characteristics.

The claw tufts are much more restricted in Mygalomorphae than in Araneomorphae. They are present only in Theraphosidae and Barychaelidae with a parallelism in *Melloina* (Paratropididae), but in this last case without scopula (Raven 1985; Pérez-Miles et al. 2017). According to Pérez-Miles et al. (2017) and Wolff et al. (2013), claw tufts seem to be acquired twice in Mygalomorphae, just in cursorial/burrower spiders, and they were derived from scopula.

References

Autumn K, Niewiarowski PH, Puthoff JB (2014) Gecko adhesion as a model system for integrative biology, interdisciplinary science, and bioinspired engineering. Ann Rev Ecol Evol Syst 45:445–470

Barnes W, Jon P (2011) Functional morphology and design constraints of smooth adhesive pads. MRS Bull 32:479–485

Beutel RG, Gorb SN (2001) Ultrastructure of attachment specializations of hexapods (Arthropoda): evolutionary patterns inferred from a revised ordinal phylogeny. J Zool Syst Evol Res 39:177–207

Beutel RG, Gorb SN (2006) A revised interpretation of the evolution of attachment structures in Hexapoda with special emphasis on Mantophasmatodea. Arthropod Syst Phylo 64:3–25

Bond JE, Opell BD (2002) Phylogeny and taxonomy of the genera of south-western North American Euctenizinae trapdoor spiders and their relatives (Araneae: Mygalomorphae, Cyrtaucheniidae). Zool J Linn Soc Lond 136:487–534

Bond JE, Hendrixson BE, Hamilton CA, Hedin M (2012) A reconsideration of the classification of the spider infraorder Mygalomorphae (Arachnida: Araneae) based on three nuclear genes and morphology. PLoS One 7(6):e38753. https://doi.org/10.1371/journal.pone.0038753

Bullock JM, Drechsler P, Federle W (2008) Comparison of smooth and hairy attachment pads in insects: friction, adhesion and mechanisms for direction-dependence. J Exp Biol 211:3333–3343

Coddington JA, Agnarsson I, Hamilton CA, Bond JE (2019) Spiders did not repeatedly gain, but repeatedly lost, foraging webs. PeerJ 7:e6703. https://doi.org/10.7717/peerj.6703

Dirks JH, Federle W (2011a) Fluid-based adhesion in insects—principles and challenges. Soft Matter 7:11047

Dirks JH, Federle W (2011b) Mechanisms of fluid production in smooth adhesive pads of insects. J R Soc Interface 8:952–960

Dunlop JA (1994) Movements of scopulate claw tufts at the tarsus tip of a tarantula spider. Neth J Zool 45:513–520

Eggs B, Wolff JW, Khun-Nentwig L, Gorb SN, Nentwig W (2015) Hunting without a web: how lycosoid spiders subdue their prey. Ethology 121:1166–1177

Federle W (2006) Why are so many adhesive pads hairy? J Exp Biol 209:2611–2621

Ferretti N, Pompozzi G, Copperi S, Wehitt A, Galíndez E, González A, Pérez-Miles F (2017) A comparative morphological study of the epiandrous apparatus inmygalomorph spiders (Araneae, Mygalomorphae). Micron 93:9–19

Foelix RF (1970) Structure and function of tarsal sensilla in the spider *Araneus diadematus*. J Exp Zool 175:99–124

Foelix RF (2011) Biology of spiders, 3rd edn. Oxford University Press, New York, pp 188–217

Foelix RF, Chu-Wang IW (1975) The structure of scopula hairs in spiders. In: Proc 6th Int Congr Arachnol, Amsterdam, pp 56–58

Foelix R, Erb B, Rast B (2013) Alleged silk spigots on tarantula feet: Electron microscopy reveals sensory innervation, no silk. Arthropod Structure and Development 42:209–217

Foelix RF, Jackson RR, Henksmeyer A, Hallas S (1984) Tarsal hairs specialized for prey capture in the salticid Portia. Rev Arachnol 5:329–334

Foelix RF, Rast B, Erb B (2012a) Hafthaare bei Vogelspinnen: Vergleich einer bodenlebenden Brachypelma mit einer baumlebenden *Poecilotheria*. Arachne 17:16–23

Foelix R, Rast B, Peattie AM (2012b) Silk secretion from tarantula feet revisited: alleged spigots are probably chemoreceptors. J Exp Biol 215:1084–1089

Gerschman de Pikelin BS, Schiapelli RD (1973) La subfamilia "Ischnocolinae" (Araneae: Theraphosidae). Rev Mus Arg C Nat "Bernardino Rivadavia", Entomologia 4:43–77

Gorb SN (1998) The design of the fly adhesive pad: distal tenent setae are adapted to the delivery of an adhesive secretion. Proc R Soc Lond B 265:747–752

Gorb SN (2001) Attachment devices of insect cuticle. Kluwer Academic, Dordrecht

Gorb SN, Niederegger S, Hayashi CY, Summers AP, Vötsch W, Walther P (2006) Silk-like secretion from tarantula feet. Nature 443:407

Gorb SN, Sinha M, Peressadko A, Daltorio KA, Quinn RD (2007) Insects did it first: a micropatterned adhesive tape for robotic applications. Bioinspir Biomim 2:117–125

Guadanucci JPL (2005) Tarsal scopula significance in Ischnocolinae phylogenetics (Araneae, Mygalomorphae, Theraphosidae). J Arachnol 33:456–467

Hanna G, Barnes WJP (1991) Adhesion and detachment of the toe pads of tree frogs. J Exp Biol 155:103–125

Harris DJ, Mill PJ (1973) The ultrastructure of chemoreceptor sensilla in *Ciniflo* (Araneida, Arachnida). Tissue Cell 5:679–689

Hill DE (2010) Jumping spider feet (Araneae, Salticidae). Peckhamia 85:1–41

Homann H (1957) Haften Spinnen an einer Wasserhaut? Naturwissenschaften 44:318–319

Huber G, Mantz H, Spolenak R, Mecke K, Jacobs K, Gorb SN, Arzt E (2005) Evidence for capillarity contributions to gecko adhesion from single spatula nanomechanical measurements. Proc Natl Acad Sci USA 102:16293–16296

Irschick DJ, Austin CC, Petren K, Fisher RN, Losos JB, Ellers O (1996) A comparative analysis of clinging ability among pad-bearing lizards. Biol J Linn Soc 59:21–35

IUPAC (2006) Compendium of chemical terminology, 2nd ed. (the "Gold Book"). Compiled by A. D. McNaught and A. Wilkinson. Blackwell Scientific Publications, Oxford (1997). XML on-line corrected version: http://goldbook.iupac.org. Created by M. Nic, J. Jirat, B. Kosata; updates compiled by A. Jenkins. ISBN 0-9678550-9-8. https://doi.org/10.1351/goldbook

Jagota A, Hui CY (2011) Adhesion, friction, and compliance of bio-mimetic and bio-inspired structured interfaces. Mater Sci Eng Rep 72:253

Kesel AB, Martin A, Seidl T (2003) Adhesion measurements on the attachment devices of the jumping spider *Evarcha arcuata*. J Exp Biol 206:2733–2738

Kesel AB, Martin A, Seidl T (2004) Getting a grip on spider attachment: an AFM approach to microstructure adhesion in arthropods. Smart Mater Struct 13:512–518

Labarque FM, Wolff JO, Michalik P, Griswold CE, Ramírez MJ (2017) The evolution and function of spider feet (Araneae: Arachnida): multiple acquisitions of distal articulations. Zool J Linnean Soc 20:1–34

Labonte D, Clemente CJ, Dittrich A, Kuo CY, Crosby AJ, Irschick DJ, Federle W (2016) Extreme positive allometry of animal adhesive pads and the size limits of adhesion-based climbing. Proc Natl Acad Sci 113:1297–1302

Lapinski W, Walther P, Tschapka M (2015) Morphology reflects microhabitat preferences in an assemblage of neotropical wandering spiders. Zoomorphology 134:219–236

Marples BJ (1967) The spinnerets and epiandrous glands of spiders. Zool J Linn Soc 46:209–223

Miller GL, Miller PR, Brady AR (1988) Adhesive hairs in lycosid spiders of various life styles, including the occurrence of claw tufts in *Lycosa hentzi* banks. Bull Br Arachnol Soc 7:213–216

Millot J (1930) Glandes venimeuses et glandes sèricigènes chez les Sicariides. Bull Soc Zool Fr 55:150–174

Monterroso B (1928) Observazioni preliminar sella del genere *Scytodes* (Walk) (Araneae Verae: Sicariidae). Rend Reale Accad Naz Lincei Ser 6(6):171–174

Niederegger S (2013) Functional aspects of spider scopulae. In: Nentwig W (ed) Spider ecophysiology. Springer, Berlin, pp 57–66

Niederegger S, Gorb SN (2006) Friction and adhesion in the tarsal and metatarsal scopulae of spiders. J Comp Physiol A 192:1223–1232

Peattie AM, Full JR (2007) Phylogenetic analysis of the scaling of wet and dry biological fibrillar adhesives. Proc Natl Acad Sci 104:18595–18600

Peattie AM, Dirks JH, Henriques S, Federle W (2011) Arachnids secrete a fluid over their adhesive pads. PLoS One 6(5):e20485. https://doi.org/10.1371/journal.pone.0020485

Pekar S, Sobotnık J, Lubin Y (2011) Armoured spiderman: morphological and behavioural adaptations of a specialised araneophagous predator (Araneae: Palpimanidae). Naturwissenschaften 98:593–603

Penney D, Selden PA (2011) Fossil spiders: the evolutionary history of a megadiverse order, Monograph series, volume 1. Siri Scientific Press, Manchester, p 128

Pérez-Miles F (1994) Tarsal scopula division in Theraphosinae (Araneae, Theraphosidae): its systematic significance. J Arachnol 22:46–53

Pérez-Miles F, Ortíz-Villatoro D (2012) Tarantulas do not shoot silk from their legs: evidence in four species of New World tarantulas. J Exp Biol 215:1749–1752

Pérez-Miles F, Weinmann D (2010) *Agnostopelma*: a new genus of tarantula without a scopula on leg IV (Araneae: Theraphosidae: Theraphosinae). J Arachnol 38:104–112

Pérez-Miles F, Panzera A, Ortíz-Villatoro D, Perdomo C (2009) Silk production from tarantula feet questioned. Nature 461:E9–E10

Pérez-Miles F, Perafán C, Santamaría L (2015) Tarantulas (Araneae: Theraphosidae) use different adhesive pads complementary during climbing on smooth surfaces: experimental approach in eight arboreal and burrower species. Biol Open 4:1643–1648

Pérez-Miles F, Guadanucci JPL, Jurgilas JP, Becco R, Perafán C (2017) Morphology and evolution of scopula, pseudoscopula and claw tufts in Mygalomorphae (Araneae). Zoomorphology 136:435–459

Pocock RI (1897) On the spiders of the suborder Mygalomorphae from the Ethiopian Region, contained in the collection of the British Museum. Proc Zool Soc London 1897:724–774

Ramírez MJ (2014) The morphology and phylogeny of dionychan spiders (Araneae, Areneomorphae). Bull Am Mus Nat Hist 390:1–374

Raven R (1985) The spider infraorder Mygalomorphae (Araneae): cladistics and systematics. Bull Am Mus Nat Hist 182:1–180

Richards AG, Richards PA (1979) The cuticular protuberances of insects. Int J Insect Morphol Embryol 8:143–157

Rind FC, Birkett CL, Duncan BJ, Ranken AJ (2011) Tarantulas cling to smooth vertical surfaces by secreting silk from their feet. J Exp Biol 214:1874–1879

Riskin DK, Racey PA (2010) How do sucker-footed bats hold on, and why do they roost head-up? Biol J Linn Soc 99:233–240

Roscoe DT, Walker G (1991) The adhesion of spiders to smooth surfaces. Bull Br Arachnol Soc 8:224–226

Rovner JS (1978) Adhesive hairs in spiders: behavioral functions and hydraulically mediated movement. Symp Zool Soc Lond 42:99–108

Rovner JS (1980) Morphological and ethological adaptations for prey capture in wolf spiders (Araneae, Lycosidae). J Arachnol 8:201–215

Shultz JW (1989) Morphology of locomotor appendages in Arachnida: evolutionary trends and phylogenetic implications. Zool J Linnean Soc 97:1–56

Simon E (1892) Histoire Naturelle des Araigneés, vol 1. Roret, Paris, p 256

Speck J, Barth FG (1982) Vibration sensitivity of pretarsal slit sensilla in the spider leg. J Comp Physiol A 148:187–194

Vollrath F, Knight DP (2001) Liquid crystalline spinning of spider silk. Nature 410:541–548

Von Byern J, Grunwald I (eds) (2010) Biological adhesive systems: from nature to technical and medical application. Springer, Wien

Williams EE, Peterson JA (1982) Convergent and alternative designs in the digital adhesive pads of Scincid lizards. Science 215:1509–1511

Wohlfart E, Wolff JO, Arzt E, Gorb SN (2014) The whole is more than the sum of all its parts: collective effect of spider attachment organs. J Exp Biol 217:222–224

Wolff J, Seiter M, Gorb SN (2015) Functional anatomy of the pretarsus in whip spiders (Arachnida, Amblypygi). Arthropod Structure and Development 44:524–540

Wolff JO, Gorb SN (2012a) Comparative morphology of pretarsal scopulae in eleven spider families. Arthropod Struct Dev 41:419–433

Wolff JO, Gorb SN (2012b) The influence of humidity on the attachment ability of the spider *Philodromus dispar* (Araneae, Philodromidae). Proc R Soc B Biol Sci 279:139–143

Wolff JO, Gorb SN (2012c) Surface roughness effects on attachment ability of the spider *Philodromus dispar* (Araneae, Philodromidae). J Exp Biol 215:179–184

Wolff JO, Gorb SN (2013) Radial arrangement of Janus-like setae permits friction control in spiders. Sci Rep 3:1101

Wolff JO, Gorb SN (2015) Adhesive foot pads: an adaptation to climbing? An ecological surveyin hunting spiders. Zoology 118:1–7

Wolff JO, Gorb SN (2016) Attachment structures and adhesive secretions in arachnids. Springer International Publishing, Cham, p 184. https://doi.org/10.1007/978-3-319-45713-0

Wolff JO, Nentwig W, Gorb SN (2013) The great silk alternative: multiple co-evolution of web loss and sticky hairs in spiders. PLoS One 8(5):e62682. https://doi.org/10.1371/journal.pone.0062682

Chapter 13
Biomechanics of Locomotion in Tarantulas

Carlo M. Biancardi and Valentina Silva-Pereyra

Abstract Terrestrial multi-legged locomotion is an energetically demanding activity. The limbs need to exert force on the ground to support and move the body weight and negotiate uneven surfaces. The locomotor performances of Theraphosidae are limited by their poor aerobic capacities. The smaller body size and longer legs of the more active sex (males) are considered results of an optimisation to reduce the high metabolic cost of locomotion.

A large fraction of the mechanical work is done against gravity, to lift the body centre of mass with each step. Both horizontal work (to push the centre of mass forward) and internal work (done to move the limbs with respect to the centre of mass) represent a small part of the total work.

Unlike other spiders, Theraphosidae employ all of their limbs for locomotion. The first described stepping pattern was an alternating tetrapod gait, in which the odd limbs on one side move together with the contralateral even limbs. Nevertheless, we are able to discriminate different quadruped-similar gait patterns, such as lateral and diagonal walking and trotting. Unlike quadrupedal vertebrates, the highest speeds are reached mainly by increasing stride frequency, while stride length remains roughly constant.

13.1 Introduction

Theraphosidae are large terrestrial spiders, which live in different habitats, from deserts to cloud forests (see Chap. 7). As burrowing predators, they ambush to capture their prey in trees or on the ground (see Chap. 8). The efficiency of the cardio-respiratory system of these spiders is generally low, and aerobic metabolism can

C. M. Biancardi
Biomechanics Laboratory, Department of Biological Sciences, CenUR Litoral Norte, Universidad de la República, Paysandú, Uruguay

V. Silva-Pereyra (✉)
Department of Biophysics, Faculty of Medicine, Universidad de la República, Montevideo, Uruguay

F. Pérez-Miles (ed.), *New World Tarantulas*, Zoological Monographs 6,
https://doi.org/10.1007/978-3-030-48644-0_13

sustain only slow locomotor movements above the basal activities (Anderson and Prestwich 1985; Paul et al. 1989) (see Chap. 10). Skeletal muscle work is one of the most demanding activities in terms of energetic cost. Therefore, the development of locomotion in terrestrial spiders—considering their physiological limits and the structural and physical constraints derived from their anatomy, their behaviours and the environments in which they live—is of particular interest.

In this chapter we will explore different aspects of Theraphosidae locomotion: the gait patterns, the metabolic costs, the cruise and maximum speed ranges, the mechanics and efficiency of locomotion, and the particular hydraulic system that allows for limb extension.

13.2 Gait Analysis

In the terrestrial environment, displacement with legs is not an easy locomotive action. It requires enough force to move the body forward while supporting the body weight. It also requires continuous replacement of legs on the substrate, which generates a cyclic locomotive action (Saibene and Minetti 2003). This continuous rearrangement of the limbs must be carefully coordinated to avoid trips and falls, overcome obstacles, turn, change speed, climb and descend slopes, and, in some cases, even walk upside down. The action of the legs must meet such demands and adjust to the needs of the species relative to the environment that surrounds them (Seyfarth 1985). To fulfill these demands, not just any combination of limb action sequences is possible or efficient (Wilson 1967); they should at least provide dynamic stability, which is one of the main functions of locomotion (Alexander 2003). Static stability of a standing-legged animal is achieved when the vertical projection of its body centre of mass (b_{COM}) lies within the convex polygon formed by all of the supporting feet (McGhee and Iswandhi 1979). Therefore, the minimum requirement for static stability is three simultaneously supporting feet (Ting et al. 1994). During locomotion, the vertical projection of b_{COM} can approach the edges of, and fall outside, the support polygon, or the polygon itself can disappear when fewer than three feet are in contact with the ground. In these cases, dynamic adjustments to avoid a fall are needed (Alexander 2003). The time for a gravitationally induced failure scales inversely with b_{COM} height; therefore, smaller animals need more supporting legs to achieve quasi-static stability during locomotion (Ting et al. 1994; Wilshin et al. 2018).

In natural environments, each step cycle is slightly different from the next, as each limb should adjust its movement to avoid any obstacles previously detected by the sensory organs. This information will be integrated by the central nervous system, together with proprioceptive information, to generate coordinated movement through motor pattern generators (Seyfarth 1985; Nishikawa et al. 2007). Motor pattern generators command the repetitive and rhythmic limb action in each cycle of locomotion (Seyfarth 1985).

A step cycle (stride) is a time interval that is initiated with the action of one leg (e.g. the first instant of foot contact with the ground) and ends the instant before the same action occurs again. The limbs with locomotive function during a step cycle have two phases: a support phase (also called the stance phase), in which the foot is in contact with the substrate, and a swing phase, in which the relocation of the leg occurs without contact with the ground (Cavagna et al. 1976). It is during the stance phase that the foot is fixed, with zero displacement speed, and exerts force against the ground to move the centre of body mass (Cavagna 2017).

Booster et al. (2015) observed how legs work to move the body of *Aphonopelma hentzi* (Girard, 1852) forward, studying the femur–patella and tibia–metatarsus joints of the first and fourth pairs of legs. They reported that the forelegs and hind legs move in opposite phases but work together when in contact with horizontal surfaces. During the support phase, the joints of the forelegs flex, while the joints of the hind legs extend. However, in the swing phase, the joints of the forelegs extend and the joints of the hind legs flex.

Nevertheless, the stance phase is composed of two functionally important subphases: one in which the limbs generate a braking force to receive the body weight, followed by one in which the limbs generate a propulsive force to move the body centre of mass in the direction of the displacement (Griffin et al. 2004). An important feature of locomotion for Theraphosidae are the adhesive setae on the tarsi (Pérez-Miles 1994; Pérez-Miles et al. 2015, 2017) (see Chap. 12), which are found in most Theraphosidae species (Wolff et al. 2013). The distal setae (claw tufts) produce adhesion when the legs pull on the surface, while the proximal setae (scopulae) produce adhesion when the legs push (Wohlfart et al. 2014; Pérez-Miles et al. 2015, 2017). When a tarantula climbs on vertical or positively inclined surfaces, the forelegs pull with the distal setae and the hind legs push with the proximal setae (Pérez-Miles et al. 2015, 2017). Silva-Pereyra et al. (2019) proposed a dynamic interpretation that considers the braking and propulsive subphases of contact. On a substrate without an incline, to produce a braking force, the forelegs make contact with the proximal part of the tarsus and exert push adhesion, while the hind legs exert pull adhesion with the distal setae. To produce a propulsive force, the forelegs make contact with the distal setae and exert pull adhesion at the same time that the hind legs push with the proximal adhesive setae. During locomotion on positively inclined surfaces, extra force is required to advance and overcome gravity (Gabaldón et al. 2004; Birn-Jeffery and Higham 2014). For this purpose, the forelegs should maximise pull adhesion, which is probably achieved by an increase in the time of contact with the distal part of the tarsi, and simultaneously, the hind legs should increase the time of contact with the proximal part to maximise push adhesion—both without changing the total contact time. However, experiments that demonstrate these hypotheses are necessary to prove that there are braking and propulsion forces, and that all legs simultaneously contribute to body displacement.

13.3 Stride Length and Frequency

The movements of the limbs respond to the needs of the spiders—expressed, for example, during fleeing to escape predators or in a slow walk. In legged animals, a change in the speed of displacement can be achieved by modifying the number of strides in time (i.e. stride frequency) and/or by adjusting the distance travelled with each stride (i.e. stride length) (Cavagna et al. 1988).

Analyses performed in different species show that modification of stride frequency is the main strategy found in Theraphosidae, in both males (Booster et al. 2015; Silva-Pereyra et al. 2019) and females (Anderson and Prestwich 1985; Biancardi et al. 2011), not modification of stride length. Still, Biancardi et al. (2011) found that at a slow speed, stride length also contributes significantly to speed changes. As shown in Fig. 13.1, stride frequency rises linearly with velocity, while stride length increases linearly at low speeds and remains constant at high speeds (Anderson and Prestwich 1985; Biancardi et al. 2011; Booster et al. 2015; Silva-Pereyra et al. 2019). Araneomorph spiders also modify both stride length and stride frequency (Spagna et al. 2011).

Changes in stride frequency to modify speed have also been reported in other species of hexapod arthropods—for example, beetles (Evans 1977) and cockroaches (Ting et al. 1994). It seems that changing stride frequency is the preferred strategy used by animals with exoskeletons, since the hardness of the components limits changes to stride length through the intervention of other body parts, whereas in vertebrates, the spine has an important role in changes to stride length (Griffin et al. 2004).

Both stride frequency and stride length are useful to describe limb actions but are not enough to define whether an animal is walking, trotting or performing another gait. The duty factor is another important parameter, defined as the fraction of cycle time that a limb is in contact with the surface (Alexander 2003). An increase in speed leads to a decrease in contact time and, therefore, a decrease in the duty factor (Birn-Jeffery and Higham 2014; Maes et al. 2008).

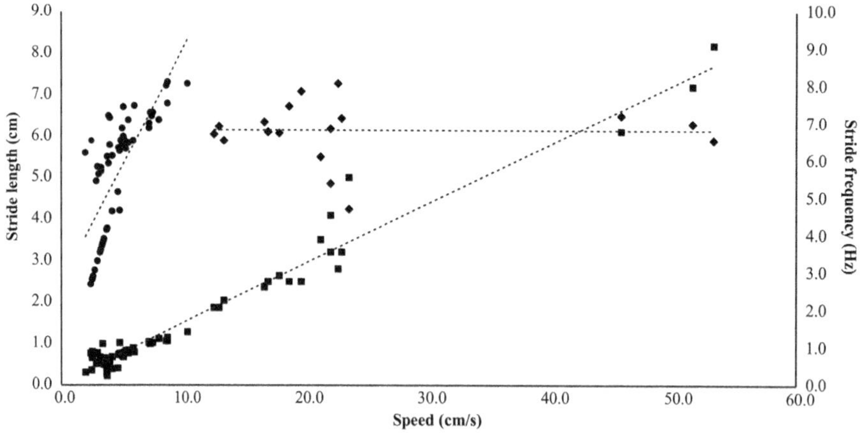

Fig. 13.1 Stride length and stride frequency versus speed in Theraphosidae (*circles*: stride length at slow speeds; *diamonds*: stride length at fast speeds; *squares*: stride frequency at all speeds; *dotted lines*: linear regression lines for the four species together). (Data from Anderson and Prestwich (1985), Biancardi et al. (2011), Booster et al. (2015) and Silva-Pereyra et al. (2019))

Vaulting gaits (e.g. walking) have been distinguished from bouncing gaits (e.g. bipedal running or the quadrupedal trot or gallop) through the phases of the locomotion cycle. In vaulting gaits, at least one foot is always in contact with the ground (Cavagna et al. 1976; Griffin et al. 2004). Bouncing gaits are characterised by the presence of a flight phase during the locomotive cycle, in which no limb is in contact with the ground (McMahon and Cheng 1990). In this case, the duty factor is a useful parameter. The fraction of the time that a foot is in contact with the ground in a vaulting gait is always greater than 50%, whereas in a gait with a flight phase, it is always less than 50% (Alexander 2003). This criterion is not always accurate in differentiating the gaits of spiders or other arthropods, because the increasing number of locomotive limbs reduce the possibility of detecting a flight phase, even at a relatively high speed (Ting et al. 1994; Kram et al. 1997; Biancardi et al. 2011). Nevertheless, in small to medium-sized araneomorph spiders, aerial phases have been reported (Spagna et al. 2011). In tarantulas, the duty factor is always greater than 50% in spite of changes in speed and/or gaits; therefore, they are always walking (Biancardi et al. 2011; Silva-Pereyra et al. 2019). Despite this, the duty factor allows for the classification of locomotive gaits into two categories: symmetric and asymmetric. The symmetric gaits are those in which the duty factor of each pair of legs is equal. Conversely, in the asymmetric gaits, the duty factor of at least one pair of legs is different from that of the others (Hildebrand 1966; Abourachid 2003; Abourachid et al. 2007; Maes et al. 2008). The duty factor, combined with the relative phasing of the leg pairs, is important for defining the gait patterns (Hildebrand 1989; Abourachid 2003).

13.4 Gait Patterns

To distinguish between gaits, it is useful to plot a gait diagram, in which all of the limbs are observed together during a step cycle. In a gait diagram, the stance period of each foot is marked with a black line, whereas the swing period is left unmarked.

The first work that analysed the gait pattern of Theraphosidae was conducted by Wilson (1967), probably in the species *Aphonopelma hentzi*, which was identified in that publication by its former name, *Dugesiella hentzi*. To study the action sequence of the limbs, this work used a metachronal model—a theoretical model based on the fact that the swing sequence of the limb occurs from the back part of the body toward the front part (caudo-rostral). The model predicted some of the locomotion patterns of the studied species but not all of the experimentally observed gait sequences, including one of the most frequently observed patterns. This pattern was characterised by the following unilateral sequence: leg IV, leg II, leg III and leg I, with leg I being the foreleg pair and leg IV the hind leg pair. This movement was close to that predicted by the alternate tetrapods model, in which bilaterally opposite legs and legs of the adjacent segments tend to move out of phase (Wilson 1967).

Hildebrand (1989) described the different quadrupedal gaits by plotting the duty factor against the time lag between the footfalls of hind feet and forefeet on the same side of the body. More recently, Abourachid (2003) proposed another quantitative method to analyse the coordination of limb movements and understand the complex pattern used in locomotion: the anteroposterior sequence of movement (APS). This method is not yet widely used to analyse the footfall sequences in Theraphosidae or other spiders. Proposed by Abourachid (2003) to understand the gait patterns of quadruped animals, it has some assumptions that can be mostly extrapolated to tarantulas. There is a morphological and functional similarity in a pair of limbs but not necessarily between the pairs, which is very accepted in tarantulas. The second assumption is that the forelimbs are the first to detect the environment and, therefore, redress the movement (Abourachid 2003; Abourachid et al. 2007). Regarding this, in tarantulas, the first pair of legs has both locomotive and exploratory functions (Anderson and Prestwich 1985; Foelix 2011). This suggests rostro-caudal activation by the central nervous system (Abourachid 2003; Abourachid et al. 2007), which, in the case of tarantulas, corresponds to the suboesophageal ganglion (Seyfarth 1985; Foelix 2011). In this way, locomotion depends on a basic coordination pattern that triggers all of the gaits and controls the movement of the limb pairs and the movement of the pairs between them (Abourachid 2003; Abourachid et al. 2007).

Gait patterns are characterised by both spatial and temporal coordination (Abourachid et al. 2007; Maes et al. 2008). Spatial interlimb coordination is the way in which limbs distribute the space during footfalls, and is reflected by the temporal coordination regardless of speed, at least in the pairs of legs (Maes et al. 2008); this is justified by the relation between the distance travelled by a pair of legs with speed and time, as the limb pairs are connected through a rigid structure. However, this relationship was not observed for the coordination between the pairs, because there is no rigid structure connecting the pairs in the animals in which the method was developed. In fact, it was found that the parameter that measures this coordination is specific for each gait (Maes et al. 2008).

The method allows us to distinguish between symmetric and asymmetric gait patterns, as well as to differentiate patterns within this classification and observe the transition between gaits (Abourachid 2003; Abourachid et al. 2007). It also has the potential to observe locomotion with disturbances and the implication of morphology in the coordination of different species (Abourachid et al. 2007). To quantify the coordination, it uses a set of temporal parameters (Table 13.1) that are considered enough to identify the gait. The parameters quantify the time lag between

Table 13.1 Theoretical parameters of the anteroposterior sequence of movement (Abourachid 2003; Maes et al. 2008)

Gait		FL (%)	HL (%)	PL (%)
Symmetric	Lateral walk	≅50	≅50	>50
	Diagonal walk	≅50	≅50	<50
	Trot	≅50	≅50	Dependent speed
Asymmetric	Half bound	<50	≅0	Dependent speed
	Bound	≅0	≅0	Dependent speed
	Transverse gallop	<50	<50	Dependent speed

FL time lag between forelegs, *HL* time lag between hind legs, *PL* time lag between pairs of legs

forelegs (FL), the time lag between hind legs (HL) and the time lag between pairs (PL) (Abourachid 2003; Maes et al. 2008). Among the symmetric gaits are the trot, the lateral walk and the diagonal walk. They are characterised by uniformity in time with respect to the distribution of the footfall within the pairs of legs (Abourachid 2003; Abourachid et al. 2007). The FL and HL are close to 50%, and the PL has different values for each gait (Maes et al. 2008). The asymmetric gaits, such as the gallop and half bound, do not show a uniform distribution of the footfall in at least one of the pairs (Abourachid 2003; Abourachid et al. 2007). In those gaits, FL and HL deviate from 50% with a quantity proportional to their speed (Maes et al. 2008).

In tarantulas, the parameters that quantify the time lag between two successive foot contacts (Fig. 13.2) have been defined for the four pairs of legs, as well as for the ipsilateral limbs between legs I and II, and between legs III and IV. The pattern can be analysed as two synchronised quadrupeds: the anterior (composed of legs I and II) and the posterior (integrated by legs III and IV) (Wilson 1967; Biancardi et al. 2011; Wilshin et al. 2018; Silva-Pereyra et al. 2019). Therefore, a time lag between the two groups of four limbs (anterior and posterior quadrupeds) can be defined as the quadruped lag (QL), calculated as the time lag among two homologous limbs—for example, R1 and R3 (Biancardi et al. 2011).

Biancardi et al. (2011) focused on female *Grammostola anthracina* (C.L. Koch 1842) (Araneae, Theraphosidae), formerly known as *Grammostola mollicoma*. During free displacement, they observed gait according to two speed categories: slow and fast (Fig. 13.3), with a cut-off speed of 11 cm.s^{-1}. The mean of the slow speed was 5.6 cm s^{-1} and that of the fast speed was 18.9 cm s^{-1}.

At slow speed, *G. anthracina* moves with an alternating tetrapod gait: R1, R3, L2, L4 to L1, L3, R2, R4. The four pairs of legs have the same pattern, with FL, HL and PL values (Table 13.2) that correspond to a quadruped walking in general terms,

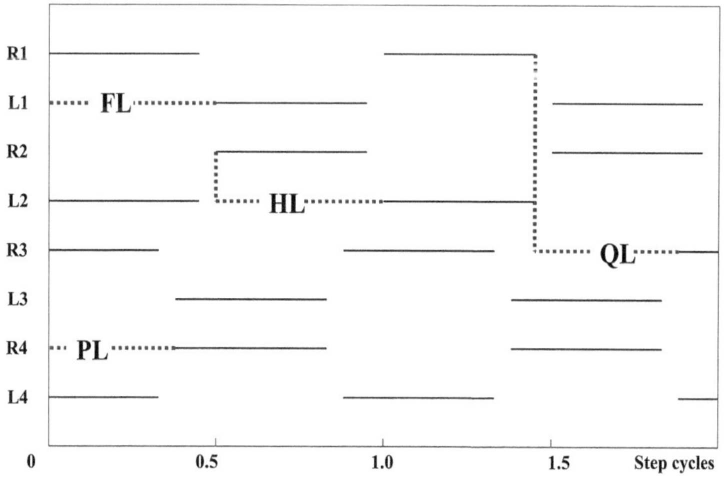

Fig. 13.2 Theoretical trot pattern using the anteroposterior sequence in Theraphosidae. Temporal coordination parameters: time lag between forelegs, R1–L1 and R3–L3 (FL); time lag between hind legs, R2–L2 and R4–L4 (HL); time lag between ipsilateral legs, R1–R2 and R3–R4 (PL); time lag between homologous legs, R1–R3 (QL)

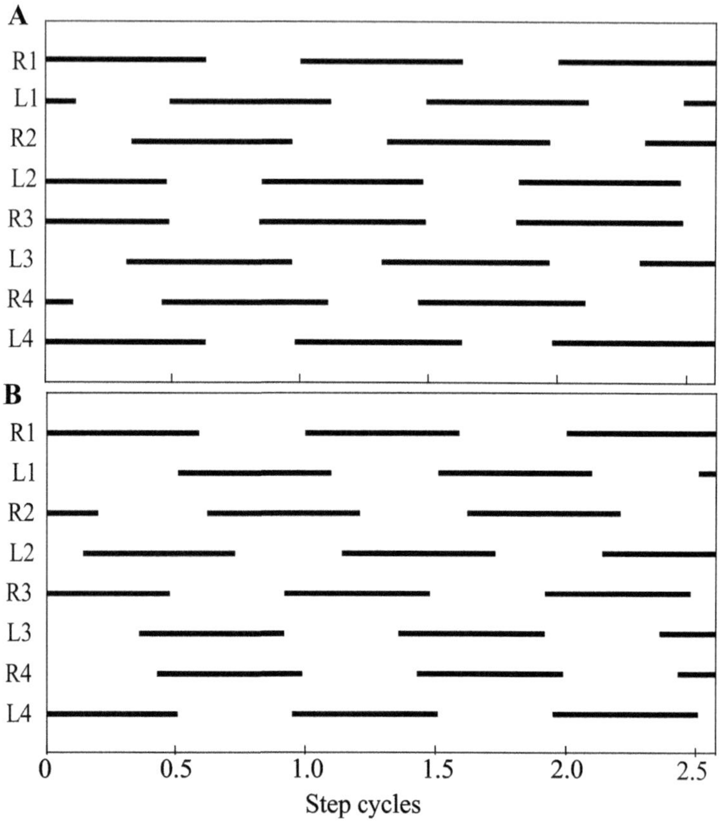

Fig. 13.3 Gait patterns of *Grammostola anthracina* at a slow speed of 5.6 cm s^{-1} (**a**) and at a fast speed of 18.9 cm s^{-1} (**b**). (Data from Biancardi et al. (2011))

Table 13.2 Experimental data on time lags in Theraphosidae (Biancardi et al. 2011; Silva-Pereyra et al. 2019)

	Anterior quadruped			Posterior quadruped			
	FL (%)	HL (%)	PL (%)	FL (%)	HL (%)	PL (%)	QL (%)
Grammostola anthracina							
Slow speed	49	51	34	48	52	62	84
Fast speed	51	56	35	44	52	51	92
Eupalaestrus weijenberghi							
Walk–trot	42	51	49	44	51	43	
Walk–bound	14	23	44				

FL time lag between forelegs, *HL* time lag between hind legs, *PL* time lag between ipsilateral legs, *QL* time lag between homologous legs

with a 16% anticipation of the posterior quadruped with respect to the anterior (Biancardi et al. 2011). This was also observed in males of the species *Eupalaestrus weijenberghi* (Thorell, 1894) (Silva-Pereyra et al. 2019) and in other arachnids, araneomorphae and scorpions (Spagna and Peattie 2012; Wilshin et al. 2018), as well as in other arthropods: crabs (Blickhan and Full 1987) and cockroaches (Ting et al. 1994). It is similar to the symmetric trot of the four-legged vertebrates, with a longer support time and without a flight phase, in which diagonal limbs move in phase (Biancardi et al. 2011). At fast speeds, the pattern of *G. anthracina* was also similar to the alternation of the legs of tetrapods, although more variable, moving away from the general model of tetrapod alternation. The parameters FL, HL and PL correspond to a variant of a quadruped walk, with an 8% anticipation of the posterior quadruped with respect to the anterior. Contrary to slow speed, which was a symmetric pattern, at fast speed, the pairs of legs I and II were different from the pairs III and IV; pairs I and II were more asymmetric than pairs III and IV (Biancardi et al. 2011). Therefore, during faster gaits, the patterns are different in both groups of legs. The major changes with speed that were observed in *G. anthracina* were the time lag between ipsilateral legs (PL) in the posterior quadruped and the time lag between homologous legs, one in the anterior quadruped and the other in the posterior quadruped (QL) (Table 13.2). This suggests changes in the coordination between the pairs of legs when the speed changes.

Biancardi et al. (2011) observed a diagonal walk pattern in the anterior quadruped of *G. anthracina.* In the diagonal walk pattern, the legs move out of phase and a posterior leg slightly anticipates the diagonal anterior leg. In *E. weijenberghi*, two walking patterns were observed in the anterior quadruped, mostly the trot-like walk and, in a few records, a pattern that was called walk–bound; this last pattern is similar to a bound in quadrupeds and has a brief time lag between pairs (Silva-Pereyra et al. 2019). In both species, the high trajectory and greater phase of swing of the first pair of legs was interpreted as the possible cause of these patterns, associated with the explorative function (Blickhan and Barth 1985; Anderson and Prestwich 1985; Foelix 2011; Biancardi et al. 2011; Silva-Pereyra et al. 2019). In *G. anthracina*, the posterior quadruped showed a symmetric pattern (Biancardi et al. 2011); the same was found in *E. weijenberghi,* which showed a trot-like pattern (Silva-Pereyra et al. 2019).

The analysis of time lags is sufficient to define the gaits (Abourachid 2003). However, the duty factor provides information about the versatility of the quadruped (Silva-Pereyra et al. 2019). In *E. weijenberghi*, the reported mean duty factors are 51% for legs I, 57% for legs II, 66% for the third pair and 67% for the fourth pair (Silva-Pereyra et al. 2019).

Wilshin et al. (2018) speculated that spiders would adapt their gaits to maximise the overall static stability, and indicated the symmetric gaits (alternating gaits, lateral walking, trotting) as more stable (Wilshin et al. 2017, 2018). The high values of the duty factor in the posterior quadruped (with respect to the anterior one) and the symmetric pattern are related to the location of the centre of mass, behind the fovea and between legs III and IV (Biancardi et al. 2011). These legs must support the body weight and are more involved in maintaining stability. In addition, legs IV also have the important role of pushing (Biancardi et al. 2011; Maes et al. 2008). The low duty factor of legs I indicates that they can move with more versatility (being able to explore the environment) as well as giving direction to the displacement (Silva-Pereyra et al. 2019). In tarantulas, the posterior quadruped shows more symmetric patterns, while the anterior quadruped can have both types of patterns, depending on the behaviour that the tarantula is performing (Silva-Pereyra et al. 2019).

13.5 Cost of Locomotion

Locomotion requires energy, but how can we measure or estimate the amount of energy necessary to maintain a body in movement? One approach would be to measure the total work required to move an animal body per unit of distance. When divided by the animal's body mass, this measure is called the mechanical cost of transport (Alexander 2003). However, the mechanical cost of transport does not account for the total metabolic energy expenditure of the muscles employed during locomotion. Metabolic energy is converted into mechanical work with variable efficiency, which depends, inter alia, on the actual velocity, frequency and efficiency of muscle contraction, and the work done by antagonist muscles (co-contractions). The metabolic cost of transport is the metabolic energy required to move one unit of the mass of the animal one unit of distance (Schmidt-Nielsen 1972; Alexander 2003).

Assuming that locomotion is basically an aerobic exercise, the metabolic energy expenditure is estimated through a measure of the rate of oxygen consumption or, sometimes, through the rate of carbon dioxide production (Nagy 1989). The relationship between the rate of oxygen consumption (VO_2) and the rate of carbon dioxide production (VCO_2) is called the respiratory coefficient ($RC = VCO_2/VO_2$), a dimensionless number that depends on the kind and relative quantity of metabolised substrates. The RC ranges from 0.7 (100% lipids metabolised) to 1.0 (100% carbohydrates metabolised) (Di Prampero 1985).

The cost of transport (CoT) is calculated by dividing the net rate of gas exchange (active rate of gas exchange – resting metabolic rate) by the velocity of locomotion. The minimum cost of transport (C_{min}) (Taylor et al. 1970), which scales inversely to body mass, is defined as the slope of the regression line relating VO_2 (or VCO_2) to speed. C_{min} is proportional to (body mass)$^{-0.32}$ (Alexander 2003, 2005) and is used to compare the locomotor performances of different species (e.g. see Biancardi et al. (2011)).

The cost of transport is usually expressed (using the joule as the energy unit) as $J\ kg^{-1}\ m^{-1}$ and this is the form we will use in this chapter. However, CoT can be specified as the volume of gas per unit of distance per unit of mass (e.g. $ml\ O_2\ kg^{-1}\ m^{-1}$). In this case, the value should be multiplied by the energetic equivalent of the measured gas (Walsberg and Wolf 1995). The energy produced by one unit of volume of oxygen depends on the substrate that is catabolised, which can be estimated through the respiratory coefficient. The RC is not always available, but the energetic equivalent of consumed oxygen differs by only around 10% between pure lipid ($19.616\ J\ ml\ O_2^{-1}$) and pure carbohydrate ($21.127\ J\ ml\ O_2^{-1}$) (Di Prampero 1985). Therefore, the error made by assuming an intermediate value is generally accepted (Walsberg and Wolf 1995).

The standard metabolic rates of arachnids at 20 °C are generally lower than those of other poikilotherm species of a similar size (Anderson 1970; Canals et al. 2007). In Theraphosidae, the metabolic rate ranges between 30% and 35% of the predicted values for poikilotherms (Anderson 1970) (see Chap. 10). It has been proposed that the low metabolic rate of arachnids could be related to their evolutionary history and their role as 'sit-and-wait' predators, foraging in a way that requires little energy but also depends on a 'bonanza' style of eating, with few big meals stretched over a long period of time (Anderson 1970).

Herreid (1981) analysed seven unspecified theraphosinae spiders and established that they increase their oxygen consumption rate to five times the resting metabolic rate during locomotion. However, the kinetics of their oxygen consumption differ considerably from those of other invertebrate and vertebrate species. In fact, the 'steady-state' rate, usually reached after 2–3 minutes of aerobic exercise (Di Prampero and Margaria 1968), appears only after 15 minutes of locomotion (Herreid and Full 1980). The recovery period after exercise was also abnormally in comparison with those of other taxa (Herreid 1981).

The energy cost during locomotion has been measured in a few species of Theraphosidae, whose body mass ranges from 6.91 to 30 g (see Table 13.3 for details and references). Because of the difference in their sizes, the metabolic rate and the energy expenditure during locomotion of male and female Theraphosidae are usually considered separately. The four mentioned investigations spanned a wide temporal range, from 1980 to 2016. Despite this long time interval, the experimental set-ups were quite similar. In all cases, a treadmill was employed to control the speed during the locomotor trials. To collect the respiratory gases, a flow-through respiratory mask was built by Anderson and Prestwich (1985), while in the other cases, a cabinet or plexiglass chamber was used. Carbon dioxide production was measured in the more recent works (Shillington and Peterson 2002; Grossi et al. 2016), and the net oxygen consumption was calculated accordingly. Only Grossi et al. (2016) provided the results (as the carbon dioxide flow rate) at different speeds. Shillington and Peterson (2002) and Grossi et al. (2016) calculated C_{min} as the slope of the regression of the flow rate against the speed. For the other works, it was assumed that the provided results were measured at the maximum aerobic speed and were therefore very close to C_{min}.

In Fig. 13.4, the minimum cost of transport of Theraphosidae is plotted against body mass. The dashed line indicates the predicted values of C_{min} according to the regression equation ($C_{min} = 10.7\ M^{-0.32}$, where the body mass (M) is in kilograms), calculated over an extended range of vertebrate and invertebrate species (Alexander 2003). The C_{min} values for *Aphonopelma anax* males and females are slightly higher than, but close to, the predicted values for other species with the same body mass (Shillington and Peterson 2002). Anderson and Prestwich (1985) reported an average steady-state VO_2 of *Brachypelma smithi*, with small variations among three different speeds (Table 13.3). This suggests the implication of anaerobic metabolism because in aerobic conditions, the oxygen flow rate should increase with speed. The cost at a higher speed is obviously underestimated, and a more realistic estimation

Table 13.3 Cost of locomotion at different speeds in some species of Theraphosidae

Species	Number of specimens	Mass (g)	Net VO_2 (ml O_2 g^{-1} h^{-1})	Speed (km h^{-1})	CoT (J kg^{-1} m^{-1})	Sex	Reference
Aphonopelma anax	6	6.91		0.063	62.93[a]	Males	Shillington and Peterson (2002)
Grammostola rosea	25	10.1	0.076	0.025	63.14	Males	Grossi et al. (2016)
Grammostola rosea	25	10.1	0.127	0.050	52.78	Males	Grossi et al. (2016)
Grammostola rosea	25	10.1	0.155	0.072	44.60	Males	Grossi et al. (2016)
Grammostola rosea	25	10.1	0.197	0.122	33.38	Males	Grossi et al. (2016)
Grammostola rosea	25	10.1			34.06[a]	Males	Grossi et al. (2016)
Theraphosinae sp.	7	12.7	0.225	0.209	22.31[a]		Herreid and Full (1980)
Aphonopelma anax	6	15.72		0.059	45.54[a]	Females	Shillington and Peterson (2002)
Grammostola rosea	25	16.7	0.045	0.025	37.43	Females	Grossi et al. (2016)
Grammostola rosea	25	16.7	0.078	0.050	32.37	Females	Grossi et al. (2016)
Grammostola rosea	25	16.7	0.111	0.072	32.06	Females	Grossi et al. (2016)
Grammostola rosea	25	16.7	0.127	0.122	21.56[b]	Females	Grossi et al. (2016)
Grammostola rosea	25	16.7			8.75[a]	Females	Grossi et al. (2016)

Table 13.3 (continued)

Species	Number of specimens	Mass (g)	Net VO_2 (ml O_2 g^{-1} h^{-1})	Speed (km h^{-1})	CoT (J kg^{-1} m^{-1})	Sex	Reference
Brachypelma smithi	5	30	0.170	0.072 0.180	33.60[c]	Females	Anderson and Prestwich (1985)

Net oxygen consumption during locomotion (net VO_2) = measured oxygen consumption (VO_{2tot}) − resting metabolic rate (VO_{2rest}). Data from Grossi et al. (2016) have been converted using a respiratory coefficient of 0.92 (Shillington et al. 2002), which corresponds to an energetic equivalent of 20.7 joules. This energetic equivalent value has been used to calculate the cost of transport (CoT).
[a]Minimum cost of transport (C_{min}).
[b]According to the authors, this speed is above the aerobic capacity of females; therefore, the CoT value lies below the C_{min} threshold (see the explanation in the text).
[c]C_{min} estimated at an average speed

of C_{min} could be derived considering the measured flow rate at an average speed. This estimation is in agreement with the predicted value (Fig. 13.4).

The low C_{min} reported by Herreid (1981) for an undefined theraphosinae species (51% of the predicted value) was probably underestimated because at the reported speed, part of the energy was supplied by anaerobic metabolism (see Herreid (1981; this was also reported by Shillington and Peterson (2002)). The minimum cost for

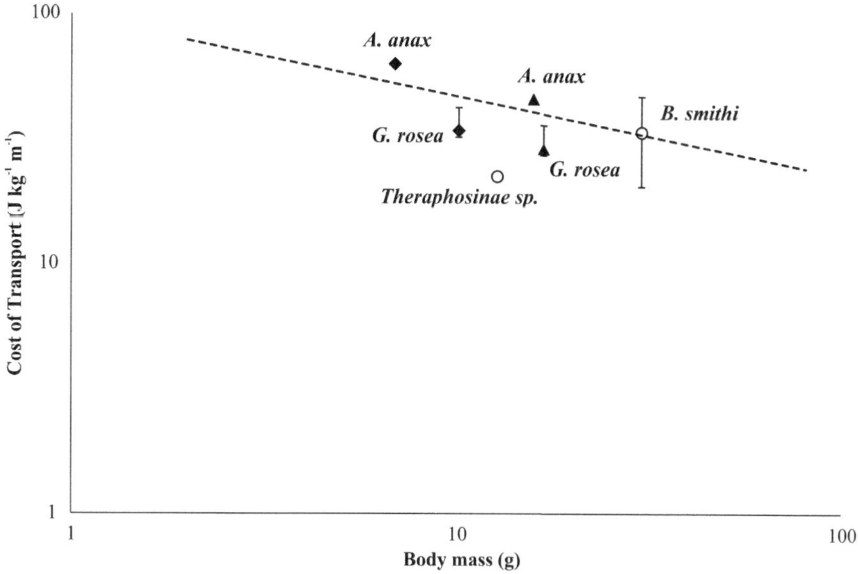

Fig. 13.4 Minimum cost of transport (C_{min}) versus body mass in Theraphosidae (*diamonds*: males; *triangles*: females; *open circles*: undefined; *error bars*: range (minimum–maximum) of the estimation; *dashed line*: predicted values of C_{min}). (Data from Herreid and Full (1980), Anderson and Prestwich (1985), Shillington and Peterson (2002) and Grossi et al. (2016))

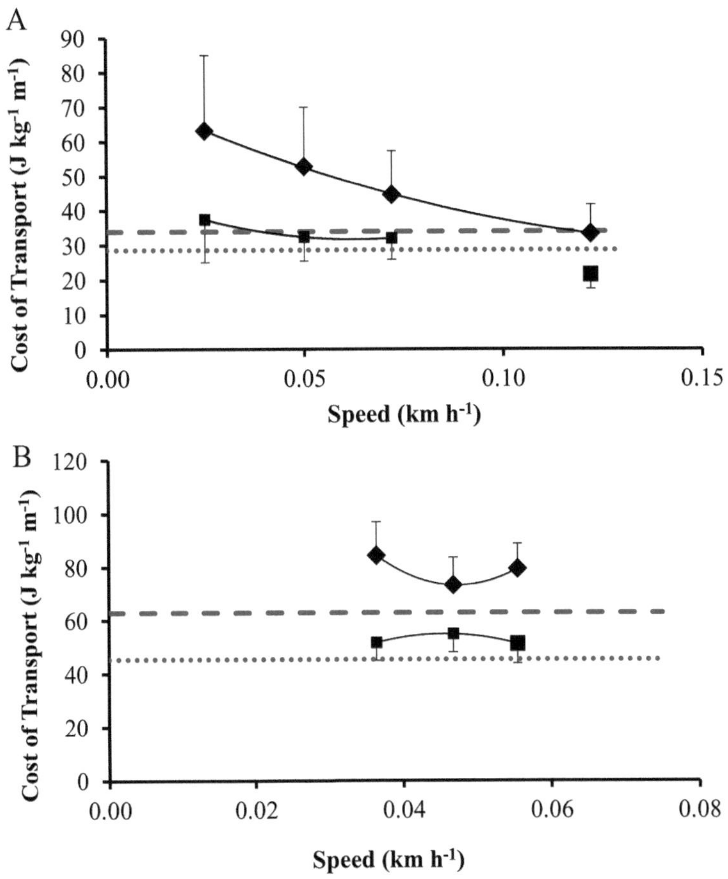

Fig. 13.5 Cost of transport (C) versus speed (*diamonds*: males; *squares*: females; *error bars*: standard deviation; *dashed line*: C_{min} of males; *dotted line*: C_{min} of females). (**a**) *Grammostola rosea* (data from Grossi et al. (2016)). (**b**) *Aphonopelma anax* (data from Shillington and Peterson (2002))

Grammostola rosea females and males displayed values slightly lower than the predicted values (Grossi et al. 2016). The authors performed an accurate estimation of the minimum cost as the slope of the regression line of VCO_2 versus speed; however, the conversion from the carbon dioxide flow rate to joules is affected by the respiratory coefficient. If the RC was close to 1 (100% carbohydrate metabolised), the C_{min} was around 73% of the predicted value. The closer the RC was to 100% lipid (0.7), the closer C_{min} was to the predicted value (Fig. 13.4).

Shillington and Peterson (2002) and Grossi et al. (2016) provided the VCO_2 at different speeds; therefore, it was possible to plot the CoT versus speed (Fig. 13.5a, b). From the results for *G. rosea*, it appears clear that the higher speed probably represents the higher aerobic speed of males, while females reach this limit at a lower speed. The CoT of females, at 0.12 km h^{-1}, is affected by the contribution of anaerobic metabolism (Grossi et al. 2016). Despite their smaller size, and thanks to their longer legs, males can aerobically sustain higher speeds than females, as an adaptation to their more active behaviour (Grossi et al. 2016). The same does not appear from the data for *A. anax*, in which the CoT is almost constant in the range of considered speeds, which (according to the authors) is within the aerobic range of the species (Shillington and Peterson 2002).

13.6 Size and Speed

According to the dynamic similarity hypothesis (Alexander and Jayes 1983), the movements of two animals of different sizes are dynamically similar when they move at the same Froude number (*Fr*), a dimensionless quantity defined as:

$$Fr = v^2 \ g^{-1} \ l^{-1}$$

where v is the progression speed, g is the gravity acceleration and l is a characteristic measure of length. In bipeds and cursorial mammals, the proposed characteristic measure was the leg length (Alexander and Jayes 1983), whereas in animals with semi-erect or sprawling postures, l would be the height of the hip joint from the ground during a normal standing posture (Alexander 2003; Irschick and Jayne 2000).

Comparisons of velocity-related features between dimorphic males and females (e.g. Fig. 13.5a, b) could change when plotted against the relative speed (Froude number) instead of the 'real' absolute speed. However, the figures would change dramatically if we considered the actual leg length rather than the height of the coxa. So far, no attempt to compare males and females has been made. Biancardi et al. (2011) calculated *Fr* using the coxa height, and Silva-Pereyra et al. (2019) did the same to compare the two ranges of speeds, concluding that the speeds of *E. weijenberghi* were within the 'slow' range of speeds of *G. anthracina*.

Comparing the maximum aerobic speeds of *G. rosea* (Grossi et al. 2016) with the velocities recorded during free displacements of *G. anthracina* (Biancardi et al. 2011)—two species of comparable size—one could infer that during normal activities, these spiders must employ a large amount of energy derived from anaerobic metabolism.

13.7 Mechanical Work of Locomotion

Mechanical work is the amount of energy transferred by a force to produce the displacement of a mass. As stated at the beginning of Sect. 13.5, the total work required to move an animal body is related to the metabolic energy expenditure of those movements. Actually, the mechanical efficiency of a movement is defined as the

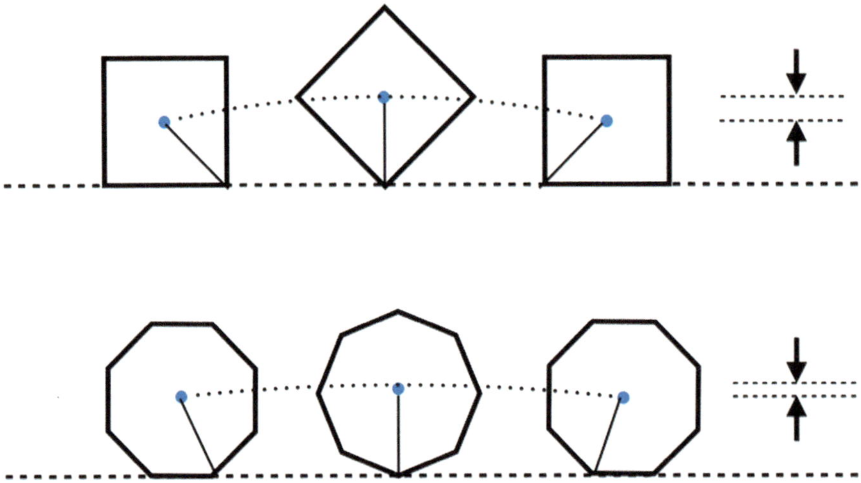

Fig. 13.6 Different polygonal wheels and their effects on the vertical displacements of the centre of mass

ratio between the total work done (or the power delivered) by a system and the total energy employed (or the power supplied) to perform the movement.

In terrestrial legged locomotion, the body centre of mass is subjected, at each step of the gait cycle, to a speed change (Cavagna 2017). Every time a foot makes contact with the ground, it loses velocity and needs to be accelerated to maintain a constant progression of speed. In walking gaits, the centre of mass behaves as though it is in a 'square wheel' (Cavagna 2017). In polygonal wheels, the vertical displacements of the centre of mass decrease as the number of sides increases, approaching zero as the polygon approaches the shape of a circle (Fig. 13.6). The vertical displacements of the body centre of mass (b_{COM}) in walking can be affected in the same way as the number of locomotor limbs increases.

As bouncing gaits—that is, those with an aerial phase, such as bipedal running or quadrupedal trotting and galloping—have never been observed in theraphosid spiders, we will focus only on the mechanics of walking-like gaits. The oscillations in b_{COM} during walking gaits have been modelled as the movement of an 'inverted pendulum' (Alexander 2003). In a theoretical ideal pendulum, the oscillation is maintained by a continuous and complete interchange between potential and kinetic energy through each period, without any energy supply. In real pendula, both the friction at the pivot and the air resistance would cause an amount of energy loss during each cycle, which should be supplied to maintain the movement. In real walking, muscles should supply the energy to overcome friction and air resistance, and to re-accelerate the limbs at each stride. In a simple bipedal model, at midstance, b_{COM} is in the highest position (Fig. 13.6, central position), then part of the potential energy is converted to kinetic energy as the body gains speed falling forward, until the contact of the next foot (Fig. 13.6, side positions). The increase in kinetic energy (E_K) is recovered to gravitational potential energy (E_P) as b_{COM} rises to the next

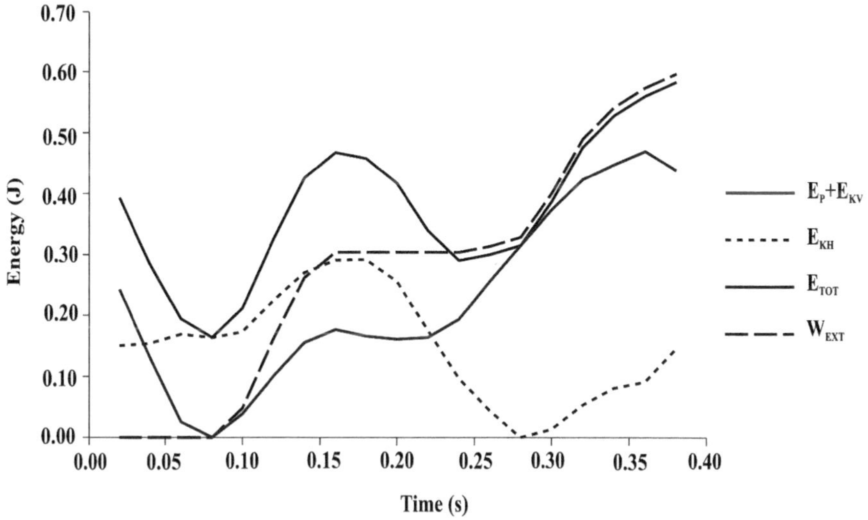

Fig. 13.7 Energy changes and external work (W_{EXT}) during one stride of *G. anthracina* at 0.6 km h^{-1}. During this stride, the recovery was 23% and W_{EXT} = 0.5 J kg^{-1} m^{-1}. E_{KH} horizontal component of kinetic energy, E_{KV} vertical component of kinetic energy, E_P gravitational potential energy, E_{TOT} total energy. (Data from Biancardi et al. (2011))

midstance position. The total energy of b_{COM} is therefore the algebraic sum of the kinetic and potential energies ($E_{TOT} = E_K + E_P$).

The work done to raise and accelerate b_{COM} is defined as external work (W_{EXT}); according to the work–energy theorem, it is given by the sum of the positive increments of the total energy. In an ideal pendulum, as stated above, the total energy is constant and W_{EXT} is zero. However, we can separately calculate the vertical work done to raise b_{COM} (W_V), as the sum of the positive increments of the vertical energy (i.e. $E_P + E_{KV}$, the vertical component of the kinetic energy), and the horizontal work done to move the body forward (W_H), as the sum of the positive increment of E_{KV}, the horizontal component of the kinetic energy (Cavagna 1975, 2017). The amount of energy recovered at each cycle is defined as 'recovery' (Cavagna et al. 1977) and is usually expressed as a percentage, defined as:

$$Recovery = \frac{|W_V| + |W_H| - |W_{EXT}|}{|W_V| + |W_H|}$$

Energy recovery tends to approach 1 (100%) as W_{EXT} tends to approach zero, as in an ideal pendulum. Figure 13.7 provides an example of energy interchanges during one stride of a theraphosid species (*G. anthracina*).

The importance of pendulum mechanics in the morphological evolution of spiders, both hanging and wandering species, has been underlined by the results of different studies (Moya-Laraño et al. 2008; Grossi and Canals 2015). Pendulum mechanics can also explain the sexual size dimorphism (SSD) exhibited by different Theraphosidae species, as the morphological divergence is coupled to differences in behaviour and locomotor necessities and performances. For instance, a smaller body size and relatively longer legs can increase locomotor performance (Grossi and Canals 2010, 2015). However, few investigations have focused on these mechanical aspects of spider locomotion (e.g. Sensenig and Shultz (2006) and Weihmann (2013)), and only two have focused on theraphosid species (Biancardi et al. 2011; Silva-Pereyra et al. 2019).

Despite its involvement in large wandering spider locomotion, the main beneficial mechanism of pendula—the interchange between potential and kinetic energy to save metabolic energy expenditure—seems to work poorly in these species. Mechanical energy recovery can reach the considerable value of 70–80% in bipeds (Griffin et al. 2004), 30–65% in quadrupeds (Griffin et al. 2004) and 15.7% in cockroaches (Exapoda) (Full and Tu 1990). In large wandering spiders, the reported average values are 8% (Brüssel 1987), 17–18.6% (Biancardi et al. 2011) and 5.3% (Silva-Pereyra et al. 2019). There is an inverse relationship between the number of locomotor limbs and the value of mechanical energy recovery. One exception is the ghost crab, a functional octopod, with an average value of 31% (Blickhan and Full 1987); however, in their lateral locomotion, the right and left side limbs behave like hind limbs or forelimbs, with gaits more similar to those of quadrupeds.

Low energy recovery seems to be unrelated to the reduced vertical oscillation of b_{COM} (Fig. 13.6); rather, it is related to the inequality between energy values. Especially at lower walking speeds, the forward kinetic energy values are much lower than the vertical ones (gravitational potential and vertical kinetic energies), and a great part of W_{EXT} accounts for the vertical work. Therefore, as shown by Biancardi et al. (2011), the energy recovery increases with speed as the pendulum interchange mechanism becomes more efficient.

External work is often expressed in units of distance and units of body mass—namely, in the same units as the cost of transport (J kg^{-1} m^{-1}). However, unlike CoT, W_{EXT} is relatively independent of the body mass and is less influenced by speed (Full and Tu 1990; Biancardi et al. 2011; Silva-Pereyra et al. 2019). Data are only available for *E. weijenberghi* males, with free short displacements on different surfaces and gradients (Silva-Pereyra et al. 2019), and for *G. anthracina* females, with free short displacements on undefined surfaces (Biancardi et al. 2011). The average W_{EXT} of *E. weijenberghi* on glass (0.568 J kg^{-1} m^{-1}, standard deviation 0.098) was consistent with that of *G. anthracina* (0.452–0.643 J kg^{-1} m^{-1}). On a Teflon surface, in the same range of speeds, the external work showed an average twofold increase (1.202 J kg^{-1} m^{-1}, standard deviation 1.038) but only for some stride sequences, as supported by the large standard deviation. In general, for arthropods (cockroaches and crabs), birds and mammals, the work necessary to move 1 kilogram of an animal 1 metre is approximately 1 joule (Full and Tu 1990). Theraphosid data point to a lower average external work percentage (50–60%) than the reported values for other species (Biancardi et al. 2011; Silva-Pereyra et al. 2019). As these spiders are the first truly octopedal species to be measured, their lower W_{EXT} can be related to reduced vertical oscillations of b_{COM} due to the increased number of locomotor limbs (Fig. 13.6).

Theraphosid spiders can move up severe gradients (Pérez-Miles et al. 2015). Moving uphill, the work against gravity increases and so does the external work. In *E. weijenberghi* males, W_{EXT} rose to 2.135 J kg^{-1} m^{-1} (standard deviation 0.128) on glass and 2.578 J kg^{-1} m^{-1} (standard deviation 0.721) on Teflon, both at a gradient of 21% (12° inclination) (Silva-Pereyra et al. 2019).

Although W_{EXT} represents most of the total mechanical work of locomotion, it is not all of it. To complete the information, we need to consider the work done to accelerate and reposition the limb segments with respect to the body centre of mass (i.e. the kinetic internal work (W_{INT})) (Cavagna 2017). Still considering W_{INT}, there is a missing part of the mechanical work done by the muscles, which includes the energy necessary to overcome the frictional resistance of joints and ligaments and the concomitant contractions of agonist and antagonist muscles; this generally neglected part represents the viscous internal work (Minetti 2011).

Kinetic internal work is obviously related to the number of limb segments involved and their relative mass. There is a trade-off between the opposite necessities of providing muscular power to the limbs and restraining their inertial parameters; this explains why cursorial ungulates have thin limbs with a muscular mass concentrated in the proximal segments (Alexander 2003). Fedak et al. (1982) calculated an allometric equation that related W_{INT} (in J kg^{-1} m^{-1}) to the square root of speed (allometric exponent = 0.53). However, more recently, Minetti (1998) proposed a model equation to predict W_{INT}, proving a linear relationship to speed. The results, published by Biancardi et al. (2011), which calculated the kinetic internal work of *G. anthracina* over a wide range of speeds, showed W_{INT} increasing linearly with speed, in agreement with the model created by Minetti (1998).

Despite the large number of locomotor limbs, and their non-negligible relative masses, the kinetic internal work in theraphosid spiders is considerably lower than that in vertebrates, ranging from 1% to 17% of the total mechanical work (22% in horses (Minetti et al. 1999) and 25% in humans (Saibene and Minetti 2003; Biancardi et al. 2011; Silva-Pereyra et al. 2019)).

13.8 Efficiency and Maximum Power

Mechanical efficiency is defined as the ratio between the mechanical output and the energetic input, and is usually expressed as a percentage. The two quantities must be in the same unit of measure; therefore, when the metabolic energy expenditure is expressed as CoT (J kg^{-1} m^{-1}), the efficiency is computed using the total mechanical work (W_{EXT} + W_{INT}), as shown in Sect. 13.7. Mechanical efficiency increases with body mass: larger animals are more efficient than smaller ones. This is due to the almost constant mechanical work necessary to move one unit of mass one unit of distance (Full and Tu 1990), versus the CoT, which decreases with body mass (Alexander 2003).

Biancardi et al. (2011) estimated a mechanical efficiency of 2.6–3.8% in *G. anthracina* on the basis of the probably underestimated C_{MIN} reported by Herreid and Full (1980). Using the more recent data reported by Grossi et al. (2016), measured in females of the similar species *G. rosea*, the mechanical efficiency of

G. anthracina should be set around 2%; this value is lower than the estimated efficiency of vertebrates of a similar mass (Alexander 2003, 2005).

Because of their poor aerobic capacities, spiders, for their movements, rely on high-energy storage compounds in the muscles and anaerobic metabolism. One of the indicators of the energy available in the muscles is analysis of explosive or maximum power (Alexander 2003).

The peak mechanical power in Theraphosidae was estimated in *G. anthracina*, during maximum accelerations from rest (Biancardi et al. 2011). The resulting value (8.3 W kg^{-1}) was very low in comparison with the maximum power of other species: around 40–50 W kg^{-1} in humans, around 80 W kg^{-1} in insects and more than 300 W kg^{-1} in frogs (Ellington 1985; Lutz and Rome 1994).

13.9 Hydraulic Joint Movements

Unlike other terrestrial species, spiders extend some of their leg segments hydraulically (Kropf 2013). Both the femur–patella joint and the tibia–metatarsus joint extensions occur when haemolymph is pumped into the legs by the contraction of muscles in the prosoma (Anderson and Prestwich 1975). Haemolymph flows back to the prosoma during leg flexion. During locomotion, this back-and-forth flow occurs at each stride (Sensenig and Shultz 2003).

The impact of this hydraulic system on the metabolic cost of transport is unknown. With this device, spiders do not need to activate extensor muscles at these joints, and this could be an energy-saving benefit. On the other hand, haemolymph is always pumped toward the eight legs. Thus, the temporal shift of extension that normally occurs during locomotion is regulated by the increased work of the antagonist flexor muscles. Moreover, the high pressure necessary to extend the leg segments (more than 50 times the resting pressure) would require a large amount of muscular work (Stewart and Martin 1974). So far, the real benefit of the hydraulic system is the space saved by the absence of extensor muscles, enabling slimmer leg segments with more space available for flexor muscles (Kropf 2013).

Speed, more than temperature, may affect the efficiency of the hydraulic joint movement. Booster et al. (2015) have shown that at high speeds, the increased stride frequency limits the time available for haemolymph to properly flow through all of the leg segments. On the other hand, because of temperature variations, haemolymph viscosity changes do not seem to influence the movements (Booster et al. 2015).

13.10 Conclusion

Although spiders are part of one of the largest groups of terrestrial species, the locomotion of spiders, and theraphosids in particular, is still poorly understood. Therefore, approaching the end of this review, we feel the need to evaluate the status of our knowledge and underline some future directions of research:

1. In terrestrial locomotion, from small insects to large vertebrates, the purpose of the legs is to raise and displace the whole body (in biomechanics, we would say b_{COM}) at each step. Indeed, we observe the same basic mechanical models and patterns of leg coordination across the different taxa. In spiders, the most commonly employed gait is an alternating tetrapod, similar to the alternating tripod of hexapods. However, the peculiar characteristics of Theraphosidae—that is, their sprawled posture, the particular arrangement of the leg around the prosoma, the presence of adhesive setae on their tarsi and their lack of extensor muscles in some leg segments—determine deviations from the basic gait patterns that require more investigations.
2. The role of elastic structures in spider locomotion is still poorly understood. On the one hand, the rigid exoskeleton of the prosoma does not permit flexion movements, as in the vertebrate trunk. On the other hand, the 'bow' shape of the legs could allow elastic oscillations. Elastic structures could play a double function, helping to increase stride length and storing potential elastic energy, which could be reused during locomotion.
3. In general, too few data are available on the mechanical energy of locomotion. In particular, for the internal work—the mechanical energy associated with the limb movements—contradictory data have been found so far. More research is needed to understand the relative importance of internal versus external work in theraphosid locomotion.
4. The Froude number has been widely used to normalise the speed of different species according to the dynamic similarity hypothesis (Alexander and Jayes 1983). In bipeds and quadrupeds, the 'normal'—or self-selected—walking speed generally ranges around Froude 0.25 (Alexander 2003). However, when applied to tarantulas (e.g. Biancardi et al. (2011)), the Froude values are much lower than expected. The suitability of the Froude number out of the vertebrate world, and its relationship to the number of legs or the body size, are open questions.
5. The minimum cost of transport, which has been determined for few species, is associated with very low speeds on a treadmill. When measured during free displacements, spiders move at higher speeds, hinting that they usually proceed above their aerobic threshold. The low aerobic capacities of spiders and their relationships to locomotor performances and with muscle physiology need more investigation. Although the protein composition of the skeletal muscles of theraphosids indicates a mix of muscle fibre types (Zhu et al. 2009), the characteristics of such fibre types (aerobic, anaerobic, fast twitch, slow twitch…) are still unknown.

Acknowledgements The authors are grateful to Fernando Pérez-Miles for his invitation to contribute to this book, and to Santiago Fernandez for his help with the English editing of this chapter.

References

Abourachid A (2003) A new way of analysing symmetrical and asymmetric gaits in quadrupeds. C R Biol 326:625–630
Abourachid A, Herbin M, Hackert R, Maes L, Martin V (2007) Experimental study of coordination patterns during unsteady locomotion in mammals. J Exp Biol 210:366–372
Alexander RMN (2003) Principles of animal locomotion. Princeton University Press, Princeton
Alexander RMN (2005) Models and the scaling of energy costs for locomotion. J Exp Biol 208:1645–1652
Alexander RMN, Jayes AS (1983) A dynamic similarity hypothesis for the gaits of quadrupedal mammals. J Zool 201:135–152
Anderson JF (1970) Metabolic rates of spiders. Comp Biochem Physiol A Physiol 33:51–72
Anderson JF, Prestwich KN (1975) The fluid pressure pumps of spiders (Chelicerata, Araneae). Zoomorphology 81:257–277
Anderson JF, Prestwich KN (1985) The physiology of exercise at and above maximal aerobic capacity in a theraphosid (tarantula) spider, *Brachypelma smithi* (F.O. Pickard-Cambridge). J Comp Physiol B155:529–539
Biancardi CM, Fabrica CG, Polero P, Loss JF, Minetti AE (2011) Biomechanics of octopedal locomotion: kinematic and kinetic analysis of the spider *Grammostola mollicoma*. J Exp Biol 14:3433–3442
Birn-Jeffery AV, Higham TE (2014) The scaling of uphill and downhill locomotion in legged animals. Integr Comp Biol 54:1159–1172
Blickhan R, Barth FG (1985) Strains in the exoskeleton of spiders. J Comp Physiol A 157:115–147
Blickhan R, Full RJ (1987) Locomotion energetics of the ghost crab: II. Mechanics of the centre of mass during walking and running. J Exp Biol 130:155–174
Booster NA, Su FY, Adolph SC, Ahn AN (2015) Effect of temperature on leg kinematics in sprinting tarantulas (*Aphonopelma hentzi*): high speed may limit hydraulic joint actuation. J Exp Biol 218:977–982
Brüssel A (1987) Belastungen und Dehnungen im Spinnenskelett unter natürlichen Verhaltensbedingungen. Thesis (unpublished), Goethe University, Frankfurt
Canals M, Salazar MJ, Durán C, Figueroa D, Veloso C (2007) Respiratory refinements in the mygalomorph spider *Grammostola rosea* Walckenaer 1837 (Araneae, Theraphosidae). J Arachnol 35:481–486
Cavagna GA (1975) Force platforms as ergometers. J Appl Physiol 39:174–179
Cavagna GA (2017) External, internal and total mechanical work done during locomotion. In: Cavagna GA (ed) Physiological aspects of legged terrestrial locomotion: motor and the machine. Springer, Cham, pp 129–138
Cavagna GA, Thys H, Zamboni A (1976) The sources of external work in level walking and running. J Physiol 262:639–657
Cavagna GA, Heglund NC, Taylor CR (1977) Mechanical work in terrestrial locomotion: two basic mechanisms for minimizing energy expenditure. Am J Phys Regul Integr Comp Phys 233:R243–R261
Cavagna GA, Franzetti P, Heglund NC, Willems P (1988) The determinants of the step frequency in running, trotting and hopping in man and other vertebrates. J Physiol 399:81–92
Di Prampero PE (1985) La locomozione umana su terra, in acqua, in aria: fatti e teorie. Edi. Ermes, Milan

Di Prampero PE, Margaria R (1968) Relationship between O_2 consumption, high energy phosphates and the kinetics of the O_2 debt in exercise. Pflugers Arch 304:11–19
Ellington CP (1985) Power and efficiency of insect flight muscle. J Exp Biol 115:293–304
Evans MEG (1977) Locomotion in the Coleoptera Adephaga, especially Carabidae. J Zool 181:189–266
Fedak MA, Heglund NC, Taylor CR (1982) Energetics and mechanics of terrestrial locomotion. II. Kinetic energy changes of the limbs and body as a function of speed and body size in birds and mammals. J Exp Biol 97:23–40
Foelix R (2011) Biology of spiders. Oxford University Press, New York
Full RJ, Tu MS (1990) Mechanics of six-legged runners. J Exp Biol 148:129–146
Gabaldón AM, Nelson FE, Roberts TJ (2004) Mechanical function of two ankle extensors in wild turkeys: shifts from energy production to energy absorption during incline versus decline running. J Exp Biol 207:2277–2288
Griffin TM, Main RP, Farley CT (2004) Biomechanics of quadrupedal walking: how do four-legged animals achieve inverted pendulum-like movements? J Exp Biol 207:3545–3558
Grossi B, Canals M (2010) Comparison of the morphology of the limbs of juvenile and adult horses (*Equus caballus*) and their implications on the locomotor biomechanics. J Exp Zool A Ecol Genet Physiol 313:292–300
Grossi B, Canals M (2015) Energetics, scaling and sexual size dimorphism of spiders. Acta Biotheor 63:71–81
Grossi B, Solis R, Veloso C, Canals M (2016) Consequences of sexual size dimorphism on energetics and locomotor performance of *Grammostola rosea* (Araneae; Teraphosidae). Physiol Entomol 41:281–288
Herreid CF (1981) Energetics of pedestrian arthropods. In: Herried CF, Fourtner CR (eds) Locomotion and energetics in arthropods. Plenum, New York, pp 491–526
Herreid CF, Full RJ (1980) Energetics of running tarantulas. Physiologist 23:40
Hildebrand M (1966) Analysis of the symmetrical gaits of tetrapods. Folia Biotheor 6:9–22
Hildebrand M (1989) The quadrupedal gaits of vertebrates. Bioscience 39:766–776
Irschick D, Jayne B (2000) Size matters: ontogenetic variation in the three-dimensional kinematics of steady-speed locomotion in the lizard *Dipsosaurus dorsalis*. J Exp Biol 203:2133–2148
Kram R, Wong B, Full RJ (1997) Three-dimensional kinematics and limb kinetic energy of running cockroaches. J Exp Biol 200:1919–1929
Kropf C (2013) Hydraulic system of locomotion. In: Nentwig W (ed) Spider ecophysiology. Springer, Berlin, pp 43–56
Lutz GJ, Rome LC (1994) Built for jumping: the design of the frog muscular system. Science 263:370–372
Maes LD, Herbin M, Hackert R et al. (2008) Steady locomotion in dogs: temporal and associated spatial coordination patterns and the effect of speed. J Exp Biol 211:138–149
McGhee RB, Iswandhi GI (1979) Adaptive locomotion of a multilegged robot over rough terrain. IEEE Trans Syst Man Cybern 9:176–182
McMahon TA, Cheng GC (1990) The mechanics of running: how does stiffness couple with speed? J Biomech 23:65–78
Minetti AE (1998) A model equation for the prediction of mechanical internal work of terrestrial locomotion. J Biomechan 31:463–468
Minetti AE (2011) Bioenergetics and biomechanics of cycling: the role of 'internal work'. Eur J Appl Physiol 111:323–329
Minetti AE, Ardigo LP, Reinach E, Saibene F (1999) The relationship between mechanical work and energy expenditure of locomotion in horses. J Exp Biol 202:2329–2338
Moya-Laraño J, Vinković D, De Mas E, Corcobado G, Moreno E (2008) Morphological evolution of spiders predicted by pendulum mechanics. PLoS One 3:e1841
Nagy KA (1989) Doubly-labeled water studies of vertebrate physiological ecology. In: Rundel PW, Ehleringer JR, Nagy KA (eds) Stable isotopes in ecological research. Ecological Studies (Analysis and Synthesis), vol 68. Springer, New York, pp 270–287

Nishikawa K, Biewener AA, Aerts P, Ahn AN, Chiel HJ, Daley MA, Daniel TL, Full RJ, Hale ME, Hedrick TL, Lappin AK, Nichols AP, Quinn NR, Satterlie RA, Szymik B (2007) Neuromechanics: an integrative approach for understanding motor control. Integr Comp Biol 47:16–54

Paul R, Fincke T, Linzen B (1989) Book lung function in arachnids. J Comp Physiol B 159:409–418

Pérez-Miles F (1994) Tarsal scopula division in Therephosinae (Araneae, Theraphosidae): its systematic significance. J Arachnol 22:46–53

Pérez-Miles F, Perafán C, Santamaría L (2015) Tarantulas (Araneae: Theraphosidae) use different adhesive pads complementarily during climbing on smooth surfaces: experimental approach in eight arboreal and burrower species. Biol Open 4(12):1643–1648

Pérez-Miles F, Guadanucci JPL, Jurgilas JP, Becco R, Perafán C (2017) Morphology and evolution of scopula, pseudoscopula and claw tufts in Mygalomorphae (Araneae). Zoomorphology 136:435–459

Saibene F, Minetti AE (2003) Biomechanical and physiological aspects of legged locomotion in humans. Eur J Appl Physiol 88:297–316

Schmidt-Nielsen K (1972) Locomotion: energy cost of swimming, flying, and running. Science 177:222–228

Sensenig AT, Shultz JW (2003) Mechanics of cuticular elastic energy storage in leg joints lacking extensor muscles in arachnids. J Exp Biol 206:771–784

Sensenig AT, Shultz JW (2006) Mechanical energy oscillations during locomotion in the harvestman *Leiobunum vittatum* (Opiliones). J Arachnol 34:627–633

Seyfarth EA (1985) Spider proprioception: receptors, reflexes, and control of locomotion. In: Barth FG (ed) Neurobiology of arachnids. Springer, Berlin, pp 337–347

Shillington C, Peterson CC (2002) Energy metabolism of male and female tarantulas (*Aphonopelma anax*) during locomotion. J Exp Biol 205:2909–2914

Silva-Pereyra V, Fábrica CG, Biancardi CM, Pérez-Miles F (2019) Kinematics of males *Eupalaestrus weijenberghi* (Araneae, Theraphosidae) locomotion on different substrates and inclines (No. e27520v1). PeerJ 7:e7748. https://doi.org/10.7717/peerj.7748

Spagna JC, Peattie AM (2012) Terrestrial locomotion in arachnids. J Insect Physiol 58:599–606

Spagna JC, Valdivia EA, Mohan V (2011) Gait characteristics of two fast-running spider species (*Hololena adnexa* and *Hololena curta*), including an aerial phase (Araneae: Agelenidae). J Arachnol 39(1):84–92

Stewart DM, Martin AW (1974) Blood pressure in the tarantula, *Dugesiella hentzi*. J Comp Physiol 88:141–172

Taylor CR, Schmidt-Nielsen K, Raab JL (1970) Scaling of energetic cost of running to body size in mammals. Am J Physiol Legacy Cont 219:1104–1107

Ting LH, Blickhan R, Full RJ (1994) Dynamic and static stability in hexapedal runners. J Exp Biol 269:251–269

Walsberg G, Wolf B (1995) Variation in the respiratory quotient of birds and implications for indirect calorimetry using measurements of carbon dioxide production. J Exp Biol 198:213–219

Weihmann T (2013) Crawling at high speeds: steady level locomotion in the spider *Cupiennius salei*—global kinematics and implications for centre of mass dynamics. PLoS One 8:e65788

Wilshin S, Reeve MA, Haynes GC, Revzen S, Koditschek DE, Spence AJ (2017) Longitudinal quasi-static stability predicts changes in dog gait on rough terrain. J Exp Biol 220(10):1864–1874

Wilshin S, Shamble PS, Hovey KJ, Harris R, Spence AJ, Hsieh ST (2018) Limping following limb loss increases locomotor stability. J Exp Biol 221(18):jeb174268

Wilson DM (1967) Stepping patterns in tarantula spiders. J Exp Biol 47:133–151

Wohlfart E, Wolff JO, Arzt E, Gorb SN (2014) The whole is more than the sum of all its parts: collective effect of spider attachment organs. J Exp Biol 217:222–224

Wolff JO, Nentwig W, Gorb SN (2013) The great silk alternative: multiple co-evolution of web loss and sticky hairs in spiders. PLoS One 8:e62682

Zhu J, Sun Y, Zhao FQ, Yu J, Craig R, Hu S (2009) Analysis of tarantula skeletal muscle protein sequences and identification of transcriptional isoforms. BMC Genomics 10:117

Chapter 14
Communication and Reproductive Biology of Tarantulas

Nelson E. Ferretti

Abstract Theraphosidae spiders (commonly known as tarantulas) comprise some of the largest known spiders. It is the most diverse family among Mygalomorphae and Theraphosinae, endemic to the Neotropics, is the richest subfamily. However, the knowledge on some aspects of their reproductive biology is still unknown. Usually, the sexual behavior of tarantulas has been considered as "simple," that is, males just walk searching for females and when they randomly find one, the mating should occur. This point of view is changing as the number of studies has grown during the last 20 years, suggesting that the sexual behavior of Theraphosidae is far from simple. Such complexity may be represented by specific searching and courtship behavior of males and active roles of females, leading to an intricate intersexual communication before mating, the occurrence of male copulatory, and complex patterns of palpal insertions. This chapter describes the ways of communication of tarantulas mainly during the sexual encounters. The patterns of courtship and copulation of representatives of most subfamilies will be described based on a bibliographic review. Finally, some general strategies of mating and reproduction of tarantulas will be discussed and topics for future research are presented.

14.1 Communication in Tarantulas: From Sounds to Smells

Communication is typically an exchange of information between individuals through one or more signals (acoustic, vibratory, chemical, or visual) sent from one animal to another (Uetz and Stratton 1983).

The poor vision that characterize most of theraphosid taxa (Dippenaar-Schoeman 2002), probably related to their nocturnal habits, could increase pressures of selection that may have favored the use of vibratory and chemical signals during communication between two individuals of this group. Moreover, most burrow-living

N. E. Ferretti (✉)
Centro de Recursos Naturales Renovables de la Zona Semiárida (CERZOS-CONICET), UNS, Bahía Blanca, Buenos Aires, Argentina
e-mail: nferretti@conicet.gov.ar

F. Pérez-Miles (ed.), *New World Tarantulas*, Zoological Monographs 6,
https://doi.org/10.1007/978-3-030-48644-0_14

tarantulas are sit-and-wait predators and prey is usually detected by vibration receptors on the palps and legs of these spiders. The receptors may consist of trichobothria that detect prey-generated air currents or silk- or soil-vibration detectors such as the slit sensilla or club-shaped trichobothria. An advantage of these acoustic or seismic signals is that they are relatively independent of environmental conditions (light, temperature, humidity) for efficiency signal propagation (Krafft 1982; Redondo 1994; Foelix 2010). Another advantage of acoustic or seismic signals is their temporal characteristic, which can be modified quickly according to the motivational state of the animal (e.g., from the response of a female). Disadvantages include the short temporal persistence of the signal and the high cost of production. The advantages could explain why acoustic/vibratory signals are so widespread in tarantulas.

Acoustic/vibratory signals in tarantulas can be produced by stridulation and percussion/vibration. Stridulation can be defined as the process of sound production by friction of one rigid body part across a second part (Uetz and Stratton 1983). The stridulatory organs are widely distributed among theraphosids (see Chap. 11). The evidence suggests that the stridulatory setae are related only with a defensive display (Pérez-Miles et al. 2005; Bertani et al. 2008a). When disturbed, some tarantulas stridulate, showing a great diversity in volume and sound production. The sound is produced by the specialized setae mainly located on proximal segments of front appendages. Regarding vibratory signals, they are important in the communication systems of tarantulas. Indeed, many males are capable of making sounds and vibrations by drumming or vibrating the whole body of some appendages. Theraphosid spiders are known to use percussion of palps or legs to produce sound and vibrations on the ground surface or even over the silk threads surrounding the female burrow. Some authors (Prentice 1992, 1997; Quirici and Costa 2005) found that compared to acoustic signals, seismic signals have the advantage of propagating through long distances and twice as faster than the acoustic signals, depending on the type of soil and degree of soil moisture. Arguably, taking into account the advantages of seismic signals and the widespread occurrences of vibration behavior in theraphosids, it is probable that seismic signals are the main communicatory channel used by burrowing tarantulas, mainly during courtship.

On the other hand, chemically mediated communication is common in spiders but has been poorly studied in tarantulas. A study that aimed to determine whether chemical cues influence the behavior of females of *Tliltocatl vagans* (Ausserer, 1875) during encounters with previously inhabited burrows, found that tarantulas are strongly attracted to the cues left by other conspecific females (Dor et al. 2008). However, it was demonstrated that when *T. vagans* females encounter each other, aggressive interactions normally occur, thus authors suggested that chemical cues are not signals that are deliberately released by tarantulas, but rather these compounds may escape inadvertently from the burrow-inhabiting females and cannot be easily suppressed (Dor et al. 2008).

Chemical signals called pheromones are substances secreted by one individual and received by another which elicit a specific response (for example, sex pheromones secreted by females stimulate courtship behavior in males). Theraphosids are known to use contact pheromones involved in courtship behavior that is important for intra-specific recognition prior to mating (Costa and Pérez-Miles 2002). There is also evidence that several groups of spiders use airborne pheromones (see Gaskett 2007 for a review); however, the emission of airborne pheromones has not been reported in Theraphosidae yet.

14.2 Prepared for Sex: Sperm Induction

When a male tarantula reaches sexual maturity, he charges his palpal organs with sperm, a process known as sperm induction (Foelix 2010). For example, males of *Aphonopelma chalcodes* Chamberlin, 1940 make their first sperm induction about 10 days after maturity (Minch 1979). Sperm induction is the first stage of the reproductive process of a tarantula and usually is characterized by the construction of a large and dense sperm web attached to rocks, logs, or shrubs and inclined with respect to the soil. A small area of this web is reinforced and the sperm droplet is deposited there. To achieve this, the male placed the genital opening under the reinforced area of the web where he then deposits a sperm droplet. During this deposition the prosoma remains out of the web (Costa and Pérez-Miles 2002).

Curiously, male tarantulas, like other spiders, have additional silk glands (epigastric glands) which are not associated with the spinnerets occurring on the anterior margin of the genital furrow. Specialized glands and associated spigots (known as epiandrous apparatus) are used during the construction of the sperm web (Marples 1967; Costa and Pérez-Miles 2002; Ferretti et al. 2017). Although the specific function of those glands has not been confirmed, it is possible that the silk produced may reinforce the sperm web or be involved in maintaining the sperm droplet attached to the web since epiandrous silk is mostly limited to the zone where the sperm drop is deposited. In a recent study, it was found that many tarantula species have epiandrous spigots mixed with plumose setae arranged in an evident half-moon shape above the genital opening (Ferretti et al. 2017). This feature is notorious before the male reaches adulthood, allowing a quick and easy way to identify sub-adult males in Theraphosidae (Ferretti et al. 2017).

The construction of the sperm web takes a few hours in some tarantula species (Costa and Pérez-Miles 2002). Costa and Pérez-Miles (2002) described in detail this behavior. In a first step, the male digs a shallow depression removing the soil. Then, he begins the spinning of a dense web. Later, with its ventral surface toward the web, the male moves under the sperm web producing a notch in an edge and separating the sperm web from the substrate. Afterward, the web remains hanging loose and fixed by the other edges. Then the male moves from under the web and climbs on the web for tightening the notch. Next, the male moves under the web again, with the venter upward, and spins especially in an area close to the notch. This may be the instance when male uses the epiandrous apparatus to spin this special area of the sperm web (Fig. 14.1). The male orients himself by placing the hairs of the coxae of the fourth pair of legs against the notch edge. The genital opening is

Fig. 14.1 Male of *Grammostola vachoni* before sperm deposition. Note the reinforced area (dense web), where the sperm droplet will be deposited (photo: G. Pompozzi)

placed under the reinforced area of the web where he deposits a sperm droplet. Then the male stands on the web with the venter oriented downward, searches for the web edge and seeks the droplet with his palps. Once found, the male starts alternate rapid palpal movements contacting the tip of the palpal bulb with the droplet, which is absorbed by capillarity. Finally, the male cleans his palps and usually the web is consumed (Costa and Pérez-Miles 2002).

Theraphosid males spend a long time in sperm induction in comparison to other spiders. Moreover, it was observed that males of species that live in rocky environments showed higher sperm charge durations in relation to tarantulas species that inhabit strictly in open fields (Costa and Pérez-Miles 2002). The short duration of sperm charge on those species could be related to the scarcity of protected sites for this conspicuous event. Moreover, some burrowing species such as *Aphonopelma joshua* Prentice 1997 or *Eupalaestrus weijenberghi* (Thorell, 1894) are able to perform the sperm induction inside the burrow if it is sufficiently wide (Prentice 1997; Costa and Pérez-Miles 2002; Pérez-Miles et al. 2005). The same male recharges his palp organs several times. For example, Baerg (1958) reported a record of more than 17 sperm inductions in 6 weeks. The sperm charged during one sperm induction probably is not enough to inseminate several females what could occur when the sex ratio is biased toward females, or maybe there is some selective pressure to avoid old sperm in the palps (Costa and Pérez-Miles 2002). The first interpretation seems to be the more adequate since it was observed in *Eupalaestrus weijenberghi* that one sperm induction is enough for at least two copulations and copulating males recharge palpal organs after mating and more frequently than males that do not copulate (Pérez-Miles et al. 2007).

Sperm induction is a costly behavior in tarantulas due to the time spent (hours) and the energy consumed, thus this behavior could reflect a good health and indirectly reinforce also their attractiveness. However, little is known about which facts stimulate males to perform sperm induction or if there is a relation of this behavior and male performance during the mating, since most of the studies completely ignore this step or assume it as an innate behavior of mature males. After the process of sperm induction, the male is ready for mate search, an activity valuable in terms of reproductive effort invested by males, acting as a trade-off against other activities such as foraging.

14.3 Send Me a Sign: Male Courtship Behavior and Female Sexual Responses

A general statement about theraphosid behavior until the early 1980s was that they showed simple repertoires and inconspicuous courtship behaviors (Baerg 1928; Gerhardt 1929; Platnick 1971). For example, courtship of tarantulas has been catalogued as *Level I*, in which males respond only to direct contact with the female and mating then usually occurs (Baerg 1928; Platnick 1971). However, this scenario has changed and an increasing number of studies on sexual behavior of tarantulas during the last 25 years led to a better understanding of the intricate and complex mechanisms of communication displayed by theraphosids (Ferretti et al. 2013).

After a male tarantula reaches maturity, the primary goal is reproduction, and to achieve a successful mating, pre-copulatory isolating mechanisms need to be overcome. Moreover, the male must suppress the female's predatory behavior as well as "convince" her to mating. The function of courtship in general has been the subject of considerable debate. Under a traditional view, courtship among spiders is considered to have three main functions: (1) partner recognition (species and sex); (2) orientation of individuals and synchronization of mating; and (3) as said above, suppression of nonreproductive behavioral patterns (e.g., prey capture behavior) (Foelix 2010). The female tarantula is anything but a passive partner, and in fact it has been shown that she can be the first to promote direct contact by the use of pheromones as sexual attractants.

Contact pheromones are widespread in the sexual communication of theraphosids and are usually associated with silk threads released by females. Males have chemoreceptors on the pedipalps and forelegs and after they contact female silk, they frequently groom their own legs or pedipalps and start courtship, suggesting that contacting with female silk promotes male sexual stimulation (Gaskett 2007). Recent studies suggest that chemical sexual communication via sex pheromones is the rule among tarantulas (Pérez-Miles et al. 2005, 2007; Almeida-Silva et al. 2008; Ferretti and Ferrero 2008; Copperi et al. 2012; Ferretti et al. 2013; Copperi 2018), but how does it actually work under natural conditions? A recent finding indicates that female cues (sex pheromones) are characterized by a long-term persistence on the field. For example, Costa et al. (2015) observed in *Eupalaestrus weijenberghi* that female cues elicited male courtship up to at least 46 days under different climate conditions. However, male courtship responses decline slowly over time and this is in relation to the decrease of attractiveness of female cues over time, mainly caused by the damage of the silk threads exposed to wind or rain (Costa et al. 2015). Once a male contacts female cues the courtship behavior is elicited. The sexual behavior of male theraphosids is characterized by alternation of periods of activity and inactivity, where activity usually consists of body vibrations and palpal drumming (Quirici and Costa 2005).

The subfamilies Aviculariinae (Simon 1874), Theraphosinae (Thorell 1870), and Ischnocolinae (Simon 1892) are the more diverse in the Neotropical Region (See Chap. 1) and by far include the most studied species regarding sexual behavior. Although the courtship behavior exhibited is species specific, some behaviors are displayed and shared by many males of different species. Also, different behaviors can be split between the stages before and after contacting with the female. Thus, pre-contact male courtship usually involves the following behaviors:

Palpal drumming: Up and down alternating movements of pedipalps against the substrate or on female's silk threads.

Leg tapping: Legs I, legs II, or both of them, hit vigorously against the substrate, the female's silk threads or the female's body.

Body vibration: High-frequency movement of the body or legs usually caused by the inward contractions of legs III with femora positioned almost vertically.

These behaviors are usually displayed before contacting the female. Regarding palpal drumming, it seems to act as long- and also short-distance communication since the frequency of this behavior clearly increases at the entrance of the female burrow or shelter, or when the male gets close to the female. Leg tapping and body vibration are involved in transmitting the species-specific signals through the substrate or the silk threads. Body vibration comprises a key behavioral unit in the courtship of male tarantulas and the third leg seems to be essential to elicit female response and sexual receptivity. According to Baruffaldi et al. (2013), the body vibration is a stereotyped behavior that could be displayed with all pairs of legs, and when spiders had the third pair of legs tied (according to an experimental approach), their body vibrations were less frequent than spiders with legs II tied, suggesting the importance of the third leg for increasing body vibrations and also for a complete female sexual response. Moreover, legs III are in the best position and seem to be the best candidates from transmitting vibrations from the top to the bottom of the male's body. This would allow transferring the mechanical energy produced by leg muscle contractions to the substrate by vibrating their bodies (Baruffaldi et al. 2013).

Seismic sexual male signals are able to propagate for at least 1.10 m through the substrate and the characteristics of vibrations bouts seem to maintain the reproductive isolation between sympatric and synchronic tarantula species (Quirici and Costa 2007). Although strong reproductive barriers are expected between similar species that co-occur in time and space, a noticeable case of two sympatric and synchronic tarantula species, *Acanthoscurria cordubensis* Thorell, 1894 and *Eupalaestrus weijenberghi* was reported as males courted heterospecific female cues (Costa et al. 2013). Moreover, authors observed that females of one species seem to be more receptive or attracted to heterospecific males than conspecific ones in early courtship stages. Although precopulatory confusion occurs, no heterospecific copulations were recorded, so probably, species recognition occurred through chemotactic and other short-distance mechanisms, as occurs in many spider species (Costa et al. 2013).

If signals from male courtship are received by females, they may respond to them. This behavior has been reported in species of the genera *Acanthoscurria* Ausserer 1871, *Aphonopelma* Pocock 1901, *Avicularia* Lamarck 1818, *Eupalaestrus* Pocock 1901, *Grammostola* Simon 1892, *Homoeomma* Ausserer 1871, and *Sickius* Soares and Camargo 1948. Females usually respond to male courtship by tapping with forelegs and palps against the substrate indicating a sexually receptive state (Stradling 1994; Quirici and Costa 2005; Pérez-Miles et al. 2007; Copperi et al. 2012). Moreover, female sexual responses also help male orientation toward her location from long distances, for example, in species that inhabit crevices or under stones (Copperi et al. 2012). Another female response to male courtship is called "piston" behavior (Costa and Pérez-Miles 2002; Quirici and Costa 2005; Pérez-Miles et al. 2007). This behavior comprises a ritualized agonistic response because after perceiving the signal the male immediately escapes. This threatening behavior is observed in *Eupalaestrus*, *Grammostola* and occurs inside the burrow, generating air currents or vibrations through the burrow entrance. It could also generate seismic signals (Pérez-Miles et al. 2007; Copperi 2018). The complex sexual communication with special acceptance or rejecting signals reduce the time males spend outside the female burrow, since courting males constitute a cost for females because they could attract predators or just compete with other important biological female activities such as feeding, moulting, and/or eggsac care.

After spiders make contact with each other, they adopt an elevated position reaching a characteristic front-to-front posture (Fig. 14.2). Most theraphosid species have specialized tibial apophyses to clasp female chelicerae following the inner curvature of his palp. This clasp improves male security during copulation and also provides stability to the female, so that the male can reach the female genital area.

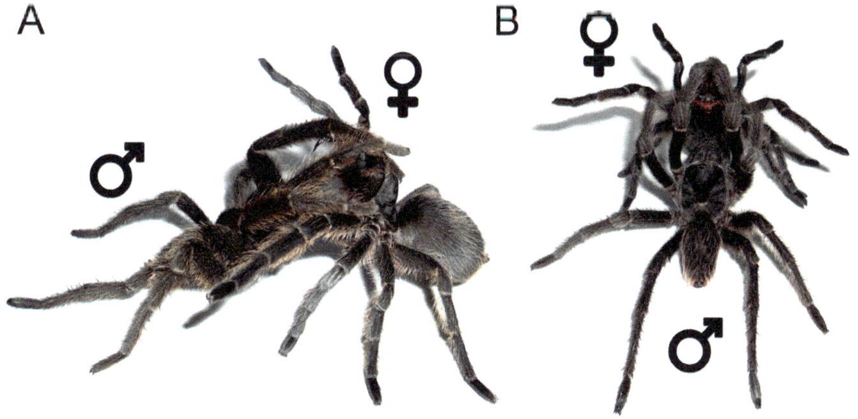

Fig. 14.2 Mating position of the theraphosid *Grammostola doeringi*. (**a**) Early stage of mating with the male clasping female's fangs. (**b**) Late stage of mating with the male already clasped with the female's fangs and making palpal insertions (photos: G. Pompozzi)

One exception is that reported for males of *Sickius longibulbi* Soares and Camargo, 1948 (Subfamily Schismatothelinae) which clasp the female on her first pair of legs (Bertani et al. 2008b).

During the contact phase, males of theraphosids can perform the following behaviors:

Bite fangs: Male and female interweave forelegs and open chelicerae and fangs; then the male presses the female fangs with his own.

Palpal boxing: Up and down alternating movements of palps against the female sternum.

Spasmodic beats: Male extends legs II or III and make vigorous backward and forward movements against the legs of female.

The male bites have only been reported for *Grammostola iheringi* and comprise a conspicuous and ritualized behavior performed in face-to-face position, with no damage for either sex (Postiglioni and Costa 2006). This frontal bite behavior is part of the male courtship and has probably evolved from an aggressive context and was later incorporated to the species-specific ritualized sexual behavioral patterns (Postiglioni and Costa 2006). The palpal boxing and spasmodic bets are widespread among Neotropical genera of Theraphosidae and is always displayed before palpal insertions. It has been reported for *Avicularia* (Aviculariinae), *Acanthoscurria, Aphonopelma, Brachypelma, Eupalaestrus, Grammostola, Plesiopelma,* (Theraphosinae) and in *Catumiri* and *Dolichothele* (Ischnocolinae) (Yañez et al. 1999; Costa and Pérez-Miles 2002; Ferretti et al. 2013; Teixeira 2017). The male spasmodic beats with legs II over the female was thought to be an exclusive behavior displayed by the genus *Grammostola* (Costa and Pérez-Miles 2002; Pérez-Miles et al. 2007; Ferretti and Ferrero 2008), but it was more recently recorded in the genus *Dolichothele* (Teixeira 2017) and even in the family Nemesiidae (Ferretti et al. 2013). This behavior is displayed after contacting the female and its function seems to be the relaxation of female fangs, considering that this behavior is performed during clasping and unclasping (Ferretti et al. 2013).

14.4 Mating Behavior and Sperm Transfer

The characteristic mating position adopted by most Theraphosidae is with the male pushing up the female body and raising her to reach a position forming an angle of about 60–80° between carapace and abdomen (Ferretti et al. 2013) (Fig. 14.2). A notable exception to this typical mating position is that adopted by males and females of *Sickius longibulbi* (Schismatothelinae). Male of this species leans the female down by pushing her body until she lays on her back and then the male locates himself at an angle of 90° with the reclining female (Bertani et al. 2008b). This male behavior demands a lot of energy and one possibility for this unusual mating position is that the male could be obtaining a more efficient sperm transfer because while in this posture the extremely long embolus can deposit more sperm into the female oviduct, where she retains the sperm because she does not have spermathecae (Bertani and da-Silva 2002; Bertani et al. 2008b).

In arboreal species of Theraphosidae (See Chap. 4), for example, those of the genus *Avicularia*, the copulation takes place outside the female retreat. In the case of *Avicularia avicularia* males approach occupied retreats and after courtship receptive females emerge to mate (Stradling 1994). However, in the case of burrowing theraphosids the copulation always takes place at the entrance (Costa and Pérez-Miles 2002; Ferretti et al. 2013). This location is probably due to space limitations inside the female burrow, but also avoids the risk of the male of being seized by the female inside the burrow. On the other hand, the couple suffers predation risk while mating at an exposed site. However, this risk is minimized by the brief copulation duration. Usually the mating just lasts from one or less minutes to about 15 (Costa and Pérez-Miles 1992, 2002; Pérez-Miles and Costa 1992; Punzo and Henderson 1999; Pérez-Miles et al. 2007; Ferretti and Ferrero 2008; Copperi 2018). An exceptional long mating was reported just during one encounter in *Sickius longibulbi*, lasting 45 min (Bertani et al. 2008b). Sometimes, the mating duration extends due to a high number of failed palpal insertions, which are frequent in some species.

During the mating position achieved by the couple, the male initiates insemination with either palps and may switch one palp to the other in initial insertions in consecutive matings. An ipsilateral insertion pattern was observed by Minch (1979) in *Aphonopelma chalcodes*, in which the male's right palp entered the female's right opening to the spermathecae, while the left palp entered the female's left opening to the spermathecae in an alternating palpal insertions. However, a given palp is able to inseminate either or both spermathecal receptacles (Costa et al. 2000). Moreover, there is no evidence of morphological or ethological constraints which prevent a palp from delivering sperm to either receptacle. The sperm is deposited by the embolus deep into the spermathecal receptacle (Costa et al. 2000). During inseminations, both palps are employed but a perfect alternating pattern usually is not performed. Moreover, in some cases just the same palp is used many times or even just one palp is inserted one time during mating (Minch 1979; Pérez-Miles et al. 2007).

Usually, most theraphosid species show a low number of brief palpal insertions (ranging from 1 to 15). Once the male and the female separate from each other, the male rapidly moves away from the female range. Females remain immobile in a quiescent or cataleptic state for some minutes after mating. Typically, the female stays static with their first pair of legs and palps raised. The sexual cannibalism seems to be occasional and rare in theraphosids, with some scarce cases of female attacks toward males during or after mating (Prentice 1997; Shillington and Verrell 1997; Yañez et al. 1999; Costa and Pérez-Miles 2002; Postiglioni and Costa 2006; Pérez-Miles et al. 2007; Bertani et al. 2008b; Ferretti and Ferrero 2008). However, sexual cannibalism in theraphosids has been reported by some authors (Brazil and Vellard 1926; Bücherl 1952; Punzo and Henderson 1999).

14.5 Sexual Dimorphism and Mating Systems

In general, spiders are sexually dimorphic in size, with males being smaller than females (Hormiga et al. 2000). Although sexual dimorphism is extreme in web-building spiders, non-web-building spiders usually have a lower degree of sexual size dimorphism (Walker and Rypstra 2001). In Theraphosidae, males and females appear of similar size (Hénaut et al. 2015), but it is known that they show sexual dimorphism considering metabolic rates (Shillington 2005; Copperi 2018). Shillington (2002, 2005) reported the significance of life history traits in sexual dimorphism of *Aphonopelma anax* (Chamberlin, 1940). In this case, mature males have a very active life that implies searching for females for reproduction, which leads to higher energy demands and a higher resting metabolic rate than females (Hénaut et al. 2015).

Males of *Aphonopelma anax* have a smaller abdomen and longer legs than females (Shillington and Peterson 2002). In *Grammostola anthracina* (Koch, 1842), a variation in somatic characters between males and females was observed, particularly for legs and prosoma (Pérez-Miles 1989). A clear evidence of sexual size dimorphism was reported for *Tliltocatl vagans* (Hénaut et al. 2015). In that case, males have longer tibia than females, probably as a result of selective pressures related to the reproductive role of wandering males. In many tarantula species sexually mature males abandon their burrows and actively search for females over a large geographical area (Shillington 2002; Hénaut et al. 2015), so their body is adapted to be fast and efficient searchers. Another explanation for the sexual dimorphism observed in this species could be that during clasping access to female genitalia could be favored by longer legs in males. Moreover, male larger leg size may also be an adaptation for protection against females that may not predate organisms of an equivalent or larger size (Hénaut et al. 2015, see also Chap. 8).

Copperi (2018) found sexual size dimorphism in two tarantula species, *Grammostola vachoni* and *Plesiopelma longisternale*, with females larger than males. Arguably, a more recent study involving the tarantula *Grammostola rosea* found that males have a higher aerobic speed, average speed, distance travelled, and critical angle of climbing than females, indicating better performance than females (Grossi et al. 2016). This study also reported that males have lower costs of transport than females. In conclusion, sexual dimorphism in wandering spiders with active males such as tarantulas, which are characterized by smaller body size and longer legs than females, could be associated with low transport costs, high speed, and better locomotor skills (Grossi et al. 2016).

Male and female tarantulas probably reach adulthood with a 1:1 sex ratio but due to the longer lifespan of females (from about 7 years to 30 years in some species) the sex ratio could be strongly biased toward females (Costa and Pérez-Miles 2002; Ibler et al. 2013). For this reason, males copulate several times along one reproductive period (Buskirk et al. 1984; Costa and Pérez-Miles 2002; Ferretti et al. 2013; Copperi 2018). For example, in the genus *Aphonopelma*, a male was observed to copulate 12 times with females, thus an estimation of the actual sex ratio during the reproductive season should be one male to six or seven females (Baerg 1958). Also, in some species of this genus, it was reported that females could copulate between five and seven times (Shillington and Verrell 1997). In the burrowing tarantula *Grammostola doeringi*, a maximum of five copulations was observed for a male (Copperi 2013). In the case of the arboreal *Avicularia avicularia*, a male can copulate about five times (Stradling 1994).

Given the abundance and availability of females in the field, intrasexual direct competition among males (male-male fighting or mate guarding) does not occur in theraphosids. Moreover, male competition could be restricted to be the first in finding females. In this scenario, in which neither females nor other resources are easily monopolized by a single male, males avoid aggressive interactions with each other and instead canalize their mating effort into searching for receptive females (Shillington and Verrell 1997). Searchers will be selected, for example, males that are able to locate any or the most females as is possible (Duvall and Schuett 1997; Shillington and Verrell 1997). In addition, this is related with the "*cul de sac*" spermathecae morphology of the Theraphosidae, which rewards the last male in mating (Austad 1984).

Theraphosid females continue molting after adulthood; consequently, old spermathecae and remaining sperm are lost during molting. In this context, males clearly increase their reproductive success by mating multiple times with different females (Davies et al. 2012; Costa-Schmidt et al. 2017). The benefits for a female under this system could be the gain in genetic diversity, female choice, and also avoiding costs of maintaining sperm for long periods (Peretti and Aisenberg 2015; Costa-Schmidt et al. 2017). Additionally, cryptic female choice may come into play (Eberhard 1996; Peretti and Aisenberg 2015; Costa-Schmidt et al. 2017). Thus, the mating system known for most theraphosid species comprises a polygyny/polyandry strategy, with females and males copulating several times.

Pérez-Miles et al. (2007) reported a polygynic/monandry strategy in the tarantula *Eupalaestrus weijenberghi*, with females copulating once and immediately becoming sexually reluctant to further matings. In this species matings are extremely brief and involve a single palpal insertion (Pérez-Miles et al. 2007). It is important to note that because females live many years, they will copulate with several males during their lifetime; so they are considered as sequential monandrics (Pérez-Miles et al. 2007). If females do not remate, they are renouncing to several possible direct and indirect reproductive benefits, as it was previously mentioned (Simmons 2005). One explanation could be that monogamy with brief copulations including a single palpal insertion minimizes the risk of infections through sexual transmission of parasites and pathogens (Pérez-Miles et al. 2007). However, monogamy could also reduce or exclude sperm competition and cryptic female choice.

Because females of this species are reluctant after a single mating, the available number of receptive females declines rapidly and, in consequence, a change in the operational sex ratio takes place as the reproductive period advances. This scenario generates a strong selective pressure on males for locating and finding a female as soon as possible. Finally, assuming rapid and dramatic changes in the operational sex ratio, it is expected to find sexy males and choosy females in this species. Thus, females select males and sexual selection seems to shape male mate search (Pérez-Miles et al. 2007).

14.6 Phenology, Eggsac, Progeny, and Dispersal

A few studies have been made to try to understand the phenology of tarantulas at the field along the different habitats they occupy. Most reports about this are based on captive breeding and development. Although no generalizations could be made about the reproductive periods of most tarantula species, it is well known that reproductive activities of theraphosids inhabiting temperate regions coincide with seasonal changes (Pérez-Miles et al. 2005) and those species that live in tropical areas show a reproductive period that matches the beginning of rainy season (Stradling 1994; Marechal et al. 2009) (See also Chaps. 7 and 10).

Most tarantula species from temperate regions show restricted sexual periods, usually of approximately 2 months (Costa and Pérez-Miles 2002; Pérez-Miles et al. 2005). Sexual periods can occur during spring, summer, or early autumn, coinciding with warmer months (Pérez-Miles et al. 2005). However, some species are known to have their sexual period during colder months and this is related to an ecological strategy to avoid predation (Pérez-Miles et al. 1993; Costa and Pérez-Miles 2002). Seasonal mass movement of males is a widespread phenomenon in tarantulas that has been considered as a form of migration by some authors, and is usually related to weather conditions (Baerg 1958; Magnusson 1985). Walking males of some species seem to be synchronized with meteorological conditions such as cloudy days, with high temperature and low atmospheric pressure (Pérez-Miles et al. 2005).

The movement of male tarantulas has been subject of speculations and interest of arachnologists for many years (Smith 1994; Janowski-Bell and Horner 1999). Unfortunately, just one study has extensively documented the movement of male tarantulas through telemetry in *Aphonopelma hentzi* (Girard 1852) (Janowski-Bell and Horner 1999). It is unlikely that tarantulas use visual environmental cues when searching for female burrows. Male tarantulas conduct systematic searches to locate females within a patch and the movement of the males comprises a combination of random walks and straight line movements (Janowski-Bell and Horner 1999). Males are able to move large distances; for example, it was reported that a male of *Aphonopelma hentzi* covered about 1300 m during a period of time of about 18 days, while searching for females (Janowski-Bell and Horner 1999). The large size and especially long legs of males favor male locomotion in open field during the reproductive season. Large body size in males could also help reduce predation, in combination with conspicuous defensive behaviors exhibited by some species (Pérez-Miles et al. 2005). Adult males do not feed, or feed scarcely under natural conditions, but invest in walking and mating. Consequently, males by the end of the reproductive period are scarce and show body deterioration.

Females can store sperm for several months since copulation and oviposition usually take place during the warmer months. Most theraphosid species seem to produce only one sac per year and per female along just a short period of time (usually 2 months), and in some tropical species the occurrences of egg sacs by females are in a prolonged period but anyway with just one clutch per reproductive period (spring–summer). The egg sac construction by most theraphosids is characterized by the complete covering of the walls of the shelter by a dense web spinned from side to side in circular ways (Fig. 14.3a). This process can take several hours, since the initial silk deposition to the final oviposition (Melchers 1964). In some species egg sac care by the female involves positioning the egg sac in the entrance of the burrow and maintaining it under her body (Fig. 14.3b), what comprises a free and mobile cocoon (Table 14.1). In other species, such as some Ischnocolinae, females construct silk tubes with dense walls and the egg sac is flattened, discoid and remains fixed inside the tube wall (Costa and Pérez-Miles 2002).

Some theraphosines incorporate urticating setae during egg sac construction (Marshall and Uetz 1990; Bertani and Guadanucci 2013). The function seems to be to protect the egg sac against phorid fly larvae. Egg sacs usually contain a few to several hundred eggs, which in conjunction with the relatively larger size of the eggs compared with those of insects, renders them as potentially valuable food resource for any predator (Austin 1985). *Acanthoscurria*, *Avicularia*, *Lasiodora* Koch 1850, *Megaphobema* Pocock 1901, *Prohapalopus* Mello-Leitão, 1923, *Theraphosa* Thorell 1870, and *Vitalius* Lucas, Silva and Bertani, 1993 (Marshall and Uetz 1990; Pérez-Miles and Costa 1994; Bertani et al. 2003; Bertani and Guadanucci 2013) are some of the Neotropical genera that incorporate urticating hairs to their egg sacs. Astonishingly, it was found that these setae not only are effective against fly larvae but can also protect them from ant attacks (Bertani and Guadanucci 2013).

Clutch sizes of Neotropical theraphosid spiders can be quite variable (Table 14.1). Smaller species such as *Grammostola burzaquensis* Ibarra-Grasso 1946, *Hapalotremus vilcanota* Ferretti, Cavalllo, Chaparro, Ríos-Tamayo, Seimon and West, 2018, *Homoeomma uruguayense* (Mello-Leitão, 1946), or *Catumiri parvum* (Keyserling, 1878) and several species of *Avicularia* are characterized by relatively small clutch sizes (12–120 eggs/egg sac) (Ibarra-Grasso 1961; Charpentier 1992;

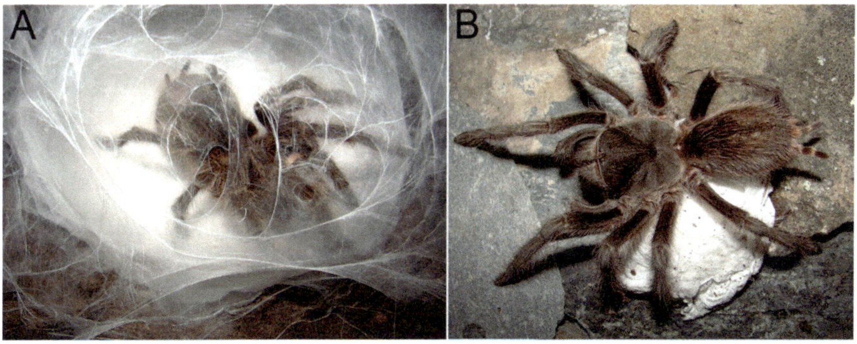

Fig. 14.3 Egg sac of *Grammostola doeringi*. (**a**) Dense web spun by the female before the deposition of eggs. (**b**) Female holding the mobile egg sac under her legs (photos: G. Pompozzi)

Table 14.1 Reports of clutch sizes and egg sac type (mobile or fixed to the burrow or web shelter) in Neotropical Theraphosidae. Data was obtained from the literature or personal observations

Subfamily	Genus/species	Egg sac type	Clutch size	References
Aviculariinae	*Avicularia taunayi*	Mobile	29	Bertani and Motta (2013)
	Avicularia avicularia	Mobile	103–154	Stradling (1994)
	Avicularia metallica	Mobile	178–182	Charpentier (1992)
	Antillena rickwestii	Mobile	15	Bertani and Huff (2013)
	Caribena versicolor	Mobile	12–221	Charpentier (1992), Marechal et al. (2009)
	Pachistopelma rufonigrum	Mobile	30	Dias and Brescovit (2003)
Ischnocolinae	*Catumiri parvum*	Fixed	71	Costa and Pérez-Miles (2002)
	Catumiri argentinense	Fixed	28	Ferretti pers. obs.
Theraphosinae	*Acanthoscurria paulensis*	Mobile	1600–2000	Lourenço (1978)
	Acanthoscurria musculosa	Mobile	600	Ibarra-Grasso (1961)
	Brachypelma smithii	Mobile	> 700	Clarke (1991)
	Tliltocatl vagans	Mobile	≈ 200	Dor and Hénaut (2012)
	Eupalaestrus weijenberghi	Mobile	167–334	Pérez-Miles et al. (2007)
	Grammostola anthracina	Mobile	≈ 90	Panzera et al. (2009)
	Grammostola burzaquensis	Mobile	100–120	Ibarra-Grasso (1961)
	Grammostola doeringi	Mobile	176–246	Ferretti et al. (2012)
	Grammostola vachoni	Mobile	248–434	Ferretti et al. (2012)
	Hapalotremus vilcanota	Mobile	27	Ferretti et al. (2018)
	Homoeomma chilensis	Mobile	130	Montenegro et al. (2018)
	Homoeomma uruguayense	Mobile	40	Ferretti et al. (2019)
	Magnacarina primaverensis	Mobile	25	Mendoza-Marroquín et al. (2016)
	Pamphobeteus roseus	Mobile	1200	Ibarra-Grasso (1961)
	Pamphobeteus sp.	Mobile	1200–2000	Bücherl (1951)
	Plesiopelma longisternale	Mobile	16–111	Costa and Pérez-Miles (1992, 2002)
	Theraphosa blondi	Mobile	36–78	Lambert and Dupre (1992), Marshall and Uetz (1993)
	Vitalius nondescriptus	Mobile	52	Bertani et al. (2012)

Costa and Pérez-Miles 2002; Ferretti et al. 2018). Although some large species as those of the genus *Pamphobeteus* Pocock 1901 can lay over 1500 eggs (Bücherl 1951), the large theraphosid *Theraphosa blondi* (Latreille 1804) (weighting up to 70 g, with a leg span of 25 cm), one of the largest spiders in the world, lays a surprisingly small number of eggs (about 78 eggs per egg sac) (Lambert and Dupre 1992; Marshall and Uetz 1993). However, the size of the eggs of *T. blondi* is considerable larger than that of other theraphosids (about 200 mg each) (Marshall and Uetz 1993). Apparently having larger spiderlings developing from larger eggs is one of the primary mechanisms by which this spider achieves its formidable size (Punzo and Henderson 1999). Interestingly, the development of low clutch size with large eggs and larger spiderlings is achieved by a much smaller species of Theraphosidae, *Hapalotremus vilcanota* (Ferretti et al. 2018). However, in that case, this is not related with the size of adults (it is a very small tarantula of just 2.5 cm in total length); instead it may be an adaptation to the extreme habitat conditions of this tarantula inhabiting altitudes of more than 4000-m high (See Chap. 7), so when the spiderlings develop and emerge they are large and have higher chances for survival (Ferretti et al. 2018).

The time required for egg production and incubation is quite variable among species. In many tropical species, females lay eggs within a few weeks after mating whereas temperate species may wait for several months. For example, it is well known that some temperate tarantulas actually store sperm for 7–9 months before oviposition (Pérez-Miles et al. 2005). The eggs mature internally and the time at which they are deposited depends on female age, the time since the last molt, nutritional factors, and temperature (Schaeffer 1987; Punzo and Henderson 1999).

The duration of egg sac development (gestation) is also quite variable for many theraphosid species. For example, gestation periods range from 29 to 80 days in some tropical species of *Avicularia* (Stradling 1994; Bertani and Motta 2013). In some temperate and small-sized species, such as *Plesiopelma longisternale* (Schiapelli and Gerschman, 1942), the gestation period is about 49 days (Costa and Pérez-Miles 1992). Comparable gestation periods have been reported for *Catumiri parvum* (37–41 days) (Costa and Pérez-Miles 2002), *Homoeomma chilensis* Montenegro and Aguilera, 2018 (56 days) (Montenegro et al. 2018) and *Grammostola burzaquensis* (51 days) (Ibarra-Grasso 1961). Slightly longer incubation periods have been reported for medium- to large-sized species such as *Eupalaestrus weijenberghi* (83–87 days) (Pérez-Miles et al. 2007), *Grammostola doeringi* (Holmberg, 1881) (62–76 days) (Ferretti et al. 2012), *Grammostola vachoni* Schiapelli and Gerschman, 1961 (75 days) (Ferretti et al. 2012), and *Theraphosa blondi* (77 days) (Marshall and Uetz 1993). The percentage of spiderlings that hatch is also variable, ranging from 15 to 96% (Punzo and Henderson 1999).

Unfortunately, the factors of the discrepancy between the number of laid eggs and the final number of spiderlings are unclear. In some cases, the mortality of spiderlings is high due to abnormal molts resulting in defective neonates, probably by the critical high sensitivity of spiderlings to humidity levels (Costa and Pérez-Miles 2002; Panzera et al. 2009; Ferretti et al. 2012).

Once spiderlings are ready to emerge from the cocoon, they may have the ability to emerge without the assistance of their mother, as was reported by Panzera et al. (2009) in *Grammostola anthracina* (Koch 1842). The natural perforations usually found in some cocoons indicate that spiderlings can also use preexisting holes in the egg sac (Panzera et al. 2009). Moreover, in *G. anthracina* copulation during egg sac care was observed, thus the ability of spiderlings to emerge without assistance comprises an important trait that allows the mother to be sexually receptive and mate without threatening the success of her spiderlings (Panzera et al. 2009). In some theraphosid species specialized structures on spiderlings such as bifurcated tips on cheliceral fangs are related with the opening of cocoon (Galiano 1969). However, in others the absence of such special structures related to cocoon opening suggests that they open the cocoon with their chelicerae (Panzera et al. 2009).

After emerging from the cocoon and molting, spiderlings are ready to leave the maternal burrow and begin the dispersal. This process does not usually occur immediately. It was observed in *Tliltocatl vagans* (Ausserer, 1875) that spiderlings remain inside and around the maternal burrow for several weeks and even undergo at least one molt after emerging from the egg sac (Shillington and McEwen 2006). Spiderlings are able to move easily through the silk covering the burrow entrance. The silk network around the burrow provides chemotactic cues for orientation of the juveniles (Minch 1978; Shillington and McEwen 2006). Silk deposition by spiderlings during emergence from the mother's burrow was first reported by Dor and Hénaut (2012) in the tarantula *Tliltocatl vagans*. Authors confirmed that spiderlings produce and lay silk during its gregarious stage and at least during the first instance of dispersal (Dor and Hénaut 2012). Also, it was observed that spiderlings are able to construct superficial burrows near the mother refuge, what could be a possible indicator of territorial behavior in tarantulas (Dor and Hénaut 2012).

The dispersal process in spiders is well known in some araneomorph families and even in some mygalomorphs. However, none of these studies included Theraphosidae, but later, juvenile behaviors associated with dispersal were described (Reichling 2000). However, some sporadic observations or assumptions led to think that spiderlings disperse in mass by walking out of the maternal burrow and moving in all directions (randomly) (Gertsch 1949). Since spiderlings are much too large to balloon away on silken lines (dispersal phenomenon known for araneomorphs and some mygalomorphs as ballooning), they usually settle down in the neighborhood of the burrow, hiding under stones or logs for a while and then occupying small burrows in the ground or constructing arboreal retreats depending on the species (Gertsch 1949; Stradling 1994).

Spiderlings of *Tliltocatl vagans* disperse from their maternal burrow in a novel way (Reichling 2000; Shillington and McEwen 2006; Dor and Hénaut 2012). In this case, groups of spiderlings walk in a single line that slowly snakes its way, resembling a column of "ants" (Fig. 14.4). This line can reach more than one meter in length (Reichling 2000). The spiderlings maintain close proximity to one another while walking, often slightly touching the abdomen of the individual ahead of them with their front legs (Reichling 2000). The juvenile occupying the front of the line changes direction frequently and the spider behind moves ahead and substitutes the

Fig. 14.4 A row of spiderlings of *Tliltocatl vagans* at dispersion stage (photo: Ariane Dor)

previous leader, which reinserts farther back in line (Reichling 2000). During this stage of dispersal, some silk-threads are produced by juveniles (Dor and Hénaut 2012). Spiderlings may produce silk as they walk, and the "ant column" dispersal pattern allows the adhesion of many silk threads of the successive spiderlings that pass that way. According to Shillington and McEwen (2006), not all spiderlings of the same maternal burrow disperse on the same time, thus the remaining juveniles disperse along the same path as the column of their siblings. Moreover, spiderlings may use the silk as a support for their chemical cues and to orient themselves (Dor and Hénaut 2012; Dor et al. 2008).

14.7 Final Considerations

For many years the study of the sexual behavior of mygalomorph spiders, mainly in the family Theraphosidae, was scarce in relation to araneomorphs. One true fact is that this group of spiders was traditionally considered relatively simple regarding their genitalia, courtship, and copulatory behavior. Fortunately, currently we are beginning to understand and elucidate the intricate mechanisms of communication, sexual behavior, and mating systems of tarantulas, opening exciting fields for future research. New studies on sexual behavior of theraphosids are focusing on female behavior, usually not recorded in previous studies, and on male and female communication before, during and after copulation. Future studies on sexual behavior of Neotropical theraphosid spiders should include detailed observations of male behavior before contacting the female in order to evaluate the occurrence of pheromones associated with female silk. Also, females should be exposed in similar conditions to nature, allowing burrow construction, thus eliciting natural behavioral responses to male courtship. Finally, careful observations and recording of individuals' behaviors during sexual interactions will allow new interpretations of the selective pressures driving their origin and maintenance.

Acknowledgments The author thanks the National Council for Technical and Scientific Research (CONICET). Special thanks to Gabriel Pompozzi and Ariane Dor for the photos.

References

Almeida-Silva LM, Camacho A, Brescovit AD, Lucas SM, Brazil TK (2008) Redescription and notes on the natural history of the arboreal tarantula *Iridopelma seladonium* (Araneae: Theraphosidae: Aviculariinae). Rev Bras Zool 25(4):728–736

Austad SN (1984) Evolution of sperm priority patterns in spiders. In: Smith RL (ed) Sperm competition and the evolution of animal mating systems. Academic, San Diego, CA, pp 223–249

Austin AD (1985) The function of spider egg sacs in relation to parasitoids and predators, with special reference to the Australian fauna. J Nat Hist 19:359–376

Baerg WJ (1928) The life cycle and and mating habits of the male tarantula. Q Rev Biol 3:109–116

Baerg WJ (1958) The Tarantula. University of Kansas Press, Lawrence, KS

Baruffaldi L, Perafán C, Perdomo C, Montes de Oca L, Pérez-Miles F, Costa FG (2013) Third pair of legs is a key feature for eliciting female receptivity in the road tarantula spider *Eupalaestrus weijenberghi* (Araneae: Theraphosidae). J Arachnol 41:213–214

Bertani R, da-Silva JPI (2002) The first mygalomorph spider without spermathecae: *Sickius longibulbi*, with a revalidation of *Sickius* (Araneae, Theraphosidae, Ischnocolinae). J Arachnol 30:519–526

Bertani R, Guadanucci JPL (2013) Morphology, evolution and usage of urticating setae by tarantulas (Araneae: Theraphosidae). Zoologia 30(4):403–418

Bertani R, Huff J (2013) *Avicularia rickwestii* sp. nov., a remarkable new species of *Avicularia* (Theraphosidae: Aviculariinae). Zoologia 30:333–337

Bertani R, Motta PC (2013) Redescription of *Avicularia taunayi* and notes on its habitat and geographical distribution (Araneae: Theraphosidae: Aviculariinae). Zoologia 30:107–114

Bertani R, Boston T, Evenou Y, Guadanucci JPL (2003) Release of urticating hairs by *Avicularia versicolor* (Walckenaer, 1837) (Araneae, Theraphosidae). Bull Br Arachnol Soc 12:395–398

Bertani R, Fukushima CS, da Silva-Júnior PI (2008a) Two new species of *Pamphobeteus* Pocock 1901 (Araneae: Mygalomorphae: Theraphosidae) from Brazil, with a new type of stridulatory organ. Zootaxa 1826:45–58

Bertani R, Fukushima CS, da-Silva JPI (2008b) Mating behavior of *Sickius longibulbi* (Araneae, Theraphosidae, Ischnocolinae), a spider that lacks spermathecae. J Arachnol 36:331–335

Bertani R, Nagahama RH, Fukushima CS (2012) *Vitalius nondescriptus* comb. nov. (Araneae: Theraphosidae: Theraphosinae): an example of theraphosid taxonomic chaos. Zoologia 29:467–473

Brazil V, Vellard J (1926) Contribuicao ao estudo do veneno das aranhas. II. Mem Inst Butantan 3:243–295

Bücherl W (1951) Estudos sobre a biologia e a sistemática do género *Grammostola* Simon, 1892. Monogr Inst Butantan 1:1–203

Bücherl W (1952) Instintos maternais nas aranhas brasileiras. Dusenia 3:57–74

Buskirk RE, Frohlich C, Ross KG (1984) The natural selection of sexual cannibalism. Am Nat 123:612–625

Charpentier P (1992) The genus *Avicularia*. Exothermae 1:1–41

Clarke D (1991) Captive-breeding programme for the red-kneed bird-eating spider *Euathlus smithii* at London Zoo. Int Zoo Yearb 30:68–75

Copperi S (2013) Comportamiento reproductivo de *Grammostola doeringi* (Holmberg 1881) (Araneae: Theraphosidae) en condiciones de laboratorio y a campo. Thesis (unpublished), Universidad Nacional del Sur, Argentina

Copperi S (2018) Biología reproductiva de arañas migalomorfas (Araneae: Mygalomorphae): patrones y mecanismos de selección sexual. Thesis (unpublished), Universidad Nacional del Sur, Argentina

Copperi S, Ferretti N, Pompozzi G, Pérez-Miles F (2012) Can't you find me? Female sexual response in an Argentinean tarantula (Araneae, Theraphosidae). Rev Colomb Entomol 38:164–166

Costa FG, Pérez-Miles F (1992) Notes on mating and reproductive success of *Ceropelma longisternalis* (Araneae, Theraphosidae) in captivity. J Arachnol 20:129–133

Costa FG, Pérez-Miles F (2002) Reproductive biology of Uruguayan theraphosids (Araneae, Theraphosidae). J Arachnol 30:571–587

Costa FG, Pérez-Miles F, Corte S (2000) Which spermatheca is inseminated by each palp in Theraphosidae spiders?: a study of *Oligoxystre argentinense* (Ischnocolinae). J Arachnol 28:131–132

Costa FG, Baruffaldi L, Perafán C, Perdomo C, Panzera A, Montes de Oca L, Pérez-Miles F (2013) Confusion and isolation in the courtship of two sympatric and synchronic tarantula spiders (Mygalomorphae, Theraphosidae). Arachnology 16:106–109

Costa FG, Curbelo B, Pérez-Miles F (2015) Long-term persistence and water resistance of female sex cues in the tarantula *Eupalaestrus weijenberghi* (Araneae: Theraphosidae). Arachnology 16:311–313

Costa-Schmidt LE, Albo MJ, Bollatti F, Cargnelutti F, Calbacho-Rosa L, Copperi S, Ferretti N, Izquierdo M, Peretti AV (2017) Sexual selection in Neotropical spiders: examples from selected groups. In: Viera C, Gonzaga MO (eds) Behaviour and ecology of spiders: contributions from the Neotropical region. Springer, Cham, pp 303–350

Davies NB, Krebs JR, West SA (2012) An Introduction to Behavioural Ecology, 4th edn. Wiley-Blackwell, London

Dias SC, Brescovit AD (2003) Notes on the behavior of *Pachistopelma rufonigrum* Pocock (Araneae, Theraphosidae, Aviculariinae). Rev Bras Zool 20:13–17

Dippenaar-Schoeman AS (2002) Baboon and Trapdoor spiders of Southern Africa: an identification manual. Agricultural Research Council, Pretoria

Dor A, Hénaut Y (2012) Silk use and spiderling behavior in the tarantula *Brachypelma vagans* (Araneae: Theraphosidae). Acta Zool Mex 28:1–12

Dor A, Machkour-M'Rabet S, Legal L, Williams T, Hénaut Y (2008) Chemically mediated burrow in the Mexican tarantula *Brachypelma vagans* female. Naturwissenschaften 95:1189–1193

Duvall D, Schuett GW (1997) Straight-line movement and competitive mate searching in praire rattlesnakes, *Crotalus viridis viridis*. Anim Behav 54:329–334

Eberhard WG (1996) Female control: sexual selection by cryptic female choice. Princeton University Press, Princeton, NJ

Ferretti N, Ferrero A (2008) Courtship and mating behavior of *Grammostola schulzei* (Schmidt 1994) a burrowing tarantula from Argentina. J Arachnol 36:480–483

Ferretti N, Pompozzi G, Copperi S (2012) Notes on egg-sacs and spiderlings of two species of *Grammostola* (Araneae: Theraphosidae) from central Argentina. J Br Tarantula Soc 27:53–61

Ferretti N, Pompozzi G, Copperi S, González A, Pérez-Miles F (2013) Sexual behaviour of mygalomorph spiders: when simplicity becomes complex; an update of the last 21 years. Arachnology 16:85–93

Ferretti N, Pompozzi G, Copperi S, Wehitt A, Galíndez E, González A, Pérez-Miles F (2017) A comparative morphological study of the epiandrous apparatus in mygalomorph spiders (Araneae: Mygalomorphae). Micron 93:9–19

Ferretti N, Cavallo P, Chaparro JC, Ríos-Tamayo D, Seimon TA, West R (2018) The Neotropical genus *Hapalotremus* Simon, 1903 (Araneae: Theraphosidae), with the description of seven new species and the highest altitude record for the family. J Nat Hist 52:1927–1984

Ferretti N, Copperi S, Pompozzi G (2019) Discovery of an isolated population of the dwarf tarantula *Homoeomma uruguayense* (Araneae, Theraphosidae) in central Argentina. Caldasia 41(2):436–441

Foelix RF (2010) Biology of spiders. Harvard University Press, Cambridge, MA

Galiano ME (1969) El desarrollo postembrionario larval de *Grammostola pulchripes* (Simon, 1891) (Araneae, Theraphosidae). Physis 29:73–90

Gaskett AC (2007) Spider sex pheromones: emission, reception, structures, and functions. Biol Rev 82:27–48

Gerhardt U (1929) Zur vergleichenden Sexualbiologie primitiver Spinnen, insbesondere der Tetrapneumonen. Z Morphol Ökol Tiere 14:699–764

Gertsch WJ (1949) American spiders. Van Nostrand, New York

Grossi B, Solís R, Veloso C, Canals M (2016) Consequences of sexual size dimorphism on energetics and locomotor performance of *Grammostola rosea* (Araneae; Theraphosidae). Physiol Entomol 41:281–288

Hénaut Y, Machkour-M'Rabet S, Weissenberger H, Rojo R (2015) Dimorphism and population size of the Mexican redrump tarantula, *Brachypelma vagans* (Araneae: Theraphosidae), in Southeast Mexico. Rev Mex Biodivers 86:737–743

Hormiga G, Scharff N, Coddington J (2000) The phylogenetic basis of sexual size dimorphism in orb-weaving spiders (Araneae Orbiculariae). Syst Biol 49:435–462

Ibarra-Grasso A (1961) Datos biológicos sobre *Grammostola buzarquensis* Ibarra-Grasso 1946 y su distribución geográfica (Araneae Theraphosidae). Neotrópica 7:7–12

Ibler B, Michalik P, Ficher K (2013) Factors affecting lifespan in bird-eating spiders (Arachnida: Mygalomorphae: Theraphosidae)—a multi-species approach. Zool Anz 253:126–136

Janowski-Bell ME, Horner NV (1999) Movement of the male brown tarantula, *Aphonopelma hentzi* (Araneae, Theraphosidae), using radio telemetry. J Arachnol 27:503–512

Krafft B (1982) The significance and complexity of communication in spiders. In: Witt PN, Rovner JS (eds) Spider communication: mechanisms and ecological significance. Princeton University Press, Princeton, NJ, pp 13–66

Lambert N, Dupre G (1992) Note preliminaire sur le developpement postembryonaire de *Theraphosa leblondi* (Thorelli, 1804) (Theraphosidae, Theraphosinae). Arachnides 12:7–8

Lourenço WR (1978) Notas sobre a biologia de *Acanthoscurria atrox* Vellard 1924 (Araneae, Theraphosidae). Rev Bras Biol 38:161–164

Magnusson WE (1985) Group movement by male mygalomorph spider. Biotropica 17:56

Marechal P, Rollard C, Leguin EA, Monpiou L (2009) Étude sur l'état des populations de *Avicularia versicolor*, mygale endémique de la Martinique. Muséum National d'Histoire Naturelle, Paris

Marples BJ (1967) The spinnerets and epiandrous glands of spiders. Zool J Linn Soc 46:209–223

Marshall SD, Uetz GW (1990) Incorporation of urticating hairs into silk: a novel defense mechanism in two Neotropical tarantulas (Araneae, Theraphosidae). J Arachnol 18:143–149

Marshall SD, Uetz GW (1993) The growth and maturation of a giant spider: Theraphosa blondi (Latrielle, 1804) (Araneae, Theraphosidae). Rev Arachnol 10:93–103.

Melchers M (1964) Zur biologie der vogelspinnen (Fam. Aviculariidae). Z Morphol Okol Tiere 53:517–536

Mendoza-Marroquín JI, Locht A, Kaderka R, Medina F, Pérez-Miles F (2016) A new genus of theraphosid spider from Mexico, with a particular palpal bulb structure (Araneae, Theraphosidae, Theraphosinae). Eur J Taxon 232:1–28

Minch EW (1978) Daily activity patterns in the tarantula *Aphonopelma chalcodes* Chamberlin. Bull Br Arachnol Soc 4:231–237

Minch EW (1979) Reproductive behaviour of the tarantula *Aphonopelma chalcodes* Chamberlin (Araneae, Theraphosidae). Bull Br Arachnol Soc 4:416–420

Montenegro R, Aguilera MA, Casanueva MA (2018) First record of *Homoeomma* Ausserer, 1871 in Chile and description of two new species. Spixiana 41:13–25

Panzera A, Perdomo C, Pérez-Miles F (2009) Spiderling emergence in the tarantula *Grammostola mollicoma* (Ausserer 1875): an experimental approach (Araneae, Theraphosidae). J Arachnol 37:92–96

Peretti AV, Aisenberg A (2015) Cryptic female choice in arthropods. Springer, Cham

Pérez-Miles F (1989) Variación relativa de caracteres somáticos y genitales en *Grammostola mollicoma* (Araneae, Theraphosidae). J Arachnol 17:263–274

Pérez-Miles F, Costa FG (1992) Interacciones intra e intersexuales en *Grammostola mollicoma* (Araneae, Theraphosidae) en condiciones experimentales. Bol Soc Zool Uruguay 7:71–72

Pérez-Miles F, Costa FG (1994) *Acanthoscurria atrox* incorporates urticating hairs into its shedding mat. Forum Am Tarantula Soc 3:63–64

Pérez-Miles F, Costa FG, Gudynas E (1993) Ecología de una comunidad de Mygalomorphae criptozoicas de Sierra de las Animas, Uruguay (Arachnida, Araneae). Aracnología 17:1–22

Pérez-Miles F, Costa FG, Toscano-Gadea C, Mignone A (2005) Ecology and behaviour of the "road tarantulas" *Eupalaestrus weijenberghi* and *Acanthoscurria suina* (Araneae, Theraphosidae) from Uruguay. J Nat Hist 39:483–498

Pérez-Miles F, Postiglioni R, Montes de Oca L, Baruffaldi L, Costa FG (2007) Mating system in the tarantula spider *Eupalaestrus weijenberghi* (Thorell, 1894): evidences of monandry and polygyny. Zoology 110:253–260

Platnick N (1971) The evolution of courtship behaviour in spiders. Bull Br Arachnol Soc 2:40–70

Postiglioni R, Costa FG (2006) Reproductive isolation among three populations of the genus *Grammostola* from Uruguay (Araneae, Theraphosidae). Iheringia 96:71–74

Prentice TR (1992) A new species of North American tarantula, *Aphonopelma paloma* (Araneae, Mygalomorphae, Theraphosidae). J Arachnol 20:189–199

Prentice TR (1997) Theraphosidae of the Mojave desert west and north of the Colorado River (Araneae, Mygalomorphae, Theraphosidae). J Arachnol 25:137–176

Punzo F, Henderson L (1999) Aspects of the natural history and behavioural ecology of the tarantula spider *Aphonopelma hentzi* (Girard, 1854) (Orthognatha, Theraphosidae). Bull Br Arachnol Soc 11:121–128

Quirici V, Costa FG (2005) Seismic communication during courtship in two burrowing tarantula spiders: an experimental study on *Eupalaestrus weijenberghi* and *Acanthoscurria suina*. J Arachnol 33:199–208

Quirici V, Costa FG (2007) Seismic sexual signal design of two sympatric burrowing tarantula spiders from meadows of Uruguay: *Eupalaestrus weijenberghi* and *Acanthoscurria suina* (Araneae, Theraphosidae). J Arachnol 35:38–45

Redondo T (1994) Comunicación: teoría y evolución de las señales. In: Carranza J (ed) Etología: Introducción a la Ciencia del Comportamiento. Universidad de Extremadura, Cáceres, España, pp 255–297

Reichling SB (2000) Group dispersal in juvenile *Brachypelma vagans* (Araneae, Theraphosidae). J Arachnol 28:248–250

Schaeffer M (1987) Life cycles and diapause. In: Nentwig W (ed) Ecophysiology of spiders. Springer, Heidelberg, pp 331–347

Shillington C (2002) Thermal ecology of male tarantulas (*Aphonopelma anax*) during the mating season. Can J Zool 80:251–259

Shillington C (2005) Inter-sexual differences in resting metabolic rates in the Texas tarantula, *Aphonopelma anax*. Comp Biochem Physiol 142:439–445

Shillington C, McEwen B (2006) Activity of juvenile tarantulas in and around the maternal burrow. J Arachnol 34:261–265

Shillington C, Peterson CC (2002) Energy metabolism of male and female tarantulas (*Aphonopelma anax*) during locomotion. J Exp Biol 205:2909–2914

Shillington C, Verrell P (1997) Sexual strategies of a North American "Tarantula" (Araneae, Theraphosidae). Ethology 103:588–598

Simmons LW (2005) The evolution of polyandry: sperm competition, sperm selection and offspring viability. Annu Rev Ecol Evol Syst 36:125–146

Smith AM (1994) Theraphosid spiders of the new world. Vol. 2. Tarantulas of the U.S.A. and Mexico. Fitzgerald, London

Stradling DJ (1994) Distribution and behavioral ecology of an arboreal 'tarantula' spider in Trinidad. Biotropica 26:84–97

Teixeira JAA (2017) Estudo do comportamento de corte e cópula de *Dolichothele exilis*– (Mygalomorphae, Theraposidae) em um contexto comparativo com espécies congenéricas. Thesis (unpublished). Universidade Estadual Paulista, Instituto de Biociências de Rio Claro, Brazil

Uetz GW, Stratton GE (1983) Communication in spiders. Endeavour 7:13–18

Walker SE, Rypstra AL (2001) Sexual dimorphism in functional response and trophic morphology in *Rabidosa rabida* (Araneae: Lycosidae). Am Midl Nat 146:161–170

Yañez M, Locht A, Macías-Ordóñez R (1999) Courtship and mating behaviour of *Brachypelma klassi* (Araneae, Theraphosidae). J Arachnol 27:165–170

Chapter 15
Tarantulas in Captivity: Raising and Breeding

Laura Montes de Oca and Jorge Mendoza

Abstract Tarantulas are animals that you either love or find disgusting. Some people are fascinated with them, whereas others fear them because of ignorance or aversion to what they consider dangerous on the basis of their appearance. Despite their reputation, many people study, maintain and use these spiders. The interest in keeping and propagating tarantulas has increased worldwide in recent decades. The aim of this chapter is to discuss the basic and ethical considerations involved in keeping these spiders in captivity (whether for research or as an enthusiast), as well as being aware of the origin and conservation status of the most common commercial species. As more has been learned about tarantula biology, there has been greater interest in how to breed them in captivity. However, few studies have addressed the reproductive biology of tarantulas. Despite this, many species are propagated successfully by enthusiasts, and the conditions for their reproduction are empirically known. We approach this chapter by dividing it into five sections: (1) natural history; (2) keeping and breeding tarantulas in captivity; (3) common ailments; (4) tarantulas' popularity as pets and their husbandry; and (5) traffic, endangered species and responsible breeding facilities.

L. Montes de Oca (✉)
Departamento de Ecología y Biología Evolutiva, Instituto de Investigaciones Biológicas Clemente Estable, Montevideo, Uruguay
e-mail: lmontesdeoca@iibce.edu.uy

J. Mendoza
Colección Nacional de Arácnidos, Módulo D Planta Baja, Departamento de Zoología, Instituto de Biología, Universidad Nacional Autónoma de México, Coyoacán, Mexico City, DF, Mexico

F. Pérez-Miles (ed.), *New World Tarantulas*, Zoological Monographs 6,
https://doi.org/10.1007/978-3-030-48644-0_15

15.1 Natural History

15.1.1 Common Names and History

Tarantulas are mygalomorph spiders belonging to the family Theraphosidae. Currently, 985 species belonging to this family are recognized, and more than half are from the New World (World Spider Catalog 2020). Theraphosid spiders are known for being hairy, large and long lived in comparison with other spiders. Although many people consider them dangerous and nasty, mostly because of lack of information, they have become very popular within others. Because of their popularity, these spiders are known by several common names. Theraphosids are known as baboon spiders in Africa, earth tigers in Asia, whistling or barking spiders in Australia and bird-eater spiders or tarantulas in the rest of the world. Indeed, the Committee on Common Names of Arachnids of the American Arachnid Society has developed a list of common names principally for those arachnids that are very common in houses, museum or zoo exhibitions, or the pet trade, in accordance with the following criteria: medical or agricultural importance (for their venom, as pests, or as predators of arthropod pests), endangered or threatened status, and abundance or conspicuousness. According to these criteria spiders from the family Theraphosidae can be named tarantulas (in lowercase) and common species will have an associated common name; for example, the (Goliath) bird-eater tarantula is the common name of *Theraphosa blondi* (Latreille, 1804) (Breene 1998). However, the common names are not universal and can vary between different countries and different communities. Hence, the words "tarantula" and "mygalomorph" are technically wrong (Psaila 2005; Pizzi 2012) and their use can provoke confusion. When people consult arachnologists about "big tarantulas" in their gardens, they often turn out to be wolf spiders. This is because the common name "tarantula" originates from the Araneomorphae spider *Lycosa tarantula* (Linnaeus, 1758), a wolf spider distributed in southeastern Europe, the Mediterranean and the Near East (World Spider Catalog 2020). History tells us that in the seventeenth and eighteenth centuries in Italy, the bites of these spiders (and other insects) were combatted with a frenzied dance for a couple of days to remove the "tarantism," accompanied by music known nowadays as tarantella. Also, there is a whip spider (Amblypygi) genus known as *Tarantula* Fabricius, 1793. Another name is *Mygale*, which means "field mouse," and was first used in the fourteenth and fifteenth centuries to name ermines (Mustelidae) (Sedinova 2015).

15.1.2 Natural History

Tarantulas can be found in tropical and subtropical regions, and are adapted to inhabit a wide range of habitats, from the rainforest canopy to desert soils. They are shy and typically solitary, spending most of the time inside their retreats. As ectotherms, tarantulas are sensitive to temperature. In nature, they construct their retreats in order to maintain constant humidity and temperature, avoiding exposure to high

Fig. 15.1 Burrows. (**a**) *Bistriopelma* sp. from Peru. (**b**) *Pamphobeteus insignis* from Peru. (**c**) *Aphonopelma seemanni*. (**d**) *Caribena* sp. from Puerto Rico. (Photos (**a**) and (**b**): L. Montes de Oca; photos (**c**) and (**d**): J. Richards)

or low temperature conditions (Punzo and Henderson 1999; Hardin and Sincage 2015). Burrow construction will vary depending on the habitat and the species. Terrestrial tarantulas use hollows or dig a vertical tunnel into the soil with or without chambers, or use sac-like chambers under rocks or logs; arboreal ones use holes or tree bark (Fig. 15.1). For example, the skeleton tarantula, *Ephebopus murinus* (Walckenaer, 1837), constructs two kinds of retreat: a silken tube mixed with leaves in standing vegetation, or a fossorial J-shaped vertical burrow (a tunnel) with a final chamber, whose entrance is ornamented with a wide funnel of silk and debris (Marshall and West 2008). The latter is similar to that of the white-collared *Eupalaestrus weijenberghi* (Thorell, 1894), but the entrance is smoothly surrounded by silk (Pérez-Miles et al. 2005, 2007). The Mexican redrump, *Tliltocatl vagans* (Ausserer, 1875), constructs a vertical and then a horizontal tunnel (20 cm long) leading from the entrance, with one to four chambers, using the last ones for leaving prey remains and molts (Locht et al. 1999; M'rabet et al. 2007). The female's burrow entrance is covered with more silk during the reproductive season (Locht et al. 1999). Species of the genus *Grammostola* Simon, 1892 can construct vertical tunnels in an open field or live under rocks in sac-like chambers (Ferretti et al. 2012; Costa and Pérez-Miles 2002). The arboreal pinktoe, *Avicularia avicularia* (Linnaeus, 1758), constructs its retreat in a hole, behind tree bark or on a tree branch (Cloudsley-Thompson and Constantinou 1985; Stradling 1994). Some tarantulas build only one burrow and enlarge it as they grow (Dippenaar-Schoeman 2002), whereas others present ontogenic habitat shifts. For example, some terrestrial juveniles construct retreats off the ground, avoiding the burrows of bigger juveniles or adult females (Marshall and West 2008), or look for retreats to occupy instead of digging. Arboreal juveniles construct retreats within the leaves of climbing plants on trees and, as they grow, exchange those retreats for higher ones behind loose bark or in hollow branches without using leaves (Stradling 1994; Marshall and West 2008). Besides providing a suitable microhabitat, retreats are essential for protection while spiders are molting, for safety during the cold season, while spiders are taking care of egg sacs and also for protection against parasites and predators. For example, tarantulas have to beware of their biggest enemy, the tarantula hawk, *Pepsis* (Pompilidae) (Punzo 1994; Costa et al. 2004; Punzo 2005; Rego et al. 2004; Kurczewski 2010; Kurczewski and Edwards 2012). Indeed, some species such as the white-collared tarantula (*Eupalaestrus weijenberghi*) construct an extra small chamber at the end of their burrow, annexed to the main chamber, in order to prevent the wasp reaching them.

Most of the tarantulas in America have a special kind of hairs for their defense, called urticating hairs (setae), located in the dorsal part of the abdomen. These setae cause irritation on contact with skin. When a tarantula is disturbed, it throws the stinging hairs at the aggressor, which is why some tarantulas remain "bald" on the abdomen. If this is not enough, it raises its front legs, showing its fangs (Gallon 2000).

Tarantulas have two remarkable characteristics, the first being their low vagility. They spent most of their lives inside their retreats, emerging only to catch prey or when attracted by the courtship of a male. As they cannot balloon (as some other spiders do) to disperse, their big moment for dispersal is when they leave their mother's burrow as spiderlings. Reichling (2000) published the first description of the juvenile dispersion of *Tliltocatl vagans*. He found three groups of 72, 76 and 135 spiderlings walking at night, forming a line, and every 7–10 cm, the leader changed position, allowing the following spiderling to take the leader position. On this occasion, the nearest female burrow found was approximately 50 m away. In addition, Shillington and McEwen (2006), observing the same species, reported a maximum spiderling dispersal distance of 9 m from the mother's burrow. This behavior could explain why these spiders are found in local aggregations, while in adjacent areas with similar vegetation, they are absent (Reichling 2000).

Their second remarkable characteristic is their great longevity. Tarantulas from temperate zones, which are inactive during droughts or in the winter season (with exceptions), have a slower metabolism and live even longer. For example, *Aphonopelma hentzi* (Girard, 1852) mature at 10–12 years (Baerg 1958) and *Grammostola* species at 7–13 years (Costa and Pérez-Miles 2002). In the tropics, tarantulas mature faster; for example, *Avicularia avicularia* reaches adulthood in 3–4 years (Stradling 1978). Females become mature much later than males (Stradling 1978; Pizzi 2012). This has been suggested to be a temporal barrier to prevent breeding between siblings in nature (Pizzi 2012). After reaching adulthood, females tarantulas can live for 10–20 more years; moreover, in captivity, a female's lifespan can reach 20–34 years (Baerg 1928; Costa and Pérez-Miles 2002; Foelix 2011; Montes de Oca, personal observation 2010–2020). On the other hand, males have a shorter lifespan, and after they reach adulthood (sexual maturity), they can live on for between 4 months and 6 years, depending on the species—some probably only until the end of the mating season (Baerg 1928; Millot 1943; Stradling

1978; Locht et al. 1999; Costa and Pérez-Miles 2002; Pérez-Miles et al. 2005, 2007; Schultz and Schultz 2009). Some males can become more long lived and very occasionally can present a change after their sexual maturity but lose their reproductive capacity (Verdez and Cléton 2004). Even though in captivity they can live longer, they are unable to keep growing. Males that have molted after becoming sexually mature present anomalies in their pedipalps (their bulbs cannot be regenerated) and sometimes die during the molting process.

Spider growth depends on environmental factors; for example, food can affect the size (Baerg 1928; Turnbull 1962, 1965), as can the photoperiod and temperature (Peck and Whitcomb 1970; Li and Jackson 1996). If we have to rear one, we must bear that in mind. As arthropods, tarantulas have an exoskeleton and must change it in order to become bigger, in a process called molting or ecdysis. Days before molting, the tarantula will start fasting (Pizzi 2012) (and after ward too). Then it will create a soft sheet of silk on the ground and lie down dorsally (upside down) (Fig. 15.2). This is a critical moment; you must leave it in peace and alone, without prey in the terrarium, until the whole process has ended. The process starts a week before molting, when the body starts to prepare. Epidermal cells secrete chitinases and protease enzymes into the exuvial space (the gap between the epithelium and cuticle), which start dissolving the endocuticle, and the cuticle separates from the epidermis. Hypodermal cells secrete the cuticulin (the first new layer), which is protected from digestion. All dissolved substances are reabsorbed. Once the endocuticle has been largely digested, the prosoma starts to tear in the pleural areas. Ecdysis starts with lifting of the carapace, followed by liberation of the abdomen and, lastly,

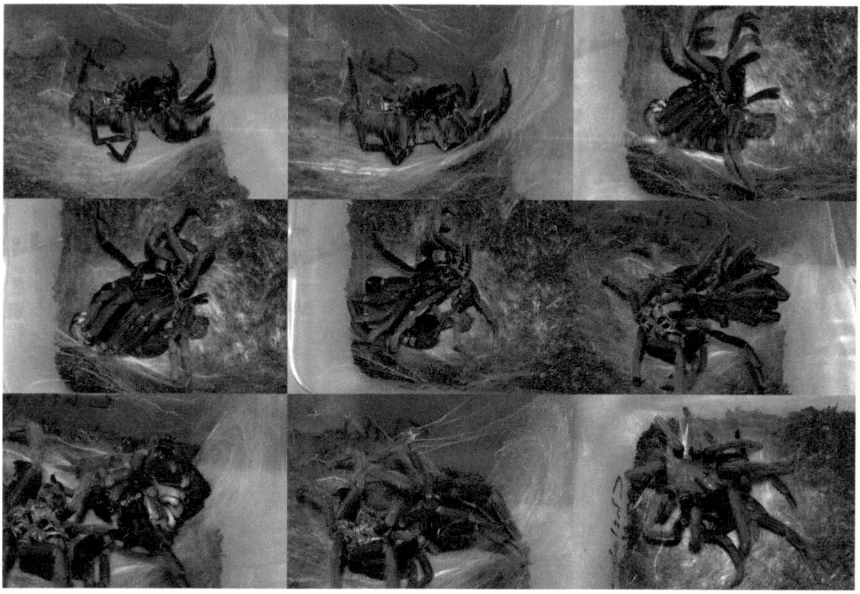

Fig. 15.2 *Chilobrachys dyscolus* molting sequence. (Photos: J. Mendoza)

extraction of the extremities. Finally, the exocuticle starts to harden. During the subsequent weeks, the exoskeleton will increase its thickness by endocuticle deposition during the weeks following the molt (Foelix 2011). Furthermore, in the molting process, the spider regenerates lost extremities and also recovers the urticating hairs and all of the setae and spines that were present originally. It is common for spiders to lose some of their extremities. Legs can be autotomized from the coxa–trochanter junction if the spider gets hurt somewhere on the appendages below the coxae or if a predator takes an extremity. Regeneration will occur slowly; first, the spider will grow a shorter leg because the new one, after autotomy, will be developed inside the coxa. After the spider's molt, the new leg will be shorter and thin at first, but then, after consecutive molts, it will resemble the original one (Baerg 1926, 1938; Foelix 2011). Molting periodicity will vary depending on the stage. Spiderlings molt weekly, young spiders will molt less frequently (up to once a month) and juveniles each 2–6 months until they become adults. After that only females will keep molting every year or two (Baerg 1938, 1958; Costa and Pérez-Miles 2002).

Spiders are mostly carnivorous "sit-and-wait" predators. This strategy requires little energy because they can use their anerobic capacity to catch prey and their venom to paralyze it (Prestwich 1983a, b). But they do not have jaws, so they have to start the digestion outside their body by regurgitating digestive fluids into the prey in order to dissolve it and produce a liquefied meal. Then, activating the pharynx and stomach muscles, the spider sucks the juice through the mouth opening and transports it to the midgut to absorb the nutrients (Foelix 2011). Tarantulas prey on invertebrates, principally arthropods, and the big ones can even feed on small vertebrates such as lizards, frogs, snakes, mice and birds (Yánez 1998 in Locht et al. 1999; Punzo and Henderson 1999; M'Rabet et al. 2007; Foelix 2011). The majority are active at night or early in the morning. They remain in the entrance of their safe retreat (Cloudsley-Thompson and Constantinou 1985; Yáñez and Floater 2000) until some prey comes within a short distance. The spider must first touch the prey to sense its size and, if it is appropriate, it will catch the prey with its front legs and bite it. This hunting technique can result in long fasting periods; fortunately, spiders are adapted to survive these conditions for several months by reducing their metabolism (Baerg 1928; Punzo 1989; Anderson 1970, 1974). Indeed, in temperate zones, spiders do not eat (much) during the cold season, and nor do females when they are taking care of their egg sacs, during which time they remain inside their warm and closed burrows (Cloudsley-Thompson and Constantinou 1985; Locht et al. 1999; Pérez-Miles et al. 2005).

15.2 Keeping and Breeding Tarantulas in Captivity

Husbandry of a tarantula is not trivial, but it is very intuitive. The most important thing is to use common sense and observe the spider's behavior. The first thing you should know is where the species lives in its natural habitat, such as its natural climate and refuge construction. You should try to reproduce the natural conditions in order to keep your spider in optimal conditions. Bengston et al. (2014) found behavioral differences in *Brachypelma smithi* (F. O. Pickard-Cambridge, 1897) between individuals housed in enrichment and nonenrichment conditions. Those who were raised in conditions similar to their natural conditions (enrichment conditions) showed fewer threat displays and more prey attacks and exploratory behavior. Sects. 15.2.1–15.2.10 give details of general care that must be borne in mind. For information about special care, see Sect. 15.4.

15.2.1 The Terrarium

Tarantulas are principally solitary, so they must be placed in separate containers. Although some exceptions exist, it is recommended to keep individuals separately. For better observation, the containers can be made of glass, acrylic or other transparent plastic. Although they are used to living in small burrows, they need space to prepare their retreat, feed, and have room while they are molting. Marshall (1996) suggests a minimum container size of 15 times the spider's leg span. For terrestrial tarantulas, you must consider a container length and height not more than twice the spider's length (e.g., 30 × 30 cm) in order to avoid injures from falling (Pizzi 2010; a top door is recommended. In contrast to terrestrial tarantulas, for arboreal tarantulas the vertical dimension is more important than the horizontal dimension, so you should consider a taller container (e.g., 30 × 50 cm), and both front and top doors are recommended (Psaila 2005; Bennie et al. 2011). Lids and doors should always be secure, do not underestimate the tarantula strength and perseverance to move them, they will try to get out from the container (Pizzi 2012; Montes de Oca, personal observation 2010–2020). Make sure the container has ventilation; small holes can be made in the lid or at the top of the walls. When thinking about the container location, the things you must avoid are (1) noise: remember that spiders are very sensitive to vibration; (2) direct sun: when it hits the container, it increases the temperature; and (3) draught: this also causes a temperature disruption. It is recommended to keep them away from chemical substances and at a height where it is warmer than near the floor.

In the case of a spiderling, it can be kept in a small plastic container at least three to four times bigger than its leg span. This container must be changed as the tarantula grows to a size at which it can live in its final terrarium. It will be fine as long as you ensure that there is some moisture in the substrate. In cold places, heat can be provided by placing it on an electrical heating pad. It is preferable to place the pad on one side of the container or put a small piece of cardboard on the heating pad and place the tarantula's container on the cardboard so that it does not receive too much heat (Hardin and Sincage 2015; Mendoza, personal observation (2008–2020)).

15.2.2 Substrates

There are a variety of options for the substrate. It is most important to select one with the ability to retain water (e.g., turf or coir) and also permit it to drain. Here, we summarize a couple of mixtures. Mix I contains three parts of sterilized peat and one part of sand, which are then mixed with water. The mixture must be moist but not soaked, and it must be able to hold the shape of a burrow. (Psaila 2005). Mix II contains two parts of coir, one part of fine coarse sand and one part of topsoil (Hardin and Sincage 2015). You can place 5–10 cm of substrate for terrestrial spiders and 3–4 cm for arboreal ones, softly compressing the surface to make it firm underfoot (Saul-Gershenz 1996; Psaila 2005; Hardin and Sincage 2015). Whatever substrate ingredients you select, ensure that they are sterile and do not contain chemicals or parasites (Marnell 2016). Avoid gravel and anything with sharp edges that could injure the spider if it falls (Hardin and Sincage 2015). A new technique is the use of "living soil," involving addition to the terrarium of topsoil, which will include microarthropods such as isopods, collembola, and microorganisms such as bacteria and fungi. These will help to decompose organic matter, such as prey remains, and feed on unwanted acarids, for example, keeping the substrate "clean." Moreover, their movement within the soil will aerate it, improving its drainage and structure.

15.2.3 Hides

Burrowing species will be fine with a deep substrate. For other species, you can use natural hides (such as logs, coconut shells or rocks) or artificial ones (such as pots, PVC pipe or plastic containers). For arboreal species, you can add a bark layer (Baxter 1995). Always leave half of the floor area clear, providing space for eating, molting and reproduction (Hardin and Sincage 2015). Remember to provide an enrichment environment to improve healthy development (Bennie et al. 2011; Bengston et al. 2014). It is recommended not to decorate it too much, in order to make it easy to clean it periodically.

15.2.4 Light

Tarantulas do not need too much light; a room where daylight comes in through a window is sufficient. Avoid direct sunlight on the terrarium. Spiders need to be able to distinguish the day from the night, so if they are in a dark room, it is recommended to use dim lights (not bright ones). During the day, you can turn on a low-wattage light, and during the night, if illumination is necessary, use low-wattage colored bulbs or black light (Psaila 2005; Bennie et al. 2011).

15.2.5 Food

As tarantulas are ambush predators and use their hairs to feel the vibrations of prey when they are nearby, they need to be fed with live food. In captivity, tarantulas can be fed with crickets (*Acheta domesticus*), mealworms (*Tenebrio molitor*) and cockroaches (*Blaberus discoidalis, Blaptica dubia, Gromphardorhina protentosa*). Mealworms should preferably be offered carefully with forceps until the tarantula takes them, so they do not fall into the substrate and become buried. Mice or other small vertebrates are not recommended, because the remains start to rot, attracting parasites. Spiderlings can be fed with fruit flies (*Drosophila melanogaster*) and worker termites. You can also try moths, grasshoppers, earthworms, waxworms or cicadas. Always be vigilant for the presence of pesticide in areas where you find the prey, as the spider can become intoxicated if it is fed prey contaminated with pesticide. The size of the prey will depend on the size of the spider; you can start with the length of the spider's cephalothorax as a reference point. As you raise the spider and take care of it, you will get to know the spider's preferences (and personality). Adult spiders can be fed every 4–7 days manually ad libitum, waiting until they capture the first prey with their fangs before you offer the next one. For example, a medium tarantula can eat up to 3–4 mealworms at a time (Montes de Oca, personal observation 2010–2020). Small vertebrates should be offered up to twice a year; more could cause obesity (Hardin and Sincage 2015). If the spider rejects the food it is recommended to remove it. The prey could stress the spider, eat or damage its egg-sac, even attack the spider during molting. Another reason is if the prey dies, it will start to decompose, attracting parasites (e.g. it is common for some prey to drown in the spider's water bowl) (Psaila 2005; Saul-Gershenz 1996; Montes de Oca, personal observation 2010–2020). Remember that a spider starts a fasting period 30–60 days before ecdysis (Philip 2006; Philip and Shillington 2010; Foelix 2011). Other factors that can influence fasting are seasonal temperature changes, the light cycle and reproductive activities. In nature, mature males are known to stop eating in order to look for females to copulate with, consequently dying by starvation after the reproductive season (Pérez-Miles et al. 2005). In captivity, they will become more active and can survive for a few years (depending on the species) if we provide water and try to feed them (De Voe 2009). Studies in araneomorph spiders have found that food limitation in juveniles reduces egg production at the adult stage, and food limitation in adults reduces the clutch size. In both cases (though the impact of food limitation at the juvenile stage is greater), food limitation reduces the life span (Kleinteich et al. 2015). Other studies have found that spiderlings with a restricted diet have reduced growth but are able to gain mass once food is more plentiful again (Jespersen and Toft 2003).

15.2.6 Heating and Temperature

Species from temperate zones can tolerate a temperature range from 21 °C to 27 °C (Baxter 1995), but tropical species need a constant temperature, which may be around 27–30 °C (Psaila 2005). High temperatures may cause terrestrial species to start climbing inside the container, increasing the risk of falls, and tarantulas can become more aggressive and increase their water consumption. More importantly, a reduction in humidity can cause dehydration (Psaila 2005). Low temperatures will slow down the spider's metabolism, making it lethargic and reducing its appetite

(Psaila 2005). If you have to provide external heat and use a heat pad or lights, they should be placed only on one side of the container in order to not overheat it and to create a temperature gradient so that the spider can move to where it is most comfortable and thermoregulate as needed (Baxter 1995; Hardin and Sincage 2015).

15.2.7 Water and Humidity

Tarantulas have two pair of book lungs, giving them a greater propensity to lose water by evaporation (Davies and Edney 1952; Figueroa et al. 2010). Subtropical species should be kept in a humidity range of 40–60% (Shultz 1998 in Psaila 2005), whereas tropical species need higher humidity of 65–90% (de Vosjoli 1991; Baxter 1995). For terrestrial spiders, always keep a shallow bowl of water on the soil without any kind of cotton or sponge, which could cause infections (Marshall 2001; De Voe 2009; Riley and Barron 2016; Cléton et al. 2015; Mendoza, personal observation 2008–2020). As small spiders or prey can fall into it and drown, it is suggested to put small stones inside it to help them escape (Hardin and Sincage 2015), or maintain a damp area of substrate in the cage for the spider to drink from. Arboreal spiders will prefer to drink water drops on their web or in their surroundings, provided with a spray bottle (Frye 1992; De Voe 2009). To maintain the humidity, you can also cover the vents with a silk screen, plastic food wrap or laminated paper; this will depend on the kind of container you have selected and the mesh or hole size you use to provide ventilation (Psaila 2005; Saul-Gershenz 1996). Another way is spraying water every day (Bennie et al. 2011; Saul-Gershenz 1996). A dry environment can cause problems during molting, making the new skin stick to the old skin, which can cause loosening of extremities and respiratory difficulties (including asphyxia) if it gets stuck in the book lungs.

15.2.8 Cleaning

Containers must be maintained and cleaned periodically. The temperature and humidity conditions can cause feces, food remains or dead bugs to become infested by fungi and parasites. To prevent disease, containers should be kept under hygienic control. The inside surfaces can be cleaned with a 3400 ppm (part-per-million) sodium hypochlorite solution (SHS). Use different forceps for feeding and removing food and for removing or replacing items. This will avoid contamination of clean items. When you have finished using them, all forceps can be autoclaved or washed with SHS for use the next time. Some spiders leave their feces on the wall; they can be removed using cotton soaked in water (Psaila 2005; Saul-Gershenz 1996). Soil sterilization can be achieved by baking it at 200–250 °C for 2 hours in a conventional oven (Saul-Gershenz 1996) or even in a microwave oven (at 625 watts and 2450 MHz); leave it for at least 30 seconds until you see steam, and then it is ready. The time needed will depend on the amount of soil used and can be between

1 and 3 minutes (e.g., 1 kg of soil for 150 seconds; see Trevors (1996) for more methods). Let it cool before removing it from the oven.

15.2.9 Handling and Transport

Tarantulas are not animals that should be manipulated; they are fragile creatures that can easily be hurt accidentally, so it is always advisable to handle them only if it is strictly necessary. Although most tarantula species are docile, they all have venom glands and, if they feel stressed or threatened, they can bite you with their big fangs. Although it is not lethal, their bite can produce pain, local or extensive swelling, fever, dysesthesia, cramps, myalgia, or shivers (Haro and Jouglard 1998). Moreover, the urticating hairs present on New World tarantulas can cause chronic conjunctivitis (Haro and Jouglard 1998). To avoid the urticating hairs, it is recommended to use latex gloves and a facial mask when moving the tarantula or cleaning the terrarium (Saul-Gershenz 1996). To catch a spider, you can use a container (such as a glass) or a goldfish net and a small stick to tap on the burrow entrance. The use of cotton or gardening gloves is also recommended. If the tarantula is inside the burrow, you can dig (very gently) with a spoon. In the case of arboreal tarantulas, if all you need to do is change the substrate, remove the spider with its whole retreat (the log) or, if possible, you can start pressing on the silk tube from the back in order to persuade the spider to go outside, and then capture it with a container. The best time to manipulate a tarantula is during the day, as they are more active during the night (Psaila 2005).

15.2.10 Breeding

By examining the molts, sexes can be determined while tarantulas are still juveniles. Females can be recognized by the presence of spermathecae on the inner side of the epigastric furrow, between the anterior book lungs (Fig. 15.3). Juvenile females will have spermathecae that are sclerotized. If such a structure is not seen, you must have

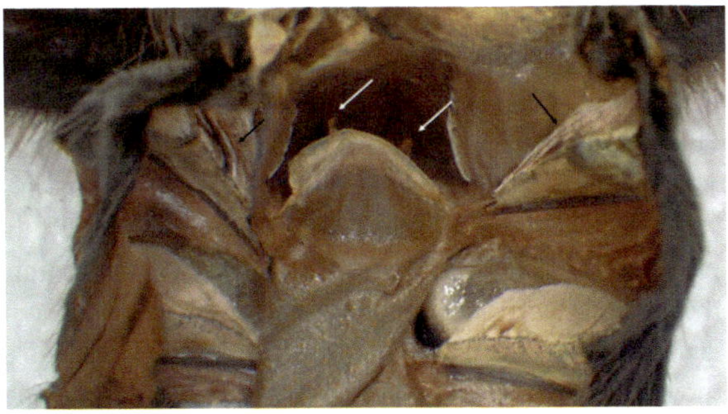

Fig. 15.3 *Grammostola anthracina* spermathecae (*white arrows*) between anterior booklungs (*black arrows*). (Photo: L. Montes de Oca)

a juvenile male. For example, in *Brachypelma* Simon, 1891, the spermatheca can be seen from the fifth instar (Hardin and Sincage 2015). Also, adult females can show an epigastric furrow that is swollen (raised) (Pizzi 2012). Males become sexually mature in their last molt, in which they develop a tibial apophysis (or spur) on legs I—a structure they use to hold onto females by their chelicera during copulation (except for the common species *Theraphosa stirmi* Rudloff & Weinmann, 2010 and Old World *Poecilotheria* spp.). Also, the tarsus on the palp is modified and appears as a bulb—a sclerotized structure that transfers sperm into the female spermathecae. As their testicles are in their abdomen, males have to transfer the sperm into their bulbs, in a process called sperm induction (Costa and Pérez-Miles 2002; Foelix 2011) (see Chap. 14). This process occurs during a 2-week period after the last molt, and then the males are ready for copulation. You can see the remains of the dense sheet of silk as a thick white string.

Preparation for mating Each species has its own reproduction period (see Sect. 15.4) and, depending on the species, females may copulate twice a year or every second year (Petrunkevitch 1911; Prentice 1997; Shillington and Verrell 1997; Costa and Pérez-Miles 2002). Females of some species may copulate several times (Petrunkevitch 1911), whereas others become reluctant after their first copulation (Pérez-Miles et al. 2007). In nature, the female remains in her retreat until a male approaches, detects her pheromones and starts his courtship. In captivity, copulation should occur in the female's container. Mature females should be well fed and within 2–3 months of their last molt. Before introducing the male, clear the terrarium, removing all objects from it to create space and avoid obstructions in case you need to separate the couple in the event that something goes wrong. Once the male is inside the terrarium and perceives the female's presence, he will start his courtship with vibration of his posterior legs and palp drumming. If the female is receptive, she will answer the male drumming with her pedipalps and approach him. Once they are close, the male will hook up his spurs on the female's chelicera in order to lift her and get access to the ventral part of her abdomen, and reach the epigastric furrow with his pedipalps to introduce the bulb and transfer the sperm into the spermathecae (Fig. 15.4). Copulation will end with the couple separating, at which point the female will return to her burrow or remain quiet for a while, and the male will move away. You can also help the male to get out in order to avoid an attack from the female. It also can happen that a male starts his courtship but the female does not respond, because she is still immature, is close to molting or is already gravid. In those cases, the male may intensify his courtship or move away. Be alert in case the female becomes upset or tries to attack the male. Depending on the species, courtship and copulation can take up to an hour. More details of the process are given in Chap. 14. If the female molts after copulating, the stored sperm will be lost; this can occur if the female is overfed (Clarke 1987).

A few weeks after copulating, the female will construct an egg sac, starting with a dense horizontal silk disk and a cylindrical wall, then she will lay her eggs and proceed with the fertilization, which takes a few minutes. After that, she will close the egg sac with a cover plate and finally wrap it with a mesh of threads (Foelix 2011). Most terrestrial tarantulas roll and silk the egg sac until it is spherical, and if they perceive that it is misshapen, they may eat it (Marshall 2001). The egg sac will

Fig. 15.4 *Megaphobema mesomelas*. (**a**) Female. (**b**) Male. (**c**) First contact. (**d**) Male hooking up the female. (Photos: J. Mendoza)

maintain its humidity and temperature, as well as providing mechanical defense. It is common for females to remain hidden in their retreats until the spiderlings emerge (Dias and Brescovit 2003; Foelix 2011) (Fig. 15.5). The developmental stages are (1) the embryonic stage, from the time of egg fertilization until the body shape of the spider is established, (2) the larval stage, which includes the prelarva and the larva, feeding on yolk; and (3) the nympho-imaginal stage, once all of the organ systems are present (nymph = juvenile, imago = adult). All stages beyond the larval stage are separated by ecdysis (Foelix 2011).

In captivity, some people remove the egg sac from the female and incubate it artificially in order to prevent cannibalism, which can occur if the female is disturbed. An interesting finding is that when eggs are incubated at 27 °C and then the juveniles are subsequently well fed, the males will be bigger (in terms of both mass and size) when they mature (Reichling and Gutzke 1998).

15.3 Common Ailments

Health problems can be avoid through control of humidity, temperature, ventilation and hygiene. Sects. 15.3.1–15.3.3 discuss the most common health problems observed in tarantulas.

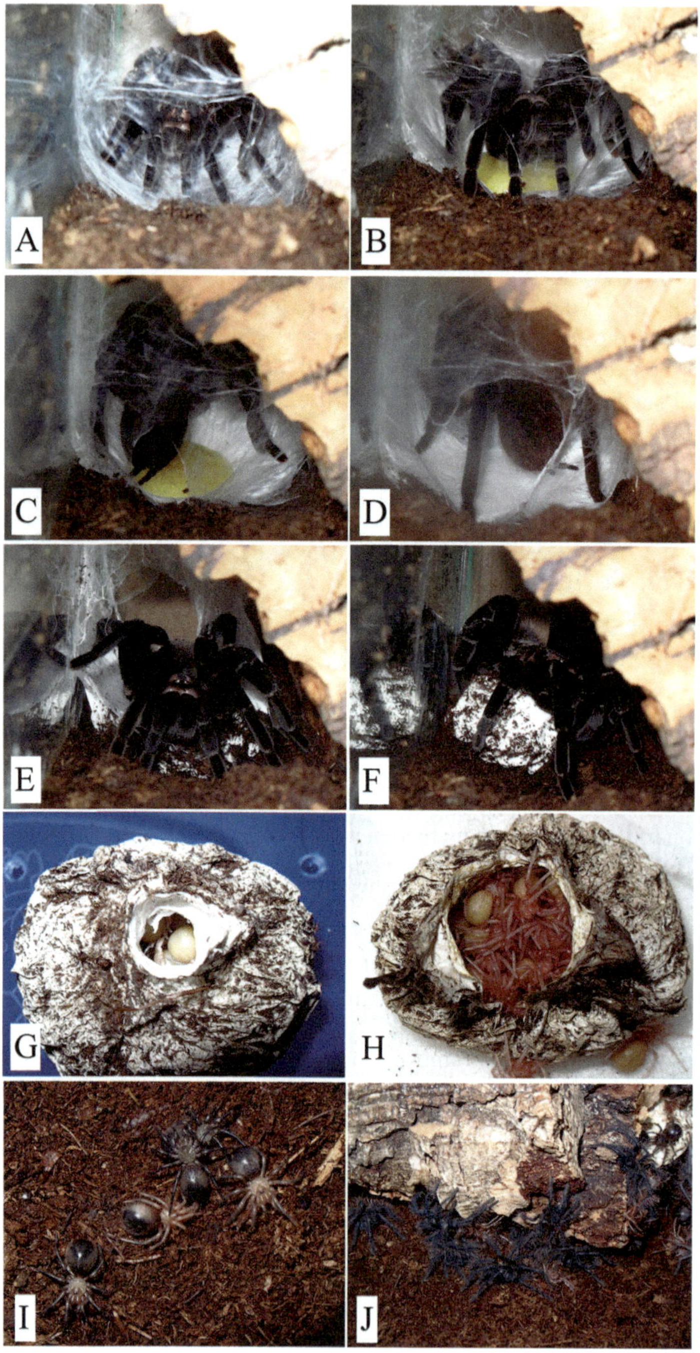

Fig. 15.5 *Xenesthis immanis* egg sac construction. (**a**) Horizontal disk silk and cylindrical wall. (**b**) Egg laying. (**c**) Fertilization. (**d–e**) Egg sac closing with a cover plate. (**f**) Wrapping and rolling with a mesh of threads. (**g**) Egg sac open at the larval stage. (**h**) Spiderling emergence. (**i**) Larval stage. (**j**) Juveniles. (Photos: J. Mendoza)

15.3.1 Dehydration

Dehydration can cause ataxia (loss of control of body movements) because, like all spiders, tarantulas use blood pressure instead of muscles to extend their legs, and lack of water lowers the blood pressure. Severe cases also can be distinguished by a shrunken abdomen (Draper and Trim 2018). Be alert to the possibility of dehydation if the spider drinks too much water, the water container is always empty or the spider sits in it. The recommended treatment is to check the water bowl and mist the terrarium. Place the spider's cephalothorax over the water container, allowing the chelicera to be in contact with the water for at least 1 hour. This should be supervised to ensure that the book lungs in the abdomen do not touch the, which could drown the spider. If the spider is still weak, repeat the procedure in a couple of days. Also, try to increase the relative humidity by closing some of the ventilation holes and adding water to the substrate (Psaila 2005).

15.3.2 Tissue Damage

Tissue damage can be caused if the spider falls from or inside the terrarium, if molting goes wrong or if the spider suffers a cut from a sharp object used for decoration. These kinds of injuries are risky because the spider can lose too much hemolymph, causing dehydration. In small injuries, loss of hemolymph can be stopped with use of wax (e.g., paraffin wax or beeswax) or cyanoacrylate adhesive. Infections can be avoided by use of antibiotic ointment (e.g., neomycinum). Never use iodine, alcohol or hydrogen peroxide, since they could poison the tarantula. It is better to avoid injuries, because most tarantulas do not survive them. Depending on the volume of hemolymph that is lost, the spider may require treatment for dehydration (Hardin and Sincage 2015; Pellett et al. 2015).

15.3.3 Parasites

One of the more lethal parasite infections found in captive tarantulas is caused by Panagrolaimidae nematodes, which have also been found in wild specimens. An infected spider will manifest anorexia, will become less mobile, and may adopt a huddled posture or appear to be "standing on the tips of its toes." You can inspect the mouth, washing the oral region between the chelicera with a saline solution, using a low-powered microscope or an endoscopic camera to try to see any mobile nematodes (which are less than 2 mm long). If the infection advances, a thick, white oral discharge can be observed, which may be mistaken for a bacterial infection. No treatment has been found, and the spider will die within weeks after the infection is first noticed. Precautions should be taken when handling specimens, because related nematodes have also been reported to infect humans, so a tarantula bite can be aggravated by nematode infection of the wound, which is difficult to treat. It is strongly recommended to euthanize the spider in order to stop dispersal of the nematode infection (Pizzi 2009).

Saprophytic mites can appear as a result of moisture or dampness, so, again, it is important to control the humidity of the tarantula's environment. When they occur in large numbers, they may settle on the tarantula, and the resulting stress may make it stop eating and become weak. The main problem is if they infest the book lungs, interfering with the spider's breathing (Breene 1998). If you notice mites, place the tarantula in a new, empty container and remove the mites with a fine paintbrush. Disinfect (or autoclave) all of the terrarium objects or replace them. In North America, you can buy the predatory mite *Hypoaspis miles* (Laelapidae) (West 1995), and the recommended dose is half a teaspoon in the terrarium (Breene 1998). Also, in the substrate, you can include woodlice, which are isopods and help to dispose of prey remains, thereby helping to avoid the appearance of saprophytic mites (Schultz and Schultz 2009). Other alternative treatments have been described by Pizzi (2009).

Scuttle flies (genus *Megaselia*) are attracted by rotting flesh or food remains. Maggots can get inside the spider through its mouth or book lungs and eat it from the inside out. The symptoms are anorexia and immobility, followed by swelling of the abdomen and eventually death (Machkour-M'Rabet et al. 2015). This parasite has been reported in natural populations of *Megaphobema robustum* (Ausserer, 1875) and *Pamphobeteus* Pocock, 1901 in Colombia (Weinmann and Disney 1997), *Theraphosa* Thorell, 1870 in French Guyana (Marshall and Uetz 1990) and *Tliltocatl vagans* (Ausserer, 1875) in Mexico (Machkour-M'Rabet et al. 2015).

Acrocerid spider fly are cosmopolitan endoparasites. The last larval stage is fatal to the spider, as the larvae burst out of the host's abdomen to pupate (Cady et al. 1993; Larrivée and Borkent 2009; Pizzi 2009; Barneche et al. 2013).

Spiders can be also be infected by bacteria, causing lethargy, fasting, weight loss and open wounds. Indeed, bacterial co-infections are very common in spiders infected with nematodes. Diagnosis can be difficult because little is known about the relevant microflora. For confirmation of the diagnosis, a hemolymph culture should be made (Mitchell and Tully 2008).

15.4 Tarantulas' Popularity as Pets and Their Husbandry

Spiders vary in size, color and specific characteristics within each group, according to their lifestyle, but a very particular group of spiders that, throughout history, have created great fear while also fascinating people are tarantulas. These great giants among spiders have impressive strength, a perception of their environment that can seem other-worldly, and ways of eating, breathing and even reproducing that may seem very strange. The large size of these spiders suggests that they are very poisonous, but tarantulas pose no danger to humans; contrary to popular belief, no tarantula has a poison capable of killing people (Mendoza 2009).

As mentioned previously, tarantulas are very long-lived animals, with a life expectancy of more than 10 years and even up to 30 years in captivity for the females of some species. It is because of this longevity that there is greater demand for females. Tarantulas come in a wide color range, which varies from black to brown, red, orange, blue and even violet. These characteristics have made them popular animals as pets. Like other types of exotic animals, they are increasingly in demand among hobbyists. More and more species are being commercialized, described and kept in captivity (Reichling 2003; Rojo 2004; West 2005).

15.4.1 Popular Pet Species

Some species of tarantula commonly kept as pets include the Mexican red-legged tarantula (*Brachypelma smithi*), the burgundy Goliath tarantula (*Theraphosa stirmi*), the Costa Rican zebra tarantula (*Aphonopelma seemanni*) (F. O. Pickard-Cambridge, 1897), the Pinktoe tarantula (*Avicularia avicularia*) and the Old World Togo starburst tarantula (*Heteroscodra maculata* Pocock, 1900). These are just some of those considered the most popular and common species among hobbyists worldwide. These species are relatively easy to get for those interested in acquiring a tarantula as a hobby. These species are preferred mainly by those who are taking up the hobby, on the basis of their docility (except for *H. maculata*), coloration, longevity or size. It should be noted that most of these species are from the New World. This could be due to the fact that the majority of American species are large, are generally docile (as opposed to Asian or African species), come in a great variety of colors and, in most cases, are easy to maintain and breed in captivity. However, there are also other tarantulas from different parts of the world that are popular as pets and are regularly bred and offered for sale as pets (Table 15.1). Not all species are suitable for people who are new to the hobby; some tarantulas require more specific maintenance and management conditions. So, it is very important to learn and read about the species that we want to keep, in order to provide them with the most appropriate conditions in captivity. In the following sections, we will describe the basic requirements and care of some tarantulas that are kept in captivity in many parts of the world.

15.4.2 *Acanthoscurria geniculata* (C. L. Koch, 1841)

Historical names *Mygale geniculata* C. L. Koch, 1841
Scurria geniculata C. L. Koch, 1850

Synonyms *Acanthoscurria transamazonica* Piza, 1972

Common names Brazilian whiteknee tarantula, Giant whiteknee tarantula

Geographic distribution Brazil (states of Rondônia, Roraima, Pará and Mato Grosso)

Table 15.1 Popular tarantulas in the pet trade

Acanthoscurria geniculata
Aphonopelma seemanni
Avicularia spp.
Brachypelma spp.
Caribena versicolor
Ceratogyrus spp.
Chilobrachys spp.
Cyriopagopus spp. (*Omothymus*)
Ephebopus cyanognathus
Ephebopus murinus
Grammostola spp.
Haplopelma spp. (*Cyriopagopus*)
Heteroscodra maculata
Lampropelma spp.
Pamphobeteus spp.
Phormictopus spp.
Poecilotheria spp.
Psalmopoeus spp.
Pterinochilus murinus
Stromatopelma calceatum
Tapinauchenius spp.
Theraphosa spp.
Xenesthis spp.

Physical description The carapace is black, typically bordered by white. The legs are black with a white ring at the joints of each segment. Typically, there are two white stripes over the patellae and tibiae, and there may also be two narrow white vertical stripes on the patellae. A small cream-white vertical stripe is visible on one third of the metatarsi. The abdomen is black with scattered long red hairs. Adult males have a black carapace and legs. The adult female has a body length of up to 8 cm, with a leg span of 20 cm. The adult male has a body length of up to 7 cm, with a leg span of 18 cm (Fig. 15.6).

Habitat Brazilian tropical forest with a warm and humid climate and marked rainy and dry seasons. As an exclusively terrestrial species, it lives in burrows located under rocks, inside fallen logs, inside living trees and in ravines on the ground (Paula et al. 2014).

Longevity In captivity, the female matures in a period of 3–4 years, while the male matures at 2 years; females can live for up to 20 years, while males can live for up to 4–5 years.

Terrarium The substrate must be at least 10 cm deep because this species is a digger and tends to make a burrow if the substrate is to its liking.

Fig. 15.6 Female *Acanthoscurria geniculata*. (*Photo: J. Mendoza*)

Temperature 25–28 °C during the day and 21–24 °C at night

Humidity The recommended humidity is 70–80%. After the female has mated, the humidity can be dropped gradually to 60%; then, after 3–4 months, it can be increased again to 80% to simulate a wet season and let the female lay an egg sac. Adults tolerate lower humidity better; for spiderlings, the humidity can be up to 90% (in a well-ventilated container). The humidity can be maintained by sprinkling water on half the substrate at least once a week (Marshall 2001; Cléton et al. 2015; Mendoza, personal observation 2008–2020).

Feeding You can offer crickets, cockroaches, mealworms (adult specimens are well received), large roaches or grasshoppers. Young spiders can feed on fruit flies (occasionally) and pinhead crickets, preferably every third day. It is a very voracious tarantula and will grow quickly, changing its skin at a frequency of around 30–40 days (Mendoza, personal observation 2008–2020).

Reproduction This is an easy species to breed. The mating season is from June to September. The female may or may not be slightly aggressive toward the male. If the male is introduced very quickly into the terrarium of the female, she will attack it most of the time. If the female is not ready for mating, she will try to escape in the first instance, but if the male is persistent, she may attack him to defend herself. After copulation, the male will try to run away quickly. During the subsequent 2 months, the female should be fed regularly; during the third month, the feeding should be more sporadic. The latency of oviposition after mating can range from 3 to 6 months. Once the egg sac is formed, the female will take care of it for approximately 3 months before the hatchlings emerge. During this period, the average temperature of the terrarium should be around 28 °C. If desired, the egg sac can be incubated artificially; in that case, it should be removed between 4 and 6 weeks after it is laid. The number of offspring can range from 800 to 2000 (Paula et al. 2014; Cléton et al. 2015; Mendoza, personal observation 2008–2020).

Handling and behavior This species has a generally docile but very unpredictable character and can bite if you put your hand inside its terrarium, but the most likely reason is that this species is very voracious and may mistake your hand for some kind of food. If you disturb it, it is most likely to throw urticating hairs. The urticating hairs of this species are particularly effective against humans and can cause a severe reaction (in some people) of burning of the skin, eyes and nose, so be careful (Cléton et al. 2015; Mendoza, personal observation 2008–2020). Its venom is not dangerous to humans, and its bite, although painful, does not cause major effects. The pain and swelling can disappear in a few hours or a couple of days (Mendoza, personal observation 2008–2020).

Conservation status There is not enough information available on this species, but it is not considered threatened.

15.4.3 Aphonopelma seemanni (F. O. Pickard-Cambridge, 1897)

Historical names *Eurypelma seemanni* F. O. Pickard-Cambridge, 1897
Rechostica seemanni Smith, 1986

Synonyms None

Common names Costa Rican zebra tarantula, Costa Rican stripe knee tarantula

Geographic distribution Costa Rica (province of Guanacaste), Nicaragua (departments of Rivas and Carazo)

Physical description The female carapace is black or dark brown, typically bordered by light brown. The male carapace is black with dark purple around the eyes, bordered by light brown. The legs are black with scattered long brownish hairs. Typically, there are two white stripes over the patellae and tibiae, but the stripes on the tibiae are less visible in adult males. A small cream-white vertical stripe is visible on one third of the metatarsi and palp tarsi. The abdomen is black with scattered long brown hairs. From the ventral view, the prosoma and opisthosoma are brown. The adult female has a body length of up to 7 cm, with a leg span of 14 cm. The adult male has a body length of up to 6 cm, with a leg span of 13 cm (Teyssié 2015; Mendoza, personal observation 2008–2020) (Fig. 15.7).

Habitat Costa Rican dry forest and grassland. It is an adaptable species and can be found frequently around rich green embankments on the front lawns of local homeowners (Smith 2013; Mendoza, personal observation 2008–2020). As an exclusively terrestrial species, it lives in burrows located in open grasslands, on hillsides or on slopes (Herrero and Valerio 1986; Mendoza, personal observation 2008–2020).

Fig. 15.7 *Aphonopelma seemanii*. (**a**) Burrow. (**b**) Female. (**c**) Male. (**d**) Mating. (Photos: J. Mendoza)

Longevity In captivity, the female matures in a period of 4–5 years, while the male matures at 3 years; females can live for up to 15 years, while males can live for up to 2–3 years.

Terrarium The substrate must be at least 10 cm deep because this species is a digger and tends to make a burrow if the substrate is to its liking.

Temperature 24–27 °C during the day and 21–23 °C at night

Humidity The recommended humidity is 70–80%. After the female has mated, the humidity can be dropped gradually to 50%; then, after 3–4 months, this can be increased to 80% to simulate a wet season and let the female lay an egg sac. Adults are more tolerant of minimum humidity. The humidity can be maintained by sprinkling water on half the substrate at least once a week.

Feeding You can offer crickets, cockroaches and mealworms. Young spiders can feed on fruit flies (occasionally) and pinhead crickets, preferably every third day. Juveniles can be fed with medium-sized crickets or *Tenebrio molitor*. Females grow more slowly than males, changing their skin at a frequency of around 40 days as juveniles (Mendoza, personal observation 2008–2020).

Reproduction This species rarely breeds in captivity. The mating season is from August to January. The female in general is not aggressive toward the male. If the female is not ready for mating, she will try to escape in the first instance, but if the male is persistent, she may attack him to defend herself. After copulation, the male will try to run away. During the subsequent months, the female should be fed regularly. The latency of oviposition after mating can range from 4 to 6 months; however, an Arachnida breeding facility keeps a female which laid an egg sac 14 months after mating (Mendoza, personal observation). Once the egg sac is formed, the female will take care of it for approximately 3 months before the hatchlings emerge. During this period, the average temperature of the terrarium should be around 26 °C. It is recommended to artificially incubate the egg sac in an incubator to avoid the risk of cannibalism. The egg sac can be removed after 4 weeks. The number of offspring can range from 200 to 500 (Cléton et al. 2015; Mendoza, personal observation 2008–2020).

Handling and behavior Many individuals are very skittish and can run very fast, so you must be careful when opening the terrarium. It is rare for this species to throw urticating hairs. (Marshall 2001; Mendoza, personal observation 2008–2020). Its venom is not dangerous to humans, and the bite does not cause a major effect. Not much is known about the effect of its venom.

Conservation status There is not enough information available on this species, but it is not considered threatened.

15.4.4 Avicularia avicularia (Linnaeus, 1758)

Historical names *Aranea avicularia* Linnaeus, 1758
Mygale avicularia Latreille, 1804
Mygale scoparia C. L. Koch, 1841
Eurypelma avicularia C. L. Koch, 1850
Avicularia avicularia Simon, 1892

Synonyms *Avicularia ancylochira* ***Mello-Leitão, 1923***
Avicularia velutina Simon, 1889
Avicularia exilis Strand, 1907
Avicularia cuminami Mello-Leitão, 1930
Avicularia nigrotaeniata Mello-Leitão, 1940

Common names Pinktoe tarantula, Common pinktoe tarantula, Guyana pinktoe tarantula

Geographic distribution Venezuela, Guyana, Suriname, French Guyana, Trinidad and Tobago, Brazil, Peru and Bolivia

Physical description Females have a brown carapace with short brown body setae with a green sheen. The carapace is bordered with long reddish brown setae with a pink sheen. The ventral region is dark brown. The legs and palps have short brown body setae with a green sheen and reddish brown guard setae with homogeneous dark coloration on the anterior legs and guard setae with a darker base and a contrasting whitish apex on the posterior legs. There are leg rings on the distal femora and tibiae, and the metatarsi are whitish. The abdomen has scattered long reddish brown hairs with a pink sheen grouped on the lateral, dorsal and anterior areas, with short black body setae. The ventral abdomen is brown. Males have a brown carapace with short brown body setae with a green and golden sheen. The carapace is bordered with long setae the same color as the short body setae on the dorsal carapace. The ventral region is dark brown. The legs and palps have short brown body hairs with a green sheen and reddish brown guard setae with homogeneous dark coloration on the anterior legs and guard setae with a darker base and a contrasting whitish apex on the posterior legs. There are leg rings on the distal femora and tibiae, and the metatarsi are whitish. The dorsal abdomen has scattered long reddish brown hairs with a pink sheen and dark short body setae. The ventral abdomen is brown (Fukushima and Bertani 2017). The adult female has a body length of up to 7 cm, with a leg span of 13 cm, and the adult male has a body length of up to 6 cm, with a leg span of 12 cm (Fig. 15.8).

Color pattern ontogeny Juveniles are brownish without a metallic sheen, and with black tarsi contrasting with other lighter parts, and a reddish dorsal abdomen, with a dorsal central longitudinal black stripe disconnected from transversal black stripes. When mature, both males and females lose this pattern (Fukushima and Bertani 2017).

Habitat Tropical forests in South America. Their retreats are found in different settings in vegetation and in human constructions. Younger juveniles are found between the leaves of low-growing plants (especially *Heliconia*), holding the edges of leaves together with silk to create a retreat. After this life stage, these spiders tend to build their retreats in more elevated structures without incorporating leaves into their retreat. The locations of their retreats include hollow stumps of the açaí palm, folded leaves of bananas, the hollow center of pineapple plants and the palm leaf thatch of native houses (Fukushima and Bertani 2017).

Fig. 15.8 *Avicularia avicularia*. (**a**) Female. (**b**) Mating. (Photos: J. Mendoza)

Longevity In captivity, the female matures in a period of 2–3 years, while the male matures at 1–2 years; females can live for up to 12 years, while males can live for up to 2–3 years.

Terrarium This should have a vertical orientation, and the substrate must be at least 5 cm deep to retain moisture. It is recommended to put in a piece of cork bark or a hollow log to let the tarantula build its silk tube web there.

Temperature 26–28 °C during the day and 22–25 °C at night

Humidity The recommended humidity is 80–90%. The humidity can be maintained by sprinkling water on half the substrate at least once a week or even around the silk nest of the spider.

Feeding You can offer crickets, cockroaches and mealworms. Young spiders can feed on fruit flies (occasionally) and a small cricket twice per week. This is a voracious tarantula and will grow quickly, changing its skin at a frequency of around 30 days (Mendoza, personal observation 2008–2020).

Reproduction This is an easy species to breed. The mating season is from July to October. The female does not exhibit aggression toward the male, and they can live together for a few days. The female will lay an egg sac 2–4 months after mating. After 1–2 months of incubation, between 100 and 200 spiderlings will emerge from the egg sac. There is no problem if the egg sac is left with the female, but, if desired, the egg sac can be incubated artificially; in that case, it should be removed 4 weeks after it is laid. During incubation, the average temperature of the terrarium should be around 28 °C (Schultz and Schultz 2009; Fukushima and Bertani 2017; Mendoza, personal observation 2008–2020).

Handling and behavior This species is generally docile but can leap toward the ground if it feels in danger. It can also shed its excreta as a defense mechanism. This species possesses urticating hairs. These are not thrown into the air but placed by direct contact with the attacker (Fukushima and Bertani 2017; Mendoza, personal observation 2008–2020). Its venom is not dangerous to humans, and its bite does not cause a major effect (Mendoza, personal observation 2008–2020).

Conservation status There is not enough information available on this species, but it is not considered threatened.

15.4.5 Brachypelma auratum Schmidt, 1992

Historical names None

Synonyms None

Common names Mexican flame knee tarantula

Geographic distribution Mexico (states of Jalisco, Michoacán, Guerrero and Estado de México)

Physical description This is a very robust tarantula with a black background color throughout the body, with both dark and whitish hairs on the legs. It presents red coloration in the area of the knee (patella) of each of the legs and palps that resembles the shape of a flame, hence its common name. Also, at the end of each segment of the legs, it has a small ring that is slightly orange. It has a few reddish hairs on its abdomen, while the carapace is typically black with a light yellow border. Its ventral region is totally black (Mendoza, personal observation 2008–2020). The adult females have a body length of up to 7 cm, with a leg span of 15 cm. Males usually tend to be a bit smaller than females, with a body length of 5–6 cm and a leg span of up to 14 cm, although there are some specimens that are even larger than many females and of a similar build, reaching a leg span of almost 16 cm (Cooper et al. 2019) (Fig. 15.9).

Habitat Regions of subwarm and subhumid temperate climate with mixed forest vegetation of deciduous plants and thorn. As an exclusively terrestrial species, it lives in burrows and prefers mainly clay-type soils, where it digs its tunnels under rocks, in tree roots, in cracks and (because of proximity to several urban centers) under the walls of houses. Some specimens can occupy the setting of a grassland

Fig. 15.9 *Brachypelma auratum*. (**a**) Habitat. (**b**) Female. (**c**) Male. (**d**) Mating. (Photos: J. Mendoza)

field or farmland because, as a result of the destruction and fragmentation of their habitat, they have been forced to adapt to these sites at which they can find the prey they feed on (West 2005; Cooper et al. 2019).

Longevity Females live for up to 25 years in captivity, and males live for 6–10 years.

Terrarium The substrate must be at least 12 cm deep because this species is a digger and will tend to make a burrow. It is also recommended to put a slightly curved piece of cork bark over the substrate to simulate a burrow. The terrarium can be decorated but should be kept minimally ornamented to facilitate cleaning, feeding and location of the specimen inside the terrarium.

Temperature 25–29 °C during the day and 20–24 °C at night

Humidity It has been believed that because of the kind of habitat in which this species lives, it is better to keep it very dry, but we must remember that inside its burrow, the humidity is higher than in the environment, so although they can be maintained in terrariums with humidity of 50%, humidity of 70–80% is recommended. The maximum percentage is especially recommended to avoid dehydration of the eggs during the incubation period. The humidity can be maintained by sprinkling water on half the substrate at least once a week or even around the silk nest of the spider.

Feeding You can offer crickets, cockroaches and mealworms (occasionally). Young spiders can feed on small crickets or cockroaches twice per week. This species is considered a slow-growing tarantula, so it molts every 40–50 days, on average, as a spiderling; juveniles molt twice a year and adults once a year (Mendoza, personal observation 2008–2020).

Reproduction The main challenge in breeding this species is to obtain an adult male, as they are not commonly available. The mating season runs from October to January. The adult males age very fast, so it is recommended that the male should mate with the female during the first 3 months after the male matures. The female can exhibit some aggression toward the male. Sometimes the female will approach the male with her chelicerae spread. If mating goes well, the female will bend quite far backward when the male introduces the male palpal bulb; after this, the male can run away. The female will lay an egg sac 4–8 months after mating. After 3 months of incubation, between 400 and 800 spiderlings will emerge from the egg sac. The egg sac can be incubated artificially; in that case, it should be removed 4–5 weeks after it is laid. During the incubation time, the average temperature of the terrarium should be around 28 °C (Mendoza, personal observation 2008–2020).

Handling and behavior This species is generally docile but has a nervous character. It can be manipulated manually but is very likely to throw urticating hairs. It is rare to come across specimens that do not hesitate to bite in order to defend themselves, but it is always better to try to understand the kind of character that the animal we keep has in particular. This species has urticating hairs on its abdomen, which can cause a severe reaction (in some people) of burning of the skin, eyes and nose, so be careful. Its venom is not dangerous to humans, and its bite, although painful, does not cause more major effects than local pain and slight swelling, which disappear in a few hours.

Conservation status All species in the *Brachypelma* genus are listed in the Convention on International Trade in Endangered Species of Wild Fauna and Flora (CITES) Appendix II, so their international trade is strictly regulated. This species is very exploited in Mexico for its illegal trade as a pet, and there are already problems in its wild populations because of the extraction rate and the fragmentation that its habitat is suffering. Because of the lack of population studies, we cannot know exactly where a relatively abundant population lives (Fukushima et al. 2018; Cooper et al. 2019).

15.4.6 Brachypelma hamorii Tesmoingt, Cléton & Verdez, 1997

Historical names None

Synonyms None

Common names Mexican orange knee tarantula

Geographic distribution Mexico (states of Jalisco, Colima and Michoacán)

Note This species was misidentified for years as *Brachypelma smithi*. Most of the specimens the pet trade has erroneously sold as *B. smithi* have, in fact, been specimens of *B. hamorii*.

Physical description Adult females have two carapace patterns: (1) brownish pink around the border and black dorsomedially (juveniles and subadults have the same pattern); and (2) pale orange-yellow around the border and behind the fovea, with a starburst black pattern from the fovea to the caput. The dorsal chelicerae are light bluish gray with two brownish pink stripes (not all specimens have clearly visible stripes; they are more readily seen in recently molted specimens). The ventral prosoma is brownish black. The abdomen is black with light reddish brown setae dorsally and brownish black ventrally. The legs and palps have black femora, patellae with a proximal dorsomedian deep orange flame-shaped area, a pale orange-yellow distodorsal paramedian area (specimens of some populations only present pale orange-yellow on the distal half of the patella) and brownish pink setae laterally.

Fig. 15.10 *Brachypelma harmorii*. (**a**) Chelicera. (**b**) Female. (Photos: J. Mendoza)

The tibiae are proximally half black with brownish pink setae, and the distal half is light orange-yellow with brownish pink setae. The metatarsi are black with brownish pink setae and a yellowish white ring at the terminal end. The tarsi are black with a few dorsal brownish pink setae. The adult females have a body length of up to 7 cm, with a leg span of 14–15 cm. Males usually tend to be a bit smaller than females, with a body length of 5–6 cm and a leg span of up to 14 cm (Mendoza and Francke 2017) (Fig. 15.10).

Habitat Thorn and deciduous secondary forests. As an exclusively terrestrial species, its modified or self-excavated burrows can be found under fallen logs or large rocks, among large tree roots or thorny brush, or in tall grass thickets. The burrows do not have any silk around the entrance (West 2005; Mendoza and Francke 2017).

Longevity Females live for up to 25 years in captivity; males live for 6–10 years.

Terrarium This species can be maintained in a terrarium like the one used for *Brachypelma auratum*.

Temperature 25–29 °C during the day and 22–24 °C at night

Humidity This is the same as that used for *Brachypelma smithi*.

Feeding You can offer crickets, cockroaches and mealworms (occasionally), although some specimens do not like to eat cockroaches or mealworms (Mendoza, personal observation 2008–2020). Young spiders can feed on small crickets or newborn cockroaches twice per week. It is considered a slow- to medium-growing tarantula, depending on the maintenance conditions. Spiderlings will molt every 40–50 days on average; juveniles molt three times a year and adults once a year (Mendoza, personal observation 2008–2020).

Reproduction This is similar to that of *Brachypelma smithi*.

Handling and behavior These are the same as those specified for *Brachypelma auratum*.

Conservation status All species in the *Brachypelma* genus are listed in CITES Appendix II, so their international trade is strictly regulated. Not only is this species exploited in Mexico for its illegal trade as a pet, but also they live in a very fragmented area. Because of the lack of population studies, it is not possible to know exactly where a relatively abundant population lives and how threatened they are (Mendoza and Francke 2017). This species is listed in the International Union for Conservation of Nature (IUCN) Red List and considered vulnerable (Fukushima et al. 2018).

15.4.7 Brachypelma smithi (F. O. Pickard-Cambridge, 1897)

Historical names *Eurypelma smithi* F. O. Pickard-Cambridge, 1897
Brachypelma smithi Pocock, 1903

Synonyms *Brachypelma annitha* Tesmoingt, Cléton & Verdez, 1997

Common names Mexican redknee tarantula

Geographic distribution Mexico (state of Guerrero)

Physical description In live specimens, adult females have three carapace patterns: (1) grayish yellowish pink around the border and behind the fovea, with a bluish black starburst pattern from the fovea to the caput; (2) light brown around the border and bluish black dorsomedially; and (3) grayish yellowish pink on almost all of the carapace, except for two longitudinal black patches on the caput. juveniles and subadults have light brown coloration around the border, with brownish black coloration dorsomedially. The dorsal chelicerae are dark grayish blue. The ventral coxae, labium, maxillae and sternum are brownish black. The abdomen is black dorsally with light orange setae and brownish black ventrally. The legs and palps have bluish black femora, and the patellae have a dark reddish orange flame-shaped proximal dorsomedian area and a light yellowish pink distodorsal paramedian area, with light orange setae laterally. The tibiae have a bluish black proximal half with light orange setae and a pale yellowish pink distal half. The metatarsi are bluish black with light orange setae and a pinkish white ring at the terminal end. The tarsi are bluish black (Mendoza and Francke 2017). A females have a body length of up to 7 cm, with a leg span of 15–16 cm. Males usually are smaller than females, with a body length of 5–6 cm and a leg span of up to 14 cm, although there are some specimens that are even larger than many females and of a similar build, reaching a leg span of almost 17 cm (Mendoza, personal observation 2008–2020) (Fig. 15.11).

Habitat This species is known to inhabit the Pacific coast of Guerrero in dense thickets or vegetation of dry thorn forests and deciduous plants. Burrows are found under large rocks or tree roots; there are no traces of silk at the burrow entrance, and

Fig. 15.11 *Brachypelma smithi*. (**a**) Habitat. (**b**) Chelicera. (**c**) Female. (**d**) Male. (Photos: J. Mendoza)

the interior is often multitunneled. Some specimens can occupy the surroundings of grassland fields or farmland because the destruction and fragmentation of their habitat have forced them to adapt to these sites at which they can find the prey they feed on (West 2005; Mendoza and Francke 2017).

Longevity Females live for up to 25 years in captivity; males live for 5–10 years.

Terrarium This species can be maintained in a terrarium like the one used for *Brachypelma auratum*.

Temperature 26–30 °C during the day and 20–22 °C at night

Humidity The recommended humidity is 75–80%. During the incubation period, it is possible to raise it to 85%. The humidity can be maintained by sprinkling water on half the substrate at least once a week.

Feeding You can offer crickets, cockroaches and mealworms (occasionally), although there are some specimens who do not like to feed on cockroaches or mealworms (Mendoza, personal observation 2008–2020). Young spiders can feed on small crickets or newborn cockroaches twice per week. It is considered a slow- to medium-growing tarantula, depending on the maintenance conditions. Spiderlings molt every 40–50 days on average; juveniles molt three times a year and adults once a year (Mendoza, personal observation 2008–2020).

Reproduction The main difficulty in breeding this species is obtaining an adult male; they are not so common among hobbyists. Also, because it looks similar to *B. hamorii*, one must be careful to not try and pair two specimens of these different species. The mating season is from November to January. The adult male ages very fast, so it is recommended to mate it with a female during the first 4 months after the male matures. The female can exhibit some aggression toward the male. Sometimes the female will approach the male with her chelicerae spread. If the mating goes well, the female will bend quite far backward when the male introduces the palpal bulb; after this, the male will go away. After the mating, it is recommended to drop the humidity to 60% and lower the temperature to 22–24 °C. Then, after 1 or 2 months under these conditions, the humidity can be increased to 85% and the temperature to 28–30 °C to trigger the female to elaborate the egg sac. It will be laid between 4 and 8 months after the mating, depending on the environmental conditions. After 3 months of incubation, between 450 and 1000 spiderlings will emerge from the egg sac. The egg sac can be incubated artificially; in that case, it should be removed 4 weeks after it is laid (Mendoza, personal observation 2008–2020).

Handling and behavior These are the same as those specified for *Brachypelma auratum*.

Conservation status All species in the *Brachypelma* genus are listed in CITES Appendix II, so their international trade is strictly regulated. This species is not only exploited in Mexico for its illegal trade as a pet but also killed by local people who think they are dangerous. Because of the lack of population studies, it is not possible to know exactly where a relatively abundant population lives and how threatened they are (Mendoza and Francke 2017). This species is listed in the IUCN Red List and considered near threatened (Fukushima et al. 2018). It is important to mentioned that this species is protected by Mexican law and listed as a threatened species.

15.4.8 Caribena versicolor (Walckenaer, 1837)

Historical names *Mygale versicolor* Walckenaer, 1837
Avicularia versicolor Simon, 1892
Caribena *versicolor* Fukushima & Bertani, 2017

Synonyms *Avicularia rutilans* Ausserer, 1875

Common names Martinique pinktoe tarantula, Antilles pinktoe tarantula, Martinique red tree spider

Geographic distribution Martinique

Physical description Females have a brown carapace with short golden body setae with a very intense green sheen. The carapace is bordered with long setae the same color as the carapace. The ventral region is light brown. The legs and palps have short gold body hairs with a green sheen and long brown guard setae with a very intense iridescent sheen. The leg rings on the distal femora, tibiae and metatarsi are the same color as the rest of the segment. The dorsal abdomen has scattered long red iridescent hairs and short black body setae. The ventral abdomen is brown. Urticating hairs form a very distinctive small bronze patch on the dorsoposteriad area of the abdomen. The male's carapace is brown with short golden body setae with a very intense green sheen. The carapace is bordered with long setae the same color as the dorsal carapace short body setae. The ventral region is light brown. The legs and palps have short gold body hairs with a green sheen and long brown guard setae with a very intense iridescent sheen. The leg rings on the distal femora, tibiae and metatarsi are the same color as the rest of the segment. The dorsal abdomen has scattered long vivid red hairs with a very intense iridescent sheen that is homogeneously distributed, and short dark body setae. The ventral abdomen is brown. Urticating hairs form a very distinctive small bronze patch on the dorsoposteriad area of the abdomen. Two color forms are known: one with specimens that have leg and palp hairs in bright red and the other with specimens that have darker hairs on the legs and palps (Fukushima and Bertani 2017). The adult female has a body length of up to 6 cm, with a leg span of 12 cm. The adult male has a body length of up to 5 cm, with a leg span of 10–11 cm (Fig. 15.12).

Color pattern ontogeny Spiderlings have a metallic blue sheen, and their abdomen has a dark fish-bone pattern on it. Juveniles have a metallic sheen; all parts are the same blackish color, and the dorsal abdomen has a central longitudinal black stripe connected to all transverse black stripes. When mature, they lose this pattern (Fukushima and Bertani 2017).

Habitat Cloud forests of Martinique; their retreats are found in bromeliad leaves, between tree branches, in bamboo, in tree hollows and also in people's houses (Marechal et al. 2009).

Longevity In captivity, the female matures in a period of 2 years, while the male matures at 1 year; females can live for up to 10 years, while males can live for up to 2–3 years.

Terrarium This should have a vertical orientation, and the substrate must be at least 5 cm deep to retain moisture. It is recommended to put in a piece of cork bark or a hollow log to let the tarantula build its silk burrow there.

Temperature 26–28 °C during the day and 22–24 °C at night

Fig. 15.12 *Caribena versicolor*. (**a**) Female. (**b**) Spiderling. (**c**) Mating. (Photos: J. Mendoza)

Humidity The recommended humidity is 80%. The humidity can be maintained by sprinkling water on half the substrate at least once a week or even around the silk nest of the spider.

Feeding You can offer crickets, cockroaches and mealworms. Young spiders can feed on fruit flies (occasionally) and a quarter-inch (6 mm) cricket twice per week. It is a voracious tarantula and will grow quickly, changing its skin at a frequency of around 30–40 days (Mendoza, personal observation 2008–2020).

Reproduction This is a relatively easy species to breed in captivity. The female can be a little aggressive toward the male sometimes. The female will lay an egg sac 2–4 months after mating. After 1–2 months of incubation, between 100 and 200 spiderlings will emerge from the egg sac. There is no problem if the egg sac is left with female, but, if desired, the egg sac can be incubated artificially; in that case, it should be removed 3 weeks after it is laid. During the incubation time, the average temperature of the terrarium should be around 27–28 °C. Spiderlings can be separated in individual containers about 4 cm high, but some authors also recommend separating them into small groups (e.g., groups of 10 individuals) and raising them that way, after noticing instances of cannibalism. It is necessary to provide a suitable small surface on which the spiderlings can build their silk nests in the upper area of their container (Cléton et al. 2015; Mendoza, personal observation 2008–2020).

Handling and behavior This species is generally docile but can leap toward the ground if it feels in danger, and if you persist in disturbing it, it can use its urticating hairs. It can also shed its excreta as a defense mechanism. Its venom is not dangerous to humans, and its bite does not cause a major effect (Mendoza, personal observation 2008–2020).

Conservation status There is not enough information available on the wild populations of this species, but it is not considered threatened.

15.4.9 Psalmopoeus irminia Saager, 1994

Historical names None

Synonyms None

Common names Venezuelan suntiger tarantula

Geographic distribution Venezuela, Guyana and Brazil

Physical description The female has a greenish carapace with a metallic sheen, bordered with olive-green setae. The ventral region is dark brown. The legs are black with an orange stripe at the end of the tarsi and a larger longitudinal stripe on the metatarsi. The abdomen has a black background with eight orange oval marks in the dorsal region. The ventral abdomen is light brown. The male is smaller and less striking than the female; when mature, it is gray with a feathery appearance. The adult male also has orange stripes on its metatarsi and tarsi (Mendoza, personal observation 2008–2020) (Fig. 15.13).

Fig. 15.13 Female *Psalmopoeus irminia*. (Photo: J. Mendoza)

Color pattern ontogeny Spiderlings have an orange carapace. The abdomen is black with eight oval whitish spots on the dorsum. The legs are black, but all of the metatarsi are white. The juveniles have the same coloration as the adults but paler. The adult female has a body length of up to 7 cm, with a leg span of 15 cm. The adult male has a body length of up to 5 cm, with a leg span of 13 cm (Mendoza, personal observation 2008–2020).

Habitat This species inhabits rainforests. Their retreats are found between tree branches, inside tree hollows, under bark or even close to large tree roots (Mendoza, personal observation 2008–2020).

Longevity In captivity, the female matures in a period of 2 years, while the male matures at 1 year; females can live for up to 12 years, while males can live for up to 2–3 years.

Terrarium This should have a vertical orientation, and the substrate must be at least 8 cm deep to retain moisture, but also because in captivity, this species will use some of the substrate to build its retreat. It is recommended to put in a piece of cork bark or a hollow log to let the tarantula build its silk retreat there. If possible, try to locate the water plate up high, close to the retreat's entrance.

Temperature 26–28 °C during the day and 20–22 °C at night

Humidity The recommended humidity is 80%, but this can be lowered to 60% after mating. The humidity can be maintained by sprinkling water on half the substrate at least once a week or even around the silk nest of the spider.

Feeding You can offer crickets, cockroaches and mealworms (occasionally). Young spiders can feed on fruit flies (occasionally) and a quarter-inch (6 mm) cricket twice per week. This is a voracious tarantula and will grow quickly, changing its skin at a frequency of around 25–35 days (Mendoza, personal observation 2008–2020).

Reproduction This is an easy species to breed in captivity. The mating season is from July to October. The female can be a little aggressive toward the male sometimes. The female will lay an egg sac 2–4 months after mating. After 1–2 months of incubation, between 100 and 200 spiderlings will emerge from the egg sac. There is no problem if the egg sac is left with the female, but, if desired, the egg sac can be incubated artificially; in that case, it should be removed 3 weeks after it is laid. During the incubation time, the average temperature of the terrarium should be around 27–28° C. This species can lay a second egg sac 4–5 months after the first one (Cléton et al. 2015; Mendoza, personal observation 2008–2020).

Spiderlings can be kept in individual containers approximately 4 cm high. It is necessary to provide a suitable small surface on which the spiderling can build its silk nest in the upper area of its container.

Handling and behavior This is a very fast and fearful species when it is young; first, it will tend to flee before facing its aggressor. However, as it grows, it tends to be more defensive. If it feels threatened, it will adopt a defensive position, showing its fangs, and if you persist in bothering it, it may bite. This species possesses a stridulating organ, which it uses as a warning, and lacks urticating hairs. Its venom is not dangerous to humans but can cause some pain for hours, depending on the sensitivity of the person (Mendoza, personal observation 2008–2020).

Conservation status This species is widely bred in captivity. There is not enough information available on the wild populations of the species, but it is not considered threatened.

15.4.10 Pseudoclamoris gigas (Caporiacco, 1954)

Historical names *Tapinauchenius gigas* Caporiacco, 1954
Pseudoclamoris gigas Hüsser, 2018

Synonyms None

Common names Orange tree spider, orange chevron tarantula

Geographic distribution French Guyana

Physical description Females have a greenish carapace bordered with red-orange setae. The ventral region is dark. The abdomen and legs are a red-orange color, with a slightly marked spike pattern over the abdomen. The ventral abdomen is light orange. The male is smaller and less striking than the female; when mature, it is a grayish brown color with a feathery appearance and presents a light orange color on the metatarsi and tarsi (Mendoza, personal observation 2008–2020). The adult female has a body length of up to 6 cm, with a leg span of 13 cm. The adult male has a body length of up to 5 cm, with a leg span of 10–11 cm (Fig. 15.14).

Color pattern ontogeny Spiderlings have a black carapace. The abdomen is black with eight oval orange spots on the dorsum (four on each side). The legs are orange and the tarsi are all black. Juveniles have the same coloration as adult females but paler.

Habitat Tropical rainforest; their retreats are found between tree branches, inside tree hollows, under bark or even close to large tree roots (Mendoza, personal observation 2008–2020).

Longevity In captivity, the female matures in a period of 2 years, while the male matures at 1 year; females can live for up to 12 years, while males can live for up to 2–3 years.

Fig. 15.14 *Pseudoclamoris gigas*. (**a**) Female. (**b**) Mating. (Photos: J. Mendoza)

Terrarium This species can be kept in the same conditions as *Psalmopoeus irminia*.

Temperature 27–30 °C during the day and 24–26 °C at night

Humidity The recommended humidity is 75%, but it can be lowered to 60% after mating. The humidity can be maintained by sprinkling water on half the substrate at least once a week or even around the silk nest of the spider.

Feeding This species can be fed in the same way as *Psalmopoeus irminia*.

Reproduction This is similar to that of *Psalmopoeus irminia*.

Handling and behavior This is a very fast and fearful species when it is young. First, it will tend to flee before facing its aggressor. However, as it grows, it tends to be more defensive and, if it feels threatened, it will adopt a defensive position by showing its fangs. If you persist in bothering it, it may bite. Its venom is not dangerous to humans but can cause some pain for hours, depending on the sensitivity of the person (Mendoza, personal observation 2008–2020).

Conservation status This species is widely bred in captivity. There is not enough information available on the wild populations of the species, but it is not considered threatened.

15.4.11 Theraphosa stirmi Rudloff & Weinmann, 2010

Historical names None

Synonyms None

Common names Burgundy Goliath bird-eater

Fig. 15.15 *Theraphosa stirmi*. (**a**, **b**) Female. (**c**) Egg sac. (Photos: J. Mendoza)

Geographic distribution Guyana (Takutu)

Physical description Both males and females have a carapace and legs that are mud-brown in color. They typically have two stripes over the patellae. The abdomen is the same color as the carapace and legs, but darker, with scattered long pale brown hairs. The adult female has a body length of up to 10 cm, with a leg span of 27 cm. The adult male has a body length of up to 9 cm, with a leg span of 26–28 cm. The female is more robust than the male, but the male can have larger legs and therefore can have a greater leg span (Fig. 15.15).

Habitat Tropical forest with a warm and humid temperate climate and with marked rainy and dry seasons. As a species that is both terrestrial and fossorial, it lives in areas near water, where it digs burrows under the roots of large trees and even between the roots. It also takes over existing burrows of small mammals or reptiles (Mendoza, personal observation 2008–2020).

Longevity Females can live for up to 20 years, while males can live for up to 4–5 years.

Terrarium The substrate must be at least 15 cm deep because this species will tend to make a burrow. A deep substrate is particularly useful for breeding this species. The terrarium can be decorated but should be kept minimally ornamented to facilitate cleaning, feeding and location of the specimen inside the terrarium.

Temperature 20–22 °C during the day and 17–18 °C at night. This species is sensitive to high temperatures so must be kept in accordance with these recommendations.

Humidity The recommended humidity is 80–85%. After the female has mated, the humidity can be dropped gradually to 65%; then, after 2–3 months, it can be increased again to 85% to trigger the female to lay an egg sac. This species is sensitive to fungal infections, so there must be excellent ventilation in the terrarium. The humidity can be maintained by sprinkling water on half the substrate at least once a week. After the substrate has been left to almost dry out after mating, it is better to sprinkle water in the terrarium once a day (Mendoza and Matias, personal observation 2008–2020).

Feeding You can offer crickets, cockroaches, mealworms (adult specimens are well received), large roaches or grasshoppers. Young spiders of a large size (e.g., those with a leg span of 3 cm) can be fed medium-sized crickets, preferably every third day. This is a very voracious tarantula and will grow quickly, changing its skin at a frequency of around 30–40 days (Mendoza, personal observation 2008–2020).

Reproduction Breeding in captivity is moderately difficult. The mating season is from June to September. The female may or may not be slightly aggressive toward the male, so you must be careful to separate them if necessary. If the male is introduced very quickly into the terrarium of the female, she can turn aggressive toward the male. If the female is not ready for mating, she will try to warn the male by tapping on the substrate with the first pair of her legs. After the mating is finished, the male will try to run away quickly. During the subsequent 1 or 2 months, the female should be fed regularly, but be careful not to overfeed her. The size of her abdomen should be slightly bigger than her carapace. The substrate should be allowed to dry out almost completely, and the temperature should be kept at 16–18 °C. During the second and third month, the feeding should be more sporadic as the abdomen of the female increases in size. Then the humidity must be raised again to 85% by spraying the terrarium (do not flood the substrate), and the temperature must be kept at 20–22 °C. Once the egg sac is formed, the female will take care of it for approximately 3 months before the hatchlings emerge. During this period, the average temperature of the terrarium should be around 20 °C. If desired, the egg sac can be incubated artificially; in that case, it should be removed between 4 and 6 weeks after it is laid. The number of offspring can range from 50 to 80.

The spiderlings are big (2–3 cm) and must be kept in individual containers according to their size (Mendoza, personal observation 2008–2020).

Handling and behavior This species is very nervous and will not hesitate to use its urticating hairs, which are particularly effective against mammals, including humans. Urticating hairs can cause a severe reaction (in some people) of burning of the skin, eyes and nose, so be careful (Cléton et al. 2015; Mendoza, personal observation 2008–2020). If the use of urticating hairs has not been enough to dissuade the attacker, then the spider will bite. The bite is particularly painful because of the size of the chelicera (approximately 2 cm). The venom is not dangerous to humans, and its bite, although painful, does not cause a major effect. The pain and swelling can disappear in a few hours or a couple of days (Mendoza, personal observation 2008–2020).

Conservation status There is not enough information available on this species, but it is not considered threatened. For years, this species was collected and sold as

Theraphosa blondi. However, some people suspected that it was a different species, which was confirmed after its description in 2010. Of the three known Goliath tarantulas, this is the one most commonly found in trade. Also, because there has been more success in breeding this species than in breeding *T. blondi* and *T. apophysis*, it is the least expensive and most common species of the three on the market.

15.4.12 Xenesthis immanis (Ausserer, 1875)

Historical names *Lasiodora immanis* Ausserer, 1875
Xenesthis colombiana Simon, 1891
Xenesthis immanis Pocock, 1901

Synonyms None

Common names Colombian lesserblack tarantula

Geographic distribution Panama, Colombia and Venezuela

Physical description The carapace is black with a bright pink radial pattern, which is more evident in the ocular area. The carapace is typically bordered by large brownish setae. The legs are black. The abdomen is black with scattered long reddish hairs. Adult males have a black carapace with an iridescent pinkish or bronze radial pattern, which, unlike that observed in females, extends throughout the carapace and its border. The legs are black and the femora are a wine-like color. The adult female has a body length of up to 8 cm, with a leg span of 21 cm. The adult male has a body length of up to 7 cm, with a leg span of 18–20 cm. Some males can be as long as females, but females are more robust (Fig. 15.16).

Habitat Tropical forests with a warm and humid temperate climate and with marked rainy and dry seasons. As an exclusively terrestrial species, it lives in bur-

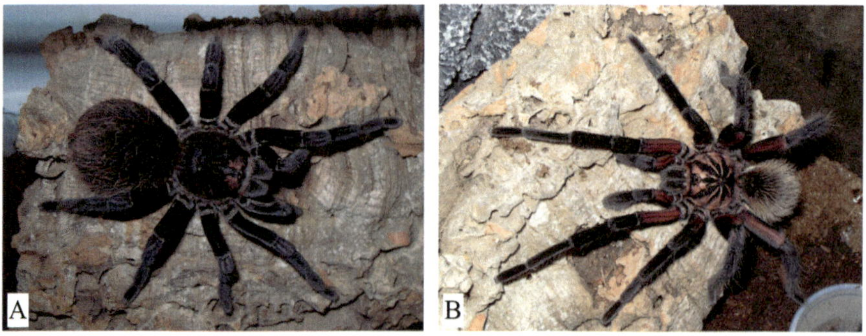

Fig. 15.16 *Xenesthis immanis*. (**a**) Female. (**b**) Male. (Photos: J. Mendoza)

rows located under fallen logs or among tree roots; some can make burrows on hillsides and at ground level (Mendoza, personal observation 2008–2020).

Longevity In captivity, the female matures in a period of 3–4 years, while the male matures at 2–3 years; females can live for up to 15 years, while males can live for up to 4–5 years.

Terrarium The substrate must be at least 10 cm deep because this species is a digger and tends to make a burrow. It is also possible to place a flat piece of cork bark between the rear wall of the terrarium and the substrate, creating a refuge that the tarantula can use as its burrow and a nice place for it to lay its egg sac.

Temperature 25–28 °C during the day and 20–22 °C at night

Humidity The recommended humidity is 80–85%. After the female has mated, the humidity can be dropped gradually to 65%; then, after 3–4 months, it can be increased again to 85% by wetting the substrate to simulate a wet season, which will trigger the female to lay an egg sac. Adults can better tolerate minimum humidity; for spiderlings, it should be at least 80% (in a well-ventilated container). The humidity can be maintained by sprinkling water on half the substrate at least once a week.

Feeding You can offer crickets, cockroaches, mealworms (adult specimens are well received), the largest cockroaches or grasshoppers. Young spiders can feed on *Tenebrio molitor* or medium-sized crickets, preferably every third day. This is a very voracious tarantula and will grow relatively quickly, changing its skin at a frequency of around 40 days (Mendoza, personal observation 2008–2020).

Reproduction This species is not so easy to breed and requires a dry and humid season for successful breeding. The mating season is from June to September. The female may or may not be slightly aggressive toward the male. If the male is introduced very quickly into the terrarium of the female, he will be attacked. If the female is not ready for mating, she will try to escape in the first instance, but if the male is persistent, she could attack him to defend herself and use her rear legs to prevent the male from inserting his palpal bulb. After copulation, the male will try to run away quickly. During the subsequent 3 months, the female should be fed regularly, but during the third month, the feeding should be more sporadic. The humidity and temperature should be lowered to the recommended minimum levels. After the third or fourth month, the humidity and temperature must be increased. The latency of oviposition after mating can last from 3 to 8 months. Once the egg sac is formed, the female will take care of it for approximately 3 months before the hatchlings emerge. During this period, the average temperature of the terrarium should be around 28 °C. If desired, the egg sac can be incubated artificially; in that case, it should be removed between 4 and 6 weeks after it is laid. The number of offspring can range from 60 to 100. Before molting into spiderlings, the larvae will climb the walls of the female's burrow, setting themselves as high as possible to facilitate their molt. During artificial incubation, two pieces of gauze can be placed to allow the tarantulas to climb and be attached properly to molt (Cléton et al. 2015; Mendoza, personal observation 2008–2020).

Handling and behavior This species is very nervous; when disturbed, it is almost certain to throw urticating hairs. These urticating hairs are particularly effective against humans and can cause a severe reaction (in some people) of burning of the skin, eyes and nose, so be careful (Cléton et al. 2015; Mendoza, personal observation 2008–2020). Its venom is not dangerous to humans, and its bite, although painful, does not cause a major effect. The pain and swelling can disappear in a few hours or a couple of days (Mendoza, personal observation 2008–2020).

Conservation status There is not enough information available on this species, but it is not considered threatened.

15.5 Traffic, Endangered Species and Responsible Breeding Facilities

Over the years, in many countries tarantulas have been offered in markets, pet stores, reptile shows or even tarantula shows. A lot of large specimens from distant places are offered, and it seems that no one cares about the origin of these spiders. No one asks where they come from, if they are wild-caught and, if so, whether they were collected with appropriate permission from the country of origin—or these issues are simply ignored. It is uncertain how many tarantula hobbyists ask or care about the condition of the natural habitat of the tarantulas they keep. There are arguments for and against collection of wild-caught tarantulas. At first, all specimens of the species kept by hobbyists were wild-caught, and this was necessary in order to start a breeding process. However, the lack of regulation of spider collection can have consequences, as was shown with the Mexican redknee tarantula (*Brachypelma smithi*) in the 1980s. This species was collected in its thousands for the pet trade and sent mainly to the USA and Europe. Because of this, large colonies of *B. smithi* declined, population sizes became affected and the species was declared threatened. In September 1985, *B. smithi* was officially listed in Appendix II of the Convention on International Trade in Endangered Species (CITES) (Smith 1994; Schultz and Schultz 2009).

CITES is the most important international convention for regulation of the wildlife trade. Its fundamental purpose is to protect and control the international trade of organisms that may be threatened or endangered by any of the signatory countries (Schultz and Schultz 2009). However, despite the increased recognition of this issue, these regulatory measures do not necessarily benefit wild species. While there have been numerous revisions to the listing criteria, they still focus almost exclusively on the biological and trade status of the species and scarcely touch on whether the listing will benefit the conservation status of the species (Dickinson 2002). Unfortunately, the conservation status of the vast majority of tarantulas is unknown. The Red List of the International Union for Conservation of Nature (IUCN) includes only some species of the genera *Brachypelma* Simon, 1891, *Chilobrachys* Karsch, 1892, *Grammostola* Simon, 1892, *Haploclastus* Simon, 1892, *Poecilotheria* Simon, 1885 and *Thrigmopoeus* Pocock, 1899 (Molur et al. 2008; Ferretti and Popozzi 2012; Fukushima et al. 2018), and in CITES, only the genus *Brachypelma* is listed.

According to Reichling (2003), one negative result of the *Brachypelma* CITES listing has been a decrease in the supply of some species, leading to increased desirability and thus demand. This has encouraged the development of a black market, which smuggles protected species both locally and internationally. On the other hand, lack of supply and availability of some species lead hobbyists and local dealers to breed their own specimens. Captive breeding allows production and sale (trade) of spiderlings between breeders and hobbyists. However, there are some inherent issues with this activity. For example, the genetic diversity of a captive population is finite, and without control, inbreeding will occur (CEC 2017). That is one reason why tarantulas well established in captivity are continuously smuggled. It is common to see, at least once a year in the news, that a person or some luggage has been intercepted at an airport, with smuggled tarantulas (Phillips 2009; Cooper 2011; Agencia EFE 2016; Carranza 2018).

Some of the most commonly smuggled tarantulas include *Brachypelma* spp., *Poecilotheria* spp., *Grammostola* spp., *Avicularia* spp., *Haplopelma* spp., *Pamphobeteus* spp., *Xenesthis* spp., *Theraphosa* spp., and many undescribed species from different subfamilies. Unfortunately, not all countries have records or policies on the tarantula trade and its implications for their conservation status. The most extensively documented cases of this involve *Brachypelma* in Mexico and *Poecilotheria* in India and Sri Lanka (Smith 1994; Rojo 2004; Molur et al. 2008; Benjamin et al. 2012; Mendoza and Francke 2017; CEC 2017). Moreover, it has been documented that different species are smuggled from Ecuador, Brazil and Uruguay. In the case of Mexico, traffickers pay the local people who collect them around US\$2–3 per tarantula, which can then be sold for up to US\$200 on the international market (Inecc 2012). Large tracts of land are pitted from excavation of burrows. Further habitat loss also occurs during the rainy season, as a result of landslides. The principal method of illegal exportation known within the international tarantula trade community is the "brown box". Large numbers of live tarantulas are individually sealed in padded containers and put together in a brown box. This box is airmailed out of the country of origin with documentation declaring that its contents are a gift, a costume, or some other nonliving content that will avoid suspicion from postal service employees or regulatory authorities. The principal destinations are the European Union and Asia, and sometimes the USA (CEC 2017). In some cases, spiders are hidden in photographic film containers, small plastic tubes or plastic bags inside toys, electronic devices or handicrafts (BWPM 2014). The result of that kind of packing is that animals are crammed in with each other, and many dies as a result of dehydration, insufficient space during molting or suffocation in their own exuviae (Rojo 2004). In Mexico, some of these packages are intercepted and reviewed by the authorities, and the live specimens are turned over to legal breeders or zoos rather than being destroyed. Confiscated adult specimens are then used as additional breeding stock for renewed genetic diversity in breeding facilities (Mendoza, personal observation 2008–2020).

With the increase in illegal trade (not only of tarantulas but also of other organisms), several countries have closed their borders to trade of their wildlife, prohibiting its export and increasing fines for its illicit trade. Brazil, Costa Rica, Venezuela, Australia and Sri Lanka are some of the countries that do not allow export of their tarantulas in any way. This, in the first instance, could be observed as something favorable for these and other organisms. However, illegal trade is only one part of the risks faced by all wildlife. Habitat loss poses the greatest threat to species. Without a strong plan to create protected terrestrial and marine areas, important ecological habitats will continue to be lost (Laurance 2010). At sustainable levels of consumption, both wildlife and people benefit from trade. Granting people an economic stake in wildlife provides the best incentive for careful stewardship of species and habitats (Carey 1999). It is certainly not the removal of a few tarantulas, with the intention of breeding them, that will lead to the extinction of a species (Cléton et al. 2015); the greater damage is the removal of large numbers of spiders for illegal export, only a few of which are expected to reach their final destination alive. On the other hand, the collection of tarantulas with purely commercial intentions is usually very harmful because it is done by extracting as many specimens as possible from some areas of the natural distribution of tarantulas. This practice can damage the habitat even more because the collection techniques are destructive, removing large numbers of stones, digging up soil and plants, and even using substances such as gasoline to force the tarantulas to leave their burrows more quickly (Verdez and Cléton 2004; West 2005).

As a group, spiders are one of the most proficient colonizers, but tarantulas are an exception in that they lack any ability to travel farther than their legs will carry them. The tendency of tarantulas to keep their populations together and to be common only in small areas throughout their distribution makes them vulnerable to both natural and anthropogenic factors that directly affect their habitat, such as drought or transformation of native vegetation in areas of cultivation (Reichling 2003). Some arachnids such as tarantulas have low mobility, limited mechanisms, of dispersion and sedentary habits. Depending on each group of arachnids, these characteristics can greatly influence their dispersal capacity (Ferretti et al. 2014). That is why many species of tarantula are vulnerable to destruction of their natural habitats. Despite this, laws that prohibit sustainable use of natural resources have some contradictory issues. On the one hand, there is extreme protection of wildlife against being traded or even bred in captivity; on the other hand, there is the fact that their habitat continues to disappear because of a lack of interest in direct actions that can benefit the tarantulas and the people who live together at their distribution sites. Hectares of land continue to be devastated, and loss of biodiversity continues unabated. We must understand that we can take advantage of resources sustainably and contribute to their care and conservation. For this, there are some models that can be implemented for legal rearing and trade of native tarantulas from each region, allowing collection of a certain number of individuals in order to breed them in captivity and release part of the progeny with a planned reintegration strategy, thereby competing legally with the illegal trafficking of them.

One example of this is the case of Mexico, where there is a permanent program that integrates all people interested in sustainable use of wildlife in the country in a system called the UMA (Unidad de Manejo para la Conservación de la Vida Silvestre) (DOF 2000). UMAs can be properties or property owners who voluntarily participate in sustainable use of the wild species that live there. They also respond to the need to conserve biodiversity and boost production and socioeconomic development of the country (Robles de Benito 2009). The general wildlife law (Ley General de Vida Silvestre (LGVS)) and its respective UMA regulate sustainable use, conservation and management of native tarantulas in Mexico. The UMA program allows qualified persons to collect a limited number of wild tarantulas in order to keep and breed them in captivity, and only the resulting offspring can be sold domestically or exported (CEC 2017). It should be noted that these breeding facilities can sell only those specimens that have been bred in captivity and not the wild-caught specimens they have been allowed to collect, since their only permitted use is as initial breeding stock for captive breeding (Mendoza, personal observation 2008–2020). This is a substantial difference in the use of tarantulas, since although their use is allowed, this has limitations, and only a few specimens are allowed to be collected in order to breed them in captivity to supply the existing demand for some protected species or even other species within the country. These actions are intended to remove pressure on native populations of tarantulas, giving people the opportunity to legally buy specimens bred in captivity that can be maintained or exhibited without legal problems, rather than those that have been illegally removed from their natural habitat (Zuñiga, personal communication 2019). As these specimens bred in captivity can also be exported to other countries, their illegal collection for international trafficking can be reduced. In addition, when the authorities confiscate tarantulas whose origin can not be determined and for any reason they cannot be returned to their country of origin (or, in the case of Mexican tarantulas, reintegrated into their habitat), they are given to any of these legal breeding facilities in order that they can be kept in the best possible living conditions. In this way, if it is possible, tarantula-breeding facilities can reproduce these species and trade the offspring. Legal importation of tarantulas into Mexico is also allowed, so any person with requirements for legal importation of animals can introduce exotic species into the country, on the condition that they are kept exclusively in captivity (DOF 2000).

Mexico is among the countries considered megadiverse and is considered the country with the second-largest number of known species of tarantula, after Brazil (Locht 2008; Candia-Ramírez and Francke 2017). Therefore, we consider that the way Mexico has implemented its legislation to be able to regulate and enable sustainable use of and trade in tarantulas is a good example of how other countries could allow responsible captive breeding of and trade in tarantulas. While it is true that around the world, there are a large number of people and breeding sites that reproduce tarantulas as part of a hobby, many of them are located in areas where there is little or no diversity of native tarantulas. So, it is important to recognize that large breeding facilities in Europe, Canada and the USA contribute to an important part of the indirect conservation of tarantulas by offering thousands of captive-bred individuals to supply the growing market demand. However, it should be noted that many of these tarantulas bred in captivity today were once illegally harvested from their places of origin. Thus, if there could be a legal alternative allowing trading and supply of tarantulas to the international market by the countries where these organisms are distributed naturally, it could reach a point where breeding facilities similar to those found in Mexico could be opened. Local breeders could not only benefit from this activity but also promote the importance and conservation of tarantulas in their native habitat, thereby managing to curb an important part of the illegal trafficking and collection of these organisms. On the other hand, importers of tarantulas in places such as Europe, Canada, and the USA could access specimens bred in captivity and exported legally, with the assurance that they were the product of sustainable use. In such a way, with the passage of time, they would preferentially acquire tarantulas legally from breeders in different countries and bring more income to families dedicated to the breeding of tarantulas in captivity.

References

Agencia EFE (2016) Un envío fallido de tarántulas de Uruguay a Dominicana destapa el tráfico ilegal de esta especie, 23 January. https://www.efe.com/efe/america/cono-sur/un-envio-fallido-de-tarantulas-uruguay-a-dominicana-destapa-el-trafico-ilegal-esta-especie/50000553-2818821. Accessed 5 Nov 2018

Anderson JF (1970) Metabolic rates of spiders. Comp Biochem Physiol 33(1):51–72

Anderson JF (1974) Responses to starvation in the spiders *Lycosa lenta* Hentz and *Filistata hibernalis* (Hentz). Ecology 55(3):576–585

Baerg WJ (1926) Regeneration of appendages in the tarantula *Eurypelma californica* Ausserer. Ann Entomol Soc Am 19(4):512–513

Baerg WJ (1928) The life cycle and mating habits of the male tarantula. Q Rev Biol 3(1):109–116

Baerg WJ (1938) Tarantula studies. J N Y Entomol Soc 46(1):31–43

Baerg WJ (1958) The tarantula. University of Kansas Press, Lawrence

Barneche JA, Gillung JP, Gonzalez A (2013) Description and host interactions of a new species of *Exetasis* Walker (Diptera: Acroceridae), with a key to species of the genus. Zootaxa 3664(4):525–536

Baxter RN (1995) Keeping and breeding tarantulas, 2nd edn. Chudleigh, Ilford

Bengston SE, Pruitt JN, Riechert SE (2014) Differences in environmental enrichment generate contrasting behavioural syndromes in a basal spider lineage. Anim Behav 93:105–110

Benjamin SP, Nanayakkara RP, Dayananda SK (2012) The taxonomy and conservation status of the spiders (Arachnida: Araneae) in Sri Lanka. In: Weerakoon DK, Wijesundara S (eds) The national red list 2012 of Sri Lanka; conservation status of the fauna and flora. Ministry of Environment, Colombo, pp 42–45

Bennie M, Loaring C, Trim S (2011) Laboratory husbandry of arboreal tarantulas (Theraphosidae) and evaluation of environmental enrichment. Anim Technol Welfare 10(3):163

Breene RG (1998) The ATS arthropod medical manual: diagnoses and treatment. American Tarantula Society, Carlsbad, NM, p 32

BWPM (2014) Códigos de barras de la vida silvestre México. http://bwp-mex.blogspot.mx/. Accessed 11 Aug 2018

Cady A, Leech R, Sorkin L, Stratton G, Caldwell M (1993) Acrocerid (Insecta: Diptera) life histories, behaviors, host spiders (Arachnida: Araneida), and distribution records. Caad Entomol 125(5):931–944

Candia-Ramírez DT, Francke OF (2017) Taxonomic revision of the genus *Crassicrus* Reichling & West, 1996 (Araneae: Theraphosidae: Theraphosinae), with the description of additional keels on the embolus. J Arachnol 45(1):67–98

Carey JE (1999) Improving the efficacy of CITES by providing the proper incentives to protect endangered species. Wash U L Rev 77:1291–1322

Carranza D (2018) Incautan en Colombia 23 tarántulas que iban al comercio ilegal en Seúl. Anandolu Agency. https://www.aa.com.tr/es/mundo/incautan-en-colombia-23-tarántulas-que-iban-al-comercio-ilegal-en-seúl/1271224. Accessed 2 Nov 2018

CEC (2017) Sustainable trade in tarantulas: action plan for North America. Commission for Environmental Cooperation, Montreal

Clarke D (1987) The captive breeding of arachnids. In: Proceedings of a Symposium on Spiders and Their Allies, London, pp 71–76

Cléton F, Sigwalt Y, Verdez JM (2015) Tarantulas, breeding experience & wildlife. Edition Chimaira, Frankfurt

Cloudsley-Thompson JL, Constantinou C (1985) Diurnal rhythm of activity in the arboreal tarantula *Avicularia avicularia* (L.) (Mygalomorphae: Theraphosidae). Biol Rhythm Res 16(2):113–116

Cooper R (2011) Smuggler caught at airport with 261 Mexican tarantulas shoved into his luggage. Daily Mail. https://www.dailymail.co.uk/news/article-2051210/Smuggler-caught-airport-261-Mexican-tarantulas-shoved-luggage.html. Accessed 5 Nov 2018

Cooper EWT, West R, Mendoza J (2019) Identification of CITES-listed tarantulas: *Aphonopelma*, *Brachypelma* and *Sericopelma* species. Commission for Environmental Cooperation, Montreal

Costa FG, Pérez-Miles F (2002) Reproductive biology of uruguayan theraphosids (Araneae, Mygalomorphae). J Arachnol 30(3):571–588

Costa FG, Pérez-Miles F, Mignone A (2004) Pompilid wasp interactions with burrowing tarantulas: *Pepsis cupripennis* versus *Eupalaestrus weijenberghi* and *Acanthoscurria suina* (Araneae, Theraphosidae). Stud Neotrop Fauna E 39(1):37–43

Davies ME, Edney EB (1952) The evaporation of water from spiders. J Exp Biol 29(4):571–582

De Voe RS (2009) Captive invertebrate nutrition. Vet Clin North Am Exot Anim Pract 12(2):349–360

de Vosjoli P (1991) Arachnomania: general care and maintenance of tarantulas and scorpions. Advanced Vivarium Systems, Lakeside

Dias SC, Brescovit AD (2003) Notes on the behavior of *Pachistopelma rufonigrum* Pocock (Araneae, Theraphosidae, Aviculariinae). Rev Bras Zool 20(1):13–17

Dickinson B (2002) International conservation treaties, poverty and development: the case of CITES. Nat Resour Perspect 74:1–4

Dippenaar-Schoeman AS (2002) Baboon and trapdoor spiders of southern Africa: an identification manual. Plant Protection Research Institute handbook no. 13. Agric Res Council, Pretoria

DOF (Diario Oficial de la Federación) (2000) Ley general de vida silvestre. Diario Oficial de la Federación. http://www.diputados.gob.mx/LeyesBiblio/ref/lgvs/LGVS_orig_03jul00.pdf. Accessed 3 July 2018

Draper E, Trim SA (2018) Dyskinetic syndrome in tarantula spiders (Theraphosidae). Vet Nurs J 33(8):230–232

Ferretti NE, Pompozzi G (2012) *Grammostola vachoni*. The IUCN red list of threatened species 2012: e.T18217620A18217711. https://doi.org/10.2305/IUCN.UK.2012.RLTS.T18217620A18217711.en. Accessed 5 Nov 2018

Ferretti N, Pompozzi G, Copperi S (2012) Notes on egg-sacs and spiderlings of two species of *Grammostola* (Araneae: Theraphosidae) from Central Argentina. J Br Tarantula Soc 27(2):53–61

Ferretti N, Gonzalez A, Pérez-Miles F (2014) Identification of priority areas for conservation in Argentina: quantitative biogeography insights from mygalomorph spider (Araneae: Mygalomorphae). J Insect Conserv 18(6):1087–1096

Figueroa DP, Sabat P, Torres-Contreras H, Veloso C, Canals M (2010) Participation of book lungs in evaporative water loss in *Paraphysa parvula*, a migalomorph spider from Chilean Andes. J Insect Physiol 56(7):731–735

Foelix R (2011) Biology of spiders. Oxford University Press, New York

Frye FL (1992) "Tarantulas" and other spiders. In: Captive invertebrates a guide to their biology and husbandry. Krieger, Malabar, pp 7–17

Fukushima CS, Bertani R (2017) Taxonomic revision and cladistic analysis of *Avicularia* Lamarck, 1818 (Araneae, Theraphosidae, Aviculariinae) with description of three new Aviculariinae genera. ZooKeys 659:1–185

Fukushima C, Mendoza J, West R, Longhorn S, Rivera-Téllez E, Henriques S, Cardoso P (2018) *Brachypelma*. The IUCN red list of threatened species 2018: e.T8152A21416539. https://doi.org/10.2305/IUCN.UK.2018-2.RLTS.T8152A21416539.en. Accessed 5 Nov 2018

Gallon RC (2000) The natural history of tarantula spiders. J Br Tarantula Soc. http://www.bts.ndirect.co.uk/natural.htm. Accessed 5 Oct 2018

Hardin T, Sincage J (2015) AZA Mexican red kneed tarantula (*Brachypelma smithi*) species survival plan. Mexican red kneed tarantula care manual. Association of Zoos and Aquariums, Silver Spring. http://digitalcommons.unl.edu/entodistmasters/1. Accessed 27 Oct 2018

Haro L, Jouglard J (1998) The dangers of pet tarantulas: experiences of the Marseilles Poison Centre. J Toxicol Clin Toxicol 36(1&2):51–53

Herrero MV, Valerio CE (1986) Análisis de 1ª actividad diaria de *Aphonopelma seemanni* (Araneae, Theraphosidae) en Costa Rica. J Arachnol 14:79–82

Inecc (2012) Tráfico ilegal de especies silvestres y sus impactos. http://www.inecc.gob.mx/descargas/dgioece/2012_sem_trafico_pon01_alow.pdf. Accessed 21 Nov 2018

Jespersen LB, Toft S (2003) Compensatory growth following early nutritional stress in the wolf spider *Pardosa prativaga*. Funct Ecol 17:737–746

Kleinteich A, Wilder SM, Schneider JM (2015) Contributions of juvenile and adult diet to the lifetime reproductive success and lifespan of a spider. Oikos 124(2):130–138

Kurczewski FE (2010) Prey and nesting behavior of some North American spider wasps (Hymenoptera: Pompilidae). Northeast Nat 1:115–124

Kurczewski FE, Edwards GB (2012) Hosts, nesting behavior, and ecology of some North American spider wasps (Hymenoptera: Pompilidae). Southeast Nat 11(m4):1–72

Larrivée M, Borkent CJ (2009) New spider host associations for three acrocerid fly species (Diptera, Acroceridae). J Arachnol 37(2):241–242

Laurance WF (2010) Habitat destruction: death by a thousand cuts. In: Sodhi NS, Ehrlich PR (eds) Conservation biology for all. Oxford University Press, Oxford, pp 73–87

Li D, Jackson RR (1996) How temperature affects development and reproduction in spiders: a review. J Therm Biol 21(4):245–274

Locht A (2008) Estudio sobre la sistemática y distribución de la familia Theraphosidae (Arachnida, Araneae) en México. Dissertation, Facultad de Ciencias, Universidad Nacional Autónoma de México

Locht A, Yáñez M, Vázquez I (1999) Distribution and natural history of Mexican species of *Brachypelma* and *Brachypelmides* (Theraphosidae, Theraphosinae) with morphological evidence for their synonymy. J Arachnol 1:196–200

M'Rabet SM, Hénaut Y, Sepúlveda A, Rojo R, Calmé S, Geissen V (2007) Soil preference and burrow structure of an endangered tarantula, *Brachypelma vagans* (Mygalomorphae: Theraphosidae). J Nat Hist 41(17-20):1025–1033

Machkour-M'Rabet S, Ariane D, Hénaut Y (2015) *Megaselia scalaris* (Diptera: Phoridae): an opportunistic endoparasitoid of the endangered Mexican redrump tarantula, Brachypelma vagans (Araneae: Theraphosidae). J Arachnol 43(1):115–119

Marechal P, Rolland C, Leguin EA, Monpiou L (2009) Etude sur l'etat des populations de *Avicularia versicolor*, mygale endemique de la Martinique, rapport de synthese, missions 2007–2008. Museum national d'Histoire naturelle, Paris. http://www.side.developpement-durable.gouv.fr/EXPLOITATION/DEFAULT/Infodoc/ged/viewportalpublished.ashx?eid=IFD_FICJOINT_0006180&search=. Accessed 10 Oct 2019

Marnell C (2016) Tarantula and hermit crab emergency care. Vet Clin North Am Exot Anim Pract 19(2):627–646

Marshall SD (1996) Tarantulas and other arachnids: a complete pet owner's manual. Barron's, New York

Marshall SD (2001) Tarantulas and other arachnids. Barron's Educational Series, Hauppauge

Marshall SD, Uetz GW (1990) Incorporation of urticating hairs into silk: a novel defense mechanism in two Neotropical tarantulas (Araneae, Theraphosidae). J Arachnol 18(2):143–149

Marshall SD, West R (2008) An ontogenetic shift in habitat use by the Neotropical tarantula *Ephebopus murinus* (Araneae, Theraphosidae, Aviculariinae). Arachnology 14(6):280–285

Mendoza J (2009) La ciencia de las tarántulas: anatomía y fisiología. In: Memorias del 2° Curso de Manejo de Tarántulas y Escorpiones en Cautiverio. Universidad Nacional Autonóma de México, Facultad de Medicina Veterinaria y Zootecnía

Mendoza JI, Francke OF (2017) Systematic revision of *Brachypelma* red-kneed tarantulas (Araneae: Theraphosidae), and the use of DNA barcodes to assist in the identification and conservation of CITES-listed species. Invertebr Syst 31(2):157–179

Millot J (1943) Une mygale vivante de l'Uruguay. Bull Soc de Pathol Exotique 36:130–133

Mitchell M, Tully TN (2008) Manual of exotic pet practice. Saunders-Elsevier, St. Louis

Molur S, Daniel BA, Siliwal M (2008) *Poecilotheria*. The IUCN red list of threatened species. e.T63563A12681959. https://doi.org/10.2305/IUCN.UK.2008.RLTS.T63562A12681695.en. Accessed 4 Nov 2018

Paula FS, Gabriel R, Indicatti RP, Brescovit AD, Lucas SM (2014) On the Brazilian Amazonian species of *Acanthoscurria* (Araneae: Theraphosidae). Zoologia (Curitiba) 31(1):63–80

Peck WB, Whitcomb WH (1970) Studies on the biology of a spider, *Chiracanthium inclusum* (Hentz). Bull Ark Exp Stn 753:1–76

Pellett S, Bushell M, Trim SA (2015) Tarantula husbandry and critical care. Companion Anim 20(2):54–60

Pérez-Miles F, Costa FG, Toscano-Gadea C, Mignone A (2005) Ecology and behaviour of the 'road tarantulas' *Eupalaestrus weijenberghi* and *Acanthoscurria suina* (Araneae, Theraphosidae) from Uruguay. J Nat Hist 39(6):483–498

Pérez-Miles F, Postiglioni R, Montes de Oca L, Baruffaldi L, Costa FG (2007) Mating system in the tarantula spider *Eupalaestrus weijenberghi* (Thorell, 1894): evidences of monandry and polygyny. Zoology 110:253–260

Petrunkevitch A (1911) Sense of sight, courtship and mating in *Dugesiella hentzi* (Girard), a theraphosid spider from Texas. Zoologische Jahrbücher (Systematik) 31:355–376

Philip BN (2006) The effect of prey availability on metabolism and activity in the tarantula *Phormictopus cancerides* (Araneae: Theraphosidae). Dissertation, Eastern Michigan University. http://commons.emich.edu/theses/82

Philip BN, Shillington C (2010) The effect of prey availability on metabolism and activity in the tarantula *Phormictopus cancerides*. Can J Zool 88(1):90–98

Phillips T (2009) London petshop owner try to smuggle 1000 spiders. The Guardian. https://www.theguardian.com/world/2009/nov/12/pet-shop-owner-spider-smuggling. Accessed 27 Oct 2018

Pizzi R (2009) Parasites of tarantulas (Theraphosidae). J Exot Pet Med 18(4):283–288

Pizzi R (2010) Invertebrates. In: Meredith A, Johnson-Delaney CA (eds) BSAVA manual of exotic pets, 5th edn. BSAVA, Gloucester, pp 373–385

Pizzi R (2012) Spiders. In: Lewbart G (ed) Invertebrate medicine, 2nd edn. Blackwell, Chichester pp 187–221

Prentice TR (1997) Theraphosidae of the Mojave Desert west and north of the Colorado River (Araneae, Mygalomorphae, Theraphosidae). J Arachnol:137–176

Prestwich KN (1983a) Anaerobic metabolism in spiders. Physiol Zool 56:112–121

Prestwich KN (1983b) The roles of aerobic and anaerobic metabolism in active spiders. Physiol Zool 56:122–132

Psaila N (2005) Husbandry manual for exotic tarantulas. http://nswfmpa.org/Husbandry%20Manuals/Published%20Manuals/Invertebrata/Exotic%20Tarantulas.pdf. Accessed 2 July 2018

Punzo F (1989) Effects of hunger on prey capture and ingestion in *Dugesiella echina* Chamberlin (Orthognatha, Theraphosidae). Bull Br Arachnol Soc 8(3):72–79

Punzo F (1994) The biology of the spider wasp, *Pepsis thisbe* (Hymenoptera: Pompilidae) from Trans Pecos, Texas. II. Temporal patterns of activity and hunting behavior with special reference to the effects of experience. Psyche 101(3–4):243–256

Punzo F (2005) Experience affects hunting behavior of the wasp, *Pepsis mildei* Stål (Hymenoptera: Pompilidae). J N Y Entomol Soc 113(3):222–229

Punzo F, Henderson L (1999) Aspects of the natural history and behavioral ecology of the tarantula spider *Aphonopelma hentzi* (Girard, 1854) (Orthognatha, Theraphosidae). Bull Br Arachol Soc 11(4):121–128

Rego FN, Rheims CA, Venticinque EM (2004) Notes on the predation of an *Aviculariine* spider (Araneae: Theraphosidae, *Avicularia* sp.) by *Pepsis frivaldszkyi* (Hymenoptera: Pompilidae) in Brazilian Amazonia. J Br Tarantula Soc 20(1):18–25

Reichling SB (2000) Group dispersal in juvenile *Brachypelma vagans* (Araneae, Theraphosidae). J Arachnol 28(2):248–251

Reichling SB (2003) Tarantulas of Belize. Krieger, Malabar

Reichling SB, Gutzke WH (1998) Phenotypic consequences of incubation temperature and feeding regimen in captive-bred tarantulas. Zoo Biol 17(5):405–414

Riley J, Barron H (2016) Wildlife emergency and critical care. Vet Clin North Am Exot Anim Pract 19(2):613–626

Robles de Benito R (2009) Las unidades de manejo para la conservación de la vida silvestre y el corredor biológico Mesoamericano México. CONABIO, Mexico City

Rojo R (2004) Las tarántulas de México: pequeños gigantes incomprendidos. Biodiversitas 56:7–11

Saul-Gershenz L (1996) Laboratory culture techniques for the Goliath tarantula *Theraphosa blondi* (Latreille, 1804) and the Mexican red knee tarantula, *Brachypelma smithi* (Araneae: Theraphosidae). In: American Association of Zoological Parks and Aquariums (AAZPA) Regional Conference Proceedings, pp 773–777

Schultz SA, Schultz MJ (2009) The Tarantula Keeper's guide: comprehensive information on care, housing, and feeding. Baron's Educational series, Hauppauge, NY, p 384

Sedinova H (2015) The shrew or ermine? New meaning of Aristotele's term mygale in the middle ages. Liaty Filologicke 138(1-2):119–146

Shillington C, McEwen B (2006) Activity of juvenile tarantulas in and around the maternal burrow. J Arachnol 34(1):261–266

Shillington C, Verrell P (1997) Sexual strategies of a North American 'tarantula' (Araneae: Theraphosidae). Ethology 103(7):588–598

Smith AM (1994) Tarantula spiders: tarantulas of the USA and Mexico. Fitzgerald, London

Smith A (2013) The story of the Costa Rican zebra tarantula, *Aphonopelma seemanni*. http://www.lovetarantulas.com/costaricanzebratarantula.pdf. Accessed 14 Aug 2018

Stradling DJ (1978) The growth and maturation of the "tarantula" *Avicularia avicularia* L. Zool J Linn Soc 62:291–303

Stradling DJ (1994) Distribution and behavioral ecology of an arboreal tarantula spider in Trinidad. Biotropica 1:84–97

Teyssié F (2015) Tarantulas of the world. NAP Editions, Verrières-le-Buisson

Trevors JT (1996) Sterilization and inhibition of microbial activity in soil. J Microbiol Methods 26(1–2):53–59

Turnbull AL (1962) Quantitative studies of the food of *Linyphia triangularis* Clerck (Araneae; Linyphiidae). Can Entomol 94:1233–1245

Turnbull AL (1965) Effects of prey abundance on the development of the spider *Agelenopsis potteri* (Blackwell) (Araneae: Agelenidae). Can Entomol 97(2):141–147

Verdez JM, Cléton F (2004) Tarántulas. Introducción, guía y reproducción. Reptilia Ediciones, Barcelona

Weinmann D, Disney RH (1997) Two new species of Phoridae (Diptera) whose larvae associate with large spiders (Araneae: Theraphosidae). J Zool 243(2):319–328

West RC (1995) Mighty mites. J Br Tarantula Soc 10(3):86–88
West RC (2005) The *Brachypelma* of Mexico. J Br Tarantula Soc 20:108–119
World Spider Catalog (2020) World Spider Catalog. Version 21.5. Natural History Museum Bern. http://wsc.nmbe.ch. Accessed 11 Sept 2020. https://doi.org/10.24436/2
Yáñez M, Floater G (2000) Spatial distribution and habitat preference of the endangered tarantula, *Brachypelma klaasi* (Araneae: Theraphosidae) in Mexico. Biodivers Conserv 9(6):795–810

Chapter 16
Situation and Conservation of Tarantulas in the Americas

Jorge Mendoza

Abstract Tarantulas are the world's largest spiders and they are more abundant in the Americas than in any other place. They are primarily found in tropical, semi-tropical, and arid regions over the continent; some of them even have adapted to anthropogenic habitats. As other spiders, tarantulas play many important roles in ecosystems as predators and sources of food for other creatures. Although tarantulas are an integral part of biodiversity whose conservation is justified, they are not usually included in conservation strategies. It is evident that regardless the scientific and technical arguments for their protection, the cultural weight awash with prejudice and undervaluation can ruin the best reasons. Because of this, the sustainable use and management of tarantulas native to the Americas depend on the knowledge of its main threats, its relevance to the environment, and creation of public policies that involve society for management, protection, and conservation actions.

16.1 Social and Ecological Importance of Tarantulas

Currently, there are 147 genera and 985 described species of the family Theraphosidae in the world (WSC 2020). Tarantulas are the world's largest spiders and are more abundant in the Americas than in any other place. They are primarily found in tropical, semi-tropical, and arid regions over the continent; some of them even adapted to anthropogenic habitat (West 2005; Rojo 2004). In the American continent, species of the subfamily Theraphosinae are the most predominant and are characterized by having in the dorsal area of the posterior part of their body (the opisthosoma) a patch of urticating setae that are their first defense mechanism against predators. These could be called the "typical tarantulas," as these are the large, hairy, ground-dwelling spiders that enter most of the people minds when they think about tarantulas (Reichling 2003). Of all the tarantulas, one of the most famous in the world is *Brachypelma smithi* (F. O. Pickard-Cambridge, 1897), the so-called "Mexican

J. Mendoza (✉)
Colección Nacional de Arácnidos, Módulo D planta baja, Departamento de Zoología, Instituto de Biología, Universidad Nacional Autónoma de México, Mexico City, Mexico

F. Pérez-Miles (ed.), *New World Tarantulas*, Zoological Monographs 6,
https://doi.org/10.1007/978-3-030-48644-0_16

Fig. 16.1 Mexican redknee tarantula (*Brachypelma smithi*) in habitat. Photo: Jorge Mendoza

Fig. 16.2 Tarantulas from illegal trade that were confiscated and subsequently placed under guard in conditions suitable for their recovery. Photo: Jorge Mendoza

red-kneed" (Fig. 16.1). Due to its bright coloration, docile behavior, longevity, and long tradition in the world of exotic pets, this species has been illegally shipped out of Mexico in large numbers through the years (Rojo 2004; Mendoza and Francke 2017; Cooper et al. 2019). Because of this, population sizes have been affected and the species was declared threatened by Mexican government in 1994 (SEMARNAP 1995). In August 1985, *B. smithi* was officially placed on Appendix II of the Convention on International Trade in Endangered Species (CITES) (CITES 1985). Despite these restrictions, illegal trade continues. There are records of some shipments with individuals of the genus *Brachypelma* that have been confiscated mainly at European airports (Mendoza and Francke 2017). However, beside species protected by CITES, other species from South America, Asia, or Africa are taken without permits from their countries of origin. Generally, these animals are shipped in poor conditions and many of them die due to dehydration or to the conditions to which they are subjected when being transported in plastic bags, tightened with each other (Fig. 16.2). Sometimes they are sent inside plastic containers, but either because they are confined in a small space or because they may be in the process of moving; they do not find enough space and succumb to suffocation (Rojo 2004; CEC 2017).

16.1.1 Ecological Significance

Spiders are generalist terrestrial predators which provide arthropod-mediated ecosystem services; they are sensitive to changes in the environment, and they can be grouped into different functional guilds (Theron 2017). This diverse group of predators occupies a wide range of different niches within the environment (Cardoso et al. 2011), thereby providing important ecosystem services (Sunderland and Samu 2000). They are highly adapted to thrive within their specific niche, and these so-called functional guilds allow spiders to exploit a variety of different resources within the environment (Foelix 1996). Guilds refer to groups of species that share similar resources, although do not occur in the same or similar niches (Wilson 1999). Niche partitioning refers to the process by which ecologically similar, sympatric species may avoid competitive exclusion by different patterns of resource use (Cloyed and Eason 2017). This niche partitioning allowed spiders to occupy almost every part of the world (Cardoso et al. 2011). While araneomorphs generally have high vagility, mygalomorphs usually have little dispersal ability (Coyle 1985; Raven 2010; Ferretti et al. 2014). They are habitat specialists, and thus juveniles and females are mainly sedentary, adult males increase their motility during mate search in the breeding season (Main 1987; Coyle and Icenogle 1994). Females and juveniles usually leave their burrows for short distances for prey capture and for discarding prey debris (Pérez-Miles and Perafán 2017) (Figs. 16.3 and 16.4). Some mygalomorphs extend the prey detection area by using silk threads with twigs,

Fig. 16.3 An adult female of *Crassicrus tochtli* at the entrance of her burrow waiting for prey. Photo: Jorge Mendoza

Fig. 16.4 A *Tliltocatl vagans* excavate its burrow under a big rock where is protected from potential predators. Photo: Jorge Mendoza

grass, or debris around the burrow entrances (Main 1978). Trap-door mygalomorphs do not abandon the burrow to hunt (Dippenaar-Schoeman 2002).

Most of the ecological studies of Theraphosidae deal with ecotype description, microhabitat use and home range determination (Dias and Brescovit 2003). Tarantulas as other Mygalomorphs have low vagility that can promote geographic fragmentation over space and time and small geographic distribution (Bond et al. 2006). This small geographic distribution facilitated the preservation of historical biogeographic signal, enabling the study of species evolutionary history and the geological processes in its environment (Hedin et al. 2013). Low vagility, combined with high habitat specificity usually makes tarantulas excellent models for biogeographical research (Bond et al. 2001; Bond and Stockman 2008; Bertani 2012; Ferretti et al. 2012; West et al. 2012; Hedin et al. 2013). Wilson et al. (2012) suggested that tarantulas should be used as ecosystem health indicator species due to their characteristics as group. They are long-lived, are habitat specialists, and show high local endemicity (Coyle and Icenogle 1994). Despite the poor ecological knowledge and cryptic habits, theraphosids can be considered threatened species in red lists since they are very popular in pet trade, having a commercial value; as also due to habitat loss and their restricted geographical occurrence (Molur et al. 2008; Ferretti and Pompozzi 2012; Fukushima et al. 2018).

Tarantulas are top predators among arthropods (Wilson et al. 2012). These animals can prey mainly on insects such as ants, beetles (Pérez-Miles et al. 2004), cicadas, cockroaches, grasshoppers (Fig. 16.5), crickets, termites, Lepidoptera (mostly Saturniidae and Sphingidae), and Hymenoptera (Baerg 1958; DeWet 1991; Coyle 1995; Paulsen 1999; Dippenaar-Schoeman 2002). Likewise, they can feed on other arachnids as other spiders, solifugids, scorpions and even occasionally other tarantulas from its own or different species (Paulsen 1999; Dor et al. 2011; Dor and Hénaut 2011). Some can even predate other invertebrates such as millipedes (Coyle 1995), centipedes (Mendoza pers. obs. 2018) (Fig. 16.6), slugs (Rudloff 2008), and earth worms (Nyffeler et al. 2001). It is not known with which frequency tarantulas

Fig. 16.5 Remains of a grasshopper at the entrance of the burrow of an *Aphonopelma* sp. Photo: Jorge Mendoza

Fig. 16.6 Centipede Scolopendra polymorpha recently hunted by a North America cobalt blue tarantula (*Aphonopelma mooreae*). Photo: Jorge Mendoza

can feed on a variety of small vertebrates, but there are reports of predation over small lizards (Coyle 1995, Streicher et al. 2011; Vieira et al. 2012), amphibians (Menin et al. 2005; Toledo 2005; Ramírez-Castaño et al. 2014; Marín-Martínez and Rojas-Morales 2016), snakes like *Bothrops asper* (Garman, 1884) (Silvano pers. obs. 2016), *Micrurus ibiboboca* (Merrem, 1820) (Nunes et al. 2010), *Ninia sebae* (Duméril, Bibron & Duméril, 1854) (Aguilar-López et al. 2014) (Fig. 16.7), *Erythrolamprus almadensis* (Wagler, 1824) (Borges et al. 2016). There are also reports of some tarantulas feeding on bats (Das et al. 2012; Nyffeler and Knörnschild 2013), birds (Campos and Almeida 2016), fishes (Horstkottet et al. 2010), and small mammals (Azevedo and Smith 2004).

Due to their large size, it can be thought that tarantulas have no enemies that can attack them. However, precisely because of their size they are considered an excellent source of food for other organisms. Tarantula spiders can be attacked by a wide variety of predators, parasitoids, and parasites, some of which take advantage of the sedentary habits of tarantulas (Pérez-Miles and Perafán 2017, see also Chap. 8). The coati *Nasua* spp. is known to be an avid predator of tarantulas (Hirsch 2009; Markovics 2013). Some toads as *Rhinella* spp. also feed on them (Reagan and Waide 1996). In addition, large quantities of organisms such as insects, arachnids,

Fig. 16.7 Snake *Ninia sebae* being predated by a tarantula of the genus *Acentropelma*. Photo: José Luis Aguilar

lizards, frogs, birds, centipedes, insectivorous mammals (honey badgers, shrews, bats, mice, and baboons), and other arachnids such as vinegaroons, solifuges, scorpions or other spiders, prey also on tarantulas (Verdez and Cléton 2004; Dor et al. 2011; Machkour-M'Rabet et al. 2012; Pérez-Miles and Perafán 2017). One of the most dangerous enemies of the tarantulas are the wasps of the family Pompilidae (*Pepsis* spp and *Hemipepsis* spp), which specialize in hunting tarantulas. If the wasp finds a spider, it will lure the spider out of its hole. The wasp tries sting the spider in its central nerve ganglion and paralyzing it. If success, the wasp drags the spider to a suitable place and digs a hole or use the tarantula burrow. It lays one egg on the hapless spider and buries it in an underground chamber. The wasp larva eats the spider, starting with the nonessential body parts to keep the meat fresh. After a few weeks, the larva pupates and it emerges to repeat the cycle (Kingsley 2016) (Fig. 16.8). There are other organisms besides Pompilidae wasps that parasitize the tarantulas and use them as sustenance. Ichneumonidae wasps lay their eggs on the spider's legs or abdomen but do not paralyze the spider as do Pompilidae wasps. The larva develops as an ectoparasite, ingesting the tarantula's hemolymph. Some ichneumonid larvae also parasitize egg sacs (Pizzi 2006). Sphecidae mud daubers are solitary wasps. Similar to pompilid wasps, they paralyze spiders for the larvae to feed on (Breene 1998; Pizzi 2006). Acroceridae spider flies are endoparasites. Larvae slowly crawl on the tarantula's body to the book lungs, penetrating the

Fig. 16.8 Parasite emerging from the abdomen of a tarantula *Tliltocatl* sp. Photo: Jorge Mendoza

opisthosoma. The larva may be present for several months or even years in tarantulas. The larva consumes opisthosoma tissues before bursting through the dorsal opisthosoma to pupate (Pizzi 2006). Up to 14 larvae have been found in the opisthosoma of one tarantula (Foelix 1996), but the presence of just one larva is most common. There is also known that humpback flies (Phoridae) can lay eggs on tarantulas. Adult tarantula death appears to be caused by the parasitic larvae infecting the book lungs before penetrating the opisthosoma and internal organs (Pizzi 2006; Schultz and Schultz 2009). Oral Panagrolaimidae nematodes parasite has been seen in a wide variety of Theraphosidae species from Africa, Americas, and Asia (Pizzi 2006). These parasites are lethal for tarantulas; after infection, spiders manifest anorexia and a gradually increasing lethargy that progresses and finish in death (Pizzi et al. 2003). Parasitic mites (Acari) have been rarely seen on tarantulas, but they can attack the soft tissue of the leg joints or abdomen. As mites are uncommon parasites, it is difficult to evaluate their significance as threat of tarantulas health (Pizzi 2006). Breene (1998) proposes that mites may occlude the moist surfaces of the book lungs when present in high numbers. Although ants are not parasites, a few species have been reported to injure or even kill captive tarantulas (Baerg 1958). The main problematic species in North America are *Solenopsis* spp. (fire ants), *Linepithema humile* Mayr, 1868 (the Argentine ant), and *Monomorium pharaonis* (Linnaeus, 1758) (the pharaoh ant) (Pizzi 2006). Tarantulas can be particularly vulnerable to predators at earlier stages. Despite many tarantula species have hundreads of spiderlings per eggsac, we have poor data of this early period of spider lives, except that their rate of mortality is very high (Schultz and Schultz 2009). Baerg (1958) states that predation is the main cause of the low ratio survival in tarantula spiderlings, mainly due to cannibalism or fierce predators such as some species of ants.

In addition to the importance of tarantulas as predators and prey, it has also been documented that some species may exhibit commensal relationships with some other organisms. Commensalism is when one organism benefits from a close relationship between two organisms, and the other organism is not harmed by the association, so it is not beneficial or detrimental to the other species (Siliwal and

Ravichandran 2008, see also Chap. 8). Hunt (1980) describes an association in which the frog *Gastrophryne olivacea* (Hallowell, 1856) inhabits burrows of the theraphosid spider *Aphonopelma hentzi* (Girard, 1852). Powell et al. (1984) reported the frog *Physalaemus pustulosus* (Cope, 1864) inside burrows of the theraphosid spider *Aphonopelma* sp. A commensal relationship of *Chiasmocleis ventrimaculata* (Andersson, 1945) and the tarantula *Xenesthis immanis* (Ausserer, 1875) was studied by Cocroft and Hambler (1989). Feeding trials suggest that *Xenesthis* readily feeds on some other species of anurans, and observations implicate the role of chemical defenses in preventing *Xenesthis* predation on *Chiasmocleis* (Garton and Mushinsky 1979). In this case, the little frog benefits from the protection of its spider-mate. The adult female tarantula will aggressively defend its burrow against many predators (Csakany 2002). Apparently, the tarantula receives no benefit from its fellow frog; however, some researchers theorize that the frog inside the burrow may prey on ants or other small predators (Rödel et al. 2013) (Figs. 16.9 and 16.10). Teyssié (2015) mentioned another possible relation of commensalism between tarantulas and phorid flies. The tiny winged flies lay their eggs on the carapace of the tarantula (close to the fovea); they become into larvae which can feed primarily

Fig. 16.9 Adult female of *Tliltocatl kahlenbergi*, a species that can be found sharing its burrow with small frogs. Photo: Eddy Hijmensen

Fig. 16.10 A small frog *Engystomops pustulosus* that was found inside a burrow of *Tliltocatl kahlenbergi*. Photo: Eddy Hijmensen

on the leftovers of uneaten preys, but also sometimes on the prey that tarantulas are currently eating. This associations have been seen in *Theraphosa blondi* (Latreille, 1804), *Megaphobema robustum* (Ausserer, 1875) and *Pamphobeteus vespertinus* (Simon, 1889). As has been observed, tarantulas present diverse relationships within their ecosystems. The majority of their ecologic aspects is still unknown; thus, it is difficult to determine the importance of the role of theraphosids in the ecosystem. Future studies can clarify the biological cycle and importance of this particular group of spiders.

16.1.2 Tarantulas as Hobby and Other Usages

Tarantulas are organisms with an appearance, habits, and behavior that arouse fascination in a few, curiosity in some, and a certain repulsion in the vast majority of people, which in extreme cases can become a pathological modality such as arachnophobia. Probably the tarantulas are the most feared spiders due to their large size and hairy appearance. Most people accept the myth that tarantulas are aggressive and sinister creatures looking for an opportunity to pounce on any human victim to inflict a deadly bite (Reichling 2003). However, tarantulas are not dangerous and there are even many people who buy them as pets. By nature, they are non-domesticable wild animals and do not like to be handled, even so there are some species of more docile character (Fig. 16.11). They do not usually attack humans although they could use their fangs if they are manipulated incorrectly or too often. Although tarantula bites do not have medical importance, they can cause respiratory problems, inflammation of the eyelids, itching, decrease in blood pressure, acceleration of heart rhythm, rashes and swelling in the lips, throat or affected area depending on the sensitivity of the person bitten (Kaiser et al. 1994; Escoubas et al. 1997; Isbister et al. 2003). Tarantulas, like other types of exotic animals, had their demand

Fig. 16.11 Handling of a *Tliltocatl verdezi* tarantula, some species of tarantulas do not tend to be aggressive so they can be taken with caution. Photo: Jorge Mendoza

increased by the pet trade. For more than 30 years, these animals have become very popular in Europe, Canada, and the United States (Copperi et al. 2011). Much of their popularity is due to its diversity of sizes, coloration, and longevity (Mendoza and Francke 2017) (Fig. 16.12). Compared to other spider families, Theraphosidae are long-lived, with maximal lifespans reaching up to 20–30 years (Baerg 1958; Rojo 2004; Schultz and Schultz 2009; Costa and Pérez-Miles 2002; Schmidt 2003; Klaas 2007). In most species, females continue to grow after they reach adulthood. The males, on the other hand, rarely survive more than a season after reaching sexual maturity and having mated (Fig. 16.13).

Essentially, the most important and obvious reason for a person to want to have a tarantula as a pet, is simply that they like or are attracted to these animals. Due to their variety of sizes and colors, they can be a source of aesthetic pleasure for their owner. Also, many people like to raise tarantulas because they consider it a challenge or a stimulating experience to get their tarantulas to reproduce (Fig. 16.14).

Fig. 16.12 A beautiful tarantula from Socotra Island *Monocentropus balfouri* which is commonly found in the pet trade. Photo: Jorge Mendoza

Fig. 16.13 Remains of an adult male of *Brachypelma emilia*, several tarantula males are at risk of dying while they are looking for females to reproduce. Photo: Jorge Mendoza

Fig. 16.14 Mating of *Brachypelma smithi* in captivity. Photo: Jorge Mendoza

Fig. 16.15 Tarantula enclosure for exhibition of arboreal tarantulas. Photo: Jorge Mendoza

In some cases, the tarantula spiderlings can be sold at a good price, although few species manage to compensate the time and expenses incurred by the owners in their care (Rankin and Walls 2008). Another advantage of tarantulas is that differently from other animals, tarantulas need much less space to be kept in captivity, even when compared to other animals that can also be maintained in a terrarium. Since in the wild these animals also have a relatively small mobility, they do not need a huge terrarium. A medium-sized spider with a body about 5-cm long needs a terrarium of 25 × 20 × 20 cm, for example (Marshall 2001; Schultz and Schultz 2009) (Fig. 16.15). Depending on the species, tarantulas can even be an easy animal for those who seek their first pet. Considering the easiness of maintenance, there are some species that are more difficult to keep because they require more specific environmental characteristics or are very fast and defensive (Fig. 16.16). An important point that should be clear to those who want to have a tarantula home is that they are not pets to be manipulating. The charm of having one of these animals lies in observing their natural habits. It is also imperative to mention that in many occasions the tarantulas are

Fig. 16.16 A psychedelic tarantula *Haploclastus devamatha* in defense position after being disturbed. Photo: Jorge Mendoza

extracted from their habitat to be sold as pets. So it should be liable if you want to acquire one and purchase only with legal traders to assure good captivity conditions to the animal.

16.1.3 Tarantula's Utilization: Food Resources, Medical Applications, and Other Applications

There is little information about the utilization of tarantulas by people differently from the current use as pets. However, there are some reports of their potential use in medicine or as a food source for people (Bristowe 1932; Smith 1990; Breet 1993; Meyer-Rochow and Changkija 1997; Escoubas and Rash 2004; Paoletti and Dufour 2005; Corzo et al. 2008; Münke 2012; Mourão et al. 2013; Yen and Ro 2013; Richards et al. 2018). In Southwest Africa, tarantulas are disturbed by people with stalks of grass until they come out from their burrows. Spiders are then collected and placed on a skewer over a fire, the time enough to singe the urticating hair. Among the bushmen such practice is viewed as an important initiation rite for young men about to engage in a hunt (Smith 1990). There is also a filmed report of the South American native Piaroa people who used *Theraphosa apophysis* (Tinter, 1991) and *Theraphosa blondi* (Latreille, 1804) for food. Legs of spiders are bent backward, then they are held with a leaf, rolled over and roasted in hot coals (Breet 1993). More recently, with increasing of tourism to Cambodia over recent years, there are many internet references to the fried tarantulas sold as food. The main market for tarantulas is the small town of Skuon, known as "spider town" to many Cambodians. Skuon had long used the local tarantulas in traditional medicine; they were thought to be good for the heart, throat, and lungs. The practice of using them as a food source started in the years of the Khmer Rouge. At that time, starvation was rife in Cambodia and people ate anything they could get their hands, including insects (Rigby 2002; Byrne 2009; DuFord 2009; Midena 2010; Serath

2011). Now, assuming that there are around 30 vendors operating in Skuon area, and each of them can sell 50 spiders per day in average, then this gives an estimate of 1500 spiders sold daily at Skuon. Even if estimates were low, there is no doubt that a lot of spiders are sold every day, and there are crucial unanswered questions about both the sustainability of this trade and the potential ecological implications of overharvesting a large invertebrate predator (Yen and Ro 2013).

In the case of medicinal uses, in Brazil it is reported the use of the Goliath bird-eating tarantula *Theraphosa* spp. to treat "erysipelas" (or "Holy fire"), fortification of teeth and asthma (Costa Neto and Resende 2004; Costa-Neto 2006). In Mexico, in state of Chiapas, is reported that the ethnic group Tzotziles and Tzetzales use a "big spider" in their medicine and that a tarantula species (possibly *Tliltocatl vagans* (Ausserer, 1875)) is used to treat tumors (Hunn 1977; Enríquez et al. 2006). The traditional use of the tarantula *Tliltocatl vagans* by medicine men in the Mayan Chol community is also carried out. The people who present an illness called "aire de tarantula" with symptoms including chest pain, coughing, and asthma, are treated with a tarantula-based beverage (Machkour-M'Rabet et al. 2011). There is also an interest of biomedical research of tarantula venoms. Because the functional role of venom is to neutralize prey or predators, tarantula venoms are naturally rich sources of (1) ion channel-acting toxins (neurotoxins) affecting particular ion conductance, (2) presynaptic toxins affecting neurotransmitter release/exocytosis, and (3) post-synaptic toxins altering binding of neurotransmitters to their cellular targets (Saez et al. 2010). Theraphosid venoms are an untapped resource for reveal new toxins many of which may prove to have clinical uses. Among others, the main potential applications of spider toxins/peptides in therapy are for treating cardiovascular diseases, microbial infection, neurological disorders, epilepsy, arrhythmia, and pain (Escoubas and Rash 2004; Saez et al. 2010; Mourão et al. 2013; Nicholson 2013; Monge-Fuentes et al. 2015; Richards et al. 2018). However, other medical applications might exist due to the variety of biological actions of spider toxins/peptides (Nicholson 2013). Apart from those potential medical applications of spider compounds in humans, it should be noted that mygalomorphs, contain many insect-selective toxins that would be "leads" for developing novel biopesticides (Corzo et al. 2008; Hardy et al. 2013). Generally, these toxins are phyla-specific and target voltage-gated ion channels which are ubiquitous among insects. The molecular basis of toxin specificity (insect vs. mammal ion channels) is still poorly understood and need to be clarified for an effective development of insecticidal toxins (or derivatives). The later would help to produce non-peptide mimetics that opened the way to foliar sprays (Nicholson 2013; Pennington et al. 2018).

An unusual use of tarantulas is observed in Indonesia: some people utilize tarantulas as an ingredient for a hair dye (Schultz and Schultz 2009). More recently it was observed that the coloration in tarantulas is produced using different types of structural mechanisms, and those colors (especially blue) are present since ancient times in evolutionary history and its presence occurred many times during the evolution of the group. Tarantulas are thus a good model system to study structural color evolution under natural selection, bringing new perspectives and insights on our understanding of structural color evolution and possibly revealing new structural mechanisms or photonic nanostructure designs (Hsiung et al. 2015). Scientists from Karlsruhe Institute of Technology (KIT), in cooperation with international colleagues, have now succeeded in replicating nanostructures that generate the same color irrespective of the viewing angle (Hsiung et al. 2017). Microscopic analysis of blue tarantulas inspires production of nanostructures with bright colors that are independent of the viewing angle for textile, packaging, and cosmetic industries. In contrast to pigments, structural colors are nontoxic, more vibrant and durable (Landgraf 2016).

16.1.4 Economic Significance

Most people do not think of tarantulas as valuable. Even fewer people consider that they can have an economic or commercial value in various areas. Several places, products, and services use the name tarantula or some tarantula motif in their business name, logo, and advertising (Schultz and Schultz 2009). In general, when a reference is made to "trade," one usually does not think in live animals being traded, and it is likely that initially people thought about products made by industries, manufactured goods, fuels, merchandise or others. However, international trade in wildlife is a huge business that involves many billions of dollars a year and hundreds of millions of individuals of plants and animals (Reuter and Mosig 2010). The number of tarantula spiders sold and kept as exotic pets is increasing throughout many countries in the world (Rojo 2004; West 2005). The presence of tarantula dealers in many different countries and continents indicates that the market for live tarantulas is increasing internationally (Fig. 16.17). The international markets include Canada, United States, EU, and Asia, with demands exceeding the legal supply. Some big tarantula dealers within Canada, the United States, the EU, and Asia re-sell and re-export specimens to either final customers or other tarantula dealers. Tarantula dealers in Canada, United States, and EU stated they also purchase spiderlings from hobbyists from their own country who occasionally breed their specimens (Fig. 16.18). Tracking the species and the number of individuals generated by breeding done by amateurs is difficult because most of them do not keep records of

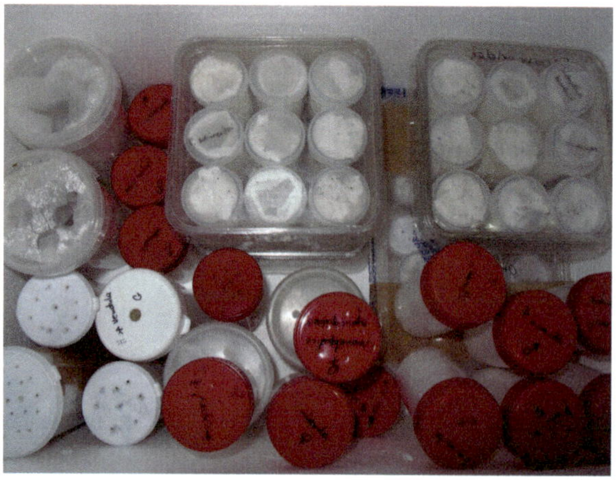

Fig. 16.17 Several tarantulas packed and ready to be sent for export. Photo: Jorge Mendoza

Fig. 16.18 Hundreds of spiderlings bred in captivity. Photo: Jorge Mendoza

their production. As spiderlings increase in size, their price rises accordingly, with adults reaching the highest prices (CEC 2017). The prices of the specimens vary accordingly to the demand in each country, the availability of the species and their ease in both their care and reproduction. However, one must be aware that these animals can also be object for illegal and unsustainable trade, so not all animals offered in pet trade are captive breed. The strengthening of capacities in reducing and controlling the flow of illegal and unsustainable tarantula trade can have benefits for a part of the local economy and the conservation of wild tarantulas and their habitat. The sustainable use and trade of tarantulas and even other arachnids will generate greater long-term benefits, preventing the local extinction of species and the degradation of their natural habitats. Illegal trade negatively affects initiatives of the local communities, can threaten wild populations in a faster way, and it does not provide any kind of incentive for local people to care about their protection.

16.2 Main Threats

16.2.1 Habitat Loss

Every year several new tarantulas' species are described and in the same way as most tropical arthropods, there are many species to be discovered or described (Reichling 2003). As theraphosids are generally long-lived, to understand the effects of harvesting on the population age structure and subsequently in population dynamics is important. For landscapes, the ongoing loss and fragmentation of the habitat have an obvious effect: when habitat is lost, dependent specimens are also likely to be lost and, populations decline. Habitat fragmentation isolates small populations in patches, which also increases the risk of extinction (Mantyka-Pringle et al. 2012) (Fig. 16.19). There is an urgency to conduct biodiversity studies focused on habitat dynamics, because burgeoning human populations worldwide and associated human

Fig. 16.19 Deforestation in the deciduous forest in the state of Michoacán, Mexico. The typical habitat of *Brachypelma baumgarteni*. Photo: Jorge Mendoza

Fig. 16.20 Land clearing for being used in agriculture, a big problem for many ecosystems. Photo: Jorge Mendoza

activities are rapidly eliminating natural habitats that sustain the biodiversity. Spiders are considered good indicators of habitat disturbance, not only for being commonly found in high numbers in almost all ecosystems, but also for depending on the habitat physiognomy for the construction of their webs or burrows (Uetz 1991). There are species of spiders associated with environments with a certain degree of modification (Shochat et al. 2004). Some spiders are related to urban environments and are indicators of anthropic impact (Desales-Lara et al. 2013). Other species benefit from human activity to expand their geographic dispersion and adapt to new environments with great ease, becoming introduced or invasive species that tend to displace the native ones, subsequently generating a loss of biodiversity in the environment (Lowe et al. 2017; Rozwalka et al. 2017). During the twenty-first century, there was a great interest on the study of spiders as environmental indicators (Taylor and Doran 2001). Tarantulas in general are sensitive to habitat loss. In the particular case of the American continent, the traditional slash and burn agriculture is being replaced by a more prejudicial practice in which the fallow periods are greatly shortened (Fig. 16.20). This eventually results in a patchwork of cleared lands not suitable for most wildlife, including tarantulas (Reichling 2003; Laurance 2010).

16.2.2 Impact of Illegal Trade

Anthropogenic activities such as agriculture, livestock farming, urbanization, hunting and poaching of wild species, have caused the modification, fragmentation and loss of natural ecosystems and its biodiversity (Sarukhán et al. 2012). Among these threats, the poaching of wildlife to meet illegal trade has a direct and irreversible impact. It is possible to observe its deleterious effects in the great decline that populations of species of high commercial value have shown in recent years (WWF 2012). The illegal traffic of wildlife is defined as the extraction, storage, transportation, commercialization, and possession of species of flora and/or wild fauna through hunting and collection in violation of national and international laws and treaties (Zimmerman 2003). In recent years, the illegal wildlife trade has increased significantly, probably because it represents an attractive illicit business with large profits and low risk (Nadal et al. 2013). Virtually all tarantula species of large sizes and/or vivid colorations are threatened due to the progressive increase in the demand of these specimens in the pet trade, or by being included in resins and used as paperweight or in other "souvenir" objects, both for the national and international market (Amat-García et al. 2007).

For years, the genus *Brachypelma* was considered one of the most trafficked tarantulas worldwide (Smith 1994; Rojo 2004; Mendoza and Francke 2017; Turner et al. 2018). Due to its bright coloration, docile behavior, longevity, and robustness in captivity, it has been shipped out of Mexico in large numbers over the years (Reichling 2003; Mendoza 2016). This problem stimulated the first step in theraphosid conservation, the inclusion of *Brachypelma smithi* in the CITES' Appendix II (CITES 1985). Posteriorly, the entire genus *Brachypelma* was included on CITES' Appendix II (CITES 1994), in order to prevent that its international trade threats wild populations (Longhorn et al. 2007). However, one negative result of listing *Brachypelma* in CITES was the decrease in the supply of some species, leading to increased desirability and thus demand. This has encouraged the development of a black market, which smuggles many other tarantula species both locally and abroad in different countries (Reichling 2003; Phillips 2009; Cooper 2011; Benjamin et al. 2012; Nanayakkara 2014; Ghione et al. 2017; Carranza 2018). The evolutionary history and taxonomy of many traded tarantulas is poorly understood. Specimens are found on pet trade often before being scientifically described and there is no formal knowledge to protect them (Turner et al. 2018). Tarantulas are collected by poor local people to help sustain their families. A skilled spider collector is able to dig up several hundred of spiders in a few months (Rigby 2002; Schultz and Schultz 2009). Unfortunately, collectors disturb the habitat and do not care about maintaining sustainable harvesting practices. Traffickers can pay to local collectors around \$2–3 US dollars per tarantula, which can be sold for \$200–500 dollars on the international market depending on the species (Inecc 2012; Mendoza pers. obs. 2014). Large tracts of land are damaged due to excavation of burrows and after rainy season parts of the habitat is lost in landslides. Smugglers have unique methods of facilitating their illicit activities, often hiding tarantula specimens in photographic film containers, small plastic tubes or plastic bags (West 2005). The animals are crammed against each other and many die due to dehydration, insufficient space during moulting, or suffocation in their own exuviae (Rojo 2004). An increasing number of countries, including Brazil, Mexico, Venezuela, Australia, South Africa, Costa Rica, have therefore banned the export of animals and plants without respective permits. Violations of these laws are punishable by substantial fines or even imprisonment (Cléton et al. 2015).

Scientists work against the clock to discover and describe new species. Ironically, in cases of commercially valuable taxa, publishing new species descriptions may inadvertently facilitate their extinctions (Stuart et al. 2006). On the other hand, the publication of imprecise information on the geographical distribution of biodiversity to avoid poaching hampers effective conservation decisions. Major efforts are underway to fill knowledge gaps, but there are increasing concerns that publishing the locations of species is dangerous. The scientific descriptions of some taxa with a potential to be exploited for the pet trade provide hobbyists or collectors with enough data and information about the location and habitat where these species are located. This can be especially damaging for those species whose distribution is very limited and or whose populations are subject to strong environmental pressures as a result of human activities and consequently the removal of individuals can have a huge impact on population maintenance. In the case of tarantulas, many species are taken out of their habitat and offered in the hobby even before they have been formally described, their natural populations status be known, their distribution or their evolutionary history (Burgman 2005; Stuart et al. 2006; Costello et al. 2013; Tulloch et al. 2018; Turner et al. 2018).

16.2.3 Natural Disasters

Very little is known about the conservation status of many tarantulas, especially those restricted to areas particularly vulnerable to overexploitation. In general, tarantulas tend to have clumped distributions in small areas with specific soil types, and are found only sporadically across their entire geographic range, making them vulnerable to natural stochastic events (Reichling 2003).

The arachnologist William Baerg (1958) reported a decline in a large population of *Aphonopelma crinitum* (Pocock, 1901) in a shrubland and mezquite area in Durango, Mexico. He observed a big aggregation of hundreds of individuals but after just 8 years, only a few tarantulas remaining. He returned to the same site 20 years later and then the whole population had disappeared. Baerg attributed the extinction of tarantula population to a chronic drought season in the area. Mendoza and Francke (2017) reported the nearly complete destruction of two *Brachypelma*

Fig. 16.21 Papagayo river flow after hurricane season in 2013. Photo: Jorge Mendoza

smithi populations during the hurricane season of 2012–2013, when the Mexican government reported serious environmental damage and changes to the flow of rivers along the Pacific coast (Fig. 16.21). The coast of the state of Guerrero was one of the most heavily affected areas, with extensive flooding of the Papagayo River (Mendoza and Francke 2017). These examples show that when local environmental conditions become untenable, tarantulas are among the least equipped of organisms to move to a better location and would probably face extinction (Reichling 2003).

16.3 Actions for Tarantula Conservation

The ways human beings benefit from the conservation of invertebrates are hard to quantify because there is so many ways in which we have benefits from ecosystem services directly or indirectly related with invertebrates and the general public is often unaware of them (Cardoso et al. 2011). Conserving wildlife and biodiversity in general cannot be, as everyone knows, a volatile cause, away from a practicable rationality. We only preserve what we appreciate and value, and this can vary accordingly to complex preferences of individuals, organizations, and institutions. The will and resources necessary for conservation can flow from the private sphere, public sector or from international sources, but only when there is shared information, identifiable benefits for the actors involved or sufficiently strong public opinion. The conservation of invertebrate species depends on a large extent on the knowledge of their biology. Without a basic knowledge it is difficult to qualify the degree of vulnerability or threat that a species is facing (Amat-García et al. 2007). The loss of predators such as spiders may cause the loss of ecosystem capacity to control pest outbreaks and the consequent loss in productivity (Landis et al. 2000; Symondson et al. 2002). The conservation of arachnids in general is only possible if the conservation of natural habitats is advocated. Measures aimed at the direct conservation of some specific species of spiders, as for example is done with some species of mammals or birds, are today unthinkable. The attention given to them, in general, is considerably lower than that given to many other life forms, both in relation to the number of people who dedicate their attention to them, and in relation to the amount of resources they are assigned in conservation priorities (Barrientos 2004). On the other hand, conservation is not an easy initiative that can be undertaken without the assistance of the governmental agencies in charge of protecting the environment. It is these national regulatory entities that would have the capacity of evaluating the conservation proposals of particular species.

16.3.1 Need of Studies With Wild Populations

Biogeography incorporates ecology, genetics, and geology to understand the history of a geographical area and its inhabitants. It also may help us to predict species responses to present and future pressures by discovering how species reacted to changes in their environment. Conservation biogeography involves the application of biogeographical principles and methods to conservation issues. The identification of areas of endemism is important for both historical biogeography and conservation (Ferretti et al. 2014). Collecting data on species distributions and habitat requirements also facilitates evaluation of a species conservation status, and of areas that should be preserved if the species is endangered (Smith 2016) (Fig. 16.22). Important conservation actions include protecting the natural habitat of tarantulas, creating management plans, and conducting systematic monitoring to inform the recovery or loss of subpopulations. It would be highly recommended start with those species that occur in areas already protected in order to help in conservation, suggesting the creation of any specific area just in case is need it for a particular unprotected species (Ferretti et al. 2012). Additionally, systematic monitoring and protection could be done in known populations. Further studies are needed to confirm population, harvest, and trade trends, and how the latter affect harvest levels. Hendrixson et al. (2014) mentioned that the most significant challenge in assessing

Fig. 16.22 Data collection of distribution and habitat of tarantulas in Mexico. Photo: Emmanuel Goyer

the need for conservation in some tarantula species is our general lack of knowledge about their natural history and population dynamics. In the case of the montane tarantulas, they concluded the paucity of samples in conjunction with a lack of long-term ecological data precludes researches from being able to fully assess the extent of these tarantulas' distributions or the changes in their population size, two of the most important criteria used by the IUCN Red List for considering a taxon 'threatened' (IUCN 2018). This is just one example of the necessity of obtaining as many information as possible from the species on field. The information generated by this kind of studies could allow the delimitation of the best areas for protection of tarantulas, establishing their relevance, endemism degree, and the frequency of taxa inside them (Ferretti et al. 2014). In addition, captive studies may also be useful to test the resilience of individuals to changes in salinity, moisture, and temperature, and perhaps a long-term captive study could assess whether there are alterations in fertility during periods of environmental change by evaluating the eggs per cocoon in females maintained under different environmental conditions. Boosted regression modelling may also be a useful approach to predicting distributions, particularly with regards to future climate change (Smith 2016).

16.3.2 Accurate Identifications

Most living species are still to be described (Mora et al. 2011). This problem is especially prevalent in invertebrates such as spiders, with researchers still far from agreeing on the possible number of species. As with several other invertebrates harvested for the exotic pet trade, the evolutionary history and taxonomic uniqueness of most tarantulas is not well known (Turner et al. 2018). For currently known and as for newly described species, accurate data on their geographic origin and natural distribution are vital to study how the species interact with their environment, how they coexist with other species, and especially, to consider what factors promote and maintain the uniqueness of an alleged new form (Longhorn 2014). Increased taxon sampling and the use of molecular characters will likely reveal new lineages and cryptic species throughout the Theraphosidae family (Hamilton et al. 2016; Montes de Oca et al. 2016; Mendoza and Francke 2017; Ortiz and Francke 2017; Turner et al. 2018). Such data will be especially important to fully understand the limits of species and the population dynamics of taxa with wide ranges and variable geographic habitats, and could have more implications for conservation (Longhorn et al. 2007). Modern taxonomy is more focused in solving high-level phylogenic questions using molecular techniques that require specialized skills and equipment than in dealing with species descriptions and diagnoses. The lack of taxonomists and the wider recognition by policymakers that to conserve biodiversity it may be important to know what biodiversity is present. Knowledge allows wise decisions and should guide priorities for best use of very restricted resources available for practical conservation (Cardoso et al. 2011). Well-crafted legislation includes mechanisms to extend protection despite taxonomic changes; initiatives such as Convention on International Trade in Endangered Species (CITES) and the

International Union for Conservation of Nature (IUCN) specialist groups already link taxonomy and its changes with conservation (Thomson et al. 2018). In the case of tarantulas, the genus *Brachypelma* is protected by CITES Appendix II to avoid overexploitation due to pet trade, and that initiative have been taken not only for their protection but for sustainable trade (CEC 2017). It is a good example of how to make conservation plans focused directly on tarantulas. With recent studies based on morphological and molecular data it has been observed that the species included in this genus actually belong to two completely different lineages, the so-called "red legs" (*Brachypelma sensu stricto*) and the one of "red rump" "(*Brachypelma sensu lato*) (Turner et al. 2018; Mendoza and Francke 2020). Thus, with the help of several specialists, assessments of the extinction risk of *Brachypelma* species were done and several species of the 'red legs' species met IUCN Red List's criteria for Vulnerable (VU) or Endangered (EN) (Fukushima et al. 2018) due to their limited geographic distribution on the Mexican Pacific coast (Locht 2008; Locht et al. 1999; West 2005; Mendoza and Francke 2017).

Changes in taxonomy do not have consistent and predictable impacts on conservation. According to Morrison III et al. (2009) there are some general trends on the impact of the taxonomic changes on conservation: (1) All of the researches where taxonomic change helped protection of species involve splitting. (2) Taxonomic change has least impact on the protection of iconic or charismatic organisms, protected areas of special status, and economically important groups. (3) Taxonomic progress can be detrimental to conservation when it involves species amalgamation (lumping) or reveals the hybrid nature of a species. Most of the problems for conservation resulting from the dynamic taxonomic process could be avoided if future conservation legislation followed the international conventions and explicitly referencing the specific taxon concept implied by a name, that is, by citing the original species description and the most recent scientific name. Taxonomists and conservation biologists should join forces to promote effective legislative mechanisms to deal with a changing taxonomy rather than engage in infighting about the proper way to do taxonomy (Thomson et al. 2018) (Fig. 16.23).

Fig. 16.23 Part of the scientific collection of tarantulas from the Oxford Museum. The data that can be obtained from the museological collections allow us to know important biological characteristics of organisms. Photo: Jorge Mendoza

16.3.3 Education and Sustainable Use

As wild flora and fauna do not respect political boundaries, an international environmental regulation is necessary. Following the launch of the World Conservation Strategy in Zoos, which will be carried out by the International Union for the Conservation of Nature (IUCN), the zoological community must devote its full potential to the conservation of species, their natural habitats and ecosystems (Rodríguez 2006). Environmental education is conceived as a strategy aimed at solving environmental problems that directly or indirectly affect the communities surrounding the areas of wildlife distribution. It must be focused on specific environmental problems, with the purpose of training the communities involved in these problems along the path of self-management and inter-institutional cooperation, fostering autonomy and civic leadership. Sustainable practices for the use of wildlife will generate greater long-term benefits, while preventing the local extinction of species and the degradation of natural environments (Robles de Benito 2009). Within these policies it is necessary to promote legal, sustainable, and traceable trade. Because tarantulas have increased their demand within the pet trade, they are a group that must be prioritized for the implementation of a sustainable use that allows to combat their illegal trade. In this context, Mexico is one of the few countries that regulates the sustainable use, conservation, and management of native tarantulas (CEC 2017). Mexican legislation gives priority to the use of tarantulas or other organisms in a sustainable manner, this can be extractive (hunting, commercial, subsistence, traditional rites and ceremonies, repopulation, and research) or non-extractive, as it happens with certain protected species or in some degree of threat (ecotourism, environmental education and research) (DOF 2000). In some cases, although the species are protected and are considered as a priority for conservation, such as the genus *Brachypelma*, commercial exploitation is possible when it is controlled, constantly monitored and with the possibility of designing compensatory measures as the reintroduction of some individual in the wild (Zamorano 2009; Mendoza and Francke 2017) (Fig. 16.24). The use of native tarantulas is carried out

Fig. 16.24 One of the largest tarantula breeding facilities in Mexico. Many Mexican species that are legally exported to other countries are breed and raised in places as it is. Photo: Jorge Mendoza

through the so-called system of Units for Management and Sustainable Exploitation of Wildlife (UMA), which are defined as registered properties and facilities that operate in accordance with an approved management plan and within which permanent monitoring is given to the state of the habitat and of the populations or specimens that are distributed there (CEC 2017) (Fig. 16.25).

A purely technical assessment of the resource management scheme that involves the construction of an UMA allows us to clearly appreciate its value for conservation, based on a concept that at first glance is simple: the owner of a specific property decides to join a program of management of one or several species in wildlife, making some profitable business such as legal trade or ecological tourism) and in return the owner has to preserve habitat portions in the area that still have suitable conditions to be occupied by populations of the species of the interest. Then, the UMA, in addition to promoting the productive reconversion of rural properties, bringing them closer to more sustainable models, tend to favor the existence of lands that contribute to the conservation of the ecosystems of the region where they develop (Reuter and Mosig 2010). Under the UMA program, qualified persons may

Fig. 16.25 Part of the exhibition of live tarantulas in the interactive museum of the UMA Aracneé. On this site they show tarantulas from different parts of the world and teach the general public why they shouldn't fear them. Photo: Jorge Mendoza

Fig. 16.26 Thousands of baby tarantulas are raised in Mexico to meet international demand. These breeding facilities compete against illegal trade so that wild populations can remain stable

be permitted to collect a limited number of wild tarantulas to keep and breed in captivity. The resulting offspring may then be sold domestically or exported (Fig. 16.26).

16.4 Public Policies for Management, Protection, and Conservation

It is considered essential that any country that proposes schemes for the sustainable use of natural resources for any purpose allocate the necessary resource to simultaneously promote the mechanisms, tools and control and monitoring capacity that ensure respect and application of the laws. Otherwise, there is a risk that a percentage of the legally established exploitation and management initiatives will become suitable establishments for the laundering of wild fauna and flora (Reuter and Mosig 2010). Like other spiders, tarantulas play many important roles in ecosystems as predators and food sources for other creatures, so their conservation is justified. However, although tarantulas are an integral part of biodiversity, they are not easily contemplated within this perspective. It is clear that, regardless of the scientific and technical arguments for their protection, the cultural weight flooded with prejudices and undervaluations can ruin the best reasons (Reichling 2003). Because of this, the sustainable use and management of tarantulas native to the Americas it will depend greatly on the knowledge of the main threats to its populations and the creation of public policies that involves the whole society for management, protection, and conservation actions. It is necessary to establish and develop national and regional initiatives that integrate, in a strategic manner, the scattered interests and efforts, in order to exponentiate the impacts in favor of conservation and the sustainable use of wild resources such as tarantulas. This will only be possible if there is a legal framework that allows it and is consistent with the initiatives, policies, and plans related to the subject. At the international level, instruments such as CITES can also be very useful tools, but only if they are applied appropriately. It is essential that any country that proposes schemes for the sustainable use of any type of natural resources allocate the necessary and sufficient funds to simultaneously promote the mechanisms, tools, the control and monitoring capacity that guarantee the respect and application of the relevant laws. In addition, strengthening capacities to reduce and control the flow of illegal and unsustainable wildlife trade will benefit local economies, their levels of well-being, and the conservation of wild species and their habitat.

References

Aguilar-López JL, Pineda E, Luría-Manzano R (2014) Depredación de tres especies de herpetozoos por arañas en la región tropical de Veracruz, México. Rev Mex Biodiv 85:965–968

Amat-García G, Andrade MG, Amat-García E (2007) Libro Rojo de los Invertebrados terrestres de Colombia. Instituto de Ciencias Naturales- Universidad Nacional de Colombia, Conservación Internacional Colombia Instituto Alexander von Humboldt, Ministerio de Ambiente, Vivienda y Crédito Territorial, Bogotá 204 pp

Azevedo CS, Smith M (2004) Araneofauna na região do Seringalzinho. In: Janelas para diversidade no Parque Nacional do Jaú. Uma estratégia para o estudo da biodiversidade na Amazônia, pp 135–142

Baerg WJ (1958) The Tarantula. University of Kansas Press, Lawrence, 85 pp

Barrientos JA (2004) Conservar y proteger a los arácnidos ¿cuales, cómo y por qué? Munibe 21:138–153

Benjamin SP, Nanayakkara RP, Dayananda SK (2012) The taxonomy and conservation status of the spiders (Arachnida: Araneae) in Sri Lanka. In: Weerakoon DK, Wijesundara S (eds) The National Red List 2012 of Sri Lanka; conservation status of the fauna and flora. Ministry of Environment, Colombo, pp 42–45

Bertani R (2012) Revision, cladistic analysis and biogeography of *Typhochlaena* C. L. Koch, 1850, *Pachistopelma* Pocock, 1901 and *Iridopelma* Pocock, 1901 (Araneae, Theraphosidae, Aviculariinae). ZooKeys 230:1–94

Bond JE, Stockman AK (2008) An integrative method for delimiting cohesion species: finding the population-species interface in a group of Californian trapdoor spiders with extreme genetic divergence and geographic structuring. Syst Biol 57:628–646

Bond JE, Hedin MC, Ramirez MG, Opell BD (2001) Deep molecular divergence in the absence of morphological and ecological change in the Californian coastal dune endemic trapdoor spider *Aptostichus simus*. Mol Ecol 10:899–910

Bond JE, Beamer DA, Lamb T, Hedin M (2006) Combining genetic and geospatial analyses to infer population extinction in mygalomorph spiders endemic to the Los Angeles region. Anim Conserv 9:145–157

Borges LM, Da Rosa CM, Franzoi Dri G, Bertani R (2016) Predation of the snake *Erythrolamprus almadensis* (Wagler,1824) by the tarantula *Grammostola quirogai* Montes De Oca, D'Elía and Pérez-Miles, 2016. Herpetol Notes 9:321–322

Breene RG (1998) The ATS arthropod medical manual: diagnoses and treatment. American Tarantula Society, Carlsbad, NM, 32 pp

Breet C (1993) Documentary: giant spiders of the lost world, Survival Anglia, Summit Television

Bristowe WS (1932) Insects and other invertebrates for human consumption in Siam. Trans R Entomol Soc Lond 80:387–404

Burgman MA (2005) Risks and decisions for conservation and environmental management. Cambridge University Press, New York, 488 pp

Byrne R (2009) Cambodian spiders under threat. Khmer Media. http://mediakh.net/business-and-eco/cambodian-spiders-under-threat. Accessed 8 Aug 2018

Campos JV, Almeida F (2016) A small homage to Maria Sibylla Merian, and new records of spiders (Araneae: Theraphosidae) preying on birds. Rev Bras Ornitol 24:30–33

Cardoso P, Erwin TL, Borges PAV, New TR (2011) The seven impediments in invertebrate conservation and how to overcome them. Biol Conserv 144:2647–2655

Carranza D (2018) Incautan en Colombia 23 tarántulas que iban al comercio ilegal en Seúl. Anandolu Agency. Retrieved from https://www.aa.com.tr/es/

CEC (2017) Sustainable trade in Tarantulas: action plan for North America. Commission for Environmental Cooperation, Montreal, 52 pp

CITES (1985) Amendments to appendices I and II of the convention. Cites Secretariat, Geneva

CITES (1994) Other proposals: to include all species (and subspecies) of the genus Brachypelma (also known as Euathus) in Appendix II. Cites Secretariat, Geneva

Cléton F, Sigwalt Y, Verdez JM (2015) Tarantulas, breeding experience & wildlife. Frankfurt am Main, Edition Chimaira, 231 pp

Cloyed CS, Eason PK (2017) Niche partitioning and the role of intraspecific niche variation in structuring a guild of generalist anurans. R Soc Open Sci 4:170060

Cocroft RB, Hambler K (1989) Observations of a commensal relationship of the microhylid frog *Chiasmocleis ventrimaculata* and the burrowing theraphosid spider *Xenesthis immanis* in southeastern Peru. Biotropica 21:2–8

Cooper R (2011) Smuggler caught at airport with 261 Mexican tarantulas shoved into his luggage. Daily Mail. Retrieved from https://www.dailymail.co.uk

Cooper EWT, West R, Mendoza J (2019) Identificación de tarantulas Listadas en la CITES: especies de *Aphonopelma, Brachypelma y Sericopelma*. Comisión para la Cooperación Ambiental, Montreal, 80 pp

Copperi S, Ferretti N, González A, Pérez-Miles F, Pompozzi G (2011) Arañas migalomorfas: ¿simpáticas mascotas o aterradores peligros? Ciencia Hoy 21:36–43

Corzo G, Diego-García E, Clement H, Peigneur S, Odell G, Tytgat J, Possani LD, Alagón A (2008) An insecticidal peptide from the theraposid *Brachypelma smithi* spider venom reveals common molecular features among spider species from different genera. Peptides 29:1901–1908

Costa FG, Pérez-Miles F (2002) Reproductive biology of uruguayan theraphosids (Araneae, Mygalomorphae). J Arachnol 30:571–588

Costa Neto EM, Resende JJ (2004) A percepção de animais como "insetos" e sua utilização como recursos medicinais na cidade de Feira de Santana, estado da Bahia, Brasil. Acta Sci 26:143–149

Costa-Neto EM (2006) Bird-spiders (Arachnida, Mygalomorphae) as perceived by the inhabitants of the village of Pedra Branca, Bahia State, Brazil. J Ethnobiol Ethnomed 2:50

Costello MJ, May RM, Stork NE (2013) Can we name Earth's species before they go extinct? Science 339:413–416

Coyle FA (1985) Ballooning behavior of *Ummidia* spiderlings (Araneae, Ctenizidae). J Arachnol 13:137–138

Coyle FA (1995) A revision of the funnelweb Mygalomorph spider subfamily Ischnothelinae (Araneae, Dipluridae). Bull Am Mus Nat Hist 226:1–133

Coyle FA, Icenogle WR (1994) Natural history of the Californian trapdoor spider genus *Aliatypus* (Araneae, Antrodiaetidae). J Arachnol 22:225–255

Csakany J (2002) Study on the chemical communication between the Microhylid frog, *Chiasmocleis ventrimaculatata*, and a Theraphosid Spider involved in a commensal relationship. Masters dissertation report submitted to State University of New York, 13 pp

Das KSA, Sreekala LK, Abdurahima O (2012) Predation on the Kelaart's Pipistrelle Bat, *Pipistrellus ceylonicus* Kelaart (Chiroptera: Vespertilionidae), by the Reddish Parachute Tarantula, P*oecilotheria rufilata* Pocock (Araneae: Theraphosidae), in Chinnar Wildlife Sanctuary, Kerala, India. Trop Nat Hist 12:257–260

Desales-Lara MA, Francke OF, Sánchez-Nava P (2013) Diversidad de arañas (Arachnida: Araneae) en hábitats antropogénicos. Rev Mex Biodiv 84:291–305

DeWet JI (1991) A revision of the genus *Ceratogyrus* Pocock (Araneae: Theraphosidae). MSc thesis, Rand Afrikaans University, Johannesburg

Dias C, Brescovit A (2003) Notes on the behavior of Pachistopelma rufonigrum Pocock (Araneae, Theraphosidae, Aviculariinae). Rev Bras Zool 20:13–17

Dippenaar-Schoeman AS (2002) Baboon and trapdoor spiders of Southern Africa: an identification manual. Plant protection research institute handbook no. 13. Agricultural Research Council, Pretoria, p 129

DOF (Diario Oficial de la Federación) (2000) Ley general de vida silvestre. Diario Oficial de la Federación 3 July 2000. http://www.diputados.gob.mx/LeyesBiblio/ref/lgvs/LGVS_orig_03jul00.pdf

Dor A, Hénaut Y (2011) Are cannibalism and tarantula predation, factors of the spatial distribution of the wolf spider *Lycosa subfusca* (Araneae Lycosidae)? Ethol Ecol Evol 23:375–387

Dor A, Calmé S, Hénaut Y (2011) Predatory interactions between *Centruroides* scorpions and the tarantula *Brachypelma vagans*. J Arachnol 39:201–204

DuFord D (2009) How to eat fried tarantula in Cambodia. http://www.worldhum.com/features/how-to-eat-fried-tarantulas-in-cambodia-20091201. Accessed 28 June 2018

Enríquez P, Mariaca R, Retana ÓG, Naranjo EJ (2006) Uso medicinal de la fauna silvestre en los altos de Chiapas, México. Interciencia 31:491–499

Escoubas P, Rash L (2004) Tarantulas: eight-legged pharmacists and combinatorial chemists. Toxicon 43:555–574

Escoubas P, Célérier ML, Nakajima T (1997) High-performance liquid chromatography matrix-assisted laser desorption/ionization time-of-flight mass spectrometry peptide fingerprinting of tarantula venoms in the genus *Brachypelma*: chemotaxonomic and biochemical applications. Rapid Commun Mass Spectrom 11:1891–1899

Ferretti N, Gonzalez A, Pérez-Miles F (2014) Identification of priority areas for conservation in Argentina: quantitative biogeography insights from mygalomorph spider (Araneae: Mygalomorphae). J Insect Conserv 18:1087–1096

Ferretti NE, Pompozzi G (2012) *Grammostola vachoni*. The IUCN red list of threatened species 2012: e.T18217620A18217711. Accessed 18 Aug 2018

Ferretti N, Pompozzi G, Copperi S, Pérez-Miles F, González A (2012) Mygalomorph spider community of a natural reserve in a hilly system in central Argentina. J Insect Sci 12:1–16

Foelix RF (1996) Biology of spiders, 2nd edn. Harvard University Press, Cambridge, 330 pp

Fukushima C, Mendoza J, West R, Longhorn S, Rivera-Téllez E, Henáut Y, Henriques S, Cardoso P (2018) *Brachypelma*. The IUCN red list of threatened species 2018: e.T8152A21416539. Accessed 28 Nov 2018

Garton JD, Mushinsky HR (1979) Integumetary toxicity as an antipredator mechanism in the narrow mouthed frog, *Gastrophryne carolinensis*. Can J Zool 57:1965–1973

Ghione S, Coelho L, Costa FG, García LF, González M, Jorge C, Laborda A, Montes de Oca L, Pérez-Miles F, Postiglioni R, Simó M, Toscano-Gadea C, Viera C, Aisenberg A (2017) Arácnidos prioritarios para la conservación en Uruguay. Bol Soc Zoológica Urug 26:1–8

Hamilton CA, Hendrixson BE, Bond JE (2016) Taxonomic revision of the tarantula genus *Aphonopelma* Pocock, 1901 (Araneae, Mygalomorphae, Theraphosidae) within the United States. ZooKeys 560:1–340

Hardy MC, Daly NL, Mobli M, Morales RAV, King GF (2013) Isolation of an orally active insecticidal toxin from the venom of an Australian tarantula. PLoS One 8:e73136

Hedin M, Starrett J, Hayashi C (2013) Crossing the uncrossable: novel trans-valley biogeographic patterns revealed in the genetic history of low-dispersal mygalomorph spiders (Antrodiaetidae, Antrodiaetus) from California. Mol Ecol 22:508–526

Hendrixson BE, Guice AV, Bond JE (2014) Integrative species delimitation and conservation of tarantulas (Araneae, Mygalomorphae, Theraphosidae) from a North American biodiversity hotspot. Insect Conserv Div 8:120–131

Hirsch BT (2009) Seasonal variation in the diet of ring-tailed coatis (*Nasua nasua*) in Iguazu, Argentina. J Mammal 90:136–143

Horstkottet J, Riesch R, Plath M, Jager P (2010) Predation by three species of spiders on a cave fish in a Mexican sulphur cave. Bull Br Arachnol Soc 15:55–58

Hsiung BK, Deheyn DD, Shawkey MD, Blackledge TA (2015) Blue reflectance in tarantulas is evolutionarily conserved despite nanostructural diversity. Sci Adv 1:e1500709

Hsiung BK, Siddique RH, Jiang L, Liu Y, Lu Y, Shawkey MD, Blackledge TA (2017) Tarantula-Inspired Noniridescent Photonics with Long-Range Order. Adv Opt Mater 5:1600599

Hunn ES (1977) Tzeltal folk zoology: The classification of discontinuities in nature. Academic Press, New York

Hunt RH (1980) Frog sanctuary in a tarantula burrow. J Nat Hist 89:48–53

INECC (2012) Tráfico ilegal de especies silvestres y sus impactos. Available at http://www.inecc.gob.mx/descargas/dgioece/2012_sem_trafico_pon01_alow.pdf. Accessed 9 Sept 2018

Isbister GK, Seymourb JE, Grayc MR, Raven RJ (2003) Bites by spiders of the family Theraphosidae in humans and canines. Toxicon 41:519–524

IUCN (2018) The IUCN red list of threatened species, Version 2018.2. Available online at http://www.iucnredlist.org. Accessed 29 Oct 2018

Kaiser II, Griffin PR, Aird SD, Hudiburg S, Shabanowitz J, Francis B, John TR, Hunt DF, Odell GV (1994) Primary structures of two proteins from the venom of the Mexican red knee tarantula (*Brachypelma smithi*). Toxicon 32:1083–1093

Klaas P (2007) Vogelspinnen völlig neu bearbeitete Auflage. Datz Terrarienbücher, Ulmer, Stuttgart, 142pp

Kingsley KJ (2016) Common, conspicuous, and curious arthropods of Zion National Park. Self-published as a pdf file. Available online at https://www.entsoc.org/sites/default/files/Arthropods%20of%20Zion%20National%20Park.pdf

Landgraf M (2016) Dressing up like a Peacock: bright colors by nanotechnology. Sci Daily 144:1–3

Landis DA, Wratten SD, Gurr GM (2000) Habitat management to conserve natural enemies of arthropod pests in agriculture. Annu Rev Entomol 45:175–201

Laurance WF (2010) Habitat destruction: death by a thousand cuts. In: Sodhi NS, Ehrlich PR (eds) Conservation biology for all. Oxford University Press, Oxford, pp 73–87

Locht A (2008) Estudio sobre la sistemática y distribución de la familia Theraphosidae (Arachnida, Araneae) en México. Facultad de Ciencias, Universidad Nacional Autónoma de México, Mexico City

Locht A, Yáñez M, Vázquez I (1999) Distribution and natural history of Mexican species of *Brachypelma* and *Brachypelmides* (Theraphosidae, Theraphosinae) with morphological evidence for their synonymy. J Arachnol 27:196–200

Longhorn SJ (2014) On type localities of Mexican tarantulas, either missing for Bonnetina or misleading for Brachypelma, with appeal for accurate biogeographic data. J Brit Tarant Soc 29:16–28

Longhorn SJ, Nicholas M, Chuter J, Vogler AP (2007) The utility of molecular markers from non-lethal DNA samples of the CITES II protected "tarantula" *Brachypelma vagans* (Araneae, Theraphosidae). J Arachnol 35:278–292

Lowe EC, Wilder SM, Hochuli DF (2017) Life history of an urban-tolerant spider shows resilience to anthropogenic habitat disturbance. J Urban Ecol 3:jux004

Machkour-M'Rabet S, Hénaut Y, Winterton P, Rojo R (2011) A case of zootherapy with the tarantula Brachypelma vagans Ausserer, 1875 in traditional medicine of the Chol Mayan ethnic group in Mexico. J Ethnobiol Ethnomed 7:1–7

Machkour-M'Rabet S, Hénaut Y, Calmé S, Legal L (2012) When landscape modification is advantageous for protected species. The case of a synanthropic tarantula, *Brachypelma vagans*. J Insect Conserv 16:479–488

Main BY (1978) Biology of the arid-adapted Australian trapdoor spider *Anidiops villosus* (Rainbow). Bull Br Arachnol Soc 4:161–175

Main BY (1987) Ecological disturbance and conservation of spiders: implications for biogeographic relics in southwestern Australia. In: Majer J (ed) The role of invertebrates in conservation and biological surveys. Western Australian Department of Conservation and Land Management Report, pp 89–98

Mantyka-Pringle CS, Martin TG, Rhodes JR (2012) Interactions between climate and habitat loss effects on biodiversity: a systematic review and meta-analysis. Glob Chang Biol 18:1239–1252

Marín-Martínez M, Rojas-Morales JA (2016) Predation by a Mygalomorphic Spider *Xenesthis immanis* (Araneae: Theraphosidae) on a Stream-dwelling Frog, *Rheobates palmatus* (Anura: Aromobatidae). Reptile Amphib 23:175–177

Markovics J (2013) White-nosed Coati: Racoon's cousin. Beaport, New York, 32 pp

Marshall SD (2001) Tarantulas and other arachnids. Barron's Educational Series, Hauppauge, NY, 109 pp

Mendoza J (2016) Las tarántulas endémicas: problemática y conservación. In: Mendoza J (ed) La Biodiversidad de la Ciudad de México, vol II. Conabio/Sedema, Mexico City, pp 245–248

Mendoza JI, Francke OF (2017) Systematic revision of *Brachypelma* red-kneed tarantulas (Araneae: Theraphosidae), and the use of DNA barcodes to assist in the identification and conservation of CITES-listed species. Invertebr Syst 31:157–179

Mendoza J, Francke O (2020) Systematic revision of Mexican threatened tarantulas Brachypelma (Araneae: Theraphosidae: Theraphosinae), with a description of a new genus, and implications on the conservation. Zool J Linnean Soc 188:82–147

Menin M, Rodrigues DJ, Azevedo CS (2005) Predation on amphibians by spiders (Arachnida, Araneae) in the Neotropical region. Phyllomedusa 4:39–47
Meyer-Rochow VB, Changkija S (1997) Uses of insects as human food in Papua New Guinea, Australia, and North-East India: cross cultural considerations and cautious conclusions. Ecol Food Nutr 36:159–187
Midena K (2010) Spider hunting the latest craze in Cambodia. http://www.news.com.au/travel/news/spider-hunting-the-latest-craze/story-e6frfq80-1225880970301. Accessed 09 Aug 2018
Molur S, Daniel BA, Siliwal M (2008) *Poecilotheria*. The IUCN red list of threatened species. e.T63563A12681959. Accessed 18 Aug 2018
Monge-Fuentes V, Medeiros FM, Alves GA, Silva JC, Mayer A, Carneiro L, Coimbra J, Soares K, Mortari MR (2015) Neuroactive compounds obtained from arthropod venoms as new therapeutic platforms for the treatment of neurological disorders. J Venom Anim Toxins incl Trop Dis 21:31
Montes de Oca L, D'Elía G, Pérez-Miles F (2016) An integrative approach for species delimitation in the spider genus *Grammostola* (Theraphosidae, Mygalomorphae). Zool Scr 45:322–333
Mora C, Tittensor DP, Adl S, Simpson AGB, Worm B (2011) How many species are there on Earth and in the Ocean? PLoS Biol 9:e1001127
Morrison WR III, Lohr JL, Duchen P, Wilches R, Trujillo D, Mair M, Renner SS (2009) The impact of taxonomic change on conservation: Does it kill, can it save, or is it just irrelevant? Biol Conserv 142:3201–3206
Mourão CBF, Oliveira FN, Carvalho AC, Arenas CJ, Duque HM, Gonçalves JC, Macêdo JKA, Galante P, Schwartz CA, Mortari MR, Almeida MFM, Schwartz EF (2013) Venomic and pharmacological activity of *Acanthoscurria paulensis* (Theraphosidae) spider venom. Toxicon 61:129–138
Münke C (2012) Informal market and potential contribution to rural livelihood (preliminary findings from a field survey) Khon Khaen, Regional Symposium. Available online at http://www.fao.org/fileadmin/templates/rap/files/meetings/2012/120531_ib3.pdf
Nadal L, Carmona A, Trouyet M (2013) Illegal wildlife trafficking. Secretariat of Environment and Natural Resources, Mexico, 32 pp
Nanayakkara RP (2014) Tiger spiders "*Poecilotheria*" of Sri Lanka. Biodiversity Secretariat, Ministry of Environment & Renewable Energy, Colombo, 167 pp
Nicholson GM (2013) Spider peptides. In: Kastin AJ (ed) Handbook of biologically active peptides, 2nd edn. Academic Press, California, 2032 pp
Nunes GSS, De-Carvalho CB, Dias DM, Magina GCT, Carvalho CM (2010) *Micrurus ibiboboca* (Caatinga Coral Snake). Predation. Nat Hist Notes Herpet Rev 41:367–368
Nyffeler M, Knörnschild M (2013) Bat predation by spiders. PLoS One 8:e58120
Nyffeler M, Moor H, Foelix RF (2001) Spiders feeding on earthworms. J Arachnol 29:119–124
Ortiz D, Francke OF (2017) Reconciling morphological and molecular systematics in tarantulas (Araneae: Theraphosidae): revision of the Mexican endemic genus *Bonnetina*. Zool J Linnean Soc 180:819–886
Paoletti MG, Dufour DL (2005) Edible Invertebrates among Amazonian Indians: a critical review of disappearing knowledge. In: Paoletti MG (ed) Ecological implications of Minilivestock. Science Publishers, Enfield, NH, pp 293–342
Paulsen M (1999) A Harpactira burrow reveals the secrets of its occupant. J Br Tar Soc 15:14–23
Pennington MW, Czerwinski A, Nortonb RS (2018) Peptide therapeutics from venom: current status and potential. Bioorg Med Chem 26:2738–2758
Pérez-Miles F, Perafán C (2017) Behavior and biology of mygalomorphae. In: Viega C, Gonzaga MO (eds) Behavior and ecology of spiders. Springer, Cham, 448 pp
Pérez-Miles F, Costa FG, Toscano-Gadea C, Mignone A (2004) Ecology and behaviour of the 'road tarantulas' *Eupalaestrus weijenberghi* and *Acanthoscurria suina* (Araneae, Theraphosidae) from Uruguay. J Nat Hist 39:483–498
Phillips T (2009) London petshop owner try to smuggle 1000 spiders. The Guardian. Retrieved from: https://www.theguardian.com

Pizzi R (2006) Spiders. In: Lewbart G (ed) Invertebrate medicine, 1st edn. Wiley, Ames, IA, 327 pp
Pizzi R, Carta L, George S (2003) Oral nematode infection of tarantulas. Vet Rec 152:695
Powell R, Little LW, Smith DD (1984) Eine Wohngemeinschaft von *Physalaemuspustulosus* (Cope, 1864) (Salientia: Leptodactylidae) mit einer bodenbewohnenden Vogelspinne. Salamandra 20:273–274
Ramírez-Castaño VA, Robledo-Ospina LE, Gutiérrez-Cárdenas PDA (2014) *Hyloscirtus palmeri* (Palmer's Treefrog) Predation. Herp Rev 45:304
Rankin W, Walls JG (2008) Manuales del terrario. Tarántulas y escorpiones. Editorial Hispano Europea, 64 pp
Raven RJ (2010) A review of the Mygalomorphae: biology, morphology and systematics. Book of abstracts of the 18th international congress of arachnology, 11–17 Jul 2010, Siedle, Poland
Reagan DP, Waide RB (1996) The food web of a tropical rain forest. University of Chicago Press, Chicago, IL, 623 pp
Reichling SB (2003) Tarantulas of Belize. Krieger Publishing, Florida, p 127
Reuter A, Mosig P (2010) Comercio y aprovechamiento de especies silvestres en México: observaciones sobre la gestión, tendencias y retos relacionados. Available online at: https://www.traffic.org/site/assets/files/10139/comercio-y-aprovechamiento-de-especies-silvestres-en-mexico.pdf
Richards KL, Milligan CJ, Richardson RJ, Jancovski N, Grunnet M, Jacobson LH, Undheim EAB, Mobli M, Chow CY, Herzig V, Csoti A, Panyi G, Reid CA, King GF, Petrou S (2018) Selective NaV1.1 activation rescues Dravet syndrome mice from seizures and premature death. Proc Natl Acad Sci 115:E8077–E8085
Rigby R (2002) Cambodia: eating tarantulas with the spiderwomen of Skuon. Available at http://www.rhymer.net/tdarchive28b.htm. Accessed 16 Nov 2019
Robles de Benito R (2009) Las unidades de manejo para la conservación de la vida silvestre y el Corredor Biológico Mesoamericano México. CONABIO, Mexico, 134 pp
Rödel MO, Brede C, Hirschfeld M, Schmitt T, Favreau P, Stöcklin R, Wunder C, Mebs D (2013) Chemical camouflage, a frog's strategy to co-exist with aggressive ants. PLoS One 8:e81950
Rodríguez C (2006) Directrices generales para la Conservación ex situ de fauna silvestre en parques zoológicos y acuarios de Colombia. Ministerio de Ambiente, Vivienda y Desarrollo Territorial de Colombia, Bogotá, 80 pp
Rojo R (2004) Las Tarántulas de México: pequeños gigantes incomprendidos. Biodiversitas 56:7–11
Rozwalka R, Dawidowicz L, Wawer W (2017) Three alien spider species (Araneae: Theridiidae) newly found in Poland. Fragm Faun 60:61–66
Rudloff JP (2008) Zur Kenntnis der Gattung *Phormictopus* Pocock, 1901 – einem Faunenelement der Grossen Antillen (Mygalomorphae: Theraphosidae: Theraphosinae). Arthropoda 15:2–51
Saez NJ, Senff S, Jensen JE, Er SY, Herzig V, Rash LD, King GF (2010) Spider-venom peptides as therapeutics. Toxins 2:2851–2871
Sarukhán J, Carabias J, Koleff P, Urquiza-Haas T (2012) Capital natural de México. Acciones estratégicas para su valoración, preservación y recuperación. Comisión Nacional para el Conocimiento y Uso de la Biodiversidad, México, 95 pp
Schmidt G (2003) Die Vogelspinnen: eine weltweite Übersicht. Neue Brehm-Bücherei, Hohenwarsleben, 383pp
Schultz SA, Schultz MJ (2009) The tarantula keepers guide. Barrons Educational Series, New York, 384 pp
SEMARNAP (1995) Norma Oficial Mexicana NOM-059-ECOL-1994, que determina las especies, subespecies de flora y fauna silvestres terrestres acuáticas en peligro de extinción, amenazadas, raras y las sujetas a protección especial y que establece especificaciones para su protección. Gaceta Ecológica, México 7:8–72
Serath (2011) Fried tarantula, a popular delicacy in Cambodia. http://khmerization.blogspot.com/2011/11/fried-tarantula-popular-delicacy-in.html. Accessed 22 Aug 2018
Shochat E, Stefanov W, Whitehouse M, Faeth S (2004) Urbanization and spider diversity: influences of human modification of habitat structure and productivity. Ecol Appl 14:268–280
Siliwal M, Ravichandran B (2008) Commensalism in Microhylid frogs and mygalomorph spiders. Zoos' Print 23:13

Smith AM (1990) Baboon spiders Tarantulas of Africa and The Middle East. Fitzgerald, London, 142 pp
Smith AM (1994) Tarantula spiders: Tarantulas of the U.S.A. and Mexico. Fitzgerald, London, 411 pp
Smith VR (2016) An unexpected journey: the biogeography and conservation ecology of the trapdoor spider genus *Cantuaria* Hogg, 1902. Doctor of Philosophy thesis, Lincoln University, 253 pp
Streicher JW, Fajardo GR, Vázquez-Almazán CR (2011) *Plestiodon sumichrasti*. Predation Herp Rev 42:607–608
Stuart BL, Rhodin AGL, Grismer LL, Hansel T (2006) Scientific description can imperil species. Science 312:1137
Sunderland K, Samu F (2000) Effects of agricultural diversification on the abundance, distribution, and pest control potential of spiders: a review. Entomol Exp Appl 95:1–13
Symondson WOC, Sunderland KD, Greenstone MH (2002) Can generalist predators be effective biocontrol agents. Annu Rev Entomol 47:561–594
Taylor RJ, Doran N (2001) Use of terrestrial invertebrates as indicators of the ecological sustainability of forest management under the Montreal Process. J Insect Conserv 5:221–231
Teyssié F (2015) Tarantulas of the world. NAP Editions, Verrières-le-Buisson, p 487
Theron KJ (2017) Conservation of spider diversity within an agricultural mosaic: insights from the Greater Cape Floristic Region, biodiversity hotspot. Master Thesis, Faculty of Agri Sciences, Stellenbosch University
Thomson SA, Pyle RL, Ahyong ST, Alonso-Zarazaga M, Ammirati J, Araya JF (2018) Taxonomy based on science is necessary for global conservation. PLoS Biol 16:e2005075
Toledo LF (2005) Predation of juvenile and adult anurans by invertebrates: current knowledge and perspectives. Herp Rev 36:395–400
Tulloch AIT, Auerbach N, Avery-Gomm S, Bayraktarov E, Butt N, Dickman CR, Ehmke G, Fisher DO, Grantham H, Holden MH, Lavery TH, Leseberg NP, Nicholls M, O'Connor J, Roberson L, Smyth AK, Stone Z, Tulloch V, Turak E, Wardle GM, Watson JEM (2018) A decision tree for assessing the risks and benefits of publishing biodiversity data. Nat Ecol Evol 2:1209–1217
Turner SP, Longhorn SJ, Hamilton CA, Gabriel R, Pérez-Miles F, Vogler AP (2018) Re-evaluating conservation priorities of New World tarantulas (Araneae: Theraphosidae) in a molecular framework indicates nonmonophyly of the genera, *Aphonopelma* and *Brachypelma*. Syst Biodivers 16:89–107
Uetz GW (1991) Habitat structure and spider foraging. In: Bell SS, Mccoy ED, Mushinsky HR (eds) Habitat structure: the physical arrangement of objects in space. Chapman and Hall, London, 464 pp
Verdez JM, Cléton F (2004) Tarántulas. Introducción, Guía y Reproducción. Reptilia ediciones, España, 205 pp
Vieira WLS, Gonçalves MBR, Nóbrega RP (2012) Predation on *Tropidurus hispidus* (Squamata: Tropiduridae) by *Lasiodora klugi* (Aranea: Theraphosidae) in the semiarid caatinga region of northeastern Brazil. Biota Neotrop 12:1–3
West RC (2005) The *Brachypelma* of Mexico. J Br Tar Soc 20:108–119
West RC, Nunn SC, Hogg S (2012) A new tarantula genus, *Psednocnemis*, from west Malaysia (Araneae: Theraphosidae), with cladistic analysis and biogeography of Selenocosmiinae Simon 1889. Zootaxa 3299:1–43
Wilson JB (1999) Guilds, functional types and ecological groups. Oikos 86:507–522
Wilson JS, Gunnel CF, Wahl DB, Pitts JP (2012) Testing the species limits of the tarantulas (Araneae: Theraphosidae) endemic to California's Southern Coast Ranges, USA. Insect Conserv Div 6:365–371
World Spider Catalog (2020) World Spider Catalog Version 21.5. Natural History Museum Bern. Available online at http://wsc.nmbe.ch. Accessed 12 Sep 2020
WWF (2012) The fight against illicit wildlife trafficking: a consultation with governments. Dalberg, London

Yen AL, Ro S (2013) The sale of tarantulas in Cambodia for food or medicine: is it sustainable? J Threat Taxa 5:3548–3551

Zamorano P (2009) La flora y fauna silvestres en México y su regulación. Estudios Agrarios 15:159–167

Zimmerman ME (2003) The black market for wildlife: combating transnational organized crime in the illegal wildlife trade. J Trans Law 36:1657–1689

Chapter 17
Tarantulas, Gods and Arachnologists: An Outline of the History of the Study of New World Theraphosid Spiders

Andrew M. Smith

Abstract Tarantula spiders have repelled much of humankind since the beginning of recorded time. But not all. For a minority the spider is fascinating. This chapter is all about such individuals. Firstly from the Curiosity Cabinets of Europe's trading elite—later in the spirit rooms of the new grandiose cathedrals of science that had sprung up in the cities of both the Old World and the New World. By 1900 no museum collection could be called complete—without a tarantula spider on display. From the earliest Neolithic image of a spider daubed on an Egyptian rock face, amongst sacred symbols that no longer have meaning—to the iconic image of a copper plate engraving of a Bird Eating Spider, by the seventeenth-century naturalist and artist Madam Maria Sibylla Merian—this chapter traces the history of natural history. From the Victorian scientists of the heyday of European research, to the exciting DNA studies being undertaken today by a new generation of arachnologists—the heart of all that research remains one of the most awe-inspiring animals to ever walk the face of this planet—The Tarantula Spider.

17.1 History of the Study of New World Theraphosid Spiders

The Mayan knew them as *Chihua*, the Aztecs *Tocatl*, the Hopi *Gogyeng Sowuhti* and the first Europeans settlers: *tarantulas*. Tarantulas, Gods and Arachnologists traces the 300 year history of the study of the Theraphosidae of the New World, from the first tarantula spider *Phalangium americanum* to be described by Carolus Clusius at Leiden University in 1611, to Jorge Mendoza and Oscar Francke's, Systematic revision of *Brachypelma* red-kneed tarantulas (Aranaea: Theraphosidae) (2017) and the use of DNA barcodes to assist in identification and conservation of CITES listed species.

Spiders, particularly tarantula spiders, have fascinated and repelled humankind for thousands of years. The earliest image that we have of an arachnid is of a spider

A. M. Smith (✉)
Retired, Research Associate Natural History Museum, London, England

Bexhill-On-Sea, England

F. Pérez-Miles (ed.), *New World Tarantulas*, Zoological Monographs 6,
https://doi.org/10.1007/978-3-030-48644-0_17

and a fly from a Neolithic rock painting in the Gasulla Gorge, Castellon, Spain (Hancock and Hancock 1992). This was an image created by a hunter gatherer thousands of years ago and which reflects I suspect, the respect shown by one hunter for another hunter that transcends species. By the seventeenth century, the new copperplate printing technology had enabled the production of magnificent natural history books and the creation of the eye-catching, classic image of a tarantula spider eating a hummingbird. An image, which proved so powerful, that it quickly spawned in all of the major European languages, the term Bird-Eating Spider.

At the same time that a human hand was drawing the outline of a spider on a rock face, the human tongue was giving birth among hunter gather communities, to the earliest camp-fire Creation stories. Stories that told of Divine Spiders, which bridged the great void between the heavens and Earth with silken ladders, and whose egg sacs contained the beginnings of mankind. These stories gradually morphed into the female divinities that are the beginning of the Earth Mother cults, in which spinning, fertility, and procreation are quickly turned into spider imagery.

In the New World, The Great Goddess was to be eclipsed by the arrival of jaguar and serpent gods and the rise of an aristocratic male priesthood, which spawned in turn warlike deities such as Head-Hunting Tarantula, whose images began to appear on the stone carvings of the temple and civil complexes. All of which, probably reflects the increasingly expansionist nature of these Mesoamerican and South American native Indian cultures (Burger 1995).

And yet, it is possible that The Great Goddess still held sway in the furthest corners of the temple complex at Teotihuacan, where the stylised images of a tarantula spider appear on the stone gate posts of the temple of an agricultural fertility goddess, alongside butterflies and frogs. Karl Taube (1983) described this goddess as *"The Teotihaucan Spider Woman"*. My own observation is that the painting is of a goddess of agriculture and the spider imagery simply links her to the coming of the rains. But what is noteworthy is that although she is surrounded by araneomorph spiders and butterflies, it is a tarantula spider that performs the role of a guardian on the door posts.

In North America, the earliest folklore of the native Indian cultures of the South West initially linked Spider Woman to Creationist stories but, over time, the goddess was gradually downgraded to a trickster. At best, she is the tribes grandmother, who is the embodiment of a nomadic people's garnered wisdom, passed down through the generations in the form of humorous camp-fire stories. At worst, she is a malevolent spirit (Fig. 17.1) which waits in the wilderness to torment those who stumble upon her burrow (Talayesva 1942).

All of these cultures had their own names for the giant, earth-bound spiders, which priests and shamans linked humankind with the ghosts and spirits of the underworld. In the language of the Aztecs, the spiders were called *Tocatl*, in Mayan *Chihuo*, among the Hopi *Gogyeng Sowuhti*, but it was the name given to these spiders by the Conquistadors, who sailed from Europe, that was to become the universal world term for the theraphosid spider:—*Tarantula*.

The name hails from Taranto in Italy, but was also in common usage in Spain. In both countries, the bite of the spider was associated with hysteria and melancholic

Fig. 17.1 Tarantula Woman tricks and shames a proud young warrior into giving her not only his weapons but even his clothes. Hopi and Navajo legend. Drawing by Steven Kirk. Copyright: author's collection

trances which, it was commonly believed, could only be cured by music and dancing. So disruptive of civil life and commerce were these rowdy and often ribald processions that in its heyday in the 1560s, (Hillyard 1994) the Catholic Church was convinced that what it was witnessing was a revival of the pagan rites of Bacchus and strived hard to contain them (Smith 1994).

From the fifteenth century onwards, we witness the discovery, exploration and exploitation by Europe of the New World, initially by Spain and Portugal and then increasingly by England, France and Holland, all eager to share the fabled treasures of the Americas. As is often the way in human history, this period of exploration, imperial expansionism, military conflict and intense economic activity, allied with the growth of the printing press and religious rivalry, acted as a catalyst for the study of science and technology. This was the embryonic period of science that saw the growth of independent enquiry and the gradual move away from a slavish adherence to the works of Aristotle. This was a period, when exploration, discovery and economic expansion meant that the very building blocks of biological science were laid. The classification and cataloguing of the world's flora and fauna, began in earnest.

From the sixteenth century onwards, Spain's expeditions to the New World told of Mexican temples, the fancied gold of Eldorado, and the real gold of Peru and Mexico. But increasingly, they also returned not just with gold and silver, but with examples of the flora and fauna: potatoes, tomatoes, parrots, monkeys and even the indigenous native peoples. It is highly probable that a preserved example of a giant spider would have first appeared in the Courts of Spain and Portugal at that time. They would have been called *tarantulas*, because that is what the Conquistadors would have called these giant spiders, convinced that their large size could only mean that they were as dangerous as the fabled spiders of Taranto. Unfortunately, many of the historical archives of Madrid and Lisbon were destroyed by fires during the twentieth century, which means that we have to look further afield to northern Europe for the first records of these spiders. Here, a growing passion for curiosity cabinets among Europe's elite trading families encouraged the collection of exotic fauna and flora. And what is more exotic than a giant spider!

The first record and image that we have of a tarantula spider is in *Exoticorum Libri Decem* by Carolus Clusius (1611), which has a woodcut of a theraphosid spider. Described as *Phalangium americanum*, this name predates Linnaeus (1758) and so is not recognised. The same woodcut appeared in three other publications over the next 40 years:—Joannes de Laet (1633), Piso and Marcgrave (1648) and Worm (1655). Why the same woodcut? Because, in order to both save and make money, publishers sold woodcuts on to other publishers.

Carl Linnaeus believed them to be the same spider and listed them in 1758 in the tenth edition of *Systema Naturae* as *Aranea avicularia* (the pink/orange toe arboreal tarantula that ranges from the Amazon to Suriname). Unfortunately, closer examination of these classic books and the woodcut images would indicate that Linnaeus was wrong. The descriptions in Clusius (1611) and de Laet (1633) are of a specimen housed at the University of Leiden, but now lost. The description indicated that the collection site was *Sanctos Baya, de Todos Los,* which is a bay near the city of Salvador in the state of Bahia, Brazil. The specimen was probably collected by a Dutch mariner or slave trader while visiting one of the free trade ports that sprung up regularly along this section of the Brazilian coast during the seventeenth century, trading sugar, rum and molasses for slaves. Fukushima and Bertani (2017) speculated that the spider was of the genus *Pachistopelma* (Fig. 17.2) a common arboreal theraphosid of the region, which is plausible, because the ship, after unloading its human cargo, would have had to fill its water barrels and re-provision, which meant the spider may have turned up in firewood. It is also likely that the ship's officers would have been invited to dine at the plantation homes of wealthy local dignitaries, where they may have had the opportunity to see these large spiders in a curiosity cabinet, the keeping of which had become increasingly popular. Which, meant that they would have realised that these zoological curiosities were of commercial value. Another factor, which supports Fukushima and Bertani's suggestion, is that the outline of the woodcut does look remarkably similar to the stumpy shape of a *Pachistopelma* spider.

Nevertheless, it is unwise to look at a sixteenth-century woodcut and interpret it as an accurate illustration of the spider. This particular woodcut would have been copied by an engraver from an illustration made by Carolus Clusius or one of his students. The same woodcut was then used by Joannes de Laet 20 years later in his 1633 Latin edition of *The History of the New World.* As de Laet was the successor of Clusius at Leiden University, it is likely that he examined the same spider and for

Fig. 17.2 *Pachistopelma bromelicola* Bertani, R. 2012. This specimen was collected north of Salvador and is a common spider in the region. Alongside the west African spider, *Stromatopelma calceatum* (Fabricius, 1793) it is likely that a *Pachistopelma* spp. was one of the very first tarantula spiders to be borne by ships engaged in the slave trade, to the curiosity cabinets of northern Europe. Copyright: author's collection

Fig. 17.3 *Lasiodora parahybana* Mello-Leitao, 1917. An endemic tarantula in the Recife region of Brazil, which is capable of living for over 20 years and, although not reaching *Theraphosa blondi* proportions, can grow to a considerable size. Copyright: author's collection

the sake of economy used the same woodcut in his book. But this cannot be said of the classic publication *Historia Naturalis Brasiliae* (Piso and Marcgrave 1648), where, once again, the very same woodcut turns up.

We know from three independent images (Eckhout 1641; Jacob Marrell 1645; Zacharius Wagenaer circa 1640) that the spider in this publication is probably the large terrestrial theraphosid spider *Lasiodora parahybana* Mello-Leitao 1917 (Fig. 17.3). Nevertheless, probably for reasons of economy, the publisher decided to use the earlier woodcut instead of commissioning a new woodcut or a more expensive copper-plate engraving.

The second image which we have of a tarantula spider, appeared in 1641, in the corner of Albert Eckhout's famous painting (displayed in the National Museum of Copenhagen) of a Tapuya Indian with spears. Albert Eckhout (1604–1679), alongside William Piso (1611–1678) and George Marcgrave (1610–1644) were members of a small group of artists and scientists, who accompanied Johan-Maurits (1604–1679), count of Nassau-Siegen, when he was appointed governor-general of the short-lived Dutch occupation of Brazil. It has always been presumed that the image is of the large tarantula spider, *Theraphosa blondi* (Latreille 1804), but the common terrestrial theraphosid of the region, is *Lasiodora parahybana*, which is also capable of growing to a very large size. This is partly confirmed by the existence of a contemporary journal by Zacharius Wagenaer (1614–1668), which contained copies of Eckhouts field sketches, including an image of a tarantula spider. Wagenaer was a professional soldier, who may have acted as military escort to the scientists and artists and thus had ample opportunity to observe and copy their work. Nevertheless, it cannot be discounted that it is likely that specimens of *Theraphosa blondi,* from Browns Berg in Suriname, (Fig. 17.4) would have begun to reach Holland's curiosity cabinets by this period, as the colonies sugar plantations had expanded far up the River Surinam and were within easy reach of this collection site. This observation is important, as there are a number of art historians who believe that Eckhout completed the final painting in Holland, from preparations sketches undertaken in Brazil (Buvelot 2004:32). In which case, the spider that was chosen by Eckhout to appear in painting could have easily have come from a Dutch or German curiosity cabinet.

William Piso (1611–1678), the co-author of *Historia Naturalis Brasiliae* (1648), may have been led to believe by his sponsor, Johan-Maurits, count of Nassau-Siegen, that the publication was to have had copper plates, not inferior woodcuts. This may be surmised by the fact that we have a fascinating sketch in Maria Sibylla Merian's St Petersburg *Studienbuch* of a tarantula spider, which was drawn, not by

Fig. 17.4 *Theraphosa blondi* (Latreille, 1804). It is likely that specimens of this spider, collected in Dutch Suriname and French Guyana would have appeared very early on in the European collections. Holland, in 1640, also had trading station at Para, on the Amazon, which would have been another source of these giant spiders. Copyright: author's collection

her, but by her stepfather, the artist and engraver Jacob Marrel (see Merian 1976). Dated 1645, 3 years before Merian was born, the preparatory sketch was commissioned by Joannes de Laet, who was acting as the editor for Piso & Marcgrave's, *Historia Naturlalis Brasiliae*. The Dutch text, beneath the illustration (translated here) also reveals that the spider was brought back to Holland as a live specimen: "*Jacob Marrel made in 1645 in Leiden, this spin has been eating bread for 2 years. General Erdtschocktzky has donated her Mr Johan de Lait*". Note: This was possibly General Arciszewski/Artichofski (a senior administrator in Dutch Brazil) who had returned to Europe by this date to take up a new appointment in Poland in 1646. Which begs the question: were live tarantula spiders being shipped into Europe in the seventeenth century alongside parrots, monkeys and other exotic animals? Up to now, it has always been presumed that only preserved arachnid material was being brought back to Europe for the university and curiosity cabinet collections. The spider in the image is black, with red hairs, which means it was not *Lasiodora parahybana* but possibly *L. klugi* (C. L. Koch 1841) (Fig. 17.5).

The woodcut appeared for its final time in Ole Worm's (1655) catalogue of his Museum Wormianum, in Copenhagen. In this case, Worm's publisher was simply seeking a cheap woodcut of a theraphosid spider to illustrate a library catalogue.

By 1666, the Clusius woodcut had finally had its day, and was replaced by a new woodcut of a tarantula spider, commissioned by Frederick III, Duke of Holstein-Gottorp, for Adam Oleurius's (1666) catalogue of the of the Gotorp library. Under the supervision of Adam Oleurius, Frederick's cabinet had grown considerably larger with the acquisition in 1651 of the collection of the Dutch scholar and physician Bernardus Paludanus. It is likely that the spider illustrated in the 1666 catalogue, hails from this collection. The spider illustrated in this woodcut is a terrestrial theraphosid, and, if it was from a Dutch collection then it would have been collected in Dutch Brazil in the region of Recife, or at the short lived Dutch trading post at

Fig. 17.5 *Lasiodora klugi* (C. L. Koch, 1841). Although not described until the mid-nineteenth century, the tarantula illustrated by both Jacob Marrel and Zacharius Wagener does not appear to be *L. parahybana*, but possibly another large theraphosid spider, which is common in the region *L. klugi*. Copyright: author's collection

Belem. If not Brazil, then in Dutch Suriname. We know the likely provenance of the spider but what we cannot tell from the image is its identity.

This explosion of interest in curiosity cabinets, in which the new, wealthy trading classes could display not only preserved exotic flora and fauna, but also their own education and intellectual prowess, meant that there soon sprang up an industry to supply this lucrative market with both specimens and richly illustrated publications.

One of the best known figures of the period, associated with the boom in curiosity cabinets, was the Dutch/German publisher and illustrator Maria Sibylla Merian. She is primarily known for her 1705 publication, *Metamorphosis Insectorum Surinamensium,* which contained wonderful copper-plate engravings of the butterflies that she had discovered over the 2 years she had spent in the Dutch colony of Suriname. It also contained the classic—one could even say iconic image of an *Avicularia avicularia* spider raiding the nest of a hummingbird (See Chap. 4). The illustration shows one spider emerging from a silk tube/sock web and seizing a large ant, and a second spider in a nest with a dead hummingbird in its grasp. The text accompanying the illustration and separate text found as notes in *Studienbuch* (now housed in St Petersburg) are the first observations that we have of a tarantula spider made by a naturalist in the field (Fig. 17.6).

> "They do not spin long webs as some travellers would have us believe. They are covered in hair all over and are supplied with sharp teeth, with which they give deep dangerous bites, at the same time injecting a fluid into the wound. Their habitual food and prey is ants, who find it difficult to escape them as they move over the trees. These spiders have eight eyes; with two they see upwards, with two downwards, with two to the right and two to the left. When they fail to find ants they take small birds from their nests and suck all the blood from their bodies (Merian 1705, legend to plate 18.; see Merian 1976).

> These large spiders make a tissue [she does not say web] like caterpillars on Couyabe trees. When you rip up the tissue they get out. It seems a harmful animals due to its sharp teeth".

She was also the first naturalist to observe that spiders, *"moult from time to time like caterpillars"*.

It is from this period that the term Bird-Eating Spider began to be used increasingly in Protestant Europe, gradually replacing the southern European name tarantula.

A century later, both Merian and this image fell out of favour in the Age of Reason, which followed the French Revolution. In this new, perceived era of radicalism and reasoned logic, the image was held to be the fanciful fabrication of a woman traveller prone to hysteria. Despite the fact that a second traveller Palisot de Beavois witnessed a similar incident in 1805, Count Langsdorf in his 1812 publication *Expedition into the Interior of Brazil* stated that he *"disputed its truth"* (Savory 1961). We then have the remarkable scenario, over the next century, of various authorities striving to prove that Merian's observation was a fiction.

In 1834, the Australian naturalist W. S. Macleay published the following observation: "*...the genus Mygale, of which several and enormous species exist in Cuba, cannot possibly catch birds because it spins no net; because it lives during the day*

Fig. 17.6 *Avicularia avicularia* (Linnaeus, 1758). This image illustrates the thick silk tube webs that this arboreal tarantula spins on the side of trees. This spider is endemic throughout Suriname and is common in even the parks of the country's capital:—Paramaribo. Copyright: author's collection

in holes, or in tubes sometimes three foot deep in the earth, which generally open up under stones and where certainly no Humming-bird can get at it; and finally, because Mygale is itself too inactive in its motions and humbly keeps too close to its mother earth to be able to get near a Hummingbird" (Macleay 1834:191).

In 1842, he added in a letter published *in The* Annals of the Magazine of Natural History—"*I will even go so far as to add my utter disbelief in the existence of any bird-catching spider. In short, my conviction is, that Madam Merian has painted a falsehood: and that her naïve followers have to hastily placed confidence in her idle tales"* (Macleay 1842: 325; McKeown 1963).

Eventually Merian's (1705) description was corroborated by the observation of the field collector and naturalist Henry Bates in his classic book *The Naturalist on the River* Amazons (1863). Here, he recorded that at Cameta, *"I chanced to verify a fact relating to the habits of a large hairy spider of the genus Mygale, in a manner worth recording. The species was Mygale avicularia, or one very closely allied to it; the individual was nearly two inches in length of the body but the legs expanded seven inches and the entire body was covered in coarse grey and reddish hairs, I was attracted by a movement of the monster on a tree trunk: it was close beneath a deep crevice in the tree, across which was stretched a dense white web. The lower part of the web was broken and two small birds, finches were entangled in the pieces; they were about the size of an English siskin. One of them was quite dead, the other lay under the body of the spider not quite dead and was smeared by the liquor or saliva. I drove away the spider and took the birds, but the second one soon died*" (Bates 1863: 86).

As an aside, Bates swiftly followed up the first observation with a second, which makes him I believe, the first naturalist to record the presence of urticating hairs on a theraphosid spider. "*The hairs with which they are clothed come off if touched, and cause a peculiar and almost maddening irritation. The first specimen that I killed and prepared was handled incautiously, and I suffered terribly for three days afterwards*" (See Chap. 9).

But what is not commonly known is that Madam Merian was one of the leading traders in exotic fauna in Europe during this period. Her daughter and husband supplied her with material from Suriname, while a second son-in-law supplied further material From Ceylon/Sri Lanka. This trade is illustrated in business correspondence between Merian and Mr Volckamer of Amsterdam dated October 8, 1702 (Wettengl 1998: 264):

> "I have also brought with me all the animals comprised within this work, dried and well preserved in boxes, so that they can be seen by all. I currently still have jars with liquid containing one crocodile and many kinds of snakes and other animals as well as about twenty round boxes of butterflies, beetles, humming-birds, lantern flies and other animals which are for sale. If the gentleman desires to have them he need only so order. I also have other people in America who catch such animals and send them to me to sell and I hope to receive things from the Spanish West Indies, as soon as the way has opened to allow ships to there."

She also added to the October 8th letter:

> "I will soon be in a position to send East Indian fauna, for my youngest daughters husband has gone there as chief surgeon and will do his utmost to find as many things as possible".

From this source, she procured from Sri Lanka the first known examples of the beautiful Indian and Sri Lankan tarantula spider, *Poecilotheria* Simon 1885 and was probably the source of the specimen illustrated in Albertus Seba's (1734) Locupletissimi rerum naturalium *thesauri*. [Republished 2001 by Taschen as a facsimile edition under the title, *Cabinet of Natural Curiosities*]. When examined, the layout of the plate (Tome 1. Tabula 69) was obviously heavily influenced by Merian's bird-eating spider illustration.

Another letter, simply dated October, gives some idea of her method of preservation:

> I take it you wish to have 5 boxes of West Indien insects at 3f per box, which I am enclosing here. I have coated them and sealed them with turpentine oil in the presence of Mr Schey

The Albertus Seba collection, which was purchased by Tsar Peter the Great in 1718, contained specimens of both the Suriname and Sri Lankan spiders *Avicularia*

avicularia and *Poecilotheria vittata*, all of which we can safely presume was supplied by Merian. In my view, it is likely that the majority of the royal collections of the period would have contained material supplied by Merian, whom we know from correspondence was specifically sought out by agents acting on behalf of both Sir Hans Sloane and Peter the Great (Wettengl 1998).

This period, and the Age of Reason, have often been described as the birth of science but, for biologists, it is primarily the age of classification and collection at a time when there was no effective method of preservation. It is a horrible truth that barely nothing remains from that period, unless the specimen was placed in a sealed, pest-proof container. In 1806 and 1809, huge bonfires were held in the grounds of the British Museum, whereby much of the Sloane collection was destroyed. Many private collections simply rotted away. A decade later, a large part of the eighteenth-century British Museum, reptile spirit collection (preserved in Boyles spirit of wine) was buried in the grounds of the museum, the curators of the period having discovered the spirit of wine was not an effective long-term method of preservation (Stearn 2001).

I am often asked after a lecture why Linnaeus (1707–1778) frequently listed the illustrations and engravings contained in the natural history books of the period, when describing a given animal. The explanation is simple: only by linking a scientific name and description with a published illustration was it possible to have an effective method of identification. The concept of linking a scientific description with a type specimen could only effectively take root with the development of industrial alcohol in the early nineteenth century, and later the discovery of formaldehyde by the Alexander Butlerov in 1859 (Cato and Jones 1991).

In January 2006 Ray Gabriel, Richard Gallon and myself were privileged to be able to examine two historical theraphosid spiders, labelled as *Avicularia avicularia* held in the Linnaean Society collection in London. To our astonishment, on examination, these very fragile dried specimens were identified by Gallon as *Stromatopelma calceatum* (Fabricius 1793) from the Gold Coast and *Harpactira atra* (Latreille 1832) from Cape Town. Both specimens, hailing from key trading ports (associated with either the western slave trade triangle or the eastern spice trade) would have been common in the collections of the period, but it does give the reader an understanding how one should take nothing for granted when examining these historical collections. As an aside, Jean Christian Fabricius (1745–1807), who described *S. calceatum,* was a student of Linnaeus and may have even examined these particular specimens. (Gabriel et al. 2007).

When Linnaeus described *Avicularia avicularia* in 1758, it is highly probable that it was not a physical specimen of the spider on his desk, but a copy of Madam Merian's 1705 *Metamorphosis Insectorum Surinamensium.* Fukushima and Bertani (2017) proposed that a dried historical specimen housed in the Linnaean Collection at the Museum of Evolution of Uppsala University, be designated as the lectotype of *Avicularia avicularia*. It was also tentatively identified as hailing from the Amazon. There is no doubt that Linnaeus examined this material and that it warrants a lectotype designation, but it is highly unlikely that the specimens came from Brazil as suggested in the paper. With Europe torn apart by religious conflict, it is improbable that natural history specimens would have been collected from the catholic colonies of Spain and Portugal for the curiosity cabinets of protestant northern Europe. It is much more likely that any *Avicularia* material housed in the northern European collections of this period, would have hailed from the Dutch colony of Suriname (Fig. 17.6) and was probably sourced from Merian, who was actively marketing natural history specimens.

One year after the 1904 Sixth International Congress of Zoology in Berne, the newly founded Commission on Zoological Nomenclature issued *Regles internationales de la Nomenclature zoologique,* which chose the 1758 (tenth) edition, of Linnaeus's *System Naturae* as the date from which systematic zoology using binominal nomenclature could begin. Before that date, with a handful of exceptions (such as Carl Clerck's, *Svenska Spindlar* 1756, which also used binominal nomenclature), all earlier systems of nomenclature were deemed to be redundant. This also solved the problem of striving to identify historical species dating back to the beginning of the sixteenth century when at best, there was only a primitive engraving and a brief description.

Linnaeus (1758) assigned 12 genera to the Insecta Aptera, a broad grouping which we would now call arthropods: Lepisma, Podura, Termes, Pediculus, Cancer, Monoculus, Oniscus, and Scolopendra are insects, crustaceans and myriapods. Acarus, Phalangium, Aranea and Scorpio belong to the Arachnida, which were going to have to wait another half century before being recognised as a specific grouping. Nevertheless, 1758 was the beginning of systematic arachnology.

We are now at the gates of a period in time which has come to be known as the Age of Reason. A period, which was dominated by the political and economic turmoil of the French Revolution and the Napoleonic Wars. A period that was also one of immense technological change and which has been aptly described as the birth of science (Savory 1961).

For this period, we now have to switch from Sweden, the home of Linnaeus and Clerck, to late eighteenth century France and specifically Paris. Merian had gone to Suriname in 1699 to touch the face of God and illustrate his divine creations. One hundred years later, the philosophical biologist Jean Baptiste Lamarck (1744–1829) was debating transformism and the very beginnings of organic evolution, which he presented in his 1809 publication *Zoological Philosophy*. But from an arachnologists viewpoint, it is his 1801 *System des Animaux sans Vertebres*, which is much more important. In this work, for all intents and purposes he continued the work of Linnaeus by taking the Insecta Aptera and subdividing it by creating Classe Troisieme, Les Arachnides. Unfortunately, we then have the problem that he divides this group into Palpistes and Antennistes; in other words creatures with palps and those with antennae, which is not a particularly helpful designation for those seeking a standalone arachnid grouping. Nevertheless, it was Lamarck who first created the name Arachnides, derived from the Greek name for spider: Arachne, whose origins, in turn were derived from the mythological story of Arachne and Pallas, as narrated in the Roman poet Ovid's *Metamorphoses*. During this period, Ovid, like Caesars Gallic Wars was used as a children's Latin primer, and would have been familiar to any child who had enjoyed a classical education. Lamarck was also responsible for the scientific terms; *biology, invertebrate*, and *vertebrate.* He also gave us *Avicularia*, which was to prove more problematic:—of which more anon.

If Lamarck's *System des Animaux sans Vertebrates* (1801) can be said to be a natural development of Linnaeus's *System Naturae,* then the publications of the French biologists Latreille and Walckenaer, which were to follow, defined Paris as the birthplace of Scientific arachnology. From this point we are looking at academic works which are not only striving to understand and classify the natural world, but also create diagnostic methodology.

At the end of the eighteenth century, despite the fact that a separate *"Classe: Les Arachnides"* now existed, it still only contained one spider genus:—the *Aranea* of Linnaeus. What is more, that single genus had acquired 670 species, although from the perspective of this paper only one of these was a theraphosid spider species:—*Aranea avicularia* Linnaeus 1758.

Pierre Andre Latreille (1762–1833) and Charles Athanasius Baron Walckenaer (1771–1852) were to change that. Between them, in the first half of the nineteenth century they produced an astonishing body of work, which not only created new genera but also introduced nomenclature and diagnostic terminology, which is still valid today.

It would be fascinating to know on what terms these two giants of French arachnology functioned, bearing in mind that they were living in the same city at the same time and whose publications often duplicated each other. Unfortunately no correspondence is known to have survived which gives us any idea of the relationship between these two men. What we do know is that they were very different men from very different social backgrounds. Walckenaer can be essentially defined as a wealthy, gentleman naturalist and a renowned man of letters and cartography. Latreille was a professional zoologist, member of the Academie des Sciences and the first man to hold a chair of entomology, but who was so poorly paid that at one point he lodged in rooms in the museum. Yet in the world of zoology in early nineteenth century Paris, it was Latreille who dominated arachnology and only after his death in 1833, would it seem that Walckenaer regained the intellectual freedom and initiative to publish as he pleased on arachnids. As an insight into Latreille's personality, we know that when he was working under Lamarck he was unable to accept the older man's controversial theory of transformism. Like Cuvier, he believed in the immutability of species and the role of a divine and all wise life force. All of which indicates that he was deeply influenced by his theological training as a young man. Yet, it was to be Latreille who delivered Lamarck's graveside eulogy when most others had forsaken this controversial old man of French science (Burkhardt 1973).

Nevertheless, the first step in the development of arachnology as a discipline was made by Walckenaer, who proposed in a memorandum published in 1802:—*Memoire lu a la Societe sur le nouveau genre Mygale [Memoire ou l'auter separe les Mygales; je na'ai pas su le trouver],* the need to separate the Mygales from the Araignees.

By today's criteria this proposed genus is somewhat confusing in that it included both burrowers/*mineuses* and arboreal/*aviculaires* spiders. Which immediately raises the question of what actually was understood to be a genus in the early 1800s? This problem is the elephant in the room that dogged much of Walckenaer's early work. In his two volume work, *Fauna of Paris/Fauna Parisienne*, (1802) where he described 131 spiders, he commented that their differences did not seem great enough to warrant the creation of new genera (Savory 1961:54). Nevertheless, 40 years later Walckenaer's epic work—the four volume Histoire naturelle des Insectes Apteres, (1837–1847), occupies a permanent place in arachnid history because of the large number of still valid species and genera first described in it. The new world theraphosid species, *Avicularia versicolor, Ephebopus murinus, Grammostola rosea, Lasiodora saeva, Homoeomma nigrum and Tapinauchenius sanctivincenti* appear for the first time in 1837. He died in Paris in 1852.

Walckenaer essentially came into his own after Pierre Andre Latreille's death in 1833, but it was Latreille (who was primarily a systematist), who dominated the first three decades of the nineteenth century. Based on Walckenaer's 1802 publication, he created the first genera which marks the beginning of the systematic study of spiders. It was Latreille who took the name Mygale and designated it as a specific genus. What followed can best be described as a tsunami of publications. Between 1802 and 1805 we have 14 volumes of *Histoire naturelle generale et particuliere des Crustaces et des Insectes*. From 1806 to 1809 four volumes of Genera *Crustaceorum et Insectorum*. More works followed and in 1832, shortly before his death, he published a final paper that described the new world theraphosids, *Citharacantus spinicrus*, and *Cyrtopholis bartholomaei*. But among theraphosid researchers, he will be remembered as the arachnologist who described the iconic tarantula spiders *Theraphosa blondi, Phormictopus cancerides* and Sri Lankan species *Poecilotheria fasciata*. All of which, in 1832, would have been known as *Mygale blondi, Mygale cancerides* and *Mygale fasciata*.

Although the period was essentially dominated by Latreille and Walckenaer, other academic researchers were working in Paris on theraphosid spiders. The voyager, traveller and man of letters, Antoine Guillaume Oliver (1756–1814), published an article in 1811 which made an attempt at defining the new genus *Mygale*, which he concluded were ground burrowing spiders. Arboreal tarantulas such as *avicularia* reverted back to Linnaeus's *Aranea* genus. This was obviously not a satisfactory situation, which brings us back to Lamarck (1818), who in the *Araneides* volume of an 8 volume work:—*Histoire naturelle des animaux sans vertebres*, interceded by arguing that burrowers should be categorised as *Mygales* and arboreal theraphosid spiders should be placed into a new genus, which he named *Avicularia*.

Unfortunately, Lamarck's standing in the European scientific community was extremely poor and this publication was essentially ignored. One of the most powerful scientific personalities of the period, Baron Georges Cuvier (1769–1832) loathed and despised him. There is a story that at the end of his life, frail and blind he attended a lecture at which Cuvier was present and presented a short paper on blind cave fish, arguing that they had lost their eyes due to disuse. All of which fitted into his controversial and widely derided theory of transformism. Cuvier publicly taunted Lamarck with the words:—"*Perhaps your own refusal to use your eyes to look at nature properly has caused them to stop working?*" Lamarck's daughter, Cornelie is supposed to have replied, "*Have no doubts father, posterity will honour you*" (Milner 1990:264).

Although at this point, it should be noted that Walckenaer, as early as 1805, was possibly thinking along the same lines, as suggested in *Tableau des Araneides ou Caracteres essentiels des tribus genres families et races que renferme le genre Aranea de Linne*. Here, he argued *Mygale blondi* be placed in a new grouping called *Theraphosa*, while Linnaeus's *avicularia* should remain in Latreille's new genus *Mygale*. Unfortunately, Walckenaer's failure to specifically create a new genus meant that he is not recognised as the author of *Theraphosa*.

Sadly, for the science of arachnology, the next generation of arachnologists, Carl Hahn and Carl Ludwig Koch remained firmly wedded to the idea of *Mygale*, as a single super genus and simply ignored the work of Walckenaer, Latreille and Lamarck. It was not until the early 1870s, with the arrival of Tord Teodor Thorell and Anton Ausserer on the European arachnological scene, that Walckenaer's *Theraphosa* and Lamarck's *Avicularia* were rediscovered, re-described and utilised in an updated format:—Thorell 1870 and Ausserer 1871.

After the publication of Walckenaer's 1837 volume of *Histoire naturelle des animaux sans vertebres*, the world of the theraphosid arachnohistorian switches from France to Germany, with the realisation of the monumental publishing project *Die Arachniden*. Despite the fact that Walckenaer lived on for another decade, he published nothing more on the theraphosidae. Why? I suspect that he had simply exhausted the limited collection of tarantula spiders that were in the French collections of the period. When you look at the material described, much of it comes from the French colonial possessions of French Guiana and Caribbean. We see a handful of material from South America and the Far East (probably collected by French traders), but with defeat of Napoleon, stiff British competition and the rise of an independent New World, French dreams of an extensive colonial empire became increasingly limited to Africa and the Far East. The New World became increasingly the fiefdom of British engineers and German coffee planters.

But we do have the lone voice of one French field naturalist:—a Mr M. Goudet, who in 1817 had newly returned from French Guiana. He would greatly assist Latreille in the writing of his excellent account of the behaviour of *Avicularia avicularia*. "*They establish their domicile in the clefts of trees, under the bark, in the fissure of rocks or between leaves of various plants. The cell of the Mygale avicularia has the form of a tube, narrowed into a point at its posterior extremity. It consists of a white web, of close, very fine texture, semi-diaphanous and resembling muslin. One of them presented to me by M. Goudet, when unrolled was about two decimetres on length and six centimetres in breadth. The cocoon of the same species was the figure and size of a large walnut. It appears that the young are hatched in it and undergo there first change of tegument there. The naturalist just mentioned that he had taken a hundred of them from a single cocoon*" (Cuvier 1817:287).

Our focus now switches from France to Germany and the astonishing arachnological publishing project by Carl Wilhelm Hahn (1786–1836) and Carl Ludwig Koch (1778–1857). The sheer scale of the project and the large number of new theraphosid spider species described in these publications gives us some idea of the wealth of new material that could now be found in both the German state and private collections. The reason for this was probably due to the fact that large numbers of well-educated Germans (having no colonial possessions themselves at this point), worked abroad as civil servants, commodity traders, engineers and agriculturists for other foreign powers. Many German agriculturists ran or owned coffee plantations throughout South America and British Ceylon. Other German's sold machinery or traded in commodities. All of which gave potential collectors the opportunity to collect.

One typical such fellow was John Nietner (1820–1874), who arrived in Ceylon/Sri Lanka in the mid-1840s. He initially worked at the Peradeniya Botanical Gardens under George Thwaites (another enthusiastic spider collector for the European museums), before he acquired his own coffee plantation and wrote books on coffee pests. Like his mentor, he also became a phenomenal field collector of theraphosid spiders (Green 1912 p. 81). Numerous other John Nietners could be found all over the world.

What is fascinating is the lengths to which they went and the dangers they faced in their bid to collect specimens. Karl Ludwig Schmarda (1819–1908) was a Ph.D. and Professor of zoology with a comfortable post at the University of Prague. Yet at the age of 34 he resigned his post and with limited funds embarked on an epic, whirl-wind tour, which had him robbed in Panama, losing his collections in a fire in Chile and being so poor that he was suffering from scurvy in South Africa. He survived and the material that he collected can be found in museum collections throughout Germany (Dictionary of German Biography 2005 8:230). Many others did not survive. Henry Walter Bates (1825–1892) wrote in a letter to the Zoologist in 1852, *"The yellow fever is still bad here: this week all of the crew of a Hamburg vessel in port, died one by one—the captain, two mates, cook and men: worse than Africa"* (Bates 1852:353).

Coffee plantations need more than just agriculturists and one such man was Theodor Cordua (1796–1857), civil servant, coffee machinery salesman and commodities seller, who operated out of Suriname between the years 1819 and 1841 before going bankrupt and joining the California Gold Rush. He also added to his funds by selling natural history specimens, one of which was to become the type specimen of *Tapinauchenis plumipes*, described by Koch in 1842, now housed in the Berlin Natural History Museum (Smith 2013:86).

The description is to be found in the 1842 volume of *Die Arachniden*. A 16 volume work (2000 pages and 563 colour plates), which Pierre Bonnet described *as a marvel for its age*. The British arachnologist Theodore Savory added. *"It was the first work to bear a name which embraced the whole of the newly established class of arachnids—and may be said to mark a stage in the history of arachnology"* (Savory 1961:55).

The principal author of this great work is Carl Ludwig Koch, but he was not the catalyst. That honour goes to an eccentric, now forgotten figure, who between the years of 1820 and 1836 was responsible for a shambolic series of beautifully illustrated booklets, which when complete make up Carl Hahn's *Monographie der Spinnen*. It is so rare that it is simply known by the names of the libraries which are fortunate to possess a copy. This first volume also contains images of the four key tarantula spiders that we may deem to have been common in the European collections of the period. These were *Mygale avicularia, Mygale blondi, Mygale cancerides* and *Mygale fascicata.*

Professor Paolo Brignoli (1942–1986) was of the opinion that Hahn's pioneering work was the blueprint for *Die Arachniden* (Brignoli 1985).

We have a wonderful letter of reference from his employer, Count Friedrich von Puckler, "*This person is indeed extremely foolish, but is incidentally polite, draws exceedingly well, has a beautiful hand and has much knowledge of natural history and botany and is skilled in stuffing birds. However because of his foolishness he requires strict supervision. His other talents make him worthy of consideration*" (Hahn and Sacher 1988:115).

Unfortunately, *Monograph der Spinnen* was a publishing disaster and Hahn switched to a new publisher ZEH to embark on the new arachnid project *Die Arachniden.* A terse statement in the first instalment in August 1831 makes his position clear, *"I have explained in the advertisement for this work that my Monograph of Spiders (of which five booklets have appeared since 1820, published by the bookseller Herr J. Lecher) will no longer be produced because neither I nor the titular customer will live to experience the end of it, due to the extremely slow appearance of the booklets".*

On Hahn's early death in 1836, Carl Ludwig Koch was invited by ZEH to take over the project. Like Hahn, Koch was an artist, but unlike Hahn who was essentially an ornithologist, Koch was a botanist. Consequently there is a sudden change in the layout of the illustrations. With Hahn we have the traditional entomological approach of a straight forward, flat dorsal view, which was probably based on the need to draw a fragile dried specimen pinned in a cabinet tray. Koch ignored the traditional approach and instead opted for a perspective that is normally associated with vertebrate or botanical subjects. This was probably due to the fact that more and more specimens were being preserved in industrial standard alcohol, which meant that they could be manipulated. Hahn was probably responsible for all of the illustrations contained in the first two volumes, but the task of illustrating the remaining 14 volumes would have been enormous, and variations in the style of the illustrations would indicate that Koch could draw upon the skills of other illustrators. The colouring, done by hand, would have been undertaken by female piece workers and children. In Germany, the bookseller Salomon Schinz trained children from the local orphanage to colour *Anleitung zu der Pflanzenkenntnis.* The low wages and repetitive nature of the work did not encourage high standards. When Sir Joseph Hooker complained about poor colouring, the publisher swiftly admonished him thus, "*Print-colourists are not artists and cannot do artists work and nor will the work permit artists pay*" (Desmond 2003:22).

The final volume of *Die Arachniden* was published in 1848, but before his death 9 years later in Nuremberg, he published *Uebersicht des Arachnidensystems* (1851). This publication is important in that it demonstrates how far Koch had moved away from the idea of a super genus (*Mygale*) and begun the process of dividing up the theraphosidae into specific genera. The genus *Lasiodora* makes its first appearance in this paper. We may speculate that this revision may have been prompted by an increasing awareness of the existence of a paper published by William Macleay

(1835), which observed that the genus name *Mygale* was occupied by a rodent genus, created by Cuvier in 1800 to house *Sorex moschatus* (Macleay 1835:187). With this discovery the researchers of the day were forced to re-examine the revision work of what had gone before, and it was at this point that Lamarck's *Avicularia* genus once more came to the fore—but not in Koch's revision. Lamarck's *Avicularia* would have to wait until 1871 with the arrival of a talented young Austrian arachnologist Anton Ausserer, before being universally recognised as a valid name.

Unfortunately, despite the fact that a large number of the tarantula spiders are illustrated in the 1842 volume, only a handful of Koch's species are recognised today:—the more well-known being, *Chaetopelma olivaceum, Lasiodora klugi, Acanthoscurria geniculata* and *Tapinauchenius plumipes*. The problem being that a large number of the descriptions were based on specimens housed in private collections which are now lost, and although the illustrations and physical descriptions are far superior to anything that came before them, it is still not possible to identify many of the spiders. There is also a problem with incorrectly labelled plates; an example being the illustration of *Theraphosa blondi* (tab. CCCX1X p89), which was labelled *Selenocosmia javanensis*. As this would appear to have gone unnoticed until the late twentieth century, one can only ponder the bemusement of colonial Dutch museum officials when they consulted their Koch and discovered the world largest spider resided in Java.

By 1860, although the idea of linking a designated a type specimen to a species description was still in its infancy, the development of natural history museums and national collections meant that the concept was becoming increasing popular. Unfortunately, the muddled hand of history, war and curatorial carelessness has meant that in some cases even if a specimen was placed in a museum collection it has now gone missing.

Two such historical spiders with stories to tell, which are now missing, are *Aphonopelma hentzi* (Girard 1852) and *Brachypelma emilia* (White 1856), known in their day as *Mygale hentzi* and *Mygale emilia.*

When Captain Randolph (5th Infantry) finally returned home from the Red River Expedition in late 1852, he discovered that the Eastern newspapers had informed their readers that his expedition had been wiped out by the Comanche. Fortunately, these reports proved untrue because among the material collected were the first two tarantula spiders to be recorded from North America:—described by Charles Girard in 1853 as *Mygale hentzi*. As an aside, the spider was first described in 1853 as an appendix in the monograph—*The Exploration of the Red River of Louisiana in the year 1852 by Randolph B. Marcy (Captain 5th Infantry US Army).* The date 1854, which is often linked to this spider, were reprints published for the Senate and House.

The Red River Expedition was made up of 90 men who rode out of Fort Smith in March 1852, with an objective reminiscent of Star Trek. To "*collect and report everything that may be useful and interesting*". Hundreds of zoological specimens were trapped and collected, including 20 new species of mammal, 10 reptiles and numerous entomological specimens, most of whom were secured by Marcy's second in command, Captain George B. McClellan and the expeditions surgeon,

George Shumard. Marcy wrote of his future son-in-law George McClellan, "*An interesting collection of reptiles and other specimens in alcohol were also made under his superintendence and put onto the hands of Professors Baird and Girard*". Having collected North America's first tarantula spider, McClellan, was to go onto even greater things, eventually taking command of the Army of the Potomac and standing against Abraham Lincoln in the presidential election of 1864. As an aside, it has always been presumed that the type material of *hentzi* was collected in present day Oklahoma, Smith (1994) and Warriner (2008), but new research by Smith (2015) would indicate that the male was collected in the vicinity of Amarillo in Texas.

It has always fascinated me how adventurous and exciting were the lives of many of the individuals who were drawn to collect or work on theraphosid spiders. Charles Girard (1822–1895), having trained under Lois Agassiz in Switzerland followed the great man to Harvard, where to Agassiz's fury Girard left him to work under Professor Spencer Baird at the Smithsonian. At that time the Smithsonian was the primary beneficiary of the extensive field collections made by the government survey parties of the period, which meant that Girard described numerous herpetological and arthropod species, including North America's first description of a tarantula spider, *Mygale hentzi.* During the civil war, Girard's sympathies lay with the South, and he was appointed The Commissioner of Medical Supplies. This entailed organising the blockade runners, who smuggled munitions and medical supplies through the union naval blockade of the Confederacy.

The spider was named after Nicholas Hertz (1797–1856), author of *Spiders of the United States* (1875). Hentz, described by Comstock as the father of American arachnology, was noted by his contemporaries for his deep sense of spirituality. He painted the All Seeing Eye on his study door and it was observed that without warning, in the middle of a conversation he would drop to his knees and pray fervently. After a few minutes he would stop and return back to the conversation. Described by his doctors as having a nervous disposition brought on by his mother having to flee the French Revolution while he was still in the womb, he was prescribed morphine:—an addiction responsible for his early death (Hanley 1977:119).

We know that both a male and female specimen of *Mygale hentzi* were described by Girard, and it is presumed that it was placed into the Smithsonian collection. Unfortunately both specimens are now deemed lost. The same fate was to befall the type specimen of *Brachypelma emilia* (White 1856).

Charles Darwin (1809–1882) was to make his name as a young man, with the publication of the third volume of the H.M.S. Beagle series, *The Narrative of the Surveying Voyages of H.M.S. Adventure and H.M.S Beagle 1826 and 1836.* Likewise John MacGillivray (1821–1867), with his account of the voyage of H.M.S Rattlesnake to the Pacific in 1846. Both men understood that it was great honour to be appointed as the ships official naturalist on a Royal Navy surveying expedition (Goodman 2005:33). And yet today, one of the most popular accounts from that period:—*The Narrative of the Voyage of H.M.S. Herald 1845–1851,* by Berthold Seemann (1827–1871), is largely forgotten.

Botanist, naturalist and field collector, Seemann was appointed in July 1846, (on the recommendation of Sir W. J. Hooker at Kew Gardens), to replace the first ships naturalist, Thomas Edmonston (1825–1846), who had been killed in a firearms accident the previous year. Seemann was immediately despatched to Panama, only to find that H.M.S. Herald had put to sea. He spent the remainder of the year collecting on the isthmus before finally joining the ship in January 1847. For much of that year the ship surveyed the Pacific coast of Mexico and Central America, putting into sheltered bays every so often to re-provision, fill the water barrels and exercise the crew. One such bay was Bahia Culebra in Costa Rica, and it was there that Seemann was able to collect the well-known Zebra legged tarantula, *Aphonoplema seemanni* (Pickard–Cambridge FO 1897). Later on in the voyage, it is commonly believed that it was Seemann who collected the strikingly beautiful Mexican tarantula *Brachypelma emilia* (White 1856), somewhere in the region of Durango.

H.M.S. Herald finally returned to England in June 1851, and the two spiders found their way first to the spirit room of the British Museum, and then in 1882 to the new Natural History Museum in South Kensington. In fact, so many boxes of specimens were coming back from these Royal Navy survey expeditions that the museum workers of the day had considerable difficulty in keeping up with them. *Aphonopelma seemanni*, for example, had to wait 50 years before it was described in the *Biologia of Centrali-America*, but *Brachypelma emilia* was such an attractive spider that it would appear to have caught the eye of Adam White (1817–1878) and in doing so, it became the first tarantula spider to be described in Britain.

The paper was titled *Description of Mygale emilia. A spider form Panama,—Hitherto apparently unrecorded* (1856). The paper, from the perspective of an arachnohistorian, is richly illuminating in that it is packed with interesting information mostly garnered it seems from Adam White's world-wide correspondence. One such revelation is the discovery that live tarantulas were being shipped to enthusiastic collectors from abroad during this period. Adam White notes:—*I have but once seen a Mygale alive; the specimen was sent to the late John Doubleday by post. The day after its arrival he gave it cockroaches. They were put into the small box with the Mygale. It apparent at first did not see them running around its legs—then the great spider drew itself up and darted its chelicera into one of them, tearing its intestines with its fearful armed hook. The blatta was soon devoured.*

It is likely that the recipient of the spider, John Doubleday mentioned in the text, was the renowned restorer, who Adam White would have worked alongside in the British Museum and who died in the year that the paper was published.

But what is truly intriguing about this paper is the title, and the belief that the spider was found in Panama. It would seem that Seemann had no recollection of where he collected one of the magnificent spiders in the world, which has led me to the conclusion that he did not collect it. I believe that it was collected 4 years earlier in Mexico by either the ships surgeon, or more likely the two ships officers, a Mr Romaine and a Mr McNamara, who left the ship and rode up to the British consulate in Tepic to collect fresh orders form the admiralty. The ship's log informs us that Edmonston was killed when a musket was accidently discharged by a member of the ship's crew, when returning back to the ship after collecting botanical specimens. As Seemann had no idea of where the specimen had been collected, we can only presume that in the confusion after Edmonston's death the specimen was never correctly labelled and sat in a jar of alcohol on board ship, until relocated. This theory is further strengthened with the discovery that in Panama (where Seemann was forced to spend an extended stay), there is a *Sericopelma* species with a similar distinctive pyramid marking on the carapace (Ray Gabriel: personal correspondence). Seemann probably found the preserved specimen and concluded that it was the same spider that he had seen in Panama. Fortunately, the inclusion of a wonderful coloured plate by a Miss Spooner of Kentish Town, means that the matter can be settled by examining the prominent tibial spur of the male illustrated in the engraving. The males of the genus *Sericopelma* do not have a tibial spur.

With Adam White's paper we see the arrival of a new breed of arachnologists. The professional, increasingly university educated, museum curator. Men such as Theodor Thorell (1830–1901), Ferdinand Karsch (1853–1936) and Reginald Pocock (1863–1947), who were to become the high priests of theraphosid spider research in the new cathedrals of science that were now to be found in many large European cities by the 1880s. Nevertheless, arachnology still had room for the gentleman, amateur scientist, who had presided over earlier generations. Men like Graf Eugen von Keyserling (1833–1889), agriculturist and landowner, Anton Ausserer (1843–1889), school teacher, Octavius Pickard-Cambridge (1828–1917), clergyman and his nephew and Frederick Octavius Pickard-Cambridge (1860–1905), clergymen, natural history illustrator and radical socialist. We may only surmise how much these fellows must have envied Eugene Simon (1848–1924), whose inherited private fortune enabled him to devote his entire life to travel and the study of spiders.

For American researchers, Graf Eugen von Keyserling is forever linked with the first major publication on the arachnids of the United States:—*Die Spinnen Amerikas,* the first two volumes, of which, appeared between 1880 and 1886. After Keyserling's death in 1889, volumes three and four were finished by George Marx (1838–1895), who had emigrated to America from Germany in 1850, served in the Union Army during the Civil War and worked in the Washington Museum until his early death from cancer in 1895. Marx would later greatly assist Simon's work on North American theraphosids and was responsible in 1890 for the first reliable list of the spiders of the United States (Savory 1961:77). He was also responsible for describing in 1888, the second theraphosid to be described in the United States, *Aphonopelma rileyi* from Santa Barbara, near San Francisco. Unfortunately, Marx is now chiefly remembered as the somewhat eccentric curator of the national collection, whose chaotic custodianship has not boded well for posterity. The end result being that Simon's 1890 paper on American tarantulas, *List des esp de la fam. des Aviculariidae qui hab. L'Amerique du Nord*, was fatally flawed by the inclusion of specimens which had been incorrectly labelled (Smith 1994:5).

One of Keyserling's greatest achievements was to act as a mentor for Anton Ausserer (1843–1889), who between 1871 and 1875 produced two papers on the Theraphosidae which were to set new standards for the arachnological papers of the period. The 1871 paper, Beiträge zur Kenntniss der Arachniden-Familie der

Territelariae Thorell (Mygalidae Autor) is one of the most important papers published on the Theraphosidae during this period. Not only did it set new standards for species descriptions (with illustrations of the male genitalia), but the confident and sweeping suggestions for systematic change would indicate that this 28-year-old school teacher from Austria, (who, in 1871 was still 3 years away from gaining his doctorate) had a firm grasp of his subject matter. But all of this would have meant nothing if Ausserer had not had access to a large number of preserved specimens, in the Austro-Hungarian and German collections.

These surprisingly rich overseas natural history collections, in an empire which had no overseas possessions, was the result of the marriage in 1817 of the Brazilian Crown Prince Dom Pedro and Leopoldina, the daughter of the Emperor Francis. In that year the Austrian Government took the opportunity to dispatch a "Scientific Mission" to Brazil, which was to prove so successful that the Hofburg Palace Museum proved too small to contain the acquisitions that flowed back from South America. Many of these scientists died of fever, but one collector Johann Natterer (1787–1843) was to remain on the continent for 18 years and was responsible for much of the biological and ethnological material sent back to Vienna. His collection alone contained 60,000 insects. The Austrian Navy also took this golden opportunity to survey the Amazon, and was later responsible for sending the frigate Novara on a scientific expedition which from 1857 to 1859 circumnavigated the world. All of which meant that Ausserer had access to a vast collection of biological material (Kollman and Schultz 1989:59). An additional collector was Franz Steindachner (1834–1919), curator of Herpetology at the Naturhistorisches Museum in Vienna. Having access to a private income, Steindachner travelled the world collecting specimens for the museum (Adler 1989:54). Ausserer named *Aphonopelma steindachneri* (Ausserer 1875) in his honour.

In the winter of 1870/1871, Ausserer, supported by a research grant from the Innsbruck Education Authority, embarked on a trip first to Vienna, and then to Munich, to examine the Keyserling collection. The end result was the 1871 paper which could be described as the first modern paper on the Theraphosidae. In this paper we see five new genera (including such familiar names as *Tapinauchenius, Selenocosmia, Acanthoscurria, Ischnocolus* and *Chaetopelma*) and 12 new species. In 1875, he added five additional genera, *Cyclosternum, Euathlus, Hapalopus, Homoeomma* and the Far Eastern genus *Selenocosmia.* Species first described by him include such classic names as *Brachypelma vagans, Avicularia metallica, Grammostola mollicoma, Xenesthis immanis, Megaphobema immansis, Sericopelma rubronitens, Pamphobetues ferox and Aphonopelma steindachneri.* In May 1888, his health always poor, Ausserer contracted "*violent bronchitis*" and died in the arms of his young wife in July 1889 (Maurer 1890:36).

Tord Tamerlan Teodor Thorell (1830–1901) and Ferdinand Karsch (1853–1936) need not concern us greatly as both tended to publish primarily on the theraphosids of Africa and the Far East, with an occasional foray into New World spiders. The Costa Rican genus *Sphaerobothria*, described by Karsch in 1876, is interesting in that the single species assigned to the genus, *hoffmanni* has an unusual horn on the carapace. The spider was named after the doctor, field collector and naturalist Karl

Hoffmann (1823–1859), who died nursing typhoid victims. As an aside, Karsch's arachnological publications tend to decrease by the late 1890s, as his interests increasingly turn to ethnographic studies, particularly homosexuality in tribal societies. Karsch from this period became closely associated with a number of the leading figures in Berlins academic gay circles. His last publication on homosexuality was in 1933, the year that Adolf Hitler came to power (Aldrich 2001).

Which brings us to the two key movers and shakers in the world of theraphosid taxonomy at the end of the nineteenth century:—Eugene Simon in Paris and Reginald Innes Pocock at the British Museum (Natural History) in London. Simon was essentially a wealthy gentleman scientist, who dominated the study of spiders in Europe during this period. Pocock, on the other hand, was a professional museum taxonomist/curator who was responsible for producing a steady stream of highly influential arachnological papers. I was astonished to discover that in reality, he only worked on the group for 18 years and was so desperate to escape from the world of arthropods and work on mammals that in the end, he resigned from the museum and became the Superintendent at the Zoological Society of London.

Simon was born in Paris in April 1848, and read science at the Sorbonne, but later in life credited his father, a keen naturalist, for his lifelong passion for arachnology. Whilst still in his teens, after reading the works of Walckenaer, he realised the rich opportunity that the study of such an imperfectly known group of animals proffered. This young savant set to work without delay and in 1864 at the age of 16 he published the first edition of *Histoire Naturelle des Araignees*. This was essentially a summary of the work of his predecessors, but it enabled him to study and absorb the existing literature as a springboard to his own research. As such, his life was taken up with two great projects:—*Les Arachnids of France*, which need not concern us and further volumes of *Histoire Naturelle des Araignees,* which was essentially a systematic study the world's arachnids.

Early on, Simon realised his biggest problem was going to be acquiring enough specimens to study, which he overcame by embarking on a series of field trips:—Sicily (1864), Spain (1865 and 1868), Corsica (1869), Morocco (1871), Tunis and Algeria (1875) and then after an epic bout of writing, Venezuela (1887–1888), Suez and Aden (1889–1890), Philippines (1890), Ceylon/Sri Lanka (1892), and South Africa in 1893. Nobody had ever tackled a zoological field of study, in such a systematic fashion, especially one as obscure as the study of spiders. These collections resulted in the following genera, many of which are classic names in theraphosid literature: *Brachypelma, Coremiocnemis, Cyriocosmus, Cyriopagopus, Cyrtopholis, Encyocrates, Ephobopus, Grammostola, Hapalotremus, Haplocastus, Haplopelma, Hemiercus, Hemirrhagus, Hysterocrates, Lampropelma, Loxomphalia, Loxoptygus, Myostola, Nesiergus, Orphnaecus, Ozopactus, Paraphysa, Phlogius, Poecilotheria, Tmesiphantes and Xenesthis.*

His obituary described him as an engaging travel companion and great friend who inspired the warmest affection in the hearts of his colleagues. As Simon grew older he held court at his home:—16 Villa Sais, near the Bois de Boulogne. The French arachnologist Lucien Berland (1888–1962) was to write. "*All those who had the honour to approach him will preserve a touching remembrance—where behind his work table, accompanied by his inseparable pipe, the savant arachnologist always reserved for them so affectionate greeting"* (Berland 1925:224).

Simon was probably the greatest physical collector of spiders in the history of arachnology and yet the collection that he bequeathed to the Natural History Museum (Paris) remains underworked because of the sheer difficulty of accessing and examining the eccentrically catalogued specimens. *"The collections for the most part are still undetermined, sorted into geographical regions, waiting for active araneologists. But the specimens, kept in 6x22mm, vials have only the minimum of information"* (Levi 1964:17).

Like Simon, Reginald Innes Pocock (1863–1947) produced a body of work which has embedded itself into the heart of theraphosid taxonomy. Born in 1863 in Bristol, he was educated in Oxford where he studied under the renowned zoologist E. B. Poulton (1856–1943). He joined the British Museum (Natural History) in 1885, where he was to remain until 1903. In that year, following the death of his beloved daughter Natalie from meningitis, it would seem the sheer frustration of having to remain in the arthropod department when his real interest was mammals, triggered his resignation and departure from the world of arachnids (Hindle 1947).

As with Karsch and Thorell, Pocock is primarily associated with the arachnid fauna of Africa, India and the Far East, but he did publish in the Annals of the Magazine of Natural History in 1901 and 1903, two very influential papers on the theraphosids of the New World. Again we see names which have become classic genera. *Aphonopelma, Dugesiella, Citharacanthus, Megaphobema, Phormictopus, Pamphobeteus, Pachistopelma, Iridopelma* and *Psalmopoeus.*

Pocock was in a unique position in that he sat at the heart of an empire that encompassed a third of the world. All of which meant that a steady stream of specimens from all over the British Empire flowed into the museum. Many came from the engineers, planters, missionaries, doctors, army officers and civil servants, who when not burdened by the responsibility of empire, were, in their spare time, natural history enthusiasts. But some of the most important specimens were also purchased from professional collectors, such as W.F.H Rosenberg (1868–1957), of 48 Charing Cross Road, who in the 1890s sent his catalogues to many of the European museums of the period. Rosenberg was a traveller, naturalist, soldier and adventurer. He served in the Queens Westminster Rifles, in South Africa, as a roughrider and then was off to Columbia, where in 1894 he collected in the Cauca Valley, before moving on to Ecuador between 1996 and 1897 (Anonymous 1957:108). It was Rosenberg, who collected the large Columbian theraphosids, which Pocock was to describe as *Megaphobema* Pocock 1901.

Correspondence between Rosenberg and Pocock also sheds an interesting light on the relationship between the commercial dealer and the museum curator, which, it seems, if successful, hinged on immense patience on the part of the dealer. Having waited 9 months for a reply, one senses Rosenberg's exasperation in a letter to Pocock in April 1898 *"Would you kindly let me know whether you have selected the 30 spiders from the Ecuadorian collection, in order that I may take the remainder away?"* Payment, it would seem, also depended on the curator, not the dealer's estimation of a specimen's value. *"I enclose a bill as requested and accept your valuation of the specimens selected*". In the same letter, he reminds Pocock that a bill for the sum of £1-15s-6p was still outstanding for the Columbian material the curator had purchased the previous year (Pocock box. Rosenberg file. BMNH).

The man, who was expected to replace Pocock at the BMNH, when he suddenly resigned in 1903 and took up a post at London Zoo, was Frederick Octavius Pickard-Cambridge. Unfortunately, F.O.P (as he was known to family and friends), shortly before he was to join the museum, blew out his brains with his brother's service revolver in the rented rooms that they shared in Wimbledon. Why, remains unclear, but on doing so, for all intents and purposes the study of New World theraphosid spiders ceased in Britain for the next 80 years. This was to remain so until renewed interest in them came about in 1986, with the publication of Andrew Smith's, *The Tarantula Classification and Identification Guide*. The one exception, being the research undertaken by David John Stradling (1939–2012) during the early 1970s in Trinidad, on the growth and maturation of the arboreal tarantula *Avicularia avicularia* (Stradling 1978).

Frederick Octavius Pickard-Cambridge (1860–1905) was educated at Sherborne and Exeter College Oxford where he followed both his father and his uncle (the renowned arachnologist Octavius Pickard-Cambridge) into the church where he become a curate in Carlisle. He was an outstanding naturalist and a gifted illustrator, but he was also a driven man whose massive output of work was marred by his "own personal limitations" (Savory 1961:129). Nevertheless, his enormous contribution to the *Biologia Centrali-Americana* means that his role in British arachnology should not be underestimated. This epic 42 volume publishing project was the inspiration of two wealthy British academics, Osbert Salvin (1835–1898) and Frederick Godman (1834–1919). Each volume was lavishly illustrated, which in the case of the two arachnid volumes was the responsibility of F.O.P.-Cambridge, as was all of the text for the 1897 volume. Much of the text for the 1892 volume, which includes the description of the Costa Rican tarantula *Megaphobema mesomelas,* officially was undertaken by his uncle Octavius, although the latter's advanced age does raise questions about the validity of this assumption. In the 1897 volume, Frederick described *Brachypelma smithi*, which he named after the American collector Herbert Huntingdon Smith (1851–1919), one of the great tarantula field collectors of the nineteenth century and the man responsible for collecting the beautiful Mexican Tarantula *Brachypelma smithi* (F.0.P.-Cambridge 1897) (Fig. 17.7). Another field collector who richly deserves a mention and who contributed greatly to the project was George Champion (1851–1927), a coffee planter, who collected extensively in Guatemala and Panama. Champion's papers and field journal are in the Hope archive in the University Museum in Oxford and are richly illustrated by watercolours of the spiders that he collected.

Although F.O.P.-Cambridge is essentially associated with *Biologia Centrali-Americana*, what is not commonly known is that he was invited by the German engineer Alexander Siemens (1847–1928), to join him on S.S. Faraday, which was laying a telegraph cable along the Amazon. New species descriptions resulting from the field trip were published in 1896 in the Proceedings of the Zoological Society London, which is one of the finest, stand-alone papers published on the Theraphosidae

Fig. 17.7 *Brachypelma smithi* (F.0.P.-Cambridge 1897). The classic pet shop red knee tarantula, photographed at the type site, where Herbert Huntingdon Smith first collected them at Dos Arroyas, north of Acapulco in Mexico. Copyright: author's collection

during this period. As an aside, I have a letter sent by F.O.P.-Cambridge to Reginald Pocock, which would indicate that F.O.P was not Siemens first choice for this expedition. It would seem that Pocock was forced to withdraw from the fieldtrip when his wife who was in poor health, lost her child (Smith Collection).

While in Brazil, F.O.P-Cambridge made the acquaintance of Emil Goeldi (1859–1917), a Swiss national who had been invited to reorganise the Para Museum of Natural History. He informs us that Goeldi was responsible for a series of handbooks, issued by the museum, one of which published in 1894, contained a brief summary of the tarantula spiders of the region. This would make it the first faunal list published outside of Europe on the South American Theraphosidae.

With the exception of Embrik Strand (1897–1947) in Germany, and Lodovico di Caporiacco (1900–1951) in Italy, following F.O.P.-Cambridge's death, research into New World theraphosid spiders, for all intents and purposes, switched for the next 50 years to the indigenous countries where tarantula spiders are to be found:—specifically Brazil and Costa Rica. Little was to emerge from Europe until the early 1980s, when Gunter Schmidt (1926–2016) in Germany and Andrew Smith in London began to publish extensively on the group. The catalyst being the development of a pet trade in these large spiders, which created renewed interest in the group.

Caporiacco need not concern us greatly, in that although an outstanding arachnologist (who tragically died of cancer at an early age), his research only briefly touched upon the tarantula spiders of French Guyana and Venezuela. Embrik Strand on the other hand, aged 19, published in 1919 a list of his zoological publications which already numbered 1200. Born in Norway, he primarily worked in Germany at the museums of Senckenberg and Berlin. Between 1907 and 1922, he published close on a hundred papers on arachnids. Unfortunately, he was obsessed with multiplying the number of genera and species, which his name was appended:—to the point that Savory described his actions as "*moral tarantulism*" (Savory 1961:168). Many tarantula species have his name attached, but like his contemporary Karsch, he published mainly on African and Far Eastern material. The papers are horribly dated with few illustrations, which has meant that it is impossible to identify many of the species he described, where the type material was lost to allied bombing during the last war.

Although Eduardo Holmberg (1852–1937) is generally deemed to be the father of South American arachnology, he only published two accounts of theraphosid spiders, the most interesting being *Roca's Expedicion al Rio Negro (Patagonia)* in 1881. A much earlier paper by Hercule Nicolet (1801–1872) published in Gay's *Historia de Chile,* (1849), featured the description of a spider which Nicolet named *Mygale rubiginosa.* Although, this species is now believed to be synonymous with *Grammostola rosea* (Walckenaer 1837), it still has the honour to be the first New World tarantula described outside of Europe.

Nevertheless, despite his many faults Candido Firmino de Mello-Leitao (1886–1948) must be deemed to be the father of South American theraphosid research, in that he devoted his life to the study of arachnids and published extensively between the wars on the Theraphosidae. Unfortunately, although Mello-Leitao collected in abundance and published with much enthusiasm, he failed to ensure that his collections were curated to a standard that would ensure that his work would not become a trial for those who followed.

On Mello-Leitao's death in 1948, although Wolfgang Bucherl (1911–1985), Salvador de Toledo Piza (1898–1988), Benedicto Abilo Monteiro Soares (1914–1985) and Helio Ferraz de Almeida Camargo (1922–2006) continued a steady stream of arachnological papers, these essentially encompassed toxicology and behaviour, with only a handful of papers on theraphosid taxonomy.

Outside of Brazil, Pelegrin Franganillo-Balboa (1873–1955) worked on the spiders of Cuba, but unfortunately his papers were described by Brignoli, as "*amongst the worst published since 1870*" (Brignoli 1983:3). Nevertheless, Brignoli also noted, "*Part of Franganillo-Balboa's collection is still in existence and some of his taxa appear to be valid*". So for Cuban researchers, all is not lost. More successful, in terms of posterity, were Addalberto Ibarra Grasso (1910–1993) and Maria Elena Galiano (1928–2000) in Argentina, who added a number of tarantulas to their faunal lists:—although, their research was inevitably hampered by not having access to the European museum collections where much of their countries type material was stored.

In North America, two figures stand out in the first four decades of the twentieth century: Alexander Ivanovitch Petrunkevitch (1875–1964) and Ralph Vary Chamberlin (1879–1967). Both men have left behind a mixed legacy in the eyes of many of today's American arachnologists. Petrunkevitch's outstanding work must be his *System Aranearum* (1928), which predated the catalogues of C. F. Roewer, *Katalog der Araneae* (1942, 1954) and P. Bonnet, *Bibliographia Araneorum* (1945). This was followed up by his *Catalogue of American Spiders* (1939), which, unfortunately Roewer did not have access, when drawing up his own *Katalog der Araneae* (1942).

Only a small percentage of Petrunkevitch's work touched upon theraphosids, but while at Yale he published in 1911 a fascinating paper of the mating behaviour of *Aphonopelma hentzi*, which was illustrated by a wonderful set of images taken by the author himself. He would also appear to have been fond of the Cuban tarantula

Phormictopus cancerides (Latreille 1806), which figured extensively in his physiological studies. In the late 1930s, he had as many as 180 specimens housed in his laboratory and it was this spider, the British arachnologist John Parker was to later recall, which figured extensively in a lecture that Petrunkevitch, aged 74, delivered in 1949 at Kings College, Newcastle-Upon-Tyne (Smith 1994:7). At this point it is interesting to note that *Phormictopus cancerides, Dugesiella hentzi* and *Aphonopelma chalcodes* are the three names which appear regularly in pre- and post- war laboratory catalogues and papers, which would indicate that they were the favoured spiders of biological supply companies. John Steinbeck's character, Doc in his Cannery Row series, is very much based on the sort of individuals who collected natural history specimens for these supply companies (See Chap. 15).

Ralph Chamberlin, alongside his student and research assistant, Wilton Ivie (1907–1969) were the two leading American researchers, working on tarantula spiders during the interwar period, their last paper being 1940; during which time they described many of North America's theraphosid species. Although Chamberlin is the lead author, it is commonly believed in American arachnological circles that from 1937, it was Ivie who did much of the "*work collecting, identifying, describing and illustrating the new species. To keep Ivie in his laboratory Chamberlin condescended to add him as a junior author to the publications*" (Vogel 2011:4). Ivie, also had a formidable reputation as a first class field collector, despite being restricted by childhood polio (Murphy, J. Private correspondence). Ivie left Chamberlin in 1947, and was tragically killed in a road accident while driving to New York to take up post at the American Museum of Natural History, where he was to join his friend Willis Gertsch (1906–1997).

During the same period, we also have William J. Baerg (1985–1980), a college lecturer and tarantula enthusiast who was to become a legendary figure among tarantula hobbyists (See Chap. 15). Baerg saw his first tarantula when he took up a post at the University of Arkansas in 1918, an event which he was to describe as *enchantment-at-first-sight* and which was to last all of his long life. The fruit of this fieldwork research culminated with the publication of the classic book *The Tarantula* (1958). According to William B. Peck (1920–2001) (Smith Private correspondence), he became increasingly frustrated and eventually exasperated by the vagaries of North American tarantula taxonomy and expressed the belief that no researcher had any real hope of identifying a tarantula spider at a species level (Smith 1994:8). It is said that Baerg like to test the mettle of his student freshmen, by asking each to hold a tarantula spider. Only one person was ever bitten, he averred, and many a character was strengthened (Baerg 1997:2a).

Which brings us to the 1960s and the remarkable papers published by Berta S. Gerschman (1905–1977) and Rita Delia Schiapelli (1906–1976). Both ladies hailed from Argentina, and between the years 1945 and 1979, published some of the most outstanding arachnological papers on the Theraphosidae to have come out of South America. In fact they were of such a high standard, we may include Europe into that category as well. This achievement is made even more special, when one remembers that before any research could take place, it was necessary for these two ladies to embark on expensive trips to London and Paris and examine the historical collections. The end result was a series of richly illustrated papers that introduced new standards in the layout of arachnological publications. Nelson Ferretti has since picked up the mantle and continued their fine work in Argentina.

The 1970s was to be the decade, where students of arachnology were to see the formation of two professional arachnological societies:—The British Arachnological Society (1969) and the American Arachnological Society (1972). For the first time in the history of arachnology, spider enthusiasts now had their own specialist journals to send papers. Journals, which in turn were quickly responsible for publishing a wave of original research on all aspects of arachnology that in the past would have been scattered through a myriad of different publications.

In 1985, Robert Raven published his highly influential, ground breaking publication *The Spider Infraorder Mygalomorphae (Araneae) Cladistics and Systematics.* Today it is difficult to understand how important this work was. What Raven did, was take a large number disparate genera published over two hundred years with numerous authors and even more numerous synonymies and apply a brutal cladistic and systematic analysis. In doing so, he challenged the next generation of arachnologists to re-examine the group from a fresh perspective. What followed over the next 30 years has been a tsunami of revision work.

By the early 1980s, we also have the fascinating phenomenon that these large spiders were now being kept as pets in both America and Europe, which meant that there was an increasing interest in their identification (See Chap. 15). In Europe, with its history of university educated amateur naturalists, this new found interest in the taxonomy of theraphosids was to act as a catalyst for a revival of the privately funded, amateur arachnologist. The first of these individuals were Andrew M. Smith (school teacher) and Dr. Gunter Schmidt (a retired chemist). Smith, using the catalogues of Roewer/Bonnet/Petrunkevitch, published in 1986, *The Classification & Identification Guide*, which was a literature based survey of the Theraphosidae. During the same period, Schmidt began to publish, and continued to publish, a stream of taxonomic papers up until death at the age of 90. Although there are issues with much of his later work, nevertheless he was responsible for a number of important papers on New World tarantulas. It is widely felt that his best work was in the 1990s, and during this period he was responsible for describing a large number of well-known tarantula spiders.

Smith published primarily on the theraphosid spiders of Africa and Far East, but in 1994 he published his ground-breaking *Tarantula Spiders—Tarantulas of the USA and Mexico,* which was the first major revision of North American theraphosid spiders, since Chamberlin and Ivie in the late 1930s. Greatly assisted by the American field collector Michael Sullivan (1956–2012) Smith described a number of new species, but more importantly re-described and illustrated type material widely dispersed across museum collections on two continents.

Another important figure in the mid-1990s, working on North American tarantulas was Tom R. Prentice, who was responsible for the exciting discovery of micro tarantulas, which had been overlooked by previous collectors. The quality of his new species descriptions raised the bar of what was now expected of such publications and his field work greatly added to our knowledge of North American theraphosid spiders.

At the same time, Schmidt and Smith had first begun to publish what was to be a steady stream of publications over a period of 20 years, there appeared in Europe and America, a new generation of tarantula field collectors:—Vincent Hull-Williams, Paul Carpenter, Peter Kirk, Peter Klaas, Andreas Tinter, Jean-Michel Verdez, John and Kathleen Hancock, Fabian Vol, Andre Leetz, Andre Braunhausen, Philip Charpentier, Guy Tansley, Ray Gabriel, Mark Pennel, Marc Baumgarten, Thorsten Kroes, Thomas Maerklin, Nicolai Pedersen, Henrik Wessel-Frank and in America, Rick West and Michael Sullivan. These individuals had begun to travel the world and actively collect tarantula spiders, many of which were to prove new to science. In many cases, these spiders were also collected and captive bred for the European and American pet trade. At the same time we also see the emergence of a number of tarantula societies in both Europe and America that catered for this growing hobby, and which had the financial resources to produce printed journals:—the most influential being the Journal of the British Tarantula Society which, under the editorial ship of Peter Kirk has been responsible for pioneering the captive breeding of theraphosid spiders (See Chap. 15). These publications primarily published papers on captive breeding, but they also became vehicles for new species descriptions outside of traditional scientific journals. By the mid-1990s, a second wave of European and American researchers (many amateur, but among them professional biologists) begin to publish, some of whom were now attached to institutions and publishing in both academic and amateur journals. Peter Kirk, Richard Gallon, Dr Stuart Longhorn, Rick West, Michael Jacobi, Ray Gabriel, Fabian Vol, Marc Tesmoingt, Heinz-Josef Peters, Jan Peter Rudloff, Siegfried Huber, Jean-Michel Verdez, R. Struchen, Eddy Hijmensen, Dirk Weinmann and more recently Daniella Sherwood.

Although, the collecting, captive breeding and new species descriptions was essentially dominated by Europeans, a number of figures stand out in North America, as having been influential in our understanding of these spiders. The first was the veteran field collector and photographer Rick C. West, who later in life began to publish cladistics and systematic papers with the Australian researcher Steven C. Nunn. West has also collected extensively throughout Mexico for the Mexican national collection and has described a number of new Mexican theraphosids. The second was Robert Gale Breene (1952–2009), a doctor of entomology, a prolific writer and populariser of science. For over two decades Breene was responsible for editing The Journal of the American Tarantula Society and did much to encourage amateur research on that continent. A third figure was the tarantula field collector Michael Sullivan (1956–2012), who for four summers in a row worked alongside the author in the preparation of the reference work Tarantula Spiders—Tarantulas of the USA & Mexico (Smith 1994). Alongside Marc Baumgarten, who was collecting for Smith in Mexico, his knowledge of the desert and his physical contribution to the publishing project was to prove invaluable. Which brings us to our last name in this group, Michael Jacobi, a breeder, traveller and natural history writer, whose innovative and influential articles on captive breeding and husbandry made him a popular guest speaker in Europe.

But what of the research on the Theraphosidae, by indigenous biologists in the New World?

After the deaths of Rita Delia Schiapelli (1906-1976) and Berta S Gerschman de Pikelin (1905-1977) in the late 1970s, two Argentinian female arachnologists whose publications were highly influential during this period, little takes place until 1980, when we see the first of two major papers by the Costa Rican arachnologist Carlos Valerio. These papers not only re-examined the historical material of the region, but also added to numerous new species.

In Brazil, at the Institute Butantan, Sylvia Marlene Lucas, a student of Wolfgang Bucherl, re-described in 1972 Strand's *Dryptopelmides* and added a new species to the genus. Raven was to synonymise the genus in 1985, but Lucas has published extensively on the Theraphosidae and her place in its history is assured. One of her students, Rogerio Bertani has also gone on to become Brazil's leading authority on the family. He in turn has encouraged a new generation of young arachnologist to publish on the group, one of the most exciting being Caroline Sayuri Fukushima, who, alongside Bertani, is the key author of a sweeping and much needed revision of the genus *Avicularia* (2017). Also in Brazil, Flavio Uemori Yamamoto and Jose Paulo Leite Guadanucci have added to the countries list of theraphosid spiders.

Another author, who has done sterling work on the little known theraphosid fauna of Belize, was Steven B. Reichling, who published a flurry of papers on the region, in the latter part of the last century.

In Mexico in the late 1990s, building on access to the type material illustrated in Andrew Smith's *Tarantulas of the USA and Mexico,* a new wave of Mexican researchers began to publish on the tarantula spiders of Mexico. Early on, Arturo Locht, Martha Yanez and Vazquez, I (1999) produced an important paper on the genus *Brachypelma*, but it was only when the veteran arachnologist, Oscar Francke became curator of the National Arachnid Collection (UNAM) that we see the study of spiders in Mexico really take off. The same thing was to take place in Brazil, with Lucas/Bertani and in Uruguay with Fernando Perez-Miles. Essentially, if you have an enthusiastic, senior academic, with a passion for a particular subject supervising PhDs, this can often act as a catalyst, which attracts and encourages students to excel in that particular field (Vogel 2011:14). Under his auspices we have seen Carlos Perafán (Columbia), David Ortiz (Cuba) and Jorge Ivan Mendoza Marroquin (Mexico) publish not only on Mexican theraphosids, but progress, in the case of Perafán and Ortiz, to publish on the tarantula spiders of their own countries.

Along with Rogerio Bertani in Brazil, it has been Fernando Perez-Miles in Uruguay who has picked up the mantle of the early European arachnologists and published extensively on the Theraphosidae. His first paper in 1992 was a revision of the genus *Eupalaestrus*, followed in 1996 (while still a very young man), by the astonishingly ambitious *Systematic revision and cladistics analysis of the Theraphosinae* (Perez-Miles 1996). All of which was only the beginning of what has been almost two decades of publications.

Which brings us back to North America and the next chapter in the history of theraphosid research. Since Raven (1985), the emphasis among professional new world arachnologists has been increasingly on large-scale systematic and cladistics studies, usually at a genus level. Many of these would have involved computer modelling, but on a taxonomic level there has been nothing published that could not have been understood by a Victorian arachnologist. In 2014 that changed with the first of a flurry of molecular and phylogenetic papers (DNA), which inevitably will become a steady stream of molecular systematic publications (Hamilton et al. 2014). This new discipline, combined with traditional taxonomy and cladistics systematics has, for the first time in the history of arachnology enabled researchers to punch through the glass ceiling that has plagued theraphosid taxonomy since the nineteenth century. That problem being the lack of clear, stable, diagnostic taxonomic features in the New World Theraphosidae, which make it possible to readily identify tarantula spiders at a species level. The initial new practitioners of this new age in arachnology have been Chris A. Hamilton, Brent E. Hendrixson, Jason E. Bond, in the United States and Stuart Longhorn and Jorge Ivan Mendoza Marroquin in Mexico (See Chap. 2).

The downside to this new illuminating science is that for the last 300 years arachnology has always been a level playing field for both amateur and professional arachnologists. That may have to change as "*preparing and analysing DNA sequencing is a time consuming and expensive business. A university may be able to afford one shared molecular lab, but not one for every biologist*" (Winston 1999:447). The future may see amateur arachnologists as just that popularisers of science, while major scientific taxonomic work is increasingly confined to the university laboratory. In which case the danger may be a succession of academic technical papers published by students, who have only a fleeting interest in spiders and even less interest in what has passed before.

Acknowledgements I am indebted to Paul Hillyard, Robert Peck, Herb Levi, Sylvia Lucas, Rogerio Bertani, Fernando Pérez-Miles, Robert Raven, Stuart Longhorn, Dieter Scholz, Michael Ziegler, Ray Gabriel, Rick West, Eddy Hijmenson, Richard Gallon, Peter Kirk, Peter Klaas, Michael Sullivan, and Gunter Schmidt. I am grateful to the anonymous reviewer whose wise and helpful advice helped immensely in the redrafting of my first draft. I am also grateful to Carl Portman, for taking time to look at my even more idiosyncratic grammar.

References

Adler K (1989) Contributions to the history of herpetology, vol 1. SSAR, New York, pp 1–202

Aldrich R (ed) (2001) Gay & lesbian history. Routledge, London, p 390

Anonymous (1957) Obituary – W.F.H. Rosenberg. The Entomologist 90:108

Ausserer A (1871) Beiträge zur Kenntniss der Arachniden-Familie der Territelariae Thorell (Mygalidae Autor). Verhandl K K Zool Bot Gesell Wien 21:117–224

Baerg WJ (1958) The Tarantula. University of Kansas Press, Lawrence, p 88

Baerg WJ (1997) In: Smith A (ed) The Tarantula: biography. Fitzgerald, London, p 96

Bates HW (1852) Extracts from the correspondence of Mr. Bates. Zoologist 10:353

Bates HW (1863) The naturalist on the river Amazons: a record of the adventures, habits of animals, sketches of Brazilian and Indian life, and aspects of nature under the equator, during eleven years of travel. Murray, London, p 407
Berland L (1925) Notice necrologique sur Eugene Simon. Ann Soc Ent Fr 94:73–100
Bonnet P (1945) Bibliographia Araneorum, vol 1. Toulouse, p 832
Brignoli P (1983) A catalogue of the Araneae described between 1940 and 1981. Manchester University Press, Manchester, p 755
Brignoli P (1985) Correct dates of publication of the arachnid tax described in some work by C. W. Hahn and C. L. Koch. Bull Br Arachnol Soc 6:414–416
Burger R (1995) Chavin and the origins of Andean civilization. Thames & Hudson, London, p 96
Burkhardt RW (1973) Lamarck – dictionary of scientific biography, vol 7, pp 190–204
Buvelot Q (2004) Albert Eckhout: a Dutch artist in Brazil. Wanders, Zwolle, p 159
Cato P, Jones C (eds) (1991) Natural history museums: direction for growth. Texas Tech Univ Press, Lubbock, TX, p 252
Clusius C (1611) Curae posteriors, seu plurimarum non ante cognitarum, aut descrptarum stirpium, peregrinorumque aliquot animalium novae descriptions. Officina Plantiniana Raphelengii, Leiden & Antwerp
Cuvier G (1817) The Animal Kingdom Arranged in Conformity with its Organization. Preface. (Latreille). Preface Tome 3, Paris
de Laet J (1633) Novus Orbis seu descriptionis Indiae Occidentalis Libri XVIII authore Joanne de Laet Antverp. Novis talulis geopgraphicus et variss animantium. Plantarum Fructuumque iconibus illustrata, Antwerp
Desmond R (2003) Great natural history books & their creators. British Library/Oak Knoll Press, London/New Castle, DE. 176 pp
Fukushima C, Bertani R (2017) Taxonomic revision and cladistics analysis of *Avicularia* Lamarck 1818 (Araneae, Theraphosida, Aviculariinae) with description of three new aviculariinae genera. Zookeys 659:1–185
Gabriel R, Gallon R, Smith A (2007) Examination of the *Avicularia avicularia* specimens from the Linnaen Collection. J Br Tarantula Soc 23(1):22–30
Goodman J (2005) The Rattlesnake. Faber, London, p 355
Green E (1912) The pioneers of the Ceylon natural history. Spolia Zeylanica 13:81
Hahn C (1831–1836) Monographie der Spinnen. Zentralantiquariat Leipzig, Leipzig, p 143
Hahn C, Sacher P (1988) Carl Wilhelm Hahn Monographie der Spinnen (1820–1836). Zentralantiquariat Leipzig, Leipzig, p 143
Hamilton CB, Hendrixson BE, Brewer MS, Bond JS (2014) An evaluation of sampling effects on multiple DNA barcoding methods leads to an integrative approach for delimiting species: A case study of the North American genus *Aphonopelma* (Araneae, Mygalomorphae, Theraphosidae). Mol Phylogenet Evol 71:79–93
Hancock K, Hancock J (1992) Tarantulas: keeping & breeding in captivity. Serpents Tale, pp 2
Hanley W (1977) Natural history in America. Quadrangle, pp 339
Hillyard P (1994) The book of the Spider: from Arachnophobia to the love of Spiders. Hutchinson, London, p 196
Hindle E (1947) Reginald Innes Pocock. Obit Nat Roy Soc 6:189
Koch CL (1836–1848) Die Arachniden. Part 3–16. Hahn Parts 1–2. BMNH Library, Nurnberg
Koch CL (1851) Uebersicht des Arachnidensystems. Nurnberg
Kollman H, Schultz O (1989) The museum of natural history Vienna. Verlag Museum, Vienna, p 64
Lamarck JB (1801) Systeme des animaux sans vertebres, ou tableau general des orders et des genres de ces animaux, vol 8. Deterville, Paris, pp 1–432
Lamarck JB (1818) Histoire naturelle des Animaux sans vertebres. Deterville, Paris
Latreille P (1806) Genera crustaceeorum et insectorum. Koenig, Paris
Levi H (1964) South American Collectors. Unpublished manuscript
Macleay WS (1834) A few remarks tending to illustrate the natural history of two annulose genera, viz. Urania of Fabricius, and Mygale of Walckenaer. Trans Zool Soc London 1:179–194

Macleay WS (1842) On doubts respecting the existance of bird-catching spiders. Ann Mag Nat Hist 8:324–325
Maurer F (1890) Nachruf an Dr. Anton Ausserer. Programm 1. Staatsgymnasium Graz. 33–40.
McKeown K (1963) Australian Spiders. Sirius Books, Sydney, p 288
Mendoza J, Francke OF (2017) Systematic revision of *Brachypelma* red kneed tarantulas (Araneae: Theraphosidae) and the use of barcodes to assist in the identification and conservation of CITES listed species. Invertebr Syst 31:157–179
Merian, MS (1705) Metamophosis insectorum Surinamensium. Amsterdam. [Facsimile 2009 edition published by Taschen with a commentary by Katharina Schmidt-Loske Based on the copy owned by the University of Basel], Koln, pp 190
Merian MS (1976) Schmetterlinge, Käfer und andere Insekten. Leningrader Studienbuch. Wienheim. [Facsimile edition of unpublished notes and illustrations, with commentary by Irina Lebedeva and Wolf –Dieter Beer], Wienheim
Milner R (1990) The encyclopedia of evolution. Facts on File, New York, p 481
Pérez-Miles F, Lucas SM, da Silva P, Bertani R (1996) Systematic revision and cladistic analysis of Theraphosinae (Araneae: Theraphosidae). Mygalomorph 1(3):33–62
Pickard–Cambridge FO (1896) On the theraphosidae of the lower Amazons: being an account of the new genera and species of this group of spiders discovered during the expedition of the Steamship Faraday up the River Amazons. Proc Zool Soc London 64(3):716–766
Piso W, Marcgrave G (1648) Historia naturalis Brasiliae. Johemmes de Alte, Antwerp
Raven R (1985) The spider infraorder Mygalomorphae (Araneae): cladistics and systematics. Bull Am Mus Nat Hist 182:1–180
Roewer CF (1942) Katalog der Araneae 1 Natura Bremen, 1040 pp
Roewer CF (1954) Katalog der Araneae 11 Instit royal des Sciences naturelles de Belgique Bruxelles, 1751 pp
Savory T (1961) Spiders, Scorpions & Men. Univ. London Press, London, p 52
Seba, A (1734) A Cabinet of Natural Curiosities. Taschen. Facsimile edition 2001. Commentary Irmgard Musche, Rainer Willmann, Jes Rust. Based on a copy owned by Koninkelijke Bibliotheek, Hague
Simon E (1890) Liste des esp. de la fam. des Aviculariidae qui hab. l'Amerique du Nord. Act Soc Linn Bordeaux 44:307–326
Smith A (1994) Tarantula Spiders: Tarantulas of the USA and Mexico. Fitzgerald, London, p 196
Smith A (2013) From Times Past and Far Flung Lands. J Br Tarantula Soc 28:86
Smith A (2015) *Aphonopelma hentzi* – named after a very spiritual Spider Man. J Br Tarantula Soc 30:18
Stearn W (2001) The Natural History Museum of South Kensington, vol 17. The Natural History Museum, London
Talayesva D (1942) Sun Chief. Yale University, Yale, p 407
Taube K (1983) The Teotihuacan Spider Woman. J Latin-Am Lore 9:107–189
Vogel B (2011) AAS History. http://www.americanarachnology.org/about_AAS/history.html
Volckamer J (1702). Correspondence. Univ. Erlangen, Trew-Bibliothek. Johan (32) Volckamer. Oct 1702. Ms 1834
Walckenaer C (1837) Histoire naturelle des Insectes Apteres 1 Paris, 682 pp
Warriner M (2008) Distribution and taxonomic status of tarantulas in Arkansas. J Acad Sci 62:107–114
Wettengl K (ed) (1998) Maria Sibylla Merian. 1647–1717. Art Nat Verlag Gerd, Hatje, p 265
White A (1856) Description of *Mygale emilia*, a spider from Panama. Proc Zool Soc London 24:183–185
Winston EJ (1999) Describing species. Columbia, New York, p 518
Worm O (1655) Museum Wormianum. Seu historia rerum rariorum, tam naturalium, quam artificialum, tam domesticarum, quam exoticarum, quae Hafniae Danorum in aedibus authoris servantur. Elzevier, Leiden

Index

F. Pérez-Miles (ed.), *New World Tarantulas*, Zoological Monographs 6,
https://doi.org/10.1007/978-3-030-48644-0

D

E

W

X

Y